POLYMERIC GENE DELIVERY

Principles and Applications

T0172653

POLYMERIC GENE DELIVERY

Principles and Applications

EDITED BY

Mansoor M. Amiji

CRC PRESS

Boca Raton   London   New York   Washington, D.C.

Cover image illustrates the expression of green fluorescent protein in NIH-3T3 mouse fibroblast cells transfected with EGFP-N1 plasmid DNA using poly(beta-amino ester) nanoparticles.

Library of Congress Cataloging-in-Publication Data

Polymeric gene delivery : principles and applications / editor, Mansoor M. Amiji.
        p. cm.
Includes bibliographical references and index.
ISBN 0-8493-1934-X (alk. paper)
1. Gene therapy. 2. Polymeric drug delivery systems. I. Amiji, Mansoor M.

RB155.8.P656 2004
615.8'95—dc22   2004049668

This book contains information obtained from authentic and highly regarded sources. Reprinted material is quoted with permission, and sources are indicated. A wide variety of references are listed. Reasonable efforts have been made to publish reliable data and information, but the author and the publisher cannot assume responsibility for the validity of all materials or for the consequences of their use.

Neither this book nor any part may be reproduced or transmitted in any form or by any means, electronic or mechanical, including photocopying, microfilming, and recording, or by any information storage or retrieval system, without prior permission in writing from the publisher.

All rights reserved. Authorization to photocopy items for internal or personal use, or the personal or internal use of specific clients, may be granted by CRC Press LLC, provided that $1.50 per page photocopied is paid directly to Copyright Clearance Center, 222 Rosewood Drive, Danvers, MA 01923 USA. The fee code for users of the Transactional Reporting Service is ISBN 0-8493-1934-X/05/$0.00+$1.50. The fee is subject to change without notice. For organizations that have been granted a photocopy license by the CCC, a separate system of payment has been arranged.

The consent of CRC Press LLC does not extend to copying for general distribution, for promotion, for creating new works, or for resale. Specific permission must be obtained in writing from CRC Press LLC for such copying.

Direct all inquiries to CRC Press LLC, 2000 N.W. Corporate Blvd., Boca Raton, Florida 33431.

Trademark Notice: Product or corporate names may be trademarks or registered trademarks, and are used only for identification and explanation, without intent to infringe.

Visit the CRC Press Web site at www.crcpress.com

© 2005 by CRC Press LLC

No claim to original U.S. Government works
International Standard Book Number 0-8493-1934-X
Library of Congress Card Number 2004049668
Printed in the United States of America 1 2 3 4 5 6 7 8 9 0
Printed on acid-free paper

Dedication

This book is dedicated to my lovely wife, Tusneem, and to my daughters — Zahra, Anisa, and Salima

Preface

The primary goals of gene therapy are to correct the genetic defects that underlie a disease process and to provide supplemental therapeutic modality through genetic engineering. In order to correct genetic deficiency or treat diseases by gene therapy approaches, it is necessary to formulate and evaluate vector systems that can deliver the genes to the appropriate tissues and cells in the body in a specific and efficient manner. Over 75% of current gene therapy is performed using viruses as gene delivery vehicles. However, with viruses, there are serious concerns over issues of toxicity, immunogenicity, payload gene size limitations, and difficulty in scale-up for industrial production.

Nonviral vectors, therefore, have attracted attention from academic and industrial scientists. Among the nonviral vectors, polymeric systems offer several important advantages. First, polymers are tremendously versatile and can provide physical, chemical, and biological properties that are necessary for gene delivery applications. Second, polymers can be synthesized in a parallel synthesis pathway for high-throughput screening of biocompatibility and transfection efficiency. Third, various formulations, designs, and geometries can be made from polymeric materials for specific types of gene delivery application. Lastly, the surface of polymeric carriers can be easily modified with biological ligands for specific recognition in the body.

Polymeric Gene Delivery: Principles and Applications is intended to serve as an up-to-date guide and to promote further research by encouraging more scientists to contribute to this exciting field. In the 40 chapters of *Polymeric Gene Delivery: Principles and Applications*, academic and industrial scientists from around the world who are conducting research in the area of polymeric gene delivery systems are united to provide the most comprehensive treatment of the subject. The authors of each chapter are leaders in their respective fields and have contributed to cutting-edge research. Starting from the introductory chapter by Bob Langer, the book is divided into five sections that deal with challenges and opportunities in gene delivery (Part I), condensing polymeric systems (Part II), non-condensing polymeric systems (Part III), microspheres and nanospheres (Part IV), and specialized delivery systems (Part V).

First of all, I am deeply indebted to all the chapter contributors, who have done a superb job in writing for this book. I am especially grateful to Bob Langer for initiating my interest in polymeric gene delivery when I did my sabbatical work in his laboratory at MIT in 2000, and to Dave Lynn, who was a post-doc in Bob's lab (now at the University of Wisconsin, Madison), for allowing me to follow along as he marvelously synthesized and tested many different types of poly(β-amino ester)s. Special thanks are due to my colleagues and friends at Northeastern University and in the Boston area with whom I have collaborated on many different projects. Most importantly, I am grateful to the wonderful people at CRC Press, including Stephen Zollo, Barbara Uetrecht-Pierre, Patricia Roberson, and Susan Fox, for transforming the concept of a polymeric gene delivery book into a reality.

The editor would greatly appreciate receiving comments and constructive criticisms of the book. They may be sent by email to m.amiji@neu.edu.

Mansoor M. Amiji
Boston, Massachusetts

Contributors

Akin Akinc
Massachusetts Institute of Technology
Cambridge, Massachusetts

Valery Yu. Alakhov
Supratek Pharma, Inc.
Dorval, Quebec, Canada

Maria Jose Alonso
University of Santiago de Compostela
Santiago de Compostela, Spain

Mansoor M. Amiji
Northeastern University
Boston, Massachusetts

Daniel G. Anderson
Massachusetts Institute of Technology
Cambridge, Massachusetts

Alexei A. Antipov
Max-Planck Institute of Colloids and Interfaces
Potsdam/Golm, Germany

Tony Azzam
The Hebrew University – Hadassah Medical
  School
Jerusalem, Israel

Zain Bengali
Northwestern University
Evanston, Illinois

James C. Birchall
Cardiff University
Cardiff, Wales, United Kingdom

Maytal Bivas-Benita
Leiden University
Leiden, The Netherlands

Gerrit Borchard
Leiden University
Leiden, The Netherlands

Gianluca Carlesso
Department of Veteran Affairs Medical Center
Nashville, Tennessee

Polly Chang
SRI International
Menlo Park, California

Daniel M. Checkla
Clear Solutions Biotech, Inc.
Stony Brook, New York

Weiliam Chen
State University of New York at Stony Brook
Stony Brook, New York

Shih-Jiuan Chiu
The Ohio State University
Columbus, Ohio

Joon Sig Choi
Chungnam National University
Daejeon, Korea

Patrick Couvreur
University of Paris-Sud
Paris, France

Dan J.A. Crommelin
Utrecht Institute for Pharmaceutical Sciences
Utrecht, The Netherlands

Noemi Csaba
University of Santiago de Compostela
Santiago de Compostela, Spain

Jeffrey M. Davidson
Department of Veteran Affairs Medical
  Center
Nashville, Tennessee

Mark E. Davis
California Institute of Technology
Pasadena, California

Philip Dehazya
Clear Solutions Biotech, Inc.
Stony Brook, New York

Charles L. Densmore
Baylor College of Medicine
Houston, Texas

Sujatha Dokka
Isis Pharmaceuticals
Carlsbad, California

Abraham J. Domb
The Hebrew University – Hadassah Medical
  School
Jerusalem, Israel

Elias Fattal
University of Paris-Sud
Paris, France

Paolo Ferruti
University of Milan
Milan, Italy

Ilia Fishbein
University of Pennsylvania Medical Center and
  Children's Hospital of Philadelphia
Philadelphia, Pennsylvania

Jacopo Franchini
University of Milan
Milan, Italy

Hamidreza Ghandehari
University of Maryland
Baltimore, Maryland

Mitsuru Hashida
Kyoto University
Kyoto, Japan

Mary Lynne Hedley
Zycos, Inc.
Lexington, Massachusetts

Wim E. Hennink
Utrecht Institute for Pharmaceutical Sciences
Utrecht, The Netherlands

Suzie Hwang Pun
University of Washington
Seattle, Washington

Yong S. Jong
Spherics Inc.
Lincoln, Rhode Island

Alexander V. Kabanov
University of Nebraska Medical Center
Omaha, Nebraska

Yasufumi Kaneda
Osaka University
Osaka, Japan

Goldie Kaul
Northeastern University
Boston, Massachusetts

Berma M. Kinsey
Baylor College of Medicine
Houston, Texas

Angela Kim
Clear Solutions Biotech, Inc.
Stony Brook, New York

Ji-Seon Kim
University of Iowa
Iowa City, Iowa

Sung Wan Kim
University of Utah
Salt Lake City, Utah

Tae-il Kim
Seoul National University
Seoul, Korea

Ralf Kircheis
Igeneon Immunotherapy of Cancer
Vienna, Austria

Lori A. Kubasiak
Dendritic Nanotechnologies, Inc.
Mt. Pleasant, Missouri

Vinod Labhasetwar
University of Nebraska Medical Center
Omaha, Nebraska

Robert Langer
Massachusetts Institute of Technology
Cambridge, Massachusetts

Minhyung Lee
University of Utah
Salt Lake City, Utah
and
Inha University Medical School
Inchon, Korea

Robert J. Lee
The Ohio State University
Columbus, Ohio

Yan Lee
Seoul National University
Seoul, Korea

Robert Levy
University of Pennsylvania Medical Center and
  Children's Hospital of Philadelphia
Philadelphia, Pennsylvania

Yong-beom Lim
Seoul National University
Seoul, Korea

Dijie Liu
University of Iowa
Iowa City, Iowa

David M. Lynn
University of Wisconsin
Madison, Wisconsin

Wenxue Ma
University of Nebraska Medical Center
Omaha, Nebraska

Zaki Mageed
University of Maryland
Baltimore, Maryland

Ram I. Mahato
University of Tennessee Health Science
  Center
Memphis, Tennessee

Edith Mathiowitz
Brown University
Providence, Rhode Island

Randall J. Mrsny
University of Wales
Wales, United Kingdom

Mark Newman
Epimmune, Inc.
San Diego, California

Michael Nicolaou
Epimmune, Inc.
San Diego, California

Makiya Nishikawa
Kyoto University
Kyoto, Japan

Manfred Ogris
Centre of Drug Research
LMU Munich, Germany

Frank M. Orson
Baylor College of Medicine
Houston, Texas

David Oupicky
Wayne State University
Detroit, Michigan

Jong-Sang Park
Seoul National University
Seoul, Korea

Celso Perez
University of Santiago de Compostela
Santiago de Compostela, Spain

Itay Perlstein
University of Pennsylvania Medical Center and
  Children's Hospital of Philadelphia
Philadelphia, Pennsylvania

Swayam Prabha
University of Nebraska Medical Center
Omaha, Nebraska

Ales Prokop
Vanderbilt University
Nashville, Tennessee

David Putnam
Cornell University
Ithaca, New York

Kevin G. Rice
University of Iowa
Iowa City, Iowa

Christopher B. Rives
Northwestern University
Evanston, Illinois

Yon Rojanasakul
West Virginia University
Morgantown, West Virginia

Krishnendu Roy
University of Texas
Austin, Texas

Alejandro Sanchez
University of Santiago de Compostela
Santiago de Compostela, Spain

Camilla A. Santos
Spherics Inc.
Lincoln, Rhode Island

Lonnie D. Shea
Northwestern University
Evanston, Illinois

Dinesh B. Shenoy
Max-Planck Institute of Colloids and Interfaces
Potsdam/Golm, Germany

Srikanth Sriadibhatla
University of Nebraska Medical Center
Omaha, Nebraska

Gert Storm
Utrecht Institute for Pharmaceutical
  Sciences
Utrecht, The Netherlands

Gleb B. Sukhorukov
Max-Planck Institute of Colloids and
  Interfaces
Potsdam/Golm, Germany

Yoshinobu Takakura
Kyoto University
Kyoto, Japan

Donald A. Tomalia
Dendritic Nanotechnologies, Inc.
Mt. Pleasant, Missouri

Ferry J. Verbaan
Utrecht Institute for Pharmaceutical
  Sciences
Utrecht, The Netherlands

Ernst Wagner
Ludwig-Maximilians University
Munich, Germany

Don Wen
Clear Solutions Biotech, Inc.
Stony Brook, New York

Sharon Wong
Cornell University
Ithaca, New York

Yang H. Yun
State University of New York at
  Stony Brook
Stony Brook, New York

Table of Contents

Part V
Specialized Delivery Systems

CHAPTER 1

Introduction

Robert Langer

Gene therapy holds enormous promise for correcting genetic defects and treating countless diseases. However, perhaps the single biggest challenge that many scientists face today in gene therapy is delivery. Two general approaches have been used for gene delivery. The first is viral vectors, which are highly effective. In this type of approach the viruses can act so as to sneak foreign genes into cells. Unfortunately, viruses, even disabled ones, can cause serious side effects. The death of gene therapy trial volunteer Jesse Gelsinger in 1999 is a striking example of this.

The second approach involves using a nonviral vector such as a polymer, lipid, or liposome. While potentially safer, these synthetic systems are not as effective as viral vectors. Thus, it is critically important that efforts toward creating synthetic gene therapy vectors such as polymers be developed. This book discusses, in broad terms, polymeric delivery systems of all types as well as the significant challenges that one must face in gene therapy delivery. Specifically, gene therapy delivery involves major design initiatives. To succeed, polymers must be able to condense or package DNA into small sizes so that it can be taken up by cells, stabilize the DNA before and after cellular uptake, bypass or escape the cell's endocytotic pathways, deliver the DNA to the cell's nucleus, and unpackage DNA by releasing it in active form. In Part I of this book, a number of authors cover these important issues, discussing biological barriers, cellular uptake, trafficking, and even ways of targeting genes to specific cells and tissues.

Clearly a major challenge in trying to develop a polymeric gene therapy system is the polymer itself. Part II of this book discusses condensing polymer systems. Various examples of nondegradable polymers such as polylysine, polyethylene imine, polyethylene glycol conjugates, and novel cyclodextrins are discussed in the first section of Part II. This is followed by a second set of chapters on degradable polymer systems where novel polymers such as poly(β-amino ester)s, poly(amidoamine)s, polyimidazoles, various polysaccharides, and chitosans are discussed.

Part III of this book involves noncondensing polymer systems. Here poly(ethylene oxide)/poly(propylene oxide) copolymers are discussed, as are various poly(N-vinyl pyrrolidone) systems and hydroxypropyl methacrylate copolymers.

Part IV of the book discusses polymeric nanospheres and microspheres. These types of systems might provide controlled release or could be taken up by cells. Biodegradable polymers such as poly(ethylene oxide) and poly(alkylcyanoacrylate) nanoparticles are discussed. In addition, this section covers protein nanoparticles and adenovirus-based systems. Section B of Part IV discusses poly(lactide-coglycolide) microspheres, polyanhydrides, and hyaluronic acid systems.

Finally, the book examines various specialized systems that may be useful for gene therapy delivery. These include protein constructs and glycosylated polymers, as well as systems that could

0-8493-1934-X/05/$0.00+$1.50
© 2005 by CRC Press LLC

be targeted to different receptor systems, such as folate or transferrin. The book then discusses approaches for delivering genes to or through different places in the body. Specifically, ways of delivering genes to the central nervous systems, to the lungs, and through the skin are discussed. Important issues in wound repair and gene delivery by tissue engineering are also examined. The last two chapters discuss ways of using gene therapy in polymers to effect restenosis and biologically relevant micro-electro-mechanical systems (MEMs) systems to deliver genes.

The chapters in this book provide a comprehensive overview of the many ways in which scientists are trying to deliver genes using novel polymeric approaches. With the enormous amount of effort going into this area, it is hoped that gene therapy will reach its full promise in some day addressing the countless problems that it has the potential to solve.

PART I

Gene Delivery: Challenges and Opportunities

CHAPTER 2

Tissue- and Cell-Specific Targeting for the Delivery of Genetic Information

Randall J. Mrsny

CONTENTS

2.1 INTRODUCTION

Delivery of genetic information has been proposed to provide a plethora of novel therapeutic and prophylactic treatments. Although this field has made rather remarkable progress in the 50 years since the structure of DNA was initially reported,[1] it is sobering to consider that gene therapy is far from a standard of medical practice.[2] Ten years ago, experts in this field[3] projected that gene therapy would become commonplace if three technical breakthroughs could be made:

1. Development of a gene transfer vector that could be taken off the shelf and used to target a desired tissue or cell type
2. Identification of methods to integrate delivered genetic material safely in a homologous recombinant fashion with the genetic target it is intended to replace or modulate
3. Introduction of genetic information where it would be responsive to appropriate physiological stimuli or signals

0-8493-1934-X/05/$0.00+$1.50
© 2005 by CRC Press LLC

Although progress has been made in all three of these areas, there are still more problems than solutions for each.

These three technological advances focus on concerns related primarily to gene replacement strategies. A variety of alternative methods have been examined, however, for their effectiveness in altering the biological outcomes of a defective gene, including a wide variety of gene-modulating approaches involving either antisense oligonucleotides[4] or ribozymes[5] or interfering RNA[6] to diminish the function of selected genes. Delivery of a *trans*-splicing ribozyme has been shown to repair defective messenger RNA.[7] In some ways, these approaches are in opposition to the "classical" concept of gene therapy, where a defective cellular function is rectified by the delivery of a genetic material that results in the expression of a desired gene product. These lateral efforts using ribozymes, interfering RNA, and so on have recently come to the fore with the hope that some of these approaches would have fewer technological hurdles than those observed for gene therapy. Unfortunately, application of both gene therapy and these gene-modulating technologies is compromised by the absence of the simple, off-the-shelf delivery system mentioned above, one that would act specifically to target a desired cell or tissue.

The delivery strategy for both gene therapy and gene modulation typically involves targeting the cytoplasm of target cells, where these materials can either act on cytoplasmic components (e.g., messenger RNA) or access the nucleus, where they might achieve its optimal action(s). Although it has been suggested that in some cases these nucleic acid-based drugs only require direct interaction with cell surfaces to induce a biological outcome,[8] by accessing endogenous intracellular transport trafficking pathways to the nucleus. Indeed, oligonucleotides tend to collect in the nucleus following direct injection into the cytoplasm of a cell (as discussed by Stull and Szoka[9]), and cytoplasmic delivery of DNA complexes seem to have a similar fate. In consideration of this, the steps in a successful, optimized targeted gene delivery would be as follows:

1. Successful transport or access to a particular tissue of the body
2. Evasion of the immune system
3. Binding to a cell-specific structure
4. Internalization through a mechanism that culminates in entry into the cell's cytoplasm
5. Targeting to the nucleus (or mitochondria, another potentially important gene therapy target)
6. Unfolding or relaxation of the delivered genetic material for optimal transcription/integration events[10]

Since there are so many ways for these systems to become inactivated or misdirected, analytical tools are required to expedite the process of identifying successfully targeted gene therapy and gene modulation systems. Advances in achieving these goals have involved, and will continue to require, accurate methods to assess steps one through six above and not simply the final outcome. Further, novel methods of improved characterization of *in vivo* events, such as the use of positron emission tomography to follow the fate of oligonucleotides[11–13] will be important in evaluating the success of various approaches. This chapter will discuss some current methods for targeting certain tissues and cells selectively for the delivery of genetic material and will relate such targeting issues and approaches to the six steps noted above for successful delivery of nucleotide-based therapies.

2.2 BACKGROUND ISSUES

Any successful strategy for targeted delivery of genetic material must acknowledge and incorporate concerns related to the basic physical and chemical characteristics of oligonucleotides. The types of oligonucleotide molecules used in these therapeutic efforts are generally large (sometimes extremely large), highly charged (usually one phosphate group per residue) and bulky (due to charge-charge repulsion of the extensive negative charge profile). Some nucleotide-based therapies have been modified to reduce some of these concerns, e.g., size minimization and charge neutralization. Any method used to target genetic material to a specific cell or tissue must consider the physical and chemical properties of oligonucleotides that affect their stability since nucleotide-based therapies must be delivered intact for an efficacious outcome. There are concerns of physical damage; very large macromolecules can break merely by handling, such as passage through a pipette tip or syringe needle. Potential enzymatic degradation is a constant concern; RNAses are ubiquitous. Thus, there are many challenges in any approach to delivery of nucleotide-based therapies.

Although it seems as though there is a nearly infinite list of potential targets for nucleotide-based therapies, the reality is that only a limited set are likely to be considered for gene therapy or genetic manipulation in the near future. Many genetic defects have been experimentally associated with unsuccessful fetal development. These targets would require not only early *in utero* intervention, but also extremely early detection. Although some inroads have been made into fetal gene therapy intervention,[14] the events required to achieve this goal extend beyond our current practical and ethical understanding. There are also fewer genes in the human genome than initially expected,[15] supporting the suggestion that protein interactions may be modulated through the temporal and spatial expression of an individual protein and that this acts as an additional layer of complexity in genome function.[16] Introduction of a "corrected" gene with unclear functions or under conditions where it might not be properly regulated by appropriate physiological signals could lead to unanticipated systemic outcomes. To further complicate these outcomes, it appears that genetically regulated events do not occur only through modulation of expressed, functional proteins. A recent analysis concluded that approximately 200 to 255 human genes (nearly 1% of the human genome) encode short strands of RNA, termed microRNA, that may produce biological outcomes through regulating the actions of individual genes or sets of genes.[17] Thus, of the genetic defects that clearly correlate with a known condition or disease, we understand only a few sufficiently to provide some assurance that using a nucleotide-based therapy will have acceptable long-term safety. At present, inherited diseases such as cystic fibrosis and adenosine deaminase deficiency and acquired diseases such as AIDS and cancer are the prominent areas being examined for gene therapies.

Typically, the delivery of genes or genetic material is considered a therapeutic intervention. There are, however, potential applications in other areas such as vaccination through the induced expression of antigens or adjuvants (agents that stimulate or accentuate immune responses). Although this topic will not be specifically addressed, it brings up the issue of transient versus sustained transfection events. The original concept of gene therapy was for the permanent correction of a defective cellular function — a sustained outcome achieved using transfection strategies that target cells having either a long-term life expectancy or where genetic information is integrated into a cell that retains the ability to replicate. In the case of vaccination, establishment of potentially sustained transfection would result in a transient event, since cells expressing the introduced antigen will ultimately be targeted for destruction by the immune system. Similarly, there is concern that individuals exposed for the first time to the protein product of a previously defective gene could raise an immune response as a consequence of non-self antigen recognition. This concern is magnified since the "new" protein would be expressed for the first time in a cell that may also carry viral markers residual from the delivery system used for transfection. Thus, for many safety-related reasons, it may be more desirable to establish transient transfections in some cases. Most importantly, it is critical that the delivery system selected provide the desired outcome, whether it is sustained or transient.

Many topics related to successful gene therapy following cell or tissue targeting will not be discussed in this chapter, such as mechanisms of intracellular targeting and ways of overcoming limitations of expression outcomes. There are a number of other factors that could affect the outcomes of tissue- and cell-specific targeting schemes (e.g., methods of stabilizing nucleic acids, intracellular targeting strategies, etc.) that will only be briefly discussed since these are the topics of other chapters in this text. In advance it must also be stated that the number of studies performed to date in the areas of tissue- and cell-specific targeting related to the delivery of gene modulating structures is vast. Thus, recent reviews have been frequently cited to provide the reader with archives of information and additional perspectives that might be useful in explaining broad, general principles. In most cases, no reference has been made to the many excellent references that describe original, seminal findings in these areas.

2.3 GENERAL FACTORS TO CONSIDER

2.3.1 Viral Systems

A survey of published gene therapy will reveal two main strategies — those involving viral delivery systems and those utilizing condensed DNA complexes (nonviral systems). The goal of both approaches is to achieve efficient uptake and expression of genetic information. In the case of viral delivery systems, aspects of how the virus packages genetic information and the number (and size) of genes required for the synthesis of the mature virus dictate the amount of exogenous genetic information that can be incorporated (Figure 2.1). In general, viruses carry the minimal amount of genetic material essential for the next round of replication, leaving little room to place a large amount of additional genetic information. This is particularly true if that genetic material is encapsulated within a constrained three-dimensional capsid (nonenveloped) structure. Enveloped viruses typically have more flexibility in the amount of additional genetic material they can incorporate (Table 2.1). One way to increase available room is to delete some of the endogenous genetic material of the virus.[18] This frequently works out well since it can be advantageous to remove certain viral genes associated with pathogenic function.[10] It may also be desirable to remove genetic information that dictates viral tropism, since this natural targeting may compromise the intended, modified targeting scheme of a viral gene delivery vehicle. Since viral tropism is optimized for entry into and actions within a particular cell type, viral delivery systems with modified or unnatural targeting components may have reduced potential use for gene therapy. An example of this would be the observation that modification of the tropism of murine leukemia virus (a retrovirus) through alteration of envelope proteins does not alter its requirement for a replicating cell host,[19] thus limiting potential applications of this gene delivery platform.

Viral particles typically have a defined tropism that dictates an optimized scheme of infection, one that is coordinated with successful dissemination and reinfection (Table 2.1). Some of these principles are rather obvious; others are more subtle. For example, cytomegalovirus uses the epidermal growth factor receptor for entry into target cells.[20] Retroviruses are more promiscuous in their cell surface tropism but integrate genetic information into the host genome; for this integration to occur, a mitotic division is required. Thus, it would not make sense for a retrovirus to be modified to have a tropism for a cell, such as a neuron, that is essentially postmitotic. Interestingly, some of the most aggressively dividing cells can be those of a tumor, making retroviruses not only a good vector for targeting these cells but also an opportunist for infecting them should they arise in a patient treated for some other reason. Ebola and Marburg are members of the Filoviridae family of viruses. These viruses are similar to retroviruses (Table 2.1) in that they express a surface glycoprotein that mediates receptor binding with specific host cells. Exchange of these glycoproteins between lentiviruses and filoviruses has been used to retarget these potential viral gene therapy delivery vectors.[21] Although such modifications can lead to a retargeting of these viral vectors, the receptor through which they target is sometimes poorly understood.[22,23] It is also possible to focus a delivery therapy based upon functional aspects of the target cell population. In the case of cancer cells expressing an activated Ras/Raf/mitogen-activated protein kinase pathway, regulated replication of a viral gene delivery system can be used to provide selective expression

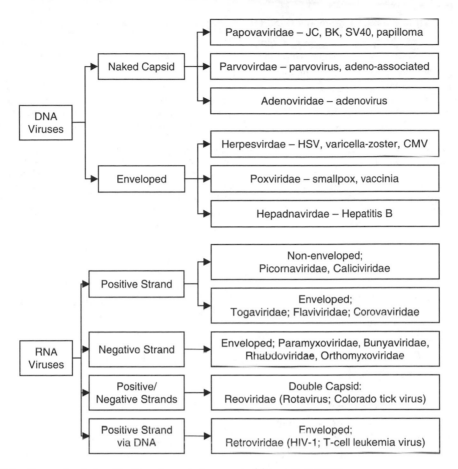

Figure 2.1   Generalized classification of some viruses used for gene therapy.

of delivered genetic information.[24] Thus, prior to using the targeting capacity of viral particles, it would be important to understand their tropisms as completely as possible.

One concern is that the tropism of a virus may differ due to the route of entry into the body. Most viruses infect individuals via a mucosal exposure. As an example, both the Coxsackievirus and the adenovirus bind to a receptor on the surface of epithelial cells located at the tight junction structures that hold adjacent cells together;[25] this same receptor is distributed differently on non-polarized cells. Additionally, adenovirus has been suggested to target hematopoietic stem cells,[26] making the route of entry of this vector critical to its potential targeting. Rarely do viruses infect in the manner (by injection) or in the numbers (as much as 10^{11} particle forming units [PFUs]) that are used in most gene therapy protocols. A few viruses do produce natural infection by injection, and their delivery by injection would emulate their natural route of infection. The Dengue virus infects via an injection from an infected mosquito.[27] Once injected into the body, the Dengue virus attaches to the host cell and enters cells via the major envelope glycoprotein E, a protein capable of locally disrupting host cell membranes, allowing entry of genetic information.[27] As stated above, many other viruses infect initially at a mucosal surface and not by an injected route. Tropism of a viral particle may be different — somewhat less selective — when delivered in an unnatural fashion and in very high numbers. Additionally, many viral vectors have been modified to reduce or eliminate a particular pathogenic characteristic (as discussed above), and this may also affect the outcome of their capacity to target to a specific cell type or tissue.

Table 2.1   Virus Families Commonly Used for Delivery of Genetic Information

Virus	Packaging Capacity	Cell/Tissue Application(s)	Outcome of Gene Therapy
Poxviridae (e.g., vaccinia, variola, avipox)	25 kb	Epidermal, respiratory, various epithelia	Transient expression, highly immunogenic
Adenoviridae (e.g., adenovirus — more than 50 serotypes)	< 7.5 kb	Broad: hepatocytes, endothelial cells, smooth muscle cells, airway cells, ocular tissues, neurons	Transient expression, highly immunogenic
Herpesviridae (e.g., simplex, cytomegalovirus, EHV)	> 30 kb	Broad: stem cells, neurons, muscles	Durable expression, low toxicity with some mutant strains.
Retrovirus (e.g., murine leukemia virus, human immunodeficiency virus-1, equine infectious anemia virus, Maloney murine leukemia virus)	8 kb	Mostly restricted to dividing cells: tumor cells, lymphocytes, hepatocytes, bone marrow cells	Genome integration, durable expression, safety concerns
Parvovirus (e.g., adeno-associated virus-2)	< 4 kb	Broad: infects both dividing and nondividing cells	Slow expression onset, can be both integrating and nonintegrating, durable expression
Alphavirus (e.g., Semliki forest virus, Sindbis virus, Venezuelan equine encephalitis virus)	< 7.5 kb	Broad, and there are neuron- and glial-specific strains	Transient, high expression with low immunogenicity

Abbreviation: kb = kilobases (kilonucleotides), EHV = equine herpesvirus 1.

2.3.2   Nonviral Systems

Nonviral gene delivery systems do not typically have limitations in the size of the genetic material that may be delivered, although other physical aspects related to manipulating large DNA structures may provide an upper limit. While viruses have the ability to package DNA carefully and selectively, condensation of DNA for nonviral delivery systems, particularly in the large-scale commercial processes required for an industry, may be much less refined. Also in contrast to viral delivery systems, nonviral delivery approaches do not have the inherent selectivity of cell surface binding and internalization (tropism). Additionally, the intracellular fate of the delivered genetic material is more efficiently directed (targeted to specific intracellular compartments with break-out mechanisms to reach the cytoplasm) compared to nonviral vectors. Because of these differences, nonviral systems can typically be used in a more universal fashion for gene delivery than viral systems. However, these differences lead to the generalized rule that viral vectors are more efficient than nonviral vectors. Although nonviral vectors are perceived to be less immunogenic than viral vectors, nonviral systems can incite a significant, dose-dependent inflammatory response that can limit both safety and efficacy.[28]

Nonviral systems have been investigated even more aggressively since the death of a patient in a virus-based gene therapy trial[29] and the occurrence of leukemia following gene therapy of children with X-linked severe combined immune deficiency using a retroviral gene therapy vector.[30] Previous efforts focused primarily on cationic lipid-DNA complexes frequently composed of combinations of diolexyoxy-propyl-triethylammunium chloride (DOTMA), dimethylaminoethane-carbamoyl cholesterol (DC-Chol), and/or dioleoyl phosphatidylethanolamine (DOPE). These complexes stabilize incorporated DNA against physical and enzymatic damage. Numerous agents (chitosan, polyethyleneimine, poly[L-lysine], dendrimers, gelatin, etc.) have now been extensively investigated as nonviral gene delivery systems.[31] Studies have been performed to compare a number of these nonviral vectors.[32] The success of these agents is directly correlated with their ability to overcome issues of low efficiency and inconsistent preparation that have plagued previous nonviral delivery systems. High levels of these agents can produce remarkable levels of transfection, but come with the concern of significant toxicity relative to viral vectors.[33] What appears to be coming out of

Table 2.2 Some Cell-Membrane-Penetrating Peptides

Peptide	Sequence*	Natural Source
pVEC	LLIILRRRIRKQAHAHSK	Murine vascular endothelial-cadherin (VEC)
Penetratin	RQIKIWFQNRRMKWKK	*Antennapedia* homeodomain
48-60 Tat peptide fragment	GRKKRRQRRRPPQ	HIV-1 nuclear transcription activator
Transportan	GWTLSNAGYLLKINLKALAALAKKIL	Galanin and Mastoparan peptides

* Standard single amino acid designations are used.

these current efforts is that a combination approach, where several of these agents are used simultaneously, can provide efficient and durable transfection. Combinations of nonviral approaches may best emulate the delivery capabilities of viral vectors.

Nonviral gene therapy systems, in comparison with viral vectors, have a poor success rate of DNA delivery to the cytoplasm (and to the nucleus) of a cell following uptake at the plasma membrane. Fusogenic peptides (some examples are shown in Table 2.2) can be used to deliver oligonucleotides directly to the cytoplasm of cells,[34] presumably bypassing endocytosis processes. More recent studies have suggested that at least some of these fusogenic peptides may follow an endocytosis pathway.[35] Fusogenic peptides have also been used for gene delivery,[36] some causing disruption of the liposome-based delivery vector into which they are entrapped.[37] Genetic information delivered in this way, directly to the cytoplasm, avoids the fate of genetic material taken up via endocytosis that is destroyed in lysosomes.[38] Several methods have been identified to improve the efficiency of transfection following endocytosis. Improved delivery to endosomes can be achieved using pH-sensitive liposomes[39] and pH-responsive polymeric carriers.[40] Enhanced release from endosomes of genetic information delivered using a nonviral vector, reducing its destruction in following fusion with lysosomes, can be achieved *in vivo* by treatment with chloroquine.[41] Although this approach does work, it is likely to be limited in its practical applications, and other methods of endosomal escape are being explored.[42] Many of the most successful of these involve polymers that induce membrane rupture when activated by a drop in pH,[43] such as occurs in the maturing endosome prior to its fusion with a lysosome.[44] Other approaches have described the use of peptide-based scaffolds.[45] Although genetic material presented to the cytoplasm can be directed to the nucleus, incorporation of nuclear targeting signals[46,47] into these systems may further improve the efficiency of genetic material being delivered to the nucleus following endosomal escape.[48] Alternately, cytoplasmic expression systems may provide a possible alternative for nonviral vector delivery of genetic information to the cytoplasm rather than the nucleus.[49]

Nonviral delivery of genetic material represents a rather remarkable feat of engineering. In order to emulate a virus, a nonviral delivery system must have a component to hold the genetic information in a stable form, a means of directing it to a specific location, and some strategy for getting the genetic material into the cytoplasm once it has been taken up by a cell.[50,51] Viral targeting reflects an optimization of these functions that has occurred through many, many generations of replication under intense selective pressure. The result is that viruses have become exquisitely efficient in stimulating their own uptake, delivery of genetic material critical for the next round of replication, and dispersal for reinfection. In order to improve their efficiency, nonviral systems can be modified to emulate some functions of a virus — certain aspects of tropism, DNA uptake, or gene incorporation. Targeting by nonviral systems is inefficient compared to targeting by viruses, which have acquired elaborate targeting and delivery schemes to focus their tropism and enhance their uptake. However, there are certain drawbacks that come with that improved targeting and delivery. Surface components used by viral particles for selective tissue and cellular targeting make them more easily recognized by the immune system. In response, viruses have acquired additional characteristics that allow them to evade the immune system or to manipulate it to their own benefit. In doing so, these viruses have become so aggressive that they usurp or modify the normal function of the host cell to the extent that their infections result in some associated pathological condition.

2.3.3 Immune Concerns

Any virus used as a vehicle for the delivery of genetic material can incite an intense immune response following its administration.[52] Indeed, some viral vectors, such as Ankara virus, are particularly effective at inducing an immune response in humans.[53] Pre-existing immunity can dramatically reduce the efficacy (potentially compromising the safety) of a viral delivery system, and adaptive immune responses are most prominent after the second or third administration of the vector.[54] Immediate responses to a viral delivery vector, through the actions of the innate immune system, can occur after the first exposure and can also result in a decreased clinical efficacy. Viral delivery vectors that require replication to function optimally provide an additional concern of response by both the innate and adaptive immune systems. Broad antiviral defense mechanisms are still being identified that could affect the outcome of viral-based gene therapies.[55] Genetic alteration of some viral vectors, such as adenovirus, can effectively be used to alter tropism as well as immunogenicity of the delivery vector.[56] Covalent modification of viruses with molecules such as polyethylene glycol (PEG) can also diminish their overt immunogenicity by reducing their recognition by the immune system. Many viruses have specialized gene products that act to stymie the actions of the adaptive or the innate immune system. These functions ensure survival of the parent virus. In some cases these evasive actions may benefit the outcome of the desired gene therapy (e.g., gene replacement) while in other cases their actions may thwart the desired outcome (e.g., cancer vaccination). Thus, removal of these gene products to make room for the genetic information to be delivered must be carefully considered. Additionally, loss of these natural mechanisms to evade immune components may lead to modified or to more complex responses at the cellular, tissue, and systemic level, following their introduction as a delivery vehicle.

Toll-like receptors associated with initial innate immune responses are of particular concern since they recognize the molecular pattern associated with the double-stranded RNA found in some viruses.[57] Nonviral delivery systems are not as readily observed by the immune system, although it is possible that some of the polymeric structures used could also have a molecular pattern recognized by some toll-like receptor. An individual's immune response to even a few viral particles can be quite striking,[58] but can be catastrophic when someone is exposed to a huge bolus of viral particles all at once; it is not uncommon to find viral doses as high as 10^{11} PFUs in a clinical trial protocol.[59] To date there has been one unfortunate death from a high dose of an adenovirus-based therapy,[60] and some severe combined immuno-deficient patients treated using a retroviral vector have developed a leukemia-like condition.[61] In the first case, the immune response against this viral vector was probably a factor involved in the clinical events that followed exposure. In the second, outcomes from this viral delivery system may not be recognized sufficiently for an individual to maintain growth regulation of transfected cells. Thus, the melding of aspects of viral delivery components with nonviral systems appears to represent a particularly promising approach for long-term success in gene delivery, but only with the caveat that utilization of some viral components may not only increase efficiency but may also bring along negative aspects such as increased immunogenicity or other potentially significant clinical issues.

2.3.4 Targeting Issues

Selection of a viral delivery vector because of its natural tropism optimizes the potential for delivered genetic information to target discrete tissues and organs specifically and selectively. For example, the fiber modification of adenovirus has been used to redirect this viral vector to cell surface receptors for folate,[62] epidermal growth factor (EGF), or fibroblast growth factor.[63] A substantial number of ligand-receptor interactions used by viruses to bind to target cells have now been elucidated (Table 2.3). Alteration of a natural tropism, however, can lead to unexpected cell and tissue targeting complexities for a viral vector. Introduction of a novel tropism into a viral vector is complicated and the tissue and organ delivery outcomes are often not completely anticipated in advance. One must

Table 2.3   Cell Surface Recognition Components for Potential Viral Gene Delivery Vehicles

Virus	Viral Component	Cell Component	Target Cells
Epstein-Barr	gp350, gp220	C3 complement receptor CR2 (CD21)	B lymphocytes
HIV-1	gp160	CD4 and chemokine coreceptors	Helper T lymphocytes
Rhinovirus	Virus proteins 1–3	Intercellular adhesion molecule-1	Epithelial cells
Rabies virus	G-protein	Acetylcholine receptor	Neurons
Influenza virus	Hemagglutinin	Sialic acid	Epithelial cells
Parvovirus (B19)		Erythrocyte P-antigen precursors	Erythroid cells
Adenovirus	Fiber protein	Cocksackievirus and adenovirus receptor	Epithelial cells
Reovirus	Sigma-1	Reovirus receptor (gp67?)	Intestinal cells

remember that attempting to isolate, standardize, or reproduce any one aspect of a viral vector, or trying to mix the targeting component of one virus with the packaging capacity of another, can result in the loss of much of the inherent capacity of the virus to evade the immune system and effectively deliver genetic information. A recent review of viral vectors for gene therapy has outlined some of these concerns.[64] Because safety must ultimately have precedent over efficacy, the manipulation of any virus so that it may be used as a gene therapy vector must be carefully considered.

Novel methods of targeting gene delivery to specific tissues and cells are constantly being explored. Attempts to emulate the actions of certain envelope viruses as gene therapy targeting and delivery systems have been examined by recreating these structures, known as virosomes.[65] Attenuated *Salmonella typhimurium* has been used as a gene delivery vector that targets tumors.[66] Others have explored the idea of using cells as carriers to chaperone gene therapy vectors to selected sites.[67] Filamentous phage, typically used for bacterial transformation, can be modified for the delivery of genetic information by the introduction of a surface-expressed cell-targeting ligand.[68,69] These approaches represent the types of alternative methods being explored for the targeted delivery of genetic material. The majority of methods that have been extensively examined for specific cell- and tissue-targeting gene therapy, however, involve either viruses or lipid-based nonviral vectors.

Targeting strategies with nonviral vectors commonly utilize an agent that selectively recognizes a cell-surface component. In many cases, this targeting agent is an antibody that can bind to an overexpressed antigen on the surface of a cancer cell[70,71] or uniquely distributed on selected cells within the body. In other instances, a two-step targeting–docking method can be used to target genetic information.[72] Following interaction with some cell surface component, the viral or nonviral vector must be internalized in order to deliver genetic information effectively. This is commonly achieved through internalization of the newly established receptor–ligand complex, involving the vector through a mechanism that entails receptor activation (commonly via a phosphorylation event). This commonly occurs during the internalization process of viral vectors. Retroviruses, such as human T-lymphocyte virus, can accelerate host cell division. This type of targeting, involving cell-surface receptor activation, can also be a component of nonviral vectors. An example of this using EGF to target cancer cells overexpressing EGF receptors has been described.[161] An important aspect of this targeting approach is that introduction of these vectors through such a targeting/uptake pathway would lead not only to entry of the vector, but to activation of the target cell.[8] An appreciation of outcomes associated with events resulting from such activation might want to be considered in relation to the desired outcome of the gene therapy.

2.4 TARGETING CELLS AND TISSUES

Cell-specific targeting is different from tissue-specific targeting in that the former identifies a singular cell type that may be within several organs or tissues of the body, while the latter may either select a cell type specific for a discrete organ or several cell types within that organ or tissue. Regional targeting to a tissue can be achieved by some expressed or functional aspect of that tissue with targeting strategies that utilize a localized protease activity to activate the targeting capability of some gene therapy vectors.[73] Oligonucleotides coupled to an asialoglycoprotein will target the liver through fairly selective interactions with hepatocytes,[74] while administration of lentiviral vectors results in transduction of both parenchymal and nonparenchymal cells of the liver.[75,76] Anionic liposomes and lactosylated low-density lipoproteins (LDL) have been successively used for the targeted delivery of antisense oligonucleotides to Kupffer cells.[77] In both cases, the liver was targeted and the outcomes, depending upon what was delivered, might be comparable. Which is more effective? The answer must typically be determined through experimentation. Since several factors may describe optimal outcome parameters, initial aspects such as efficiency and duration of expression may be overwhelmed by safety (toxicity) issues.

Clearly, the most efficient and selective method of cell or tissue targeting is an *ex vivo* transfer strategy. Here cells are removed, modified by transduction with a delivery vector, and returned to the host. The greatest delivery concern in this case typically involves matching the delivery vehicle with the isolated cells to be transduced. If there is sufficient specificity (tropism) in gene delivery using a viral vector, it may not be necessary to completely purify the individual cell type being targeted. If a pure population of the desired cells for targeting can be prepared, then it is quite feasible to use either viral or nonviral delivery systems to transduce cells that can be isolated, such as peripheral blood cells, stem cells, hepatocytes, skin fibroblasts, tumor cells, and tumor-infiltrating lymphocytes. Although this approach could lend itself to performing gene therapy on germ cells, the studies described to date have focused on somatic cells as targets (as discussed by Afione et al.[10]). Several studies have described isolation of a discrete population of cells from a patient's bone marrow that can be used for *ex vivo* introduction of the desired gene. This approach has been used for several hematological disorders including hemophilia, Franconi anemia, sickle cell disease, β-thalassemia, and chronic granulomatous disease.[78] Retroviral vectors have been particularly effective for this and have been used to introduce a number of genes, including the glucocerebrosidase that is defective in Gaucher disease,[79,80] the enzyme lacking in lysosomal acid lipase deficiency,[81] and the enzyme defective in globotriaosylceramide storage events associated with Fabry disease.[82] Not all of these gene therapy efforts have met with success.[80] Although retroviral delivery vehicles can be very effective, some recent studies have shown that these types of efficient viral delivery systems can retain a pathogenic nature and induce transformation events as well.[83]

Discrimination between cell-based and tissue-based targeting approaches can often be difficult; many examples of tissue-specific targeting for the delivery of genetic material are achieved through the restricted expression of cell-surface components. True targeting to tissues, based upon properties of that tissue, can be achieved by a topical application or some method that limits distribution of the gene therapy vector to a particular tissue or regional location. While most viral vectors have a discrete cellular tropism, rather than a generalized tissue targeting, some can be used in a tissue-selective manner. Information presented in Table 2.1 represents the targeting specificities of viruses based upon cellular tropisms under physiological conditions. Viral entry into cells, however, can be dramatically different from these observed tropisms when delivered at very high doses, or through a route of administration that differs from the normal route of infection used by these viruses.

2.5 TOPICAL DELIVERY APPROACHES

One of the most obvious places for topical gene therapy is the easiest to access: the skin. The barrier properties of the skin must, however, be overcome for a reasonable amount of any dose to act in a desired manner. Movement of materials into (through) the skin is difficult and occurs with only moderate efficiencies at best.[84] A number of methods have recently been identified to overcome the barrier properties of the skin for the delivery of structures such as large antisense oligonucleotides[85] and genes (as discussed by Touitou[86]). One approach in particular, that which establishes a defined series of minute openings, might be one of the most viable for gene therapy.

Since it is important to be able to define the absolute exposure to a gene therapy vector, movement of a set amount of the vector (either passively or assisted through some form of applied energy) may allow for such a controlled application across skin.[87] Introduction of minute defects in skin using a technique known as electroporation has also been used to perform skin-targeted gene delivery.[88] Other methods of producing minute defects in the skin's barrier have also been examined. Again, the critical factor here is establishing methods that provide defined dosing, something that is absolutely required to move any gene therapy effort into human studies.

Topical application of viral or nonviral delivery vectors provides one method of tissue targeting that can eliminate the need for complex targeting capabilities by the delivery vector. There are a variety of anatomical, biological, and physiological hurdles to overcome. Direct application to a mucosal surface in the respiratory, gastrointestinal, reproductive, or urinary tract could achieve a targeted delivery. Once at these mucosal surfaces, the delivery vector must effectively surmount the biological and anatomical barrier of entry into the cells lining that site. Virus particles have devised efficient strategies for uptake at these surfaces, and designers of nonviral systems have made strides to emulate these capabilities. For example, topical gene therapy of oral squamous cell carcinoma is logical, based upon its accessibility;[89] but this surface has a variety of enzymatic activities and is constantly flushed with saliva. Enhanced cell surface binding through interaction with specific receptors provides viruses with a mechanism to efficiently enter target cells. Cell-surface association of nonviral vectors can be similarly enhanced using adhering agents. One example of this is the use of lectins to increase the affinity of polymeric bioconjugates to the surface of colon cells.[90] Thus, adherence can dramatically improve transfection efficiency.

Alternately, maintenance of gene therapy vectors can be achieved by anatomical constraints. The lumen of the bladder, for example, can be readily reached via a catheter,[91] but there may be significant barriers to interaction of the gene delivery vector with the surface of the target cell. Mucus secreted onto a mucosal surface can be problematic for many topical gene therapy efforts. Adenoviral vectors have been used extensively for the delivery of genetic materials to the epithelial surfaces of the nose and lung.[59] These efforts have demonstrated the promise of this approach as well as its challenges.

Although this virus typically infects at this tissue surface, it would never be present at the high levels required for effective gene therapy. Nonviral delivery of genes into the lungs has also been examined, but these efforts have not provided particularly promising results, possibly because airway surface fluid components complicate this approach by binding and inactivating the transfection capacity of cationic polymers and lipid complexes.[92] Lung-specific gene delivery can be achieved by insufflation of cationic liposome-DNA complexes,[93] but is compromised by these vectors becoming trapped within the mucus that covers the surfaces of mucosal epithelia. Pilo-carpine administration along with a mucolytic agent appears effective at eradicating this mucus barrier and improving the prospects of topical gene transfer to epithelial cells.[94] Even removal of mucus may not be sufficient in some cases. For example, the critical site of the genetic defect in the inherited disease cystic fibrosis does not appear along the airway cells of the lung but rather in the submucosal glands (discussed by Wilson[58]); this is a particularly difficult site to reach from a topical application to the lung. Targeting cancer cells in the bladder is amenable to a topical application but may require disruption of the covering glycosaminoglycan layer prior to the administration of a viral gene delivery system.[91] In general, a variety of barriers must be overcome for successful topical delivery of gene therapy systems.

2.6 LOCAL INJECTION

One can readily envisage the application of gene therapy vectors to anatomically restricted yet accessible sites via injection. Cerebrospinal fluid circulates along the spinal cord and the surfaces (both internal and external) of the brain. Administration of a gene therapy vector into this compartment could be used to target tissue layers that cover the underlying nervous system components.[95] Gene delivery across the blood–brain barrier can be achieved using cationized albumin-coupled liposomes.[96] Intracerebroventricular delivery, for example, can deliver gene therapy directly to the brain.[97] Similarly, the synovial fluid that bathes the articulating surfaces of joints can be accessed directly using arthroscopic techniques. Intraperitoneal tumors can be targeted for the delivery of genetic information using cationic liposomes that are recognized by a population of folate receptors.[98] Introduction of a viral or nonviral delivery vector could result in a focused, effective gene therapy when directly injected into a particular regional site in the body (e.g., a tumor) or applied onto a specific tissue surface (a vascular bed or mucosal surface). These examples represent only several of many opportunities for topical, focused delivery of gene therapy systems to discrete cell populations of specific tissues.

Many viruses (e.g., adenovirus, adeno-associated virus, herpes simplex virus-1, vaccinia virus, Newcastle Disease virus, parvovirus, and reovirus) have been injected directly into tumors as a form of topical delivery of genetic information.[99] Again, these are administered at very high doses, although they are invariably altered to minimize their inherent characteristics as pathogens. Similarly, nonviral vectors can be very effective for the delivery of genetic material when directly injected. Recent studies suggest that both the complex size and the charge ratio will affect the efficiency of nonviral vectors delivered in this way.[100] This kind of direct delivery provides a high dose focused at a particular site, a factor that is frequently critical for the desired level of gene therapy. Substantial gene transfer can even be achieved by a direct injection of a DNA delivery system into a target site,[101] actually eliminating the need for an elaborate viral delivery system. A nonviral envelope vector system appears to increase the efficacy of gene transfer into the central nervous system following direct injection into the hypothalamus of the brain.[102] The extent of transfection can also be further enhanced by the use of electrically assisted uptake, as has been shown in muscle gene therapy.[103]

Genetic material may be delivered into many cell types but only activated or expressed in certain ones. Replicating viruses have been provisionally identified for a variety of therapeutic approaches.[104] Although this approach can achieve cell-specific outcomes, it is extremely inefficient. Intracellular activation of some dextran conjugate structures can be directed to many cells, but only activated primarily by macrophages.[105] Use of a hypoxic-responsive promoter can be used to limit the expression of a delivered gene only to a site of low oxygen tension.[106] Promoters required for the transcription of a delivered gene can be selected in a cell-specific fashion to achieve selectivity for a particular therapy. For example, an ovarian-specific promoter can be used to direct a suicide gene therapy to ovarian cancer cells.[107] One concern of such an approach is that this same promoter is typically functioning at some level, particularly during tissue repair, in the unaffected cells from that same tissue. One intriguing approach to dealing with such an issue is to use a vector that only functions in cancer cells, an adenovirus vector that selectively replicates in cancer cells, due to the lack of a functional gene that is usually present, and acts to restrain uncontrolled growth in untransformed cells.[108] Although such a cell-targeting system should allow for a highly selective delivery to cancer cells, a number of issues, including the required systemic dose and the potential for an undesirable immune event, have led to topical application by intra-tumor injection for the clinical use of such a replication-selective system.[109]

2.7 VASCULAR TARGETING

Intravenous injection of viral and nonviral vectors can be considered another method for cell and tissue targeting for gene therapy. It really just depends upon the distribution of sites that are effectively altered by the system being delivered. Initially, one must consider potential interactions between the delivery system and blood components. Pre-existing antibodies may immediately neutralize the efficacy of a virus-based delivery vehicle. There may be extensive interactions of nonviral particles, such as those containing cationic lipid–DNA complexes, with blood cells and serum proteins.[110–112] Modification of a cationic liposome-based nonviral vector with PEG prolongs its serum half-life, presumably by decreasing its interactions with serum proteins and blood cells.[113] Gene delivery distribution can be affected by physical aspects of the vasculature, with prominent capillary beds acting as sites of collection for particles. Cationic lipid–DNA complexes target the lung vasculature rather selectively.[114] Although others have not been able to replicate this observation,[93] the selective delivery of oligonucleotides to the pulmonary epithelium using lipid-based vectors has been supported.[115] Specific tissue targeting of liposome–DNA complexes to capillary beds, such as in the lung,[62] may be dictated by the size of the complex used.[116] Inflamed tissue vasculature has increased the uptake of structures less than one micron in diameter,[117] and these vascular beds show increased uptake of nonviral gene therapy vectors, a method of targeting such sites.[118] Additionally, vascular beds associated with inflamed tissues or tumors may have increased expression of phospholipase activities[119] that would potentially act to destabilize the liposome-based gene therapy vectors locally. All of these factors, and several others, affect the pharmacokinetic profiles of both viral and nonviral vectors.[120]

Targeting the vascular beds of discrete tissues would be very useful for some gene therapy applications. The use of inserted stents at specific vascular sites is one way to focus the delivery of genetic information.[121] Another way to do this would be by direct proximal administration, as in the case of an adeno-associated viral vector being injected into the subretinal space for gene delivery to vessels involved in ocular neovascularization.[106] Isolated mesangioblasts, a class of vessel-associated stem cells, have been transduced with a lentiviral vector and injected to establish capillary bed expression of a desired gene product.[122] Alternately, an antibody directed against the transferrin receptor has been shown to target materials to an endocytosis pathway potentially useful for delivery to the brain.[123] Phage display has been used to identify potential targets of tumor vasculature for delivery of genetic information. One effort in this area has identified the amino acid sequence arginine-glycine-aspartic acid (RGD) as a potential targeting tool.[124] It is important to emphasize that the desired outcome of each delivery must be matched with the vector being used; retroviruses require one round of replication of the host cell DNA prior to their integration and stable expression of a delivered gene, while others such as herpes simplex virus-1 might be better for targeting to post-mitotic cells such as neurons.[125] Similar concerns must be kept in mind for the specific targeting of viral vectors to prostate tissue.[126]

Tumor vasculature is clearly a desirable target for gene therapy systems. Vascular endothelial cells express a number of surface markers characteristic of their immature status during periods of rapid growth; however, the expression of components on the surfaces of tumor endothelial cells is heterogeneous.[127] The integrin $\alpha_v\beta_3$ expressed by tumor vasculature has been used to augment the targeting of nanoparticles that might be used to deliver a gene therapy.[128] A peptide containing the critical RGD amino sequence that binds $\alpha_v\beta_3$ coupled to polyethylene spacers can be used similarly to target gene delivery systems composed of the cationic polymer polyethyleneimine.[129] Similarly, RGD moieties coupled to gelatin modified with ethylenediamine, spermidine, and spermine can increase the targeting and efficiency of nonviral transfections.[130] Once at the tumor vascular beds, particles having a defined size range extravasate into the paravascular space with appreciable efficiency due to the increased permeability properties at these sites.[131] Particles taken up into tumors via this phenomenon typically penetrate only short distances into the tumors, due to their high internal pressure.[132,133] Cationic liposomes and microspheres, nanospheres, dendrimers, and polyethyleneimine have all been shown to effectively function in the particle-mediated delivery of genetic material to tumor vasculature.[134–136]

2.8 CELL-BASED TARGETING

Many studies have described selective targeting to particular organs or tissues. In reality these approaches commonly target to a cell type that is only present within that particular tissue or organ.

Liver targeting provides several examples. Hepatocytes are an excellent gene delivery target, having a variety of fairly specific uptake mechanisms that are used to clear certain materials from the blood.[137] Lacosylated oligonucleotide structures target directly to hepatocytes.[138] Liposome–DNA complexes coupled to the asialglycoprotein, asialofetuin, are similarly targeted to the liver,[116] and this may be further enhanced by the inclusion of protamine in the complex.[139] Galactosylated polyethylenimines coupled to DNA also target hepatocytes.[140,141] Are any of these methods of targeting specific cells of the liver superior to the other methods? For any particular gene and for any particular desired outcome, there is probably an optimal targeting and delivery method. However, it is rare that a researcher has the luxury to perform such head-to-head comparison studies, and so the answer to this question is unlikely ever to be determined. Instead, each approach is sequentially modified through trial and error to achieve a desired outcome and not compared to other methods. The initial concept of delivery to hepatocytes may (more or less) provide an ultimate delivery method to the liver.

The multiple methods of targeting the liver listed above are useful because they represent natural methods of identifying and binding structural elements as they pass by in the blood. Similarly, other cells have targeting mechanisms for the selective trafficking of cells at certain sites. Endothelial cells have unique surface components during rapid growth, such as during tumor growth (discussed above). However, endothelial cells must also express surface markers related to normal functions involving inflammation and regeneration. For example, cells of the immune system must be able to discriminate between inflamed and noninflamed tissues so that diapidesis of these cells can occur at discrete sites and only at times of active inflammation. Additionally, lymphocytes are constantly homing in to specific tissues for immune surveillance.[142] Intestinal homing of T-cell lymphocytes occurs through the integrin $\alpha_4\beta_7$ and CCR9, the receptor for a gut-associated chemokine,[143] could provide a paradigm for targeting both viral and nonviral vectors. Similarly, circulating stem cell populations must be able to select a particular tissue vascular bed to escape from to find their way to a required site.[144,145] Similar inflammation-related targeting markers should exist for blood–brain barrier endothelial cells because of the observed movement of immune cells into the brain in conditions such as multiple sclerosis. Recently, some progress has been made in identifying potential routes of transport of genetic therapy agents across these cells following targeting with these markers.[95] There appear to be a number of potential targeting opportunities that utilize unique surface properties of endothelial cells involved in stem cell and immune cell trafficking throughout the body.

Outside of the vascular bed, there are a number of possibilities for targeting specific cells through selective interactions with surface structures. Herpes simplex virus targets dorsal root ganglion cells and can be used for efficient gene delivery to this site.[146] Adeno-associated virus serotypes 1 and 6 have shown fairly selective targeting to muscle cells.[147] This cell-directed tropism can be acquired for nonviral delivery systems by incorporation of a similar targeting element. The synthetic peptides that bind to integrin $\alpha_5\beta_1$ can be used to target polycationic DNA complexes to neuroblastoma cells.[100] It should be remembered that it is not always necessary to target the exact cell type that expresses a defective gene in order to obtain the desired outcome; in fact, in some cases it may not be possible. A number of studies have shown that low levels of transfection can be achieved in isolated human pancreatic islets using cationic lipid and polymer-based gene delivery systems.[148] However, in the case of type I diabetes, the pancreatic islet β-cells of a patient may be completely obliterated prior to gene therapy. Other cell types, such as fibroblasts, have been found to be potentially useful for somatic gene therapy and subsequently responsive to alterations in blood glucose levels.[149] Gene therapy for cystic fibrosis (CF) can also be compromised by extensive tissue damage in the lung.[52] Ideally, the secretory glands where the greatest amount of the gene product defective in CF is normally expressed would be the target for these patients (as discussed by Wilson[58]). However, cells of these secretory glands can become damaged during the progression of lung disease and those that are intact are very difficult to reach with a topical gene therapy delivery to the airway. A viral vector, human respiratory syncytial virus, which selectively infects nearby ciliated airway epithelial cells, might still provide clinical benefit.[150] In the same way, muscle myoblasts can be transduced *in vitro*, and, after reintroduction, incorporated into normal muscle, so that they express an introduced gene product.[151] In this case — highlighting the potential use of cells to act as targeting vehicles — the transduced cell itself is acting as the targeting vehicle for gene delivery.

Transport of gene therapy systems into the brain is extremely challenging due to the nearly absolute properties of the blood–brain barrier. In general, viral gene therapy systems can be used, but these are difficult in that any immunological events associated with such a delivery could cause such severe clinical concerns as brain swelling. Nonviral systems have been studied that might provide a solution. Once in the brain, adeno-associated virus-2 has been shown to target mainly neurons, while a polyethylenimine-based nonviral vector targeted both neurons and glial cells.[152] To achieve a reasonable selective delivery to the brain, nonviral vectors require several elements, each one designed to overcome a barrier associated with this delivery that viral systems have mastered. For example, pegylated liposomes of 85 nm that have been decorated with an antibody that recognizes the insulin receptor can be used to target cancer cells in the brain.[153,154] Production of such a delivery system is a difficult engineering feat.

Targeting tumor cells has become one of the most intensively studied applications for the delivery of gene therapy vectors. For example, adenoviruses, complexed with a CD40 targeting molecule, can be directed to ovarian cancer cells expressing CD40 in a manner that results in the efficient expression of a delivered gene.[155] Synthetic peptides prepared from sequences of the immunoglobulin (Ig) superfamily cell adhesion molecule from peripheral nerve myelin P(0) protein and leukocyte function-associated antigen-1 have been used to target liposomes to melanoma cells expressing intercellular adhesion molecule-1.[156] This is interesting, in that most cancer cells do not express a large amount of the native receptor for adenovirus after they have lost their normal cell–cell contact interactions. Transferrin receptors expressed at the surface of endothelial cells can also be used to target and enhance the uptake of liposomes useful for gene therapy.[157,158] Similarly, folic acid coupled to cationic lipid–DNA complexes through polyethylene glycol has been used to target the folate receptors that are enriched in tumor vasculature.[62]

2.9 CONCLUSIONS

Gene therapy has tremendous promise for a broad spectrum of clinical applications. To date, however, there has been only modest progress toward this goal. Using the words of others, there are only three problems that limit gene therapy applications: delivery, delivery, and delivery.[159] At present a wide range of viral, nonviral, and novel methods are being explored for the delivery of materials that can manipulate the activity of genes within a target cell. The materials being delivered represent not only entire genes but also nucleic acid-based materials that can act to modulate gene activity. Thus, the opportunities for delivering a variety of gene modifying agents has opened up the possibilities of cell and tissue targeting dramatically since the type of materials being delivered can be varied. However, despite our best efforts to harness or emulate viruses in their capacity to target selective cells and control cellular functions, the ability to target cells and tissues for gene therapy effectively is moderate at best.

Human immunodeficiency virus-1 is well known for its ability to target and kill T-lymphocyte populations required for a competent immune response. Lassa virus and Ebola virus have recently been shown to impair dendritic cell function as a potential mechanism of their associated pathologies.[160] These types of cellular targeting, despite their unfortunate outcomes, are perhaps the optimal examples of targeting that could be hoped for in a gene therapy approach. These viruses target a selective cell type and efficiently transform it into a cell performing tasks determined by the delivered genome, rather than its own. Can viral or nonviral delivery vectors ever hope to be able to emulate this kind of efficiency for therapeutic gene modulation? The optimistic answer is

"Of course!" — but the realistic answer will be given more slowly and with much less certainty. Although it is highly likely that selective cell and tissue gene delivery will be possible in the future, examples of these successes will likely come after many miscues and will probably not provide generalized approaches but rather highly selective systems. Safety will always be the gatekeeper in these efforts, as it should be, and efforts to emulate viral vector capabilities with nonviral vectors will become more of a reality as our understanding of viral functions increases.

REFERENCES

1. Watson, J. D. and Crick, F. H. C., Molecular structure of nucleic acids: a structure for deoxyribose nucleic acid, *Nature* 171 (4356), 737–738, 1953.
2. Blau, H. and Khavari, P., Gene therapy: progress, problems, prospects, *Nat. Med.* 3 (6), 612–613, 1997.
3. Morgan, R. A. and Anderson, W. F., Human gene therapy, *Annu. Rev. Biochem.* 62, 191–217, 1993.
4. Hughes, M. D., Hussain, M., Nawaz, Q., Sayyed, P., and Akhtar, S, The cellular delivery of antisense oligonucleotides and ribozymes, *Drug Discov. Today* 6 (6), 303–315, 2001.
5. Scanlon, K. J., Jiao, L., Funato, T., Wang, W., Tone, T., Rossi, J. J., and Kashani-Sabet, M., Ribozyme-mediated cleavage of c-fos mRNA reduces gene expression of DNA synthesis enzymes and metal-lothionein, *Proc. Nat. Acad. Sci. U.S.A.* 88 (23), 10591–10595, 1991.
6. Shuey, D. J., McCallus, D. E., and Giordano, T., RNAi: gene-silencing in therapeutic intervention, *Drug Discov. Today* 7 (20), 1040–1046, 2002.
7. Rogers, C. S., Vanoye, C. G., Sullenger, B. A., and George Jr., A. L., Functional repair of a mutant chloride channel using a *trans*-splicing ribozyme, *J. Clin. Invest.* 110 (12), 1783–1789, 2002.
8. Rockwell, P., O'Connor, W. J., King, K., Goldstein, N. I., Zhang, L. M., and Stein, C. A., Cell-surface perturbations of the epidermal growth factor and vascular endothelial growth factor receptors by phosphorothioate oligodeoxynucleotides, *Proc. Nat. Acad. Sci. U.S.A.* 94 (12), 6523–6528, 1997.
9. Stull, R. A. and Szoka, F. C., Jr., Antigene, ribozyme and aptamer nucleic acid drugs: progress and prospects, *Pharm. Res.* 12 (4), 465–483, 1995.
10. Afione, S. A., Conrad, C. K., and Flotte, T. R., Gene therapy vectors as drug delivery systems, *Clin. Pharmacokinet.* 28 (3), 181–189, 1995.
11. Tavitian, B., Terrazzino, S., Kuhnast, B., Marzabal, S., Stettler, O., Dolle, F., Deverre, J. R., Jobert, A., Hinnen, F., Bendriem, B., Crouzel, C., and Di Giamberardino, L., *In vivo* imaging of oligonucle-otides with positron emission tomography, *Nat. Med.* 4 (4), 467–471, 1998.
12. Tavitian, B., In vivo antisense imaging, *Q. J. Nucl. Med.* 44 (3), 236–255, 2000.
13. Tavitian, B., Marzabal, S., Boutet, V., Kuhnast, B., Terrazzino, S., Moynier, M., Dolle, F., Deverre, J. R., and Thierry, A. R., Characterization of a synthetic anionic vector for oligonucleotide delivery using *in vivo* whole body dynamic imaging, *Pharm. Res.* 19 (4), 367–376., 2002.
14. Gharwan, H., Wightman, L., Kircheis, R., Wagner, E., and Zatloukal, K., Nonviral gene transfer into fetal mouse livers (a comparison between the cationic polymer PEI and naked DNA), *Gene Ther.* 10 (9), 810–817, 2003.
15. Roest Crollius, H., Jaillon, O., Bernot, A., Dasilva, C., Bouneau, L., Fischer, C., Fizames, C., Wincker, P., Brottier, P., Quetier, F., Saurin, W., and Weissenbach, J., Estimate of human gene number provided by genome-wide analysis using Tetraodon nigroviridis DNA sequence, *Nat. Genet.* 25 (2), 235–238, 2000.
16. Jeffery, C. J., Multifunctional proteins: examples of gene sharing, *Ann. Med.* 35 (1), 28–35, 2003.
17. Lim, L. P., Glasner, M. E., Yekta, S., Burge, C. B., and Bartel, D. P., Vertebrate microRNA genes, *Science* 299 (5612), 1540, 2003.
18. Hu, W. S., Rhodes, T., Dang, Q., and Pathak, V., Retroviral recombination: review of genetic analyses, *Front. Biosci.* 8, D143–155, 2003.
19. Hu, W. S. and Pathak, V. K., Design of retroviral vectors and helper cells for gene therapy, *Pharmacol. Rev.* 52 (4), 493–511, 2000.
20. Wang, X., Huong, S. M., Chiu, M. L., Raab-Traub, N., and Huang, E. S., Epidermal growth factor receptor is a cellular receptor for human cytomegalovirus, *Nature* 424 (6947), 456–461, 2003.

21. Kobinger, G. P., Weiner, D. J., Yu, Q. C., and Wilson, J. M., Filovirus-pseudotyped lentiviral vector can efficiently and stably transduce airway epithelia *in vivo*, *Nat. Biotechnol.* 19 (3), 225–230, 2001.

22. Chan, S. Y., Empig, C. J., Welte, F. J., Speck, R. F., Schmaljohn, A., Kreisberg, J. F., and Goldsmith, M. A., Folate receptor-alpha is a cofactor for cellular entry by Marburg and Ebola viruses, *Cell* 106 (1), 117–126, 2001.

23. Sinn, P. L., Hickey, M. A., Staber, P. D., Dylla, D. E., Jeffers, S. A., Davidson, B. L., Sanders, D. A., and McCray, P. B., Jr., Lentivirus vectors pseudotyped with filoviral envelope glycoproteins transduce airway epithelia from the apical surface independently of folate receptor alpha, *J. Virol.* 77 (10), 5902–5910, 2003.

24. Ahmed, A., Thompson, J., Emiliusen, L., Murphy, S., Beauchamp, R. D., Suzuki, K., Alemany, R., Harrington, K., and Vile, R. G., A conditionally replicating adenovirus targeted to tumor cells through activated RAS/P-MAPK-selective mRNA stabilization, *Nat. Biotechnol.* 21 (7), 771–777, 2003.

25. Cohen, C. J., Gaetz, J., Ohman, T., and Bergelson, J. M., Multiple regions within the coxsackievirus and adenovirus receptor cytoplasmic domain are required for basolateral sorting, *J. Biol. Chem.* 276 (27), 25392–25398, 2001.

26. Marini, F. C., 3rd, Yu, Q., Wickham, T., Kovesdi, I., and Andreeff, M., Adenovirus as a gene therapy vector for hematopoietic cells, *Cancer Gene Ther.* 7 (6), 816–825, 2000.

27. Modis, Y., Ogata, S., Clements, D., and Harrison, S. C., A ligand-binding pocket in the dengue virus envelope glycoprotein, *Proc. Nat. Acad. Sci. U.S.A.* 100 (12), 6986–6991, 2003.

28. Ferrari, N., Glod, J., Lee, J., Kobiler, D., and Fine, H. A., Bone marrow-derived, endothelial progenitor-like cells as angiogenesis-selective gene-targeting vectors, *Gene Ther.* 10 (8), 647–656, 2003.

29. Ferber, D., Gene therapy: safer and virus-free? *Science* 294 (5547), 1638–1642, 2001.

30. Kohn, D. B., Sadelain, M., and Glorioso, J. C., Occurrence of leukaemia following gene therapy of X-linked SCID, *Nat. Rev. Cancer* 3 (7), 477–488, 2003.

31. Liu, F. and Huang, L., Development of non-viral vectors for systemic gene delivery, *J. Control Release* 78 (1–3), 259–266, 2002.

32. Uchida, E., Mizuguchi, H., Ishii-Watabe, A., and Hayakawa, T., Comparison of the efficiency and safety of non-viral vector-mediated gene transfer into a wide range of human cells, *Biol. Pharm. Bull.* 25 (7), 891–897, 2002.

33. Benigni, A., Tomasoni, S., and Remuzzi, G., Impediments to successful gene transfer to the kidney in the context of transplantation and how to overcome them, *Kidney Int. Suppl.* 61, Supplement 1, 115–119, 2002.

34. Astriab-Fisher, A., Sergueev, D., Fisher, M., Shaw, B. R., and Juliano, R. L., Conjugates of antisense oligonucleotides with the Tat and antennapedia cell-penetrating peptides: effects on cellular uptake, binding to target sequences, and biologic actions, *Pharm. Res.* 19 (6), 744–754, 2002.

35. Richard, J. P., Melikov, K., Vives, E., Ramos, C., Verbeure, B., Gait, M. J., Chernomordik, L. V., and Lebleu, Bl., Cell-penetrating peptides: a reevaluation of the mechanism of cellular uptake, *J. Biol. Chem.* 278 (1), 585–590, 2003.

36. Gratton, J. P., Yu, J., Griffith, J. W., Babbitt, R. W., Scotland, R. S., Hickey, R., Giordano, F. J., and Sessa, W. C., Cell-permeable peptides improve cellular uptake and therapeutic gene delivery of replication-deficient viruses in cells and *in vivo*, *Nat. Med.* 9 (3), 357–362, 2003.

37. Mastrobattista, E., Koning, G. A., van Bloois, L., Filipe, A. C., Jiskoot, W., and Storm, G., Functional characterization of an endosome-disruptive peptide and its application in cytosolic delivery of immunoliposome-entrapped proteins, *J. Biol. Chem.* 277 (30), 27135–27143, 2002.

38. Lechardeur, D. and Lukacs, G. L., Intracellular barriers to non-viral gene transfer, *Curr. Gene Ther.* 2 (2), 183–194, 2002.

39. Costin, G. E., Trif, M., Nichita, N., Dwek, R. A., and Petrescu, S. M., pH-sensitive liposomes are efficient carriers for endoplasmic reticulum-targeted drugs in mouse melanoma cells, *Biochem. Biophys. Res. Commun.* 293 (3), 918–923, 2002.

40. Murthy, N., Campbell, J., Fausto, N., Hoffman, A. S., and Stayton, P. S., Design and synthesis of pH-responsive polymeric carriers that target uptake and enhance the intracellular delivery of oligonucleotides, *J. Control Release* 89 (3), 365–374, 2003.

41. Zhang, X., Sawyer, G. J., Dong, X., Qiu, Y., Collins, L., and Fabre, J. W., The *in vivo* use of chloroquine to promote non-viral gene delivery to the liver via the portal vein and bile duct, *J. Gene Med.* 5 (3), 209–218, 2003.

42. Asokan, A. and Cho, M. J., Exploitation of intracellular pH gradients in the cellular delivery of macromolecules, *J. Pharm. Sci.* 91 (4), 903–913, 2002.

43. Duncan, R., Gac-Breton, S., Keane, R., Musila, R., Sat, Y. N., Satchi, R., and Searle, F., Polymer-drug conjugates, PDEPT and PELT: basic principles for design and transfer from the laboratory to clinic, *J. Control Release* 74 (1–3), 135–146, 2001.

44. Mostov, K., Su, T., and ter Beest, M., Polarized epithelial membrane traffic: conservation and plasticity, *Nat. Cell Biol.* 5 (4), 287–293, 2003.

45. Brokx, R. D., Bisland, S. K., and Gariepy, J., Designing peptide-based scaffolds as drug delivery vehicles, *J. Control Release* 78 (1–3), 115–123, 2002.

46. Keller, G.-A., Li, W., and Mrsny, R. J., Targeting macromolecular therapeutics to specific cell organelles, in *Controlled Drug Delivery: Designing Technologies for the Future*, ed. K. Park and R. J. Mrsny, American Chemical Society, Washington, D.C., 2000, 168–183.

47. Moghimi, S. M. and Rajabi-Siahboomi, A. R., Recent advances in cellular, sub-cellular and molecular targeting, *Adv. Drug Deliv. Rev.* 41 (2), 129–133, 2000.

48. Johnson-Saliba, M. and Jans, D. A., Gene therapy: optimising DNA delivery to the nucleus, *Curr. Drug Targets* 2 (4), 371–399, 2001.

49. Brisson, M. and Huang, L., Liposomes: conquering the nuclear barrier, *Curr. Opin. Mol. Ther.* 1 (2), 140–146, 1999.

50. Zhang, X., Collins, L., and Fabre, J. W., A powerful cooperative interaction between a fusogenic peptide and lipofectamine for the enhancement of receptor-targeted, non-viral gene delivery via integrin receptors, *J. Gene Med.* 3 (6), 560–568, 2001.

51. Zhang, X., Collins, L., Sawyer, G. J., Dong, X., Qiu, Y., and Fabre, J. W., *In vivo* gene delivery via portal vein and bile duct to individual lobes of the rat liver using a polylysine-based nonviral DNA vector in combination with chloroquine, *Hum. Gene Ther.* 12 (18), 2179–2190, 2001.

52. Ferrari, S., Griesenbach, U., Geddes, D. M., and Alton, E., Immunological hurdles to lung gene therapy, *Clin. Exp. Immunol.* 132 (1), 1–8, 2003.

53. McConkey, S. J., Reece, W. H., Moorthy, V. S., Webster, D., Dunachie, S., Butcher, G., Vuola, J. M., Blanchard, T. J., Gothard, P., Watkins, K., Hannan, C. M., Everaere, S., Brown, K., Kester, K. E., Cummings, J., Williams, J., Heppner, D. G., Pathan, A., Flanagan, K., Arulanantham, N., Roberts, M. T., Roy, M., Smith, G. L., Schneider, J., Peto, T., Sinden, R. E., Gilbert, S. C., and Hill, A. V., Enhanced T-cell immunogenicity of plasmid DNA vaccines boosted by recombinant modified vaccinia virus Ankara in humans, *Nat. Med.* 9 (6), 729–735, 2003.

54. Hauck, B., Chen, L., and Xiao, W., Generation and characterization of chimeric recombinant AAV vectors, *Mol. Ther.* 7 (3), 419–425, 2003.

55. Mangeat, B., Turelli, P., Caron, G., Friedli, M., Perrin, L., and Trono, D., Broad antiretroviral defence by human APOBEC3G through lethal editing of nascent reverse transcripts, *Nature* 424 (6944), 99–103, 2003.

56. Wickham, T. J., Targeting adenovirus, *Gene Ther.* 7 (2), 110–114, 2000.

57. Beutler, B. and Rietschel, E. T., Innate immune sensing and its roots: the story of endotoxin, *Nat. Rev. Immunol.* 3 (2), 169–176, 2003.

58. Wilson, J. M., Gene therapy for cystic fibrosis: challenges and future directions, *J. Clin. Invest.* 96 (6), 2547–2554, 1995.

59. Boucher, R. C., Knowles, M. R., Johnson, L. G., Olsen, J. C., Pickles, R., Wilson, J. M., Engelhardt, J., Yang, Y., and Grossman, M., Gene therapy for cystic fibrosis using E1-deleted adenovirus: a phase I trial in the nasal cavity, *Hum. Gene Ther.* 5 (5), 615–639, 1994.

60. Check, E., A tragic setback, *Nature* 420 (6912), 116–118, 2002.

61. Dettweiler, U. and Simon, P., Points to consider for ethics committees in human gene therapy trials, *Bioethics* 15 (5–6), 491–500, 2001.

62. Hofland, H. E., Masson, C., Iginla, S., Osetinsky, I., Reddy, J. A., Leamon, C. P., Scherman, D., Bessodes, M., and Wils, P., Folate-targeted gene transfer *in vivo*, *Mol. Ther.* 5 (6), 739–744, 2002.

63. Curiel, D. T., Strategies to adapt adenoviral vectors for targeted delivery, *Ann. N.Y. Acad. Sci.* 886, 158–171, 1999.

64. Lundstrom, K., Latest development in viral vectors for gene therapy, *Trends Biotechnol.* 21 (3), 117–122, 2003.

65. Sarkar, D. P., Ramani, K., and Tyagi, S. K., Targeted gene delivery by virosomes, *Methods Mol. Biol.* 199, 163–173, 2002.

66. Mei, S., Theys, J., Landuyt, W., Anne, J., and Lambin, P., Optimization of tumor-targeted gene delivery by engineered attenuated Salmonella typhimurium, *Anticancer Res.* 22 (6A), 3261–3266, 2002.

67. Harrington, K., Alvarez-Vallina, L., Crittenden, M., Gough, M., Chong, H., Diaz, R. M., Vassaux, G., Lemoine, N., and Vile, R., Cells as vehicles for cancer gene therapy: the missing link between targeted vectors and systemic delivery? *Hum. Gene Ther.* 13 (11), 1263–1280, 2002.

68. Larocca, D., Jensen-Pergakes, K., Burg, M. A., and Baird, A., Gene transfer using targeted filamentous bacteriophage, *Methods Mol. Biol.* 185, 393–401, 2002.

69. Larocca, D., Burg, M. A., Jensen-Pergakes, K., Ravey, E. P., Gonzalez, A. M., and Baird, A., Evolving phage vectors for cell targeted gene delivery, *Curr. Pharm. Biotechnol.* 3 (1), 45–57, 2002.

70. Gonzalez, D. E., Covitz, K. M., Sadee, W., and Mrsny, R. J., An oligopeptide transporter is expressed at high levels in the pancreatic carcinoma cell lines AsPc-1 and Capan-2, *Cancer Res.* 58 (3), 519–525, 1998.

71. Mrsny, R. J., Oligopeptide transporters as putative therapeutic targets for cancer cells, *Pharm. Res.* 15 (6), 816–818, 1998.

72. Backer, M. V., Gaynutdinov, T. I., Gorshkova, II, Crouch, R. J., Hu, T., Aloise, R., Arab, M., Przekop, K., and Backer, J. M., Humanized docking system for assembly of targeting drug delivery complexes, *J. Control Release* 89 (3), 499–511, 2003.

73. Peng, K. W. and Russell, S. J., Viral vector targeting, *Curr. Opin. Biotechnol.* 10 (5), 454–457, 1999.

74. Lu, X. M., Fischman, A. J., Jyawook, S. L., Hendricks, K., Tompkins, R. G., and Yarmush, M. L., Antisense DNA delivery *in vivo*: liver targeting by receptor-mediated uptake, *J. Nucl. Med.* 35 (2), 269–275, 1994.

75. Follenzi, A., Sabatino, G., Lombardo, A., Boccaccio, C., and Naldini, L., Efficient gene delivery and targeted expression to hepatocytes *in vivo* by improved lentiviral vectors, *Hum. Gene Ther.* 13 (2), 243–260, 2002.

76. Follenzi, A. and Naldini, L., Generation of HIV-1 derived lentiviral vectors, *Methods Enzymol.* 346, 454–465, 2002.

77. Ponnappa, B. C. and Israel, Y., Targeting Kupffer cells with antisense oligonucleotides, *Front. Biosci.* 7, e223–233, 2002.

78. Herzog, R. W. and Hagstrom, J. N., Gene therapy for hereditary hematological disorders, *Am. J. Pharmacogenomics* 1 (2), 137–144, 2001.

79. Nolta, J. A., Yu, X. J., Bahner, I., and Kohn, D. B., Retroviral-mediated transfer of the human glucocerebrosidase gene into cultured Gaucher bone marrow, *J. Clin. Invest.* 90 (2), 342–348, 1992.

80. de Fost, M., Aerts, J. M., and Hollak, C. E., Gaucher disease: from fundamental research to effective therapeutic interventions, *Neth. J. Med.* 61 (1), 3–8, 2003.

81. Du, H., Heur, M., Witte, D. P., Ameis, D., and Grabowski, G. A., Lysosomal acid lipase deficiency: correction of lipid storage by adenovirus-mediated gene transfer in mice, *Hum. Gene Ther.* 13 (11), 1361–1372, 2002.

82. Park, J., Murray, G. J., Limaye, A., Quirk, J. M., Gelderman, M. P., Brady, R. O., and Qasba, P., Long-term correction of globotriaosylceramide storage in Fabry mice by recombinant adeno-associated virus-mediated gene transfer, *Proc. Nat. Acad. Sci. U.S.A.* 100 (6), 3450–3454, 2003.

83. Hacein-Bey-Abina, S., von Kalle, C., Schmidt, M., Le Deist, F., Wulffraat, N., McIntyre, E., Radford, I., Villeval, J. L., Fraser, C. C., Cavazzana-Calvo, M., and Fischer, A., A serious adverse event after successful gene therapy for X-linked severe combined immunodeficiency, *N. Engl. J. Med.* 348 (3), 255–256, 2003.

84. Green, P. G., Hinz, R. S., Kim, A., Szoka, F. C., Jr., and Guy, R. H., Iontophoretic delivery of a series of tripeptides across the skin *in vitro*, *Pharm. Res.* 8 (9), 1121–1127, 1991.

85. Brand, R. M., Topical and transdermal delivery of antisense oligonucleotides, *Curr. Opin. Mol. Ther.* 3 (3), 244–248, 2001.

86. Touitou, E., Drug delivery across the skin, *Expert Opin. Biol. Ther.* 2 (7), 723–733, 2002.

87. Mikszta, J. A., Alarcon, J. B., Brittingham, J. M., Sutter, D. E., Pettis, R. J., and Harvey, N. G., Improved genetic immunization via micromechanical disruption of skin- barrier function and targeted epidermal delivery, *Nat. Med.* 8 (4), 415–419, 2002.

88. Maruyama, H., Ataka, K., Higuchi, N., Sakamoto, F., Gejyo, F., and Miyazaki, J., Skin-targeted gene transfer using *in vivo* electroporation, *Gene Ther.* 8 (23), 1808–1812, 2001.

89. Xi, S. and Grandis, J. R., Gene therapy for the treatment of oral squamous cell carcinoma, *J. Dent. Res.* 82 (1), 11–16, 2003.

90. Lu, Z. R., Shiah, J. G., Sakuma, S., Kopeckova, P., and Kopecek, J., Design of novel bioconjugates for targeted drug delivery, *J. Control Release* 78 (1–3), 165–173, 2002.

91. Irie, A., Advances in gene therapy for bladder cancer, *Curr. Gene Ther.* 3 (1), 1–11, 2003.

92. Rosenecker, J., Naundorf, S., Gersting, S. W., Hauck, R. W., Gessner, A., Nicklaus, P., Muller, R. H., and Rudolph, C., Interaction of bronchoalveolar lavage fluid with polyplexes and lipoplexes: analysing the role of proteins and glycoproteins, *J. Gene Med.* 5 (1), 49–60, 2003.

93. Tsan, M. F., White, J. E., and Shepard, B., Lung-specific direct *in vivo* gene transfer with recombinant plasmid DNA, *Am. J. Physiol.* 268 (6 Pt. 1), L1052–1056, 1995.

94. Sandberg, J. W., Lau, C., Jacomino, M., Finegold, M., and Henning, S. J., Improving access to intestinal stem cells as a step toward intestinal gene transfer, *Hum. Gene Ther.* 5 (3), 323–329, 1994.

95. Pardridge, W. M., Drug and gene delivery to the brain: the vascular route, *Neuron* 36 (4), 555–558, 2002.

96. Thole, M., Nobmanna, S., Huwyler, J., Bartmann, A., and Fricker, G., Uptake of cationzied albumin coupled liposomes by cultured porcine brain microvessel endothelial cells and intact brain capillaries, *J. Drug Target.* 10 (4), 337–344, 2002.

97. Chauhan, N. B., Trafficking of intracerebroventricularly injected antisense oligonucleotides in the mouse brain, *Antisense Nucleic Acid Drug Dev.* 12 (5), 353–357, 2002.

98. Reddy, J. A., Abburi, C., Hofland, H., Howard, S. J., Vlahov, I., Wils, P., and Leamon, C. P., Folate-targeted, cationic liposome-mediated gene transfer into disseminated peritoneal tumors, *Gene Ther.* 9 (22), 1542–1550, 2002.

99. Kirn, D., Martuza, R. L., and Zwiebel, J., Replication-selective virotherapy for cancer: Biological principles, risk management and future directions, *Nat. Med.* 7 (7), 781–787, 2001.

100. Lee, L. K., Siapati, E. K., Jenkins, R. G., McAnulty, R. J., Hart, S. L., and Shamlou, P. A., Biophysical characterization of an integrin-targeted non-viral vector, *Med. Sci. Monit.* 9 (1), BR54–61, 2003.

101. Wolff, J. A., Malone, R. W., Williams, P., Chong, W., Acsadi, G., Jani, A., and Felgner, P. L., Direct gene transfer into mouse muscle *in vivo*, *Science* 247 (4949 Pt. 1), 1465–1468, 1990.

102. Shimamura, M., Morishita, R., Endoh, M., Oshima, K., Aoki, M., Waguri, S., Uchiyama, Y., and Kaneda, Y., HVJ-envelope vector for gene transfer into central nervous system, *Biochem. Biophys. Res. Commun.* 300 (2), 464–471, 2003.

103. Tomanin, R., Friso, A., Alba, S., Piller Puicher, F., Mennuni, C., La Monica, N., Hortelano, G., Zacchello, F., and Scarpa, M., Non-viral transfer approaches for the gene therapy of mucopolysaccharidosis type II (Hunter syndrome), *Acta Paediatr. Suppl.* 91 (439), 100–104, 2002.

104. Galanis, E., Vile, R., and Russell, S. J., Delivery systems intended for *in vivo* gene therapy of cancer: targeting and replication competent viral vectors, *Crit. Rev. Oncol. Hematol.* 38 (3), 177–192, 2001.

105. Harada, M., Imai, J., Okuno, S., and Suzuki, T., Macrophage-mediated activation of camptothecin analogue T-2513- carboxymethyl dextran conjugate (T-0128): possible cellular mechanism for anti-tumor activity, *J. Control Release* 69 (3), 389–397, 2000.

106. Mistry, A., Thrasher, A., and Ali, R., Gene therapy for ocular angiogenesis, *Clin. Sci.* 5, 5, 2003.

107. Bao, R., Selvakumaran, M., and Hamilton, T. C., Targeted gene therapy of ovarian cancer using an ovarian-specific promoter, *Gynecol. Oncol.* 84 (2), 228–234, 2002.

108. Vile, R., Ando, D., and Kirn, D., The oncolytic virotherapy treatment platform for cancer: unique biological and biosafety points to consider, *Cancer Gene Ther.* 9 (12), 1062–1067, 2002.

109. Hecht, J. R., Bedford, R., Abbruzzese, J. L., Lahoti, S., Reid, T. R., Soetikno, R. M., Kirn, D. H., and Freeman, S. M., A phase I/II trial of intratumoral endoscopic ultrasound injection of ONYX-015 with intravenous gemcitabine in unresectable pancreatic carcinoma, *Clin. Cancer Res.* 9 (2), 555–561, 2003.

110. Opanasopit, P., Nishikawa, M., and Hashida, M., Factors affecting drug and gene delivery: effects of interaction with blood components, *Crit. Rev. Ther. Drug Carrier Syst.* 19 (3), 191–233, 2002.

111. Opanasopit, P., Hyoudou, K., Nishikawa, M., Yamashita, F., and Hashida, M., Serum mannan binding protein inhibits mannosylated liposome-mediated transfection to macrophages, *Biochim. Biophys. Acta* 1570 (3), 203–209, 2002.

112. Opanasopit, P., Sakai, M., Nishikawa, M., Kawakami, S., Yamashita, F., and Hashida, M., Inhibition of liver metastasis by targeting of immunomodulators using mannosylated liposome carriers, *J. Control Release* 80 (1–3), 283–294, 2002.

113. Kim, I. S. and Kim, S. H., Development of polymeric nanoparticulate drug delivery systems: evaluation of nanoparticles based on biotinylated poly(ethylene glycol) with sugar moiety, *Int. J. Pharm.* 257 (1–2), 195–203, 2003.

114. Zhu, N., Liggitt, D., Liu, Y., and Debs, R., Systemic gene expression after intravenous DNA delivery into adult mice, *Science* 261 (5118), 209–211, 1993.

115. Ma, Z., Zhang, J., Alber, S., Dileo, J., Negishi, Y., Stolz, D., Watkins, S., Huang, L., Pitt, B., and Li, S., Lipid-mediated delivery of oligonucleotide to pulmonary endothelium, *Am. J. Respir. Cell. Mol. Biol.* 27 (2), 151–159, 2002.

116. Templeton, N. S., Lasic, D. D., Frederik, P. M., Strey, H. H., Roberts, D. D., and Pavlakis, G. N., Improved DNA: liposome complexes for increased systemic delivery and gene expression, *Nat. Biotechnol.* 15 (7), 647–652, 1997.

117. Davis, S. S., Biomedical applications of nanotechnology—implications for drug targeting and gene therapy, *Trends Biotechnol.* 15 (6), 217–224, 1997.

118. Awasthi, V. D., Goins, B., Klipper, R., and Phillips, W. T., Accumulation of PEG-liposomes in the inflamed colon of rats: potential for therapeutic and diagnostic targeting of inflammatory bowel diseases, *J. Drug Target.* 10 (5), 419–427, 2002.

119. Davidsen, J., Jorgensen, K., Andresen, T. L., and Mouritsen, O. G., Secreted phospholipase A(2) as a new enzymatic trigger mechanism for localised liposomal drug release and absorption in diseased tissue, *Biochim. Biophys. Acta* 1609 (1), 95–101, 2003.

120. Takakura, Y., Nishikawa, M., Yamashita, F., and Hashida, M., Influence of physicochemical properties on pharmacokinetics of non-viral vectors for gene delivery, *J. Drug Target.* 10 (2), 99–104, 2002.

121. Regar, E., Sianos, G., and Serruys, P. W., Stent development and local drug delivery, *Br. Med. Bull.* 59, 227–248, 2001.

122. Sampaolesi, M., Torrente, Y., Innocenzi, A., Tonlorenzi, R., D'Antona, G., Pellegrino, M. A., Barresi, R., Bresolin, N., De Angelis, M. G., Campbell, K. P., Bottinelli, R., and Cossu, G., Cell therapy of alpha-sarcoglycan null dystrophic mice through intra-arterial delivery of mesoangioblasts, *Science* 301 (5632), 487–492, 2003.

123. Qian, Z. M., Li, H., Sun, H., and Ho, K., Targeted drug delivery via the transferrin receptor-mediated endocytosis pathway, *Pharmacol. Rev.* 54 (4), 561–587, 2002.

124. Arap, W., Pasqualini, R., and Ruoslahti, E., Cancer treatment by targeted drug delivery to tumor vasculature in a mouse model, *Science* 279 (5349), 377–380, 1998.

125. Federoff, H. J., Geschwind, M. D., Geller, A. I., and Kessler, J. A., Expression of nerve growth factor *in vivo* from a defective herpes simplex virus 1 vector prevents effects of axotomy on sympathetic ganglia, *Proc. Nat. Acad. Sci. U.S.A.* 89 (5), 1636–1640, 1992.

126. Lu, Y., Viral based gene therapy for prostate cancer, *Curr. Gene Ther.* 1 (2), 183–200, 2001.

127. Chang, Y. S., di Tomaso, E., McDonald, D. M., Jones, R., Jain, R. K., and Munn, L. L., Mosaic blood vessels in tumors: frequency of cancer cells in contact with flowing blood, *Proc. Nat. Acad. Sci. U.S.A.* 97 (26), 14608–14613, 2000.

128. Reynolds, A. R., Moein Moghimi, S., and Hodivala-Dilke, K., Nanoparticle-mediated gene delivery to tumour neovasculature, *Trends Mol. Med.* 9 (1), 2–4, 2003.

129. Suh, W., Han, S. O., Yu, L., and Kim, S. W., An angiogenic, endothelial-cell-targeted polymeric gene carrier, *Mol. Ther.* 6 (5), 664–672, 2002.

130. Hosseinkhani, H. and Tabata, Y., *In vitro* gene expression by cationized derivatives of an artificial protein with repeated RGD sequences, Pronectin((R)), *J. Control Release* 86 (1), 169–182, 2003.

131. Maeda, H., The enhanced permeability and retention (EPR) effect in tumor vasculature: the key role of tumor-selective macromolecular drug targeting, *Adv. Enzyme Regul.* 41, 189–207, 2001.

132. Au, J. L., Jang, S. H., Zheng, J., Chen, C. T., Song, S., Hu, L., and Wientjes, M. G., Determinants of drug delivery and transport to solid tumors, *J. Control Release* 74 (1–3), 31–46, 2001.

133. Au, J. L., Jang, S. H., and Wientjes, M. G., Clinical aspects of drug delivery to tumors, *J. Control Release* 78 (1–3), 81–95, 2002.

134. Dass, C. R. and Su, T., Delivery of lipoplexes for genotherapy of solid tumours: role of vascular endothelial cells, *J. Pharm. Pharmacol.* 52 (11), 1301–1317, 2000.

135. Dass, C. R. and Su, T., Particle-mediated intravascular delivery of oligonucleotides to tumors: associated biology and lessons from genotherapy, *Drug Deliv.* 8 (4), 191–213, 2001.

136. Hood, J. D., Bednarski, M., Frausto, R., Guccione, S., Reisfeld, R. A., Xiang, R., and Cheresh, D. A., Tumor regression by targeted gene delivery to the neovasculature, *Science* 296 (5577), 2404–2407, 2002.

137. Wu, J., Nantz, M. H., and Zern, M. A., Targeting hepatocytes for drug and gene delivery: emerging novel approaches and applications, *Front Biosci.* 7, d717–725, 2002.

138. Kren, B. T., Bandyopadhyay, P., and Steer, C. J., *In vivo* site-directed mutagenesis of the factor IX gene by chimeric RNA/DNA oligonucleotides, *Nat. Med.* 4 (3), 285–290, 1998.

139. Arangoa, M. A., Duzgunes, N., and Tros de Ilarduya, C., Increased receptor-mediated gene delivery to the liver by protamine-enhanced-asialofetuin-lipoplexes, *Gene Ther.* 10 (1), 5–14, 2003.

140. Kunath, K., von Harpe, A., Fischer, D., and Kissel, T., Galactose-PEI-DNA complexes for targeted gene delivery: degree of substitution affects complex size and transfection efficiency, *J. Control Release* 88 (1), 159–172, 2003.

141. Kunath, K., von Harpe, A., Fischer, D., Petersen, H., Bickel, U., Voigt, K., and Kissel, T., Low-molecular-weight polyethylenimine as a non-viral vector for DNA delivery: comparison of physico-chemical properties, transfection efficiency and *in vivo* distribution with high-molecular-weight polyethylenimine, *J. Control Release* 89 (1), 113–125, 2003.

142. Campbell, J. J. and Butcher, E. C., Chemokines in tissue-specific and microenvironment-specific lymphocyte homing, *Curr. Opin. Immunol.* 12 (3), 336–341, 2000.

143. Mora, J. R., Bono, M. R., Manjunath, N., Weninger, W., Cavanagh, L. L., Rosemblatt, M., and Von Andrian, U. H., Selective imprinting of gut-homing T cells by Peyer's patch dendritic cells, *Nature* 424 (6944), 88–93, 2003.

144. Okamoto, R., Yajima, T., Yamazaki, M., Kanai, T., Mukai, M., Okamoto, S., Ikeda, Y., Hibi, T., Inazawa, J., and Watanabe, M., Damaged epithelia regenerated by bone marrow-derived cells in the human gastrointestinal tract, *Nat. Med.* 8 (9), 1011–1017, 2002.

145. Otani, A., Kinder, K., Ewalt, K., Otero, F. J., Schimmel, P., and Friedlander, M., Bone marrow-derived stem cells target retinal astrocytes and can promote or inhibit retinal angiogenesis, *Nat. Med.* 8 (9), 1004–1010, 2002.

146. Mata, M., Glorioso, J. C., and Fink, D. J., Targeted gene delivery to the nervous system using herpes simplex virus vectors, *Physiol. Behav.* 77 (4–5), 483–488, 2002.

147. Hauck, B. and Xiao, W., Characterization of tissue tropism determinants of adeno-associated virus type 1, *J. Virol.* 77 (4), 2768–2774, 2003.

148. Mahato, R. I., Henry, J., Narang, A. S., Sabek, O., Fraga, D., Kotb, M., and Gaber, A. O., Cationic lipid and polymer-based gene delivery to human pancreatic islets, *Mol. Ther.* 7 (1), 89–100, 2003.

149. Kawakami, Y., Yamaoka, T., Hirochika, R., Yamashita, K., Itakura, M., and Nakauchi, H., Somatic gene therapy for diabetes with an immunological safety system for complete removal of transplanted cells, *Diabetes* 41 (8), 956–961, 1992.

150. Zhang, L., Peeples, M. E., Boucher, R. C., Collins, P. L., and Pickles, R. J., Respiratory syncytial virus infection of human airway epithelial cells is polarized, specific to ciliated cells, and without obvious cytopathology, *J. Virol.* 76 (11), 5654–5666, 2002.

151. Dhawan, J., Pan, L. C., Pavlath, G. K., Travis, M. A., Lanctot, A. M., and Blau, H. M., Systemic delivery of human growth hormone by injection of genetically engineered myoblasts, *Science* 254 (5037), 1509–1512, 1991.

152. Hirko, A. C., Buethe, D. D., Meyer, E. M., and Hughes, J. A., Plasmid delivery in the rat brain, *Biosci. Rep.* 22 (2), 297–308, 2002.

153. Zhang, Y., Zhu, C., and Pardridge, W. M., Antisense gene therapy of brain cancer with an artificial virus gene delivery system, *Mol. Ther.* 6 (1), 67–72, 2002.

154. Zhang, Y., Jeong Lee, H., Boado, R. J., and Pardridge, W. M., Receptor-mediated delivery of an antisense gene to human brain cancer cells, *J. Gene Med.* 4 (2), 183–194, 2002.

155. Hakkarainen, T., Hemminki, A., Pereboev, A. V., Barker, S. D., Asiedu, C. K., Strong, T. V., Kanerva, A., Wahlfors, J., and Curiel, D. T., CD40 is expressed on ovarian cancer cells and can be utilized for targeting adenoviruses, *Clin. Cancer Res.* 9 (2), 619–624, 2003.

156. Jaafari, M. R. and Foldvari, M., Targeting of liposomes to melanoma cells with high levels of ICAM-1 expression through adhesive peptides from immunoglobulin domains, *J. Pharm. Sci.* 91 (2), 396–404, 2002.

157. Voinea, M., Dragomir, E., Manduteanu, I., and Simionescu, M., Binding and uptake of transferrin-bound liposomes targeted to transferrin receptors of endothelial cells, *Vascul. Pharmacol.* 39 (1–2), 13–20, 2002.

158. Voinea, M. and Simionescu, M., Designing of 'intelligent' liposomes for efficient delivery of drugs, *J. Cell Mol. Med.* 6 (4), 465–474, 2002.

159. Greco, O., Scott, S. D., Marples, B., and Dachs, G. U., Cancer gene therapy: 'delivery, delivery, delivery,' *Front Biosci.* 7, d1516–1524, 2002.

160. Mahanty, S., Hutchinson, K., Agarwal, S., McRae, M., Rollin, P. E., and Pulendran, B., Cutting edge: impairment of dendritic cells and adaptive immunity by Ebola and Lassa viruses, *J. Immunol.* 170 (6), 2797–2801, 2003.

161. Ogris, M., Walker, G., Blessing, T., Kircheis, R., Wolschek, M., and Wagner, E., Tumor-targeted gene therapy: strategies for the preparation of ligand-polyethylene glycol-polyethylenimine/DNA complexes, *J. Control Release*, 91 (1–2), 173–181, 2003.

CHAPTER 3

Biological Barriers to Gene Transfer

Yasufumi Kaneda

CONTENTS

3.1 INTRODUCTION

Currently, more than 900 clinical protocols have been approved for human gene therapy. However, in many cases no remarkable successes have been reported. In 1995, the Orkin-Motulsky report indicated the importance of vector development in human gene therapy. Numerous vectors have been developed to date. In general, viral vectors are more effective for gene transfer; the problem, however, is safety.[1] Unfortunately, in 1999, a young patient died following infusion of an adenoviral vector via the hepatic artery.[2] In 2002, two patients with X-linked severe combined immunodeficiency, treated with retroviral vector gene therapy, suffered from leukemia-like symptoms, likely induced by insertional mutagenesis of the retroviral DNA.[3] Hemophilia B gene therapy using the adeno-associated virus (AAV) vector looks promising,[4] but several potential risks still remain, such as germ line transmission. On the other hand, nonviral vectors have been evaluated for safety, but the limitation of this vector system is inefficient gene transfection.[5] Therefore, the current consensus for vector development is that highly efficient and minimally invasive vectors are the most appropriate for human gene therapy. It appears to be very difficult to satisfy both

0-8493-1934-X/05/$0.00+$1.50
© 2005 by CRC Press LLC

efficacy and safety. Before developing vector systems, we should analyze the biological barriers to gene transfer and develop methods to solve such difficulties.

All the current vector systems have their limitations and advantages.[6] Currently, there are several approaches to correct for the limitations of each vector system. For example, in adenovirus vector systems, one crucial improvement is reduced vector antigenicity, and a helper-dependent or "gutless" vector that deletes most of the adenovirus genome has been developed.[7] However, the production of the "gutless" vector was much lower than that of the classical adenovirus vector. Another approach is to overcome the limitations of one vector system by drawing upon the strengths of another. Examples of chimeric vectors include the pseudotyped retrovirus vector, in which the envelope component of the classical retrovirus vector is replaced with the vesicular stomatitis virus (VSV) G protein,[8] and a new lentivirus vector containing human immunodeficiency virus (HIV) proteins with the pseudotype retrovirus envelope-containing VSV G protein.[9] However, combinations of viral vector systems are thought to be of limited utility, because the components necessary for virus replication and packaging cannot be eliminated, and the constituent viruses may interfere with each other during vector production.

Limitations in vector systems are inevitable, as the introduction of foreign genes into cells is abnormal and is directed to disturb cellular function. Current gene therapy strategies mimic viral infection, and cells are designed to resist the gene transfer. Cells have biological barriers that protect against invasion by foreign genes and inhibit expression of the foreign gene product.[5] Protocols for *in vivo* gene transfection occur in several steps. The first step entails the approach of the vectors to the target cells from the outside environment. The second is the introduction of the transgenes to the cytoplasm. Third, the transgene must migrate to the nucleus; and finally, the transgenes must be retained and stably expressed. Each of these steps has barriers. For example, serum proteins and immune cells can attack gene transfer vectors before they reach target cells; the cell membrane can prevent transfer of foreign genes to the cytoplasm; the nuclear envelope can restrict nuclear targeting of molecules; and expression of the transgene can be inhibited by degradation of the DNA or by transcriptional silencing.

This chapter is entitled "Biological Barriers to Gene Transfer." In a narrow sense, "gene transfer" means the introduction of a transgene to the cytoplasm, and encompasses the first and second steps described above. However, it is actually pointless to refer only to the introduction of a transgene to the cells without considering the final step of gene expression. Therefore, this chapter analyzes biological barriers from the first step (approach to target cells) to the fourth step (effective expression of a transgene), and potential solutions for each barrier are discussed with respect to future views of gene transfer systems.

3.2 BIOLOGICAL BARRIERS

3.2.1 Step 1: Reaching Target Cells

To achieve tissue-specific targeting, vector systems should be developed that will achieve three goals (Figure 3.1): recognition by specific target tissues, avoidance of nonspecific uptake, and resistance to degradation in the systemic circulation.

3.2.1.1 *Recognition by Specific Target Tissues*

Targeting vectors to specific tissues has been considered an ideal method for gene delivery. Although 100% efficiency in targeting appears to be impossible, targeting systems can enhance delivery to specific tissues. Ligand-specific delivery systems have been developed for tissue-specific targeting. In one experiment, asialoglycoprotein was conjugated with DNA-poly-lysine complexes.[10] Because asialoglycoprotein binds to receptors on hepatocytes, DNA can be delivered specifically to

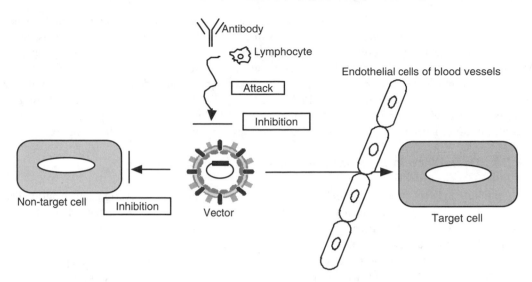

Figure 3.1 Step 1 in biological barriers to gene transfer. This step includes the approach of the vector to the target cells by passing through the vessel wall avoiding nonspecific uptake by nontarget cells and degradation by the host immune system.

hepatocytes by receptor-mediated endocytosis using this complex.[11] Transferrin has also been used for receptor-mediated delivery of DNA to cancer cells, since cancer cells express high levels of the transferrin receptor.[12] These complexes worked well in cultured cells, but the efficiency of *in vivo* gene delivery was not high. The distribution of DNA in animal tissues following peripheral administration has not been thoroughly investigated. Some viruses do, however, target specific tissues.[13] Epstein-Barr virus (EBV) infects primarily B-lymphocytes, hepatitis virus B attacks hepatocytes, and HIV targets lymphocytes and macrophages. Currently, adenovirus-based gene delivery vectors are being used for many gene therapy applications because of the broad range of hosts.[14] Cellular tropism of the adenovirus vector in tissues is regulated by fiber proteins and, based on tissue tropisms of these viruses, targeting viral vectors can be developed.[15-17] Hemagglutinating virus of Japan (HVJ), also called Sendai virus, is known to induce cell fusion, and the use of this fusion activity has been widely used for drug delivery, as described in Step 2. Two glycoproteins, HN and F, are required for cell fusion by HVJ. The HVJ envelope vector can be prepared by incorporating DNA into inactivated HVJ particles. The vector is also useful as a drug delivery system. When the HVJ envelope vector, with a diameter of 200 nm, is injected into the mouse tail vein, the vector reaches the spleen to deliver DNA into cells at the marginal zone.[18] However, when reconstituted HVJ particles containing only F protein, without HN protein, are injected into the mouse tail vein, gene expression is observed mainly in liver, because the galactose residues of the F protein are recognized by hepatocytes.[19] The spleen targeting mechanism of the HVJ envelope vector remains to be determined, but the HN protein on the envelope is likely responsible.

3.2.1.2 Avoidance of Nonspecific Uptake

Colloidal particles with a diameter more than 300 nm are trapped by the reticulendothelial system (RES) in liver, spleen and lung.[20] HVJ liposomes containing both F and HN proteins target the liver, spleen, and lung when the vector is injected into the saphenous veins of monkeys,[21] probably because the phosphatidylserine residues[22] present on the envelope are recognized by the RES, and the particle size is 470 nm in diameter. Some phospholipids such as phosphatidylserine are also recognized by the RES. The liposome-protamine sulfate-plasmid DNA (LPD) vector targets the lung, kidney, heart, liver, and spleen with the highest level of gene expression in the lung.[23,24] To avoid such nonspecific uptake, long circulating liposomes have also been developed.[25] Selecting lipid components means that the liposomes are retained in the systemic circulation much longer than other liposomes. Such long circulating liposomes are used for drug delivery to tumor tissues.[26] Liposomes conjugated with polyethylene glycol (PEG) are also of the long circulating type.[27,28] Actually, cationic lipid–DNA complexes accumulate mostly in the lung, while the "PEGylated" lipid–DNA complexes exhibit reduced lung accumulation. The folate receptor is abundant in tumor tissues. When folate is conjugated with PEGylated lipid–DNA complexes, lung accumulation of DNA is reduced and tumor-targeted DNA increases.[29] PEGylation of various types of vectors may regulate tissue-targeting when tissue-specific molecules are conjugated with the complex.

3.2.1.3 *Resistance to Degradation in the Systemic Circulation*

In the human body, protective mechanisms against viral invasion exist. Antigalactose antibodies present in human serum bind to retrovirus envelope proteins to induce complement-mediated lysis.[30] To avoid such lysis, a complement-resistant retrovirus has been developed.[31] Adenovirus is highly immunogenic and is attacked by the host immune system after repeated transfection.[32] As described above, to reduce immunogenicity, a "gutless" adenovirus vector has also been developed.[7]

Another barrier to gene transfer is the penetration and distribution of vectors in tissues. To extravasate from blood vessels to the target tissues and cells, vectors must penetrate endothelial cells. To induce gene expression in a large area of tissue, vectors must be spread diffusely across their area. Many viral vectors including adenovirus vector and adeno-associated virus vector cannot penetrate endothelial cells when administered systemically.[33] Among the nonviral gene delivery systems, cationic liposomes are the most frequently used for gene transfer. When injected into the mouse tail vein, gene expression is detected primarily in the lung.[34] However, cationic liposomes cannot penetrate past the endothelial cells when injected into the vasculature.[35] In contrast, anionic liposomes are able to penetrate endothelial cells.[36] Moreover, anionic liposomes can be distributed to broader areas of the target tissues than cationic liposomes. The mean size of liposomes is also important, because smaller liposomes are more permeable in tissue.

3.2.2   Step 2: Crossing Over the Cell Membrane

After reaching the target cells, transgene DNA must cross the cell membrane (Figure 3.2), which is a very effective barrier that excludes foreign substances from cells. There are several ways to overcome this barrier. One is to facilitate DNA transfer by utilizing endocytosis or phagocytosis. However, in this process, the foreign DNA must penetrate the membrane of the endosome or phagosome rapidly. Otherwise, the DNA will be degraded. Viruses have the ability to enter cells, and viral vectors can penetrate the cell membrane. Adenovirus can escape from the endosome by disrupting the endosomal membrane with penton fibers.[37,38] This ability has been utilized to enhance the efficiency of gene transfer by transferrin-polylysine-DNA complexes.[12] Other viruses can fuse with the cell membrane to introduce their genomes into the cytoplasm. There are two different mechanisms of virus–cell fusion: pH dependent and pH independent. Influenza virus,[39] Semliki Forest virus (SFV),[40] and VSV[41] exhibit pH-dependent fusion, whereas HVJ[42] and retrovirus[43,44] can fuse with the cell membrane at both acidic and neutral pHs. Viral fusion proteins have been identified, and synthetic vectors that express these viral fusion proteins can transfer foreign genes efficiently into the cytoplasm.[45] Because fusion proteins provide efficient gene delivery, HVJ liposomes have been constructed by fusing DNA-loaded liposomes with UV-inactivated HVJ.[46,47] The hemagglutinating (HN) protein of HVJ liposomes can bind to sialic acid receptors for HVJ. The F protein then associates with lipid molecules in the lipid bilayer of the cell membrane to induce cell–liposome fusion.[42] HVJ liposomes are useful for both *in vitro* and *in vivo* gene trans-fer.[48,49] In fact, oligodeoxynucleotides (ODN) trapped in HVJ liposomes are able to be transferred directly into the cytoplasm without undergoing lysosomal trafficking and possible degradation. By

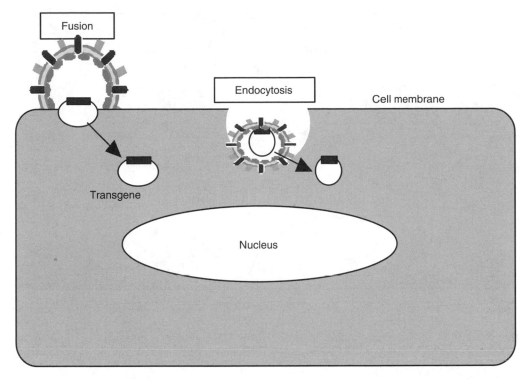

Figure 3.2 Step 2 in biological barriers to gene transfer. This step includes crossing over the cell membrane by fusion- or endocytosis-mediated transfer.

contrast, with the use of cationic liposomes (including lipofectin), the major cellular uptake mechanism of ODN is considered to be endocytosis,[50,51] creating the potential for the degradation of ODN within the lysosomes. Therefore, differences in the cellular uptake mechanisms of ODN between these two vector systems may contribute to the integrity of the ODN upon transfer into the nucleus. This hypothesis has been clearly demonstrated using fluorescence resonance energy transfer (FRET) analysis.[52]

To simplify the vector system and to develop a more effective gene delivery system than HVJ liposomes, the HVJ envelope vector has been developed.[18] In this system, macromolecules such as plasmid DNA, RNA, synthetic oligonucleotides, proteins, and peptides can be incorporated into inactivated HVJ particles quite simply by means of treatment with mild detergent and centrifugation, and can then be delivered to cells both *in vitro* and *in vivo*.

3.2.3   Step 3: Nuclear Targeting

Transport of the foreign gene to the nucleus (Figure 3.3) is required for gene expression in gene therapy. Especially in nondividing cells, DNA is not transported efficiently to the nucleus where transcription occurs. However, it should be noticed that naked plasmid DNA can be transported to the nucleus without breakdown of the nuclear envelope. The efficiency is generally 1–2% of the DNA in the cytoplasm.[53] Therefore, nuclear targeting increases gene expression, especially in tissues. Some viruses, such as adenovirus, SV40, HIV, and herpes virus, are known to induce rapid migration of their genomes to the nucleus even in nondividing cells.[54] In SV40, viral capsid proteins contain nuclear localization sequences (NLSs) that trigger translocation of the virion to the nucleus and disassembly of the virion within the nucleus.[55] Although rapid nuclear transport also occurs with adenovirus, disassembly of the viral capsid occurs in the cytoplasm, and the DNA-NLS-containing proteins are then sorted in the nucleus.[55,56] In HIV, an integrase is required for

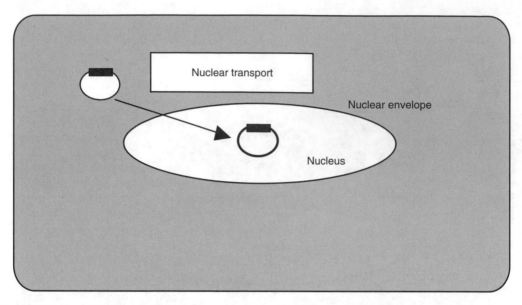

Figure 3.3   Step 3 in biological barriers to gene transfer. This step includes nuclear transport of a transgene.

nuclear migration of the viral genome.[54] However, the DNA–integrase complex, or preintegration complex, of oncoretroviruses such as Moloney leukemia virus cannot pass through the nuclear pore, and thus transport of the transgene to the nucleus does not occur in nondividing cells in these retroviral systems.[1] In nonviral vector systems, nuclear transport of the plasmid DNA occurs, but the efficiency is very low.[5] Enhancing nuclear migration of plasmid DNA is an important issue for increasing transgene expression. Complexes of plasmid DNA and nuclear proteins have been constructed to improve nuclear migration of DNA[57,58] but no successes have been reported to date. It was recently reported that conjugation of the NLS peptide derived from SV40 with a luciferase gene fragment enhanced luciferase gene expression by a factor of approximately 1000, compared with the luciferase gene without the NLS peptide.[59] NLS peptide at either the 5' or 3' end of the DNA enhanced luciferase gene expression, but incorporation of the NLS at both ends was not effective. It is thought that positioning of the NLS at one end of the gene stimulates nuclear migration of the luciferase gene, whereas incorporation of the NLS at both ends ensnares the DNA at the nuclear envelope. However, it remains unknown whether NLS conjugation improves the nuclear migration of larger DNA molecules and if NLS conjugation improves the efficiency of translocation *in vivo*. This raises the question, does only a small fraction of all the plasmid DNA go to the nucleus? Recently, several papers have suggested that nuclear migration of plasmid DNA may be sequence dependent.[60] The SV40 enhancer sequence appears to facilitate nuclear migration of DNA by binding transcription factors[61] [which are transported to the nucleus by the Ran/importin system].

RNA-based vectors have also been developed. Alphaviruses such as SFV and Sindvis virus are converted into gene expression recombinant RNA vectors,[62,63] which do not require the transport of DNA into the nucleus. In general, RNA vectors produce large amounts of proteins, and this production inhibits host protein synthesis, causing induced cell death, or apoptosis.[64] Thus, tight regulation of protein synthesis will be required for the use of gene therapy using RNA vector systems.

3.2.4   Step 4: Regulation of Gene Expression

Effective expression of a transgene is also a big issue for gene therapy. We should view this step from at least two perspectives (Figure 3.4). One is the stable retention of a transgene and the

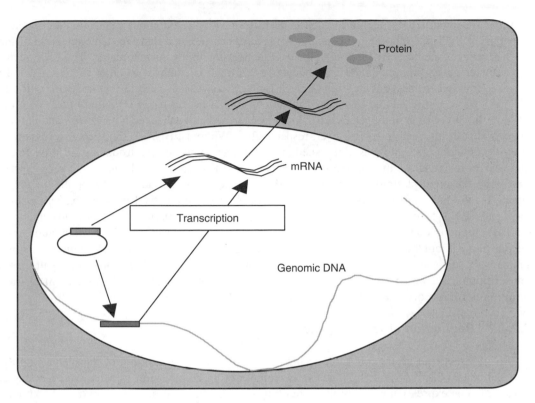

Figure 3.4   Step 4 in biological barriers to gene transfer. This step includes stable retention of a transgene and effective transcription of the transgene in the nucleus.

other is the regulation of transcription. The export of mRNA, translation, and post-translational modification of polypeptides generally affect gene expression, but they are not specific for expression of a transgene. Therefore, in this section, the events occurring before mRNA production are the focus.

3.2.4.1   *Stable Retention of a Transgene*

There are two distinct approaches to address this issue. One is the integration of the transgene into the host genome. Retroviruses can integrate into random sites of the host chromosomes, but integration occurs only in dividing cells.[1] HIV inserts its genome into the host chromosomes even in nondividing cells. A lentivirus vector system containing the recombinant HIV genome along with the therapeutic gene and the vesicular stomatitis virus G protein (VSVG) gene was used to generate long-term gene expression in mouse neurons.[9] The mechanism of insertion of the viral genome into host chromosomes has been investigated. However, reconstitution of the viral machinery that confers the ability for genomes to be inserted into nonviral gene delivery systems has not yet been successful. In 1997, *sleeping beauty*, a fish transposon/transposase system, was developed.[65] This system inserts the neomycin resistance gene into the HeLa cell genome with a very high efficiency. Application of this system to gene transfer is currently under investigation. However, insertion of the transgene into the host genome occurs at random. Although adenoassociated virus (AAV) integrates specifically into the AAVS1 region (19q13-qter) of human chromosome 19, the recombinant AAV vector is unable to do so because it does not express Rep 65 protein.[66,67]

Another approach is the stable retention of an episomal piece of DNA. Epstein-Barr virus (EBV) has been analyzed thoroughly in terms of latent infection. The *cis*-acting oriP (the latent viral DNA replication origin) sequence and the *trans*-acting EBV nuclear antigen-1 (EBNA-1) gene are required for latent EBV infection, which is characterized by autonomous replication and nuclear retention of the EBV genome in host cells.[68-70] It has been reported that the stable retention of the EBV genome is mediated by binding of the oriP sequence with the nuclear matrix. The EBV vector can transfer genes primarily to B lymphocytes, with minimal transfer to other somatic cells. To improve long-term transgene expression, an EBV replicon-based plasmid that contains oriP and EBNA-1 was constructed and transferred to various animal tissues using HVJ liposomes. With the EBV replicon vector system, gene expression was sustained and enhanced both *in vitro* and *in vivo*, in several cell lines and in mouse liver, respectively.[71] Without EBNA-1, transgene expression of an oriP-containing plasmid was diminished in mouse liver. However, reintroducing EBNA-1 to the mouse liver reactivated transgene expression.[72] This indicates that EBNA-1 can restore gene expression of an episomal plasmid containing an oriP sequence within the nucleus. When the luciferase gene was introduced into a mouse melanoma cell line, with or without the EBV replicon components (oriP and EBNA-1), effective and long-term gene expression was obtained in the plasmid containing the EBV replicon components. However, Southern blot analysis revealed that the copy number of the plasmid without the EBV components was much higher than that with the EBV replicon plasmid. This result suggests that effective gene expression by the EBV replicon plasmid results from transcriptional activation, not from stable retention of the plasmid DNA. The regulation of transcription is further discussed in the next item.

3.2.4.2 *Regulation of Transcription*

Silencing of transgene expression occurs in host cells, despite insertion of transgenes into the host genome.[73] Similar silencing has also been reported in certain viral infections. For example, retroviral gene expression was inhibited after retrovirus infection of mouse embryos.[74,75] Although low or silenced transgene expression after transfection has been a major problem in human gene therapy,[76] the mechanism(s) by which transgene expression is regulated have not yet been elucidated. Since it will be essential to control transgene expression in human gene therapy, basic research on transgene expression in human gene therapy should be pursued further.

Although transcription factors have been investigated extensively,[77] prior studies have not fully clarified the mechanism(s) of the regulation of transcription in higher eukaryotes. Recent studies have shown that chromatin remodeling factors such as the switch/sucrose nonfermenting (SWI/SNF) family and the imitation switch (ISWI) family[78] are involved in the regulation of transcription. It is thought that histone acetylase and deacetylase also regulate transcription by modifying chromatin.[79,80] Acetylation and deacetylation of histones and the other proteins associated with DNA have been shown to be important in the epigenetics of regulating gene expression.[81-83] Inhibitors of histone deacetylation have been identified including trichostatin A (TSA), trapoxin (TPX), and n-butyrate.[84] Both reagents induce transcriptional activation of the HIV-type 1 promoter through disruption of a single nucleosome located at the transcription start site, which results in markedly increased virus production.[85] These agents have been shown to reactivate chromosome-integrated transgenes[86] and amplify expression of transgenes encoded by the recombinant adenovirus vector.[87] Recently, amplification of transgene expression in animals treated with a novel histone deacetylase inhibitor was reported.[88] This amplification was more effective for transgene expression than for endogenous gene expression. These data suggest that transgene expression can be regulated by chromatin modifications, such as histone acetylation/deacetylation. Methylated DNA-binding protein MeCP2 is a component of the complex that contains histone deacetylase.[89] This suggests that DNA methylation status may dictate the state of histone acetylation/deacetylation in chromatin. To overcome the silencing of transgene expression and to sustain transgene expression, *cis*-elements such as "insulators"[90-92] that are not affected by chromatin modification may be applied for transgene expression in gene therapy.

3.3 CONCLUSION

Many significant barriers to gene transfer have been elucidated. If omnipotent vector systems are eventually developed, however, gene transfer may harm patients; this is far from therapeutic. An all-encompassing vector system may not be feasible, and each vector system must be evaluated with respect to the target disease. In the case of some diseases, such as arteriosclerosis obliterans, transient gene expression would be sufficient to cure the disease, and sustained gene expression would be detrimental. With other diseases, such as SCID-X1 and hemophilia B, it may be necessary to induce the transformation of only a small population of cells. Cancer gene therapy occupies more than 60% of current human gene therapy approaches, and no remarkable successes have been reported in this area to date. When we consider why gene therapy is necessary for cancer treatment, we must conclude that gene therapy should be used to inhibit tumor metastasis and suppress the recurrence of cancers. For these purposes, investigations in basic cancer cell biology and technological developments should be promoted much more than at present. The issues I have described in this chapter present a road map for the future technological developments for cancer gene therapy. Thus, technologies in gene therapy should be developed on the basis of analysis of the target tissues, target cells, cellular components, and molecules which are relevant to the biology.

REFERENCES

1. Mulligan, R.C., The basic science of gene therapy, *Science* 260, 926–932, 1993.
2. Marshall, E., Gene therapy death prompts review of adenovirus vector, *Science* 286, 2244–2245, 1999.
3. Kaiser, J., Gene therapy:seeking the cause of induced leukemias in X-SCID trial, *Science* 299, 495, 2003.
4. Ragni, M.V., Safe passage: a plea for safety in hemophilia gene therapy, *Mol. Ther.* 6, 436–440, 2002.
5. Ledley, F.D., Nonviral gene therapy: the promise of genes as pharmaceutical products, *Hum. Gene Ther.* 6, 1129–1144, 1995.
6. Kaneda, Y., Gene therapy: a battle against biological barriers, *Curr. Mol. Med.* 1, 493–499, 2001.
7. Morral, N., O'Neal, W., Rice, K., Leland, M., Kaplan, J., Piedra, P.A., Zhou, H., Parks, R.J., Velji, R., Aguilar-Cordova, F. et al., Administration of helper-dependent adenoviral vectors and sequential delivery of different vector serotype for long-term liver-directed gene transfer in baboons, *Proc. Nat. Acad. Sci. U.S.A.* 96, 12816–12821, 1999.
8. Burns, J.C., Friedmann, T., Driever, W., Burrascano, M., and Yee, J.K., Vesicular stomatitis virus G glycoprotein pseudotyped retroviral vectors: concentration to very high titer and efficient gene transfer into mammalian and nonmammalian cells, *Proc. Nat. Aca. Sci. U.S.A.* 90, 8033–8037, 1993.
9. Naldini, L., Blomer, U., Gallay, P., Ory, D., Mulligan, R., Gage, F.H., Verma, I.M., and Trono, D., *In vivo* gene delivery and stable transduction of nondividing cells by a lentiviral vector, *Science* 272, 263–267, 1996.
10. Wu, G.Y. and Wu, C.H., Receptor-mediated *in vitro* gene transformation by a soluble DNA carrier system, *J. Biol. Chem.* 262, 4429–4432, 1987.
11. Wu, G.Y. and Wu, C.H., Receptor-mediated gene delivery and expression *in vivo*, J. Biol. Chem. 263, 14621–14624, 1988.
12. Wagner, E., Zatloukal, K., Cotten, M., Kirlappos, H., Mechtler, K., Curiel, D.T., and Birnstiel, M.L., Coupling of adenovirus to transferrin-polylysine/DNA complexes greatly enhances receptor-mediated gene delivery and expression of transfected genes, *Proc. Nat. Acad. Sci. U.S.A.* 89, 6099–6103, 1992.
13. Flint, S.J., Enquist, L.W., Krug, R.M., Racaniello, V.R., and Skalka, A.M. (Eds.), *Principles of Virology*, ASM Press, Washington, D.C., 2000.
14. Gall, J., Kass-Eisler, A., Leinwand, L., and Falck-Pedersen, E., Adenovirus type 5 and 7 capsid chimera: fiber replacement alters receptor tropism without affecting primary immune neutralization epitopes, *J. Virol.* 70, 2116–2123, 1996.
15. Krasnykh, V.N., Mikheeva, G.V., Douglas, J.T., and Curiel, D.T., Generation of recombinant adenovirus vectors with modified fibers for altering viral tropism. *J. Virol.* 70, 6839–6846, 1996.

16. Zabner, J., Chillon, M., Grunst, T., Moninger, T.O., Davidson, B.L., Gregory, R., and Armentano, D., A chimeric type 2 adenovirus vector with a type 17 fiber enhances gene transfer to human airway epithelia, *J. Virol.* 73, 8689–8695, 1999.

17. Von Seggern, D.J., Huang, S., Fleck, S.K., Stevenson, S.C., and Nemerow, G.R., Adenovirus vector pseudotyping in fiber-expressing cell lines: improved transduction of Epstein-Barr virus-transformed B cells, *J. Virol.* 74, 354–362, 2000.

18. Kaneda, Y., Nakajima, T., Nishikawa, T., Yamamoto, S., Ikegami, H., Suzuki, N., Nakamura, H., Morishita, R., and Kotani, H., Hemagglutinating virus of Japan (HVJ) envelope vector as a versatile gene delivery system, *Mol. Ther.* 6, 219–226, 2002.

19. Ramani, K., Hassan, Q., Venkaiah, B., Hasnain, S.E., and Sarkar, D.P., Site-specific gene delivery *in vivo* through engineered Sendai viral envelopes, *Proc. Nat. Acad. Sci. U.S.A.* 95, 11886–11890, 1998.

20. Ikomi, F., Hanna, G.K., and Schmid-Schonbein, G.W., Mechanism of colloidal particle uptake into the lymphatic system: basic study with percutaneous lymphography, *Radiology* 196, 107–113, 1995.

21. Tsuboniwa, N., Morishita, R., Hirano, T., Fujimoto, J., Furukawa, S., Kikumori, M., Okuyama, A., and Kaneda, Y., Safety evaluation of hemagglutinating virus of Japan — artificial viral envelope liposomes in nonhuman primates, *Hum. Gene Ther.* 12, 469–487, 2001.

22. Allen, T.M., Williamson, P., and Schlegel, R.A., Phosphatidylserine as a determinant of reticuloen-dothelial recognition of liposome models of the erythrocyte surface, *Proc. Nat. Acad. Sci. U.S.A.* 85, 8067–8071, 1988.

23. Li, S. and Huang, L., *In vivo* gene transfer via intravenous administration of cationic lipid-protamine-DNA (LPD) complexes, *Gene Ther.* 4, 891–900, 1997.

24. Li, S., Rizzo, M.A., Bhattacharya, S., and Huang, L., Characterization of cationic lipid-protamine-DNA (LPD) complexes for intravenous gene delivery, *Gene Ther.* 5, 930–937, 1998.

25. Huang, S.K., Lee, K.D., Hong, K., Friend, D.S., and Papahadjopoulos, D., Microscopic localization of sterically stabilized liposomes in colon carcinoma-bearing mice, *Cancer Res.* 52, 5135–5143, 1992.

26. Litzinger, D.C., Buiting, A.M., van Rooijen, N., and Huang, L., Effect of liposome size on the circulation time and intraorgan distribution of amphipathic poly(ethylene glycol)-containing liposomes, *Biochim. Biophys. Acta* 1190, 99–107, 1994.

27. Blessing, T., Kursa, M., Holzhauser, R., Kircheis, R., and Wagner, E., Different strategies for formation of pegylated EGF-conjugated PEI/DNA complexes for targeted gene delivery, *Bioconjug. Chem.* 12, 529–537, 2001.

28. Choi, Y.H., Liu, F., Choi, J.S., Kim, S.W., and Park, J.S., Characterization of a targeted gene carrier, lactose-polyethylene glycol-grafted poly-L-lysine and its complex with plasmid DNA, *Hum. Gene Ther.* 10, 2657–2665, 1999.

29. Reddy, J.A., Abburi, C., Hofland, H., Howard, S.J., Vlahov, I., Wils, P., and Leamon, C.P., Folate-targeted, cationic liposome-mediated gene transfer into disseminated peritoneal tumors, *Gene Ther.* 9, 1542–1550, 2002.

30. Oroszlan, S. and Nowinski, R.C., Lysis of retroviruses with monoclonal antibodies against viral envelope proteins, *Virology* 101, 296–299, 1980.

31. Pensiero, M.N., Wysocki, C.A., Nader, K., and Kikuchi, G.E., Development of amphotropic murine retrovirus vectors resistant to inactivation by human serum, *Hum. Gene Ther.* 7, 1095–1101, 1996.

32. Dickson, G., *Molecular and Cell Biology of Human Gene Therapeutics*, Chapman & Hall, London, 1995.

33. Eslami, M.H., Gangadharan, S.P., Sui, X., Rhynhart, K.K., Snyder, R.O., and Conte, M.S., Gene delivery to *in situ* veins: differential effects of adenovirus and adeno-associated viral vectors, *J. Vasc. Surg.* 31, 1149–1159, 2000.

34. Wheeler, C.J., Felgner, P.L., Tsai, Y.J., Marshall, J., Sukhu, L., Doh, S.G., Hartikka, J., Nietupski, J., Manthorpe, M., Nichols, M. et al., A novel cationic lipid greatly enhances plasmid DNA delivery and expression in mouse lung, *Proc. Nat. Acad. Sci. U.S.A.* 93, 11454–11459, 1996.

35. Sawa, Y., Suzuki, K., Bai, H.Z., Shirakura, R., Morishita, R., Kaneda, Y., and Matsuda, H., Efficiency of *in vivo* gene transfection into transplanted rat heart by coronary infusion of HVJ liposome, *Circulation* 92, II479–482, 1995.

36. Yonemitsu, Y., Kaneda, Y., Morishita, R., Nakagawa, K., Nakashima, Y., and Sueishi, K., Characterization of *in vivo* gene transfer into the arterial wall mediated by the Sendai virus (hemagglutinating virus of Japan) liposomes: an effective tool for the *in vivo* study of arterial diseases, *Lab. Invest.* 75, 313–323, 1996.

37. Seth, P., Mechanism of adenovirus-mediated endosome lysis: role of the intact adenovirus capsid structure, *Biochem. Biophys. Res. Commun.* 205, 1318–1324, 1994.

38. Greber, U.F., Webster, P., Weber, J., and Helenius, A., The role of the adenovirus protease on virus entry into cells, *EMBO J.* 15, 1766–1777, 1996.

39. Maeda, T. and Ohnishi, S., Activation of influenza virus by acidic media causes hemolysis and fusion of erythrocytes, *FEBS Lett.* 122, 283–287, 1980.

40. Marsh, M., Bolzau, E., and Helenius, A., Penetration of Semliki Forest virus from acidic prelysosomal vacuoles, *Cell* 32, 931–940, 1983.

41. Blumenthal, R., Bali-Puri, A., Walter, A., Covell, D., and Eidelman, O., pH-dependent fusion of vesicular stomatitis virus with Vero cells. Measurement by dequenching of octadecyl rhodamine fluorescence, *J. Biol. Chem.* 262, 13614–13619, 1987.

42. Okada, Y., Sendai virus-induced cell fusion, *Methods Enzymol.* 221, 18–41, 1993.

43. McClure, M.O., Sommerfelt, M.A., Marsh, M., and Weiss, R.A., The pH independence of mammalian retrovirus infection, *J. Gen. Virol.* 71 (Pt. 4), 767–773, 1990.

44. McClure, M.O., Marsh, M., and Weiss, R.A., Human immunodeficiency virus infection of CD4-bearing cells occurs by a pH-independent mechanism, *EMBO J.* 7, 513–518, 1988.

45. Wagner, E., Plank, C., Zatloukal, K., Cotten, M., and Birnstiel, M.L., Influenza virus hemagglutinin HA-2 N-terminal fusogenic peptides augment gene transfer by transferrin-polylysine-DNA complexes: toward a synthetic virus-like gene transfer vehicle, *Proc. Nat. Acad. Sci. U.S.A.* 89:7934–7938, 1992.

46. Kaneda, Y., Iwai, K., and Uchida, T., Increased expression of DNA cointroduced with nuclear protein in adult rat liver, *Science* 243, 375–378, 1989.

47. Saeki, Y., Matsumoto, N., Nakano, Y., Mori, M., Awai, K., and Kaneda, Y., Development and characterization of cationic liposomes conjugated with HVJ (Sendai virus): reciprocal effect of cationic lipid for *in vitro* and *in vivo* gene transfer, *Hum. Gene Ther.* 8, 2133–2141, 1997.

48. Dzau, V.J., Mann, M.J., Morishita, R., and Kaneda, Y., Fusigenic viral liposome for gene therapy in cardiovascular diseases, *Proc. Nat. Acad. Sci. U.S.A.* 93, 11421–11425, 1996.

49. Kaneda, Y., Saeki, Y., and Morishita, R., Gene therapy using HVJ-liposomes: the best of both worlds? *Mol. Med. Today* 5, 298–303, 1999.

50. Bennett, C.F., Chiang, M.Y., Chan, H., Shoemaker, J.E., and Mirabelli, C.K., Cationic lipids enhance cellular uptake and activity of phosphorothioate antisense oligonucleotides, *Mol. Pharmacol.* 41, 1023–1033, 1992.

51. Ollikainen, H., Lappalainen, K., Jaaskelainen, I., Syrjanen, S., and Pulkki, K., Liposomal targeting of bcl-2 antisense oligonucleotides with enhanced stability into human myeloma cell lines, *Leuk. Lymphoma* 24, 165–174, 1996.

52. Nakamura, N., Hart, D.A., Frank, C.B., Marchuk, L.L., Shrive, N.G., Ota, N., Taira, K., Yoshikawa, H., and Kaneda, Y., Efficient transfer of intact oligonucleotides into the nucleus of ligament scar fibroblasts by HVJ-cationic liposomes is correlated with effective antisense gene inhibition, *J. Biochem.* (Tokyo) 129, 755–759, 2001.

53. Loyter, A., Scangos, G.A., and Ruddle, F.H., Mechanisms of DNA uptake by mammalian cells: fate of exogenously added DNA monitored by the use of fluorescent dyes, *Proc. Nat. Acad. Sci. U.S.A.* 79, 422–426, 1982.

54. Izaurralde, E., Kann, M., Pante, N., Sodeik, B., and Hohn, T., Viruses, microorganisms and scientists meet the nuclear pore, Leysin, VD, Switzerland, February 26–March 1, 1998, *EMBO J.* 18, 289–296, 1999.

55. Greber, U.F. and Kasamatsu, H., Nuclear targeting of SV40 and adenovirus, *Trends Cell Biol.* 6, 189–195, 1996.

56. Greber, U.F., Suomalainen, M., Stidwill, R.P., Boucke, K., Ebersold, M.W., and Helenius, A., The role of the nuclear pore complex in adenovirus DNA entry, *EMBO J.* 16, 5998–6007, 1997.

57. Fritz, J.D., Herweijer, H., Zhang, G., and Wolff, J.A., Gene transfer into mammalian cells using histone-condensed plasmid DNA, *Hum. Gene Ther.* 7, 1395–1404, 1996.

58. Sebestyen, M.G., Ludtke, J.J., Bassik, M.C., Zhang, G., Budker, V., Lukhtanov, E.A., Hagstrom, J.E., and Wolff, J.A., DNA vector chemistry: the covalent attachment of signal peptides to plasmid DNA, *Nat. Biotechnol.* 16, 80–85, 1998.

59. Zanta, M.A., Belguise-Valladier, P., and Behr, J.P., Gene delivery: a single nuclear localization signal peptide is sufficient to carry DNA to the cell nucleus, *Proc. Nat. Acad. Sci. U.S.A.* 96, 91–96, 1999.

60. Dean, D.A., Import of plasmid DNA into the nucleus is sequence specific, *Exp. Cell Res.* 230, 293–302, 1997.

61. Wilson, G.L., Dean, B.S., Wang, G., and Dean, D.A., Nuclear import of plasmid DNA in digitonin-permeabilized cells requires both cytoplasmic factors and specific DNA sequences, *J. Biol. Chem.* 274, 22025–22032, 1999.

62. Smerdou, C. and Liljestrom, P., Alphavirus vectors: from protein production to gene therapy, *Gene Ther. Reg.* 1, 33–64, 2000.

63. Bredenbeek, P.J., Frolov, I., Rice, C.M., and Schlesinger, S., Sindbis virus expression vectors: packaging of RNA replicons by using defective helper RNAs, *J. Virol.* 67, 6439–6446, 1993.

64. Liljeström, P. and Garoff, H., A new generation of animal cell expression vectors based on the Semliki Forest virus replicon, *Biotechnology* (NY) 9, 1356–1361, 1991.

65. Ivics, Z., Hackett, P.B., Plasterk, R.H., and Izsvak, Z., Molecular reconstruction of Sleeping Beauty, a Tc1-like transposon from fish, and its transposition in human cells, *Cell* 91, 501–510, 1997.

66. Summerford, C., Bartlett, J.S., and Samulski, R.J., Adeno-associated viral vectors and successful gene therapy, the gap is closing, *Gene Ther. Reg.* 1, 9–32, 2000.

67. Rutledge, E.A. and Russell, D.W., Adeno-associated virus vector integration junctions, *J. Virol.* 71, 8429–8436, 1997.

68. Lupton, S. and Levine, A.J., Mapping genetic elements of Epstein-Barr virus that facilitate extrachromosomal persistence of Epstein-Barr virus-derived plasmids in human cells, *Mol. Cell Biol.* 5, 2533–2542, 1985.

69. Yates, J.L., Warren, N., and Sugden, B., Stable replication of plasmids derived from Epstein-Barr virus in various mammalian cells, *Nature* 313, 812–815, 1985.

70. Jankelevich, S., Kolman, J.L., Bodnar, J.W., and Miller, G., A nuclear matrix attachment region organizes the Epstein-Barr viral plasmid in Raji cells into a single DNA domain, *EMBO J.* 11, 1165–1176, 1992.

71. Saeki, Y., Wataya-Kaneda, M., Tanaka, K., and Kaneda, Y., Sustained transgene expression *in vitro* and *in vivo* using an Epstein-Barr virus replicon vector system combined with HVJ liposomes, *Gene Ther.* 5, 1031–1037, 1998.

72. Kaneda, Y., Saeki, Y., Nakabayashi, M., Zhou, W.Z., Kaneda, M.W., and Morishita, R., Enhancement of transgene expression by cotransfection of oriP plasmid with EBNA-1 expression vector, *Hum. Gene Ther.* 11, 471–479, 2000.

73. Palmer, T.D., Rosman, G.J., Osborne, W.R., and Miller, A.D., Genetically modified skin fibroblasts persist long after transplantation but gradually inactivate introduced genes, *Proc. Nat. Acad. Sci. U.S.A.* 88, 1330–1334, 1991.

74. Harbers, K., Jahner, D., and Jaenisch, R., Microinjection of cloned retroviral genomes into mouse zygotes: integration and expression in the animal, *Nature* 293, 540–542, 1981.

75. Jahner, D., Stuhlmann, H., Stewart, C.L., Harbers, K., Lohler, J., Simon, I., and Jaenisch, R., *De novo* methylation and expression of retroviral genomes during mouse embryogenesis, *Nature* 298, 623–628, 1982.

76. Grossman, M., Raper, S.E., Kozarsky, K., Stein, E.A., Engelhardt, J.F., Muller, D., Lupien, P.J., and Wilson, J.M., Successful *ex vivo* gene therapy directed to liver in a patient with familial hypercholesterolaemia, *Nat. Genet.* 6, 335–341, 1994.

77. Roeder, R.G., The role of general initiation factors in transcription by RNA polymerase II, *Trends Biochem. Sci.* 21, 327–335, 1996.

78. Tyler, J.K. and Kadonaga, J.T., The "dark side" of chromatin remodeling: repressive effects on transcription, *Cell* 99, 443–446, 1999.

79. Luo, R.X. and Dean, D.C., Chromatin remodeling and transcriptional regulation, *J. Nat. Cancer Inst.* 91, 1288–1294, 1999.

80. Kuo, M.H. and Allis, C.D., Roles of histone acetyltransferases and deacetylases in gene regulation, *Bioessays* 20, 615–626, 1998.

81. Wolffe, A.P. and Pruss, D., Targeting chromatin disruption: transcription regulators that acetylate histones, *Cell* 84, 817–819, 1996.

82. Grunstein, M., Histone acetylation in chromatin structure and transcription, *Nature* 389, 349–352, 1997.

83. Pazin, M.J. and Kadonaga, J.T., What's up and down with histone deacetylation and transcription? *Cell* 89, 325–328, 1997.

84. Yoshida, M., Kijima, M., Akita, M., and Beppu, T., Potent and specific inhibition of mammalian histone deacetylase both *in vivo* and *in vitro* by trichostatin A, *J. Biol. Chem.* 265, 17174–17179, 1990.

85. Van Lint, C., Emiliani, S., Ott, M., and Verdin, E., Transcriptional activation and chromatin remodeling of the HIV-1 promoter in response to histone acetylation, *EMBO J.* 15, 1112–1120, 1996.

86. Chen, W.Y., Bailey, E.C., McCune, S.L., Dong, J.Y., and Townes, T.M., Reactivation of silenced, virally transduced genes by inhibitors of histone deacetylase, *Proc. Nat. Acad. Sci. U.S.A.* 94, 5798–5803, 1997.

87. Dion, L.D., Goldsmith, K.T., Tang, D.C., Engler, J.A., Yoshida, M., and Garver, R.I., Jr., Amplification of recombinant adenoviral transgene products occurs by inhibition of histone deacetylase, *Virology* 231, 201–209, 1997.

88. Yamano, T., Ura, K., Morishita, R., Nakajima, H., Monden, M., and Kaneda, Y., Amplification of transgene expression *in vitro* and *in vivo* using a novel inhibitor of histone deacetylase, *Mol. Ther.* 1, 574–580, 2000.

89. Nan, X., Ng, H.H., Johnson, C.A., Laherty, C.D., Turner, B.M., Eisenman, R.N., and Bird, A., Transcriptional repression by the methyl-CpG-binding protein MeCP2 involves a histone deacetylase complex, *Nature* 393, 386–389, 1998.

90. Wolffe, A.P., Transcriptional control: imprinting insulation, *Curr. Biol.* 10, R463–465, 2000.

91. Steinwaerder, D.S., and Lieber, A., Insulation from viral transcriptional regulatory elements improves inducible transgene expression from adenovirus vectors *in vitro* and *in vivo*, *Gene Ther.* 7, 556–567, 2000.

92. Muller, J., Transcriptional control: the benefits of selective insulation, *Curr. Biol.* 10, R241–244, 2000.

CHAPTER **4**

Cellular Uptake and Trafficking

Sujatha Dokka and Yon Rojanasakul

CONTENTS

0-8493-1934-X/05/$0.00+$1.50
© 2005 by CRC Press LLC

4.1 INTRODUCTION

The basis for gene therapy is to change the expression of some genes in an attempt to treat, cure, and ultimately prevent disease. One of the principal barriers to effective gene therapy is the development of efficient vectors that can deliver the therapeutic gene to its target site, the nucleus. Viral vectors have evolved into the most efficient vectors for introducing foreign genes into cells. However, safety concerns and the immunological profile of viral vectors have steered research toward the development of efficient nonviral systems.[1]

Knowledge of the transport barrier and cellular delivery of polynucleotides (plasmid DNA [pDNA], antisense oligonucleotide [AS-ON], small interfering RNA [siRNA], etc.) is of importance in the development of therapeutically effective gene delivery. If the drug molecule does not reach its intended target, it cannot exert any therapeutic effect. Because the target for pDNA is the nucleus, its activity would therefore depend on how well the molecule can penetrate the cell membrane and reach the nuclear target.

Plasmid DNA is a double-stranded supercoiled DNA molecule. Plasmids used in mammalian cell transfection studies range from 3 to 15 kb (2000 to 10000 kDa). AS-ONs are single-stranded polynucleotides that are typically 10 to 25 nucleotides long and thus have molecular weights that range from 3000 to 7500 daltons; siRNAs are small RNA duplexes, slightly larger than AS-ONs. These polynucleotide molecules are large, hydrophilic macromolecules with a net negative surface charge. These characteristics prevent them from crossing biological membranes efficiently. Although AS-ONs and siRNAs are much smaller than pDNA, the problem of cellular delivery exists with these molecules as well. Most of the discussion throughout this chapter pertains to polynucleotides in general, but with a focus on pDNA.

A number of cellular barriers restrict efficient gene delivery.[2] These include transfer across the plasma membrane, movement through the cytoskeletal network within the cytoplasm, and transfer across the nuclear envelope. The low efficiency of DNA delivery is a natural consequence of this process, because at each step of the journey the number of transferring DNA molecules decreases. Identifying and overcoming each hurdle along the DNA entry pathways can improve DNA delivery, and hence overall transfection efficiency.

While the focus of this chapter is cellular uptake of DNA, it is important to know that the problem of gene delivery involves not only cellular uptake but also a number of other obstacles. These include biological barriers, such as the immunogenicity of the vector itself[3,4] and immune stimulation by certain DNA sequences (e.g., CpG motifs)[5] in the molecule. In addition, several other hurdles also need to be addressed, including cytotoxicity, DNA compaction, serum stability, cell-specific targeting, route of administration, endosomolysis, cytoplasmic stability, and nuclear transport. These hurdles have been discussed extensively in other reviews.[2]

The objective of this chapter is to summarize the critical steps in intracellular trafficking of polynucleotides and current approaches for overcoming these hurdles. A good place to start would be to study the most efficient delivery vectors: viruses.

4.2 VIRUSES: WHY ARE THEY SUCH EFFICIENT DELIVERY SYSTEMS?

Viruses have a natural ability to infect host cells, and hence they offer an excellent means of introducing foreign DNA sequences into cells for gene therapy.[6] It has been shown that binding of a single virus particle to the cell membrane is sufficient to infect the host cell.[7] This is exactly the reason why more than 70% of current clinical trials for gene therapy employ viral vectors. Studying the viral mechanism of entry into DNA can give us a good insight into the strategies for breaking down various cellular barriers. Viruses currently in use include retroviruses, adenoviruses, adeno-associated viruses, the herpes simplex virus, the papilloma virus, and others.

Viruses are large complexes (up to 100 nm). Because of their large size, it is unlikely that nuclear delivery of viral DNA occurs by passive diffusion.[8] Viruses differ in the way they infect and eventually deliver DNA into cells. Generally, the first step involves the binding of the virus to cell surface receptors on the cell membrane. These receptors may consist of sialic acid containing cell surface glycoproteins and glycolipids (in case of influenza virus)[9] or integrins on the cell surface (in case of adenovirus).[10] Internalization of the virus takes place either by endocytosis (e.g., adenovirus) or by fusion with plasma membrane (e.g., herpes simplex virus).[11] Viruses have evolved to escape or bypass endocytosis with extremely high efficiency. After internalization into endosomes, the viruses enter an acidic environment. This acidic environment triggers conformational changes in viral coat proteins, resulting in membrane permeabilization, which allows endosomal escape of the viral DNA.

The next step is to traverse the cytoplasm to reach the final destination, the nucleus. This transport is highly efficient and it has been speculated that viruses efficiently use the cytoskeletal network, including microtubules across the cytoplasm.[6,11] The final step is nuclear entry of the viral DNA. Viral DNA can range anywhere from 2.5 to 150 kb in size, and this enormous size presents a formidable barrier for nuclear delivery.[8] However, viruses have evolved and have overcome this barrier successfully. Some viruses (such as adenovirus) use the nuclear pore complex (NPC) machinery to gain entry into the nucleus by having a nuclear localization sequence (NLS).[12,13] An NLS on viral DNA is exposed some time after endosomal escape into cytoplasm, facilitating the entry. The influenza virus has approached the problem of size limits by splitting its genome into eight small segments to facilitate nuclear entry.[14,15] Simian virus 40, HIV, and herpes viruses are known to induce rapid migration of genes to the nucleus even in nondividing cells.[16] However, some viruses are effective only in dividing cells (e.g., retroviruses) because they take advantage of the breakdown of the nuclear membrane during mitosis. Among the different classes, adenovirus has been the most extensively studied.

4.3 CELLULAR BARRIERS TO DNA DELIVERY

4.3.1 Stability in Extracellular Compartments

There are a number of nucleases in the extracellular environment, and the plasmids are very susceptible to nuclease-mediated degradation. This degradation can be partly overcome by condensing/binding with polycations or polymers.[17,18]

4.3.2 Association of DNA with the Cell Membrane

As a general rule, the plasma membrane of living cells is relatively lipophilic and negatively charged, which restricts the transport of large and charged molecules. Therefore, it might be expected that this membrane would act as a barrier for polynucleotides. DNA by itself associates very poorly with the cell membrane because of charge repulsion. This problem has been addressed by the use of polycations as delivery systems. They mask the negative charge of DNA and hence improve binding to the cell membrane. The extent of binding to the cell membrane depends on the physico-chemical properties of the DNA–delivery system complex, which need to be characterized, because there may be significant variations. Efficiency and specificity can be further increased by incorporation of ligands that recognize cell surface receptors.[19-21] *In vitro* cell culture studies have shown that larger aggregates often have a greater degree of cell association.[22,23] This could partly be because of the ability of these aggregates to sediment, and hence facilitate greater cell contact in tissue culture. However, this effect is seen *in vitro* and may not be very relevant to *in vivo* conditions.

The role of molecules or receptors on the cell surface that facilitate binding has not been studied thoroughly. However, proteoglycans (PG) on the cell membrane appear to play a role in cationic lipid mediated transfer because cells deficient in PG synthesis are more difficult to transfect.[24,25] In our labs we have found routinely that transfection efficiency varies greatly between cell types.[26]

4.3.3   Cellular Internalization

Endocytosis seems to be the most predominant pathway for uptake of either naked DNA or DNA complexes.[27] The size and composition of the DNA complex and the cell type, however, play a major role in determining the mechanism of internalization. The best known form of endocytosis is initiated by clathrin-coated pit formation; other forms of endocytosis include caveolea-mediated and clathrin-independent internalizations.[28,29] Some cell types are capable of internalizing extracellular fluid via macropinocytosis and large particles via phagocytosis.[30] It has been demonstrated in several instances that escape of DNA from the endocytic compartments is one of the major barriers to efficient gene delivery. Endosomes can associate with both actin and microtubule (MT) motor proteins, and the intracellular trafficking and processing of endocytic vesicles are regulated by the cytoskeleton network.[30]

Viruses have proteins that undergo conformational change at the low pH of the endosomes. This conformational change to an amphipathic α-helical structure leads to membrane fusion and contents are released into the cytoplasm. Indeed, studies have shown that incorporation of a shortened version of hemagglutinin (HA2 from influenza virus) protein into DNA complexes increases gene transfection efficiency.[31]

A number of approaches have been used to overcome the problem of endocytic entrapment of DNA. Some of them are discussed below and are summarized in Table 4.1.

4.3.3.1 Proton Sponge

Buffering due to high concentrations of protonating groups leads to osmotic swelling and subsequent membrane lysis and, hence, DNA release. This concept has been tested by adding compounds such as sucrose, PVP, and glycerol, with some success.[32-34] The efficacy of the cationic polyethylenimine (PEI) has been related to its extensive buffering capacity, provoking the swelling and disruption of endosomes.[35]

4.3.3.2 pH-Sensitive Liposomes

These liposomes are formulated to destabilize under low pH conditions.[36] Mixing the pH-sensitive neutral lipid dioleylphosphatidylethanolamine (DOPE) with a cationic lipid has been shown to increase the efficiency of gene transfer. This fusogenic lipid promotes the fusion of lipid/DNA particles with the endosomal membrane, thus facilitating membrane disruption and increasing the amount of plasmid molecules released into the cytoplasm.[37] Unfortunately, most of these systems are not stable *in vivo*.

Table 4.1   Endolytic Strategies for Nonviral Delivery

Agent	Mechanism
pH-sensitive liposomes	Destabilization of membrane at low pH
Fusogenic peptides	Membrane lytic activity (e.g., HA2 from influenza virus)
DMI-2	Specific inhibitor of DNase I, inhibits lysosomal nucleases
Chloroquine	Lysosomotropic agent
Glycerol, PVP, sucrose, PEI	Proton sponge, osmotic swelling

4.3.3.3 Fusogenic Peptides

This approach utilizes peptides that trigger endosomal lysis or fusion, e.g., hemagglutinin or synthetic analogs designed to have membrane lytic activity. Although these systems have shown promise *in vitro*, their *in vivo* use may be limited due to their potential to stimulate a host immune response.[38-42] Fusogenic peptides are usually water soluble and coil randomly in conformation at physiological pH, but undergo transition to an amphipathic α-helix when pH is lowered, facilitating the process of endosomal disruption and hence increasing transfection. Other pharmacological agents that have been used for their endolytic activity include chloroquine[27] and DMI-2, a specific inhibitor of DNase I.[43]

4.3.4 Cytosolic Transport of DNA

Passive diffusion of large molecules like pDNA is highly inefficient in the cytoplasm. The cytoplasm consists of a filamentous cytoskeleton network embedded in an aqueous gel-like matrix. The matrix is composed of a concentrated mix of soluble proteins and organelles that are linked to the cytoskeletal fibers.[44-47] The cytoplasm is composed of a network of microfilament and microtubule systems and a variety of subcellular organelles. The role of the actin filaments and microtubules is to maintain intracellular distribution of organelles and to facilitate trafficking between organelles.[48] Further understanding of these systems may help us design systems that promote the transport of DNA to the nucleus. The mesh-like structure of the cytoskeleton, the presence of organelles and the high protein concentration (up to 100 mg/ml) impose an intensive molecular crowding of the cytoplasm, which limits the diffusion of large-sized molecules.[49] It is estimated that the viscosity of cytoplasm is similar to that of 13% dextran.[50] A study done by Luby-Phelps et al. with varying sizes of inert particles showed that diffusion in the cytoplasm is hindered in a size-dependent manner and that particles with a radial dimension greater than 260 Å are practically immobile in the cytoplasm.[51] Intracytoplasmic diffusion of proteins is usually slower than that of inert particles (e.g., ficolls or dextrans) due to binding interactions with intracellular components.

Dowty et al. concluded that pDNA was practically immobile in the cytoplasm of microinjected myotubes because pDNA remained at the site of microinjection.[52] The size cutoff for diffusion into the nucleus seemed to be ~250 bp and hence ONs did not seem to have a problem.[53,54] This was further confirmed by the fact that microinjection of pDNA into the proximity of the nucleus or decreasing the size of the expression vector led to significant enhancement of the transfection efficiency.[52,55]

4.3.4.1 Metabolic Instability of DNA in the Cytosol

The cytoplasm contains numerous nucleases that can digest DNA. Thus, metabolic instability of DNA in the cytoplasm represents another hurdle that needs to be considered. Lichardeur et al. microinjected pDNA into the cytoplasm and quantitatively measured the decay kinetics by single-cell video image analysis in HeLa and COS-1 cells.[56] They found that 50% of the DNA was eliminated in 1 to 2 hours and that this degradation was independent of the copy number or the conformation of the plasmid. Similar results were obtained in cell cycle arrested cells; they concluded that cytosolic elimination of pDNA could not be attributed to cell division.[56] The identity of the cytosolic nucleases responsible for pDNA degradation remains to be further elucidated. Such information will be useful for future design of effective gene delivery systems.

4.3.5 Nuclear Localization of Plasmid DNA

Nuclear trafficking of pDNA is the next hurdle to efficient transfection. The mechanism of DNA nuclear translocation and the question of whether the DNA is still associated with the delivery system are not fully understood, but appear to depend on the type of delivery system employed. Overall, there seem to be three possible routes for DNA entry into the nucleus: moving through the nuclear pores; entering during mitosis when the nuclear envelope breaks down; and moving physically across the nuclear membrane. The first two routes seem to be the most likely. No matter what the route, DNA has to reach the nucleus to have any therapeutic effect.

Studying the nucleocytoplasmic trafficking of proteins and viral nucleic acids has given us an insight into the transport mechanisms of macromolecules across the nuclear envelope. In this section, we summarize key findings on the nucleocytoplasmic transport of macromolecules in general, the nuclear import of pDNA, and attempts that have been made so far to increase the nuclear uptake of DNA. Nuclear transport has been shown unequivocally to be the rate-limiting step toward efficient DNA delivery.

4.3.5.1 Nucleocytoplasmic Transport

In eukaryotic cells, the nucleus is separated from the cytoplasm by a double-layered membrane system known as the nuclear envelope.[57] The nuclear membrane separates the nucleoplasm from the cytoplasm and thereby offers a means of regulation of gene expression that is unavailable to prokaryotes.[58] Nuclear pore complexes (NPCs) are the sites of exchange of macromolecules between the two compartments. The NPCs have a mass of about 125 megadaltons in higher eukaryotes and are estimated to contain roughly 100 different polypeptides.[59] The NPC constitutes a passive diffusion channel about 9 nm in diameter.[60,61] Typically, a mammalian cell nucleus possesses about 2000 NPCs/μm^2. Nuclear pores exist in at least two conformational states. The closed state permits passive diffusion of molecules of less than 9 nm in diameter, whereas the open state facilitates transport of particles less than 26 nm.[62] Proteins above the size limit for passive diffusion can enter the nucleus only in an active way. However, even small nuclear proteins, such as histones, generally enter the nucleus actively rather than by passive diffusion.[63]

Transport across the nuclear pore occurs in both directions and involves various substrates.[61] Current models for transport through the NPC rest on three major premises: first, that distinct types of cargo contain specific molecular signals. Second, that signal-dependent translocation of cargo through the NPC is mediated by mobile (shuttling) carriers. And third, that most transport processes require the small GTPase Ran.[57] In essence, it is thought that soluble carriers bind cargo in one compartment, escort it through the NPC, release it in the other compartment, and shuttle back to the first compartment.

Nuclear import of NLS-bearing proteins is a process involving several steps, mediated by several cytoplasmic factors.[58] The best characterized import pathway is the one employing importin-α and importin-β. First, the nucleophilic protein binds the karyopherin α/β heterodimer in the cytoplasm. Importin-α binds to NLS proteins, whereas importin-β strengthens the affinity of the complex for the NLS and mediates the docking of the cargo–carrier complex to the cytoplasmic face of the NPC.[64] Binding to the nuclear envelope does not require ATP or GTP and can occur at 0°C. GTPase Ran mediates the subsequent energy-dependent translocation of the docked transport substrate through the pore complex.[65] This process is energy and signal dependent. It is saturable and thus carrier mediated.

The role of signal sequences for protein import into the nucleus is well documented.[66] NLS sequences fit no consensus but fall in general into two classes: short basic sequences of four to seven amino acids, and longer bipartite sequences consisting of two stretches of basic amino acids separated by ten less-conserved amino acids. Proteins larger than 60 kDa are excluded from the nucleus unless they harbor the NLS. However, not every nuclear protein should necessarily be expected to possess its own NLS, as in some cases nuclear entry may take place by interaction with NLS-containing partners (piggyback transport).[67,68]

Among the different NLSs, the NLS of the simian virus 40 (SV40) large T antigen has been extensively studied. This short peptide possesses five basic amino acids and a single amino acid substitution abolishes its NLS function, resulting in a complete cytoplasmic localization. It has been shown that synthetic NLSs, notably those homologous to that of SV40 T antigen, direct nuclear import of a wide range of nonkaryophilic proteins.[69]

4.3.5.2 Nuclear Import of DNA

Nuclear import of exogenous DNA is a critical step in the efficiency of cell transfection. The nucleocytoplasmic transport of pDNA is relatively inefficient in comparison to the nuclear uptake of karyophilic proteins and viral genes. It has been estimated that less than 1% of plasmid molecules introduced into the cytoplasm eventually reach the nucleus.[70] The mechanism by which nucleic acids (such as pDNA) are imported into nuclei is still not completely understood. Nuclear uptake of DNA is dependent on the size of DNA because short DNA fragments such as ONs are easily transported across the nuclear pores by diffusion, whereas the plasmids remain in the cytoplasm. The size limit for passive diffusion of DNA was found to be between 200 and 310 bp.[71]

Studies by Capecchi et al.[72] have shown that plasmids microinjected into the cytoplasm of mouse fibroblasts deficient in thymidine kinase (TK) did not result in any detectable TK activity, whereas plasmids microinjected into the nucleus resulted in TK expression in 50–100% of the cells. Similar results were obtained in human skin fibroblasts[73] and rat embryo fibroblasts.[74] Thus, it seems that the major barrier to efficient transgene expression is the final transport step, when the DNA must reach the nucleus for transcription.

The genome of a virus can range from 2.5 to 200 kbp and can present a formidable barrier for movement through the cytoplasm and entry into the nucleus. Viruses have developed efficient methods for entering the nucleus of the host cell. A good start to enhancing DNA nuclear uptake would be to study the nuclear uptake of viral nucleic acids. There is evidence that import of viral nucleic acids into the nuclei of host cells utilizes pathways similar to those of nuclear protein import. The entry of influenza virus RNA into the nucleus of permeabilized cells is mediated by an associated nucleoprotein (NP) containing an NLS.[75] Conjugating NP to nonviral RNA also induces nuclear import.[76] Similarly, the matrix protein of HIV-1 virus contains an NLS likely to function in nuclear import.[77]

In vitro studies have shown that DNA entry into mammalian nuclei occurs through the NPC, is energy dependent, and can be inhibited by wheat germ agglutinin (WGA).[78] DNA import was also prevented by inhibitors of transcription,[79] suggesting that nuclear import of exogenous DNA is dependent on import of transcription factors. This would argue that, as for viral DNA, nuclear import of pDNA necessitates its association with NLS-containing proteins. However, since pDNA does not have an NLS, the efficiency of nuclear uptake is very low. Thus, it would be reasonable to assume that attaching a nuclear targeting moiety to pDNA would enhance nuclear uptake.

Another mechanism by which DNA is thought to gain access to the nucleus is by association with nuclear material on breakdown of the nuclear envelope during mitosis. This association has been observed for both cationic lipid and polymer-mediated gene delivery by comparing transfection efficiencies for cells at various stages in the cell cycle.[40,80-83]

4.3.5.3 Attempts to Increase the Efficiency of Nuclear DNA Uptake

Numerous efforts have been made to improve the entry of pDNA into the nucleus. It is well established that the attachment of NLS to proteins not normally karyophilic enables their efficient nuclear targeting. This has inspired similar efforts to increase DNA nuclear uptake by providing an NLS to the DNA, which can be achieved by noncovalently or covalently complexing the NLS-containing molecules with DNA. The resulting DNA–NLS complexes can be recognized as nuclear import substrates by specific intracellular receptor proteins, allowing for nuclear translocation.

4.3.5.3.1 Noncovalent Complexes

Nuclear proteins are by definition destined to the nucleus; hence, improvement in the efficiency of gene transfer into the cell nucleus initially involves the coupling of pDNA to nuclear proteins. Binding of DNA to the core protein VII of adenovirus type 2, or to salmon sperm protamine, has been shown to increase the nuclear trafficking of DNA and improve transgene expression.[84] Similarly, pDNA complexed to the nonhistone chromosomal protein HMG-1 is rapidly transported into the nucleus of cultured Ltk⁻ cells.[85] The successful use of DNA–nuclear protein complexes for gene transfer in mammalian cells has provided the basis for a recent series of investigations involving the coupling of pDNA to short synthetic peptides carrying the NLS.

Collas et al.[86-89] reported that NLS peptides complexed to pDNA facilitated rapid uptake of the DNA into embryonic nuclei. They used synthetic NLS peptides analogous to that of SV40 T antigen and a plasmid carrying a firefly luciferase reporter gene. The NLS peptides were bound to pDNA by ionic interactions. Cytoplasmic injection of the pDNA–NLS complexes was shown to induce transgene expression in zebrafish embryos, indicating transfer of the exogenous DNA into nuclei.[88] In contrast, naked DNA or plasmid complexed to a reverse NLS peptide deficient in nuclear import was not readily imported into the nuclei. These investigators also studied the nuclear import of pDNA mediated by an NLS in a cell-free extract.[86] They demonstrated that nuclear import of DNA–NLS complexes is a two-step process involving binding to, and translocation across, the nuclear envelope. Peptides are also likely to mediate active transport of DNA–NLS complexes across the nuclear envelope. Binding is ATP independent, occurs at 0°C, and is Ca²⁺ independent. By contrast, translocation requires ATP hydrolysis and Ca²⁺, is temperature dependent, and is blocked by the lectin WGA. The investigators finally concluded that the requirements for NLS-mediated nuclear import of pDNA are similar to those for nuclear import of protein–NLS conjugates in permeabilized cells.

In another study, Aronsohn et al.[90] added NLS peptides to the cationic liposome/DNA complexes and studied their transgene expression. They reported a threefold increase in luciferase gene expression with the addition of NLS peptides. However, they did not investigate how the NLS enhanced nuclear uptake of pDNA and did not address the question of whether the NLS still attaches to the DNA after endosomal release. The dissociation of pDNA from the liposomal complex in the endosome has been proposed as a possible mechanism of pDNA escape from the endosome via cationic liposome-mediated gene transfer.[91] It is thus more than likely that the NLS would dissociate from the plasmid before this complex reaches the nuclear envelope.

4.3.5.3.2 Covalent Modification of Plasmid DNA

Noncovalent association of NLS sequences to DNA results in a weak association. Thus, there is always a possibility that the NLS will dissociate from DNA before reaching and interacting with the nuclear pore complex. This could explain why these systems have not always been successful.[92] This has led to an interest in covalently attaching NLS peptides to pDNA.[92-95]

To determine whether cross-linking of NLS to double-stranded DNA (dsDNA) would result in robust nuclear uptake comparable to that of NLS-conjugated proteins, Sebestyen et al.[92] developed a technique that uses the DNA alkylating moiety, cyclopropapyrroloindole (CPI), to attach cationic peptides to dsDNA. They reported a significant increase in nuclear uptake of NLS pDNA, which was dependent on the number of NLS peptides in digitonin-permeabilized cells. Attachment of similar peptides without a functional NLS sequence did not increase nuclear uptake, suggesting that the increased nuclear uptake was due to the NLS. The investigators also suggested that the NLS peptide enhanced the nuclear uptake of DNA via the classical pathway for the nuclear transport of karyophilic proteins, based on the effects of various inhibitors on the nuclear transport. Although this study showed the potential of NLS to enhance nuclear uptake, it did so at the expense of transgene expression. The major disadvantage of this nonspecific covalent approach was that transcription was abolished due to the high number of covalent modifications on the plasmid. Also, sequence-specific conjugation was not possible, because the covalent adducts resulting from the reaction were randomly positioned in the nucleic acid.

Ciolina et al.[93] used another chemical strategy to covalently attach cationic NLS peptides to pDNA. They attached a cationic peptide, NLS from SV40 large T antigen, to a photoactive tetrafluorophenylazido-containing molecule. This peptide was then covalently conjugated to pDNA by photoactivation. This strategy allowed the investigators to control the number of NLSs attached to the peptide. By decreasing the covalent modifications, they seemed to maintain some biological function. They also showed that NLS peptides attached to dsDNA interacted with the NLS-binding protein importin-α in a sequence-specific manner, suggesting that the NLS peptides are recognized despite plasmid electrostatic properties. However, even using this approach they could not specify the sites on the plasmid for covalent modification.

Neves et al.[96] developed a strategy to associate a cationic NLS peptide covalently with pDNA at a specific site on the plasmid by triple helix formation. They coupled a purified ON modified with a thiol residue to an NLS-bearing peptide containing a maleimide group. A triple helix was then formed between the ON–NLS conjugate and the pDNA. They found that the NLS on pDNA was recognized by importin-α, but there was no increase in gene expression over pDNA control.

More recently, peptide nucleic acids (PNAs) have been used successfully to link molecules to DNA in a sequence-specific manner.[97-99] PNAs are uncharged nucleic acid mimics consisting of repetitive units of 2-aminoethyl-glycine. This pseudo-peptide backbone provides biological stability and access to a variety of chemical modifications, making PNA an ideal linker molecule.[98,100,101]

Some of the most encouraging *in vivo* data so far is the work published by Dasi et al.[102] In this study they conjugated NLS peptides covalently to asialofetuin (ASF) targeted and nontargeted liposome/transgene complex via a helper lipid, DPPE. The NLS peptide-conjugated ASF-targeted liposomes showed greater transfection efficacy and longer duration of expression compared to the nontargeted NLS-conjugated liposomes.[102]

Finally, it is important to note the potential pitfalls and limitations of the NLS approach. First, the use of this approach may trigger an immune response. Second, the interaction of DNA with the NLS is highly possible, leading to the masking of NLS and rendering the NLS ineffective. Third, it was found that an optimal degree of labeling with the NLS is required because too many signals per DNA copy may actually inhibit translocation by simultaneously interacting with multiple nuclear pores.

4.3.5.3.3 Alternate Approaches

To circumvent the problem of inefficient nuclear delivery of pDNA, cytoplasmic expression vectors have been developed.[80,103] T7 polymerase introduced into the cytoplasm could be utilized to drive cytoplasmic transcription of the pDNA, thus avoiding the need for nuclear entry of the plasmid. However, the immune response to the polymerase may ultimately lead to loss of transgene expression, and these systems need to be further developed before they can become a clinical reality.

Another approach is to include in the DNA some nucleotide sequences that have an affinity for nuclear transport proteins such as transcription factors. The potential utility of this approach was demonstrated by incorporating the promoter for smooth muscle gamma actin into pDNA. It was found that pDNA was effectively transported into the nucleus of smooth muscle cells.[104] A similar concept was tested by Mesika et al.[105] They designed a vector that contained repetitive binding sites for the induction of transcription factor NF-κB. This system exhibited a 12-fold increase in nuclear delivery and expression of the pDNA.[105]

4.4 CURRENT APPROACHES FOR ENHANCING GENE DELIVERY

Gene delivery systems can be broadly divided into viral vectors, chemical/synthetic vectors (broadly classified as nonviral vectors), and physical methods.

4.4.1 Viral Vectors

These are by far the most efficient vectors, as discussed above in Section 2, and hence 70% of clinical trials utilize these vectors. Currently, the viral vectors in clinical trials include retroviruses, adenovirus, pox virus, adeno-associated virus, and herpes simplex virus. These vectors have been used for a wide range of indications from cancer to infectious diseases. Unfortunately, these vectors have a number of disadvantages, including limited DNA carrying capacity, lack of specificity, production and packaging problems, replication and recombination potential, and high cost. For these reasons, nonviral systems, especially synthetic DNA delivery systems, have become increasingly desirable. Table 4.2 lists the advantages and disadvantages of current viral vectors. These vectors have been described in much more detail elsewhere (www.wiley.co.uk/genmed).

Table 4.2 Viral Gene Delivery Systems

Vector	Advantages	Disadvantages
Retrovirus	Integration into host genome 1–30% transduction efficiency	Low transduction efficiency Packaging cell line required No targeting Replication competence Insert size 9–12 kb Infects only dividing cells Low titer and unstable in blood Random DNA insertion
Adenovirus	High transduction efficiency Infection of many cell types Infection does not require cell division	No integration/temporary effect Packaging cell line required Safety/toxicity/immunogenicity Replication competence No targeting Insert size 4–5 kb
Adeno-associated virus	Integration into host genome Infection does not require cell division	No targeting Packaging cell line required Safety Insert size 5 kb Produced in low titers Good for small scale only
Herpes simplex virus	Infects wide range of cell types Large insert size, 40–50 kb Relatively prolonged expression Very high titers	No targeting Packaging cell line required Toxicity Difficult to develop due to complexity
Vaccinia virus	Large insert size, 25 kb	Immunogenicity/toxicity/safety No targeting
Avipox virus	Infection does not require cell division Large insert size, >4 kb	Immunogenicity/toxicity/safety No targeting

4.4.2 Chemical/Synthetic Nonviral Vectors

Nonviral delivery systems have considerable advantages over viral counterparts because of greater control of their molecular composition for simplified manufacturing and analysis, flexibility in the size of the transgene to be delivered, and relatively lower immunogenicity. For these reasons, much work is being done in this area. However, the biggest disadvantage of these systems is their relative inefficiency of transfection compared to the viral vectors. Here we summarize two major classes of nonviral vectors: liposomes or lipid conjugates and DNA–protein conjugates. Synthetic nonviral delivery systems have been extensively reviewed elsewhere.[22,106-115]

4.4.2.1 Liposomes

Liposomes are classified as either anionic or cationic, based on their net negative or positive charge, respectively. The encapsulation efficiency of anionic liposomes is low, thus limiting the applicability of these liposomes for gene delivery. Felgner et al. synthesized several cationic liposomes and demonstrated that they could avidly and efficiently bind nucleic acids (which are anionic) by electrostatic interactions upon simple incubation of the liposomes with nucleic acids.[22] After their first reported use, this area has improved greatly. Cationic lipids are typically a mixture of amphipathic cationic lipid (e.g., DOTMA) and fusogenic lipid (e.g., DOPE) and are used as dispersions of small unilamellar or multilamellar liposomes. The cationic lipids differ markedly and may contain single or multiple charges, cationic detergent, or polylysine. The three basic parts of a cationic lipid are a hydrophobic lipid anchor group, which helps in forming liposomes and can interact with cell membranes; a linker group; and a positively charged headgroup that interacts with plasmid, leading to its condensation.[114] The physicochemical properties of plasmid/lipid complexes are strongly influenced by the relative proportions of each component and the structure of the headgroup. Cationic lipids with multivalent headgroups have been shown, in general, to be more effective than their monovalent counterparts.

One of the disadvantages of liposome-mediated gene transfer is its low transduction efficiency compared to viral vectors. This could be attributed to the mode of uptake of liposome–DNA complexes. Liposomes are taken up by cells by means of endocytosis, and unless an endolytic agent is added to the formulation, release of DNA from endosomes is very inefficient.[116] Free plasmids, and not plasmid–lipid complexes, injected into the nucleus are expressed.[116] This suggests that plasmids need to be released from the plasmid–lipid complexes prior to entering the nucleus for expression to occur.

Liposome-mediated gene delivery has been targeted to a number of diseases such as cancer, cystic fibrosis, and α_1-trypsin deficiency. Fusogenic liposomes[117] and pH-sensitive liposomes[118] have also been used with some success.

4.4.2.2 DNA-Protein Conjugates

These are cell-specific DNA delivery systems that utilize unique cell surface receptors on the target cell. Attaching the ligand recognized by such a receptor to the transgene DNA causes the DNA–ligand complex to become selectively bound and internalized into the target cell.[19,20] They offer a potential cell-specific gene transfer without the problems associated with viral vectors, such as replication, immunogenic viral proteins, and recombination potential. Poly(L-lysine) (PLL), a polycation, has been routinely used as it can be easily coupled to a variety of protein ligands by chemical cross-linking methods. When the PLL–ligand adduct is mixed with pDNA, macromolecular complexes form in which the DNA is electrostatically bound to the PLL–ligand molecules. These structures present ligands to the cell surface receptor that are efficiently endocytosed. The transferring receptor, the asialoorosomucoid receptor, and cell surface carbohydrates have been used to demonstrate the potential of ligand-mediated gene delivery.

To utilize the endosomal escape functions of viruses, investigators have constructed physically linked complexes between the adenovirus and the DNA–ligand adduct.[21] Adenovirus greatly enhanced the penetration of protein taken up by endocytosis into the cytoplasm, possibly by destruction of the endosomal membrane.

4.4.3 Physical Methods of DNA Delivery

4.4.3.1 Electroporation

Electroporation or electropermeabilization involves the use of an electric field to open up pores in the cell. During the time that the pores are open, DNA can enter the cell directly into the cytoplasm and ultimately into the nucleus. This process is a physical process, not dependent upon special characteristics of the cell, and therefore can be used with virtually any cell type. The optimal amplitude and length of pulse will vary for each cell type, so this procedure should be optimized for each cell type. Increasing the voltage would increase the amount of DNA per cell due to the induction of a larger pore size; however, the toxicity would also be higher. This method has been used to successfully transfer genes *in vivo* and *in vitro*. Recent data by Golzio et al.[119] sheds some light on the mechanism by which electroporation seems to enhance DNA delivery. Based on their data, the investigators outline the following steps in electroporation-mediated transfer of DNA. First, the electric field permeabilizes the plasma membrane, and negatively charged DNA migrates electrophoretically toward the plasma membrane on the cathode side. Second, plasmids interact with the plasma membrane. Third, translocation into cytosol and consequently to the nucleus occurs. Tissues targeted for gene expression *in vivo* by electroporation have been skin,[120-121] liver,[122,123] tumors,[124–126] and muscle.[127] Some initial work done by Widera et al.[128] suggests that electroporation may be effective for human vaccination.

4.4.3.2 Direct DNA Injection and Microinjection

As implied, direct DNA injection involves purified DNA directly injected into the desired tissue without any special carrier molecules. Conceptually this is a very simple, inexpensive, and nontoxic procedure with a potential to carry large DNA constructs. Currently the major application of gene introduction by direct injection of pure DNA is immunization against foreign antigens where a low level of gene expression is often sufficient to achieve an immunological response.

Microinjection is a process by which DNA can be directly injected into the cytoplasm or nucleus of an individual cell. Cells are cultured on glass coverslips on an inverted microscope and each cell is injected with the DNA using glass capillary pipettes prepared using a needle puller. The advantages of this procedure are as follows:

- It is independent of the length and sequence of the DNA
- DNA is applied directly to the site of action
- Culture environment can be controlled
- The amount of DNA reaching the target site can be controlled
- It is highly efficient

The disadvantages of this procedure are as follows:

- It requires specialized equipment
- It is technically demanding and hence not reproducible all the time
- It is a monocellular technique (it can be performed only in a single cell)
- It has limited applications *in vivo*

This method is hence suited either for mechanistic studies or for studies where *in vivo* regulated processes are to be examined in a single cell.

4.4.3.3 *Bombardment of DNA-Coated Particles*

This is a method of introducing DNA using metal particles as a carrier. It has been shown that particle bombardment is an effective means of gene transfer both *in vitro* and *in vivo*. In this method, pDNA is first coated onto the surface of gold particles (approximately 1 micron in diameter) and then propelled using a gene gun to accelerate the DNA-coated particles into superficial cells of the skin or into skin tumors.[129,130] Gene transfer mediated by particle bombardment has an advantage in that it can be applied to various tissue cells and cancer cells *in vivo* with relatively high efficiency. It has been speculated that the physical force of impact overcomes the cell membrane barrier. Disadvantages of particle-bombardment-mediated gene transfer are that it requires specific equipment and that only the surface portion of organs can be transfected. Gene expression lasts only for a few days because the pDNA does not integrate into the host cell genome. Because of the limited depth of DNA penetration, this technique is limited to surface cells that can be accessed directly. Gene-gun delivery is ideally suited for gene-mediated immunization, where only brief expression of the antigen is necessary to achieve an immune response. Since the epidermal layers of the skin are rich in antigen-presenting cells, they are the preferred targets for vaccination. The simplicity, safety and technical ease of preparation of this DNA transfer system make its large-scale application more feasible than available viral DNA delivery systems.

4.4.4 Other Physical Approaches

A number of other methods are still in the early stages of development, including shock wave-mediated delivery,[131] laser-induced pressure transients,[132] microseeding/microneedles,[133] ultrasound,[134,135] and photochemical transfection.[136] Some of these systems have great potential while others are far away from being therapeutically relevant at this point in time.

4.5 CONCLUSIONS

The ultimate success of gene therapy will be determined by the precision, efficiency, and safety by which gene transfer can be accomplished. Clearly, the field of genetic therapy is just in its infancy. Thus far the trials have not shown convincingly that gene therapy is effective in treating diseases in humans. However, these results reflect the limitations of gene transfer and not the limitations of gene therapy approach itself.

The ideal DNA delivery system should possess the following properties: ease of assembly; efficient delivery leading to total transfection; stabilization of DNA before and after uptake; capability of bypassing or escaping from endocytic pathways; efficient nuclear targeting; and high, persistent, and adjustable expression of therapeutic levels of proteins. Lessons from viruses have already improved synthetic delivery systems. Understanding and incorporating the extremely efficient mechanisms of infection by viruses will help to make future DNA delivery systems virus-like in function, though not necessarily in shape.

Introduction of foreign genes into cells is not natural, and hence cells appear to have biological barriers that protect against invasion by foreign genes. A better understanding of the cellular mechanisms by which vectors transfer to and traffic in cells should help in the design of more efficient vectors. Crucial to the success of DNA as a therapeutic modality is transfection efficiency. In the past few years, there has been significant progress in this direction. The studies reviewed above demonstrate the potential utility of various gene delivery systems and describe some of their advantages and disadvantages. The practical use of these systems will require further evaluations of their *in vivo* safety and efficacy profiles. Although much remains to be established, significant progress has clearly been made in this area that is proving to be an important advance in gene therapeutics.

REFERENCES

1. Roth, J.A. and Cristiano, R.J., Gene therapy for cancer: what have we done and where are we going? *J. Nat. Cancer Inst.* 89(1), 21–39, 1997.
2. Bally, M.B., Harvie, P., Wong, F.M., Kong, S., Wasan, E.K., and Reimer, D.L., Biological barriers to cellular delivery of lipid-based DNA carriers, *Adv. Drug Deliv. Rev.* 38(3), 291–315, 1999.
3. Dokka, S., Toledo, D., Shi, X., Castranova, V., and Rojanasakul, Y., Oxygen radical-mediated pulmonary toxicity induced by some cationic liposomes, *Pharm. Res.* 17(5), 521–525, 2000.
4. Ruiz, F.E., Clancy, J.P., Perricone, M.A., Bebok, Z., Hong, J.S., Cheng, S.H., Meeker, D.P., Young, K.R., Schoumacher, R.A., Weatherly, M.R., Wing, L., Morris, J.E., Sindel, L., Rosenberg, M., van Ginkel, F.W., McGhee, J.R., Kelly, D., Lyrene, R.K., and Sorscher, E.J., A clinical inflammatory syndrome attributable to aerosolized lipid-DNA administration in cystic fibrosis, *Hum. Gene Ther.* 12(7), 751–761, 2001.
5. Scheule, R.K., The role of CpG motifs in immunostimulation and gene therapy, *Adv. Drug Deliv. Rev.,* 44(2–3), 119–134, 2000.
6. Sodeik, B., Mechanisms of viral transport in the cytoplasm, *Trends Microbiol.* 8 (10), 465–472, 2000.
7. Seisenberger, G., Ried, M.U., Endress, T., Buning, H., Hallek, M., and Brauchle, C., Real-time single-molecule imaging of the infection pathway of an adeno-associated virus, *Science* 294(5548), 1929–1932, 2001.
8. Kasamatsu, H. and Nakanishi, A., How do animal DNA viruses get to the nucleus? *Annu. Rev. Microbiol.* 52, 627–686, 1998.
9. Marsh, M. and Helenius, A., Virus entry into animal cells, *Adv. Virus Res.* 36, 107–151, 1989.
10. Wickham, T.J., Mathias, P., Cheresh, D.A., and Nemerow, G.R., Integrins αv $\beta 3$ and αv $\beta 5$ promote adenovirus internalization but not virus attachment, *Cell* 73(2), 309–319, 1993.
11. Sodeik, B., Ebersold, M.W., and Helenius, A., Microtubule-mediated transport of incoming herpes simplex virus 1 capsids to the nucleus, *J. Cell Biol.* 136(5), 1007–1021, 1997.
12. Greber, U.F., Webster, P., Weber, J., and Helenius, A., The role of the adenovirus protease on virus entry into cells, *EMBO J.* 15(8), 1766–1777, 1996.
13. Greber, U.F., Suomalainen, M., Stidwill, R.P., Boucke, K., Ebersold, M.W., and Helenius, A., The role of the nuclear pore complex in adenovirus DNA entry, *EMBO J.* 16(19), 5998–6007, 1997.
14. Jennings, P.A., Finch, J.T., Winter, G., and Robertson, J.S., Does the higher order structure of the influenza virus ribonucleoprotein guide sequence rearrangements in influenza viral RNA? *Cell* 34(2), 619–627, 1983.
15. Compans, R.W., Content, J., and Duesberg, P.H., Structure of the ribonucleoprotein of influenza virus, *J. Virol.* 10(4), 795–800, 1972.
16. Izaurralde, E., Kann, M., Pante, N., Sodeik, B., and Hohn, T., Viruses, microorganisms and scientists meet the nuclear pore, *EMBO J.* 18(2), 289–296, 1999.
17. Rolland, A.P. and Mumper, R.J., Plasmid delivery to muscle: recent advances in polymer delivery systems, *Adv. Drug Deliv. Rev.* 30(1–3), 151–172, 1998.
18. Li, S., Tseng, W.C., Stolz, D.B., Wu, S.P., Watkins, S.C., and Huang, L., Dynamic changes in the characteristics of cationic lipidic vectors after exposure to mouse serum: implications for intravenous lipofection, *Gene Ther.* 6(4), 585–594, 1999.
19. Wu, G.Y. and Wu, C.H., Receptor-mediated gene delivery and expression *in vivo*, *J. Biol. Chem.* 263(29), 14621–14624, 1988.
20. Wu, G.Y. and Wu, C.H., Receptor-mediated *in vitro* gene transformation by a soluble DNA carrier system, *J. Biol. Chem.* 262(10), 4429–4432, 1987.
21. Wagner, E., Zatloukal, K., Cotten, M., Kirlappos, H., Mechtler, K., Curiel, D.T., and Birnstiel, M.L., Coupling of adenovirus to transferrin-polylysine/DNA complexes greatly enhances receptor-mediated gene delivery and expression of transfected genes, *Proc. Nat. Acad. Sci. U.S.A.* 89(13), 6099–6103, 1992.

22. Felgner, P.L., Gadek, T.R., Holm, M., Roman, R., Chan, H.W., Wenz, M., Northrop, J.P., Ringold, G.M., and Danielsen, M., Lipofection: a highly efficient, lipid-mediated DNA-transfection procedure, *Proc. Nat. Acad. Sci. U.S.A.* 84(21), 7413–7417, 1987.

23. Yagi, K., Noda, H., Kurono, M., and Ohishi, N., Efficient gene transfer with less cytotoxicity by means of cationic multilamellar liposomes, *Biochem. Biophys. Res. Commun.* 196(3), 1042–1048, 1993.

24. Mislick, K.A. and Baldeschwieler, J.D., Evidence for the role of proteoglycans in cation-mediated gene transfer, *Proc. Nat. Acad. Sci. U.S.A.* 93(22), 12349–12354, 1996.

25. Mounkes, L.C., Zhong, W., Cipres-Palacin, G., Heath, T.D., and Debs, R.J., Proteoglycans mediate cationic liposome-DNA complex-based gene delivery *in vitro* and *in vivo*, *J. Biol. Chem.* 273(40), 26164–26170, 1998.

26. Dokka, S., Toledo, D., Shi, X., Ye, J., and Rojanasakul, Y., High-efficiency gene transfection of macrophages by lipoplexes, *Int. J. Pharm.* 206(1–2), 97–104, 2000.

27. Guy, J., Drabek, D., and Antoniou, M., Delivery of DNA into mammalian cells by receptor-mediated endocytosis and gene therapy, *Mol. Biotechnol.* 3(3), 237–248, 1995.

28. Mukherjee, S., Ghosh, R.N., and Maxfield, F.R., Endocytosis, *Physiol. Rev.* 77(3), 759–803, 1997.

29. Nichols, B.J. and Lippincott-Schwartz, J., Endocytosis without clathrin coats, *Trends Cell Biol.* 11(10), 406–12, 2001.

30. Apodaca, G., Endocytic traffic in polarized epithelial cells: role of the actin and microtubule cytoskeleton, *Traffic* 2(3), 149–159, 2001.

31. Plank, C., Oberhauser, B., Mechtler, K., Koch, C., and Wagner, E., The influence of endosome-disruptive peptides on gene transfer using synthetic virus-like gene transfer systems, *J. Biol. Chem.* 269(17), 12918–12924, 1994.

32. Pauner, W., Kichler, A., Schmidt, W., Sinski, A., and Wagner, E., Glycerol enhancement of ligand-polylysine/DNA transfection, *Biotechniques* 20(5), 905–913, 1996.

33. Zauner, W., Kichler, A., Schmidt, W., Mechtler, K., and Wagner, E., Glycerol and polylysine synergize in their ability to rupture vesicular membranes: a mechanism for increased transferrin-polylysine-mediated gene transfer, *Exp. Cell Res.* 232(1), 137–145, 1997.

34. Ciftci, K. and Levy, R.J., Enhanced plasmid DNA transfection with lysosomotropic agents in cultured fibroblasts, *Int. J. Pharm.* 218(1–2), 81–92, 2001.

35. Boussif, O., Lezoualc'h, F., Zanta, M.A., Mergny, M.D., Scherman, D., Demeneix, B., and Behr, J.P., A versatile vector for gene and oligonucleotide transfer into cells in culture and *in vivo*: polyethylenimine, *Proc. Nat. Acad. Sci. U.S.A.* 92(16), 7297–7301, 1995.

36. Huang, L., Connor, J., and Wang, C.Y., pH-sensitive immunoliposomes, *Methods Enzymol.* 149, 88–99, 1987.

37. Litzinger, D.C. and Huang, L., Phosphatidylethanolamine liposomes: drug delivery, gene transfer and immunodiagnostic applications, *Biochim. Biophys. Acta* 1113(2), 201–227, 1992.

38. Morris, M.C., Vidal, P., Chaloin, L., Heitz, F., and Divita, G., A new peptide vector for efficient delivery of oligonucleotides into mammalian cells, *Nucleic Acids Res.* 25(14), 2730–2736, 1997.

39. Scheule, R.K., Bagley, R.G., Erickson, A.L., Wang, K.X., Fang, S.L., Vaccaro, C., O'Riordan, C.R., Cheng, S.H., and Smith, A.E., Delivery of purified, functional CFTR to epithelial cells *in vitro* using influenza hemagglutinin, *Am. J. Respir. Cell Mol. Biol.* 13(3), 330–343, 1995.

40. Wilke, M., Fortunati, E., van den Broek, M., Hoogeveen, A.T., and Scholte, B.J., Efficacy of a peptide-based gene delivery system depends on mitotic activity, *Gene Ther.* 3(12), 1133–1142, 1996.

41. Plank, C., Zauner, W., and Wagner, E., Application of membrane-active peptides for drug and gene delivery across cellular membranes, *Adv. Drug Deliv. Rev.* 34(1), 21–35, 1998.

42. Wagner, E., Effects of membrane-active agents in gene delivery, *J. Control Release* 53(1–3), 155–8, 1998.

43. Ross, G.F., Bruno, M.D., Uyeda, M., Suzuki, K., Nagao, K., Whitsett, J.A., and Korfhagen, T.R., Enhanced reporter gene expression in cells transfected in the presence of DMI-2, an acid nuclease inhibitor, *Gene Ther.* 5(9), 1244–1250, 1998.

44. Wolosewick, J.J. and Porter, K.R., Microtrabecular lattice of the cytoplasmic ground substance: artifact or reality, *J. Cell Biol.* 82(1), 114–39, 1979.

45. Schliwa, M. and van Blerkom, J., Structural interaction of cytoskeletal components, *J. Cell Biol.* 90(1), 222–35, 1981.

46. Heuser, J.E. and Kirschner, M.W., Filament organization revealed in platinum replicas of freeze-dried cytoskeletons, *J. Cell Biol.* 86(1), 212–34, 1980.

47. Porter, K.R., The cytomatrix: a short history of its study, *J. Cell Biol.* 99(1 Pt. 2), 3s–12s, 1984.

48. Cole, N.B. and Lippincott-Schwartz, J., Organization of organelles and membrane traffic by micro-tubules, *Curr. Opin. Cell Biol.* 7(1), 55–64, 1995.

49. Luby-Phelps, K., Taylor, D.L., and Lanni, F., Probing the structure of cytoplasm, *J. Cell Biol.* 102(6), 2015–2022, 1986.

50. Kao, C.Y. and Sharon, J., Chimeric antibodies with anti-dextran-derived complementarity-determining regions and anti-*p*-azophenylarsonate-derived framework regions, *J. Immunol.* 151(4), 1968–1978, 1993.

51. Luby-Phelps, K., Castle, P.E., Taylor, D.L., and Lanni, F., Hindered diffusion of inert tracer particles in the cytoplasm of mouse 3T3 cells, *Proc. Nat. Acad. Sci. U.S.A.* 84(14), 4910–4913, 1987.

52. Dowty, M.E., Williams, P., Zhang, G., Hagstrom, J.E., and Wolff, J.A., Plasmid DNA entry into postmitotic nuclei of primary rat myotubes, *Proc. Nat. Acad. Sci. U.S.A.* 92(10), 4572–4576, 1995.

53. Lukacs, G.L., Haggie, P., Seksek, O., Lechardeur, D., Freedman, N., and Verkman, A.S., Size-dependent DNA mobility in cytoplasm and nucleus, *J. Biol. Chem.* 275(3), 1625–1629, 2000.

54. Leonetti, J.P., Mechti, N., Degols, G., Gagnor, C., and Lebleu, B., Intracellular distribution of micro-injected antisense oligonucleotides, *Proc. Nat. Acad. Sci. U.S.A.* 88(7), 2702–2706, 1991.

55. Darquet, A.M., Rangara, R., Kreiss, P., Schwartz, B., Naimi, S., Delaere, P., Crouzet, J., and Scherman, D., Minicircle: an improved DNA molecule for *in vitro* and *in vivo* gene transfer, *Gene Ther.* 6(2), 209–218, 1999.

56. Lechardeur, D., Sohn, K.J., Haardt, M., Joshi, P.B., Monck, M., Graham, R.W., Beatty, B., Squire, J., O'Brodovich, H., and Lukacs, G.L., Metabolic instability of plasmid DNA in the cytosol: a potential barrier to gene transfer, *Gene Ther.* 6(4), 482–497, 1999.

57. Nigg, E.A., Nucleocytoplasmic transport: signals, mechanisms and regulation, *Nature* 386(6627), 779–787, 1997.

58. Gorlich, D. and Mattaj, I.W., Nucleocytoplasmic transport, *Science* 271(5255), 1513–1518, 1996.

59. Bastos, R., Pante, N., and Burke, B., Nuclear pore complex proteins, *Int. Rev. Cytol.* 162B, 257–302, 1995.

60. Pante, N. and Aebi, U., Molecular dissection of the nuclear pore complex, *Crit. Rev. Biochem. Mol. Biol.* 31 (2), 153–199, 1996.

61. Feldherr, C.M. and Akin, D., EM visualization of nucleocytoplasmic transport processes, *Electron Microsc. Rev.* 3(1), 73–86, 1990.

62. Ryan, K.J. and Wente, S.R., The nuclear pore complex: a protein machine bridging the nucleus and cytoplasm, *Curr. Opin. Cell Biol.* 12(3), 361–371, 2000.

63. Breeuwer, M. and Goldfarb, D.S., Facilitated nuclear transport of histone H1 and other small nucleo-philic proteins, *Cell* 60(6), 999–1008, 1990.

64. Gerace, L., Molecular trafficking across the nuclear pore complex, *Curr. Opin. Cell Biol.* 4(4), 637–645, 1992.

65. Koepp, D.M. and Silver, P.A., A GTPase controlling nuclear trafficking: running the right way or walking randomly?, *Cell* 87(1), 1–4, 1996.

66. Powers, M.A. and Forbes, D.J., Cytosolic factors in nuclear transport: what's importin? *Cell* 79(6), 931–934, 1994.

67. Miller, M., Park, M.K., and Hanover, J.A., Nuclear pore complex: structure, function, and regulation, *Physiol. Rev.* 71(3), 909–949, 1991.

68. Jans, D.A., Nuclear signaling pathways for polypeptide ligands and their membrane receptors? *FASEB J.* 8 (11), 841–7, 1994.

69. Lanford, R.E., Kanda, P., and Kennedy, R.C., Induction of nuclear transport with a synthetic peptide homologous to the SV40 T antigen transport signal, *Cell* 46(4), 575–82, 1986.

70. Boutorine, A.S. and Kostina, E.V., Reversible covalent attachment of cholesterol to oligodeoxyribo-nucleotides for studies of the mechanisms of their penetration into eucaryotic cells, *Biochimie* 75(1–2), 35–41, 1993.

71. Ludtke, J.J., Zhang, G., Sebestyen, M.G., and Wolff, J.A., A nuclear localization signal can enhance both the nuclear transport and expression of 1 kb DNA, *J. Cell Sci.* 112(Pt. 12), 2033–2041, 1999.

72. Capecchi, M.R., High efficiency transformation by direct microinjection of DNA into cultured mammalian cells, *Cell* 22(2 Pt. 2), 479–488, 1980.

73. Mirzayans, R., Aubin, R.A., and Paterson, M.C., Differential expression and stability of foreign genes introduced into human fibroblasts by nuclear versus cytoplasmic microinjection, *Mutat. Res.* 281(2), 115–122, 1992.

74. Thorburn, A.M. and Alberts, A.S., Efficient expression of miniprep plasmid DNA after needle microinjection into somatic cells, *Biotechniques* 14(3), 356–358, 1993.

75. O'Neill, R.E., Jaskunas, R., Blobel, G., Palese, P., and Moroianu, J., Nuclear import of influenza virus RNA can be mediated by viral nucleoprotein and transport factors required for protein import, *J. Biol. Chem.* 270(39), 22701–22704, 1995.

76. O'Neill, R.E., Talon, J., and Palese, P., The influenza virus NEP (NS2 protein) mediates the nuclear export of viral ribonucleoproteins, *EMBO J.* 17(1), 288–296, 1998.

77. Gallay, P., Swingler, S., Aiken, C., and Trono, D., HIV-1 infection of nondividing cells: C-terminal tyrosine phosphorylation of the viral matrix protein is a key regulator, *Cell* 80(3), 379–388, 1995.

78. Hagstrom, J.E., Ludtke, J.J., Bassik, M.C., Sebestyen, M.G., Adam, S.A., and Wolff, J.A., Nuclear import of DNA in digitonin-permeabilized cells, *J. Cell Sci.* 110(Pt. 18), 2323–2331, 1997.

79. Dean, D.A., Import of plasmid DNA into the nucleus is sequence specific, *Exp. Cell Res.* 230(2), 293–302, 1997.

80. Brisson, M., Tseng, W.C., Almonte, C., Watkins, S., and Huang, L., Subcellular trafficking of the cytoplasmic expression system, *Hum. Gene Ther.* 10(16), 2601–2613, 1999.

81. Brunner, S., Sauer, T., Carotta, S., Cotten, M., Saltik, M., and Wagner, E., Cell cycle dependence of gene transfer by lipoplex, polyplex and recombinant adenovirus, *Gene Ther.* 7(5), 401–407, 2000.

82. Fasbender, A., Zabner, J., Zeiher, B.G., and Welsh, M.J., A low rate of cell proliferation and reduced DNA uptake limit cationic lipid-mediated gene transfer to primary cultures of ciliated human airway epithelia, *Gene Ther.* 4(11), 1173–1180, 1997.

83. Tseng, W.C., Haselton, F.R., and Giorgio, T.D., Mitosis enhances transgene expression of plasmid delivered by cationic liposomes, *Biochim. Biophys. Acta* 1445(1), 53–64, 1999.

84. Wienhues, U., Hosokawa, K., Hoveler, A., Siegmann, B., and Doerfler, W., A novel method for transfection and expression of reconstituted DNA-protein complexes in eukaryotic cells, *DNA* 6(1), 81–89, 1987.

85. Kaneda, Y., Iwai, K., and Uchida, T., Increased expression of DNA cointroduced with nuclear protein in adult rat liver, *Science* 243(4889), 375–378, 1989.

86. Collas, P. and Alestrom, P., Nuclear localization signal of SV40 T antigen directs import of plasmid DNA into sea urchin male pronuclei *in vitro*, *Mol. Reprod. Dev.* 45(4), 431–438, 1996.

87. Collas, P. and Alestrom, P., Nuclear localization signals: a driving force for nuclear transport of plasmid DNA in zebrafish, *Biochem. Cell. Biol.* 75(5), 633–640, 1997.

88. Collas, P. and Alestrom, P., Rapid targeting of plasmid DNA to zebrafish embryo nuclei by the nuclear localization signal of SV40 T antigen, *Mol. Mar. Biol. Biotechnol.* 6(1), 48–58, 1997.

89. Collas, P. and Alestrom, P., Nuclear localization signals enhance germline transmission of a transgene in zebrafish, *Transgenic Res.* 7(4), 303–309, 1998.

90. Aronsohn, A.I. and Hughes, J.A., Nuclear localization signal peptides enhance cationic liposome-mediated gene therapy, *J. Drug Target.* 5(3), 163–169, 1998.

91. Xu, Y. and Szoka, F.C., Jr., Mechanism of DNA release from cationic liposome/DNA complexes used in cell transfection, *Biochemistry* 35(18), 5616–5623, 1996.

92. Sebestyen, M.G., Ludtke, J.J., Bassik, M.C., Zhang, G., Budker, V., Lukhtanov, E.A., Hagstrom, J.E., and Wolff, J.A., DNA vector chemistry: the covalent attachment of signal peptides to plasmid DNA, *Nat. Biotechnol.* 16(1), 80–85, 1998.

93. Ciolina, C., Byk, G., Blanche, F., Thuillier, V., Scherman, D., and Wils, P., Coupling of nuclear localization signals to plasmid DNA and specific interaction of the conjugates with importin α, *Bioconjug. Chem.* 10(1), 49–55, 1999.

94. Neves, C., Escriou, V., Byk, G., Scherman, D., and Wils, P., Intracellular fate and nuclear targeting of plasmid DNA, *Cell. Biol. Toxicol.* 15(3), 193–202, 1999.

95. Cimino, G.D., Gamper, H.B., Isaacs, S.T., and Hearst, J.E., Psoralens as photoactive probes of nucleic acid structure and function: organic chemistry, photochemistry, and biochemistry, *Annu. Rev. Biochem.* 54, 1151–1193, 1985.

96. Neves, C., Byk, G., Scherman, D., and Wils, P., Coupling of a targeting peptide to plasmid DNA by covalent triple helix formation, *FEBS Lett.* 453(1–2), 41–5, 1999.

97. Branden, L.J., Christensson, B., and Smith, C.I., *In vivo* nuclear delivery of oligonucleotides via hybridizing bifunctional peptides, *Gene Ther.* 8(1), 84–87, 2001.

98. Branden, L.J., Mohamed, A.J., and Smith, C.I., A peptide nucleic acid-nuclear localization signal fusion that mediates nuclear transport of DNA, *Nat. Biotechnol.* 17(8), 784–787, 1999.

99. Cutrona, G., Carpaneto, E.M., Ulivi, M., Roncella, S., Landt, O., Ferrarini, M., and Boffa, L.C., Effects in live cells of a c-myc anti-gene PNA linked to a nuclear localization signal, *Nat. Biotechnol.* 18(3), 300–303, 2000.

100. Uhlmann, E., Peptide nucleic acids (PNA) and PNA-DNA chimeras: from high binding affinity towards biological function, *Biol. Chem.* 379(8–9), 1045–1052, 1998.

101. Tung, C.H. and Stein, S., Preparation and applications of peptide-oligonucleotide conjugates, *Bioconjug. Chem.* 11(5), 605–618, 2000.

102. Dasi, F., Benet, M., Crespo, J., Crespo, A., and Alino, S.F., Asialofetuin liposome-mediated human α1-antitrypsin gene transfer *in vivo* results in stationary long-term gene expression, *J. Mol. Med.* 79(4), 205–212, 2001.

103. Brisson, M., He, Y., Li, S., Yang, J.P., and Huang, L., A novel T7 RNA polymerase autogene for efficient cytoplasmic expression of target genes, *Gene Ther.* 6(2), 263–270, 1999.

104. Vacik, J., Dean, B.S., Zimmer, W.E., and Dean, D.A., Cell-specific nuclear import of plasmid DNA, *Gene Ther.* 6(6), 1006–1014, 1999.

105. Mesika, A., Grigoreva, I., Zohar, M., and Reich, Z., A regulated, NFκB-assisted import of plasmid DNA into mammalian cell nuclei, *Mol. Ther.* 3(5 Pt. 1), 653–657, 2001.

106. Pouton, C.W. and Seymour, L.W., Key issues in non-viral gene delivery, *Adv. Drug Deliv. Rev.* 46(1–3), 187–203, 2001.

107. Brown, M.D., Schatzlein, A.G., and Uchegbu, I.F., Gene delivery with synthetic (non viral) carriers, *Int. J. Pharm.* 229(1–2), 1–21, 2001.

108. de Lima, M.C., Simoes, S., Pires, P., Gaspar, R., Slepushkin, V., and Duzgunes, N., Gene delivery mediated by cationic liposomes: from biophysical aspects to enhancement of transfection, *Mol. Membr. Biol.* 16(1), 103–109, 1999.

109. Escriou, V., Ciolina, C., Helbling-Leclerc, A., Wils, P., and Scherman, D., Cationic lipid-mediated gene transfer: analysis of cellular uptake and nuclear import of plasmid DNA, *Cell. Biol. Toxicol.* 14(2), 95–104, 1998.

110. Ledley, F.D., Nonviral gene therapy: the promise of genes as pharmaceutical products, *Hum. Gene Ther.* 6(9), 1129–1144, 1995.

111. Li, S. and Ma, Z., Nonviral gene therapy, *Curr. Gene Ther.* 1(2), 201–226, 2001.

112. Ma, H. and Diamond, S.L., Nonviral gene therapy and its delivery systems, *Curr. Pharm. Biotechnol.* 2(1), 1–17, 2001.

113. Mahat, R.I., Monera, O.D., Smith, L.C., and Rolland, A., Peptide-based gene delivery, *Curr. Opin. Mol. Ther.* 1(2), 226–243, 1999.

114. Mahato, R.I., Rolland, A., and Tomlinson, E., Cationic lipid-based gene delivery systems: pharmaceutical perspectives, *Pharm. Res.* 14(7), 853–859, 1997.

115. Mahato, R.I., Non-viral peptide-based approaches to gene delivery, *J. Drug Target.* 7(4), 249–268, 1999.

116. Zabner, J., Fasbender, A.J., Moninger, T., Poellinger, K.A., and Welsh, M.J., Cellular and molecular barriers to gene transfer by a cationic lipid, *J. Biol. Chem.* 270(32), 18997–19007, 1995.

117. Nakanishi, M., Mizuguchia, H., Ashihara, K., Senda, T., Akuta, T., Okabe, J., Nagoshi, E., Masago, A., Eguchi, A., Suzuki, Y., Inokuchi, H., Watabe, A., Ueda, S., Hayakawa, T., and Mayumi, T., Gene transfer vectors based on Sendai virus, *J. Control Release* 54(1), 61–68, 1998.

118. Legendre, J.Y. and Szoka, F.C., Jr., Delivery of plasmid DNA into mammalian cell lines using pH-sensitive liposomes: comparison with cationic liposomes, *Pharm. Res.* 9(10), 1235–1242, 1992.

119. Golzio, M., Teissie, J., and Rols, M.P., Direct visualization at the single-cell level of electrically mediated gene delivery, *Proc. Nat. Acad. Sci. U.S.A.* 99(3), 1292–1297, 2002.

120. Titomirov, A.V., Sukharev, S., and Kistanova, E., *In vivo* electroporation and stable transformation of skin cells of newborn mice by plasmid DNA, *Biochim. Biophys. Acta* 1088(1), 131–134, 1991.

121. Nomura, M., Nakata, Y., Inoue, T., Uzawa, A., Itamura, S., Nerome, K., Akashi, M., and Suzuki, G., *In vivo* induction of cytotoxic T lymphocytes specific for a single epitope introduced into an unrelated molecule, *J. Immunol. Methods* 193(1), 41–49, 1996.

122. Suzuki, T., Shin, B.C., Fujikura, K., Matsuzaki, T., and Takata, K., Direct gene transfer into rat liver cells by *in vivo* electroporation, *FEBS Lett.* 425(3), 436–440, 1998.

123. Heller, R., Jaroszeski, M., Atkin, A., Moradpour, D., Gilbert, R., Wands, J., and Nicolau, C., *In vivo* gene electroinjection and expression in rat liver, *FEBS Lett.* 389(3), 225–228, 1996.

124. Nishi, T., Yoshizato, K., Yamashiro, S., Takeshima, H., Sato, K., Hamada, K., Kitamura, I., Yoshimura, T., Saya, H., Kuratsu, J., and Ushio, Y., High-efficiency *in vivo* gene transfer using intraarterial plasmid DNA injection following *in vivo* electroporation, *Cancer Res.* 56(5), 1050–1055, 1996.

125. Nishi, T., Dev, S.B., Yoshizato, K., Kuratsu, J., and Ushio, Y., Treatment of cancer using pulsed electric field in combination with chemotherapeutic agents or genes, *Hum. Cell* 10(1), 81–86, 1997.

126. Rols, M.P., Delteil, C., Golzio, M., Dumond, P., Cros, S., and Teissie, J., *In vivo* electrically mediated protein and gene transfer in murine melanoma, *Nat. Biotechnol.* 16(2), 168–171, 1998.

127. Aihara, H. and Miyazaki, J., Gene transfer into muscle by electroporation *in vivo*, *Nat. Biotechnol.* 16(9), 867–870, 1998.

128. Widera, G., Austin, M., Rabussay, D., Goldbeck, C., Barnett, S.W., Chen, M., Leung, L., Otten, G.R., Thudium, K., Selby, M.J., and Ulmer, J.B., Increased DNA vaccine delivery and immunogenicity by electroporation *in vivo*, *J. Immunol.* 164(9), 4635–4640, 2000.

129. Cooper, M.J., Noninfectious gene transfer and expression systems for cancer gene therapy, *Semin. Oncol.* 23(1), 172–187, 1996.

130. Nakanishi, M., Gene introduction into animal tissues, *Crit. Rev. Ther. Drug Carrier Syst.* 12(4), 263–310, 1995.

131. Kodama, T., Doukas, A.G., and Hamblin, M.R., Shock wave-mediated molecular delivery into cells, *Biochim. Biophys. Acta* 1542(1–3), 186–194, 2002.

132. Lin, T.Y., McAuliffe, D.J., Michaud, N., Zhang, H., Lee, S., Doukas, A.G., and Flotte, T.J., Nuclear transport by laser-induced pressure transients, *Pharm. Res.* 20(6), 879–883, 2003.

133. Eriksson, E., Yao, F., Svensjo, T., Winkler, T., Slama, J., Macklin, M.D., Andree, C., McGregor, M., Hinshaw, V., and Swain, W.F., *In vivo* gene transfer to skin and wound by microseeding, *J. Surg. Res.* 78(2), 85–91, 1998.

134. Miller, D.L. and Quddus, J., Diagnostic ultrasound activation of contrast agent gas bodies induces capillary rupture in mice, *Proc. Nat. Acad. Sci. U.S.A.* 97(18), 10179–10184, 2000.

135. Hosseinkhani, H., Aoyama, T., Ogawa, O., and Tabata, Y., Ultrasound enhances the transfection of plasmid DNA by non-viral vectors, *Curr. Pharm. Biotechnol.* 4(2), 109–122, 2003.

136. Hogset, A., Prasmickaite, L., Hellum, M., Engesaeter, B.O., Olsen, V.M., Tjelle, T.E., Wheeler, C.J., and Berg, K., Photochemical transfection: a technology for efficient light-directed gene delivery, *Somat. Cell Mol. Genet.* 27(1–6), 97–113, 2002.

CHAPTER **5**

Pharmacokinetics of Polymer–Plasmid DNA Complex

Makiya Nishikawa, Yoshinobu Takakura, and Mitsuru Hashida

CONTENTS

5.1 INTRODUCTION

The *in vivo* gene transfer profile that is required for effective gene therapy depends on the target disease. It includes the target cell-specificity of gene transfer, the efficiency and duration of transgene expression, and the number of transfected cells.[1] For example, gene therapy designed for hemophiliacs or other patients who have a deficiency of any plasma protein requires a prolonged production of transgene, but not necessarily cell-specific transduction. Therefore, intramuscular gene transfer of human factor IX by adeno-associated viral vector showed significant therapeutic benefits in hemophiliac patients.[2] In marked contrast, the dystrophin gene needs to be definitely transduced into the affected muscle cells, because dystrophin gene transfer to other types of cells will not produce any therapeutic benefit in patients with Duchenne muscular dystrophy. In addition, the number of transfected cells is a very important factor for the efficacy of gene transfer aimed at the treatment of muscular dystrophies.[3]

0-8493-1934-X/05/$0.00+$1.50
© 2005 by CRC Press LLC

An effective gene transfer approach, therefore, needs to be developed for each target disease. This, in turn, clearly indicates that the requirements of a vector will vary depending on the target disease. If the amount of transgene product is the top priority, viral vectors are the choice. However, recent safety issues about viral vectors[4,5] have led to concern about their use in clinical situations. On the other hand, nonviral gene transfer approaches are considered to be safer and more convenient than viral vectors. A class of nonviral methods uses naked plasmid DNA (pDNA): direct injection into the extracellular matrix of tissues, electroporation or sonoporation after direct injection, administration using gene guns, and intravascular injection in a large volume with a high velocity. Another approach is the use of a pDNA complex, which is prepared by mixing it with cationic lipids, liposomes, or polymers. Such a complex is administered by various routes for *in vivo* gene transfer.

Gene transfer only occurs in the cells taking up the gene administered into the body. Therefore, the pharmacokinetics of genes are very important in determining the final output of transgene expression. Among various approaches, the polymer–pDNA complex has some advantages as far as the pharmacokinetic properties are concerned. Its tissue distribution after systemic administration can be controlled by adjusting the physicochemical properties and/or introducing any ligand for specific interaction with corresponding receptors. A lipid-based pDNA complex could also be used for the delivery of pDNA, but it has more problems in terms of the pharmacokinetic characteristics because of susceptibility to interaction with blood components after administration.[6]

In this chapter, we first discuss the basic characteristics of pDNA in terms of its pharmacokinetics after administration into animals. Then, the tissue distribution of the polymer–pDNA complex is summarized and challenges involving targeted gene delivery based on polymeric vectors are discussed, with an emphasis on the pharmacokinetic characteristics.

5.2 PHARMACOKINETIC CONSIDERATION OF MACROMOLECULES

The tissue distribution of macromolecules is highly dependent on their physicochemical characteristics and the anatomical and physiological characteristics of the body.[7,8] Distribution of macromolecules from capillaries to tissues is restricted by the capillary endothelium in most tissues except for the liver, spleen, bone marrow and kidney, all of which have fenestrated or discontinuous capillaries.[9,10] Most solid tumors also have vessels through which macromolecules can be easily transported.[11] These anatomical characteristics restrict the tissue distribution of macromolecules into some limited organs and tissues. In other words, macromolecules have an opportunity for selective distribution to these sites. Some macromolecules and drug-macromolecule conjugates selectively accumulate in tumor tissues,[7,8,12-14] due to the unique vasculature of these tissues.

Each macromolecule has distinctive *in vivo* distribution characteristics depending on its physicochemical properties such as the molecular weight and electric charge.[13,15,16] Therefore, macromolecules with a set of well-controlled physicochemical properties could be delivered to limited types of cells *in vivo*. Furthermore, if a proper ligand is used, macromolecules can be actively and selectively delivered to the target cell. Monoclonal antibodies, growth factors, and sugars are the major ligands that have been used for targeted drug delivery. Regardless of the modes of targeting (passive or active), a relationship between structure and tissue distribution needs to be identified in order to design an appropriate targeting system. Pharmacokinetic analysis translates the tissue distribution characteristics of macromolecules into quantitative parameters, which can be directly compared with those of others as well as with physiological parameters such as blood flow and the rate of fluid-phase endocytosis.

Tissue distribution of a macromolecule can be pharmacokinetically analyzed based on the clearance concept. Tissue uptake of a macromolecule consists of uptake from the plasma and efflux from the tissue. When the tissue uptake rate is assumed to be independent on its plasma concentration and the efflux process follows first-order rate kinetics, the change in its amount in a tissue with time can be described as follows:

$$\frac{dX_i}{dt} = CL_{app,i}C_p - k_{efflux,i}X_i \tag{5.1}$$

where X_i represents the amount of the macromolecule in tissue i after administration, C_p is its concentration in plasma, $CL_{app,i}$ expresses the apparent tissue uptake clearance from plasma to tissue i, and $k_{efflux,i}$ represents the efflux rate from tissue i. Under proper experimental conditions, the efflux process can be ignored, making it easier to analyze the distribution of the macromolecule pharmacokinetically. In this case, Equation 5.1 is simplified to

$$\frac{dX_i}{dt} = CL_{app,i}C_p \tag{5.2}$$

Integration of Equation 5.2 from time 0 to t1 gives

$$X_{i,t1} = CL_{app,i}AUC_{p,0-t1} \tag{5.3}$$

where $AUC_{p,0-t1}$ is the area under the plasma concentration-time curve of the macromolecule from time 0 to t1. According to Equation 5.3, $CL_{app,i}$ can be calculated from the slope when the amounts in a tissue are plotted against AUC_p.

$CL_{app,i}$ is a hybrid parameter of the plasma flow rate (Q_i) to tissue i and the intrinsic uptake clearance ($CL_{int,i}$) of the tissue and is expressed as

$$CL_{app,i} = \frac{CL_{int,i}Q_i}{CL_{int,i} + Q_i} \tag{5.4}$$

When $CL_{int,i}$ is much larger than Q, $CL_{app,i}$ approaches the value of Q and this value (plasma flow rate) is the upper limit of $CL_{app,I}$, whatever specific and rapid uptake mechanisms are involved in the tissue uptake.

The total body clearance (CL_{total}) of the macromolecule can be calculated using AUC_p for infinite time ($AUC_{p,\infty}$) and administered dose (D) as follows:

$$CL_{total} = \frac{D}{AUC_{p,\infty}} \tag{5.5}$$

Since CL_{total} is the sum of the tissue uptake clearances and urinary clearance, CL_{total} is also expressed as

$$CL_{total} = CL_{liver} + CL_{kidney} + CL_{lung} + \cdots\cdots + CL_{urine} \tag{5.6}$$

$$= CL_{target} + CL_{non-target} \tag{5.7}$$

where CL_{target} denotes the uptake clearance of a target tissue and $CL_{non-target}$ is the sum of the clearances except for CL_{target}. After administration, the fraction of macromolecule delivered to the target (F_{target}) can be calculated as

$$F_{target} = \frac{CL_{target}}{CL_{target} + CL_{non-target}}$$ (5.8)

Therefore, the potential of the targeted delivery of drugs using polymeric carriers can be quantitatively explained by the two parameters CL_{target} and $CL_{non-target}$. If inefficient delivery of a macromolecular drug to the target is due to its rapid elimination from the systemic circulation, a reduction in $CL_{non-target}$ will improve its delivery to the target. When the drug circulates in the plasma for a long time because of a small CL_{total}, the introduction of a ligand specifically recognized by a target, which increases CL_{target}, can enhance the delivery.

Figure 5.1 shows the plots of the hepatic uptake clearance (CL_{liver}) and urinary clearance (CL_{urine}) of model macromolecules after intravenous injection into mice.[13] Inulin (average molecular weight [mol wt] 5,000); Dex (T-10) (dextran with average mol wt 10,000); dextran sulfate (average mol wt 5,000); and apoprotein of neocarzinostatin (apoNCS) (average mol wt 12,000) have large CL_{urine} values that are comparable with the glomerular filtration rate of mice. The renal glomerular capillary wall functions as a size- and charge-selective filter in the filtration of macromolecules[17,18] and these macromolecules will pass through the filter without significant restriction. On the other hand, macromolecules with molecular weights larger than 40,000 have a much smaller CL_{urine}. The electric

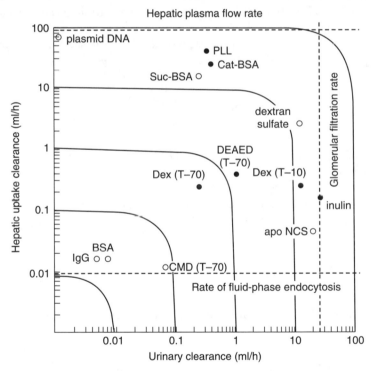

Figure 5.1 Hepatic uptake and urinary clearances of macromolecules after intravenous injection in mice. Apoprotein of neocarzinostatin (apoNCS) (weakly anionic, mol wt 14,000); bovine serum albumin (BSA) (weakly anionic, mol wt 67,000); bovine immunoglobulin G (IgG) (weakly anionic, mol wt 150,000); cationized BSA (Cat-BSA) (cationic, mol wt 70,000); succinylated BSA (Suc-BSA) (strongly anionic, mol wt 70,000); poly(L-lysine) (PLL) (cationic, average mol wt 40,000); inulin (neutral, average mol wt 5,000); Dex(T-10) or Dex(T-70), dextran (neutral, average mol wt 10,000 and 70,000, respectively); carboxymethyl-dextran (CMD) (T-70), (weakly anionic, average mol wt 70,000); diethylaminoethyl-dextran (DEAED) (T-70) (cationic, average mol wt 70,000); dextran sulfate (strongly anionic, average mol wt 5,000), plasmid DNA (strongly anionic, mol wt about 4,000,000).

charge plays an important role in the hepatic uptake of macromolecules. Cationic macromolecules, such as DEAED (T-70) (diethylaminoethyl-dextran derived from dextran with average mol wt 70,000); cationized bovine serum albumin (Cat-BSA) (average mol wt 70,000); and poly(L-lysine) (PLL) (average mol wt 40,000) have a large CL_{liver}. Weakly anionic ones such as CMD (T-70) (carboxymethyl-dextran derived from dextran with average mol wt 70,000); apoNCS; IgG (bovine immunoglobulin G); and BSA have small CL_{liver} values that are close to the rate of fluid-phase endocytosis of the liver, suggesting that they hardly interact with liver cells. The effect of electric charge on the hepatic uptake of macromolecules is explained by an electrostatic interaction between cationic macromolecules and liver cells.[19,20] On the other hand, highly anionic macromolecules like dextran sulfate, succinylated or maleylated albumins, and pDNA are rapidly taken up by liver, especially by the nonparenchymal cells,[21,22] so their disposition characteristics cannot be classified only by the electrostatic interaction with cells. When the target for delivery is any tissue other than the liver, CL_{urine} and CL_{liver} represent a great part of $CL_{non-target}$. According to Equation 5.8, macromolecules having small CL_{urine} and CL_{liver} could be good carriers for such delivery.

Thus, tissue uptake clearances are useful parameters for characterizing the tissue distribution of macromolecules because they can be directly compared with physiological parameters such as blood flow and the rate of fluid-phase endocytosis.

5.3 BASIC PHARMACOKINETIC PROPERTIES OF NAKED PLASMID DNA

Generally speaking, pDNA is a huge macromolecule with a strong negative charge. This negative charge means that DNA is in a class of polyanions, most of which are ligands for scavenger receptors. Upon administration to the body, pDNA or its complex with any vector interacts with cells, proteins, and the extracellular matrix. The summation of these interactions determines the pharmacokinetics of pDNA and its complex, which, in turn, determine the location and extent of gene transfer.

5.3.1 Cellular Uptake Mechanism

The characteristics of the cellular uptake of pDNA have been extensively examined using cultured cells. Various cells have shown the ability to take up pDNA in a concentration- and temperature-dependent manner. One example involves the brain microvessel endothelial cells,[23] which constitute the blood-brain barrier. However, the major cells determining the distribution of pDNA are immune cells such as macrophages. In an actual fact, Kupffer cells, the residential liver macrophages, make a large contribution to the hepatic clearance of pDNA administered to the blood circulation.

pDNA is efficiently taken up by cultured mouse peritoneal macrophages.[24] The profile of uptake inhibition is similar to that observed in brain microvessel endothelial cells. Another population of immune cells is dendritic cells (DCs), and these are very important as far as both innate and acquired immunity are concerned. The DC cell line, DC2.4, exhibits extensive uptake and degradation of pDNA.[25] In these cells, the uptake of pDNA is inhibited by calf thymus DNA, polyinosinic acid, dextran sulfate, and heparin, but not by polycytidylic acid and chondroitin sulfate. These results strongly suggest that any class of scavenger receptors is responsible for the cellular uptake of pDNA. However, no specific receptors have been identified as being involved in the uptake; we excluded the possibility of class A scavenger receptors (SRAs) being involved in the uptake, based on tissue distribution and uptake experiments using cultured macrophages from SRA-knockout mice.[26] The involvement of Toll-like receptor-9 (TLR-9) in the cellular response against bacterial DNA has been reported, in which some unmethylated CpG sequences are recognized as nonself, leading to the activation of an immune response.[27,28] However, the TLR-9 is not located on the surface of cells, so it may not be involved in the first step in the uptake of pDNA.

5.3.2 Tissue Distribution after Intravenous Injection

After entering the blood circulation, pDNA encounters various cells, including blood cells, endothelial cells, and macrophages. The tissue distribution after intravenous injection of single- or double-stranded DNA or oligonucleotide has been examined, and it was shown that the liver is the main organ responsible for the rapid clearance of these forms of DNA from the circulation.[29-31] However, the uptake mechanism and the cell type contributing to this hepatic uptake remained to be elucidated.

We have studied the tissue distribution of pDNA radiolabeled with ^{32}P following intravenous injection into mice.[32,33] After intravenous injection, naked pDNA rapidly disappeared from the blood circulation and was predominantly taken up by the liver. Although the brain microvessel endothelial cells show the presence of specific uptake mechanisms for pDNA, their contribution to its tissue distribution was minor. Fractionation of the liver cells revealed that the nonparenchymal cells such as liver sinusoidal endothelial cells and Kupffer cells contribute more to the uptake than hepatocytes. The hepatic uptake of pDNA was dependent on the concentration and decreased on increasing the dose, suggesting the existence of specific uptake mechanisms. The profile of the inhibition of the hepatic uptake was quite similar to that observed in cultured peritoneal macrophages and DCs.[32-34]

Tissue distribution of pDNA is also quantitatively discussed based on the tissue uptake clearances. After the tissue uptake of ^{32}P-pDNA, however, the amount of radioactivity in various tissues decreased with time, compared with the case of ^{111}In-labeled proteins or ^{14}C-labeled polysaccharides. Therefore, in our published papers, the tissue uptake clearances of pDNA were calculated using data sampled over short periods of time (several minutes) after injection when the decrease in radioactivity was theoretically minimal. Table 5.1 summarizes the clearances of pDNA after intravenous injection into mice at different doses. These clearances clearly indicate the pharmacokinetic characteristics of pDNA; at the lowest dose the uptake clearance of the liver was comparable with the hepatic plasma flow rate (85 ml/h for a 25 g mouse), but it decreased on increasing the dose to 12 ml/h at 1 mg/kg. The CL_{liver} accounts for the largest fraction of the CL_{total}, and it was much greater than those of bovine serum albumin and dextrans (Figure 5.1). Other polyanions such as succinylated bovine serum albumins and dextran sulfate possess similar patterns of tissue uptake clearances to that of pDNA.[22]

Generally speaking, following cellular uptake via endocytosis, pDNA enters the endosomes and is sorted to the lysosomes and degraded within the organella. When ^{32}P-pDNA is used, radioactive metabolites are released from cells, which is the cause of the decrease in the radioactivity in the tissues after intravenous injection in mice. Proteins that are endocytosed are also able to undergo lysosomal degradation, and radioactive metabolites are generated within the cells. The release of the metabolites makes it difficult to analyze the pharmacokinetics, especially the distribution process of any macromolecule. Therefore, a hydrophilic bifunctional chelate-radioactive metal complex or radioiodination using a sugar-containing spacer has been used as a residualizing radiolabel for protein drugs.[35-37] In these approaches, relatively large, hydrophilic molecules are

Table 5.1 Clearances of Plasmid DNA and Polymer–Plasmid DNA Complex after Intravenous Injection into Mice

Compound	Dose (mg/kg)	Total	Liver	Kidney	Spleen	Lung
^{32}P-plasmid DNA	0.1	69.9	68.6	2.9	1.2	1.2
^{32}P-plasmid DNA	0.5	36.3	26.1	0.2	0.7	1.0
^{32}P-plasmid DNA	1.0	22.8	11.6	0	0.5	0.3
$PLL_{13000}/^{32}$P-plasmid DNA	0.02	45.5	30.1	1.9	2.0	1.4
Gal_5-$PLL_{13000}/^{32}$P-plasmid DNA	0.02	47.0	35.5	3.0	2.3	0.9
Gal_{13}-$PLL_{13000}/^{32}$P-plasmid DNA	0.02	61.1	47.6	1.4	1.3	0.4

(Clearance (ml/h) spans Total, Liver, Kidney, Spleen, Lung)

incorporated into the structure of the radiolabeled adducts. After catabolism, the radioactive metabolites that are hydrophilic with a high molecular weight are hardly able to pass through the plasma and lysosomal membranes,[38] allowing the retention of radioactivity within the cells that have taken up the radiolabeled compound. Recently, we have developed a residualizing radiolabeling method for pDNA, in which a similar chemistry to the radiolabeling of proteins was applied.[39] Namely, 4-(p-azidosalicylamido)butylamine (ASBA) was coupled with diethylenetriaminepentaacetic acid (DTPA) anhydride to obtain ASBA-DTPA conjugate. Here, DTPA is a chelator of metal ions such as ^{111}In and ^{99m}Tc, and ASBA has a photoreactive group that is highly reactive toward nucleophiles, especially amines. The ASBA-DTPA conjugate was then coupled to pDNA by photoreaction, followed by labeling with [^{111}In]InCl$_3$ to obtain ^{111}In-pDNA (pDNA-ASBA-DTPA-^{111}In). This procedure of labeling hardly altered the structure of pDNA and preserved the transcriptional activity from 40–98%, depending on the degree of modification. After intravenous injection of ^{111}In-pDNA into mice, the amount of radioactivity in the liver was almost identical to that of ^{32}P-radioactivity after the injection of ^{32}P-pDNA within the first 1 h. Thereafter, the amount of ^{111}In-radioactivity remained within the liver, whereas that of ^{32}P-radioactivity decreased with time. The amount of ^{111}In-radioactivity in the lung following the administration of polyethyleneimine/^{111}In-pDNA complexes correlates well with the transgene expression, indicating that the residualizing radiolabel can offer tissue distribution data, allowing a better approach to the design of *in vivo* gene delivery.

5.4 PHARMACOKINETICS OF POLYMER–PLASMID DNA COMPLEX

Cationic polymers are a class of nonviral vectors that can be used to increase gene delivery and transfer to target cells *in vivo*. Various types of polymers have been examined with respect to their ability to protect from nuclease degradation, to deliver target cells, and to increase the transfection efficiency of pDNA. The polymer–pDNA complex is believed to be taken up by cells via an endocytotic pathway, so its transfection efficiency depends on the release of pDNA into the cytoplasm after cellular uptake. Therefore, polyplex-mediated transfection can be enhanced by the application of polymers possessing buffering ability[40] or fusogenic peptides creating pores on membranes,[41] both of which can disrupt the endosomes and the lysosomes and release the DNA complex within these vesicles into the cytoplasm.

5.4.1 Cellular Uptake

Due to the negative charge of the cellular surface, anionic macromolecules hardly interact with cells unless such macromolecules are not recognized by specific receptors. Naked pDNA, therefore, cannot result in an effective transfection to various types of cultured cells. Although a few types of cells such as macrophages take up pDNA through a scavenger receptor-like mechanism, no significant transgene expression is obtained. The important characteristic of the polymer–pDNA complex, which is quite different from naked pDNA, is its electric charge. Generally speaking, the polymer–pDNA complex is designed to have a positive charge, which is the driving force for the complex to interact with cells. Various cationic polymers have been developed so far as reagents for *in vitro* transfection.[42]

As shown with many different macromolecules (Figure 5.1), the electric charge is one of the most important factors in determining the pharmacokinetics. The cationic nature of the pDNA complex can increase the binding affinity for cells, but the binding is not specific to any type of cells. Therefore, the cationic polymer–pDNA complex has little transfection specificity as far as the type of cells is concerned.

In addition to the electric charge, the effective radius, or particle size, of the polymer–pDNA is another feature that is important for cellular uptake as well as tissue distribution. As discussed above, the effective sizes of fenestrae and intercellular gaps of endothelial cells are small, and particles of about 200 nm or greater are hardly able to pass.

5.4.2   Interaction with Biological Components

Serum proteins[43] and blood cells[44] have been reported to affect the tissue distribution of intravascularly administered pDNA complex. Interaction with various components in the body fluids, such as serum proteins, can affect the properties of the polymer–pDNA complex, which may lead to altered tissue distribution as well as reduced transgene expression.[6] Ogris et al.[45] demonstrated that when incubated with plasma, the polyethylenimine (PEI)–pDNA complex underwent aggregation. They found that various molecules, such as fibronectin, fibrinogen, complement factor C3, and IgM bind to the complex.

5.4.3   Tissue Distribution

Formation of a cationic complex greatly increases the interaction of pDNA with various cells. When injected intravenously, the interaction with the lung endothelial cells is most marked,[46,47] because the lung is the first-pass organ. Then, transgene expression is also high in these cells. When large aggregates are formed after intravenous injection of polymer–pDNA complex, they would also be trapped by the lung. In some cases, the mixing of pDNA and a cationic carrier *in vitro* resulted in very large aggregates with a size close to that of the capillary.[46,48,49] To prevent aggregation, the surface of the complex can be modified with hydrophilic compounds such as polyethylene glycol.[45,50] However, it also shields the cationic charge of the complex that is the driving force for interaction with target cells. Therefore, a surface ligand may be required for efficient delivery of such a complex.

PEI is one of the most extensively studied polymers because it has a relatively high transfection efficiency. PEI is believed to enter the cell via an endocytotic route, and to possess a buffering capacity and the ability to swell when protonated.[51,52] Therefore, at low pH values, it is believed that PEI prevents acidification of the endosome and induces a large inflow of ions and water, subsequently leading to rupture of the endosomal or lysosomal membrane so that the PEI/pDNA complex is delivered to the cytoplasmic space.[53] PEI is also reported to undergo nuclear localization while retaining an ordered structure, once endocytosed.[52] However, the transfection efficiency of PEI depends on its molecular weight and structure.[54,55] The amount of transgene expression depends on a number of variables including the nature of the polymer used and the amount of pDNA delivered. Recently, we clearly demonstrated that the amount of ^{111}In-plasmid DNA delivered to the lung was proportional to the transgene expression in the organ when the PEI/pDNA was injected into mice using a complex prepared so that it had different N/P ratios.[39]

5.5   PHARMACOKINETICS OF TARGETED POLYMER–PLASMID DNA COMPLEX

5.5.1   Fundamentals of Targeted Delivery

Targeted delivery of pDNA can be achieved by incorporating a ligand-receptor or antigen-antibody interaction into the vector system. From a theoretical point of view, the use of such a recognition system means an increase in CL_{target} in Equation 5.8. Various ligands have been used for the targeted delivery of pDNA, including asialoglycoproteins,[56] galactose,[48,55,57-59] mannose,[60-62] transferrin,[63] antibodies,[64,65] and lung surfactant proteins[66] (Table 5.2). The attached ligand can increase the affinity for the target cell through a receptor-ligand interaction. When designed properly, about 80% of a polyplex can be successfully delivered to the target organ after systemic administration.[48] However, some of these targetable polymers have only been examined using cultured

Table 5.2   Target Receptors/Molecules Employed for Cell-Specific Gene Delivery Using a Ligand–Polymer–Plasmid DNA Complex

Target Receptor/Molecule	Ligand-Polymer	Target Cell	Ref.
Asialoglycoprotein receptor	Asialoorosomucoid-PLL	Hepatocytes	56
	Lactosylated PLL	Hepatoma cells	57
	Galactosylated PLL	Hepatocytes	48, 58
	Galactosylated poly(L-ornithine)	Hepatocytes	59
	Galactosylated PEI	Hepatocytes	55
Mannose receptor	Mannosylated PLL	Macrophages	60, 62
	Mannosylated PEI	Dendritic cells	61
Transferrin receptor	Transferrin-PLL	Various cells	63
Polymeric immunoglobulin	Antibody-PLL	Respiratory epithelial cells	64
Platelet endothelial cell adhesion molecule	Antibody-PEI	Pulmonary endothelial cells	65
Surfactant protein A receptor	Surfactant protein A-PLL	Airway cells	66

cells. Even though the uptake of pDNA complex by the target cell is highly enhanced by the use of any ligand, the cellular uptake *in vivo* is limited by the flow rate and some physiological barriers. Therefore, the use of a specific ligand does not guarantee enhanced transgene expression after administration.

5.5.2   Sugar-Mediated Targeting

Sugars such as galactose and mannose are recognized by their corresponding receptors, and they have been used as ligands for targeted drug delivery. Asialoglycoproteins and galactose have been widely applied for hepatocyte-targeted delivery of drugs, proteins, and genes,[48,55,56,58,59,67,68] because hepatocytes are the only cells that express the asialoglycoprotein receptors that recognize galactose.[69] Hepatocytes are major targets for gene therapy, because there are many associated with fetal metabolic diseases resulting from a defect or deficiency in hepatocyte-derived gene products. In addition, hepatocytes can be considered as platforms to produce various proteins secreted into blood because the liver has a rich blood flow. Therefore, targetable vectors that can deliver therapeutic genes selectively to hepatocytes *in vivo* followed by efficient transgene expression within the cells are extremely valuable. On the other hand, mannose-containing carriers are used for drug or gene delivery to various macrophages, DCs, and liver endothelial cells based on their recognition by mannose receptors.[62,67,70,71]

When properly designed, the hepatic uptake clearance of galactosylated polymers becomes close to the plasma flow rate to the liver, which is the limit of the clearance as shown in Equation 5.4. Based on the pharmacokinetic behaviors of galactosylated proteins with different physicochemical properties, we have concluded that an effective density of galactose on the surface of a protein determines its affinity for the asialoglycoprotein receptors.[72] Figure 5.2 shows the correlation between CL_{liver} and the surface density of the galactose of galactosylated proteins. This figure clearly indicates that CL_{liver} is proportionally increased with an increase in the density of galactose. The recognition mechanism of naturally occurring sugar chains by the receptor was investigated *in vitro* and the importance of a precise geometry of the sugar chains was suggested using di-, tri-, and tetra-antennary oligosaccharides.[73,74] When the degree of galactosylation increases, the probability of the ligand binding to the receptor will increase, leading to a rapid uptake by the liver.

We have developed a series of galactosylated cationic polymers to deliver pDNA selectively to hepatocytes after intravenous injection.[48,55,59] PLL with various degrees of polymerization was modified with 2-imino-2-methoxyethyl 1-thiogalactoside to obtain galactosylated PLL (Gal-PLL). When intravenously injected into mice, the Gal-PLL/pDNA complex showed different tissue distribution properties. pDNA complexes with Gal-PLL having a number of galactose moieties showed

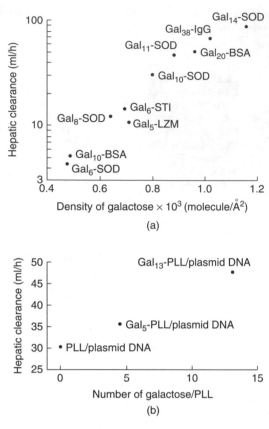

Figure 5.2 Effect of the degree of galactose modification on the hepatic clearances of galactosylated proteins (A) and Gal-PLL–plasmid DNA complex (B) after intravenous injection into mice.

a greater CL_{liver} (Figure 5.2). Using these complexes, we reported that the tissue disposition and subsequent transfection efficiency of the Gal-PLL/pDNA complex were dependent on the physicochemical properties of Gal-PLL and the complex formed. Increasing the number of galactose units per polymer would increase the affinity for the receptors, but too many galactose residues might reduce the ability of the polymer to interact with pDNA.[48] Therefore, a very precise design is required to develop an effective vector that achieves target cell-selective *in vivo* gene transfer. After intravenous injection of Gal_{13}-PLL_{13000} (PLL with an average mol wt of 13,000 and 13 galactose moieties conjugated)/^{32}P-pDNA and Gal_{26}-PLL_{29000}/^{32}P-pDNA, almost 80% of the radioactivity accumulated in the liver. Hepatocyte-specific localization in the liver was also confirmed by isolating the liver cells by collagenase. On the other hand, Gal_5-PLL_{13000}/^{32}P-pDNA resulted in insufficient delivery to the liver.

Receptor-mediated delivery of pDNA, however, has the clear drawback of ineffective transgene expression because the process normally involves sorting of the ligand to the lysosomes where it is enzymatically degraded. Therefore, membrane-destabilizing molecules have been used to increase the transgene expression by receptor-mediated gene transfer.[75] We have demonstrated that a peptide, which mimics the amino-terminal of influenza virus hemagglutinin subunit HA-2, can greatly increase transgene expression in the liver after intravenous injection of hepatocyte-targeted pDNA complex.[59] About a 70-fold increase in transgene expression in the liver was obtained by covalent binding of the peptide to a galactosylated poly(L-ornithine).

5.6 CONCLUSION

There are still lots of hurdles to be overcome for a therapeutically effective gene transfer using a polymer–pDNA complex. Low transgene expression activity and unfavorable tissue distribution characteristics are the major problems of any polymer–pDNA complex. Although very effective polymers have been reported in terms of transgene expression in cultured cells, they have not generally been proved effective *in vivo*. Even although a targetable ligand is introduced onto the polymer used, it often does not work at all after *in vivo* administration, simply because the ligand–polymer–pDNA complex cannot have access to the target cell: this is a pharmacokinetic problem. Therefore, the development of better polymers for *in vivo* gene transfer should be achieved through a detailed pharmacokinetic analysis of the polymer–pDNA complex.

REFERENCES

1. Nishikawa, M. and Hashida, M., Nonviral approaches satisfying various requirements for effective *in vivo* gene therapy, *Biol. Pharm. Bull.* 25, 275–283, 2002.
2. Kay, M.A. et al., Evidence for gene transfer and expression of factor IX in haemophilia B patients treated with an AAV vector, *Nat. Genet.* 24, 257–261, 2000.
3. Phelps, S.F. et al., Expression of full-length and truncated dystrophin mini-genes in transgenic mdx mice, *Hum. Mol. Genet.* 4, 1251–1258, 1995.
4. Marshall, E., Gene therapy death prompts review of adenovirus vector, *Science* 286, 2244–2245, 1999.
5. Hacein-Bey-Abina, S. et al., A serious adverse event after successful gene therapy for X-linked severe combined immunodeficiency, *N. Engl. J. Med.* 348, 255–256, 2003.
6. Opanasopit, P., Nishikawa, M., and Hashida, M., Factors affecting drug and gene delivery: effects of interaction with blood components, *Crit. Rev. Ther. Drug Carrier Syst.* 19, 191–233, 2002.
7. Sezaki, H. and Hashida, M., Macromolecule-drug conjugates in targeted cancer chemotherapy, *Crit. Rev. Ther. Drug Carrier Syst.* 1, 1–38, 1984.
8. Hashida, M. and Takakura, Y., Pharmacokinetics in design of polymeric drug delivery systems, *J. Control. Release* 31, 163–171, 1994.
9. Simionescu, N., Cellular aspects of transcapillary exchange, *Physiol. Rev.* 63, 1536–1579, 1983.
10. Taylor, A.E. and Granger, D.N., Exchange of macromolecules across the microcirculation, in *Handbook of Physiology: The Cardiovascular System IV*, Renkin, E.M. and Michel, C.C., Eds., American Physiological Society, Bethesda, MD, 1984, 467–520.
11. Yuan, F. et al., Vascular permeability in a human tumor xenograft: molecular size dependence and cutoff size, *Cancer Res.* 55, 3752–3756, 1995.
12. Matsumura, Y. and Maeda, H., A new concept for macromolecular therapeutics in cancer chemotherapy: mechanism of tumoritropic accumulation of proteins and the antitumor agent smancs, *Cancer Res.* 46, 6387–6392, 1986.
13. Takakura, Y. et al., Disposition characteristics of macromolecules in tumor-bearing mice, *Pharm. Res.* 7, 339–346, 1990.
14. Seymour, L.W., Passive tumor targeting of soluble macromolecules and drug conjugates, *Crit. Rev. Ther. Drug Carrier Syst.* 9, 135–187, 1992.
15. Yamaoka, T., Tabata, Y., and Ikada, Y., Fate of water-soluble polymers administered via different routes, *J. Pharm. Sci.* 84, 349–354, 1995.
16. Seymour, L.W. et al., Influence of molecular weight on passive tumour accumulation of a soluble macromolecular drug carrier, *Eur. J. Cancer* 31, 766–770, 1995.
17. Brenner, B.M., Hostetter, T.H., and Humes, H.D., Glomerular permselectivity: barrier function based on discrimination of molecular size and charge *Am. J. Physiol.* 234, F455–F460, 1978.
18. Takakura, Y., Mihara, K., and Hashida, M., Control of the disposition profiles of proteins in the kidney via chemical modification, *J. Control. Release* 28, 111–119, 1994.

19. Nakane, S. et al., The accumulation mechanism of cationic mitomycin C-dextran conjugates in the liver: *in vivo* cellular localization and *in vitro* interaction with hepatocytes, *J. Pharm. Pharmacol.* 40, 1–6, 1988.

20. Nishida, K. et al., Hepatic disposition characteristics of electrically charged macromolecules in rat *in vivo* and in the perfused liver, *Pharm. Res.* 8, 437–443, 1991.

21. Krieger, M. et al., Molecular flypaper, host defence and atherosclerosis: structure, binding properties and functions of macrophage scavenger receptors, *J. Biol. Chem.* 268, 4569–4572, 1993.

22. Yamasaki, Y. et al., Pharmacokinetic analysis of *in vivo* disposition of succinylated proteins targeted to liver nonparenchymal cells via scavenger receptors: importance of molecular size and negative charge density for *in vivo* recognition by receptors, *J. Pharmacol. Exp. Ther.* 301, 467–77, 2002.

23. Nakamura, M. et al., Uptake and gene expression of naked plasmid DNA in cultured brain microvessel endothelial cells, *Biochem. Biophys. Res. Commun.* 245, 235–239, 1998.

24. Takagi, T. et al., Involvement of specific mechanism in plasmid DNA uptake by mouse peritoneal macrophages, *Biochem. Biophys. Res. Commun.* 245, 729–733, 1998.

25. Yoshinaga, T. et al., Efficient uptake and rapid degradation of plasmid DNA by murine dendritic cells via a specific mechanism, *Biochem. Biophys. Res. Commun.* 299, 389–394, 2002.

26. Takakura, Y. et al., Characterization of plasmid DNA binding and uptake by peritoneal macrophages from class A scavenger receptor knockout mice, *Pharm Res.* 16, 503–508, 1999.

27. Hemmi, H.O. et al., A Toll-like receptor recognizes bacterial DNA, *Nature* 408, 740–745, 2000.

28. Krieg, A.M., CpG motifs in bacterial DNA and their immune effects, *Annu. Rev. Immunol.* 20, 709–760, 2002.

29. Emlen, W. and Mannik, M., Kinetics and mechanisms for removal of circulating single-stranded DNA in mice, *J. Exp. Med.* 147, 684–699, 1978.

30. Emlen, W. and Mannik, M., Effect of DNA size and strandedness on the *in vivo* clearance and organ localization of DNA, *Clin. Exp. Immunol.* 56, 185–192, 1984.

31. Du Clos, T.W. et al., Chromatin clearance in C57Bl/10 mice: interaction with heparan sulphate proteoglycans and receptors on Kupffer cells, *Clin. Exp. Immunol.* 117, 403–411, 1999.

32. Kawabata, K., Takakura, Y., and Hashida, M., The fate of plasmid DNA after intravenous injection in mice: involvement of scavenger receptors in its hepatic uptake, *Pharm. Res.* 12, 825–830, 1995.

33. Kobayashi, N. et al., Hepatic uptake and gene expression mechanisms following intravenous administration of plasmid DNA by conventional and hydrodynamics-based procedures, *J. Pharmacol. Exp. Ther.* 297, 853–860, 2001.

34. Yoshida, M. et al., Disposition characteristics of plasmid DNA in the single-pass rat liver perfusion system, *Pharm. Res.* 13, 599–603, 1996.

35. Ali, S.A. et al., Synthesis and radioiodination of tyramine cellobiose for labeling monoclonal antibodies, *Int. J. Rad. Appl. Instrum. B.* 15, 557–561, 1988.

36. Deshpande, S.V. et al., Metabolism of indium chelates attached to monoclonal antibody: minimal transchelation of indium from benzyl-EDTA chelate *in vivo*, *J. Nucl. Med.* 31, 218–224, 1990.

37. Thorpe, S.R., Baynes, J.W., and Chroneos, Z.C., The design and application of residualizing labels for studies of protein catabolism, *FASEB J.* 7, 399–405, 1993.

38. Duncan, J.R. and Welch, M.J., Intracellular metabolism of indium-111-DTPA-labeled receptor targeted proteins, *J. Nucl. Med.* 34, 1728–1738, 1993.

39. Nishikawa, M. et al., Residualizing indium-111-radiolabel for plasmid DNA and its application to tissue distribution study. *Bioconjugate Chem.,* 14, 955–961, 2003.

40. Boussif, O. et al., A versatile vector for gene and oligonucleotide transfer into cells in culture and *in vivo*: polyethylenimine, *Proc. Nat. Acad. Sci. U.S.A.* 92, 7297–7301, 1995.

41. Wagner, E. et al., Influenza virus hemagglutinin HA-2 N-terminal fusogenic peptides augment gene transfer by transferrin-polylysine-DNA complexes: toward a synthetic virus-like gene-transfer vehicle, *Proc. Nat. Acad. Sci. U.S.A.* 89, 7934–7938, 1992.

42. Nishikawa, M. and Huang, L., Nonviral vectors in the new millennium: delivery barriers in gene transfer, *Hum. Gene Ther.* 12, 861–870, 2001.

43. Li, S. et al., Dynamic changes in the characteristics of cationic lipidic vectors after exposure to mouse serum: implications for intravenous lipofection, *Gene Ther.* 6, 585–594, 1999.

44. Sakurai, F. et al., Interaction between DNA-cationic liposome complexes and erythrocytes is an important factor in systemic gene transfer via the intravenous route in mice: the role of the neutral helper lipid, *Gene Ther.* 8, 677–686, 2001.

45. Ogris, M. et al., PEGylated DNA/transferrin-PEI complexes: reduced interaction with blood components, extended circulation in blood and potential for systemic gene delivery, *Gene Ther.* 6, 595–605, 1999.

46. Mahato, R.I. et al., Physicochemical and pharmacokinetic characteristics of plasmid DNA/cationic liposome complexes, *J. Pharm. Sci.* 84, 1267–1271, 1995.

47. Liu, Y. et al., Factors influencing the efficiency of cationic liposome-mediated intravenous gene delivery, *Nat. Biotechnol.* 15, 167–173, 1997.

48. Nishikawa, M. et al., Targeted delivery of plasmid DNA to hepatocytes *in vivo*: optimization of the pharmacokinetics of plasmid DNA/galactosylated poly(L-lysine) complexes by controlling their physicochemical properties, *J. Pharmacol. Exp. Ther.* 287, 408–415, 1998.

49. Finsinger, D. et al., Protective copolymers for nonviral gene vectors: synthesis, vector characterization and application in gene delivery, *Gene Ther.* 7, 1183–1192, 2000.

50. Plank, C. et al., Activation of the complement system by synthetic DNA complexes: a potential barrier for intravenous gene delivery, *Hum. Gene Ther.* 7, 1437–1446, 1996.

51. Boletta, A. et al., Nonviral gene delivery to the rat kidney with polyethylenimine, *Hum. Gene Ther.* 8, 1243–1251, 1997.

52. Godbey, W.T., Wu, K.K., and Mikos, A.G., Tracking the intracellular path of poly(ethylenimine)/DNA complexes for gene delivery, *Proc. Nat. Acad. Sci. U.S.A.* 96, 5177–5181, 1999.

53. Klemm, A.R., Young, D., and Lloyd, J.B., Effects of polyethyleneimine on endocytosis and lysosome stability, *Biochem Pharmacol.* 56, 41–46, 1998.

54. Ouatas, T. et al., T3-dependent physiological regulation of transcription in the Xenopus tadpole brain studied by polyethylenimine based *in vivo* gene transfer, *Int. J. Dev. Biol.* 42, 1159–1164, 1998.

55. Morimoto, K. et al., Molecular weight-dependent gene transfection activity of unmodified and galactosylated polyethyleneimine on hepatoma cells and mouse liver, *Mol Ther.* 7, 254–261, 2003.

56. Wu, G.Y. and Wu, C.H., Receptor-mediated gene delivery and expression *in vivo*, *J. Biol. Chem.* 263, 14621–12624, 1988.

57. Midoux, P. et al., Specific gene transfer mediated by lactosylated poly-L-lysine into hepatoma cells, *Nucleic Acids Res.* 21, 871–878, 1993.

58. Perales, J.C. et al., Gene transfer *in vivo*: sustained expression and regulation of genes introduced into the liver by receptor-targeted uptake, *Proc. Nat. Acad. Sci. U.S.A.* 91, 4086–4090, 1994.

59. Nishikawa, M. et al., Hepatocyte-targeted *in vivo* gene expression by intravenous injection of plasmid DNA complexed with synthetic multi-functional gene delivery system, *Gene Ther.* 7, 548–555, 2000.

60. Ferkol, T. et al., Receptor-mediated gene transfer into macrophages, *Proc. Nat. Acad. Sci. U.S.A.* 93, 101–105, 1996.

61. Diebold, S.S. et al., Mannose polyethylenimine conjugates for targeted DNA delivery into dendritic cells, *J. Biol. Chem.* 274, 19087–19094, 1999.

62. Nishikawa, M. et al., Pharmacokinetics and *in vivo* gene transfer of plasmid DNA complexed with mannosylated poly(L-lysine) in mice, *J. Drug Target.* 2000, 29–38, 2000.

63. Wagner, E. et al., Transferrin-polycation conjugates as carriers for DNA uptake into cells, *Proc. Nat. Acad. Sci. U.S.A.* 87, 3410–3414, 1990.

64. Ferkol, T., Kaetzel, C.S., and Davis, P.B., Gene transfer into respiratory epithelial cells by targeting the polymeric immunoglobulin receptor, *J. Clin. Invest.* 92, 2394–2400, 1993.

65. Li, S. et al., Targeted gene delivery to pulmonary endothelium by anti-PECAM antibody, *Am. J. Physiol.* 278, L504–L511, 2000.

66. Ross, G.F. et al., Surfactant protein A-polylysine conjugates for delivery of DNA to airway cells in culture, *Hum. Gene Ther.* 6, 31–40, 1995.

67. Nishikawa, M. et al., Synthesis and pharmacokinetics of a new liver-specific carrier, glycosylated carboxymethyl-dextran, and its application to drug targeting, *Pharm. Res.* 10, 1253–1261, 1993.

68. Akamatsu, K. et al., Development of a hepatocyte-specific prostaglandin E_1 polymeric product and its potential for preventing carbon tetrachloride-induced fulminant hepatitis in mice, *J. Pharmacol. Exp. Ther.* 290, 1242–1249, 1999.

69. Ashwell, G. and Harford, J., Carbohydrate-specific receptors of the liver, *Annu. Rev. Biochem.* 51, 531–554, 1982.

70. Dumont, S. et al., Antitumoral properties and reduced toxicity of LPS targeted to macrophages via normal or mannosylated liposomes, *Anticancer Res.* 10, 155–160, 1990.

71. Fujita, T. et al., Therapeutic effects of superoxide dismutase derivative modified with mono- or polysaccharides on hepatic injury induced by ischemia/reperfusion, *Biochem. Biophys. Res. Commun.*,189, 191–196, 1992.

72. Nishikawa, M. et al., Galactosylated proteins are recognized by the liver according to the surface density of galactose moieties, *Am. J. Physiol.* 268, G849–G856, 1995.

73. Lee, Y.C. et al., Binding of synthetic oligosaccharides to the hepatic Gal/GalNAc lectin: dependence on fine structural features, *J. Biol. Chem.* 258, 199–202, 1983.

74. Townsend, R.R. et al., Binding of N-linked bovine fetuin glycopeptides to isolated rabbit hepatocytes: Gal/GalNAc hepatic lectin discrimination between Galβ(1,4)GalNAc and Galβ(1,3)GalNAc in a triantennary structure, *Biochemistry* 25, 5716–5725, 1986.

75. Wagner, E., Application of membrane-active peptides for nonviral gene delivery, *Adv. Drug Deliv. Rev.* 38, 279–289, 1999.

PART II

Condensing Polymeric Systems

A. Non-Degradable Polymers

CHAPTER 6

Poly(L-Lysine) and Copolymers for Gene Delivery

Minhyung Lee and Sung Wan Kim

CONTENTS

6.1 INTRODUCTION

To be expressed in target cells for gene therapy, DNA should be delivered to the nucleus through physiological and cellular barriers such as the cell membrane, endosomal membrane, and nucleus membrane.[1,2] Therefore, an efficient carrier should have the following characteristics. First, the carrier should condense the DNA into a small size and mask the negative charge of DNA for

0-8493-1934-X/05/$0.00+$1.50
© 2005 by CRC Press LLC

efficient internalization. Second, the carrier should have a specific ligand for targeting DNA to a specific tissue. Third, the carrier–DNA complex should escape the endosome to avoid lysosomal degradation. Fourth, the DNA should be transferred into the nucleus for transcription. Poly(L-lysine) (PLL) has been widely used for gene delivery because of its excellent characteristics as a gene carrier. PLL has positive charges at its ε-amines, which contribute to the condensation of negatively charged DNA. PLL can protect DNA from nucleases, which is an essential characteristic for *in vivo* gene delivery. In addition, a targeting moiety can be easily coupled to PLL by chemical conjugation for targeting delivery.

PLL enhances transfection efficiency compared to naked DNA. However, the PLL–DNA complex is easily aggregated and precipitated in physiological saline. In addition, PLL exhibits toxicity to cells. Furthermore, the transfection efficiency of PLL is moderate compared to poly(ethylenimine) (PEI) or liposomes, although it has a higher transfection efficiency than naked DNA. Various efforts have been made to overcome these problems. In this chapter, we will discuss the characteristics of PLL and the conjugation of ligands to PLL.

6.2 CHARACTERISTICS OF PLL AS A GENE CARRIER

The lysine of PLL has ε-amine groups with positive charges that contribute to the condensation of the negative charged phosphate groups of DNA by charge interaction (Figure 6.1). The PLL–DNA complex was first investigated by Laemmli et al. in 1975.[3] PLL condensed DNA into small complexes with donuts (toroids) or short stem structures (rods). Toroids are 25 to 50 nm in diameter and rods are 40 to 80 nm long in water,[4] with both types of complexes equally frequent in the mixture of the complex.

Complex formation is dependent on the molecular weight of PLL. PLL residues greater than 20 will bind DNA in physiological saline. However, a smaller chain of PLL will not bind DNA in physiological saline. When the PLL–DNA charge ratio is below 0.5, there is no consistent size of complex. At more than 0.5 charge ratio, DNA condenses into a more compact form. In addition, the size of the PLL–DNA complex gradually increases in a time-dependent manner at a charge ratio of 1.0. However, above or below 1.0 charge ratio, the complex size is stable.[4]

Salt concentration can determine the size of the complex. At high NaCl concentration, the PLL–DNA complex aggregates and precipitates.[5] PLL–DNA complexes at various molecular weights (26, 57, 92, or 256 kDa) precipitate in a 1.0 M NaCl solution. In a 0.15 M NaCl solution,

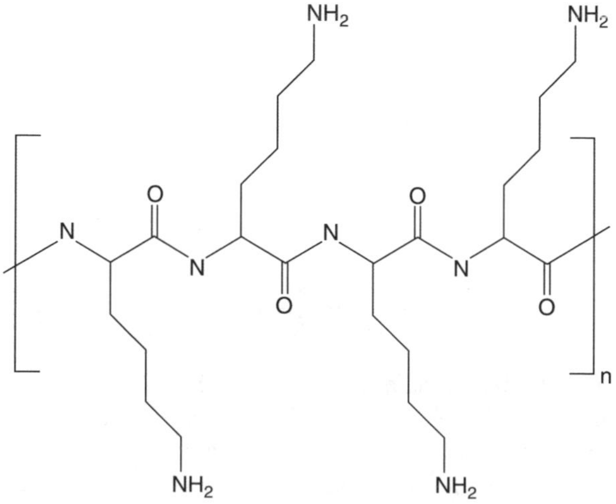

Figure 6.1 The structure of poly(L-lysine).

the complex size increases with time. Similarly, the PLL–DNA complex aggregates in physiological saline at all charge ratios. The aggregation in physiological saline can be overcome by a preparation method described by Perales et al.[6] In this method, PLL is gradually added onto DNA until the charge becomes neutral. Then, the ionic strength of the solvent is adjusted to facilitate the solubilization of the PLL–DNA complex.

The complex formation between PLL and DNA is classified into two groups, cooperative and noncooperative bindings.[5] Cooperative binding occurs in a narrow range of NaCl concentration; the concentration is dependent on the molecular weight of PLL. The cooperative and noncooperative complex formations are distinguished by the characteristics of the complex precipitation. In cooperative condensation, the DNA is fully condensed and the rest of the DNA is unbound. There are few partially condensed intermediate PLL–DNA complexes in the cooperative condensation. In the DNase I protection assay, the cooperative and noncooperative condensation showed clearly different characteristics. In the cooperative condensation, the DNA in the complex is fully protected and the nonbinding DNA completely degraded without a partial degradation product.

PLL shows the highest transfection efficiency at a 2:1 weight ratio of PLL–DNA.[7] However, the transfection efficiency of PLL is lower than that of PEI. The lower transfection efficiency of PLL compared to PEI may be due to a difference of intracellular trafficking. First, the efficiency of the endosomal escape of the PLL–DNA complex is lower than PEI. PEI has many protonation sites, which contribute to the escape from the endosomal compartment. Akinc and Langer reported the average pH environment of the plasmid delivered by PLL as 4.5, suggesting that the PLL–DNA complex may be transferred to acidic lysosomes.[8] Therefore, the plasmid delivered by PLL is broken down in the lysosomes, which is one of the reasons why PLL has a lower transfection efficiency compared to PEI. In another hypothesis, the PEI–DNA complex escapes from late endosomes, which are closer to the nucleus than early endosomes.[9] However, the PLL–DNA complexes were localized in early endosomes, which were far from the nucleus.

The cytotoxicity of PLL is relatively high. The cell viabilities decreased to 40–60% after the addition of the PLL–DNA complexes.[7,10] The cytotoxicity of the cationic particles is derived from the charge density and the shape of the particles. The PLL–DNA complex has been formulated to have a positive charge, which shows the highest transfection efficiency to various types of cells. Although the positively charged complex interacts easily with the cellular membrane, it is the positive charge that induces the cytotoxicity. It is generally accepted that the cytotoxicity of the cationic polymers is related to the transfection efficiency.[11] In addition, the low degradable rate of PLL suggests that PLL will be excreted from the body at a slow rate.

The biodistribution of the PLL–DNA complex was investigated after intravenous injection via tail vein into mice.[12] Naked DNA was rapidly cleared from the bloodstream. However, most of the naked DNA was taken up by the liver (approximately 84.9% of the injected DNA). On the contrary, the PLL–DNA complex showed a different body distribution according to the molecular weight of PLL. The circulation of the PLL–DNA complex also depends on the molecular weight of PLL. Most of the PLL 20 (20 kDa)–DNA complexes were localized in the liver (80%). In the liver, Kupffer cells took up most of the PLL–DNA complexes. Some of the complexes went to the kidney (0.8–1.9%) and lungs (1.0–2.0%). However, the PLL 211 (211 kDa)–DNA complex had long blood circulation time and approximately 47% of the complexes were taken up by the liver.

The cationic polymer–DNA complex usually interacts with the serum protein in the blood. The PLL–DNA complexes have positive charges, and therefore they can interact with negatively charged serum proteins. Therefore, in the *in vivo* situation, the transfection efficiency is dramatically reduced. The interaction with blood proteins is also dependent on the molecular weight of PLL.[12] The PLL–DNA complexes bind to blood cells *in vivo*. PLL 20–DNA complex may aggregate in the blood. However, PLL 211 and naked DNA do not aggregate.

As described above, PLL has excellent characteristics as a gene carrier. However, there are still problems that prevent PLL from being applied in *in vivo* preclinical or clinical trials. Many efforts have been made to overcome these problems and improve the characteristics of PLL. In addition,

PLL has no tissue specificity and therefore, to confer PLL with specificity and better transfection efficiency, targeting moieties have been conjugated to PLL. These PLL conjugates will be explained in the next section.

6.3 PLL CONJUGATES

6.3.1 PLL Conjugates for Low Cytotoxicity

For low cytotoxicity, two main approaches have been employed. The first is to conjugate polyethylene glycol (PEG) moiety to PLL, and the second is to introduce degradable bonds or moieties to PLL. PEG conjugation reduced the cytotoxicity of PLL and also increased the solubility of the PLL–DNA complex, which is favorable for *in vivo* gene delivery. In addition, the interaction with serum protein is prevented by PEG, which can increase the blood circulation time and the transfection efficiency in the *in vivo* situation. One of the examples of the biodegradable PLL is poly(α-[4-aminobutyl]-L-glycolic acid) (PAGA).[13] PAGA is an analog of PLL, in which the peptide bond is substituted for a degradable ester bond. A detailed explanation of PAGA is presented in Chapter 15.

6.3.1.1 PLL–PEG Block Copolymer

The block copolymer of PEG-PLL was synthesized by polymerization of the N-carboxyanhydride (NCA) of the Z-protected L-lysine, using α-methoxy-ω-amino-PEG as an initiator.[14] In the gel retardation assay, the PEG-PLL–DNA complex was completely retarded at a 2:1 charge ratio (+/-). The size of the PEG-PLL–DNA complex was over 100 nm, showing an extended shape compared to the PLL–DNA complex.[14] In another report, the PEG-PLL–DNA complex had a diameter of about 48.5 nm with a small fraction of secondary aggregates which had a diameter of 140 nm.[15] The difference between the two reports may be due to the difference in the length of PLL. Katayose and Kataoka suggested that PEG-PLL with a longer length of PLL has a larger complex size.[15] The PEG-PLL–DNA complexes have an extended form with thick linear and toroidal shapes.[14] PEG can stabilize the complex sterically.[15,16] In addition, PEG-PLL protected DNA from nuclease efficiently, showing 1.5% of degradation compared to naked DNA.[15] The protection ability of PEG-PLL increased with an increasing molecular weight of PLL.[17] The stability of the PEG-PLL–DNA complex was evaluated by means of a melting curve measurement.[15] The PEG-PLL–DNA complex showed better stability than the PLL–DNA complex and naked DNA. Analysis of the zeta potential of the PEG-PLL–DNA complex showed that PEG-PLL has a slower increase with increasing charge ratio than PLL.[14] This may be due to the charge shielding effect of PEG. The transfection efficiency of PEG-PLL was remarkably high compared to PLL. The PEG-PLL–DNA complex transfected into about 30% of 293 cells, while PLL showed 5% cell

Table 6.1   PLL-Conjugate with PEG

PLL-Conjugate	Size	Transfection Efficiency	Cell Line/Animal	Ref.
PEG-PLL block copolymer	PLL 9600 (dp = 78)	Higher transfection than PLL	A2780, 293	14
	PLL dp = 7, 19, 28, 42, 48, 78, 105, 927 PEG 4300, 12000	Higher transfection than PLL	HepG2, Balb/c mice	15, 17–19
PEG-*g*-PLL	PLL 25000 Da PEG 550 Da	Much higher transfection than PLL	HepG2, MIN6, 293, Balb/c mice	10, 20
	PLL 9600, 22400 Da PEG 5000, 12000 Da	Higher transfection than PLL	HepG2, 293	74

transfection. It was suggested that this may be due to membrane activity or the dehydrating fusogenic effect of PEG when applied at a high local concentration of PEG.[14]

Recently, polyion complex (PIC) with PEG-PLL and DNA was evaluated as a gene carrier.[18,19] In this PIC, DNA and PLL formed a hydrophobic core with the PEG shell. PIC with oligonucleotides formed a small complex size of approximately 30 nm.[18] Degradation of DNA by nuclease decreased in PIC, compared to naked oligonucleotides. The longer length of PLL protected oligonucleotides more effectively than shorter PLL.[18] PIC with plasmid DNA was evaluated in gene transfer to HepG2 cells *in vitro* and mice *in vivo*.[19] The transfection assay into HepG2 cells showed the highest transfection efficiency at a 4:1 charge ratio. At this charge ratio, PIC micelles have a small particle size, 30 to 40 nm. PIC with reporter gene was injected into mice via the tail vein. Southern blot analysis showed that PIC had longer circulation time in the blood stream than naked DNA. Like *in vitro* transfection, PIC micelles had the highest gene expression efficiency at a 4:1 charge ratio in liver. The gene expression persisted for more than 10 days with the peak expression at day 3 after the injection.

6.3.1.2 PEG-Grafted PLL

Comb-shaped PEG-grafted PLL (PEG-*g*-PLL) was first synthesized by Choi et al.[10] The PEG-*g*-PLL–DNA complex was completed retarded at a 1:1 weight ratio. PEG shields the positive charge of PLL, and therefore more PEG-*g*-PLL than PLL is required to form a completely retarded complex in a gel retardation assay. Initially, three kinds of PEG-*g*-PLLs were synthesized with different mole % of conjugated PEG: 5, 10, and 25 mole %. A dye displacement assay showed that all three kinds of PEG-*g*-PLL had a similar complex formation ability at a 3:1 weight ratio. However, at a lower weight ratio, PEG-*g*-PLL showed slightly less complex formation ability than PLL. The transfection efficiency of PEG-*g*-PLL was highest at a 3:1 weight ratio, and at this weight ratio, 10 mole % PEG-*g*-PLL had higher transfection efficiency than the other molar concentrations of PEG-*g*-PLL. PEG-*g*-PLL at 10 mole % had higher transfection efficiency than commercially available Lipofectin. In addition, PEG-*g*-PLL did not show any cytotoxicity to HepG2 cells, while Lipofectamine and PLL showed about 75% and 60% cell viabilities, respectively. PEG-*g*-PLL was evaluated *in vitro* and *in vivo* for the delivery of antisense GAD autoantigen plasmid for the prevention of autoimmune diabetes.[20] In this report, 10, 15, and 20 mole % PEG-*g*-PLL were evaluated as gene carriers. The results showed that 10 and 15 mole % PEG-*g*-PLL have higher transfection efficiencies than 20 mole % PEG-*g*-PLL. In addition, the DNA protection ability of PEG-*g*-PLL increased with decreasing the molar concentration of PEG. This result suggests that PEG shields the positive charge of PLL and that the complex with a lower molar concentration of PEG-*g*-PLL is tighter than the others. Therefore, for more efficient protection in *in vivo* administration, more polymers will be required for PEG-*g*-PLL at a higher molar concentration. The delivery of the antisense plasmid with 10 mole % PEG-*g*-PLL showed that the expression of the antisense mRNA reduced the expression of GAD autoantigen in insulinoma cells (MIN6) *in vitro*. The *in vivo* administration of the PEG-*g*-PLL–antisense plasmid complex into mice via the tail vein could deliver the antisense plasmid to the target organ, the pancreas, although PEG-*g*-PLL did not have a targeting moiety. This suggests that with targeting moiety, PEG-*g*-PLL can deliver the therapeutic plasmid to hard-to-deliver organs, such as the pancreas, more efficiently.

6.3.1.3 PLL-Based Biodegradable Polymer

Biodegradable nanoparticles with PLL or PLL-grafted polysaccharide and poly(D,L-lactic acid) were synthesized as gene carriers.[21] The nanoparticles were prepared by a solvent evaporation method or a diafiltration method. The particle size of the PLL graft copolymers determined by dynamic light scattering was about 60 nm. When the contents of the PLL graft copolymers increased, the particle size had a tendency to decrease. However, the nanoparticles with PLL did not change with the content of PLL with a 200 nm diameter. The interaction of nanoparticles with plasmid DNA was studied by a gel retardation assay. Plasmid DNA was trapped on the surface of the nanoparticles completely above a 80:1 weight ratio. The transfection efficiency of the nanoparticles is not available.

Another example of a PLL-based biodegradable polymer is PLGA-grafted PLL (PLGA-*g*-PLL).[11] PLGA-*g*-PLL forms micelles in an aqueous solution with hydrophobic PLGA as an internal core and cationic PLL as a surrounding corona. To prepare the PLGA-*g*-PLL micelles in aqueous solution, the graft copolymer was dissolved in DMSO and subjected to dialysis. The micelle size determined by dynamic light scattering was 149.6 nm with a narrow size distribution. The CMC of the micelles was 9.6 mg/l. The complete complex formation between the micelles and DNA was obtained above a charge ratio of 2.7 (+/-). The DNA protection ability of the micelles increased with an increasing charge ratio, suggesting a tighter complex formation. DNA was completely protected from nucleases at a charge ratio of 2.7, while DNA was slightly degraded along with time at a charge ratio of 1.3. The transfection efficiency of the micelles was higher than PLL and the cytotoxicity of the micelles was improved compared to PLL.

PLL-PEG multiblock copolymer with an ester bond between PEG and PLL is also a PLL-based biodegradable gene carrier (unpublished data). This multiblock copolymer is composed of low molecular weight PLL (3000 Da) and PEG. Therefore, after degradation, the multiblock copolymer produced nontoxic low molecular weight PLL and PEG. The low molecular weight PLL can be easily excreted from the body. This approach will be effective in that it will have a high molecular weight PLL in transfection and a low molecular weight PLL in cytotoxicity. Furthermore, PEG segments can stabilize the polymer–DNA complex, increase solubility of the complex in aqueous solution, reduce the cytotoxicity of the polymer carrier, shield the positive charge of PLL, and reduce the interaction with negatively charged serum protein in *in vivo* application.

6.3.2 PLL Conjugates for Receptor-Mediated Endocytosis

PLL conjugates for tissue targeting and receptor mediated endocytosis are summarized in Table 6.2. The conjugations of the ligands generally have two purposes. The first is to increase transfection efficiency of PLL using receptor-mediated endocytosis. The second is to target the PLL–DNA complex to specific tissues and organs. To achieve these goals, many kinds of specific ligands have been conjugated to PLL directly or indirectly using a spacer such as PEG.

Table 6.2 PLL-Conjugate for Receptor Mediated Endocytosis

PLL-Conjugate	PLL Size	Ligands	Target Cell	Ref.
PLL-asialoorosomucoid	4, 10, 26 kDa	Asialoorosomucoid	Hepatocytes	4, 28, 30, 31, 94, 95
PLL-transferrin	Dp = 190	Transferrin		33, 36, 37
Galactose-PLL Lactose-PLL	1.8, 7.5, 13, 15, 29, 40, 131 kDa	Galactose/lactose	Hepatocytes	40–42, 67, 96, 97
PLL-PEG-lactose	25 kDa	Lactose	Hepatocytes	44, 45
PLL-antibody	3 kDa	Rat IgG 34A and 14	Mouse lung endothelial cells	47
	20 kDa	JL1 antibody	Leukemia cells	48
PLL-folate	20 kDa	Folate	Cancer cells	50, 51
Stearyl-PLL	50 kDa	LDL (terplex system)	Artery wall cells	54–56, 61–63
PLL-PEG-AWBP	25 kDa	Artery wall binding peptide (AWBP)	Artery wall cells	64

6.3.2.1 PLL-Asialoorosomucoid

Asialoorosomucoid (ASOR) is a ligand to the asialoglycoprotein receptor.[22–27] The asialoglycoprotein receptor is mostly abundant in the membrane of hepatocyte. It is known that this receptor mediates the endocytosis of desialylated glycoproteins and functions in the removal of asialoglycoproteins from serum. Therefore, for targeting the PLL–DNA complex to hepatocytes, ASOR was conjugated to PLL in a 5:1 molar ratio using N-succinimidyl 3-(2-pyridyldithio) propionate.[28] The complex between PLL-ASOR and DNA was confirmed by a gel retardation assay. The structure of the PLL-ASOR–DNA complex was recently investigated with atomic force microscopy.[29] Like PLL, PLL-ASOR forms toroid and rod shaped complexes with DNA, with an outer diameter of 150 nm for the toroids and a width of 50.7 nm for the rods. The targeting delivery of DNA by ASOR was confirmed by the transfection into SK-Hep1 cells and HepG2 cells. Since SK-Hep1 cells do not express the asialoglycoprotein receptor, the transfection into SK-Hep1 cells was very low. However, the transfection into HepG2 cells was effective, showing a high level of the chloramphenicol acetyl transferase activity. The PLL-ASOR–DNA complex was also evaluated in *in vivo* experiments.[30] *In vivo* gene delivery showed that the reporter gene was expressed only in the liver at a detectable level, while other organs (spleen, kidney, and lungs) failed to express the reporter protein. In addition, coinjection of free ASOR prevents the reporter gene from expression in the liver. PLL-ASOR was evaluated as a gene carrier of the albumin gene using Nagase analbuminemic rat *in vivo*.[31] Nagase analbuminemic rat has a defect in the albumin gene and does not express albumin at a detectable level. Therefore, to correct this genetic disease by delivery of exogenous albumin gene, a plasmid was constructed with human serum albumin cDNA and injected into the rats using PLL-ASOR as a gene carrier. After the injection, the rats were subjected to partial hepatectomy, because partial hepatectomy increases foreign gene expression persistence *in vivo*.[32] The expression of the albumin persisted for more than 2 weeks in the liver and serum. *In vivo* application of PLL-ASOR suggests that the carrier effectively delivered the polymer–DNA complex to hepatocytes.

6.3.2.2 PLL-Transferrin

Transferrin was conjugated to PLL firstly by Wagner et al. in 1990 for receptor-mediated endocytosis.[33] Transferrin is an iron transport protein in blood. Virtually all types of cells take up iron as a transferrin-iron complex. The transferrin-iron complex binds to a specific cell-surface receptor, and then the complex is internalized by receptor-mediated endocytosis.[34,35] Transferrin was conjugated to PLL by ligation through disulfide bonds after modification with the bifunctional reagent succinimidyl 3-(2-pyridyldithio)propionate (SPDP).[33] After the conjugation, PLL-transferrin conjugates are still biologically active, binding to the receptor of the cell membrane. PLL moiety in PLL-transferrin condensed the DNA into a doughnut structure.[36] PLL-transferrin was evaluated as a gene carrier into avian HD-3 erythroblasts[33] and human erythroleukemia K-562 cells.[36–38] The transfection efficiency of PLL-transferrin was improved, compared to that of PLL. In addition, the transfection efficiency of PLL-transferrin increased after desferrioxamine pretreatment, which increased the cellular level of the transferrin receptor, suggesting that the complex was taken up by receptor-mediated endocytosis.[37]

Cotton et al. reported that the lysosomotropic agent, chloroquine, further increased the transfection efficiency of PLL-transferrin.[37] This suggested that endosome escape of the complex was a rate-limiting step in the transfection. Similarly, to increase the transfection efficiency of PLL-transferrin, influenza virus hemagglutinin HA-2 N-terminal fusogenic peptides were used for *in vitro* transfection.[38] The PLL-transferrin–DNA complex binds and interacts with serum protein, which can induce aggregation and reduce gene delivery efficiency. Therefore, the surface polymer–DNA complex was modified with poly-N-(2-hydroxypropyl)methacrylamide (pHPMA).[39] The surface modified PLL-transferrin–DNA complex was completely resistant to protein interaction and significantly decreased nonspecific uptake into cells.

6.3.2.3 Galactose-PLL and Lactose-PLL

Galactose or lactose is another ligand to the asialoglycoprotein receptor. It was previously proven that the lactosylated or glycosylated serum albumin (neoglycoproteins) were recognized and internalized by HepG2 cells. To utilize these ligands for targeting gene delivery to hepatocytes, lactose, or galactose was conjugated to PLL. Lactose-PLL (Lac-PLL) was evaluated in the transfection assay into HepG2 cells.[40] The results showed that the lactosylated PLL–DNA complex allowed a very efficient cell selective transfection. Galactose-PLL (Gal-PLL) was also synthesized and evaluated as a liver-targeting gene carrier.[41,42] PLL was reacted with 2-imino-2-methoxyethyl-1-thiogalactoside in a borate buffer. In the reporter gene transfection assay to HepG2 cells, Gal-PLL showed much higher transfection than PLL alone. *In vivo* biodistribution showed that around 60% of plasmids were localized in the liver and the level of plasmid was undetectable in other organs, showing targeted gene delivery to the liver. Physicochemical characterization of Gal-PLL–DNA complex was also identified with antisense oligonucleotides.[43] The Gal-PLL–DNA complex had a particle size of 150 nm and a zeta potential of −27.79 mV at a 1:0.6 weight ratio. The radiolabeled oligonucleotides were localized in the liver within 10 min after the injection. The protein-PLL conjugates should be prepared in a high ionic strength solution to avoid insoluble aggregation. Therefore, this approach with galactose or lactose is effective in that PLL was conjugated with a small galactose or lactose moiety, and not a protein such as ASOR, transferrin, or antibodies.

PEG was conjugated as a linker between PLL and lactose or galactose.[44-46] PEG conjugation increased the availability of the target ligand on the surface of the complex. In addition, PEG increased the solubility of the complex in aqueous solution and decreased the cytotoxicity to cells. Six, 12, 22, and 30 mole % lactose conjugated PEG-*g*-PLL was evaluated in the transfection to HepG2 cells.[44] The complete retardation was observed at a 1:1 weight ratio. The particle size of the complex was around 150 nm. Furthermore, the complex stability was similar to that of PLL. Cytotoxicity of the complex was improved compared to PLL.[44] The transfection efficiency of 30 mole % Lac-PEG-PLL showed the highest transfection efficiency. In addition, 30 mole % Lac-PEG-PLL showed higher transfection efficiency than PLL or lipofectin to HepG2 cells. The transfection assay with Gal-PEG-PLL suggests that the galactose moiety of the carrier is targeting the complex to hepatocytes efficiently.

6.3.2.4 PLL-Antibody

The specific binding of antibody to antigen has been utilized to target the gene carrier–DNA complex to specific tissue that expresses antigen.[47,48] For the conjugation of antibody to PLL, it should be confirmed that the antibody binding activity to antigen is conserved after the conjugation. Trubetskoy et al. conjugated 34A or 14 rat monoclonal antibody to PLL.[47] The conjugated PLL was localized in the spleen and had a relatively low distribution to the liver. After the PLL conjugation, the specific activity of the antibody was decreased compared to antibody alone. However, it still had substantial antigen binding affinity for the targeting delivery. In this report, the antibody was conjugated to amine groups of lysine. However, this conjugation method can often decrease antigen binding ability due to blockage of the antigen binding site. To avoid this disadvantage, the carbohydrate-directed antibody modification method was developed.[48] In this method, PLL was conjugated to the hydroxyl group of the antibody. JL-1 antibody was conjugated to PLL using the carbohydrate-directed antibody modification method as a leukemia cell targeting carrier. The specific interaction and internalization of the PLL-antibody conjugates was confirmed by confocal microscope. The PLL-anti-JL-1–DNA complex had a particle size of 280 to 300 nm.

In addition, the binding and uptake of PLL-anti-JL-1–DNA complex was confirmed in Molt 4 cells. The specificity of the antibody binding is slightly reduced when the antibody is conjugated to PLL. Specific transfection was confirmed by transfection assay on Molt 4 cells. The transfection efficiency of the PLL conjugate was higher than PLL itself and lipofectin. Therefore, this antibody-conjugated PLL demonstrated that antibody-mediated gene transfer is a very promising approach for tissue specific gene delivery.

6.3.2.5 PLL-Folate

A cancer-cell-targeting gene carrier has been developed by conjugation of ligands that are specific for cancer-cell-specific receptors.[49,50] Previously, antineoplastic drugs and nucleic acids were delivered to cancer cells using folate-mediated targeting.[49] The conjugates with folate were evaluated as gene carriers.[49-51] One of the advantages of folate is that folate conjugates enter the cytoplasm as an intact compound and avoid lysosomal degradation.[52] However, the mechanism by which folate enters cytoplasm is not clear.[53] The folate-PLL conjugate was evaluated by *in vitro* transfection into cancer cells.[50] PLL-folate showed the highest transfection efficiency at a 1.2 weight ratio (PLL-folate–DNA). At this ratio, PLL-folate had approximately 6 times higher transfection efficiency to cancer cells than PLL. Furthermore, free folate inhibited the transfection by PLL-folate, suggesting that PLL-folate–DNA complex was internalized into cells using a specific receptor. Furthermore, a simple mixture of PLL-folate–DNA did not show transfection enhancement. Therefore, the folate moiety should be conjugated to PLL to increase folate-mediated transfection. Remarkably, chloroquine increased the transfection efficiency of the complex, suggesting that endosome escape remains an important barrier for PLL-folate-mediated gene delivery.[50,52]

In vivo evaluation of PLL-folate was recently performed by Ward et al.[51] In this application, folate was conjugated directly to PLL or indirectly using PEG as a linker. The PLL-folate–DNA complex interacts with serum protein and aggregates in the bloodstream. The morphology of the complex was quite similar to that of the PLL–DNA complex. However, the zeta potential of PLL-PEG-folate reduced to zero and the serum protein association to the complex was inhibited remarkably. In addition, the solubility of the complex increased. This enhanced stability of the complex in serum increased blood circulation time effectively. Interestingly, the uptake of the PLL-folate–DNA and the PLL-PEG-folate–DNA was almost the same. However, the gene expression by the PLL-PEG-folate was improved in HeLa cells *in vitro*. This suggests that the transgene expression is not correlated to the level of internalized DNA.

6.3.2.6 The Terplex System

The terplex system was developed for receptor-mediated gene delivery,[54-56] and is composed of stearyl-PLL, low-density lipoprotein (LDL), and plasmid DNA (pDNA) (Figure 6.2). Each component in the terplex system plays an important role. PLL moiety forms complexes with plasmid DNA by charge interaction. The stearyl group interacts with LDL and integrates into LDL by hydrophobic interaction. LDL binds to the LDL receptor in the cellular membrane, and therefore LDL incorporation enhances the transfection efficiency of the system by receptor-mediated endocytosis. Stearyl-PLL was synthesized by N-alkylation of PLL (mol wt 50,000) with alkyl bromide.[54] The terplex system was completely retarded at a weight ratio of 3:3:5 of DNA:LDL:stearyl-PLL.[55,56] The zeta-potential of the terplex system was about 2 mV at a 1:1:1 weight ratio of DNA:LDL:stearyl-PLL. LDL receptors are present on the membrane of many cell types such as hepatocytes,[57] endothelial cells,[58,59] and myocytes.[60] Therefore, the terplex system showed the efficient transfection into many kinds of cells including human lung fibroblast CCD-32 Lu cells,[56] murine smooth muscle A7R5 cells,[54,56] primary smooth muscle cells, and primary endothelial cells.[61]

Pharmacokinetics and gene expression were investigated after the systemic administration with the terplex system.[62] The terplex system was injected into mice via the tail vein. The terplex system

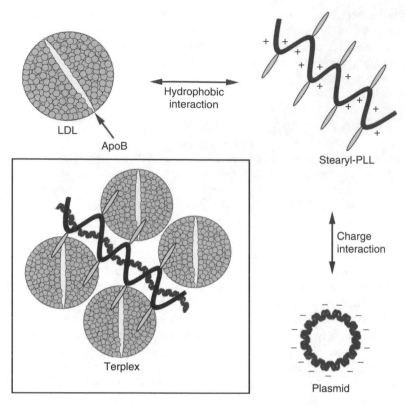

Figure 6.2   Terplex system.

showed longer circulation in the bloodstream than naked DNA. The body distribution of DNA showed that terplex complexes were localized in the lungs at earlier stages (1 h postinjection). However, terplex complexes were localized mainly in the liver at later stages (24 h postinjection). In addition, the major gene expression site was the liver after intravenous injection. The terplex system was also evaluated as a local gene delivery carrier. The terplex system with a luciferase plasmid was injected to rat or rabbit myocardium to evaluate gene expression efficiency.[63] The terplex system showed significantly higher gene expression in myocardium, compared to naked DNA. In addition, terplex complexes showed longer gene expression duration than naked DNA. The higher and longer gene expression by terplex DNA may be due to the higher gene transfection by LDL-receptor-mediated endocytosis and protection of DNA from nucleases.

6.3.2.7 Artery-Wall-Binding Peptide PEG-Grafted-PLL

Artery-wall-binding peptide (AWBP) is a core peptide of apo B-100 protein, a major protein component of LDL. Shih et al. proved that a peptide containing 1000 to 1016 amino acids of apo B-100 is the artery wall binding domain of apo B-100.[59] To target the polymer–DNA complex to the artery wall, AWBP was conjugated to PLL using PEG as a linker.[64] At a charge ratio of 1 (+/-), the PLL-PEG-AWBP–DNA complex was completely retarded in a gel retardation assay. The complex measured approximately 85.9 nm in particle size in a spherical shape. A DNase I protection assay showed that PLL-PEG-AWBP protected DNA from nucleases for more than 120 min. PLL-PEG-AWBP had about 100 times higher transfection efficiency than PLL and PEG-*g*-PLL to bovine aorta endothelial cells and aorta smooth muscle cells. This high transfection to endothelial cells and smooth cells was abolished when free AWBP was mixed with the complex as a competitor, suggesting that the transfection of PLL-PEG-AWBP was mediated by receptor-mediated endocytosis.

6.3.3 PLL Conjugates for Efficient Intracellular Trafficking

The transfection efficiency of the PLL-conjugate–DNA complex is usually enhanced by the addition of chloroquine as an endosomolytic agent.[10,40,50,65-68] In addition, the release of DNA from the endosomal compartment into the cytoplasm after transfection of the cell is a major issue for efficient gene transfer.[65,67,69-74] Therefore, the introduction of an endosomolytic moiety to PLL can increase the transfection efficiency. For this purpose, histidine has been used for PLL-mediated gene transfer. In addition, fusogenic peptide has been used to destabilize the endosomal membrane. However, in most cases, the peptides were simply mixed with the positive charged PLL–DNA complexes without covalent linkage with PLL.

6.3.3.1 Histidylated PLL

Histidine was conjugated to PLL and evaluated as a gene carrier for efficient endosomal escape.[75,76] The imidazole group of histidine has a pH around 6.0, and therefore becomes cationic in acidic solution. This means that poly(histidine) cannot form a complex with DNA at physiological pH. Therefore, the approach of the conjugation to PLL was investigated. Histidylated PLL (His-PLL) forms stable complexes with plasmid DNA. The zeta-potential of the His-PLL–DNA complex was about 25 mV.[75] In the transfection assay to HepG2 cells, His-PLL showed higher transfection efficiency than PLL with endosomolytic agent.[75] The optimum histidine conjugation percentage was 38% of the PLL ε-amino groups. Furthermore, the His-PLL–DNA complex has a low serum effect in the transfection. The transfection efficiency with 1, 5, 10, and 20% serum showed almost the same results at all serum concentrations. The His-PLL–DNA complex transfected various cells efficiently, including HepG2, HOS, COS, and HeLa.

His-PLL was evaluated as a vector *in vitro* for gene transfer into immortalized cystic fibrosis airway surface and airway gland serous cells.[76] The size of the His-PLL–DNA complex was around 100 nm in the absence of serum, 800 nm with 1% FBS, and 395 nm with 10% FBS. The transfection efficiency to SCFTE29o-cells was dependent on the concentration of serum. The transfection efficiency with 1% was almost same as without serum, but that with 10% FBS was decreased to a tenth of that without serum. The gene expression efficiency in ΣCFTE29o- and CF-KM4 cells was dependent not on the transfection time but on the dose of DNA. The gene expression persisted for more than 144 h, although gene expression was decreased along with time.

Intracellular trafficking and stability of the His-PLL–DNA complex was investigated recently.[77] Conclaves et al. compared His-PLL with His-PLL bearing lactosyl residues (His-, Lac-PLL) in terms of internalization and transfection efficiency. The His-, Lac-PLL-DNA complex was internalized more rapidly into the cells than PLL-His. However, overall gene expression efficiency of the His-PLL–DNA complex was better than that of the His-, Lac-PLL-DNA complex, which means that the His-PLL–DNA complex was more stable than the Lac-PLL–DNA complex. The complex between His-PLL and DNA had tighter condensation when pH decreased, suggesting that the His-PLL–DNA complex is more stable in the acidic endosome. In fact, the pH environment of the His-PLL–DNA complex was ~6.2, while that of naked DNA was ~6.6, suggesting the DNA in the His-PLL–DNA complex has more prolonged stability inside the endosome. Therefore, there were more undegraded plasmids in the endosome, which could increase its endosomal escape.

Another report showed that zinc could improve the transfection efficiency mediated by His-PLL.[78] With ZnCl$_2$, the transfection efficiency of His-PLL increased. However, calcium and manganese did not increase the transfection efficiency. Zinc ion can increase the fusogenic properties of a histidine-rich gene carrier.

6.3.3.2   Histidylated Oligolysines

Histidylated oligolysine (dp = 19) is not suitable for pDNA. However, highly substituted ($\geq$ 80%) histidylated oligolysines are suitable vectors for oligodeoxynucleotide (ODN) transfer.[79,80] Oligolysine (dp = 19) was substituted with histidine residues protected as $(Boc)_2$-His-OH. The cellular uptake of histidylated oligolysine–ODN complex was ten times higher than naked ODN, as determined by fluorescence intensity measurement. This effect of the lysine oligomer was dependent on cell type. In addition, the biological effect of antisense oligonucleotide delivered by histidylated oligolysine was enhanced. The ICAM-1 antisense oligonucleotide decreased the expression of the protein depending on the dose of the ODN. Similarly, the expression level of ICAM-1 mRNA was decreased by the delivery of the histidylated oligolysine–antisense ODN complex. Bafilomycin A1 is an inhibitor of the vacuolar proton pump, causing inhibition of endosome acidification.[81-83] The endosomal escape by histidylated oligolysine was dramatically reduced by bafilomycin A1, which suggests that the membrane destabilization effect of histidine enhances the endosomal escape of the complex.

6.3.3.3   Poly(histidine)-Grafted PLL

Another approach to enhancing endosomal escape is to conjugate poly(histidine) (PLH) to PLL. N-acetyl-poly(imbenzyl-L-histidine) (18-mer) was conjugated to PLL and poly(histidine)-grafted PLL (PLH-*g*-PLL) was characterized as a gene carrier.[66] The proton-buffering effect of PLH-*g*-PLL was confirmed by acid-base titration. PLH-*g*-PLL showed a higher buffering effect than PLL. The pH profile of PLH suggested that the polymer began to protonate at a pH of 6.5 to 5.0 and buffered the system throughout this range. This pH range was that of endosome pH, suggesting a physiological effect of PLH in the transfection. The complex formation between PLH-*g*-PLL and DNA was confirmed by a gel retardation assay. The transfection assay showed that the PLH-*g*-PLL–DNA complex had the highest transfection efficiency at a 10:1 weight ratio. However, the addition of chloroquine as an endosomolytic agent still enhanced the transfection efficiency, suggesting that there is still room for improvement in the polymeric gene carrier. The cytotoxicity of the PLH-*g*-PLL–DNA complex was very low, in the range of 1:1 to 20:1 weight ratio, while the PLL–DNA complex showed very high toxicity at a 20:1 weight ratio.

6.3.3.4   Fusogenic Peptides

Fusogenic peptides have been used as endosomolytic agents.[38,40,69,84,85] The simple mixture of the polymer–DNA complex and fusogenic peptides increased the transfection efficiency by enhancement of the complex from endosomal vesicles. Various kinds of natural or synthetic fusogenic peptides have been evaluated.[40,69,79,84-87] A detailed description of the peptides is beyond the scope of this chapter. A new approach with a fusogenic peptide, KALA, was investigated recently.[88] In this approach, a PEG-*g*-PLL–DNA complex was formed to have a negative charge, and then the complex surface was coated by fusogenic peptide KALA, since KALA has positive charges and the complex has negative charges. The KALA–PEG-g-PLL–DNA complex showed an improved transfection efficiency with an increasing KALA:DNA ratio.

6.3.3.5   PLL-NLS

It was previously suggested that PLL could act as a nuclear localization sequence (NLS), because PLL resembles SV40 large T antigen NLS.[89-91] However, it was proven that PLL did not function as an NLS.[92] This suggests that the conjugation of the NLS peptide to PLL can enhance the transfection efficiency by efficient trafficking of the PLL–DNA complex into the nucleus. The NLS from SV40 large T antigen was conjugated to PLL.[92] Unlike mutant-NLS-conjugated PLL or

PLL itself, NLS significantly enhanced the transfection efficiency of PLL. This enhanced transfection efficiency suggests that PLL-NLS was recognized by NLS-binding importin α/β heterodimer and the nucleus transport of the complex was facilitated. Indeed, confocal microscopy revealed that the PLL-NLS–DNA complexes accumulated rapidly in the nucleus. This NLS-mediated transfection showed the optimum transfection at a 0.4 lysine/nucleotide (Ly/Nu) ratio. At this ratio, the complex size was 60 nm, which is larger than the maximum size for nuclear pores. However, it was suggested that the complex at a 0.4 Ly/Nu ratio has the flexibility to pass through the nuclear pore.[93] This suggests that it is important to prepare the PLL-NLS–DNA complex at an optimal Ly/Nu ratio for the maximum nuclear accumulation of the complex.

6.4 CONCLUSIONS

PLL has excellent characteristics as a gene carrier. PLL condenses DNA into a small-size particle, which facilitates cellular uptake of DNA. In addition, PLL can protect DNA from endogenous nucleases. However, PLL also has disadvantages for clinical applications. First, PLL is cytotoxic to cells. In addition, PLL is excreted from the body at a slow rate, due to the low degradation rate of PLL. The PLL–DNA complex is easily aggregated and precipitated in physiological saline. The toxicity and aggregation can be overcome by PEG conjugation to PLL. In addition, biodegradable PLL analogs and copolymers enhance the excretion rate of PLL and reduce cytotoxicity. Simple conjugation of specific ligands to PLL enhance the transfection efficiency of PLL to target cells. Efficient intracellular trafficking is achieved by conjugation of histidine and NLS. Tissue targeting, endosome disrupting, nucleus transport, and low cytotoxicity should be combined to develop an effective PLL copolymer for gene therapy. In addition, the extensive *in vitro* and *in vivo* evaluation and optimization of the PLL copolymers will provide valuable information for successful and safe gene therapy applications.

REFERENCES

1. Nishikawa, M. and Huang, L., Nonviral vectors in the new millennium: delivery barriers in gene transfer, *Hum. Gene Ther.* 12, 861, 2001.
2. Wiethoff, C.M. and Middaugh, C.R., Barriers to nonviral gene delivery, *J. Pharm. Sci.* 92, 203, 2003.
3. Laemmli, U.K., Characterization of DNA condensates induced by poly(ethylene oxide) and polylysine, *Proc. Nat. Acad. Sci. U.S.A.* 72, 4288, 1975.
4. Kwoh, D.Y. et al., Stabilization of poly-L-lysine/DNA polyplexes for *in vivo* gene delivery to the liver, *Biochim. Biophys. Acta* 1444, 171, 1999.
5. Liu, G. et al., Biological properties of poly-L-lysine-DNA complexes generated by cooperative binding of the polycation, *J. Biol. Chem.* 276, 34379, 2001.
6. Perales, J.C. et al., Biochemical and functional characterization of DNA complexes capable of targeting genes to hepatocytes via the asialoglycoprotein receptor, *J. Biol. Chem.* 272, 7398, 1997.
7. Lee, M. et al., Water-soluble and low molecular weight chitosan-based plasmid DNA delivery, *Pharm. Res.* 18, 427, 2001.
8. Akinc, A. and Langer, R., Measuring the pH environment of DNA delivered using nonviral vectors: implications for lysosomal trafficking, *Biotechnol. Bioeng.* 78, 503, 2002.
9. Forrest, M.L. and Pack, D.W., On the kinetics of polyplex endocytic trafficking: implications for gene delivery vector design, *Mol. Ther.* 6, 57, 2002.
10. Choi, Y.H. et al., Polyethylene glycol-grafted poly-L-lysine as polymeric gene carrier, *J. Control. Release* 54, 39, 1998.
11. Jeong, J.H. and Park, T.G., Poly(L-lysine)-g-poly(D,L-lactic-co-glycolic acid) micelles for low cytotoxic biodegradable gene delivery carriers, *J. Control. Release* 82, 159, 2002.

12. Ward, C.M., Read, M.L., and Seymour, L.W., Systemic circulation of poly(L-lysine)/DNA vectors is influenced by polycation molecular weight and type of DNA: differential circulation in mice and rats and the implications for human gene therapy, *Blood* 97, 2221, 2001.

13. Lim, Y.B. et al., Biodegradable polyester, poly[alpha-(4-aminobutyl)-L-glycolic acid], as a non-toxic gene carrier, *Pharm. Res.* 17, 811, 2000.

14. Wolfert, M.A. et al., Characterization of vectors for gene therapy formed by self-assembly of DNA with synthetic block co-polymers, *Hum. Gene Ther.* 7, 2123, 1996.

15. Katayose, S. and Kataoka, K., Water-soluble polyion complex associates of DNA and poly(ethylene glycol)-poly(L-lysine) block copolymer, *Bioconjug. Chem.* 8, 702, 1997.

16. Mannisto, M. et al., Structure-activity relationships of poly(L-lysines): effects of pegylation and molecular shape on physicochemical and biological properties in gene delivery, *J. Control. Release* 83, 169, 2002.

17. Katayose, S. and Kataoka, K., Remarkable increase in nuclease resistance of plasmid DNA through supramolecular assembly with poly(ethylene glycol)-poly(L-lysine) block copolymer, *J. Pharm. Sci.* 87, 160, 1998.

18. Harada, A., Togawa, H., and Kataoka, K., Physicochemical properties and nuclease resistance of antisense-oligodeoxynucleotides entrapped in the core of polyion complex micelles composed of poly(ethylene glycol)-poly(L-lysine) block copolymers, *Eur. J. Pharm. Sci.* 13, 35, 2001.

19. Harada-Shiba, M. et al., Polyion complex micelles as vectors in gene therapy—pharmacokinetics and *in vivo* gene transfer, *Gene Ther.* 9, 407, 2002.

20. Lee, M. et al., Repression of GAD autoantigen expression in pancreas beta-Cells by delivery of antisense plasmid/PEG-g-PLL complex, *Mol. Ther.* 4, 339, 2001.

21. Maruyama, A. et al., Nanoparticle DNA carrier with poly(L-lysine) grafted polysaccharide copolymer and poly(D,L-lactic acid), *Bioconjug. Chem.* 8, 735, 1997.

22. Steer, C. J. and Ashwell, G., Studies on a mammalian hepatic binding protein specific for asialoglycoproteins. Evidence for receptor recycling in isolated rat hepatocytes, *J. Biol. Chem.* 255, 3008, 1980.

23. Hudgin, R. L. et al., The isolation and properties of a rabbit liver binding protein specific for asialoglycoproteins, *J. Biol. Chem.* 249, 5536, 1974.

24. Kawasaki, T. and Ashwell, G., Chemical and physical properties of an hepatic membrane protein that specifically binds asialoglycoproteins, *J. Biol. Chem.* 251, 1296, 1976.

25. Krantz, M. J. et al., Attachment of thioglycosides to proteins: enhancement of liver membrane binding, *Biochemistry* 15, 3963, 1976.

26. Stowell, C. P. and Lee, Y. C., The binding of d-glucosyl-neoglycoproteins to the hepatic asialoglycoprotein receptor, *J. Biol. Chem.* 253, 6107, 1978.

27. Tanabe, T., Pricer, W. E., Jr., and Ashwell, G., Subcellular membrane topology and turnover of a rat hepatic binding protein specific for asialoglycoproteins, *J. Biol. Chem.* 254, 1038, 1979.

28. Wu, G. Y. and Wu, C. H., Receptor-mediated *in vitro* gene transformation by a soluble DNA carrier system, *J. Biol. Chem.* 262, 4429, 1987.

29. Golan, R. et al., DNA toroids: stages in condensation, *Biochemistry* 38, 14069, 1999.

30. Wu, G. Y. and Wu, C. H., Receptor-mediated gene delivery and expression *in vivo*, *J. Biol. Chem.* 263, 14621, 1988.

31. Wu, G. Y. et al., Receptor-mediated gene delivery *in vivo*. Partial correction of genetic analbuminemia in Nagase rats, *J. Biol. Chem.* 266, 14338, 1991.

32. Wu, C. H., Wilson, J. M., and Wu, G. Y., Targeting genes: delivery and persistent expression of a foreign gene driven by mammalian regulatory elements *in vivo*, *J. Biol. Chem.* 264, 16985, 1989.

33. Wagner, E. et al., Transferrin-polycation conjugates as carriers for DNA uptake into cells, *Proc. Nat. Acad. Sci. U.S.A.* 87, 3410, 1990.

34. Karin, M. and Mintz, B., Receptor-mediated endocytosis of transferrin in developmentally totipotent mouse teratocarcinoma stem cells, *J. Biol. Chem.* 256, 3245, 1981.

35. Ciechanover, A. et al., Kinetics of internalization and recycling of transferrin and the transferrin receptor in a human hepatoma cell line. Effect of lysosomotropic agents, *J. Biol. Chem.* 258, 9681, 1983.

36. Wagner, E. et al., Transferrin-polycation-DNA complexes: the effect of polycations on the structure of the complex and DNA delivery to cells, *Proc. Nat. Acad. Sci. U.S.A.* 88, 4255, 1991.

37. Cotten, M. et al., Transferrin-polycation-mediated introduction of DNA into human leukemic cells: stimulation by agents that affect the survival of transfected DNA or modulate transferrin receptor levels, *Proc. Nat. Acad. Sci. U.S.A.* 87, 4033, 1990.

38. Wagner, E. et al., Influenza virus hemagglutinin HA-2 N-terminal fusogenic peptides augment gene transfer by transferrin-polylysine-DNA complexes: toward a synthetic virus-like gene-transfer vehicle, *Proc. Nat. Acad. Sci. U.S.A.* 89, 7934, 1992.

39. Dash, P. R. et al., Decreased binding to proteins and cells of polymeric gene delivery vectors surface modified with a multivalent hydrophilic polymer and retargeting through attachment of transferrin, *J. Biol. Chem.* 275, 3793, 2000.

40. Midoux, P. et al., Specific gene transfer mediated by lactosylated poly-L-lysine into hepatoma cells, *Nucleic Acids Res.* 21, 871, 1993.

41. Hashida, M. et al., Targeted delivery of plasmid DNA complexed with galactosylated poly(L-lysine), *J. Control. Release* 53, 301, 1998.

42. Nishikawa, M. et al., Targeted delivery of plasmid DNA to hepatocytes *in vivo*: optimization of the pharmacokinetics of plasmid DNA/galactosylated poly(L-lysine) complexes by controlling their physicochemical properties, *J. Pharmacol. Exp. Ther.* 287, 408, 1998.

43. Mahato, R. I. et al., Physicochemical and disposition characteristics of antisense oligonucleotides complexed with glycosylated poly(L-lysine), *Biochem. Pharmacol.* 53, 887, 1997.

44. Choi, Y. H. et al., Lactose-poly(ethylene glycol)-grafted poly-L-lysine as hepatoma cell-tapgeted gene carrier, *Bioconjug. Chem.* 9, 708, 1998.

45. Choi, Y. H. et al., Characterization of a targeted gene carrier, lactose-polyethylene glycol-grafted poly-L-lysine and its complex with plasmid DNA, *Hum. Gene Ther.* 10, 2657, 1999.

46. Sagara, K. and Kim, S. W., A new synthesis of galactose-poly(ethylene glycol)-polyethylenimine for gene delivery to hepatocytes, *J. Control. Release* 79, 271, 2002.

47. Trubetskoy, V. S. et al., Use of N-terminal modified poly(L-lysine)-antibody conjugate as a carrier for targeted gene delivery in mouse lung endothelial cells, *Bioconjug. Chem.* 3, 323, 1992.

48. Suh, W. et al., Anti-JL1 antibody-conjugated poly (L-lysine) for targeted gene delivery to leukemia T cells, *J. Control. Release* 72, 171, 2001.

49. Wang, S. and Low, P. S., Folate-mediated targeting of antineoplastic drugs, imaging agents, and nucleic acids to cancer cells, *J. Control. Release* 53, 39, 1998.

50. Mislick, K. A. et al., Transfection of folate-polylysine DNA complexes: evidence for lysosomal delivery, *Bioconjug. Chem.* 6, 512, 1995.

51. Ward, C. M. et al., Modification of PLL–DNA complexes with a multivalent hydrophilic polymer permits folate mediated targeting *in vitro* and prolonged plasma circulation *in vivo*, *J. Gene Med.* 4, 536, 2002.

52. Leamon, C. P. and Low, P. S., Cytotoxicity of momordin-folate conjugates in cultured human cells, *J. Biol. Chem.* 267, 24966, 1992.

53. Turek, J. J., Leamon, C. P., and Low, P. S., Endocytosis of folate-protein conjugates: ultrastructural localization in KB cells, *J. Cell Sci.* 106 (Pt 1), 423, 1993.

54. Kim, J. S. et al., *In vitro* gene expression on smooth muscle cells using a TerplexDNA delivery system, *J. Control. Release* 47, 51, 1997.

55. Kim, J. S. et al., Terplex DNA delivery system as a gene carrier, *Pharm. Res.* 15, 116, 1998.

56. Kim, J. S. et al., A new non-viral DNA delivery vector: the terplex system, *J. Control. Release* 53, 175, 1998.

57. Havel, R. J., Receptor and non-receptor mediated uptake of chylomicron remnants by the liver, *Atherosclerosis* 141 Suppl. 1, S1, 1998.

58. Sawamura, T. et al., An endothelial receptor for oxidized low-density lipoprotein, *Nature* 386, 73, 1997.

59. Shih, I. L. et al., Focal accumulation of an apolipoprotein B-based synthetic oligopeptide in the healing rabbit arterial wall, *Proc. Nat. Acad. Sci. U.S.A.* 87, 1436, 1990.

60. Hein, M. et al., Gene transfer into rat heart-derived endothelial cells, *Eur. J. Cardiothorac. Surg.* 13, 460, 1998.

61. Yu, L. et al., TerplexDNA gene carrier system targeting artery wall cells, *J. Control Release* 72, 179, 2001.

62. Yu, L. et al., Systemic administration of TerplexDNA system: pharmacokinetics and gene expression, *Pharm. Res.* 18, 1277, 2001.

63. Affleck, D. G. et al., Augmentation of myocardial transfection using TerplexDNA: a novel gene delivery system, *Gene Ther.* 8, 349, 2001.

64. Nah, J. W. et al., Artery wall binding peptide-poly(ethylene glycol)-grafted-poly(L-lysine)-based gene delivery to artery wall cells, *J. Control. Release* 78, 273, 2002.

65. Wolfert, M. A. and Seymour, L. W., Chloroquine and amphipathic peptide helices show synergistic transfection *in vitro*, *Gene Ther.* 5, 409, 1998.

66. Benns, J. M. et al., pH-sensitive cationic polymer gene delivery vehicle: N-Ac-poly(L-histidine)-graft-poly(L-lysine) comb shaped polymer, *Bioconjug. Chem.* 11, 637, 2000.

67. Erbacher, P. et al., Putative role of chloroquine in gene transfer into a human hepatoma cell line by DNA/lactosylated polylysine complexes, *Exp. Cell Res.* 225, 186, 1996.

68. Kollen, W. J. et al., Gluconoylated and glycosylated polylysines as vectors for gene transfer into cystic fibrosis airway epithelial cells, *Hum. Gene Ther.* 7, 1577, 1996.

69. Ogris, M. et al., The size of DNA/transferrin-PEI complexes is an important factor for gene expression in cultured cells, *Gene Ther.* 5, 1425, 1998.

70. Legendre, J. Y. and Szoka, F. C., Jr., Delivery of plasmid DNA into mammalian cell lines using pH-sensitive liposomes: comparison with cationic liposomes, *Pharm. Res.* 9, 1235, 1992.

71. Zauner, W. et al., Glycerol and polylysine synergize in their ability to rupture vesicular membranes: a mechanism for increased transferrin-polylysine-mediated gene transfer, *Exp. Cell Res.* 232, 137, 1997.

72. Budker, V. et al., pH-sensitive, cationic liposomes: a new synthetic virus-like vector, *Nat. Biotechnol.* 14, 760, 1996.

73. Niidome, T. et al., Binding of cationic alpha-helical peptides to plasmid DNA and their gene transfer abilities into cells, *J. Biol. Chem.* 272, 15307, 1997.

74. Toncheva, V. et al., Novel vectors for gene delivery formed by self-assembly of DNA with poly(L-lysine) grafted with hydrophilic polymers, *Biochim. Biophys. Acta* 1380, 354, 1998.

75. Midoux, P. and Monsigny, M., Efficient gene transfer by histidylated polylysine/pDNA complexes, *Bioconjug. Chem.* 10, 406, 1999.

76. Fajac, I. et al., Histidylated polylysine as a synthetic vector for gene transfer into immortalized cystic fibrosis airway surface and airway gland serous cells, *J. Gene Med.* 2, 368, 2000.

77. Goncalves, C. et al., Intracellular processing and stability of DNA complexed with histidylated polylysine conjugates, *J. Gene Med.* 4, 271, 2002.

78. Pichon, C. et al., Zinc improves gene transfer mediated by DNA/cationic polymer complexes, *J. Gene Med.* 4, 548, 2002.

79. Pichon, C., Goncalves, C., and Midoux, P., Histidine-rich peptides and polymers for nucleic acids delivery, *Adv. Drug Deliv. Rev.* 53, 75, 2001.

80. Pichon, C. et al., Histidylated oligolysines increase the transmembrane passage and the biological activity of antisense oligonucleotides, *Nucleic Acids Res.* 28, 504, 2000.

81. Johnson, L. S. et al., Endosome acidification and receptor trafficking: bafilomycin A1 slows receptor externalization by a mechanism involving the receptor's internalization motif, *Mol. Biol. Cell* 4, 1251, 1993.

82. Presley, J. F. et al., Bafilomycin A1 treatment retards transferrin receptor recycling more than bulk membrane recycling, *J. Biol. Chem.* 272, 13929, 1997.

83. Bowman, E. J., Siebers, A., and Altendorf, K., Bafilomycins: a class of inhibitors of membrane ATPases from microorganisms, animal cells, and plant cells, *Proc. Nat. Acad. Sci. U.S.A.* 85, 7972, 1988.

84. Kichler, A. et al., Glycofection in the presence of anionic fusogenic peptides: a study of the parameters affecting the peptide-mediated enhancement of the transfection efficiency, *J. Gene Med.* 1, 134, 1999.

85. Wagner, E., Ogris, M., and Zauner, W., Polylysine-based transfection systems utilizing receptor-mediated delivery, *Adv. Drug Deliv. Rev.* 30, 97, 1998.

86. Wyman, T. B. et al., Design, synthesis, and characterization of a cationic peptide that binds to nucleic acids and permeabilizes bilayers, *Biochemistry* 36, 3008, 1997.

87. Plank, C. et al., The influence of endosome-disruptive peptides on gene transfer using synthetic virus-like gene transfer systems, *J. Biol. Chem.* 269, 12918, 1994.

88. Lee, H., Jeong, J. H., and Park, T. G., PEG grafted polylysine with fusogenic peptide for gene delivery: high transfection efficiency with low cytotoxicity, *J. Control. Release* 79, 283, 2002.

89. Perales, J. C. et al., An evaluation of receptor-mediated gene transfer using synthetic DNA-ligand complexes, *Eur. J. Biochem.* 226, 255, 1994.

90. Hart, S. L. et al., Gene delivery and expression mediated by an integrin-binding peptide, *Gene Ther.* 2, 552, 1995.

91. Vitiello, L. et al., Condensation of plasmid DNA with polylysine improves liposome-mediated gene transfer into established and primary muscle cells, *Gene Ther.* 3, 396, 1996.

92. Chan, C. K. and Jans, D. A., Enhancement of polylysine-mediated transferrinfection by nuclear localization sequences: polylysine does not function as a nuclear localization sequence, *Hum. Gene Ther.* 10, 1695, 1999.

93. Chan, C. K., Senden, T., and Jans, D. A., Supramolecular structure and nuclear targeting efficiency determine the enhancement of transfection by modified polylysines, *Gene Ther.* 7, 1690, 2000.

94. Wu, G. Y. and Wu, C. H., Evidence for targeted gene delivery to Hep G2 hepatoma cells *in vitro*, *Biochemistry* 27, 887, 1988.

95. McKee, T. D. et al., Preparation of asialoorosomucoid-polylysine conjugates, *Bioconjug. Chem.* 5, 306, 1994.

96. Erbacher, P. et al., Glycosylated polylysine/DNA complexes: gene transfer efficiency in relation with the size and the sugar substitution level of glycosylated polylysines and with the plasmid size, *Bioconjug. Chem.* 6, 401, 1995.

97. Erbacher, P. et al., Gene transfer by DNA/glycosylated polylysine complexes into human blood monocyte-derived macrophages, *Hum. Gene Ther.* 7, 721, 1996.

CHAPTER 7

Gene Delivery Using Polyethylenimine and Copolymers

Manfred Ogris

CONTENTS

7.1 INTRODUCTION

Polyethylenimine (PEI) is well known in the chemical industry as a polymer with high positive charge density. It has been used in water purification, the paper industry, biotechnology (to remove nucleic acids from protein solutions), and also in the food industry. In 1995, the group of Jean-Paul Behr made the first use of this polymer for the delivery of DNA and oligonucleotides.[1] Since then, over 300 publications (as of June 2003) have appeared describing the use of PEI as a gene transfer agent. This chapter will give an overview of PEI and its functional derivatives in gene delivery, from its chemistry and transfection properties to its use in preclinical gene therapy protocols.

0-8493-1934-X/05/$0.00+$1.50
© 2005 by CRC Press LLC

7.2 POLYETHYLENIMINE: CHEMISTRY AND DERIVATIVES

PEI exists in two principal forms, branched and linear. Synthesis of branched PEI starts with aziridine monomers, which polymerize under acidic conditions to form branched polymers. The ratio of primary to secondary to tertiary amines is 1:2:1 and is independent of the polymer molecular weight.[2] At a pH of 7 approximately 10% of nitrogen atoms are protonated; at a pH of 5 this value increases to 50%.[3] Due to this effect, PEI has a strong buffering effect over a broad range of pH (this is called the proton sponge effect; see Section 7.7). Linear PEI (LPEI) can be obtained by two different methods. Similar to branched PEI, polymerization can start with aziridine monomers, but at lower reaction temperatures (for a review, see Suh et al.[4]). Alternatively, poly(2-propyl-2-oxazoline) is hydrolyzed under highly acidic conditions over several hours.[5]

7.3 COUPLING STRATEGIES FOR PEI CONJUGATES

PEI allows for a convenient coupling strategy via the amine nitrogens. Several types of conjugates have been described, including proteins, peptides, sugar molecules, and hydrophilic polymers.

7.3.1 Protein-PEI and Peptide-PEI Conjugates

Glycoproteins can be directly coupled to the primary amino groups in PEI after oxidizing the sugar residues to aldehydes and subsequent reductive amination.[6] Coupling can be performed with heterobifunctional crosslinkers, where the crosslinker usually reacts with the amines in PEI and the activated thiol group reacts with thiolated proteins or peptides.[7] Succinimidyl 3-(2-pyridyldithio)propionate (SPDP) results in disulfide-linked, reducible conjugates, whereas N-succinimidyl-[4-vinylsulfonyl]benzoate) (SVSB) forms nonreducible thioether conjugates.[8] The coupling of sugar molecules, e.g., galactose or mannose derivatives, can be carried out in a similar fashion as that described for glycoproteins. Alternatively, sugar derivates (or other compounds) containing isothiocyanate groups form thiourea derivates, and carbodiimide activated carboxyl groups lead to amide bond formation. Hydrophilic polymers like polyethylene glycol (PEG) are coupled to PEI to reduce unspecific interactions of the resulting polyplexes with nontarget cells and to increase their solubility.[9,10] PEG derivates containing amine reactive groups (succinimidyl esters) are coupled directly to PEI, whereas sulfhydryl-selective PEG reacts with thiolated PEI (or vice versa) to obtain reducible disulfide bonds or thioether bonds.[9] Heterobifunctional PEG derivates are most convenient for introducing ligands via a PEG spacer.[11,12] This can be important in the case of relatively small ligands, which are otherwise not accessible at the complex surface.

Purification of PEI conjugates is mostly carried out by chromatography. Size exclusion chromatography is the method of choice to remove small molecules (sugars, short peptides, fluorescent dyes) from higher molecular weight PEI conjugates.[13] Cation exchange chromatography separates molecules of different or equal size but different charge from PEI-conjugates. The resulting conjugates can be characterized and quantified by measuring the amine content of PEI, and/or absorption spectra of coupled ligands. PEI can also be quantified with a copper-based assay. Divalent copper ions form dark blue cuprammonium complexes with PEI;[14] this assay works for both linear and branched PEI.

7.4 BIOPHYSICS OF NUCLEIC ACID–PEI INTERACTION

Electrostatic interaction between the positively charged amino groups in PEI and the negatively charged phosphate in nucleic acids leads to subsequent condensation and formation of polyplexes. To give a measure for the ratio of PEI and nucleic acid used for complex formation, the molar ratio of nitrogen in PEI and phosphate in the nucleic acid is indicated (this is termed the N/P ratio). The condensation process can be monitored by several methods, including agarose gel electrophoresis, ethidium bromide exclusion assay,[15] and electron microscopy.[16] LDNA, supercoiled plasmid DNA, and also oligonucleotides and mRNA can be condensed with PEI. Whereas for the delivery of plasmid DNA, PEI of almost any available molecular weight can be used, short nucleic acids like mRNA are preferentially complexed with low-molecular-weight PEI[17] (LMW PEI) (see below). The N/P ratio and salt concentration can strongly influence the condensation process. Initial condensation of DNA with PEI is observed at low N/P ratios (~0.5). Individual DNA strands are packed together in a core structure, whereas uncondensed DNA loops surround the condensation nuclei.[16] Spherical particles in the range of 50 nm are formed after mixing DNA and PEI at higher N/P ratios.[15] We and other groups have observed a correlation between N/P ratio, salt concentration, and particle size. Small particles are formed at low salt concentrations or charge ratios above N/P 6, whereas aggregation of complexes is observed in the presence of salt (e.g., NaCl > 50 mM) at lower N/P ratios.[18,19] The net positive surface charge (expressed as the zeta potential) and an excess of free PEI during complex formation can prevent aggregation by repulsion of positive charges; an increase of salt concentration reduces the hydrated layer around the particles and promotes their aggregation. At N/P ratios greater than 2, free PEI is present in the complex solution, and values between 50% and 86% for free PEI have been reported.[20,21] Our own observations indicate that about two-thirds of PEI is unbound when mixing complexes at N/P 6.[81]

7.5 PEI POLYPLEXES FOR NUCLEIC ACID DELIVERY

For the delivery of DNA into cells a broad range of molecular weights of PEI can be used. Early studies have used branched PEI at 800 kDa,[1] but branched PEI at 25 kDa is also an efficient transfection reagent exhibiting reduced toxicity *in vivo*.[19] When further decreasing the molecular weight of branched PEI used for transfection, both cellular toxicity and transfection efficiency are decreased.[22] PEI polyplexes used for gene transfer experiments usually bear a net positive surface charge. Due to electrostatic interactions — e.g., with proteoglycanes[23] — the particles are bound to the cell surface and are subsequently internalized into intracellular vesicles by adsorptive endocytosis.[24] Cointernalized proteoglycanes are also supposed to influence intracellular routing of PEI polyplexes and the efficiency of transgene expression.[25] The size of PEI polyplexes can also influence transfection efficiency *in vitro*: aggregated complexes result in increased cellular association due to sedimentation onto cells and improved intracellular delivery (see below).[18] LPEI at 22 kDa was found to be far more efficient in gene delivery both *in vitro* and *in vivo*.[19,26,27] Apparently, polyplexes with LPEI exhibit a different biophysical behavior, resulting in the aggregation of initially small particles on the cell surface, which could be one reason for the exceptionally high transfection efficiency.

For delivery of short nucleic acids such as mRNA, only LMW PEI (2 kDa or less) can lead to an efficient biological effect of the transferred nucleic acids. Complexes with higher molecular weight PEI are too stable and do not release their payload already in the cytoplasm of cells.[17]

7.6 LIGAND-PEI CONJUGATES FOR GENE DELIVERY

Different ligands have been coupled with PEI to enhance the internalization of polyplexes into target cells (see Table 7.1). The use of targeting ligands incorporated into the polyplex can greatly increase the extent of binding and the specificity of internalization into endosomes. Transferrin-mediated internalization into transferrin receptor positive cells occurs quickly (in less than 1 h) compared to polyplexes lacking the ligand.[28] (Transferrin-targeted gene delivery systems are discussed in greater detail elsewhere in this Chapter 34.) Similar effects were observed for antibody

Table 7.1   Receptor-Mediated Targeting of Polyethylenimine Conjugate to Specific Cell Populations

Receptor	Ligand	Target Cells	Refs.
ASGP receptor	Galactosylated PEI	Hepatocytes, hepatoma	70–72
CD-3	Anti-CD-3	T-cell derived leukemia cells, activated PBMCs	6, 73
CD-71	Transferrin	Different tumor cell lines	6
EGF receptor	EGF	HepG2, HuH7, KB, A431	11, 74, 75
Folate receptor	Folate	KB, CT-26	76, 77
Mannose	Mannosylated PEI	Dendritic cells	78
PECAM	Anti-PECAM	Lung endothelial cells	79
Vitronectin-R	RGD-peptide	Angiogenic cells, different tumor cell lines	13, 29, 30, 80

targeted polyplexes (anti-CD3 on Jurkat cells) and polyplexes containing the epidermal growth factor (EGF). Short peptidic ligands can be chemically synthesized and coupled to PEI via standard chemistry (see above). The sequence Arg-Gly-Asp (RGD) can bind specifically to the vitronectin receptor $\alpha_v\beta_3$, an integrin which is highly upregulated in certain tumor cells and the tumor endothelium. RGD-PEI conjugates used for DNA transfection significantly increase transgene expression in different tumor cell lines[13,29] and angiogenic cells.[30]

7.7 INTRACELLULAR ASPECTS OF PEI-MEDIATED GENE DELIVERY

After internalization, gene transfer complexes are mostly found in intracellular compartments, which can be potentially acidified, such as endosomes or lysosomes, and their subsequent release into the cytoplasm represents a major bottleneck for gene delivery.[31,32] The protonation profile of PEI, bearing an intrinsic buffer capacity (see above), and its high transfection capacity led to the development of the "proton sponge" hypothesis: the buffering of the endosome by PEI causes proton accumulation and subsequent influx of chloride into the vesicle. Osmotic swelling by influx of water can thereafter lead to disruption of the vesicle and release of the polyplex into the cytoplasm.[2] This theory was at least partially confirmed by the finding that bafilomycin A, a potent inhibitor of the intracellular proton pump, selectively inhibits transgene expression of PEI polyplexes, whereas the efficiency of DOTAP, a monocationic lipid, was not significantly altered.[33] Nevertheless, additional mechanisms for cytoplasmic delivery of PEI polyplexes have been suggested, which include the possibility that PEI can rupture the lysosome directly by interacting with the membrane.[24] Studies based on flow cytometry revealed that endosomes are not completely buffered to neutral pH by PEI polyplexes,[7] and the resulting average pH was between pH 5.9 for branched PEI and 5.0 for linear PEI.[34] Apparently, the mechanism is not yet fully understood and will have to be further clarified.

Release of PEI polyplexes from endosomes can be significantly hampered when using either low molecular weight PEI[22] or small PEI polyplexes at low concentrations.[18] To promote endosomolysis of PEI polyplexes, inactivated adenovirus particles[35] or membrane active peptides[6,17,18,36] were incorporated into the polyplex and significantly augmented reporter gene expression.

Passive diffusion within the cytoplasm is limited for DNA above a certain size limit, which is due to a crowding effect caused by components of the cytoplasm.[37] PEI protects DNA from nuclease attack within the cytoplasm,[38] and the transport within the cytoplasm is supposed to be at least partially dependent on microtubuli,[39] although it cannot be excluded that the intracellular transport of PEI polyplexes is mediated within vesicles.

Nuclear import of PEI polyplexes is a major bottleneck for transfection, as only one out of 1000 copies of PEI condensed plasmid molecules will reach the nucleus after microinjection into the cytoplasm.[38] Brunner and colleagues demonstrated that there is an approximately 30-fold decrease in reporter gene expression transfecting cells either logarithmically growing or synchronized in the early G1 phase of the cell cycle.[27] Interestingly, polymer mediated transfection using linear PEI was far less dependent on cell cycle compared to branched PEI.[40] In accordance with this observation, large amounts of LPEI polyplexes were found within the nucleus of transfected cells after 7 hours.[19] Although the detailed mechanism of linear PEI is not completely clear, nuclear access due membrane rupture by aggregated polyplexes could add to this effect.

As in the case of endosomal release, different moieties were added to PEI polyplexes in order to improve their access to the nucleus. Adenoviral hexon protein is supposed to permit the access of viral DNA to the nucleus.[41] Hence, purified hexon protein was coupled to PEI, amplifying nuclear entry and leading to an increase in the efficiency of transgene expression.[8] The cationic, membrane active peptide melittin coupled to PEI polyplexes not only improved endosomal release, but also aided in the transfection of cells that divided slowly and cells that did not divide.[42]

7.8 *IN VIVO* APPLICATIONS

Gene delivery *in vivo* with nonviral gene delivery systems still remains a challenging task; depending on the therapeutic strategy different routes of administration have to be chosen. Local applications have the advantage of circumventing unwanted interaction of gene transfer complexes with blood components and the reticulo-endothelial system, although the need for diffusible particles within the tissue remains. Systemic administration is the most demanding, but offers the possibility of reaching the disseminated target throughout the organism, e.g., tumor metastases. We and others have developed PEI-based systems which enable systemic delivery to disseminated, well vascularized tumors (see Chapter 34) by shielding the complex surface with PEG or transferrin.

7.8.1 Local Delivery

Early studies with PEI polyplexes were conducted by stereotactic intracerebral injection into the brain.[1,43] Elevated expression levels and increased diffusibility within the brain were obtained using LPEI polyplexes generated in 5% glucose.[44] The small polyplexes were able to transfect both glia cells and neurons. Recently, repeated intrahecal injection of PEG-modified PEI polyplexes led to prolonged transgene expression.[45] In contrast, plain PEI polyplexes led to attenuated gene expression after repeated injection and also resulted in cellular damage. This indicates the importance of a nontoxic gene transfer formulation consisting of small polyplexes that enable diffusibility within the target tissue.

Local delivery to the liver can be achieved by application of gene transfer complexes via the hepatic artery or by direct injection into the tissue. Gharwan et al. compared the transfection efficiency of naked DNA and different PEI polyplexes by liver injection into fetal, neonatal, and adult mice.[46] At low DNA doses, LPEI polyplexes were more efficient than naked DNA, and maximal gene expression was achieved in neonatal mice, although relevant therapeutic levels of reporter gene expression could not be achieved. Injection of polyplexes into the renal artery of rats led to preferential reporter gene expression in proximal tubular cells,[47] but only PEI polyplexes below 100 nm in diameter were able to reach the target cells.[48]

Local administration to the lung can be achieved via instillation or nebulization of the gene transfer particles. Nebulized PEI polyplexes applied into mice lungs led to significant levels of reporter gene expression, which peaked after 24 h and was still detectable after 8 days.[49] Masking the surface of PEI polyplexes with hydrophilic polymers reduced interaction with bronchoalveolar lavage, although plain PEI polyplexes remained the most efficient.[50]

Direct intratumoral injections of plain PEI polyplexes were less efficient, presumably due to poor diffusion of the polyplexes within the tumor mass. Application via a micropump was much more successful, resulting in long-lasting reporter gene expression for more than 2 weeks.[51]

7.8.2 Systemic Application

In principle, three major targets can be reached after application of gene transfer complexes into the bloodstream. Highly positively charged particles will rapidly aggregate (partially by interaction with blood components) and end up in the lung, the first vascular bed encountered.[19,52] Protecting the particle surface with hydrophilic polymers enables the circulation of PEI polyplexes in the bloodstream and thereafter leads them to the liver[53] or enables a (passive) accumulation in implanted tumors.[10,54] Especially in the lung very high levels of reporter gene expression can be found using LPEI polyplexes, although toxic side effects can be observed when using either high molecular weight PEI and/or high N/P ratios.[10,52,55] The toxic effect is mostly due to the aggregating properties of PEI polyplexes and leads to activation of lung endothelium and liver damage.[55] Repeated administration of PEI polyplexes also caused sites of inflammation in liver tissue,[56] which could be due to the prolonged retention of PEI in the liver. PEI labeled with radioactive iodine could be still detected several days after intravenous injection, and was mainly found in the lysosomal fraction of the cells.[57] Absence of a toxic PEI-based gene delivery formulation will be a key point for their application in the clinic, and this highlights the importance of optimizing such formulations for reduced toxicity. This can be achieved by preventing the aggregation with blood components by surface shielding or reducing the N/P ratio of PEI polyplexes.[58] In this case, even without surface shielding, tumor tissue was efficiently transfected (see below).

7.9 THERAPEUTIC GENE DELIVERY WITH PEI

PEI has already been used for the delivery of therapeutic DNA, oligonucleotides, or different types of RNA.

Glycosylated PEI was used for the hepatic delivery of DNA–RNA oligonucleotides, also called chimeraplasts.[59,60] The genomic sequence was successfully modified by site-directed mutagenesis and was still measurable 2 years after treatment.[61] Intratumoral application of LMW PEI complexed ribozymes directed against the growth factor pleiotropin resulted in significant reduction in tumor growth.[62] Delivery of LPEI condensed antisense oligodeoxynuclotides directed against hepatitis virus inhibited viral replication in a duck liver infection model.[63]

PEI-based transfection systems are potentially suited for different strategies to treat malignant diseases (for review see Ogris and Wagner[64]): direct killing of tumor cells with bioactive proteins, e.g., cytokines or apoptosis inducers; suicide gene therapy by combining prodrugs with the expression of prodrug activating enzymes; or the induction of chemoprotection in combination with high dose chemotherapy. The following therapeutic genes have already been successfully delivered with PEI-based transfection systems *in vivo*.

The p53 gene is the most frequently altered gene in human cancers. Its normal function is to protect cellular DNA by coordinately blocking cell proliferation, stimulating DNA repair, and promoting apoptotic cell death. Overexpression of wild-type p53 is a promising approach to treat cancers caused by p53 mutations. Mice xenografted with head and neck cancer were treated by intratumoral injections of p53 expressing plasmid complexed with glycosylated PEI.[65] PEI polyxplexes were found to diffuse within the viable tumor mass, transfecting cells in the periphery of the tumor. A treatment schedule with injections twice a week led to apoptosis of tumor cells and inhibited tumor growth. Mice bearing experimental syngeneic melanoma metastases were treated by multiple applications of nebulized PEI polyplexes carrying the p53 gene.[66] The treatment led to significant reduction of tumor burden in the lung and increased the median survival time compared to treatment with control plasmid. A similar effect was found in a human osteosarcoma lung metastasis model in nude mice.[67] Systemic application of PEI polyplexes (at the low N/P ratio of 2.7) in an orthotopic bladder cancer model led to a 14-fold higher reporter gene expression in the tumor compared to the lung.[58] Treatment with 6 μg PEI polyplexes every 3 days for 3 weeks resulted in a 70% reduction in tumor size.

Tumor necrosis factor α (TNF-α) is a potent cytokine that induces hemorrhagic tumor necrosis and tumor regression. Kircheis et al. demonstrated that the local expression of TNF-α in tumors after systemic application of transferring-shielded PEI polyplexes reduced tumor growth, and in certain tumor models resulted in a cure rate of 60%.[68]

Somatostatin is a cyclic nonapeptide known to negatively regulate the growth of different cell types, including tumor cells. The antiproliferative effect mediated by the somatostatin receptor subtype 2 (sst2) was utilized for treatment of pancreatic cancer in an orthotopic syngeneic hamster model.[69] The therapeutic effects of LPEI polyplexes and recombinant adenovirus encoding sst2 were compared by intratumoral injection. Both recombinant adenovirus and LPEI polyplexes significantly reduced tumor growth.

7.10 CONCLUSIONS

Polyethylenimine is a transfection reagent that is suitable for a broad range of different gene transfer applications. The high content of primary amino groups enables the chemical coupling of targeting moieties or intracellular active components; the high density of positive charges in the molecule allows for a tight compaction of nucleic acids. PEI can influence different steps of the transfection process, i.e., endosomal release and nuclear entry. These processes have to be further clarified. Suitable formulations of PEI polyplexes with low toxicity have to be chosen for *in vivo* use, which will allow for multiple applications of the therapeutic gene. Several preclinical studies have already revealed the applicability of PEI-based gene therapy systems for a broad range of applications.

REFERENCES

1. Boussif, O. et al., A versatile vector for gene and oligonucleotide transfer into cells in culture and *in vivo*: polyethylenimine, *Proc. Nat. Acad. Sci. U.S.A.* 92, 7297, 1995.
2. Kichler, A., Behr, J.P., and Erbacher, P., Polyethylenimines: a family of potent polymers for nucleic acid delivery, in *Nonviral Vectors for Gene Therapy*, Huang, L., Hung, M.C., and Wagner, E. (Eds.), Academic Press, San Diego, 1999, Chapter 9.
3. Suh, J., Paik, H.J., and Hwang, P.K., Ionization of polyethylenimine and polyellylamine at various pHs, *Bioorg. Chem.* 22, 318, 1994.
4. Godbey, W.T., Wu, K.K., and Mikos, A.G., Poly(ethylenimine) and its role in gene delivery, *J. Control. Release* 60, 149, 1999.
5. Brissault, B. et al., Synthesis of linear polyethylenimine derivatives for DNA transfection, *Bioconjug. Chem.* 14, 581, 2003.
6. Kircheis, R. et al., Coupling of cell-binding ligands to polyethylenimine for targeted gene delivery, *Gene Ther.* 4, 409, 1997.
7. Ogris, M. et al., Melittin enables efficient vesicular escape and enhanced nuclear access of non-viral gene delivery vectors, *J. Biol. Chem.* 12, 12, 2001.
8. Carlisle, R.C. et al., Adenovirus hexon protein enhances nuclear delivery and increases transgene expression of polyethylenimine/plasmid DNA vectors, *Mol. Ther.* 4, 473, 2001.
9. Kursa, M. et al., Novel shielded transferrin-polyethylene glycol-polyethylenimine/DNA complexes for systemic tumor-targeted gene transfer, *Bioconjug. Chem.* 14, 222, 2003.
10. Ogris, M. et al., PEGylated DNA/transferrin-PEI complexes: reduced interaction with blood components, extended circulation in blood and potential for systemic gene delivery, *Gene Ther.* 6, 595, 1999.
11. Blessing, T. et al., Different strategies for formation of pegylated EGF-conjugated PEI/DNA complexes for targeted gene delivery, *Bioconjug. Chem.* 12, 529, 2001.

12. Erbacher, P. et al., Transfection and physical properties of various saccharide, poly(ethylene glycol), and antibody-derivatized polyethylenimines (PEI), *J. Gene Med.* 1, 210, 1999.

13. Erbacher, P., Remy, J.S., and Behr, J.P., Gene transfer with synthetic virus-like particles via the integrin-mediated endocytosis pathway. *Gene Ther.* 6, 138, 1999.

14. Ungaro, F. et al., Spectrophotometric determination of polyethylenimine in the presence of an oligo-nucleotide for the characterization of controlled release formulations, *J. Pharm. Biomed. Anal.* 31, 143, 2003.

15. Tang, M.X. and Szoka, F.C., The influence of polymer structure on the interactions of cationic polymers with DNA and morphology of the resulting complexes, *Gene Ther.* 4, 823, 1997.

16. Dunlap, D. D. et al., Nanoscopic structure of DNA condensed for gene delivery, *Nucleic Acids Res.* 25, 3095, 1997.

17. Bettinger, T. et al., Peptide-mediated RNA delivery: a novel approach for enhanced transfection of primary and post-mitotic cells, *Nucleic Acids Res.* 29, 3882, 2001.

18. Ogris, M. et al., The size of DNA/transferrin-PEI complexes is an important factor for gene expression in cultured cells, *Gene Ther.* 5, 1425, 1998.

19. Wightman, L. et al., Different behavior of branched and linear polyethylenimine for gene delivery *in vitro* and *in vivo*, *J. Gene Med.* 3, 362, 2001.

20. Finsinger, D. et al., Protective copolymers for nonviral gene vectors: synthesis, vector characterization and application in gene delivery, *Gene Ther.* 7, 1183, 2000.

21. Clamme, J.P., Azoulay, J., and Mely, Y., Monitoring of the formation and dissociation of polyethyl-enimine/DNA complexes by two photon fluorescence correlation spectroscopy, *Biophys. J.* 84, 1960, 2003.

22. Godbey, W.T., Wu, K.K., and Mikos, A.G., Poly(ethylenimine)-mediated gene delivery affects endot-helial cell function and viability, *Biomaterials* 22, 471, 2001.

23. Mislick, K.A. and Baldeschwieler, J.D., Evidence for the role of proteoglycans in cation-mediated gene transfer, *Proc. Nat. Acad. Sci. U.S.A.* 93, 12349, 1996.

24. Bieber, T. et al., Intracellular route and transcriptional competence of polyethylenimine-DNA com-plexes, *J. Control. Release* 82, 441, 2002.

25. Ruponen, M. et al., Extracellular glycosaminoglycans modify cellular trafficking of lipoplexes and polyplexes, *J. Biol. Chem.* 276, 33875, 2001.

26. Ferrari, S. et al., ExGen 500 is an efficient vector for gene delivery to lung epithelial cells *in vitro* and *in vivo*, *Gene Ther.* 4, 1100, 1997.

27. Brunner, S. et al., Cell cycle dependence of gene transfer by lipoplex, polyplex and recombinant adenovirus, *Gene Ther.* 7, 401, 2000.

28. Ogris, M. et al., DNA/polyethylenimine transfection particles: Influence of ligands, polymer size, and PEGylation on internalization and gene expression, *AAPS PharmSci.* 3, E21, 2001.

29. Kunath, K. et al., Integrin targeting using RGD-PEI conjugates for *in vitro* gene transfer, *J. Gene Med.* 5, 588, 2003.

30. Suh, W. et al., An angiogenic, endothelial-cell-targeted polymeric gene carrier, *Mol. Ther.* 6, 664, 2002.

31. Zabner, J. et al., Cellular and molecular barriers to gene transfer by a cationic lipid, *J. Biol. Chem.* 270, 18997, 1995.

32. Labat Moleur, F. et al., An electron microscopy study into the mechanism of gene transfer with lipopolyamines, *Gene Ther.* 3, 1010, 1996.

33. Kichler, A. et al., Polyethylenimine-mediated gene delivery: a mechanistic study, *J. Gene Med.* 3, 135, 2001.

34. Akinc, A. and Langer, R., Measuring the pH environment of DNA delivered using nonviral vectors: implications for lysosomal trafficking, *Biotechnol. Bioeng.* 78, 503, 2002.

35. Baker, A. et al., Polyethylenimine (PEI) is a simple, inexpensive and effective reagent for condensing and linking plasmid DNA to adenovirus for gene delivery, *Gene Ther.* 4, 773, 1997.

36. Lee, H., Jeong, J.H., and Park, T.G., A new gene delivery formulation of polyethylenimine/DNA complexes coated with PEG conjugated fusogenic peptide, *J. Control. Release* 76, 183, 2001.

37. Lukacs, G. L. et al., Size-dependent DNA mobility in cytoplasm and nucleus, *J. Biol. Chem.* 275, 1625, 2000.

38. Pollard, H. et al., Polyethylenimine but not cationic lipids promotes transgene delivery to the nucleus in mammalian cells, *J. Biol. Chem.* 273, 7507, 1998.

39. Suh, J., Wirtz, D. and Hanes, J., Efficient active transport of gene nanocarriers to the cell nucleus, *Proc. Nat. Acad. Sci. U.S.A.* 100, 3878, 2003.

40. Brunner, S. et al., Overcoming the nuclear barrier: cell cycle independent nonviral gene transfer with linear polyethylenimine or electroporation, *Mol. Ther.* 5, 80, 2002.

41. Saphire, A. C. et al., Nuclear import of adenovirus DNA *in vitro* involves the nuclear protein import pathway and hsc70, *J. Biol. Chem.* 275, 4298, 2000.

42. Ogris, M. et al., Melittin enables efficient vesicular escape and enhanced nuclear access of nonviral gene delivery vectors, *J. Biol. Chem.* 276, 47550, 2001.

43. Abdallah, B. et al., A powerful nonviral vector for *in vivo* gene transfer into the adult mammalian brain: polyethylenimine, *Hum Gene Ther.* 7, 1947, 1996.

44. Goula, D. et al., Size, diffusibility and transfection performance of linear PEI/DNA complexes in the mouse central nervous system, *Gene Ther.* 5, 712, 1998.

45. Shi, L. et al., Repeated intrathecal administration of plasmid DNA complexed with polyethylene glycol-grafted polyethylenimine led to prolonged transgene expression in the spinal cord, *Gene Ther.* 10, 1179, 2003.

46. Gharwan, H. et al., Nonviral gene transfer into fetal mouse livers (a comparison between the cationic polymer PEI and naked DNA), *Gene Ther.* 10, 810, 2003.

47. Boletta, A. et al., Nonviral gene delivery to the rat kidney with polyethylenimine, *Hum Gene Ther.* 8, 1243, 1997.

48. Foglieni, C. et al., Glomerular filtration is required for transfection of proximal tubular cells in the rat kidney following injection of DNA complexes into the renal artery, *Gene Ther.* 7, 279, 2000.

49. Gautam, A. et al., Enhanced gene expression in mouse lung after PEI-DNA aerosol delivery, *Mol. Ther.* 2, 63, 2000.

50. Rudolph, C. et al., Nonviral gene delivery to the lung with copolymer-protected and transferrin-modified polyethylenimine, *Biochim. Biophys. Acta* 1573, 75, 2002.

51. Coll, J.L. et al., *In vivo* delivery to tumors of DNA complexed with linear polyethylenimine, *Hum Gene Ther.* 10, 1659, 1999.

52. Goula, D. et al., Polyethylenimine-based intravenous delivery of transgenes to mouse lung, *Gene. Ther.* 5, 1291, 1998.

53. Nguyen, H.K. et al., Evaluation of polyether-polyethyleneimine graft copolymers as gene transfer agents, *Gene Ther.* 7, 126, 2000.

54. Oupicky, D. et al., Importance of lateral and steric stabilization of polyelectrolyte gene delivery vectors for extended systemic circulation, *Mol. Ther.* 5, 463, 2002.

55. Chollet, P. et al., Side-effects of a systemic injection of linear polyethylenimine-DNA complexes, *J. Gene Med.* 4, 84, 2002.

56. Oh, Y.K. et al., Prolonged organ retention and safety of plasmid DNA administered in polyethylenimine complexes, *Gene Ther.* 8, 1587, 2001.

57. Lecocq, M. et al., Uptake and intracellular fate of polyethylenimine *in vivo*, *Biochem. Biophys. Res. Commun.* 278, 414, 2000.

58. Sweeney, P. et al., Efficient therapeutic gene delivery after systemic administration of a novel poly-ethylenimine/DNA vector in an orthotopic bladder cancer model, *Cancer Res.* 63, 4017, 2003.

59. Kren, B.T. et al., Correction of the UDP-glucuronosyltransferase gene defect in the gunn rat model of crigler-najjar syndrome type I with a chimeric oligonucleotide, *Proc. Nat. Acad. Sci. U.S.A.* 96, 10349, 1999.

60. Kren, B.T., Bandyopadhyay, P., and Steer, C.J., *In vivo* site-directed mutagenesis of the factor IX gene by chimeric RNA/DNA oligonucleotides, *Nat. Med.* 4, 285, 1998.

61. Kren, B. T. et al., Modification of hepatic genomic DNA using RNA/DNA oligonucleotides, *Gene Ther.* 9, 686, 2002.

62. Aigner, A. et al., Delivery of unmodified bioactive ribozymes by an RNA-stabilizing polyethylenimine (LMW-PEI) efficiently down-regulates gene expression, *Gene Ther.* 9, 1700, 2002.

63. Robaczewska, M. et al., Inhibition of hepadnaviral replication by polyethylenimine-based intravenous delivery of antisense phosphodiester oligodeoxynucleotides to the liver, *Gene Ther.* 8, 874, 2001.

64. Ogris, M. and Wagner, E., Targeting tumors with non-viral gene delivery systems, *Drug Discov. Today* 7, 479, 2002.

65. Dolivet, G. et al., *In vivo* growth inhibitory effect of iterative wild-type p53 gene transfer in human head and neck carcinoma xenografts using glucosylated polyethylenimine nonviral vector, *Cancer Gene Ther.* 9, 708, 2002.

66. Gautam, A., Densmore, C.L., and Waldrep, J.C., Inhibition of experimental lung metastasis by aerosol delivery of PEI-p53 complexes, *Mol. Ther.* 2, 318, 2000.

67. Densmore, C. L. et al., Growth suppression of established human osteosarcoma lung metastases in mice by aerosol Gene Ther. with PEI-p53 complexes, *Cancer Gene Ther.* 8, 619, 2001.

68. Kircheis, R. et al., Tumor-targeted gene delivery of tumor necrosis factor-alpha induces tumor necrosis and tumor regression without systemic toxicity, *Cancer Gene Ther.* 9, 673, 2002.

69. Vernejoul, F. et al., Antitumor effect of *in vivo* somatostatin receptor subtype 2 gene transfer in primary and metastatic pancreatic cancer models, *Cancer Res.* 62, 6124, 2002.

70. Zanta, M.A. et al., *In vitro* gene delivery to hepatocytes with galactosylated polyethylenimine, *Bioconjug. Chem.* 8 (6), 839, 1997.

71. Bettinger, T., Remy, J.S., and Erbacher, P., Size reduction of galactosylated PEI/DNA complexes improves lectin-mediated gene transfer into hepatocytes, *Bioconjug. Chem.* 10, 558, 1999.

72. Morimoto, K. et al., Molecular weight-dependent gene transfection activity of unmodified and galactosylated polyethyleneimine on hepatoma cells and mouse liver, *Mol. Ther.* 7, 254, 2003.

73. O'Neill, M.M. et al., Receptor-mediated gene delivery to human peripheral blood mononuclear cells using anti-CD3 antibody coupled to polyethylenimine, *Gene Ther.* 8, 362, 2001.

74. Wolschek, M.F. et al., Specific systemic nonviral gene delivery to human hepatocellular carcinoma xenografts in SCID mice, *Hepatology* 36, 1106, 2002.

75. Lee, H., Kim, T.H.. and Park, T.G., A receptor-mediated gene delivery system using streptavidin and biotin-derivatized, pegylated epidermal growth factor, *J. Control. Release* 83, 109, 2002.

76. Guo, W. and Lee, R.J., Receptor-targeted gene delivery via folate-conjugated polyethylenimine, *AAPS PharmSci.* 1, Article 19, 1999.

77. Benns, J.M., Mahato, R.I., and Kim, S.W., Optimization of factors influencing the transfection efficiency of folate-PEG-folate-graft-polyethylenimine, *J. Control. Release* 79, 255, 2002.

78. Diebold, S.S. et al., Mannose polyethylenimine conjugates for targeted DNA delivery into dendritic cells, *J. Biol. Chem.* 274, 19087, 1999.

79. Li, S. et al., Targeted gene delivery to pulmonary endothelium by anti-PECAM antibody, *Am. J. Physiol. Lung Cell Mol. Physiol.* 278, L504–L511, 2000.

80. Muller, K. et al., Highly efficient transduction of endothelial cells by targeted artificial virus-like particles, *Cancer Gene Ther.* 8, 107, 2001.

81. Boeckle, S. et al., Purification of polyethylenimine polyplexes highlights the role of free polycations in gene transfer, *J. Gene Med.*, in press, 2004.

CHAPTER 8

Poly(2-(dimethylamino)ethyl methacrylate)-Based Polymers for the Delivery of Genes *In Vitro* and *In Vivo*

F.J. Verbaan, D.J.A. Crommelin, W.E. Hennink, and G. Storm

CONTENTS

8.1 INTRODUCTION

Advances in molecular biology have resulted in a new concept in the treatment of diseases, the concept of gene therapy. The unraveling of the human genome opens up the opportunity to treat diseases at their genetic origin rather than treating symptoms of diseases. However, gene therapy is not limited to the treatment of genetic defects. Therefore, gene therapy strategies are being devised for the treatment of other diseases as well. Plasmid DNA-based approaches to gene therapy involve administration of DNA ("naked DNA") or formulations of DNA. The fundamental challenge of gene delivery originates from the fact that DNA has a charged, colloidal nature, is very labile

0-8493-1934-X/05/$0.00+$1.50
© 2005 by CRC Press LLC

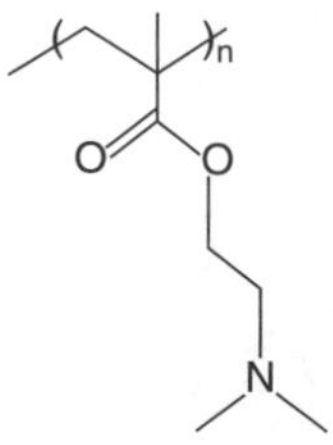

Figure 8.1 Chemical structure of pDMAEMA.

in the biological environment, and does not cross biological barriers effectively, such as an intact endothelium, the plasma membrane, or the nuclear membrane.[1] The need for an efficient and safe gene delivery system is obvious. Therefore, there is a growing interest in the development of target-cell-specific nonviral carriers for the delivery of genes.

Methacrylate polymers have been applied for the microencapsulation of cells, such as erythrocytes, fibroblasts, lymphomas, and the beta cells in the islets of Langerhans. Because of the biocompatibility with living tissue, the methacrylate polymers are also used to encapsulate cells in order to prevent tissue rejection upon cell transplantation.[2,3] The combination of their biocompatibility, their relatively easy synthesis, and the possibility of functionalizing the polymers (e.g., making targeted or PEGylated derivatives) make methacrylates interesting candidates for gene delivery studies. In the Department of Pharmaceutics at Utrecht University this led to the idea of synthesizing poly(2-[dimethylamino]ethyl methacrylate) (pDMAEMA)-based polymers as "lead compounds" for the purpose of gene delivery. The chemical structure of the cationic pDMAEMA polymer is depicted in Figure 8.1.

The purpose of this chapter is to give an overview of the different aspects of the development of pDMAEMA polymers toward an applicable gene transfectant *in vivo*. First, the *in vitro* characteristics of complexes of the pDMAEMA polymer and pDMAEMA analogs ("polyplexes") are described. Second, pharmaceutical aspects of pDMAEMA-polyplex formulations are addressed. Finally, the *in vivo* application of pDMAEMA polyplexes for tumor targeting is discussed.

8.2 *IN VITRO* TRANSFECTION EFFICIENCY OF PDMAEMA-BASED COMPLEXES: CRITICAL PARAMETERS

8.2.1 Physicochemical Properties

It has been reported that the extent of cellular uptake of the particulate system in question depends strongly on the size and charge.[4,5] Thus, the physical characteristics of pDMAEMA-polyplexes were investigated. The effect of the pDMAEMA:plasmid ratio on the size and zeta potential of the formed particles was studied by dynamic light scattering measurements and electrophoretic mobility measurements, respectively. Figure 8.2 shows that in an aqueous buffer solution (20 mM Hepes, pH 7.4), naked plasmid (5 μg/ml) has a rather large hydrodynamic size (0.3–0.4 μm). Small and stable polymer-plasmid complexes (size ~0.1–0.2 μm) could be formed under the same conditions as used for naked DNA and at polymer:plasmid ratios above 3:1 (w/w), demonstrating that the cationic polymer is able to condense plasmid DNA. Zeta potential measurements revealed that naked DNA possesses a negative zeta potential (-22 mV). After the addition of pDMAEMA, a positive zeta potential was observed which leveled off at polymer:plasmid ratios above 3:1 (w/w).[6]

The *in vitro* transfection efficiency of the pDMAEMA polyplexes is shown in Figure 8.3. The transfection efficiency shows a bell-shaped dependence on the polymer:plasmid ratio (w/w). Such dependence is observed frequently for both lipid-based[7] and cationic polymer-based transfection systems[8] and can be explained as follows. At low polymer:plasmid ratios the polyplexes have a negative

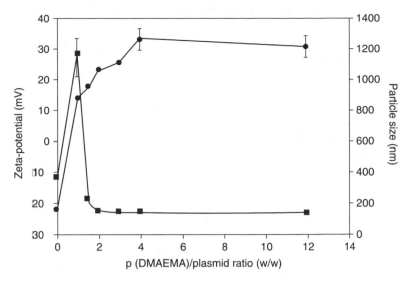

Figure 8.2   The effect of the pDMAEMA:plasmid ratio on size (as determined by DLS ■) and zeta potential (●) of pDMAEMA plasmid polyplexes. The plasmid concentration was fixed at 5 μg/ml. The results are expressed as mean values ± SD of three experiments.

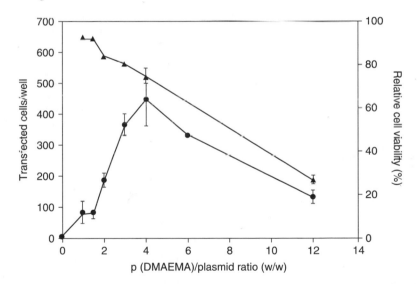

Figure 8.3   The effect of the pDMAEMA:plasmid ratio on the number of transfected cells (●) and on the relative cell viability (▲). The plasmid concentration was fixed at 5 μg/ml. The results are expressed as mean values ± SEM of three to five experiments.

zeta potential and are relatively large in size (Figure 8.2). Both characteristics are unfavorable for cellular association and internalization. At polymer:plasmid ratios above 3:1 (w/w), the polyplexes are small and have a positive zeta potential. Both properties are a favorable factor for transfection. The decreasing transfection efficiency at ratios > 3:1 can be explained by the toxicity of the free polymer present in the transfection medium at these higher ratios (see Figure 8.3).

Besides the effect of the pDMAEMA:DNA ratio, we also studied the effect of the polymer molecular weight on DNA condensation.[9] Although it has been reported that low-molecular-weight polymers are able to give small and stable polyplexes,[10] we found for pDMAEMA that the average molecular weight has to be above 100,000 g/mol to give sufficient condensation of plasmid DNA.

8.2.2 Design of pDMAEMA-Based Copolymers

A common feature of cationic polymers is their cytotoxicity.[9,11] The design of copolymers is an interesting route for optimizing the efficacy:toxicity ratio of polymeric gene transfer systems. We evaluated copolymers of pDMAEMA and hydrophobic monomers such as methyl methacrylate (MMA), or hydrophilic monomers like N-vinyl-pyrrolidone (NVP) and ethoxytriethylene glycol (triEGMA) as transfectants,[6,12] as shown in Figure 8.4. Polyplexes containing a copolymer with 20 mol % of MMA showed reduced transfection efficiency and a substantially increased cytotoxicity as compared to polyplexes formed with the pDMAEMA homopolymer of the same molecular weight. This suggests that the introduction of hydrophobic monomers is a less attractive approach to optimize the efficacy:toxicity ratio. On the other hand, polyplexes of plasmid and copolymers with a low content (20 mol %) of a hydrophilic comonomer (either NVP or triEGMA) had the same transfection and cytotoxicity properties as pDMAEMA of comparable molecular weight (around 100,000). Increasing the NVP or triEGMA content to 50 mol % resulted in less toxic polyplexes with improved transfection properties. This shows that reduction of the charge density of pDMAEMA polymers is an attractive route for optimizing the efficacy:toxicity ratio. However, a high-molecular-weight pDMAEMA (mol wt > 10^5) is still superior in mediating transfection

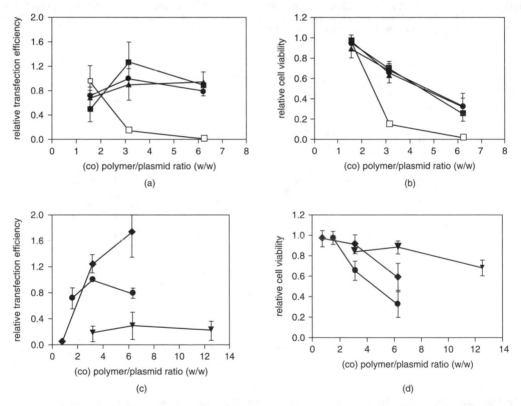

Figure 8.4   Transfection efficiency (A and C) and effect on relative cell viability (B and D) of (co)polymer–plasmid polyplexes as a function of the (co)polymer:plasmid ratio. The plasmid concentration was 5 µg/ml, the transfection time (incubation time of the complexes with OVCAR-3 cells) and the coloring time (X-Gal) were 1 and 16 h, respectively. The transfection values are normalized to the maximum number of transfected cells after incubation with plasmid complexed to pDMAEMA (M_w 100kD, polymer:plasmid ratio 3:1 [w/w]). The results are shown as mean values (± SEM) of 3–5 experiments. M_w of (co)polymers was around 100 kD. (●) pDMAEMA, (■) poly(DMAEMA-co-triEGMA) 79/21 (mol/mol), (▼) poly(DMAEMA-co-triEGMA) 52/48 (mol/mol), (▲) poly(DMAEMA-co-NVP) 86/14 (mol/mol), (◆) poly(DMAEMA-co-NVP) 46/54 (mol/mol), (□) poly(DMAEMA-co-NVP) 80/20 (mol/mol).

compared to the other evaluated polymers (data not shown), which can be ascribed to the low capability of the low-molecular-weight polymer to condense the plasmid.

The transfection capability of other water-soluble cationic polymers with structures closely related to pDMAEMA was evaluated to obtain insight into the relationship between structure and activity.[13] The following parameters were studied.

- A propyl side chain instead of an ethyl side chain with poly(3-[dimethylamino]propyl methacrylate) (pDMAPMA)
- Substitution of the ester group in pDMAEMA by an amide group with poly(2-[dimethylamino]ethyl methacrylamide) (pDMAEMAm).
- A longer side chain and an amide group with poly(3-[dimethylamino]propyl methacrylamide) (pDMAPMAm)
- Quaternary amino groups versus tertiary amino groups with poly(2-[trimethylamino]ethyl methacrylate) (pTMAEMA)
- Modification of the amino group of pDMAEMA with poly(2-[diethylamino]ethyl methacrylate) (pDEAEMA)

Almost all of the investigated cationic polymers were able to condense plasmid DNA, yielding polymer–plasmid complexes with a size of 0.1 to 0.3 μm and a positive zeta potential. However, the transfection efficiency and cytotoxicity differed widely (Figure 8.5). The highest transfection efficiency and toxicity were observed for the "parent" compound pDMAEMA. Assuming that polyplexes enter the cell via endocytosis, pDMAEMA apparently has advantageous properties to escape the endosomes. A possible explanation is that, because of its average pK_a value of 7.5, pDMAEMA is partially protonated at physiological pH and might behave as a proton sponge. This might cause a disruption of the endosomes, which results in a release of both polyplexes and cytotoxic endosomal/lysosomal enzymes in the cytosol. On the other hand, the analogs of pDMAEMA have a higher average pK_a value and have, consequently, a higher degree of protonation and a lower buffering capacity. This might be associated with a lower tendency to destabilize the endosomes, resulting in both a lower transfection efficiency and a lower cytotoxicity. Also, molecular modeling studies showed that

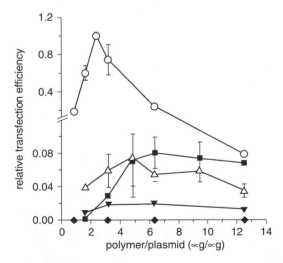

Figure 8.5 The effect of polymer/plasmid ratio on the transfection efficiency. The plasmid concentration was 5 μg/ml, the transfection time (incubation time of the polyplexes with OVCAR-3 cells) was 1 h. Transfection values are normalized to the maximum number of transfected cells after incubation with plasmid complexed to pDMAEMA. The results are shown as mean values (± SEM) of 3 experiments. (■) pDMAPMA, (▼) pDMAPMAm, (♦) pTMAEMA, (△) pDMAEMAm, (○) pDMAEMA

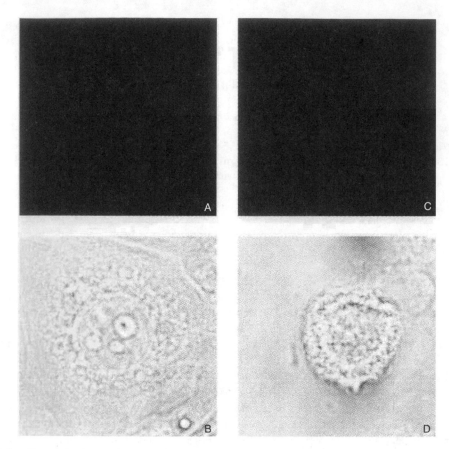

Figure 8.6   **(See color insert following page 336)** Confocal laser fluorescence microscopy (CLFM) study of OVCAR-3 cells, upon 1 h incubation at 37°C with DNA alone (A, B) and with polyplexes of pDMAEMA–DNA (w/w) at ratios of 0.9 (C, D), 0.75 (E, F), and 3 (G, H), followed by 2 days culturing at 37°C. Pictures B, D, F, and H are the direct transmitted light pictures taken simultaneously with the CLFM pictures A, C, E, and G. The plasmid was labeled with Ethd-1 and the width of a picture corresponds to 40 μm.

pDMAEMA has the lowest interaction intensity with DNA. The relative ease of dissociation was indeed demonstrated using surface plasmon resonance and gel electrophoresis.[14,15] It is, therefore, hypothesized that the superior transfection efficiency of pDMAEMA can be ascribed to an intrinsic property of pDMAEMA to destabilize the endosomes combined with relatively easy dissociation of the polyplexes once present in the cytosol or the nucleus.

8.2.3   Cellular Interactions of pDMAEMA-Based Polyplexes

The cellular interactions of pDMAEMA-based polyplexes with human ovarian cancer (OVCAR-3) cells were investigated using fluorescence-activated cell sorting (FACS), confocal laser fluorescence microscopy (CLFM), and electron microscopy (EM).[16] To gain more insight into the mode of plasmid delivery to OVCAR-3 cells, the effects of the following physicochemical characteristics on the cellular interaction of polyplexes with OVCAR-3 cells were investigated.

- Charge (manipulated by the polymer:plasmid ratio)
- Size (manipulated by the polymer molecular weight)

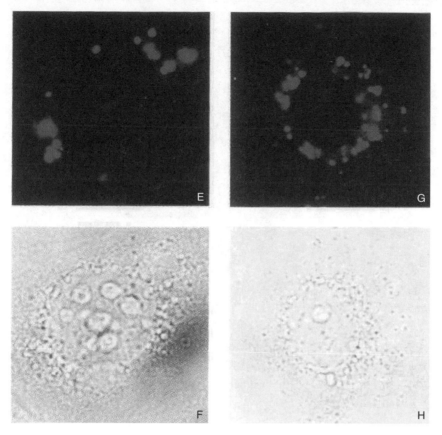

Figure 8.6 (Continued)

- Presence of tertiary versus quaternary amino groups, measured by comparing pDMAEMA with its quaternary ammonium analogue, poly(2-[trimethyl]amino ethyl methacrylate)
- Modification of polyplex surface characteristics with the hydrophilic polymer polyethylene glycol (PEG) using a graft copolymer of pDMAEMA and PEG

The effects of endocytosis inhibitors on cellular interaction and transfection activity were evaluated as well.

Cellular association and subsequent internalization only occurred when the polyplexes exhibited a positive zeta potential (i.e., a pDMAEMA:plasmid ratio of 3:1 [w/w]), as can be observed in Figures 8.6A–H. Small polyplexes have an advantage over large complexes regarding cellular entry. The combined cellular interaction and transfection results suggest that the quaternary ammonium analogue pTMAEMA does not have an intrinsic endosomal escape property, as opposed to the proton sponge effect proposed for pDMAEMA (data not shown). PEGylation of pDMAEMA effectively shielded the surface charge and yielded a notably lower degree of cellular interaction (Figures 8.7A and B). Data on the effects of the presence of endocytosis inhibitors (N-ethyl-maleimide and monensin) on cellular interaction (Figures 8.8E–H)) and transfection activity of pDMAEMA-based polyplexes confirm the hypothesis that endocytosis is the principal pathway for the intracellular delivery of plasmid. Neither the CLFM nor the EM studies revealed the presence of polyplexes or plasmid outside the endocytic vesicles or within the nucleus, suggesting that intracellular trafficking from the endosomes to the nucleus is a very inefficient process.

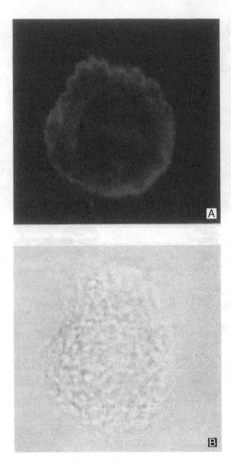

Figure 8.7   **(See color insert)** Confocal laser fluorescence microscopy (CLFM) study of OVCAR-3 cells, upon 1 h incubation at 37°C with polyplexes based on pDMAEMA-*g*-PEG–DNA at a 4:1 (w/w) ratio (A). Figure B shows the direct transmitted light pictures taken simultaneously with the CLFM picture A. The plasmid was labeled with Ethd-1 and the width of a picture corresponds to 40 μm.

8.2.4   Targeting of pDMAEMA-Based Polyplexes

Lack of cell specificity, cytotoxicity, and blood-induced aggregation are major hurdles hindering the application of pDMAEMA-based polyplexes for *in vivo* plasmid delivery. In our laboratory, two approaches are under investigation to overcome these hurdles. First, a detergent removal method to coat the cationic polyplexes with anionic lipids has been developed.[17] The intention was to create polyplexes that are negatively charged rather than positively charged, as a positive charge was supposed to be the main mediator of the above-mentioned hurdles. This method is based on the preferential formation of a lipid coat around the positively charged polyplexes upon the transition of phospholipids from the mixed micellar to the lamellar state induced by slow removal of the detergent (Figure 8.9). With this method spherical particles (referred to as lipopolyplexes) with a negative charge and a size of around 125 nm were obtained which were protected from destabilization by polyanions like hyaluronic acid. Hyaluronic acid is a component of the peritoneal ascites fluid, which probably induced a detrimental effect on the *in vivo* transfection capability of pDMAEMA polyplexes after intraperitoneal application to nude mice with OVCAR-3 cells growing in the peritoneal cavity (see Section 3.1). However, it was shown that the lipopolyplexes are much less effective in transfecting OVCAR-3 cells *in vitro* compared to positively charged pDMAEMA polyplexes, presumably due to a lower degree of cellular interaction. By conjugating Fab' fragments

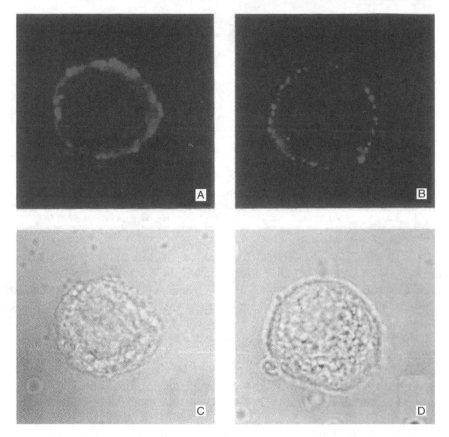

Figure 8.8 **(See color Insert)** Confocal laser fluorescence microscopy (CLFM) study of OVCAR-3 cells, upon 1 h incubation at 37°C with pDMAEMA–DNA 3:1 (w/w) polyplexes in the presence of 10 μM N-ethyl-maleimide (A, B) and in the presence of 10 μM monensin (C, D). Figures A and B are the direct transmitted light pictures taken simultaneously with the CLFM picture C and D, respectively. The plasmid was labeled with Ethd-1 and the width of a picture corresponds to 40 μm.

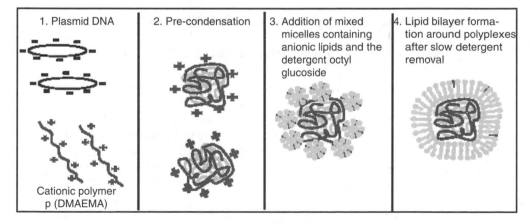

Figure 8.9 Schematic representation of lipopolyplex formation. (1) Plasmid DNA is condensed by adding the cationic polymer pDMAEMA to the DNA at a ratio of 3:1 (w/w). The formed polyplexes (2) are added to mixed micelles containing the detergent n-octyl-β-D-galactopyranoside (OG) and a total amount of 3 μM of detergent-solubilized lipids (3). Upon slow removal of detergent by adsorption to hydrophobic beads, lipid coats are preferentially formed around positively charged polyplexes due to electrostatic interactions (4).

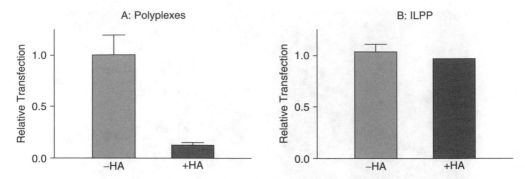

Figure 8.10   Influence of hyaluronic acid (HA) on the transfection efficiency of polyplexes versus immunoli-popolyplexes (ILPP). OVCAR-3 cells were exposed for 1 h at 37°C to (A) polyplexes or ILPP in the absence (-HA) or presence (+HA) of 2.5 mg/ml HA. Gene carriers were removed by washing and cell culture was continued for another 47 h prior to evaluation for β-galactosidase expression.

directed against the epithelial glycoprotein-2 to the lipidic surface of the lipopolyplexes, target cell specific transfection of OVCAR-3 cells could be obtained *in vitro* (Figure 8.10).

Second, we explored an active targeting strategy based on the use of folate as a targeting ligand directed against the folate receptor overexpressed on OVCAR-3 cells. Folate was coupled to preformed polyplexes with PEG as a spacer.[18] A succinimidyl-activated folate-PEG conjugate was synthesized, prepared from PEG-bisamine by subsequent conjugation with NHS-activated folate[10] and disuccinimidyl suberate (Figure 8.11A).

As the polymeric vector, a copolymer of pDMAEMA with 40% *N*-3-aminopropyl methacryla-mide (NAPMAm) was used (Figure 8.11B). Folate-mediated targeting of the polyplexes was studied with an *in vitro* transfection assay using OVCAR-3 cells. A substantial zeta potential drop occurred upon PEGylation of the polyplexes, reflecting effective shielding of the positive surface charge (Figure 8.12). The PEGylated polyplexes (nontargeted polyplexes) showed markedly decreased gene expression when compared to uncoated polyplexes. Coupling of folate to the terminal ends of the PEG chains (targeted polyplexes) restored the transfection activity to the level observed for the uncoated polyplexes (Figure 8.13).

8.3 PHARMACEUTICAL ASPECTS OF PDMAEMA-BASED COMPLEXES

8.3.1   Preparation of Highly Concentrated pDMAEMA-Polyplexes

The results presented in the earlier sections of this chapter were obtained at a relatively low plasmid concentration (5 μg/ml). At higher concentrations, however, severe aggregation of the polyplexes was sometimes observed. This is problematic, because *in vivo* studies require adminis-tration of relatively high concentrations of plasmid DNA to obtain detectable transfection levels.[19,20] Using experimental design methodology, we therefore carried out a study to gain insight into formulation parameters (pH, polymer:plasmid ratio, ionic strength, temperature, viscosity, and the presence of excipients) that might affect the polyplex size at a relatively high plasmid concentration (200 μg/ml).[21]

pDMAEMA and plasmid DNA bind to each other via electrostatic interactions. Since the charge density of the polymer depends on the pH of the aqueous solution, we selected pH values ranging from pH 5.0, where the tertiary amine side groups of the polymer are all protonated, to pH 8.0, where approximately 25% of the side-groups are protonated. At a pH of 5.0, the formation of small polyplexes was enhanced as compared to pH 7.4 and 8.0, supposedly because of strong plas-mid–polymer interactions resulting from a high charge density of the polymer.

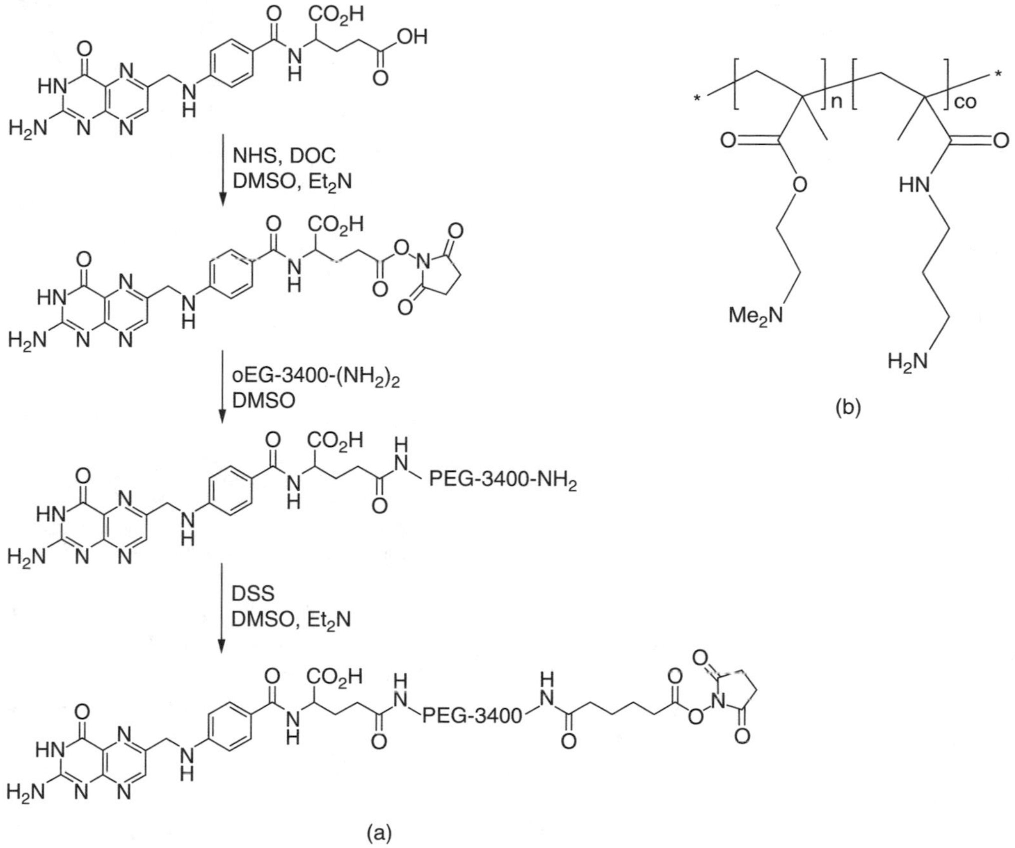

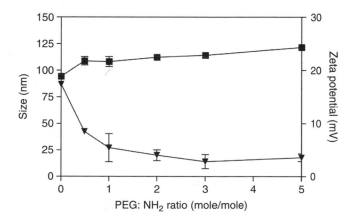

Figure 8.11 (A) Schematic representation of the synthesis of NHS-PEG-folate conjugate. (B) Structure of a copolymer of pDMAEMA with 40% N-3-aminopropyl methacrylamide (NAPMAm) (pDMAEMA-coNAPMAm)

Figure 8.12 Particle size (squares) and zeta potential (triangles) of PEG-folic-acid (FA) coated pDMAEMA-coNAPMAm–DNA (N/P = 5) complexes with PEG to primary amine ratio varying from 0 to 5 (mole/mole). Polyplexes were measured in 5 mM Hepes, pH 7.4. Data represent the means of three independently prepared formulations ± SEM.

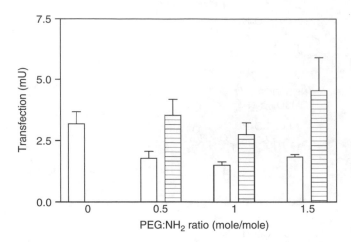

Figure 8.13 Comparison of *in vitro* transfection efficiencies of OVCAR-3 cells mediated by nontargeted and folic acid (FA)-targeted PEG polyplexes at various PEG-to-primary-amine ratios, at an N/P ratio of 10. Efficiencies were determined by measuring the β-galactosidase activity. Open bars: PEG; horizontally hatched bars: PEG-FA. Data represent the mean of three experiments ± SEM.

For pDMAEMA:plasmid ratios above 1 (w/w) and at low ionic strengths, small and stable polyplexes could be obtained. At lower ratios aggregation occurred, irrespective of the other parameters investigated.

Conditions of low ionic strength seem to favor the formation of small polyplexes, most likely because the polymer–DNA interaction increases with decreasing ionic strength.

The presence of stabilizers (PEG 600, Tween 80) as well as the temperature at which the complexes were prepared (4–40°C) had a marginal effect on the size of the polyplexes.

A preferable lyoprotectant when polyplexes are freeze-dried is sucrose. We evaluated the effect of sucrose on the size of the complexes. For concentrations of sucrose between 0 and 20% (w/v), the particle size of the polyplexes decreased. The higher viscosity of the 20% sucrose solution (compared with the viscosity of Hepes buffer) may prevent aggregation of formed polyplexes. At higher sucrose concentrations (up to 40%), the size of the formed polyplexes increased, possibly because under these conditions a viscous "barrier" hinders the interaction between the polymer and the plasmid.

8.3.2 Structural Properties of Plasmid DNA

The structure of plasmid DNA (pDNA) can be described, analogous to proteins, at four levels,[22] as summarized in Table 8.1. The primary structure is the nucleotide sequence of the plasmid, which contains the genetic information and is of major importance for its functional properties. The impact of the secondary, tertiary, and quaternary plasmid structure on the efficacy of transfection is still

Table 8.1 Structural Levels of Plasmid DNA and Detection Methods for Each Level

Structure	Description	Detection Methods
Primary	Nucleotide sequence	Enzymatic assays (restriction mapping), hybridization techniques, chemical assays, mass spectrometry
Secondary	Double-helix conformations (e.g., A, B, C form)	Circular dichroism, FTIR, UV spectroscopy
Tertiary (topology)	Supercoiled, open-circular, linear	Gel electrophoresis, anion-exchange chromatography, microscopic techniques (EM, AFM)
Quaternary	Oligomers (noncovalent)	Gel electrophoresis, light scattering, microscopic techniques (EM, AFM)Table

under investigation. Most of our work in this area has been focused on the effect of the tertiary structure — often referred to as topology — on the binding characteristics with pDMAEMA and the transfection properties *in vitro*.[23] Among the techniques available to probe the different structural levels of DNA (see Table 8.1), we selected agarose gel electrophoresis for analyzing the tertiary structure and circular dichroism (CD) spectroscopy for studying the effect of complexation with pDMAEMA on the secondary structure of the different forms of the plasmid.

Anion-exchange chromatography was used for the separation of supercoiled and open-circular plasmid from a plasmid stock solution. Linear plasmids were prepared by two different endonucleases, which cleaved the plasmid either in the promoter region or in a region that was nonspecific for expression (the ampicillin resistance region). Finally, pDNA was also heat-denatured for 6 h at 70°C, resulting in mainly open-circular and oligomeric forms. No differences in the size of the complexes or in the quenching of the DNA-intercalating fluorophore acridine orange were found as a function of the topology. However, CD spectroscopy revealed differences between the secondary structure of the different topoisomers, particularly after complexation with pDMAEMA (Figure 8.14). As compared to naked plasmid, which has a positive peak at 274 nm and a negative one at 246 nm (not shown), the peak maxima of the circular (supercoiled and open-circular) forms of plasmid complexed with pDMAEMA were red-shifted to 290 and 254 nm, respectively (Figure 8.14), indicating a change in the secondary structure from B-type to C-type.[24] Moreover, the intensity of the negative peak was increased, suggesting the presence of parallel DNA helices in the polyplexes, so-called -DNA.[25] In contrast, the CD spectra of complexed linearized forms and heat-denatured plasmid resembled those of naked plasmid. This indicates that the interactions of these topological forms and pDMAEMA were less tight and less organized as compared with supercoiled and open-circular plasmid.

The transfection efficiency in two different cell lines depended on the topology of the DNA in the order supercoiled open circular heat-denatured linear DNA prepared by cleaving in the non-specific region > linear DNA prepared by cleaving in the promoter region (Figure 8.15). Remarkably, a recent study by Bergan et al. showed that after complexation with cationic lipids, the supercoiled and the relaxed forms have the same transfection efficiency both *in vitro* and *in vivo*.[26]

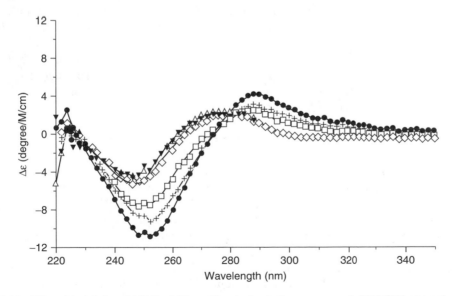

Figure 8.14 CD spectra of plasmid DNA of different topologies in the presence of pDMAEMA at a polymer:DNA 3:1 (w/w) ratio. (+) Stock (mixture of supercoiled and open circular); (□) supercoiled; (●) open circular; (◇) heat denatured; (△) linear, cleaved in promoter region; (▼) linear, cleaved in ampicillin region.

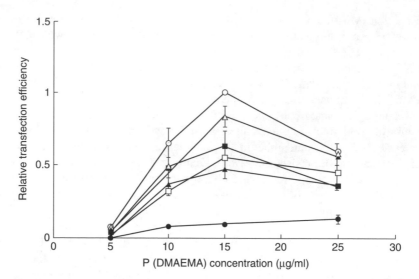

Figure 8.15   Relative transfection efficiency of polyplexes as a function of the pDMAEMA concentration at a fixed concentration of plasmid (5 µg/ml) with different topologies: (○) Stock (mixture of supercoiled and open circular); (△) supercoiled; (□) open circular; (■) heat denatured; (●) linear, cleaved in promoter region; (▲) linear, cleaved in ampicillin region. The transfection studies were performed with OVCAR-3 cells and the results are expressed as mean values ± SD of three to five experiments.

8.3.3   Stability of pDMAEMA-Polyplexes

When stored as aqueous dispersions, polyplex formulations are expected to have a limited stability. Polyplex formulations are thermodynamically unstable colloidal systems. These systems will always show aggregation depending on the charge of the particles and the ionic strength of the medium. Furthermore, in aqueous solution, DNA has a limited chemical stability. Depurination can occur and interestingly, this reaction is catalyzed by amines so that when complexed with cationic compounds (polymers or lipids) this reaction may proceed even faster.[27,28] Besides chemical degradation by hydrolysis, DNA is also susceptible to oxidative degradation.[29,30] Finally, the introduction of cationic polymers that degrade as such by hydrolysis[31,32] requires stabilization strategies. Based on what we have learned from protein-based pharmaceuticals,[33] it is logical to consider freeze-drying as an important option for the design of polyplex formulations that fulfill the stability requirements for pharmaceutical formulations.

8.3.3.1   Freeze-Drying and Freeze-Thawing of pDMAEMA-Polyplexes

pDMAEMA-plasmid polyplexes were prepared (pDMAEMA 120 µg/ml, plasmid 40 µg/ml) with sucrose concentrations varying from 0 to 5% (w/v) and the effect of freeze-thawing on size and transfection efficiency of the complexes was evaluated.[34] Figure 8.16A shows that, in the absence of sucrose, large particles with a reduced transfection efficiency were formed. On the other hand, in the presence of sucrose, both the particle size and the transfection potential of the complexes were preserved.

Polyplex formulations consisting of pDMAEMA (120 µg/ml), plasmid (40 µg/ml), and varying higher concentrations of lyoprotectant ranging from 1.25 to 20% (w/v) were prepared next. Figure 8.16B shows that after rehydration of the freeze-dried samples prepared at a low sucrose concentration (1.25%), large particles with a relatively low transfection efficiency were present. Obviously, this amount of sucrose is not enough to prevent aggregation of the polyplexes. When comparing Figures 8.16A and B, it can be concluded that, at this low sucrose concentration, damage to the

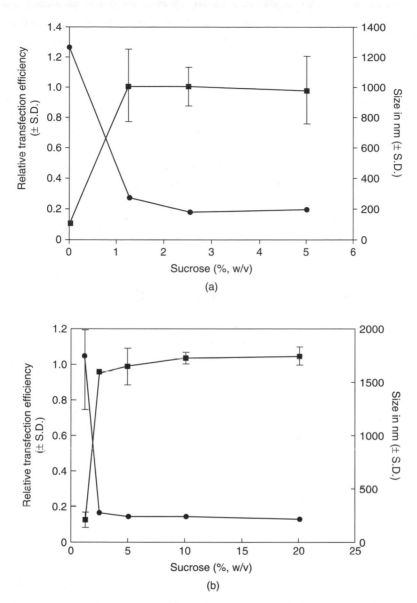

Figure 8.16 Size (●) and relative transfection efficiency (■) of polyplexes after (A) freeze-thawing and (B) freeze-drying and rehydration. The percentage of sucrose refers to the concentration present in the aqueous polymer–plasmid dispersion before the freezing step. The results are expressed as mean values ± SD of three to five experiments.

complexes occurs during the drying process. Figure 8.16B also shows that freeze-drying of polyplexes at sucrose concentrations greater than 2.5% yielded small particles after rehydration. Moreover, the transfection efficiency of these particles did not differ significantly from that of freshly prepared particles.

Knowing that sucrose is an effective lyoprotectant for polyplexes, we compared its performance with two other frequently used lyoprotectants (trehalose and maltose). Freeze-drying and freeze-thawing in the presence of 10% (w/v) of any of the investigated sugars did not affect the transfection efficiency.[34] Dynamic light scattering (DLS) experiments demonstrated that the size of the transfection complexes only slightly increased from around 0.18 μm (directly after preparation) to approximately 0.22 μm after freeze-thawing or freeze-drying. This increase in particle size was associated with a slight increase in the polydispersity index, suggesting that only limited aggregation had occurred. Interestingly, the type of lyoprotectant had no effect on either the size or the transfection efficiency of the complexes after freeze-drying and freeze-thawing.

8.3.3.2 Long-Term Stability

pDMAEMA polyplexes formulated with 10% (w/v) sucrose were aged for a period of up to 10 months at three different temperatures (4, 20, and 40°C) as aqueous dispersion as well as in the freeze-dried state.[35] The highest aging temperature (40°C) was selected because at this temperature the matrix is still in its glassy state (Tg = 50°C). Naked plasmid was aged at 40°C as a control. We evaluated the aged polyplexes for both their physico-chemical characteristics and their transfection potential. Since we demonstrated that no significant hydrolysis of the ester side-chains of pDMAEMA occurred under physiological conditions,[36] we focused our attention on changes that might occur in the secondary and tertiary structure of the plasmid both in its free form and complexed with polymer. DLS measurements revealed that after rehydration of the cake, the size of the lyophilized polyplexes remained constant over time (diameter approximately 0.15 μm; Figure 8.17). The figure also shows that, similarly, no change in particle size was observed for the polyplexes aged as aqueous dispersions at 4 and 20°C. On the other hand, the size of the liquid polyplexes aged at 40°C gradually increased with time (from 0.15 to 0.35 μm after 10 months), suggesting that limited aggregation occurred. The size of the polyplexes obtained by complexing the polymer with aged plasmid (lyophilized or aqueous) was not affected by the aging period and amounted to 0.15 μm (results not shown). No changes in the charge of the polyplexes (zeta potential 25–30 mV) were observed for the polyplexes aged under the different conditions investigated.

The secondary DNA structure of aged polyplex formulations was monitored by CD spectroscopy. Only the CD spectrum of the liquid formulations aged at 40°C changed dramatically, suggesting that the interaction between the polymer and the plasmid became stronger over time. In contrast, the CD spectra of the other aged polyplexes were similar to those of freshly prepared polyplexes (results not shown). Electrophoresis in agarose gels was used to study possible changes in the tertiary structure of plasmid DNA. Poly(aspartic acid) was used to dissociate the plasmid

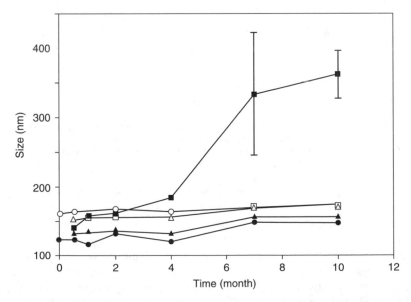

Figure 8.17 Size of polyplexes aged in 10% sucrose-aqueous solutions at 4 (●), 20 (▲), and 40°C (■) and in the freeze-dried state at 4 (○), 20 (Δ), and 40°C (□). The results are expressed as mean values ± SD of three experiments.

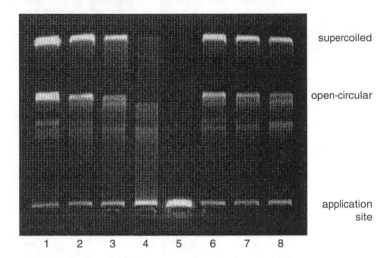

supercoiled

open-circular

application
site

1   2   3   4   5   6   7   8

Figure 8.18   Agarose gel electrophoresis of plasmid. Lane 1: naked plasmid after storage for 10 months at 4°C. Lane 2: plasmid after dissociation of freshly prepared polyplexes; plasmid after dissociation of aged polyplexes in sucrose-aqueous solutions after 10 months at 4 (lane 3), 20 (lane 4), and 40°C (lane 5). Plasmid after dissociation of aged polyplexes in freeze-dried form after 10 months at 4 (lane 6), 20 (lane 7), and 40°C (lane 8). Dissociation of the polyplexes was established by poly(aspartic acid).

from the polymer in the polyplex. The plasmid used in this study consisted of a mixture of supercoiled and open-circular DNA, with some traces of multimers (oligomers of DNA) (Figure 8.18, lane 1). The freshly prepared plasmid showed an identical pattern on agarose gel as naked plasmid after storage for 10 months at 4°C. The same pattern was observed (lane 2) when polyplexes were treated with poly(aspartic acid) shortly after preparation, which indicates that complexation of polymer with plasmid does not lead to changes in the tertiary structure of the plasmid.

In addition, when the plasmid was liberated from lyophilized polyplexes with poly(aspartic acid) after 10 months of aging, the same electrophoretic pattern was observed as for freshly prepared naked plasmid (lane 6–8). Similarly, no large changes in the tertiary structure of the plasmid were observed when the polyplexes were aged as an aqueous dispersion at 4°C (Figure 8.18, lane 3). However, supercoiled and open-circular plasmid were hardly found in the polyplexes aged in aqueous media at 20 and 40°C after 10 months, and an intense band close to the application site was observed, particularly for the sample aged at 40°C (Figure 8.18, lanes 4 and 5, respectively). Apparently, poly(aspartic acid) could not dissociate these polyplexes.

We also analyzed the tertiary structure of naked plasmid aged as an aqueous solution and in the lyophilized state. After 10 months of storage of naked plasmid in freeze-dried form at 40°C, the electrophoretic pattern did not differ significantly from that of fresh plasmid. However, naked plasmid aged in an aqueous solution was largely converted to the open-circular form after 1 month at 40°C, whereas only a minor part was still in the supercoiled form. Further aging resulted in the disappearance of the supercoiled form and the concomitant formation of open-circular, linear, and oligomeric DNA (results not shown). Figure 8.19 shows the transfection potential of polyplexes aged as aqueous dispersions containing 10% (w/v) sucrose and in the freeze-dried state. The liquid polyplex formulations almost fully preserved their transfection potential after aging at 4 and 20°C (Figure 8.19A). However, polyplexes aged at 40°C were rather unstable and lost their transfection potential (having a half-life of around 2 months). When naked plasmid DNA was aged in solution at 40°C and complexed with polymer just before the transfection experiment, a reduction in its transfection capability was also observed albeit to a lesser extent (a half-life of around 4 months). Similar results were obtained in the absence of sucrose (data not shown). After 10 months, no change in the morphology of the freeze-dried samples was observed on visual inspection. The

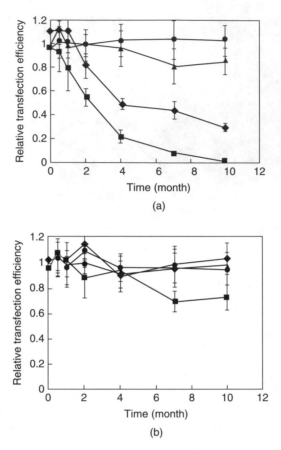

Figure 8.19   Relative transfection efficiency of aged polymer–plasmid complexes at 4 (●), 20 (▲), and 40°C
(■), and naked plasmid at 40°C (♦). The complexes were prepared in the presence of 10% (w/v)
sucrose. Aging was done (A) as a liquid dispersion or (B) in freeze-dried form. Naked plasmid
was complexed with polymer just before the transfection experiment. The results are expressed
as mean values ± SD of three experiments.

moisture content (0.5%) also remained constant over time. As observed for the polyplexes aged in
aqueous media at 4 and 20°C, freeze-dried polyplexes aged at the same temperatures preserved
their full transfection potential (Figure 8.19B). The figure also shows that, although some reduction
in transfection efficiency was observed for the freeze-dried polyplexes aged at 40°C (20–30%
reduction after 10 months), their stability was superior compared to polyplexes aged in aqueous
media. Interestingly, the naked plasmid in the freeze-dried form fully preserved its transfection
potential, which suggests that plasmid complexed with polymer in freeze-dried form also lost its
transfection potential faster than naked plasmid.

A number of recent studies demonstrate that freeze-drying is also an important method for
preserving the biophysical and transfection properties of lipoplex- and peptide-based formulations.[37-40] A future challenge is definitely to stabilize the more sophisticated nonviral systems,
which apart from a plasmid and a condensing agent (cationic polymer or lipid), also contain a
PEG-coating, a suitable homing device, and endosome disruptive and/or nuclear localization
motifs.

These data demonstrate that by rational investigation of the parameters of importance for the
characteristics of pDMAEMA-based polyplexes, stable systems can be prepared showing transfection efficiency *in vitro*. Further, we have demonstrated that freeze-drying is an excellent method
for preserving the transfectivity of pDMAEMA-based polyplexes. Besides the pharmaceutical

aspects summarized in this chapter, other items also require full attention in order to develop these systems further. These include among others the removal of impurities, the development of sterilization protocols, upscaling, and batch-to-batch consistency.

8.4 *IN VIVO* DELIVERY OF PDMAEMA-BASED COMPLEXES TO TUMORS

8.4.1 Intraperitoneal Administration

Ovarian cancer is one of the most common fatal gynecological malignancies. The OVCAR-3 human ovarian carcinoma cell line growing intraperitoneally in nude mice provides a model system suitable for studying ovarian cancer.[41] Since ovarian cancer remains confined to the peritoneal cavity throughout most of its lifetime, it is believed that ready access to the peritoneal cavity and containment of the disease progress within the peritoneal cavity favor development of anticancer gene therapy strategies. Therefore, we have defined intraperitoneally localized OVCAR-3 cells as transfection targets which should be accessible to DNA delivery systems injected directly into the peritoneal cavity.[42]

The approach taken to investigate whether OVCAR-3 cells can be transfected *in vivo* was a comparative *in vitro–ex vivo–in vivo* study utilizing similar exposure conditions of the cells to the pDMAEMA transfection complexes *in vitro* and *in vivo*. The transfection results can be summarized as follows: pDMAEMA–plasmid (pCMVLacZ) complexes can transfect OVCAR-3 cells *in vitro* with an overall transfection efficiency of 10%. Cells grown *in vivo* can be transfected *ex vivo* with pDMAEMA–plasmid complexes with an overall transfection efficiency of 1 to 2%. However, cells grown *in vivo* were very difficult to transfect *in vivo*: transfection of intraperitoneally localized OVCAR-3 cells was negligible after injection of the transfection complexes into nude mice bearing OVCAR-3 cells in the peritoneal cavity.[42]

The following reasons might explain this discrepancy.

The polyplexes may have formed aggregates induced by one or more components of the ascites fluid. We have previously observed that large-sized polyplex structures are less efficient in transfection.[9]

A potential reason for the differences found *in vitro* and *in vivo* transfection experiments may be sought in the clustering of cells growing in the peritoneal cavity. OVCAR-3 cells cultured *in vitro* grow adherently while *in vivo* cells grow in suspension in the peritoneal cavity. Clusters of cells are formed with, consequently, a reduced accessibility of a major fraction of the cells. In order to investigate whether declustering of the cells would result in improved accessibility and consequently higher transfection efficiency *ex vivo*, cells isolated from mice were treated with trypsin before incubation with the transfection complexes. Trypsin-mediated declustering did not improve transfection.

Another difference between the *in vitro* and the *in vivo* situation is the presence of body fluids, peritoneal ascites fluid in case of the particular tumor used here.

The influence of ascites fluid on the transfection activity of the polyplexes was investigated *in vitro*. In parallel, the influence of fetal calf serum (FCS) was studied in the same experiment. When ascites and FCS are absent during the experiment, the transfection optimum was observed at a polymer:plasmid ratio of 1.6 (w/w). With an increasing ascites or FCS concentration, the optimum polymer:plasmid ratio shifted to higher values. This is in good agreement with the results obtained by Yang and Huang[43] who showed that the inhibitory effect of serum on lipofection could be overcome by increasing the cationic lipid:DNA ratio. The transfection activity was increased twofold in the presence of FCS at the optimum ratio. This is possibly caused by a stimulating effect of certain FCS components on the interaction of the polyplexes with the cells. However, the *in vitro* transfection activity was strongly reduced in the presence of ascites fluid. To elucidate which component(s) of ascites had such a detrimental effect on the *in vitro* transfection activity, the influence of hyaluronic acid, which has been reported to be present in relatively high concentrations

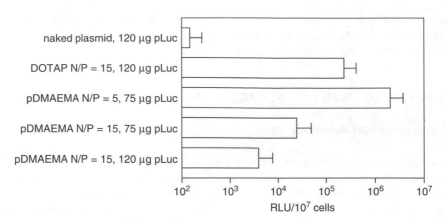

Figure 8.20   Transfection of human ovarian cancer (OVCAR-3) cells after intraperitoneal administration of pDMAEMA polyplexes and DOTAP lipoplexes. Data represent the mean ± SEM (n = 3).

in ascites[44,45] on the transfection activity, was studied. Hyaluronic acid is a polymer consisting of a regularly repeating sequence of disaccharide units (glucuronic acid and N-acetylglucosamine). Due to its polyanionic character, hyaluronic acid might have interacted with the positively charged polyplexes resulting in a reduction of the transfection activity, as has been reported for the effect of heparin on lipoplexes.[46,47] As shown in Figure 8.10, indeed, the *in vitro* transfection activity was strongly reduced in the presence of hyaluronic acid in concentrations which are in the range (up to 11 mg/ml) present in peritoneal effusions from cancer patients.[45,48] No negative effect of hyaluronic acid on cell viability was observed. This outcome suggests that one of the components of ascites fluid, hyaluronic acid, may have induced a negative effect on the transfection capability of pDMAEMA-based polyplexes.

However, relatively low DNA doses were used in this study (lower than 15 µg DNA per mouse), because of the inability to prepare polyplex dispersions with relatively high DNA concentrations. Aggregation phenomena prevented the use of dispersions with a DNA concentration higher than 40 µg/ml. Later, Cherng et al. described a procedure to prepare highly concentrated polyplex dispersions containing a DNA concentration up to 150 µg/ml.[21] pDMAEMA-polyplexes prepared according to this procedure were administered to tumor bearing mice as well (Figure 8.20). In line with expectations, administration of naked pLuc plasmid did not result in any significant gene expression. Administration of pDMAEMA-based polyplexes at a DNA dose of 120 µg and an N/P ratio of 15 resulted in low but detectable gene expression. In comparison, i.p. injection of DOTAP -based lipoplexes at the same DNA dose and N/P ratio yielded a 60-fold higher level of gene expression. Lowering the DNA dose to 75 µg and the N/P ratio to 5 yielded a 520-fold enhancement of the transfection activity of pDMAEMA-based polyplexes. The increased transfection efficiency observed upon lowering the DNA dose and N/P ratio might relate to decreased toxicity of the polyplex system under these experimental conditions.

Figure 8.21 shows that i.p. injection of pDMAEMA-based polyplexes not only leads to transfection of the cells present in the peritoneal fluid. Relatively high gene expression was also observed in several other tissues, including the spleen, diaphragm, uterus, and abdominal muscles. Low, but still significant gene expression was found in the liver, lungs, heart, colon, mesenteries, and connective tissue. In line with the tumor cell transfection results, pDMAEMA-based complexes injected at a 75 µg DNA dose and an N/P ratio of 5 yielded higher transfection levels than pDMAEMA-based polyplexes injected at a 120 µg DNA dose and an N/P ratio of 15. For the 120 µg DNA dose and an N/P ratio of 15, DOTAP-based complexes yielded higher levels of transfection in the uterus, mesenteries, and spleen when compared to pDMAEMA-based polyplexes of the same dose and N/P ratio. The transfection associated with the investigated organs after intraperitoneal administration of the transfection systems could very well be transfection of the mesothelial layer

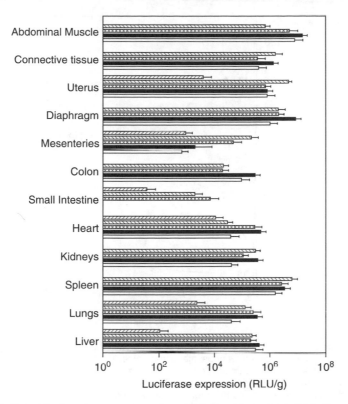

Figure 8.21 Distribution of luciferase gene expression after intraperitoneal administration of pDMAEMA poly-plexes and DOTAP lipoplexes. Horizontally hatched bars: pDMAEMA, N/P = 15, 120 µg pLuc; filled bars: pDMAEMA, N/P = 15, 75 µg pLuc; hatched bars: pDMAEMA, N/P = 5, 75 µg pLuc; left diagonally hatched bars: DOTAP, N/P = 15, 120 µg pLuc; right diagonally hatched bars: naked pLuc, 120 µg. Data represent the mean ± SEM (n = 3).

lining the peritoneal cavity and covering the tissues examined. Transfection of the mesothelial layer has also been observed after i.p. injection of viral vectors.[49]

8.4.2   Intravenous Administration

Direct injection of gene medicines in the cavity containing the tumor burden, as in the case of intraperitoneally localized ovarian carcinoma metastases, provides better chances for contact with the target tissue when compared to intravenous injection of gene medicines when the target tissue is outside the bloodstream. To determine the biodistribution of pDMAEMA-based polyplexes, pDNA (encoding for the firefly luciferase enzyme) was labeled with [α-^{32}P]-dCTP by nick trans-lation. Approximately 6-week-old female Balb/c mice received positively charged polyplexes (pDMAEMA:DNA ratio 3:1 [w/w]) labeled with trace amounts of radioactivity in an injection volume of 200 µl by tail vein injection. At various times, blood was collected from the vena cava under ether anaesthesia, and subsequently the mice were killed. Radioactivity levels in each organ were determined. It was observed that the positively charged pDMAEMA–[^{32}P]-DNA polyplexes distributed primarily to the lungs (Figure 8.22). Within minutes 80% of the injected intravenous dose was recovered from the lungs. In a second set of experiments, distribution of transfection activity was studied. Twenty-four hours after intravenous administration of pDMAEMA-based polyplexes (pDMAEMA:DNA ratio 3:1 [w/w]), luciferase levels were determined in lungs, liver, spleen, kidneys, and heart (Figure 8.23). The results showed that the gene expression profile matched the biodistribution profile of the administered positively charged polyplexes. Most of the expression

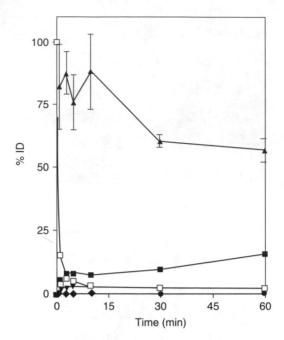

Figure 8.22   The influence of DNA dose and N/P ratio on the biodistribution and clearance of pDMAEMA–[^{32}P]-pLuc   polyplexes after intravenous injection in mice. Mice received radiolabeled DNA (complexed to pDMAEMA at an N/P ratio of 5, corresponding to a polymer:DNA ratio of 3 [w/w]) in an injection volume of 200 µl and a dose of 30 µg DNA. At the indicated time points, mice were killed and organs were excised and counted for radioactivity. (▲) Lungs, (■) Liver, (▼) Kidneys, (♦) Spleen, (●) Heart, (□) Blood. Results are expressed as the average percentage of the injected dose ± SEM (n = 3).

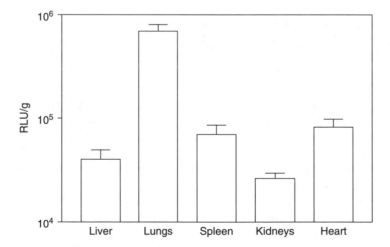

Figure 8.23   Tissue distribution of luciferase expression in mice organs after intravenous administration of pDMAEMA complexes into mice (pDMAEMA:DNA ratio 3:1 [w/w], 200 µl). Mice were injected in the tail vein with 30 µg pLuc DNA complexed to pDMAEMA. After 24 h, the mice were killed and their organs excised and homogenized, and luciferase activity was determined. Results are expressed as the average relative light units ± SEM (n = 3).

was seen in the lungs. A third set of experiments was designed to shed more light on the mechanism involved in the dominant lung uptake of polyplexes. *In vitro* turbidity experiments in serum were performed providing evidence for severe aggregation occurring upon addition of the polyplexes to the serum (data not shown). Hemagglutination experiments provided evidence that positively charged complexes induce the formation of extremely large structures upon addition to erythrocytes (Figure 8.24). If formed *in vivo*, such large aggregates can be trapped in the small capillaries of the lungs, blocking the blood flow. Another potential *in vivo* factor may be electrostatic interaction between the cationic polyplexes and the negatively charged lung endothelial cell membranes. However, incubation of polyplexes with serum albumin showed that the zeta potential of the complexes drops to negative values, making electrostatic interactions less likely.

Clearly, this "first pass" distribution severely hampers systemic tumor targeting with nonviral cationic transfection systems. To reach a tumor after intravenous administration, the gene delivery system should fulfill several important criteria. First, the interactions with biological components like erythrocytes and serum proteins should be minimal. This avoids aggregation and subsequent uptake by the lungs.[50,51] Second, if unfavorable interactions with blood constituents are avoided, the gene delivery systems should be sufficiently small (< 200 nm) and circulate long enough to be able to extravasate through the leaky tumor endothelium (via the so-called enhanced permeation and retention [EPR] effect).[52] Finally, gene delivery systems should be able to recognize and be taken up by the tumor cells. The coupling of the water-soluble polymer PEG to the surface of liposomes is a well-known strategy for increasing their circulation time via steric stabilization of their surfaces.[53,54] PEGylation has also been demonstrated to shield the surface charge of polyplex and lipoplex formulations. This results in prolonged circulation times of these colloidal systems.[55-59] Besides PEG, other water-soluble polymers (e.g., poly[N-hydroxypropyl]methacrylamide) have been successfully used to prolong the circulation of polyplexes in the bloodstream.[60]

Recent experiments yielded success with surface modification of the pDMAEMA-based polyplexes with PEG. This includes the use of either pDMAEMA-grafted PEG (pDMAEMA-*g*-PEG) or AB di-block copolymers of pDMAEMA and PEG, as well as PEGylation of preformed polyplexes (further referred to as "postPEGylation").

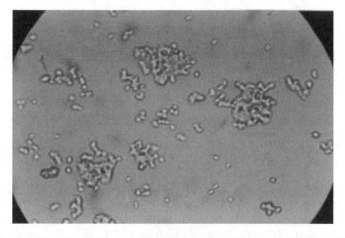

Figure 8.24 Hemagglutination of erythrocytes. Polyplexes were formed at a pDMAEMA:DNA ratio of 3:1 (w/w) and a DNA concentration of 15 µg/ml and were incubated with washed erythrocytes for 15 min.

The different types of PEG polyplexes were evaluated for their biophysical properties (size and charge) and possible interactions with blood constituents (albumin, erythrocytes). The surface charge of the pDMAEMA polyplexes was effectively shielded by two PEGylation methods (i.e., the use of pDMAEMA-g-PEG polymers and postPEGylation). The shielding effect was the highest for the postPEGylation method with PEG_{20000}, yielding polyplexes that hardly show interactions with blood components (i.e., albumin and erythrocytes). Most importantly, *in vivo* experiments showed prolonged circulation and avoidance of dominant lung localization in case of intravenous administration of PEGylated polyplexes. So far, the best results were obtained when PEG with a high molecular weight (20,000) was used. At 30 min after intravenous administration into Balb/c mice, about 50% of the injected dose was still circulating in the bloodstream, which is substantially higher when compared to the 2% of injected dose still circulating in the case of the uncoated polyplexes. Localization and gene expression in the lungs is almost absent, which is likely related to the improved colloidal stability of the complexes. For evaluating tumor targeting, we have utilized the subcutaneous C26 colon carcinoma mouse model and a subcutaneous neuro2A tumor model. The superior colloidal stability and circulation kinetics of the postPEGylated polyplexes translated into tumor accumulation which amounted to about 3.5% of the injected dose per gram tissue in a SC neuro2A tumor model and to about 4.2% of the injected dose per gram tissue in a SC C26 tumor model. Presently, experiments are being conducted in our laboratory to address the question of whether the achieved degree of tumor accumulation results in tumor transfection.

REFERENCES

1. Bally, M.B. et al., Biological barriers to cellular delivery of lipid-based DNA carriers, *Adv. Drug Del. Rev.* 38, 291–315, 1999.
2. Mallabone, C.L., Crooks, C.A., and Sefton, M.V., Microencapsulation of human diploid fibroblasts in cationic polyacrylates, *Biomaterials* 10(6), 380–386, 1989.
3. Sefton, M.V. and Stevenson, W.T.K., Microencapsulation of live animal cells using polyacrylates, *Adv. Poly. Sci.* 107, 143–197, 1993.
4. Zabner, J. et al., Cellular and molecular barriers to gene transfer by a cationic lipid, *J. Biol. Chem.* 270 (32), 18997–19007, 1995.
5. Poznanski, M.J. and Juliano, R.L., Biological approaches to the controlled delivery of drugs: a critical review, *Pharmacol. Rev.* 36, 277–336, 1984.
6. Van de Wetering, P. et al., 2-(Dimethylamino)ethyl methacrylate based (co)polymers as gene transfer agents, *J. Control. Release* 53 (1–3), 145–153, 1998.
7. Gao, X. and Huang, L., A novel cationic liposome reagent for efficient transfection of mammalian cells, *Biochem. Biophys. Res. Commun.* 179 (1), 280–285, 1991.
8. Curiel, D.T. et al., High-efficiency gene transfer mediated by adenovirus coupled to DNA-polylysine complexes, *Hum. Gene Ther.* 3 (2), 147–154, 1992.
9. Van de Wetering, P. et al., Relation between transfection efficiency and cytotoxicity of poly(2-dimethylamino)ethyl methacrylate)/plasmid complexes, *J. Control. Release* 49, 59–69, 1997.
10. Wolfert, M.A. and Seymour, L.W., Atomic force microscopic analysis of the influence of the molecular weight of poly(L)lysine on the size of polyelectrolyte complexes formed with DNA, *Gene Ther.* 3 (3), 269–273, 1996.
11. Van de Wetering, P. et al., Relation between transfection efficiency and cytotoxicity of poly(2-dimethylamino)ethyl methacrylate)/plasmid complexes, *J. Control. Release* 49, 59–69, 1997.
12. van de Wetering, P. et al., Copolymers of 2-(dimethylamino)ethyl methacrylate with ethoxytriethylene glycol methacrylate or N-vinyl-pyrrolidone as gene transfer agents, J. Control. Release 64 (1–3), 193–203, 2000.
13. Van de Wetering, P. et al., Structure-activity relationships of water-soluble cationic Methacrylate/Methacrylamide polymers for nonviral gene delivery, *Bioconjug. Chem.* 10 (4), 589–597, 1999.

14. Wink, T. et al., Interaction between plasmid DNA and cationic polymers studied by surface plasmon resonance spectrometry, *Anal. Chem.* 71 (4), 801–805, 1999.

15. Arigita, C. et al., Association and dissociation characteristics of polymer/DNA complexes used for gene delivery, *Pharm. Res.* 16 (10), 1534–1541, 1999.

16. Zuidam, N.J. et al., Effects of physicochemical characteristics of poly(2-(dimethylamino)ethyl methacrylate)-based polyplexes on cellular association and internalization, *J. Drug Target.* 8 (1), 51–66, 2000.

17. Mastrobattista, E. et al., Lipid-coated polyplexes for targeted gene delivery to ovarian carcinoma cells, *Cancer Gene Ther.* 8 (6), 405–413, 2001.

18. van Steenis, J.H. et al., Preparation and characterization of folate-targeted pEG-coated pDMAEMA-based polyplexes, J. Control. Release 87 (1–3), 167–176, 2003.

19. Keogh, M.C. et al., High efficiency reporter gene transfection of vascular tissue *in vitro* and *in vivo* using a cationic lipid-DNA complex, *Gene Therapy* 4 (2), 162–171, 1997.

20. Hofland, H.E. et al., *In vivo* gene transfer by intravenous administration of stable cationic lipid/DNA complex, *Pharm. Res.* 14 (6), 742–749, 1997.

21. Cherng, J.Y. et al., The effect of formulation parameters on the size of poly-((2-dimethylamino)ethyl methacrylate)-plasmid complexes, Eur. J. Pharm. Biopharm. 47 (3), 215–224, 1999.

22. Middaugh, C.R. et al., Analysis of plasmid DNA from a pharmaceutical perspective, *J. Pharm. Sci.* 87 (2), 130–146, 1998.

23. Cherng, J. et al., Effect of DNA topology on the transfection efficiency of poly((2-dimethylamino)ethyl methacrylate)-plasmid complexes, *J. Control. Release* 60 (2–3), 343–353, 1999.

24. Weiskopf, M. and Li, H.J., Poly(L-lysine)-DNA interactions in NaCl solutions: B to C and B to psi transitions, *Biopolymers* 16 (3), 669–684, 1977.

25. Ghirlando, R. et al., DNA packaging induced by micellar aggregates: a novel in vitro DNA condensation system, *Biochemistry* 31 (31), 7110–7119, 1992.

26. Bergan, D., Galbraith, T., and Sloane, D.L., Gene transfer *in vitro* and *in vivo* by cationic lipids is not significantly affected by levels of supercoiling of a reporter plasmid, *Pharm. Res.* 17 (8), 967–973, 2000.

27. Suzuki, T., Ohsumi, S., and Makino, K., Mechanistic studies on depurination and apurinic site chain breakage in oligodeoxyribonucleotides, *Nucleic Acids Res.* 22 (23), 4997–5003, 1994.

28. McHugh, P.J. and Knowland, J., Novel reagents for chemical cleavage at abasic sites and UV photoproducts in DNA, *Nucleic Acids Res.* 23 (10), 1664-1670, 1995.

29. Pogocki, D. and Schoneich, C., Chemical stability of nucleic acid-derived drugs, *J. Pharm. Sci.* 89 (4), 443–456, 2000.

30. Evans, R.K. et al., Evaluation of degradation pathways for plasmid DNA in pharmaceutical formulations via accelerated stability studies, *J. Pharm. Sci.* 89 (1), 76–87, 2000.

31. Lim, Y.B. et al., Biodegradable polyester, poly[alpha-(4-aminobutyl)-L-glycolic acid], as a non-toxic gene carrier, *Pharm. Res.* 17 (7), 811–816, 2000.

32. Hennink, W.E. and Bout, A., A transfection system using poly(organo)phophazenes complexes with nucleic acids, International Patent Application, 1997.

33. Arakawa, T. et al., Factors affecting short-term and long-term stabilities of proteins, *Adv. Drug Deliv. Rev.* 46 (1–3), 307–326, 2001.

34. Cherng, J.Y. et al., Freeze-drying of poly((2-dimethylamino)ethyl methacrylate)-based gene delivery systems, *Pharm. Res.* 14 (12), 1838–1841, 1997.

35. Cherng, J.Y. et al., Long term stability of poly((2-dimethylamino)ethyl methacrylate)-based gene delivery systems, *Pharm. Res.* 16 (9), 1417–1423, 1999.

36. van de Wetering, P. et al., A mechanistic study of the hydrolytic stability of poly(2-(dimethyl)amino-ethyl methacrylate), *Macromolecules* 31, 8063–8068, 1998.

37. Anchordoquy, T.J. and Koe, G.S., Physical stability of nonviral plasmid-based therapeutics, *J. Pharm. Sci.* 2000. 89(3): p. 289-96.

38. Li, B. et al., Lyophilization of cationic lipid-protamine-DNA (LPD) complexes. *J. Pharm. Sci.* 89 (3), 355–364, 2000.

39. Allison, S.D., Molina, M.C., and Anchordoquy, T.J., Stabilization of lipid/DNA complexes during the freezing step of the lyophilization process: the particle isolation hypothesis, *Biochim. Biophys. Acta* 1468 (1–2), 127–138, 2000.

40. Kwok, K.Y. et al., Strategies for maintaining the particle size of peptide DNA condensates following freeze-drying, *Int. J. Pharm.* 203 (1–2), 81–88, 2000.

41. Hamilton, T.C. et al., Characterization of a xenograft model of human ovarian carcinoma which produces ascites and intraabdominal carcinomatosis in mice, *Cancer Res.* 44 (11), 5286–5290, 1984.

42. Van de Wetering, P. et al., Comparative transfection studies of human ovarian carcinoma cells *in vitro*, *ex vivo* and *in vivo* with poly(2-(dimethylamino)ethyl methacrylate)-based polyplexes, *J. Gene Med.* 1 (3), 156–165, 1999.

43. Yang, J.P. and Huang, L., Overcoming the inhibitory effect of serum on lipofection by increasing the charge ratio of cationic liposome to DNA, *Gene Ther.* 4 (9), 950–960, 1997.

44. Veatch, A.L., Carson, L.F., and Ramakrishnan, S., Phenotypic variations and differential migration of NIH:OVCAR-3 ovarian carcinoma cells isolated from athymic mice, *Clin. Exp. Metastasis* 13 (3), 165–172, 1995.

45. Catterall, J.B. et al., Binding of ovarian cancer cells to immobilized hyaluronic acid, *Glycoconj. J.* 14 (7), 867–869, 1997.

46. Mounkes, L.C. et al., Proteoglycans mediate cationic liposome-DNA complex-based gene delivery *in vitro* and *in vivo*, *J. Biol. Chem.* 273 (40), 26164–26170, 1998.

47. Xu, Y. and Szoka, F.C., Jr., Mechanism of DNA release from cationic liposome/DNA complexes used in cell transfection, *Biochemistry* 35 (18), 5616–5623, 1996.

48. Roboz, J. et al., Hyaluronic acid content of effusions as a diagnostic aid for malignant mesothelioma, *Cancer Res.* 45 (4), 1850–1854, 1985.

49. Setoguchi, Y. et al., Intraperitoneal *in vivo* gene therapy to deliver alpha 1-antitrypsin to the systemic circulation, *Am. J. Respir. Cell Mol. Biol.* 10 (4), 369–377, 1994.

50. Kircheis, R., Wightman, L., and Wagner, E., Design and gene delivery activity of modified polyethylenimines, *Adv. Drug Del. Rev.* 53 (3), 341–358, 2001.

51. Dash, P.R. et al., Factors affecting blood clearance and *in vivo* distribution of polyelectrolyte complexes for gene delivery, *Gene Ther.* 6 (4), 643–650, 1999.

52. Nagayasu, A., Uchiyama, K., and Kiwada, H., The size of liposomes: a factor which affects their targeting efficiency to tumors and therapeutic activity of liposomal antitumor drugs, *Adv. Drug Del. Rev.* 40 (1–2), 75–87, 1999.

53. Storm, G. and Crommelin, D.J., Colloidal systems for tumor targeting, *Hybridoma* 16 (1), 119–125, 1997.

54. Moghimi, S.M., Hunter, A.C., and Murray, J.C., Long-circulating and target-specific nanoparticles: theory to practice, *Pharmacol. Rev.* 53 (2), 283–318, 2001.

55. Oupicky, D., Carlisle, R.C., and Seymour, L.W., Triggered intracellular activation of disulfide crosslinked polyelectrolyte gene delivery complexes with extended systemic circulation *in vivo*, *Gene Ther.* 8 (9), 713–724, 2001.

56. Ogris, M. et al., PEGylated DNA/transferrin-PEI complexes: reduced interaction with blood components, extended circulation in blood and potential for systemic gene delivery, *Gene Ther.* 6 (4), 595–605, 1999.

57. Fenske, D.B., MacLachlan, I., and Cullis, P.R., Long-circulating vectors for the systemic delivery of genes, *Curr. Opin. Mol. Ther.* 3 (2), 153–158, 2001.

58. Mannisto, M. et al., Structure-activity relationships of poly(L-lysines): effects of pegylation and molecular shape on physicochemical and biological properties in gene delivery, *J. Control. Release* 83 (1), 169, 2002.

59. Ward, C.M. et al., Modification of pLL/DNA complexes with a multivalent hydrophilic polymer permits folate-mediated targeting *in vitro* and prolonged plasma circulation *in vivo*, *J. Gene Med.* 4 (5), 536–547, 2002.

60. Oupicky, D. et al., Importance of lateral and steric stabilization of polyelectrolyte gene delivery vectors for extended systemic circulation, *Mol. Ther.* 5 (4), 463–472, 2002.

CHAPTER **9**

Cationic Dendrimers as Gene Transfection Vectors: *Dendri*-Poly(amidoamines) and *Dendri*-Poly(propylenimines)

Lori A. Kubasiak and Donald A. Tomalia

CONTENTS

0-8493-1934-X/05/$0.00+$1.50
© 2005 by CRC Press LLC

9.1 INTRODUCTION

Presently at least four general methods are known for the transfection of genetic materials into cells: viral vectors; nonviral vectors; naked plasmid DNA (pDNA) without a vector; and physical methods such as electroporation or the gene gun. One of the greatest challenges in the field of gene transfection and therapy is to identify viable gene delivery vectors that will elicit maximum expression with minimal toxic and immunogenicity effects and offer some potential for targeting. Until now, one of the most effective ways to transfer genes into mammalian cells has been with viral-based delivery systems.[1] Safety issues associated with viral vectors, such as recombination events and antiviral immunogenicity, have led to a new focus on the development of synthetic, nonviral vector systems.[2,3] New hybridized methods based on nonviral vector gene gun strategies are also being developed.[4]

Synthetic vectors, such as cationic polymers or block copolymers and cationic lipids, have exhibited interesting properties as gene transfection systems. Although of varying efficiencies, the low cytotoxicity, nonimmunogenicity, and ease of DNA–vector complexation observed with these nonviral vectors has promoted their interest in gene delivery.[5-6] With the exception of hyperbranched poly(ethyleneimines), all other polymeric vectors reviewed in this book belong to one of the three traditional polymeric architecture classes, namely (I) linear, (II) cross-linked, and (III) branched systems. The objective of this chapter is to focus on recent progress involving transfections with class IV dendritic architectures, as illustrated in Figure 9.1. More specifically, attention will be placed on cationic dendrimers, such as *dendri*-poly(amidoamines) (PAMAMs) and *dendri*-poly(propylene-imines) (PPIs). Other related cationic polymers such as *linear*-poly(amidoamines), poly(lysines), poly(imidazoles), and hyperbranched poly(ethylenimines) are addressed in other chapters throughout this book.

9.1.1 Dendritic Polymers: A Major New Architectural Class

Dendritic topologies are now widely recognized as the fourth major class of macromolecular architecture.[7–11] The signature for such a distinction is the unique new set of properties exhibited by this class of polymers. Presently, this architectural class consists of four dendritic subclasses: namely, (IVa) random hyperbranched polymers, (IVb) dendrigraft polymers, (IVc) dendrons, and (IVd) dendrimers (Figure 9.1). This subset order from left to right (IVa–d) reflects the relative degree of increased structural control present in each of these dendritic sublclasses.

All dendritic polymers are open, covalent nano-assemblies of branch cells.[12] They may be organized as symmetrical, monodispersed arrays, as is the case for dendrons and dendrimers, or as irregular polydispersed assemblies that typically define random hyperbranched polymers. As shown in Figures 9.1 and 9.2, these dendritic arrays of branch cells usually manifest covalent connectivity relative to some molecular reference marker (I) or core. As such, these branch cell arrays may be nonideal and polydispersed (e.g., $Mw/Mn \cong 2$–10), as observed for traditional polymer architecture types (I–III) and random hyperbranched polymers (IVa), or very ideally organized into highly controlled "core-shell" type structures as noted for dendrons and dendrimers (IVc–d): $Mw/Mn \cong 1.01$–1.0001. Dendrigraft polymers (IVb) reside between these two extremes of structural control, frequently manifesting rather narrow polydispersities of Mw/Mn 1.1–1.5 depending on their mode of preparation.

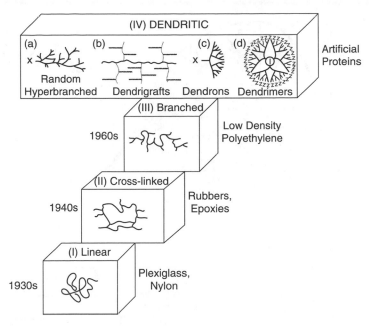

Figure 9.1   Three major traditional polymer architectural classes (I-III) leading to the fourth major class – dendritic (IV) consisting of the subclasses (a) *random hyperbranched,* (b) *dendrigrafts,* (c) *dendrons and* (d) *dendrimers.*

9.1.2   Dendrons and Dendrimers

The two most intensely investigated subsets of dendritic polymers are dendrons and dendrimers. In the past decade over 5000 literature references have appeared describing this unique class of structure-controlled dendritic polymers. The word "dendrimer" is derived from the Greek words *dendri* (branched tree-like) and *meros* (part of). The term was coined by Tomalia et al. at the Dow Chemical Company over 15 years ago, when they published the first full paper describing the "divergent synthesis" of PAMAM dendrimers.[11,13,14] Dendrons and dendrimers are nanoscale core-shell structures with architecture characterized by "branch upon branch" (dendritic) connectivity, which amplifies from the core to the surface, usually with radial symmetry. In contrast to traditional polymers, dendrimers are unique core-shell structures possessing three basic architectural components, namely (I) a core, (II) an interior of shells (generation) consisting of repetitive branch cell units, and (III) terminal functional groups (i.e., the outer shell or periphery), as illustrated in Figures 9.1 and 9.2.

Generally the first reaction sequence for a dendrimer (e.g., PAMAM) creates generation G = 0 (i.e., the core branch cell), wherein the number of arms (i.e., dendrons) anchored to the core is determined by core multiplicity (N_c) as shown in Figure 9.2. Iteration of an appropriate reaction sequence produces an amplification of terminal groups from 1 to 2 with the *in situ* creation of a branch cell (0) at the anchoring site of the dendron that constitutes G = 1. Repeating these iterative sequences (see Scheme 9.1), produces additional shells (generations) of branch cells that amplify mass and terminal groups according to the mathematical expressions described in Equation 9.2.

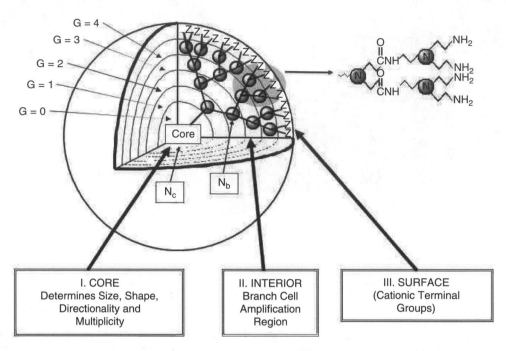

Figure 9.2   Three-dimensional projection of dendrimer core-shell architecture for G = 4 poly(amidoamine) (PAMAM) dendrimer with principal architectural components (I) core, (II) interior and (III) surface.

$$
\boxed{\begin{array}{c}\text{Number of}\\ \text{Surface}\\ \text{Groups}\end{array}} \; : \; Z \; = \; N_c N_b{}^G \qquad \boxed{\begin{array}{c}\text{Surface Group}\\ \text{Amplification/Gen.}\end{array}} \tag{9.1}
$$

$$
\boxed{\begin{array}{c}\text{Number of}\\ \text{Branch Cells}\end{array}} \; : \; BC \; = \; N_c \left[\frac{N_b{}^G - 1}{N_b - 1} \right] = \boxed{\begin{array}{c}\text{Number of Covalent}\\ \text{Bonds Formed/Generation}\end{array}} \tag{9.2}
$$

$$
\boxed{\begin{array}{c}\text{Molecular}\\ \text{Weights}\end{array}} \; : \; MW \; = \; M_c \; + \; N_c \left[M_{RU} \left(\frac{N_b{}^G - 1}{N_b - 1} \right) + M_t N_b{}^G \right] \tag{9.3}
$$

It is apparent that both the core multiplicity (N_c) and branch cell multiplicity (N_b) determine the precise number of terminal groups (Z) and mass amplification as a function of generation (G). One may view those generation sequences as quantized polymerization events. The assembly of reactive monomers, branch cells, or dendrons around atomic or molecular cores to produce dendrimers according to divergent or convergent dendritic branching principles has been well demonstrated.[12] Such systematic filling of space around cores with branch cells, as a function of generational growth stages (branch cell shells), to give discrete, quantized bundles of mass, has been shown to be mathematically predictable. Predicted molecular weights have been confirmed by mass spectroscopy and other analytical methods. Predicted numbers of branch cells, terminal groups (Z), and molecular weights as a function of generation for ethylenediamine core (EDA) ($N_c = 4$) PAMAM dendrimers are described in Table 9.1. It should be noted that the molecular weights approximately double as one progresses to generation + 1. The surface groups (Z) and branch cells (BC) amplify mathematically according to a power function, thus producing discrete, monodispersed structures with precise molecular weights as described in Table 9.1. These predicted values

Table 9.1 Calculated Values for Ethylenediamine core PAMAM Dendrimers

Generation	Surface Groups (Z)	Molecular Formula	MW	Diameter (nm)
0	4	$C_{22}H_{48}N_{10}O_4$	517	1.4
1	8	$C_{62}H_{128}N_{26}O_{12}$	1,430	1.9
2	16	$C_{142}H_{288}N_{58}O_{28}$	3,256	2.6
3	32	$C_{302}H_{608}N_{122}O_{60}$	6,909	3.6
4	64	$C_{622}H_{1248}N_{250}O_{124}$	14,215	4.4
5	128	$C_{1262}H_{2528}N_{506}O_{252}$	28,826	5.7
6	256	$C_{2542}H_{5088}N_{1018}O_{508}$	58,048	7.2
7	512	$C_{5102}H_{10208}N_{2042}O_{1020}$	116,493	8.8
8	1,024	$C_{10222}H_{20448}N_{4090}O_{2044}$	233,383	9.8
9	2,048	$C_{20462}H_{40928}N_{8186}O_{4092}$	467,162	11.4
10	4,096	$C_{40942}H_{81888}N_{16378}O_{8188}$	934,720	~13.0

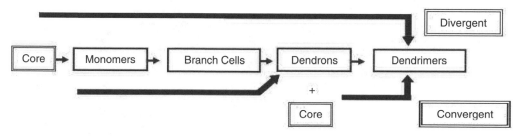

Figure 9.3 Construction components and sequences for *divergent* and *convergent* strategies to dendrimer synthesis.

can be verified by mass spectroscopy for the earlier generations; however, with divergent dendrimers, minor mass defects are often observed for higher generations as congestion-induced de Gennes dense packing begins to take effect.

In general, dendrimer syntheses involve hierarchical branch assembly strategies that require the construction components shown in Figure 9.3. Many methods for assembling these components are now known that allow the assembly of a wide variety of different core, interior, and surface compositions. Dendrimer synthesis based on the convergent method, however, is initiated from the terminal end groups.[15] First, polymeric branches are formed by reiterative coupling of branching units, followed by the anchoring of the branch assemblies (dendrons) possessing a single focal point or functional group to a core molecule. Due to steric hindrance, this method is not conducive to the formation of high generation dendrimers (with generation levels higher than G = 4), whereas the divergent method can produce dendrimers up to generation 10 (G-10) or higher. In general the convergent synthesis produces dendrimers of higher purity (isomolecularity) than those produced using the divergent method, since the dendron precursors can be purified before anchoring to the core. These divergent and convergent strategies are extensively reviewed elsewhere.[12]

9.2 POLY(AMIDOAMINE) (PAMAM) DENDRIMERS

9.2.1 Synthesis

Tomalia and coworkers while at the Dow Chemical Company reported the first divergent synthesis of poly(amidoamine) (PAMAM) dendrimers in the mid 1980s.[11,14] PAMAM dendrimers are nanoscale structures with architecture characterized by dendritic branching and radial symmetry.[16,17] Both PAMAM and poly(propylenimine) (PPI) dendrimers are synthesized using the divergent approach. This methodology involves the *in situ* branch cell construction in stepwise, iterative stages (i.e., G = 1,2,3...) around a desired core to produce mathematically defined core-shell structures. For PAMAM dendrimers, typically ethylenediamine (N_c = 4) or ammonia (N_c = 3) are used as cores and allowed to undergo reiterative two-step reaction sequences, involving exhaustive alkylation of primary amines (Michael addition) with methyl acrylate and amidation of amplified ester groups with a large excess of ethylenediamine to produce bio-friendly β-alanine repeat units and primary amine terminal groups, as illustrated in Scheme 9.1.

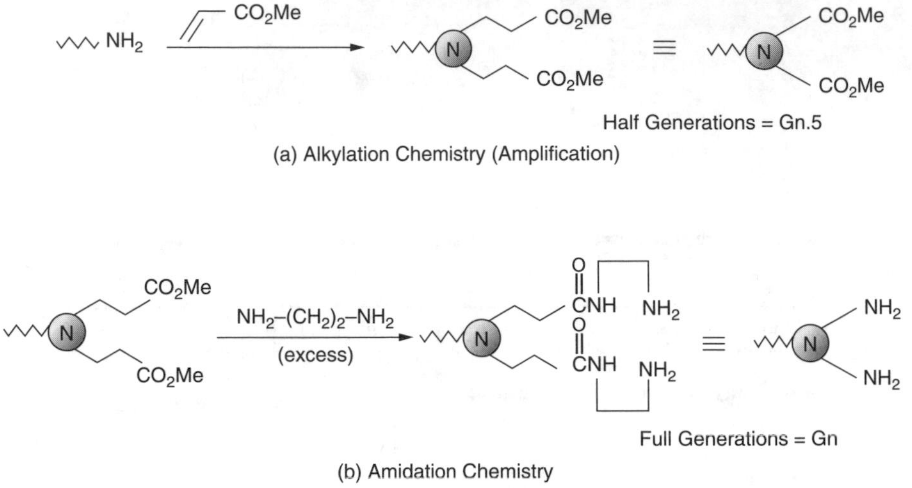

(a) Alkylation Chemistry (Amplification)

(b) Amidation Chemistry

Scheme 9.1

9.2.2   Structure

Poly(amidoamine) dendrimers (i.e., generations 0–10) range in size from 1.0 to 13.0 nm.[18] With each generation, the diameter of the dendrimer increases by ~1 nm, whereas the molecular mass approximately doubles and the number of surface groups exactly doubles (Table 9.1). As the number of surface groups amplifies, the density of the exterior branching increases and affects the dendrimer shape. Consequently, the shape observed for generations 0–4 is a more planar and flexible shape, whereas generations 5–10 exhibit more robust spherical shapes with closed surfaces due to higher branching densities at the surface.[19,20] At higher generations, de Gennes dense packing limits ideal dendrimer growth.[21] Therefore, the combination of branch and segment lengths, core and juncture multiplicities, and surface group dimensions defines a dendrimer's surface density and imposes certain limitations on the ideal dendrimer structure beyond generation 5–6 in the PAMAM series.[22] Commercially available starburst™, dendri-poly(amidoamine) (PAMAM) dendrimers are distributed by Dendritic Nano Technologies, Inc., Mt. Pleasant, MI.

9.2.3   Dendrimer Shape Changes: Nanoscale Container and Scaffolding Properties

As illustrated in Figure 9.4, dendrimers undergo congestion-induced molecular shape changes from flat, floppy conformations in the early generations to robust spheroids in the higher generations, as first predicted by Goddard and Tomalia et al.[23] Depending upon the accumulative core and branch cell multiplicities of the dendrimer family, (i.e., PAMAM dendrimers), these transitions were found to occur between G = 3 and G = 5. Ammonia core PAMAM dendrimers (N_c = 3, N_b = 2) exhibited a molecular morphogenesis break at G = 4.5, whereas the EDA core PAMAM dendrimer family (N_c = 4, N_b = 2) manifested a shape-change break around G = 3–4, and the Fréchet-type convergent

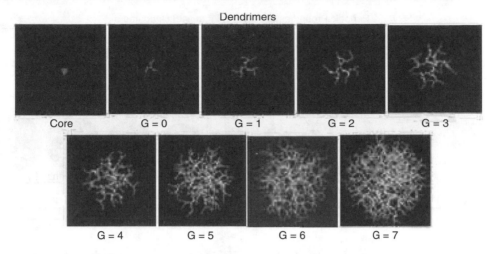

Dendrimers

Core   G = 0   G = 1   G = 2   G = 3

G = 4   G = 5   G = 6   G = 7

Figure 9.4   Molecular simulations of a PAMAM dendrimer family [NH$_3$ Core]:[G = 0–7].

dendrons (N$_b$ = 2) manifested a shape-change break around G = 4. It is readily apparent that increasing the core multiplicity from N$_c$ = 3 to N$_c$ = 4 accelerates congestion and forces a shape change at least one generation earlier. Beyond these generational transitions, one can visualize these dendrimeric shapes as nearly spheroidal or slightly ellipsoidal core-shell type architecture.

9.2.4   Nanoscale Container and Scaffolding Properties

Unimolecular container and scaffolding behavior are periodic properties that are specific to each dendrimer family or series. These properties are determined by the size, shape, and multiplicity of the construction components that are used for the core, interior, and surface of the dendrimer. Higher multiplicity components and those that contribute to tethered congestion will hasten the development of "container properties" or rigid surface scaffolding as a function of generation. Within the PAMAM dendrimer family, these periodic properties are generally manifested in three phases, as shown in Figure 9.5.

The earlier generations (i.e., G = 0–3) exhibit no well-defined interior characteristics; whereas interior development related to geometric closure is observed for the intermediate generations (i.e., G = 4–6 or 7). Accessibility and departure from the interior is determined by the size and gating properties of the surface groups. At higher generations (i.e., G 7), where de Gennes dense packing is severe, rigid scaffolding properties are observed, allowing relatively little access to the interior except for very small guest molecules.

9.2.5   Nanoscale Dimensions and Shapes Mimicking Proteins

In view of the extraordinary structure control and nanoscale dimensions observed for dendrimers, it is not surprising to find extensive interest in their use as globular protein mimics. Based on their systematic, dimensional length scaling properties (Figure 9.6) as well as electrophoretic and hydrodynamic behavior, they are referred to as artificial proteins.[7,24] Substantial effort has been focused recently on the use of dendrimers for "site isolation" mimicry of proteins and enzyme-like catalysis, as well as other biomimetic applications, drug delivery, surface engineering, and light harvesting. These fundamental properties have in fact led to their commercial use as globular protein replacements for gene therapy, immunodiagnostics, and a variety of other biological applications described throughout the chapter and elsewhere.[25]

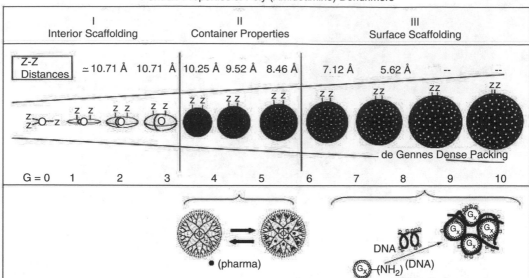

Figure 9.5   Periodic properties for poly(amidoamine) (PAMAM) dendrimers as a function of generation (G = 0–10) (I) flexible scaffolding properties (G = 0–3) (II) container properties (G = 4–6) and (III) rigid surface scaffolding properties (G = 7–10). Generation 4–6 preferred for encapsulation and controlled release of pharma; G = 7–10 preferred as scaffolding for (DNA) gene transfection.

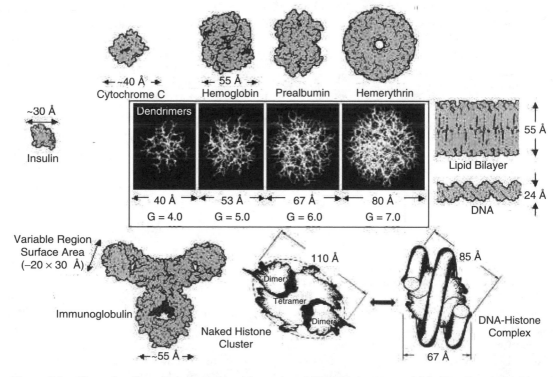

Figure 9.6   Dimensionally scaled comparison of a series of PAMAM dendrimers (NH$_3$ core; G = 4–7) with a variety of proteins, a typical lipid-bilayer membrane and DNA, indicating the closely matched size and contours of important proteins and bioassemblies.

9.2.6 Cationic (PAMAM) Dendrimer/DNA Interactions

Amine-terminated PAMAM dendrimers develop high positive charge densities at their surfaces when at physiological pHs or when they are dissolved in water. These spheroidal cationic surfaces in combination with their dimensions, which mimic in the length scale the sizes of histones, are inherent properties that favor polynucleic acid complexation. Cationic dendrimers have been shown to complex with pDNA, RNA, single-strand oligonucleotides, and various sizes of double-stranded DNA.[5,26–27] Dendrimer–DNA complex formation results from the electrostatic interactions between positively charged amino groups on the dendrimer surface and negatively charged phosphate groups on the DNA backbone. This complexation is more favored as the diameter of the dendrimer approaches the dimensions of histone octamer (i.e., ~7–8 nm), the ubiquitous DNA scaffolding found in biology. Investigations performed on double-strand DNA and amine-terminated PAMAM dendrimers (i.e., G = 2, 4, and 7) demonstrate that the DNA wrapping around G7 mimics DNA-histone wrapping, whereas such wrapping of DNA does not occur for G2 and G4 dendrimers.[28] Furthermore, the influence of pH and ionic strength were found to be critical parameters that affected electrostatic binding forces present in the DNA–dendrimer complexes. This work was in agreement with earlier findings reported by Bielinska and colleagues that suggested that the complex formation between cationic dendrimers and DNA was primarily due to charge interaction.[29] Related DNA complexation studies showed that this interaction was sufficiently strong that the DNA–dendrimer association could not be displaced by a strong DNA complexing agent such as ethidium bromide.[28] DNA–dendrimer complexation was shown to be of sufficient magnitude with higher generation PAMAM dendrimers that it offered DNA protection from nuclease degradation. It was found that its primary structure remained essentially intact, while its secondary and tertiary DNA structures were altered. Recent atomic force microscopy (AFM) studies support these findings, where it was found that the formation of DNA and G4 PAMAM dendrimer complexes protected DNA from DNase I degradation, but was dependent on dendrimer:DNA ratios and time allowed for complex formation before DNase treatment.[30]

Follow-up studies by Bielinska and colleagues defined critical parameters that affected the formation of dendrimer–DNA complexes.[31] Both soluble (low density) and insoluble (high density) complexes were analyzed. Specific dendrimer generations were demonstrated to modulate the effects of electrostatic charge. For example, generation 5 and 7 dendrimers complexed with DNA led to insoluble aggregates at lower charge ratios than those formed from G9 dendrimers. They also reported the direct effect of DNA concentration on the formation of high molecular weight and high density complexes (with nonuniform distribution) and its augmentation by increasing the dendrimer:DNA charge ratio. Increasing the dendrimer:DNA charge ratio above 20 resulted in an increase in the number of soluble (low density) complexes. It was also noted that these soluble complexes were responsible for more than 90% of the transfection, yet represented less than 20% of the total complexed DNA.

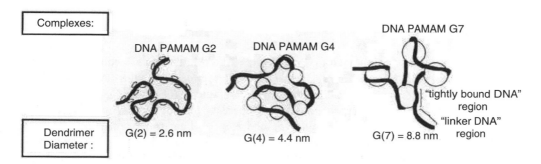

Figure 9.7 Proposed binding models for DNA-dendrimer complexes.

9.2.7   Cellular Entry

The molecular mechanism of dendrimer-mediated nucleic acid delivery into cells is not fully understood and is still under investigation.[32] It is thought that dendrimers mimic cationic liposomal delivery, wherein the complex is taken into the cell by endocytosis and nucleic acids are released from the endosome into the cytoplasm and migrate to the nucleus.[5,33-34,35] A net positive charge is necessary for DNA-dendrimer complex interaction at the cell surface. Studies using fluorescein-labeled dendrimers have shown that larger dendrimers bind more tightly to cell surfaces than lower generation dendrimers.[36] Positively charged DNA–dendrimer complexes can bind negatively charged moieties such as glycoproteins and phospholipids on the surfaces of cells, facilitating entry into the cell by endocytosis or passive transport initiated by membrane disturbance.[37] DNA–dendrimer entry is thought to be predominantly due to endocytosis,[5] and complexes not possessing substantial positive charges do not result in efficient gene transfer. Using inhibitors of endocytosis and cellular metabolism (cytocholasin B and sodium azide, respectively), Kukowska-Latallo and colleagues observed a substantial reduction in DNA–dendrimer complex entry into cells, thus illustrating the importance of endocytosis.[5]

Studies on anionic unilamellar vesicles by Zhang and Smith showed that G7 PANAM dendrimers induced significant membrane leakage when treated with standard DNA:dendrimer charge ratios used for transfection.[38] Higher generation dendrimers were significantly more effective at disrupting anionic vesicles than lower generation dendrimers or linear poly(lysines). They proposed a membrane-bending model, such that the extended radii of curvature associated with the larger spherical dendrimers forced more electrostatic distortion of the anionic membranes and induced bilayer-packing stresses, leading to lipid mixing.

Release of DNA–dendrimer complexes from endosomes is critical, and delayed release of nucleic acids into the cytosol appears to result in substantial degradation.[39] A proton sponge hypothesis has been proposed to address the issues of endosomal degradation and release. It has been hypothesized that the amine groups of the dendrimer have the capacity to buffer the endosomes, thereby inhibiting acidification and pH dependent nucleases.[40,41] Furthermore, the acidic endosomal environment has been suggested to foster polymeric swelling, thus leading to membrane disruption and consequent release of endosomal contents into the cytosol.

The complete role of dendrimers as vectors leading to gene expression is not entirely clear. Studies have shown that oligonucleotide–dendrimer complexes remain intact during uptake into vesicular compartments and subsequent entry into the nucleus.[42] Complexes of DNA and hyperbranched poly(ethylenimine) also remain intact upon entry into the nucleus.[43] This contrasts with cationic lipid delivery systems, wherein separation of nucleic acid from the vector is noted to occur in the endosome.[34,35] Overall, however, much of the molecular mechanism for DNA delivery to the nucleus remains to be elucidated.

9.2.8   *In Vitro* Transfection

9.2.8.1   Plasmids

Haensler and Szoka were the first to demonstrate PAMAM dendrimer-mediated transfection of cell cultures.[40] Using luciferase or β-galactosidase reporter plasmids with PAMAM dendrimers (G2–G10) as vectors, they investigated the transfection efficiency of both adherent and suspension cultured cells, including primary cell cultures. Adherent cell lines were represented by CV-1 (monkey fibroblast), HeLa (human carcinoma), and HepG2 (human hepatoma) cells; suspension cell cultures were represented by K-562 (human erythroleukemia), EL-4 (mouse lymphoma), and Jurkat (human T-cell) cells. Rat hepatocytes were used as a primary cell culture model. Cells from all groups could be transfected (using G6 PAMAMs); however, certain cells showed better expression than others. For example, CV-1 and K-562 cells exhibited from 30–80% and 10–30%

transfection, respectively, while EL-4 and Jurkat cells showed less than 1% transfection. This result was not surprising since most transfection systems display cell selectivity, but the molecular mechanisms for this variability remain unclear. Finally, transfection efficiency was determined to be directly related to the size of the dendrimer and the DNA:dendrimer charge ratio. Luciferase expression increased up to 3 orders of magnitude by increasing the dendrimer diameter from 4 nm to 5.4 nm (G4 to G5), and maximal expression was obtained using G6 dendrimers (6.8 nm diameter) in CV-1 cells. A dendrimer:DNA ratio of 6:1 (6 terminal amines to 1 phosphate) was shown to have optimal transfection efficiency, whereas higher ratios resulted in less efficiency.

Interestingly, it was later determined that some of the dendrimer samples used in the above studies had been inadvertently thermally degraded.[44] This spurred additional transfection studies of intentionally degraded dendrimers. Tang and colleagues performed heat treatments on PAMAM dendrimers of various generations (G5–G8) and showed that this treatment resulted in a more than 50-fold increase in transfection activity compared to the unheated control, as measured by β-galactosidase expression.[44] Lower generations required less heating time (~7 h) to reach maximal transfection activity, which was still lower than that of higher generations. Heat-activated dendrimer degradation was demonstrated using a variety of solvents, including water, 1- or 2-butanol, and 2-ethoxyethanol, wherein heating times for maximal transfection efficiency varied with their solvolytic behaviors. Thermal degradation in DMF, a nonsolvolytic solvent, had no effect on transfection efficiency, suggesting solvolysis of the dendrimer amide bonds was responsible for degradation. The heat activation of PAMAM dendrimers resulted in significant polydispersity of the typically monodisperse structure. Size exclusion chromatography and fractionation transfection data illustrated that only the high molecular weight molecules had transfection capabilities. Finally, dendrimers purposely synthesized with defective branching also enhanced transfection marginally, but did not show maximal levels such as those obtained with heat-degraded dendrimers. The model resulting from these data suggests a more flexible random hyperbranched polymer was generated *in situ* from the solvolysis and was responsible for the enhanced performance. Such thermally degraded PAMAM dendrimers, also known as "fractured" or "activated" dendrimers, are presently sold commercially by Qiagen as SuperFect™. Several studies discussed later in the chapter report the use of SuperFect™ to transfect a variety of different cells for reporter and therapeutic gene expression, as well as lentivirus production.[45]

An extensive investigation into the transfection properties of several series of intact monodispersed dendrimers was performed on a variety of cells by Kukowska-Latallo and colleagues.[5] This group used both NH_3 and EDA core PAMAM dendrimers and studied the transfection efficiencies of G0–G10 in 18 different cell lines, ranging from rat fibroblasts to human lymphoma cells. G3–G10 dendrimers were shown to form stable complexes with DNA. However, only G5–G10 exhibited significant cell transfection properties. A definite enhancement in transfection efficiency was observed by increasing generation levels from G5 to G10, with a plateau occurring after G8. Spherical shape and increase in surface charge were thought to be responsible for these effects. Overall, the PAMAM dendrimers were capable of transfecting many different cell types, including Jurkat and primary human fibroblasts, which are typically difficult to transfect, with no specific generation optimal for every type (Table 9.2).

In addition to the cell types listed in Table 9.2, PAMAM dendrimers have also been shown to be effective vectors for plasmid delivery in a number of other cell culture models. Some examples include the delivery of Epstein-Barr virus-based herpes simplex virus-1 thymidine kinase plasmids to cholangiocarcinoma cells[46] and hepatocellular carcinoma cells;[47] and the delivery of reporter plasmid β-galactosidase to cultured human chondrosarcoma-derived cells (HCS-2/8),[48] rabbit aortic smooth muscle cells, and human endothelial cells.[49] Use of PAMAM dendrimers has even been demonstrated in solid support membranes for gene therapy.[50] DNA–dendrimer complexes were found to be viable for delivering luciferase or GFP plasmids to hairless mouse skin after dissociation from solid supports or when retained on the membrane surface, even after drying. This suggests their potential use in bio-erodable membranes. In addition to plasmids, dendrimers were also found

Table 9.2   Cells Transfected with PAMAM Dendrimers

V79	Chinese hamster lung fibroblast		
CHO	Chinese hamster ovary	Jurkat	Human acute T-cell leukemia
NHBE	Human bronchial epithelium	CV-1	Monkey fibroblast
HCS-2/8	Human Chondrosarcoma-Derived	COS-1	Monkey kidney fibroblast
SW480	Human colon adenocarcinoma	COS-7	Monkey kidney fibroblast
HeLa	Human epithelial carcinoma	L-M(TK-)	Mouse connective tissue
293T	Human epithelial kidney	NIH3T3	Mouse embryonal
K-562	Human erythroleukemia	EL4	Mouse lymphoma
HF1	Human fibroblast	RAW264.7	Mouse macrophage-like
U937	Human histiocytic lymphoma	D5	Mouse melanoma
HepG2	Human hepatoma	YPE	Porcine vascular endothelial
A549	Human lung carcinoma	Rat2	Rat embryonal fibroblast
CCD-37LU	Human lung fibroblast		Rat hepatocytes
SW837	Human rectum adenocarcinoma	Clone 9	Rat liver epithelium
SAEC	Human small airway epithelium	RMI	Rat mesothelial
QS	Human synoviocyte	YB2/10	Rat myeloma

to be the most highly effective vectors compared to 20 other transfection agents evaluated for the delivery of a 60 Mb murine artificial chromosome/reporter gene to hamster lung fibroblast cells (V79-4) as well as murine connective tissue cells [L-M(TK-)].[51] The reported use of cationic dendrimers for *in vitro* transfection continues to grow as parameters affecting transfection efficiencies become better defined.

9.2.8.2 Oligonucleotides

In addition to plasmid transvections, PAMAM dendrimers were also demonstrated to be effective vectors for oligonucleotide delivery.[27,42,52,53] The efficient transfection of antisense oligonucleotides is of great interest in gene therapy due to its modulatory effects on protein expression. The concept of antisense oligonucleotide technology is centered on base pairing. Typically, a short strand of DNA is introduced that is complementary to the mRNA of the protein to be targeted. This oligonucleotide prevents the mRNA from being translated into protein by binding to the mRNA and inhibiting ribosomal interaction, thus initiating RNAse H-dependent enzymatic degradation of the mRNA.[54] Currently, the problems associated with this technology include the delivery and maintenance of high intracellular concentrations for long periods of time. Since oligonucleotides are extremely susceptible to nuclease degradation, the phosphodiester bonds are generally modified to create phosphorothioate oligonucleotides, which are more resistant to nucleases and thus exhibit longer intracellular half-lives. Delivery is currently being optimized, predominantly by increasing starting concentrations and utilizing new antisense vector systems.

Bielinska and coworkers were the first group to report dendrimers as antisense oligonucleotide transfection agents.[27] Luciferase expression in stably transfected Rat-2 fibroblasts and D5 mouse melanoma cells was maximally inhibited by ~50% using PAMAM (G7) antisense oligonucleotides complexed at a 10:1 charge ratio. Using radiolabeled oligonucleotides, the amount of radiolabeled DNA in U937 human histiocytic lymphoma, Rat-2, D5, and Jurkat cells was 300 times greater when complexed with G5, 7, and 9 dendrimers. After 24 h of transfection, PAMAM (G7) oligonucleotide-transfected cells still showed ~75% antisense inhibition of luciferase expression compared to 100% expression in uncomplexed transfected cells. Not only did dendrimers facilitate oligonucleotide delivery, but they also appeared to extend oligonucleotide intracellular effectiveness by increasing stability.

Later reports addressed issues of cellular localization of dendrimer-mediated oligonucleotide transfection. One report investigating the delivery of phosphorothioate oligonucleotides with G6 dendrimers in human astrocytoma cells (U251) showed a 50-fold increase in cell uptake of oligonucleotide.[52] Confocal microscopy illustrated enhanced intracellular localization of the dendrimer–oligonucleotide complex, in both the cytosol and the nucleus. Another report investigating the intracellular behavior of dendrimer–oligonucleotide complexes showed that the complexes were functionally active.[42] By conjugating the dendrimers and oligonucleotides with two different colored fluorescent dyes, intracellular movements of the complex or individual components could be monitored. Using fluorescence microscopy, the dendrimer and oligonucleotides were observed to be associated throughout uptake into vesicles as well as entry into the nucleus. Cellular fractionation studies revealed 5.27% of total dendrimer and 4.81% of oligonucleotides in the nucleus, implicating travel into the nucleus in the complexed form. Thus, dendrimers appear to be quite efficient at delivering oligonucleotides to the nucleus; however, studies targeting endogenous, biologically relevant proteins remain to be performed. As with pDNA, however, dendrimers are not successful transfection reagents in all systems.[55]

9.2.8.3 Enhanced Dendrimer-Mediated Transfection

Dendrimers are capable of efficient gene delivery to a number of different cell lines and several reports include reagents that enhance the dendrimer transfection system. The reagents range from surfactants to anionic oligomers, with a corresponding range of proposed mechanisms of action.

Anionic oligomers such as dextran sulfate and oligonucleotides dramatically increased plasmid transfection efficiency by PAMAM dendrimers in a study performed by Maksimenko and colleagues.[56] Dextran and dextran sulfate, a neutral molecule and anionic molecule, respectively, transfected into HeLa cells by SuperFect™ showed contrasting β-galactosidase expression. Dextran had no effect on plasmid expression, while dextran sulfate showed a substantial increase. The enhanced transfection seen with dextran sulfate was attributed to the formation of larger, low-density aggregates.

Kukowska-Latallo and coworkers reported the addition of DEAE-dextran in order to enhace transfections with dendrimers.[5] DEAE-dextran enhanced transfections were 20,000 to 40,000 times more efficient than DEAE-dextran alone, and up to 100 times more efficient than cationic lipid systems. Electron microscopy studies on G10 dendrimers revealed an inhibition of ultra-large aggregates with dextran treatment, suggesting an alteration of the DNA complex structure. Studies on lower generation dendrimers to demonstrate more effective transfection of cells were not performed.

In addition to dextran sulfate, the Maksimenko group also showed oligonucleotides to have a beneficial impact on dendrimer transfection.[56] Using several different lengths of oligonucleotides, transfection efficiency increased with oligonucleotide length up to 36 bases. Oligonucleotide sequence showed no effect, and hairpin turns slightly inhibited efficiency. Fluorescence microcopy studies demonstrated an oligonucleotide-stimulated increase in plasmid uptake into HeLa cells. Electron microscopy studies revealed morphological effects of oligonucleotides on dendrimer/plasmid complexes. Complexes formed in the presence of oligonucleotides showed much less condensed structure than the control. The authors suggested that increased protein expression with oligonucleotides was a result of forming less condensed, large structure complexes of dendrimer and DNA, which presumably increased cellular uptake.

Another category of reagents which were shown to be successful at increasing transfection efficiencies includes membrane destabilizers. Interestingly, compared to DEAE-dextran, chloroquine treatment was shown to increase reporter plasmid expression about twofold in COS-1 cells.[5] Chloroquine generally acts to disrupt endosomes, thus providing evidence for endosomal localization of DNA–dendrimer complexes. Another membrane-destabilizing peptide, GALA, has also been shown to enhance dendrimer transfection efficiency.[40] Transfection of GALA conjugated to G5 PAMAM dendrimers resulted in increased β-galactosidase expression by as much as 2 to 3 orders of magnitude. Enhanced transfection efficiency was demonstrated in a variety of cell types, including Jurkat, EL-4, K562, CV-1, HeLa, HepG2, and hepatocytes. Finally, Exosurf Neonatal™, a synthetic lung surfactant, was also demonstrated to have enhancing effects on several cell lines transfected with G9 PAMAM dendrimers.[57] Adherent cell lines CCD-37LU9 (human lung fibro-blasts), A549 (human lung carcinoma epithelial), and COS-1 (monkey kidney fibroblast), as well as nonadherent Jurkat (human T-cell) cells showed up to an 11-fold increase in luciferase expression. The most significant enhancement of luciferase expression occurred in primary cultures of normal human bronchial and tracheal epithelium (NHBE) and porcine vascular endothelial (YPE) cells, at a 41- and 25-fold increase, respectively. Tyloxapol, the nonionic surfactant component in Exosurf Neonatal™, was determined to be the active component responsible for transfection enhancement. Membrane destabilization may contribute to the effects seen, facilitating uptake of DNA–dendrimer complexes or affecting endosomal membranes.

Yet another type of reagent reported to enhance dendrimer-mediated gene delivery is cyclodex-trin. Several studies have indicated cyclodextrins to be mediators of enhanced DNA–dendrimer transfection.[58-59] Cyclodextrins are cyclic carbohydrates utilized by the pharmaceutical industry for their ability to form inclusion complexes. In transfection studies, β-cyclodextrins were initially studied in solid support-based transfection systems with G5 PAMAM dendrimers.[58] Reporter plasmid transfection yielded a 240-fold increase in transfection efficiency over the control. Later, the α- and γ-cyclodextrins were analyzed for their abilities to modulate dendrimer-based gene delivery.[59] G2 PAMAM dendrimers conjugated to α, β, or γ-cyclodextrins showed a significant increase in luciferase expression in NIH3T3 (mouse fibroblast) and RAW264.7 (mouse macrophage-like) transfected cells. Alpha-cyclodextrin conjugates exhibited the greatest transfection efficiency, with a 100-fold increase compared to dendrimer alone. Studies also revealed that specific dendrimer generations conjugated to α-cyclodextrin impacted transfection efficiency, in that G3 dendrimer conjugates resulted in higher luciferase expression. Furthermore, these data suggested that the conjugates increased both the cellular uptake of the complexes, as well as the ability to interact with endosomal membranes.

Finally, two other agents were noted to significantly enhance dendrimer-mediated transfection, namely fluorophores and adenoviruses. Investigating oligonucleotide transfer, Yoo and Juliano noted an enhanced delivery of oligonucleotides to HeLa cells with fluorophore-conjugated dendrimers, compared to unmodified dendrimers.[42] The significant increase in luciferase expression observed with Oregon green 488 conjugation was suggested to be the ability of the hydrophobic fluor to enhance disruption of endosomes. Adenoviruses, too, have been shown to enhance gene delivery through a combination of adenovirus infection and DNA–dendrimer transfection.[60] The addition of replication-deficient recombinant adenovirus to the dendrimer transfection protocol enhanced reporter gene expression in three different human cancer cell lines. Taken together, the transfection efficiency of PAMAM dendrimers has been shown to be modulated by a large variety of different agents, thus substantially expanding the number of cell types that may be useful targets of gene therapy.

9.2.8.4 In Vitro Toxicity

In order for gene transfection to have therapeutic use, the transfected cells must remain viable after DNA delivery. The most common viability assays target membrane integrity, including trypan blue exclusion, propidium iodide staining, and lactate dehydrogenase leakage into the extracellular space. Another common viability test is the MTT assay, which tests mitochondrial activity. Most studies using dendrimers as transfection agents test for cytotoxicity, but until recently, transfection-mediated cytotoxicity had not been the main focus of any research paper.

A study performed by Kukowska-Latallo and coworkers, investigating transfection efficiency of dendrimer systems in 18 different cell lines, reported that the concentrations of dendrimer and DNA eliciting maximal transfection efficiency had no effect on cell viability.[5] In this study, 10 generations of two different core dendrimers were complexed with DNA at charge ratios of 5:1 to 50:1, and cells were incubated with the complexes from 24 to 48 h.

Nonexistent or limited dendrimer-mediated transfection toxicity was also reported in several systems, including solid support-based transfection systems utilizing G5 dendrimers and COS-1 cells,[58] G5 and G7 dendrimers and rat-2 fibroblasts,[27] and SuperFect™ and either human chondrosarcoma-derived HCS-2/8 cells[48] or HeLa cells,[56] as determined primarily by trypan blue exclusion. The presence of serum in the transfection media was also shown to protect against cytotoxicity.[42]

A study performed by Haensler and Szoka had contrasting results.[40] Examination of dendrimer toxicity showed that dendrimer generation, charge ratio, and the presence of DNA affected toxicity. Assessed by the amount of protein remaining after the transfection procedure, dendrimers G5 and higher showed less protein recovery as the ratio of dendrimer to DNA increased above 6:1. Also, dendrimer-based transfection resulted in significantly higher toxicity when complexed with DNA. It should be noted, however, that these studies were only performed on CV-1 cells, and it is likely that toxicity effects vary among cell lines.

Recently, two research groups have directly addressed the cytotoxicity of PAMAM dendrimers used in transfection systems. In the first study, the *in vitro* toxicity of many gene delivery systems in L929 mouse fibroblasts and erythrocytes was assayed using MTT and lactate dehydrogenase assays, spectrophotometric hemolysis assays, and apoptosis assays, including caspase inhibition and nuclear morphology.[61] The molecular weight, charge density, and concentration of the molecules correlated with extent of cell damage, but most importantly, PAMAM dendrimers were associated with relatively low toxicity, ranking above cationized albumin and native albumin. Classical poly(ethylenimine), in contrast, exhibited maximal toxicity, followed by linear poly(lysine) and several other polymers.

The second study focused on the effects of surface modification on PAMAM cytotoxicity.[62] Using Caco-2 cells, the toxicity of G2–G4 PAMAM dendrimers was compared to that of surface-modified dendrimers. Lauroyl chain or polyethylene glycol were conjugated to the dendrimers and examined for cytotoxicity. Unmodified cationic dendrimers (full generation; amine surface) showed much higher cytotoxicity compared to anionic dendrimers (half generations; carboxylate surface). Increases in generation and concentration led to increases in toxicity for both types of dendrimer. Both the lauroyl chain and polyethylene glycol surface-modified dendrimers showed a significant decrease in toxicity, which was attributed to a shielding of the positive charge on the dendrimer surface by the surface conjugations.

9.2.8.5 Ex Vivo Gene Delivery

Another approach to gene delivery, short of *in vivo* delivery, is *ex vivo* transfection. With this approach, tissue is removed from the animal, undergoes clonal expansion followed by transfection in culture, and is then transplanted back into the animal.[63] As serum and antibiotics do not interfere with dendrimer-mediated gene transfer, cells are not deprived of these integral medium components, thereby limiting cell death with this system. To date, *ex vivo* transfection using dendrimers has been demonstrated in rabbit and human cornea models as well as murine cardiac transplant model, with great success.[64–65]

The failure rate of corneal transplantations, mainly due to graft rejection, is 26% over a 5-year time frame.[66] Transfection systems with low to no associated immunogenicity, such as dendrimers, would be a beneficial contribution to the corneal transplantation field. In the study performed by Hudde and coworkers, both rabbit and human corneas were transfected *ex vivo* with 18:1 dendrimer (SuperFect™): β-galactosidase pDNA.[64] The transfection resulted in maximal reporter expression at 3 days, decreasing to 15 days. Histochemistry revealed the transfection predominantly targeted endothelial cells, at a 10% transfection efficiency. Potential clinical use of this method was demonstrated, through successful transfer of a plasmid coding for a soluble tumor necrosis factor (TNF) receptor immunoglobulin. Again, maximal protein was detected at day 3 of the study, but the protein continued to be detected out to day 10. The production of this protein may neutralize the detrimentally high levels of TNF normally found during allograft rejection, thus enhancing the survival of grafts.[67]

Ex vivo transfection studies using murine cardiac transplant models have been equally successful. Using both vascularized and nonvascularized heart transplantation models, researchers have demonstrated effective transfection of β-galactosidase reporter plasmids by G5 dendrimers.[68,69] Both direct injection and intracoronary delivery of the DNA–dendrimer complexes resulted in efficient transfection. A DNA:dendrimer charge ratio of 1:20 was found to obtain the highest transfection efficiency, with maximal gene expression at 7 to 14 days, decreasing to 4 to 6 weeks. β-galactosidase expression was found in the epicardium, myocardium, and vascular endothelium, with predominant expression in the myocardium. Interestingly, the use of DNA–dendrimer complexes with a charge ratio of greater than 1:60 and increased concentrations of dendrimer (> 260 μg) resulted in the formation of graft thrombosis when delivered via the coronary artery. Subsequent studies by Wang and colleagues demonstrated enhanced dendrimer-mediated transfer of β-galactosidase pDNA in murine cardiac transplants using electroporation.[65] Electroporation is thought to form pores in membranes, through which DNA can be electrophoresed into cells.[70] In the study, G5 DNA–dendrimer complexes were transferred to murine cardiac grafts by immersion or intracoronary transfer, then electroporated with 20 pulses at a strength of 200 V/cm before being transplanted into recipient mice. Successful graft transfection depended on the presence of dendrimer and electroporation, with an enhanced expression of β-galactosidase from 10- to 45-fold. Indication of gene delivery was found throughout the myocardium, even penetrating to deeper cells without signs of myocardial cell death. Using the above-described electroporation protocol, transplant recipients survived the several week observation period. These results encourage subsequent studies investigating *ex vivo* gene delivery of therapeutic genes.

9.2.8.6 *In Vivo Gene Delivery*

The ultimate goal of gene transfection is to achieve safe therapeutic gene expression in the clinical setting. Successful gene delivery systems would be nontoxic and nonimmunogenic, and would result in therapeutically useful levels of gene expression.[63] Viral vectors, which show highly efficient gene transfer, have limited clinical use due to their immunogenicity. Nonviral gene delivery methods such as cationic lipids have been examined; however, inflammatory responses have been associated with their use.[71] Due to their apparent lack of cytotoxicity *in vitro*, PAMAM dendrimers have also been investigated for their utility in *in vivo* gene delivery.

Studies employing the use of PAMAM dendrimers in transfection use a variety of administration routes, including intravascular and intraperitoneal injection and intranasal delivery.[72-73] Oral availability of dendrimers has not been fully defined, so this route of administration is not commonly used.[74]

9.2.8.7 *Biodistribution*

The biodistribution of dendrimers has been investigated using ^{14}C-labeled G3, G5, and G7 PAMAM dendrimers, as well as ^{125}I-labeled G3, G4 (cationic), G2.5, G3.5, and G4.5 (anionic) dendrimers.[72,75] A study performed by Roberts and colleagues evaluated the biodistribution of radiolabeled dendrimers in male Swiss-Webster mice over 48 h.[72] Subsequent analysis of several organs was performed and data were recorded as percent injected dose (ID) per gram of tissue. Generation 3 dendrimers showed highest accumulation in the kidney, while G5 and G7 dendrimers were primarily localized to the pancreas. Interestingly, G7 also showed a high urinary excretion rate of 74% ID at 4 h. The liver showed a relatively low accumulation of dendrimer (< 10% ID).

These results contrast with results obtained in a later study performed by Malik and coworkers.[75] Using a Wistar rat model, dendrimer biodistribution in blood and liver was investigated using several different dendrimers differing in generation and surface charge, as well as intravascular and intraperitoneal administration routes. With intravascular administration, G3 and G4 dendrimers cleared from the blood much faster than G2.3, G3.5, and G5.5 dendrimers, resulting in levels less than 2% and 40% of initial dose, respectively, after 1 h. Remaining radioactive dendrimers were localized to the liver. Finally, dendrimers delivered by intraperitoneal administration entered the bloodstream within an hour and showed the same biodistribution as those dendrimers administered intravascularly. Potential reasons for discrepancy between the two studies include duration of biodistribution analysis as well as the animal model used. Additional biodistribution studies are needed, then, in order to obtain a complete understanding of the effects of surface charge and modifications, generation, and size on biodistribution in animals.

9.2.8.8 In Vivo Toxicity

In addition to distribution studies, Roberts and colleagues also performed *in vivo* toxicity experiments.[72] Male Wistar rats (5 per condition) were injected intravascularly with G3, G5, and G7 PAMAM dendrimers dissolved in phosphate-buffered saline (PBS) at doses of 5×10^{-4} to 5×10^{-6} mM/kg. Control animals were injected with PBS alone. Toxicity was evaluated by means of behavioral abnormalities and change in body weight in three different studies lasting 7 days, 30 days, and 6 months. Animals in the 7- and 30-day studies received one injection at the beginning of the study. All dendrimer treatments, regardless of generation and dose, had similar effects on body weight compared to the untreated control, which were attributed to normal animal growth over the time period. Animals in the 6 month studies were injected with 5×10^{-4} mM/kg dendrimer once a week. One animal treated with G7 dendrimer died within 24 h of administration in both the 7-day and 6-month studies. Subsequent studies utilizing a lower dose (5×10^{-5} mM/kg) established an appropriate safety margin for PAMAM G7 cationic dendrimers. At this concentration level, there was no significant difference in body weight patterns when comparing treated and untreated animals.

Interestingly, although cationic dendrimers have exhibited no serious toxicity effects in animal models at appropriate dose levels, some hemolysing effects have been reported with red blood cell (RBC) incubations.[75,76] Malik and colleagues performed an extensive study on cationic dendrimers of different generations and cores.[75] Dendrimers were added to solutions of rat RBCs prepared in PBS and incubated, shaking, for 1 h at 37°C. Free hemoglobin in the supernatant was used as an indicator of hemolysis. The amine-terminated dendrimers induced hemolysis above 1 mg/ml dendrimer concentration with the exception of PAMAM (G1). The hemolytic effects of PAMAM dendrimers were shown to be generation dependent, wherein lysis levels increased with generation. Furthermore, scanning electron microscopy (SEM) revealed significant RBC morphological changes at lower concentration (i.e., 10 μg/ml) dendrimer treatment. *In vitro* the cells appeared to exhibit aggregation tendencies; however, this *in vitro* observation appears to have no effect on animal viability. This suggests that the physiological environment may favorably alter dendrimer interactions with RBCs.[76]

Recent reports of successful *in vivo* dendrimer-based vector experiments support the potential future use of dendrimers in therapeutical applications. One study reported dendrimer-mediated gene therapy for prostate cancer.[77] Prostate-cancer-derived tumors were established in severe combined immunodeficiency mice. Intratumoral injections of dendrimer complexed with Fas ligand plasmid, a death ligand important in initiating apoptosis, resulted in the apoptosis of the tumor cells and significant growth suppression of the tumors. Another group reported the use of angiostatin and tissue inhibitor of metalloproteinase (TIMP-2) genes in an attempt to inhibit tumor growth and angiogenesis.[78] Intratumoral injection of dendrimers complexed with angiostatin or TIMP-2 plasmids significantly inhibited tumor growth by 71% and 84%, respectively, and transfection combining the two plasmids resulted in growth inhibition by 96%. These data support the viable use of dendrimer-mediated therapeutic gene delivery in animal models.

9.3 POLY(PROPYLENIMINE) (PPI) DENDRIMERS

The second class of cationic dendrimers presented in this chapter includes the commercially available compositional class referred to as *dendri*-poly(propylenimines). Much less work has been performed on these polymers related to gene transfection when compared to PAMAM dendrimers. The available literature is presented below.

9.3.1 Synthesis

Initial synthetic approaches involved the Michael addition of amines to acrylonitrile followed by the reduction of the nitriles to primary amines with various reducing agents. This synthesis was capable of producing low molecular weight dendritic polyamines (< 1.5 kilodaltons), but not higher generation dendrimers. The synthesis of these low molecular weight dendritic PPI structures was reported by Buhleier and colleagues.[79] Almost two decades later, the synthesis was dramatically improved using the Michael addition of acrylonitrile followed by catalytic hydrogenation of the nitriles with Rainey cobalt and hydrogen to primary amines.[80,81] Diaminobutane (DAB) and diaminoethane (DAE) cores are most commonly used for these PPI syntheses, with the ability to form dendrimers up to G5.

9.3.2 Structure

As shown in Figure 9.8, PPI dendrimers are characterized by dendritic branching and radial symmetry. A direct linear relationship has been demonstrated between dendrimer size and generation.[82] As with the divergent PAMAM dendrimer synthesis, retro-Michael reactions and cyclization products lead to defects in the dendrimer structure.[83] Electrospray mass spectroscopy studies show

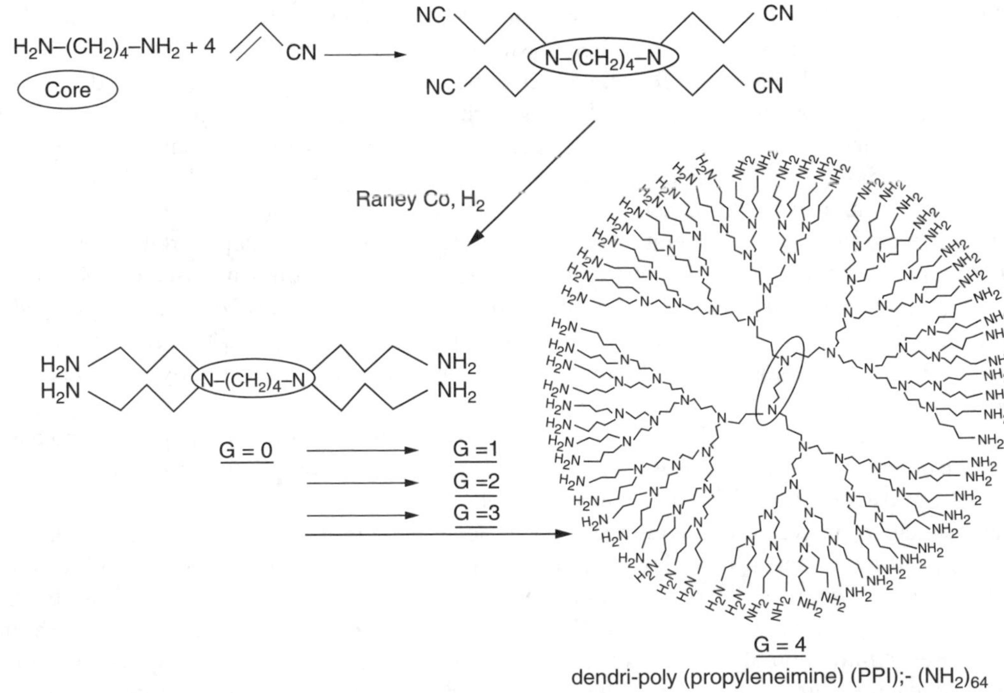

Figure 9.8 Synthetic reaction sequences to dendri-poly (propyleneimine) (PPI) dendrimers; G = 0–4.

a polydispersity of 1.002 for G5 dendrimers. Similar to PAMAM dendrimers, the outward growth of radial branching would suggest de Gennes limits on growth due to steric hindrance at the molecular surface.[21] Small-angle neutron scattering (SANS) measurements and molecular modeling studies to calculate size, radial densities, and interior volumes of PPI dendrimers support de Gennes dense-packing limitations as generations increase.[82,84]

9.3.3   DNA–Dendrimer Interactions

PPI dendrimers contain all basic protonable nitrogen atoms; thus the potential for interactions with negatively charged DNA in a gene delivery system is apparent.[85] Interactions with G5 PPI dendrimers and linear polyanions, such as DNA, have been shown using Astramol™.[86] Commercially available PPI G5 DAB core dendrimers are distributed by DSM (The Netherlands). Later studies on G1–G5 PPI dendrimers (DAB core) and DNA interactions were performed by Zinselmeyer and coworkers.[6] Dendrimer-induced DNA condensations were assessed by monitoring decreases in ethidium bromide intercalation. DNA complexation with PPI dendrimers appeared to reduce the number of ethidium bromide binding sites, thus leading to a decrease in fluorescence as measured by UV/vis spectrophotometry. Generations 1–5 were all capable of inhibiting ethidium bromide intercalation, demonstrating their ability to condense DNA. The higher generations (G3–G5) were able to condense a greater percent of DNA than lower generations (G1–G2). In addition, molecular modeling studies indicated an increase in DNA-dendrimer contact points with an increase in dendrimer generation. Studies indicated that G3 could bind an entire helical turn of DNA and the number of contact points in this complex was the minimum number for complete inhibition of ethidium bromide binding. These studies indicate that PPI dendrimers, especially G3–G5, are viable DNA complexing agents.

9.3.4   *In Vitro* Transfection

Very few reports have been published regarding the efficiency of PPI dendrimers as transfection agents. The first study performed was an evaluation of several polyplexes in gene transfer systems, including Astramol™.[87] Measured by luciferase expression, Astramol™ was shown to have very low transfection efficiency in COS-7 cells. The study then turned its focus to the more efficient transfection agents such as PAMAM dendrimers and linear polyethylenimines, evaluating these polymers in several other cell lines. The second study performed to evaluate PPI dendrimers as gene transfer agents used several more generations of the DAB core PPI dendrimers than the previous study.[6] PPI dendrimers G1–G5 were transfected into human epidermoid carcinoma cells (A431) with a β-galactosidase reporter gene, and assayed for gene expression. Generations 2 and 3 showed equivalent gene expression to DOTAP, a positive control cationic liposome transfection agent. Generation 1 PPI dendrimers were less efficient at transfecting A431 cells, resulting in an expression 40% that of DOTAP. Transfection results of G4 and G5 showed a minimal amount of transfection, at less than 10% that of DOTAP. These studies were performed at what was determined to be the optimal dose of DNA and charge ratio. Overall, very few studies have been performed using PPI dendrimers and more studies are needed in order to examine the transfection capabilities of these structures on a more extensive array of cell lines. The preliminary data suggests that G2 and G3 PPI dendrimers may be useful in gene therapy applications. To our knowledge, transfection efficiencies of PPI dendrimers have not been studied *in vivo*.

9.3.5   *In Vitro* Toxicity

As there are minimal reports in the literature regarding PPI dendrimers and transfection, there is a corresponding lack of toxicity data over a variety of cell lines. Data do exist for COS-7, A431, and rat erythrocytes, and are presented below.

In a study evaluating the transfection efficiency of several polyplexes, including Astramol™, it was determined that PPI dendrimer:DNA charge ratios exceeding 4:1 were highly toxic.[87] The 4:1 charge ratio was shown to be optimal for gene expression, although transfection efficiency was very low. Transfected into COS-7 cells, Astramol™ was shown to affect cell survival by ~20%.

A cytotoxicity analysis of G1–G5 PPI dendrimers on A431 cells was reported by Zinselmeyer and coworkers.[6] A general trend of increased toxicity with generation was noted, such that dendrimers within this family could be ranked in toxicity as G2 < G1 < G3 < G4 < G5. Both G1 and G2 were found to be less cytotoxic than cationic liposome DOTAP. Available anion binding sites appeared to be responsible for the toxicity seen, since reducing the number of sites by lowering the generation decreased cytotoxicity. With higher generations, anion binding sites were still available after the DNA bound, and increased cytotoxicity was seen.

Finally, Malik and colleagues performed studies on the toxic effects of PPI dendrimers possessing both DAB and DAE cores on RBCs.[75] The PPI dendrimers G2–G4 were added to solutions of rat RBCs prepared in PBS and incubated, shaking for 1 h at 37° C. Free hemoglobin in the supernatant was used as an indicator of hemolysis. As with PAMAM dendrimers, PPI dendrimers induced hemolysis above 1 mg/ml dendrimer concentration. DAE and DAB dendrimers were equally hemolytic, with no apparent dependence on generation. In addition, SEM revealed significant RBC clumping with 10 µg/ml dendrimer treatment. These adverse effects on erythrocytes *in vitro* need to be investigated in animal models in order to evaluate the usefulness of cationic PPI dendrimers in gene therapy systems. Additional dendrimer-based systems continue to emerge, including phosphorus-containing dendrimers[88] and dendritic poly(L-lysines),[89] the latter of which have been associated with low cytotoxicity.

9.4 TARGETED DELIVERY

The next logical step in gene transfection is targeted gene delivery. Incorporation of this localization technique would allow discriminatory gene expression in target cells. Recently, Barth and colleagues demonstrated successful targeting of epidermal growth factor (EGF)-conjugated PAMAM dendrimers to EGF receptors residing in gliomas.[90] In another instance, the concept of adding targeting functionality to dendrimer-based gene vectors via noncovalent techniques has recently been demonstrated by Lim and colleagues.[91] Such complexes between PPI dendrimers and DNA were shown to interact noncovalently with cucurbituril, a molecule with potential functionalization (targeting) capabilities. Importantly, transfection of these self-assembled complexes resulted in significantly lower cytotoxicity than cells transfected with dendrimer alone. In view of these successes, a significant demonstration of DNA–dendrimer targeting and gene expression is certain to emerge in the near future.

9.5 CONCLUSION

The search for a "magic bullet" gene delivery system continues. Although present cationic PAMAM and PPI dendrimer vectors do not satisfy all the necessary requirements for the perfect transfection system, their advantages and benefits have been well recognized and have driven intense investigations in the field. Many studies support the use of PAMAM dendrimers, especially in conjunction with appropriate expression enhancing agents. PPI dendrimers, on the other hand, will require more extensive *in vitro* as well as *in vivo* transfection characterization to support their utility in this field, although recent progress has been noted with self-assembling components that appear to offer the potential for targeting these systems.[91]

Several means of effectively delivering therapeutic genes without corresponding toxicity and immunogenicity are presently under investigation. Such work includes the use of polycationic dendrimers as coatings for nanoscale physical projectiles suitable for gene gun delivery. Recent work[4] has shown that the beneficial properties of cationic dendrimers and dendritic polymers, combined with gene gun delivery methodologies, offer completely new strategies for avoiding the complexities of chemical targeting and delivery that are now associated with more traditional approaches.

In conclusion, unique characteristics such as precise structures, well-defined functional surfaces, sizes and shapes that mimic globular proteins,[92] and nonimmunogenicity with the ability to introduce targeting functionality, continue to distinguish dendrimers as very interesting nonviral gene vector prospects.

REFERENCES

1. Wilson, J.M., Adenovirus as gene delivery vehicles. *New Engl. J. Med.* 334, 1185–1187, 1996.
2. Behr, J.P., Synthetic gene-transfer vectors. *Acc. Chem. Res.* 26, 274–278, 1993,.
3. Wagner, E. and Cotton, M., Non-viral approaches to gene therapy. *Curr. Opin. Biotechnol.* 4, 705–710.
4. Tomalia, D.A. and Balogh, L., Method and article for transfection of genetic material, U.S. Patent 6,475,994, 2002.
5. Kukowska-Latallo, J.F., Bielinska, A.U., Johnson, J., Spindler, R., Tomalia, D.A., and Baker, J.R., Jr., *Proc. Nat. Acad. Sci. U.S.A.* 93 4897–4902, 1996.
6. Zinselmeyer, B.H., Mackay, S.P., Schatzlein, A.G., and Uchegbu, I.F., The lower-generation polypropylenimine dendrimers are effective gene-transfer agents. *Pharm. Res.* 19, 960–967, 2002.
7. Tomalia, D.A., Brothers, H.M., II, Piehler, L.T., Durst, H.D., and Swanson, D.R Partial shell-filled core-shell tecto(dendrimers): A strategy to surface differentiated nano-clefts and cusps., *Proc. Nat. Acad. Sci. U.S.A.* 99 (8), 5081–5087, 2002.
8. Tomalia, D.A., Starburst dendrimers – nanoscopic supermolecules according to dendritic rules and principles. *Macromol. Symp.* 101, 243–255, 1996.
9. Tomalia, D.A., Brothers, H.M., II, Piehler, L.T., Hsu, Y., *Polym. Mater. Sci. Eng.* 73, 75, 1995.
10. Naj, A.K., Persistent inventor markets a molecule in *The Wall Street Journal*, B1, 1996.
11. Tomalia, D.A. and Fréchet, J.M.J., Discovery of dendrimers and dendritic polymers: A brief historical perspective. *J. Polymer Sci.* 40, 2719–2728, 2002.
12. Fréchet, J.M.J. and Tomalia, D.A., *Dendrimers and Other Dendritic Polymers*, J. Wiley & Sons, West Sussex, 2001.
13. Tomalia, D.A., Dewald, J.R., Hall, M.J., Martin, S.J., and Smith, P.B., *First International Conference Reprints, Kyoto, Japan, Society of Polymer Science (SPSJ)*, 1984 (August), 65.
14. Tomalia, D.A., Baker, H., Dewald, J.R., Hall, M., Kallos, G., Martin, S., Roeck, J., Ryder, J., and Smith, P., A new class of polymers: Starburst-dendritic macromolecules. *Polym. J.* 17 (117–132), 1985.
15. Hawker, C. J. and Frechet, J.M., Preparation of polymers with controlled molecular architecture. A new convergent approach to dendritic macromolecules. *J. Am. Chem. Soc.* 112, 7638–7647, 1990.
16. Tomalia, D.A., Naylor, A.N., and Goddard, W.A., III, Starburst dendrimers: Molecular level control of size, shape, surface chemistry, topology and flexibility from atoms to macroscopic matter. *Angew. Chem. Int. Ed. Engl.* 29, 138–175.
17. Fréchet, J.M., Functional polymers and dendrimers: Reactivity, molecular architecture, and interfacial energy. *Science* 263, 1710–1715, 1994.
18. Lothian-Tomalia, M.K., Hedstrand, D.M., Tomalia, D.A., Padia, A.B., and Hall, H.K., Jr., A contemporary survey of covalent connectivity and complexity. The divergent synthesis of poly(thioether) dendrimers. Amplified, genealogically directed synthesis leading to the de gennes dense packed state. *Tetrahedron* 53, 15495–15513, 1997.
19. Caminati, G., Turro, J., and Tomalia, D.A., Photophysical investigation of starburst dendrimers and their interactions with anionic and cationic surfactants. *J. Am. Chem. Soc.* 112, 8515–8522, 1990.
20. Potschke, D., Ballauff, M., Lindner, P., Fischer, M., and Vogtle, F., Analysis of the structure of dendrimers in solution by small angle neutron scattering including contrast variation. *Macromolecules* 32, 4079–4087, 1999.

21. de Gennes, P. G. and Hervet, H.J., Statistics of starburst® polymers. *J. Physique Lett. (Paris)* 44, 351, 1983.
22. Esfand, R. and Tomalia, D.A. Laboratory synthesis of poly(amidoamine) (PAMAM) dendrimers, in *Dendrimers and Other Dendritic Polymers*, Fréchet, J.M. and Tomalia, D.A., Eds., John Wiley & Sons, New York, 2001.
23. Naylor, A.M., Goddard, W.A., III, Keifer, G.E., and Tomalia, D.A., Starburst dendrimers. Molecular shape control. *J. Am. Chem. Soc.* 111, 2339–2341, 1989.
24. Esfand, R. and Tomalia, D.A., Poly(amidoamine) (pamam) dendrimers: From biomimicry to drug delivery and biomedical applications. *Drug Discov. Today* 6 8), 427–436, 2001.
25. Singh, P., Dendrimer-based biological reagents: Preparation and applications in diagnostics in *Dendrimers and Other Dendritic Polymers*, Fréchet, J.M.J. and Tomalia, D.A., Eds., John Wiley & Sons, West Sussex, 2001, 463–484.
26. Radler, J.O., Koltover, I., Salditt, T., and Safinya, C.R., Structure of DNA-cationic liposome complexes: DNA intercalation in multilamellar membranes in distinct interhelical packing regimes. *Science* 275, 810–814, 1997.
27. Bielinska, A., Kukowska-Latallo, J.F., Johnson, J., Tomalia, D.A., and Baker, J.R., Jr., Regulation of in vitro gene expression using antisense oligonucleotides or antisense expression plasmids transfected using starburst pamam dendrimers. *Nucleic Acids Res.* 24, 2176–2182, 1996.
28. Chen, W., Turro, N. J., and Tomalia, D.A., Using ethidium bromide to probe the interactions between DNA and dendrimers. *Langmuir* 16, 15–19, 2000.
29. Bielinska, A.U., Kukowska-Latallo, J.F., and Baker, J.R., Jr., The interaction of plasmid NA with polyamidoamine dendrimers: Mechanism of complex formation and analysis of alterations induced in nuclease sensitivity and transcriptional activity of the complexed NA. *Biochim. Biophys. Acta* 1353, 180–190, 1997.
30. Abdelhady, H.G., Allen, S., Davies, M.C., Roberts, C.J., Tendler, S.B., and Williams, P.M., Direct real-time molecular scale visualization of the degradation of condensed DNA complexes exposed to dnase i. *Nucleic Acids Res.* 31, 4001–4005, 2003.
31. Bielinska, A.U., Chen, C., Johnson, J., and Baker, J.R., Jr., DNA complexing with polyamidoamine dendrimers: Implications for transfection. *Bioconjugate Chem.* 10, 843–850, 1999.
32. Braun, C.S. and Middaugh, C.R., Personal Communication, Dept. of Pharmaceutical Chemistry, University of Kansas, 2003.
33. Marcusson, E.G., Bhat, B., Manoharan, M., Bennett, C.F., and Dean, N.M., Phosphorothioate oligodeoxyribonucleotides dissociate from cationic lipids before entering the nucleus. *Nucleic Acids Res.* 26, 2016–2023, 1998.
34. Zelphati, O. and Szoka, F.C., Jr., Mechanism of oligonucleotide release from cationic liposomes. *Proc. Nat. Acad. Sci. U.S.A.* 93, 11493–11498, 1996.
35. Zelphati, O. and Szoka, F.C., Jr., Intracellular distribution and mechanism of delivery of oligonucleotides mediated by cationic lipids. *Pharm. Res.* 13, 1367–1372, 1996.
36. Lai, J.C., Yuan, C., and Thomas, J.L., Single-cell measurements of polyamidoamine dendrimer binding. *Ann. Biomed. Eng.* 30, 409–416, 2002.
37. Dvornic, P.R. and Tomalia, D.A., Molecules that grow like trees. *Sci. Spectra* 5, 36–41, 1996.
38. Zhang, Z. and Smith, B.D., High generation polycationic dendrimers are unusually effective at disrupting anionic vesicles: Membrane bending model. *Bioconjugate Chem.* 11, 805–814, 2000.
39. Wattiaux, R., Laurent, N., Wattiaux-De Coninck, S., and Jadot, M., Endosomes, lysosomes: their implication in gene transfer. *Adv. Drug Deliv. Rev.* 41, 201–208, 2000.
40. Haensler, J. and Szoka, F.C., Jr., Polyamidoamine cascade polymers mediate efficient transfection of cells in culture. *Bioconjugate Chem.* 4, 372–379, 1993.
41. Boussif, O., Lezoualc'h, F., Zanta, M.A., Mergny, M.D., Scherman, D., Demeneix, B., and Behr, J.P., A versatile vector for gene and oligonucleotide transfer into cells in culture and in vivo: Polyethylenimine. *Proc. Nat. Acad. Sci. U.S.A.* 92, 7297–7301, 1995.
42. Yoo, H. and Juliano, R.L., Enhanced delivery of antisense oligonucleotides with fluorophore-conjugated pamam dendrimers. *Nucleic Acids Res.* 28, 4225–4231, 2000.
43. Godbey, W.T., Wu, K.K., and Mikos, A.G., Tracking the intracellular path of poly(ethylenimine)/DNA complexes for gene delivery. *Proc. Nat. Acad. Sci. U.S.A.* 96, 5177–5181, 1999.

44. Tang, M. X., Redemann, C.T., and Szoka, F.C., Jr., In vitro delivery by degraded polyamidoamine dendrimers. *Bioconjugate Chem.* 7, 703–714, 1996.
45. Coleman, J.E., Huentelman, M.J., Kasparov, S., Metcalfe, B.L., Paton, J.F., Katovich, M.J., Semple-Rowland, S.L., and Raizada, M.K., Efficient large-scale production and concentration of hiv-1-based lentiviral vectors for use in vivo. *Physiol. Genomics* 12, 221–228, 2003.
46. Tanaka, S.I., Harada, Y., Morikawa, T., Muramatsu, A., Mori, T., Okanoue, T., Kashima, K., Maruyama-Tabata, H., Hirai, H., Satoh, E., Imanishi, J., and Mazda, O., Targeted killing of carcino-embryonic antigen (cea)-producing cholangiocarcinoma cells by polyamidoamine dendrimer-mediated transfer of an epstein-barr virus (ebv)-based plasmid vector carrying the cea promoter. *Cancer Gene Ther.* 7, 1241–1250, 2000.
47. Harada, Y., Iwai, M., Tanaka, S., Okanoue, T., Kashima, K., Maruyama-Tabata, H., Hirai, H., Satoh, E., Imanishi, J., and Mazda, O., Highly efficient suicide gene expression in epatocellular carcinoma cells by epstein barr virus-based plasmid vectors combined with polyamidoamine dendrimer. *Cancer Gene Ther.* 7, 27–36, 2000.
48. Ohashi, S., Kuba, T., Ikeda, T., Arai, Y., Takahashi, K., Hirasawa, Y., Takigawa, M., Satoh, E., Imanishi, J., and Mazda, O., Cationic polymer-mediated genetic transduction into cultured human chondrosarcoma-derived hcs-2/8 cells. *J. Orthop. Sci.* 6, 75–81, 2001.
49. Turunen, M.P., Hiltunen, M.O., Ruponen, M., Virkamaki, L., Szoka, F.C., Jr., Urtti, A., and Yla-Herttuala, S., Efficient adventitial gene delivery to rabbit carotid artery with cationic polymer-plasmid complexes. *Gene Ther.* 6, 6–11, 1999.
50. Bielinska, A.U., Yen, A., Wu, H.L., Zahos, K.M., Sun, R., Weiner, N.D., Baker, J.R., Jr., and Roessler, B.J., Application of membrane-based dendrimer/DNA complexes for solid phase transfection in vitro and in vivo. *Biomaterials* 21, 877–887, 2000.
51. de Jong, G., Telenius, A., Vanderbyl, S., Meitz, A., and Drayer, J., Efficient in-vitro transfer of a 60-mb mammalian artificial chromosome into murine and hamster cells using cationic lipids and dendrimers. *Chromosome Res.* 9, 475–485, 2001.
52. Delong, R., Stephenson, K., Loftus, T., Fisher, M., Alahari, S., Nolting, A., and Juliano, R.L., Characterization of complexes of oligonucleotides with polyamidoamine starburst dendrimers and effects on intracellular delivery. *J. Pharm. Sci.* 86, 762–764, 1997.
53. Axel, D.I., Spyridopoulos, I., Riessin, R., Runge, H., Viebahn, R., and Karsch, K.R., Toxicity, uptake kinetics and efficacy of new transfection reagents: Increase of oligonucleotide uptake. *J. Vasc. Res.* 37, 221–234, 2000.
54. Pirollo, K. F., Rait, A., Sleer, L.S., and Chang, E.H., Antisense therapeutics: From theory to clinical practice. *Pharmacol. Ther.* 99, 55–77, 2003.
55. Jaaskelainen, I., Peltola, S., Honkakoski, P., Monkkonen, J., and Urtti, A., A lipid carrier with a membrane active component and a small complex size are required for efficient cellular delivery of anti-sense phosphorothioate oligonucleotides. *Euro. J. Pharm. Sci.* 10, 187–193, 2000.
56. Maksimenko, A.V., Mandrouguine, V., Gottikh, M.B., Bertrand, J., Majoral, J., and Malvy, C., Optimization of dendrimer-mediated gene transfer by anionic oligomers. *J. Gene Med.* 5, 61–71, 2003.
57. Kukowska-Latallo, J.F., Chen, C., Eichman, J., Bielinska, A.U., and Maker, J.R., Jr., Enhancement of dendrimer-mediated transfection using synthetic lung surfactant exosurf neonatal in vitro. *Biochem. Biophys. Res. Comm.* 264, 253–261, 1999.
58. Roessler, B. J., Bielinska, A.U., Janczak, K., Lee, I., and Baker, J.R., Jr., Substituted b-cyclodextrins interact with PAMAM dendrimer-DNA complexes and modify transfection efficiency. *Biochem. Biophys. Res. Comm.* 283, 124–129, 2001.
59. Kihara, F., Arima, H., Tsutsumi, T., Hirayama, F., and Uekama, K., Effects of structure of polyamidoamine dendrimer on gene transfer efficiency of the dendrimer conjugate with alpha-cyclodextrin. *Bioconjugate Chem.* 13, 1211–1219, 2002.
60. Dunphy, E. J., Redman, R.A., Herweijer, H., and Cripe, T.P., Reciprocal enhancement of gene transfer by combinatorial adenovirus transduction and plasmid DNA transfection in vitro and in vivo. *Hum. Gene Ther.* 10, 2407–2417, 1999.
61. Fisher, D., Li, Y., Ahlemeyer, B., Krieglstein, J., and Kissel, T., In vitro cytotoxicity testing of polycations: Influence of polymer structure on cell viability and hemolysis. *Biomaterials* 24, 1121–1131, 2003.

62. Jevprasesphant, R., Penny, J., Jalal, R., Attwood, D., McKeown, N.B., D'Emanuele, A., The influence of surface modification on the cytotoxicity of pamam dendrimers. *Int. J. Pharm.* 252, 263–266, 2003.

63. Goldspiel, B.R., Green, L., and Calis, K.A., Human gene therapy. *Clin. Pharm.* 12, 488–505, 1993.

64. Hudde, T., Rayner, S.A., Corner, R.M., Weber, M., Isaacs, J.D., Waldmann, H., Larking, D.P., and George, A.T., Activated polyamidoamine dendrimers, a non-viral vector for gene transfer to the corneal endothelium. *Gene Ther.* 6, 939–943, 1999.

65. Wang, Y., Bai, Y., Price, C., Boros, P, Qin, L., Bielinska, A.U., Kukowska-Latallo, J.F., Baker, J.R., Jr., and Bromberg, J.S., Combination of electroporations and DNA dendrimer complexes enhances gene transfer into murine cardiac transplants. *Am. J. Transplant.* 1, 334–338, 2001.

66. Williams, K.A., Muehlberg, S.M., Lewis, R.F., and Coster, D.J., Long-term outcome in corneal allotransplantation. The australian corneal graft registry. *Transplant. Proc.* 29, 983, 1997.

67. Rayner, S.A., King, W.J., Comer, R.M., Isaacs, J.D., Hale, G., George, A.J., and Larkin, D.F., Local bioactive tumour necrosis factor (tnf) in corneal allotransplantation. *Clin. Exp. Immunol.* 122, 109–116, 2000.

68. Qin, L., Pahud, D.R., Ding, Y., Bielinska, A.U., Kukowska-Latallo, J.F., Baker, J.R., Jr., and Bromberg, J.S., Efficient transfer of genes into murine cardiac grafts by starburst polyamidoamine dendrimers. *Hum. Gene Ther.* 9, 553–560, 1998.

69. Wang, Y., Boros, P., Liu, J., Qin, L., Bai, Y, Bielinska, A.U., Kukowska-Latallo, J.F., Baker, J.R., Jr., and Bromberg, J.S., DNA/dendrimer complexes mediate gene transfer into murine cardiac transplants ex vivo. *Mol. Ther.* 2, 602–608, 2000.

70. Somiari, S., Glasspool-Malone, J., Drabick, J.J., Gilbert, R.A., Heller, R., Jaroszeski, M.J., and Malone, R.W., Theory and in vivo application of electroporative gene delivery. *Mol. Ther.* 2, 178–187, 2000.

71. Logan, J.J., Bebok, Z., Walker, L.C., Peng, S., Felgner, P.L., Siegal, G.P., Frizzell, R.A., Dong, J., Howard, M., and Matalon, Preliminary biological evaluation of polyamidoamine (pamam) starburst dendrimers. *Gene Ther.* 2, 38–49, 1995.

72. Roberts, J.C., Bhalgat, M., and Zera, R.T., Preliminary biological evaluation of polyamidoamine (pamam) starburst dendrimers. *J. Biomed. Mat. Res.* 30, 53–65, 1996.

73. Kukowska-Latallo, J. F., Raczka, E., Quintana, A., Chen, C., Rymaszewski, M., and Baker, J.R., Jr., Intravascular and endobronchial DNA delivery to murine lung tissue using a novel, nonviral vector. *Hum. Gene Ther.* 11, 1385–1395, 2000.

74. Wiwattanapatapee, R., Carreno-Gomez, B., Malik, N., and Duncan, R., Anionic pamam dendrimers rapidly cross adult rat intestine in vitro: A potential oral delivery system? *Pharm. Res.* 17, 991–998, 2000.

75. Malik, N., Wiwattanapatapee, R, Klopsch, R., Lorenz, K., Frey, H., Weener, J.W., Meijer, E.W., Paulus, W., and Duncan, R., Dendrimers: Relationship between structure and biocompatibility in vitro, and preliminary studies on the biodistribution of 125i-labeled polyamidoamine dendrimers in vivo. *J. Contr. Rel.* 65, 133–148, 2000.

76. Eliyahu, H., Servel, N., Domb, A.J., and Barenholz, Y., Lipoplex-induced hemagglutination: Potential involvement in intravenous gene delivery. *Gene Ther.* 9, 850–858, 2003.

77. Nakanishi, H., Mazda, O., Satoh, E., Asada, H., Morioka, H., Kishida, T., Nakao, M., Mizutani, Y., Kawauchi, A., Kita, M., Imanishi, J., and Miki, T., Nonviral genetic transfer of fas ligand induced significant growth suppression and apoptotic tumor cell death in prostate cancer in vivo. *Gene Ther.* 10, 434–442, 2003.

78. Vincent, L., Varet, J., Pille, J.Y., Bompais, H., Opolon, P., Maksimenko, A., Malvy, C., Mirshahi, M., Lu, H., Vannier, J.P., Soria, C., and Li, H., Efficacy of dendrimer-mediated angiostatin and timp-2 gene delivery on inhibition of tumor growth and angiogenesis: In vitro and in vivo studies. *Int. J. Cancer* 105, 419–429, 2003.

79. Buhleier, E., Wehner, W., and Vogtle, F., Cascade and nonskid-chain-like synthesis of molecular cavity topologies. *Synthesis* 155, 1978.

80. de Brabander-van den Berg, E.M., and Meijer, E.W., Poly(propylene imine) dendrimers: Large scale synthesis by heterogeneously catalyzed hydrogenations. *Angew. Chem. Int. Ed. Engl.* 32, 1308, 1993.

81. Worner, C., and Muhlhaupt, R., Polynitrile- and polyamine-functional poly(trimethylene mine) dendrimers. *Angew. Chem. Int. Ed. Engl.* 32, 1306, 1993.

82. Scherrenberg, R., Coussens, B., van Vliet, P., Edouard, G., Brackman, J., and de Brabander, E., The molecular characteristics of poly(propyleneimine) dendrimers as studied with small-angle neutron scattering, viscosimetry, and molecular dynamics.*Macromolecules* 31, 456–461, 1998.

83. Hummelen, J.C., van Dongen, J.J., and Meijer, E.W., Electrospray mass spectrometry of poly(propyleneimine) dendrimers – the issue of dendritic purity or polydispersity. *Chem. Eur. J.* 3, 1489–1493, 1997.

84. de Brander-van den Berg, E.M., Brackman, J., MureMak, M., de Man, H., Hogeweg, M., Keulen, J., Scherrenberg, R., Coussens, B., Mengerink, Y., and Vanderwal, S., Polypropylenimine dendrimers: Improved synthesis and characterization. *Macromol. Symp.* 102, 9–17, 1996.

85. Van Duijvenbode, R.C., Borkovec, M., and Koper, G.M., Acid-base properties of poly(propylenimine) dendrimers. *Polymer* 39, 2657–2664, 1998.

86. Kabanov, V.A., Zezin, A.B., Rogachea, V.B., Gulyaeva, Z.G., Zansochova, M.F., Joosten, J.H., and Brackman, J., Interaction of astramol polypropylenimine dendrimers with linear polyanions. *Macromolecules* 32, 1904–1909, 1999.

87. Gebhart C.L. and Kabanov, A.V., Evaluation of polyplexes as gene transfer agents. *J. Control. Release* 73, 401–416, 2001.

88. Loup, C., Zanta, M.A., Caminde, A.M., Majoral, J.P., and Meunier, B., Preparation of water-soluble cationic phosphorus-containing dendrimers as DNA transfection agents. *Chem. Eur. J.* 5, 3644–3650, 1999.

89. Ohsaki, M., Okuda, T., Wada, A., Hirayama, T., Niidome, T., and Aoyagi, H., In vitro gene transfection using dendritic poly(l-lysine). *Bioconjugate Chem.* 13, 510–517, 2002.

90. Barth, R.F., Yang, W., Adams, D.M., Rotaru, J.H., Shukla, S., Sekido, M., Tjarks, W., Fernstermaker, R. A., Ciesieslki, M., Nawrocky, M.M., and Coderre, J.A., Molecular targeting of epidermal growth factor receptor for neutron capture therapy of giomas. *Cancer Res.* 62, 3159–3166, 2002.

91. Lim, Y.-B., Kim, T., Lee, J.W., Kim, S.-M., Kim, H.-J., Kim, K., and Park, J.-S., Self-assembled ternary complex of cationic dendrimer, cucurbituril, and DNA: Noncovalent strategy in developing a gene delivery carrier. *Bioconjugate Chem.* 13, 1181–1185, 2002.

92. Tomalia, D.A., Huang, B., Swanson, D.R., Brothers, H.M., II, and Klimash, J.W., Structure control within poly(amidoamine) dendrimers: Size, shape and regio-chemical mimicry of globular proteins. *Tetrahedron* 59, 3799–3813, 2003.

93. De Mattei, C.R., Huang, B., and Tomalia, D.A., Designed dendrimer syntheses by self-assembly of single-site, ssdna functionalized dendrons. *Nano Lett.* 4 (5), 771–777, 2004.

94. Arima, H., Kihara, F., Hirayama, F., and Uekama, K., Enhancement of gene expression by polyamidoamine dendrimer conjugates with alpha, beta, and gamma cyclodextrins. *Bioconjugate Chem.* 12, 476–484, 2001.

CHAPTER 10

Poly(ethylene glycol)-Conjugated Cationic Dendrimers

Joon Sig Choi, Tae-il Kim, and Jong-sang Park

CONTENTS

10.1 INTRODUCTION

Some representative cationic polymers, namely polyethylenimine (PEI), poly(L-lysine) (PLL), and polyamidoamine (PAMAM) dendrimer, were shown to possess cytotoxicity and form heterogeneous water-insoluble particles, leading to precipitation after complexing with DNA, and to

0-8493-1934-X/05/$0.00+$1.50
© 2005 by CRC Press LLC

unsatisfactory transfection efficiency *in vitro* and *in vivo*.[1-4] These polymers had been actively studied at an early stage of polymeric gene delivery research.

Polyethylene glycol (PEG) is one of the most important polymers in biomedical and biotechnical fields and is used in a variety of applications, including drug delivery systems.[5] Since PEG exhibits such well-known properties as high solubility in water, reduced immunogenicity, and biocompatibility, the polymer has also been competitively introduced to polymeric or liposomal gene delivery research.[6-12]

Nature is one of the best sources for stimulating our imagination. First, we were inspired by the structure of trees. At first sight, they may look like AB-type block copolymers, where the A part is composed of a main stem, boughs, limbs, and leaves, and the B part is a trunk (Figure 10.1A). However, if you consider their roots, they are in fact similar to ABA-type copolymers, where the dendritic roots constitute the other A part (Figure 10.1B). Roots help the tree stand firmly on the ground and absorb water and various nutrients, including minerals from the earth. A main stem, boughs, and limbs help the tree to withstand any physical force exerted by the wind and provide the supportive backbone, making the leaves of the tree spread out in the air to maximize the efficiency of light harvesting. This is an example of a simple natural model of multifunctional structure. Second, if you look into the structure of nerve cells (neurons), you will see that they also have an ABA-type structure (Figure 10.1C). Neurons can gather and transmit electrochemical signals. They are composed of cell body with dendrites (receiver), axon (the conducting fiber), myelin sheath (insulating layer), the node of Ranvier, and branched axon terminals (transmitters). So, we tried to mimic such natural structures and to engineer hybrid block copolymers for self-assembly with DNA that might have the potential for application to a polymeric gene delivery system. We introduced a linear PEG polymer like the trunk of a tree or the axon of a neuron, and a PLL or PAMAM dendrimer like the upper part and root of trees or the cell body with dendrites, the branched axon terminals of neurons.[13-16] We also tried to conjugate small molecular weight PEGs by surface modification of a branched PEI polymer that is very cytotoxic but is regarded as the most transfection-active material among cationic polymers, because we were interested in how to decrease the toxicity of PEI while maintaining its transfection efficiency.[17]

In this chapter, the new hybrid block copolymers developed by our group over the last few years will be reviewed and discussed in relation to their characteristics of electrostatic self-assembly with plasmid DNA forming polyionic complexes and their potential application for gene delivery *in vitro*.

10.2 SYNTHESIS AND CHARACTERIZATION OF THE HYBRID BLOCK COPOLYMERS

10.2.1 MPEG-PLLD and PLLD-PEG-PLLD (Figures 10.2A and 10.2B)

10.2.1.1 *Liquid Phase Peptide Synthesis*[18,19]

Repeated coupling and deprotection of Fmoc-Lys(Fmoc)-OH to the PEG backbone was performed. PEG was used as a polymeric supporter for liquid-phase peptide synthesis. Methoxy-PEG-NH$_2$ or NH$_2$-PEG-NH$_2$ was used and its molecular weight was 5000 and 3400 for MPEG-PLLD and PLLD-PEG-PLLD synthesis, respectively.[13,14] The coupling agents used were HOBt, HBTU, and DIPEA. After each coupling and deprotection reaction, the PEG-conjugated dendrimers were partially purified by precipitating in excess ether. The copolymer was dialyzed for 1 day against water using a Spectra/Por dialysis membrane (MWCO = 3400) (Spectrum, Los Angeles, CA) and lyophilized before use for analysis and assay.

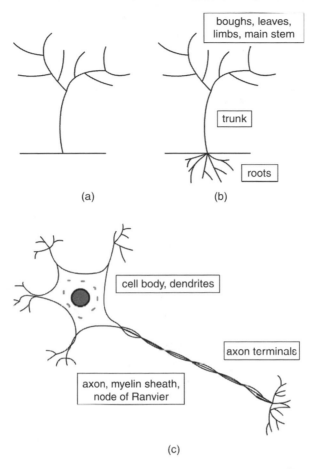

Figure 10.1 Examples of natural AB- or ABA-type structures; (A, B) trees and (C) neurons.

10.2.1.2 ¹H NMR and MALDI-TOF MS

The progress of each reaction was monitored by a ninhydrin test and confirmed by ¹H NMR and MALDI-TOF MS. MPEG-PLLD: ¹H NMR (D$_2$O): δ 1.50 (br m. -CH$_2$CH$_2$CH$_2$-), 3.00 (br m, CH$_2$-N), 3.70 (s, CH$_2$CH$_2$-O), 4.25 (br m. COCH-N). Methoxy-PEG-NH$_2$ (MPEG-NH$_2$, M$_w$ 5000) was measured by MALDI-TOF mass spectrum (M$_w$ = 5757, M$_n$ = 5697, M$_w$/M$_n$ = 1.01) and the M$_w$ and M$_n$ values of the fourth-generation copolymer were 7594 and 7553, respectively (M$_w$/M$_n$ = 1.01). The M$_w$ and M$_n$ values of the product based on its structural formula were calculated to be 7678 and 7618, respectively. PLLD-PEG-PLLD: ¹H NMR (D$_2$O): δ 1.50 (br m, -CH$_2$CH$_2$CH$_2$-), 3.00 (br m, CH$_2$-N), 3.70 (s, CH$_2$CH$_2$-O), 4.25 (br m. COCH-N). The characteristics of PLLD-PEG-PLLD are shown in Table 10.1.

10.2.2 MPEG-PAMAM and PAMAM-PEG-PAMAM (Figures 10.2C and 10.2D)

10.2.2.1 Synthesis of the Block Copolymers

MPEG-NH$_2$ (M$_w$ 5000) and NH$_2$-PEG-NH$_2$ (M$_w$ 3400) were used as polymeric supporters for AB- and ABA-type copolymers, respectively.[15,16] PAMAM dendrimers were extended from PEG by divergent stepwise synthesis.[20] The primary amine of MPEG-NH$_2$ was reacted with methyl acrylate by Michael addition and methyl ester end groups of the half-generation polymer were

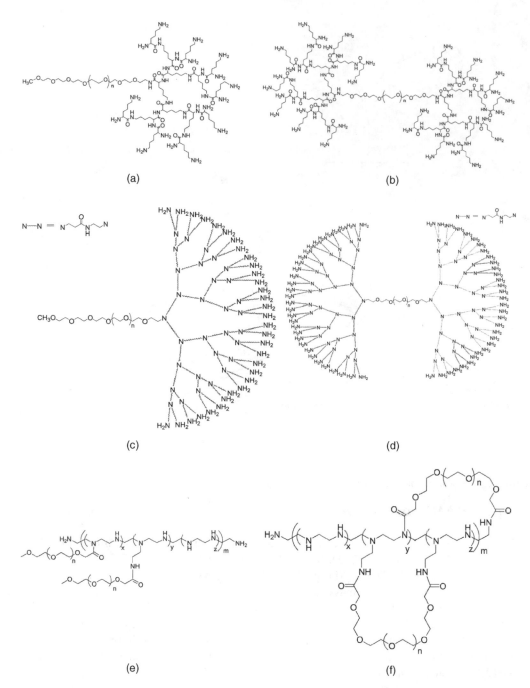

Figure 10.2 Structures of AB- or ABA-type block copolymers designed by our group. (A) MPEG-PLLD G4, (B) PLLD-PEG-PLLD G4, (C) MPEG-PAMAM G5, (D) PAMAM-PEG-PAMAM G5, (E) MPEG-PEI, and (F) PEG-PEI.

conjugated with ethylenediamine by amidation, forming the first full generation polymer which has surface primary amine groups. These steps were performed repeatedly to the desired generation. The copolymer of each step was purified by precipitating in excess ether and by dialysis against pure water. This synthetic method has the advantages of high yield and simplicity of participating reactions. For MPEG-PAMAM, the characterization results are described in Table 10.2.

Table 10.1 Characteristics of PLLD-PEG-PLLD

Polymer	Peak Intensity Ratio[a]		M_n	M_w		Number of Surface Amines
	Cal.	Exp.		Cal.	Exp.	
H_2N-PEG-NH_2	—	—	3411	—	3473	2
PLLD-PEG-PLLD G1	26.3	26.7	3682	3788	3729	4
PLLD-PEG-PLLD G2	8.77	8.75	4178	4250	4241	8
PLLD-PEG-PLLD G3	3.76	3.84	4908	5269	5265	16
PLLD-PEG-PLLD G4	1.75	1.91	7261	7317	7313	32

[a] Peak intensity ratio of the ethylene protons of PEG segment ($\delta = 3.7$ ppm) and the β-, γ-, and δ-methylene protons of PLL dendrimer ($\delta = 1.4$–1.8 ppm).

Table 10.2 The Molecular Weights of MPEG-PAMAM

Polymer	Theoretical # of α-Protons	Peak Intensity Ratio[a]		Exp. M_n	M_w		PDI
		Cal.	Exp.		Cal.	Exp.	
MPEG-PAMAM G3	28	0.056	0.054	6697	7090	6752	1.01
MPEG-PAMAM G4	60	0.120	0.117	8247	8914	8299	1.01
MPEG-PAMAM G5	128	0.256	0.250	11171	12562	11484	1.03

[a] Theoretically calculated and experimentally obtained ratio between α-protons of PAMAM backbone and PEG methylene protons of the full generation copolymers.

Each molecular weight and polydispersity index (PDI = M_w/M_n) was obtained by MALDI-TOF MS.

10.2.2.2 *¹H NMR and MALDI-TOF MS*

The synthesis of the copolymers was confirmed by ¹H NMR and MALDI-TOF MS. For example, ¹H NMR peaks of the fifth generation of PAMAM-PEG-PAMAM copolymer were as follows in D_2O: δ 2.451 (br m, -CH_2CONH-), 2.640–2.837 (br m. protons next to amines), 3.311 (br m. -$CONHCH_2$-), 3.716 (s, PEG). Peak intensity ratios of the spectra corresponded to theoretically calculated values. The number (M_n) and weight (M_w) averaged molecular weights, and polydispersity index (PDI) of the copolymer were obtained by MALDI-TOF MS. Up to the fourth generation, the MALDI-TOF MS results agreed well with theoretically expected values with slight deviations. The experimental M_w was 9289 in comparison with the theoretical M_w value, 10320. However, at the fifth generation of PAMAM-PEG-PAMAM, the MALDI-TOF result could not be obtained in various attempted conditions, presumably due to the difficulty in achieving ionization of the highly positively charged polymer.

10.2.3 MPEG-PEI and PEG-PEI (Figures 10.2E and 10.2F)

10.2.3.1 *PEG Activation and Conjugation*

MPEG (M_n 550) and PEG (M_n 600) were used.[17] MPEG was reacted with ethyl bromoacetate using NaH in THF. The ethyl MPEG acetate was dissolved in 1 N NaOH and refluxed to obtain MPEG-carboxylic acid. MPEG-carboxylic acid or PEG bis(carboxyl methyl) ether was activated to acyl chloride form using oxalylchloride and then conjugated to PEI in dichloromethane containing TEA under Ar gas. After evaporation of solvents, the viscous product was solubilized in water and dialyzed using Spectra/Por dialysis membrane (MWCO = 3400) against water and lyophilized.

10.2.3.2 *¹H NMR*

The ratio of PEG and PEI in the copolymer sample was determined from ¹H NMR spectra using integral values obtained from the number of -CH_2CH_2O- protons of PEG and -CH_2CH_2N-

Table 10.3 Characteristics of MPEG-PEI and PEG-PEI Copolymers

| Copolymer Name[a] | Reacting Polymers[b] | | Molecular Characteristics of the Synthesized Copolymers | | | | |
	PEG, Da	PEI, kDa	PEG[c]	N[d]	PEG:PEI[e]	Modification Degree[f], %	Molecular Mass[f], kDa
PEG-PEI 1:1	600	25	40	501	0.97	13.8	49.3
PEG-PEI 3:1	600	25	133	315	3.20	45.8	105
MPEG-PEI 1:1	550	25	45	536	0.99	7.75	49.7
MPEG-PEI 3:1	550	25	127	454	2.79	21.9	94.8

[a] The nomenclature for PEG and PEI copolymers accounts for the ratio of PEG:PEI. [b] Molecular masses of the reacting polymers are presented as provided by the manufacturers. [c] The number of PEG. [d] The number of amino group in PEI. [e] As determined by ^{1}H NMR analysis of the copolymer samples. [f] Calculated based on ^{1}H NMR data assuming that all polyether chains in the copolymer samples are linked to the PEI.

protons of PEI. MPEG-PEI (1:1 and 3:1) and PEG-PEI (1:1 and 3:1) were obtained. The characterization data are described in Table 10.3.

10.3 SELF-ASSEMBLY OF BLOCK COPOLYMERS WITH PLASMID DNA

10.3.1 Electrophoretic Mobility Shift Assay

Agarose gel electrophoresis was performed to assess the polyionic complex formation of the copolymers with plasmid DNA (pDNA) and the results are presented in Figure 10.3. MPEG-PLLD of generation 3 (G3) (Figure 10.3A), PLLD-PEG-PLLD G3 (Figure 10.3C), and MPEG-PAMAM G3 (Figure 10.3E) polymers were not as efficient at forming retarded complexes under agarose gel electrophoresis experiments. However, MPEG-PLLD G4 (Figure 10.3B), PLLD-PEG-PLLD G4 (Figure 10.3D), MPEG-PAMAM G4 and G5 (Figures 10.3F and 10.3G), PAMAM-PEG-PAMAM G4 and G5 (Figures 10.3H and 10.3I) copolymers of either AB or ABA type containing dendritic block(s) of G4 or higher generation showed high efficiency at complexing and complete retardation with DNA. In viewing the results, the minimum number of primary amines per dendritic block should be at least 16 to produce an electro-neutralized complex with pDNA.

For PEG-PEI and MPEG-PEI copolymers, as shown in Figures 10.3J to 10.3M, even highly substituted derivatives were still able to form polyionic complexes with pDNA. It was concluded that the conjugated PEG groups do not hamper the electrostatic interaction between the copolymer and DNA.

10.3.2 Atomic Force Microscopy (AFM)

The morphology and size distribution of PEG-conjugated DNA/dendrimers complexes was analyzed by AFM. As shown in Figures 10.4A and 10.4B, partially associated structures were observed below the charge ratio sufficient to neutralize DNA, and the MPEG-PLLD copolymer formed spherical water-soluble nanoparticles with pDNA with a mean diameter of 154.4 nm. At a diameter ranging from ~0.5 to 1 μm, pDNA was efficiently condensed into nanometer-sized particles. The formation of complexes at the nanometer scale level is generally considered to be important in polyplex-mediated gene delivery. Compared to PLLD-PEG-PLLD G3 (Figures 10.4C and 10.4D), which could not form compact discrete particles even at high charge ratios, PLLD-PEG-PLLD G4 (Figures 10.4E and 10.4F) and PAMAM-PEG-PAMAM G5 (Figures 10.4G and 10.4H) showed spherical nano-sized complexes with DNA.

The reason such a low-generation copolymer forms partially complexed polyplexes could be explained as follows. As for lower generation copolymers, some part of the plasmid DNA may

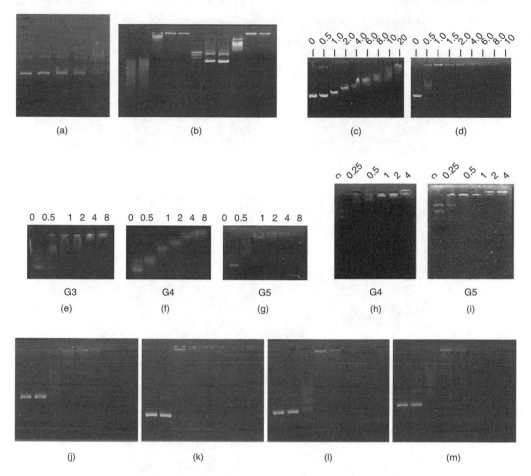

Figure 10.3 Analysis of complex formation of the block copolymers at various charge ratios by agarose gel electrophoresis. (a) MPEG-PLLD G3 1 μg of pSV-β-gal plasmid DNA only (lane 1), charge ratio of copolymer:DNA = 0.5, 1, 2, and 4 (lanes 2, 3, 4, and 5, respectively). (b) MPEG-PLLD G4 1.0 μg of sheared herring testes DNA only (lane 1),[23] charge ratio = 0.5, 1, 2, and 4 (lanes 2, 3, 4, and 5, respectively). Lane 6 is λ/Hind III cut DNA marker (23.1, 9.4, 6.5, 4.3 kbp). 1 μg of pSV-β-gal plasmid DNA only (lane 7), charge ratio of copolymer:DNA = 0.5, 1, 2, and 4 (lanes 8, 9, 10, and 11, respectively). The charge ratios are indicated at each lane (c–i). (c) PLLD-PEG-PLLD G3, (d) PLLD-PEG-PLLD G4, (e) MPEG-PAMAM G3, (f) MPEG-PAMAM G4, (g) MPEG-PAMAM G5, (h) PAMAM-PEG-PAMAM G4, (i) PAMAM-PEG-PAMAM G5. From j to m, lane 1 indicates pSV-β-gal plasmid only, and lanes 2–8 indicate polyplexes at charge ratios 0.1, 0.5, 1, 2, 4, 8, and 16, respectively. (j) MPEG-PEI 1:1, (k) MPEG-PEI 3:1, (l) PEG-PEI 1:1, and (m) PEG-PEI 3:1.

electrostatically interact with the copolymer, but more cationic charges are required to compensate fully for the excess negative phosphate anions of DNA backbones. However, at a certain concentration level, further copolymers could not be incorporated into the preformed polyplexes because PEG chains that already participated in the partially formed complexes might cause steric hindrance and prohibit additional copolymers from gaining access to the immature structure. As for higher generation copolymers, the charge density per copolymer is considered to be sufficient to fully compensate at a critical concentration level, and they could form mature polyplexes even at a lower concentration range than that of low-generation copolymers.

Based on these results, two possible self-assembly schemes between AB- or ABA-type PEG-conjugated cationic hybrid block copolymers and pDNA could be proposed, as presented in Figure 10.5. Therefore, each condensate forms electrostatically neutralized particles that are surrounded by multiple PEG chains.

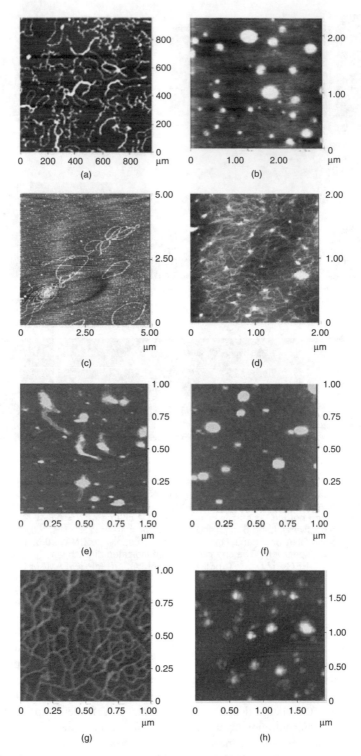

Figure 10.4   AFM images of DNA/copolymer complexes. MPEG-PLLD G4 at charge ratios 1 (a) and 2 (b). PLLD-PEG-PLLD G3 at charge ratios 4 (c) and 10 (d); G4 at charge ratios 2 (e) and 4 (f). PAMAM-PEG-PAMAM G5 at charge ratios 0 (g) and 2 (h).

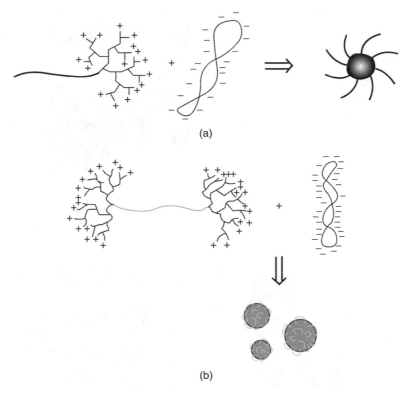

(a)

(b)

Figure 10.5 Schematic view of the formation of self-assembling complexes. The polyionic complexes are coated with hydrophilic PEG chains. (a) AB-type DNA–copolymer complexes, and (b) ABA-type DNA/copolymer complexes.

10.3.3 DNase I Protection Assay

It is interesting and important to characterize the stability of this polymeric micelle system mixed with DNA against enzymatic digestion that is present abundantly and ubiquitously *in vitro* and *in vivo* environments. In addition to maintaining the integrity of the complexed pDNA under such harsh conditions, it is also considered to be a prerequisite factor including the ability to release genetic materials at a desirable stage or site of action. It was reported that the protection efficiency of PEG-PLL copolymer was dependent on the length of linear PLL (i.e., the degree of polymerization).[21] First, we measured the difference in the UV absorbance value of DNA complexed with MPEG-PLLD′G4 after treating DNase I (Figure 10.6A). In comparison with pDNA, which had been degraded very rapidly within a few minutes by the enzyme, the DNA/polymer complex showed no change in absorbance. To further verify the integrity of the complexed DNA, it was purified from complexes treated with the enzyme and loaded onto agarose gel for electrophoresis (Figure 10.6B). As expected from the previous results, as the charge ratio increased, a greater amount of native DNA was obtained. In the case of PLLD-PEG-PLLD, G4 was more efficient in protecting DNA than G3, which results from the difference in complexing potency with DNA, as already confirmed by the agarose gel shift assay. This implies that the electrostatic interaction of G3 with DNA is weaker than that of G4 because G3 is lower in charge density than G4. Hence, the partially complexed DNAs with G3 are subject to enzymatic digestion of DNase I and show fragmented profiles in agarose gel electrophoresis data (Figure 10.6C).

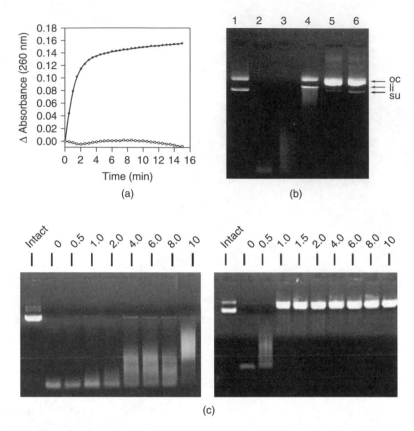

Figure 10.6 DNase I protection assay. (a) MPEG-PLLD/pSV-β-gal DNA complex. The difference in absorbance at 260 nm versus time was plotted. Plasmid DNA only (●), and complexes at charge ratio 4 (O). (b) Plasmid DNA only (lane 1), MPEG-PLLD/DNA at charge ratios 0, 0.5, 1, 2, and 4 (lanes 2, 3, 4, 5, and 6, respectively). The positions of the open circular (oc), linear (li), and supercoiled (su) forms are indicated on the right. (c) PLLD-PEG-PLLD G3/DNA (left), and G4/DNA (right). Charge ratios are indicated at each lane and the intact DNA is not treated with DNase I.

10.3.4 Zeta Potential Measurements

From the zeta potential experiments, we also expected that we might obtain important information regarding the mechanism of self-assembly between ABA-type copolymers and DNA as to whether two cationic dendritic blocks can cooperatively bind to DNA or not. If the binding is not cooperative, this means that the other cationic part which does not participate in complexing is extruded out of the polyplexes and may contribute to positive zeta potential values of the complexes. So, PLLD-PEG-PLLD G4 was used as a model polymer for performing the experiments after producing complexes with DNA at different charge ratios. The PLLD-PEG-PLLD G4 and plasmid DNA were mixed at four different charge ratios and zeta potential values were measured at physiological conditions. Interestingly, the values increased from -25 mV to around 0 with an increase of charge ratio from 0.5 up to 8.[14] It was postulated that each end of the positively charged dendritic block cooperatively interacts with pDNA, forming polyionic particulates, and that the surfaces of the complexes are shielded by cross-linked PEG strains. These results are also supportive with respect to the self-assembly between ABA-type copolymers and DNA, as already presented in Figure 10.5B.

10.4 EVALUATION OF CYTOTOXICITY *IN VITRO*

As shown in Figure 10.7, cytotoxicity was determined by comparing the amount of tetrazolium dye (MTT) reduced by cells treated with copolymers with that reduced by control cells.[22] First, a test of the time- and concentration-dependent cytotoxicity of MPEG-PLLD was performed; the results are shown in Figure 10.7A. The cells were exposed to each polymer for 4 or 24 h and the toxicity for each condition was presented. For 4 h incubation, PLL was slightly toxic to the cells, causing ~80% viability. However, when the incubation time is prolonged to 24 h, PLL caused significant toxicity to the cells even at the level of 10 µg/ml. However, the MPEG-PLLD copolymer was shown not to have any influence on cell viability for either condition even at a higher concentration level of 200 µg/ml. This is quite an outstanding characteristic of the PEG-conjugated copolymer, since there was no harm done to the cells even at highly elevated concentrations.

The *in vitro* assay was also performed for PLLD-PEG-PLLD G3 and G4 (Figure 10.7B), MPEG-PEI, and PEG-PEI (Figure 10.7C). The results indicated that all PEG-conjugated block copolymers showed much reduced or little toxicity even at high concentration levels. MPEG-PAMAM and PAMAM-PEG-PAMAM showed little toxicity profiles *in vitro*. These results are very promising characteristics of the PEG-based polymeric delivery systems for *in vivo* use, because satisfying the toxicity criteria is a prerequisite and is of great importance for clinical settings.

10.5 APPLICATION TO *IN VITRO* TRANSFECTION

The conjugated PEG chain of the block copolymer was originally considered to possess a putative fusogenic activity. However, the transient gene expression level was observed to be too low to be detected for MPEG-PLLD. The transfection efficiency of PLL and MPEG-PLLD *in vitro* was tested for 293 human embryonic kidney cells. The expressed β–galactosidase in cell lysates hydrolyzes ONPG and produces a yellow color. The absorbance at 405 nm was compared, as presented in Figure 10.8A. In comparison with PLL, MPEG-PLLD showed a very low level of transfection efficiency. In considering the results, it is more likely that the PEG caused interference in the interaction of the polyionic complex with the cell membranes, resulting in a reduced rate of endocytosis.

Interestingly, PLLD-PEG-PLLD G4 showed a very small but detectable level of gene transfection efficiency for HepG2 cells (Figure 10.8B). We were encouraged to introduce a PAMAM dendrimer, a dendritic cationic block unit that has repeating internal tertiary amino groups inside the backbone of the structure. Usually, PAMAM dendrimer showed a much higher level of transfection efficiency than PLL, and this might result from the fact that the endogenous tertiary amines also function as a buffer system in a similar way to the "endosome buffering effect" of PEI, which is one of the most potent polymeric gene carriers reported so far. As shown in Figure 10.8E, PAMAM-PEG-PAMAM G5 showed a much increased level of transfection efficiency, which is comparable to that of PEI.[16]

For PEG-PEI or MPEG-PEI, as assessed by agarose gel electrophoresis, even highly substituted PEI derivatives were still able to form polyionic complexes with DNA. However, aside from retention of the ability to condense DNA, the transfection efficiency of PEG-PEI at 4 h incubation in the absence of FBS decreased with an increase in the amount of PEG coupled (Figure 10.8C) and, as expected, PEG-PEI showed a higher level of transfection efficiency than that of native PEI at the 48 h transfection condition with a high concentration of DNA/copolymer complexes, which proved that PEG conjugation decreased the toxicity of PEI and retained transfection potency. Interestingly, the decrease was also remarkable for MPEG-PEI derivatives, resulting in a significant decrease in the transfection activity of the DNA complexes at both 1:1 and 3:1 derivatives (Figure 10.8D). In fact, the efficiency of MPEG-PEI was not detectable even at a low degree of modification,

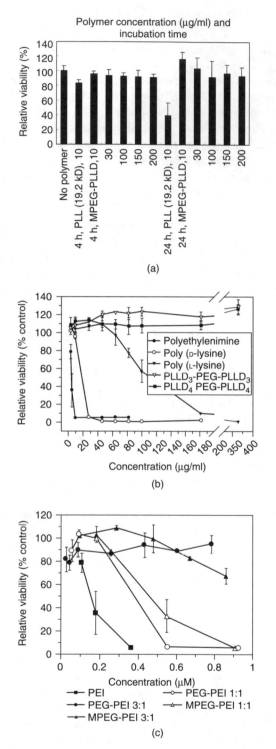

Figure 10.7 Effect of copolymers on cell viability. (a) poly(L-lysine) and MPEG-PLLD G4 on the viability of 293 cells. (b) PEI, poly(D-lysine), poly(L-lysine), PLLD-PEG-PLLD G3 and G4 on NIH3T3 cells. (c) PEI, PEG-PEI 1:1, 3:1, and MPEG-PEI 1:1 and 3:1 on NIH3T3 cells. Relative viability is expressed considering the absorbance at 570 nm of intact cells as 100%.

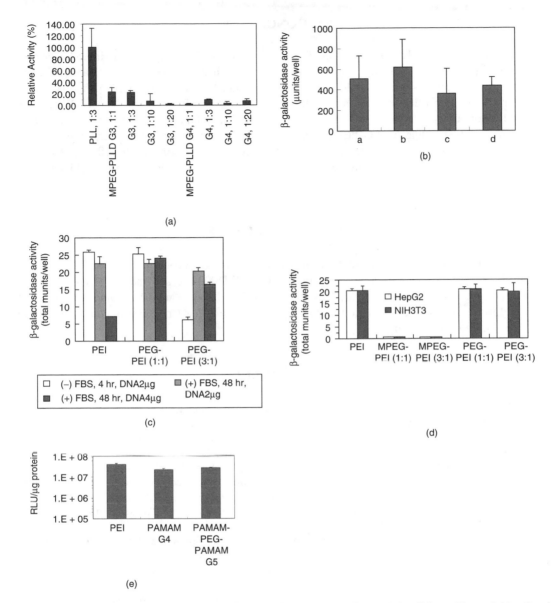

Figure 10.8 Transfection efficiency of copolymers. (a) MPEG-PLLD G4 on 293 cell lines. The weight ratio of poly(L-lysine) to DNA was 3. Various weight ratios of DNA:MPEG-PLLD (G3 and G4) were tested with 100 μM chloroquine in the absence of FBS for 4 h as indicated. (b) PLLD-PEG-PLLD G4 on NIH3T3 cells. (a) + 100 μM chloroquine, - FBS, charge ratio 4 for 4 h; (b) + 100 μM chloroquine, - FBS, charge ratio 6 for 4 h; (c) - chloroquine, + FBS, charge ratio 4 for 48 h; (d) - chloroquine, + FBS, charge ratio 6 for 48 h transfection. (c) PEG-PEI on NIH3T3 cells. (d) MPEG-PEI, PEG-PEI on HepG2 and NIH3T3 cells for 48 h transfection in the presence of serum. (e) PAMAM-PEG-PAMAM G5 at charge ratio 4 on 293 cells for 4 h in the absence of serum.

suggesting that the PEG action resulting from its type of conjugation is important for gene transfer efficiency. The results are in agreement with recent research reported by Nguyen et al.,[10] and it is presumed that the free linear PEG chains surrounding MPEG-PEI/DNA complexes may hinder the interaction between the polyplexes and cell membranes. It was found that the transfection efficiency and cytotoxicity of PEI was triggered by the mode of PEG grafting and the degree of modification.

10.6 CONCLUSION

The synthesis and polyplex formation of various hybrid block copolymers of either AB or ABA type combining linear PEG and globular dendrimers, such as PLLD, PAMAM, and PEI, was attempted. AB-type copolymers could form polymeric micelles with plasmid DNA that have core-shell structures with PEG extruding out of their surfaces. Both PEG-PEI and MPEG-PEI showed reduced cytotoxicity compared to PEI, with only PEG-PEI retaining transfection activity at the same level as PEI, while MPEG-PEI showed little efficiency. The AB-type copolymers usually showed very low or little efficiency in *in vitro* transfection. Interestingly, the ABA type showed better potency, and a relatively higher level of transfection was achieved for PAMAM-PEG-PAMAM G5.

ACKNOWLEDGMENTS

This work was supported by grants from the Basic Research Program of KOSEF (R02-2002-000-00011-0), the Korea Research Foundation (2001-15-DP0344), and the Research Center for Molecular Therapy in Sungkyunkwan University from KOSEF. The authors also wish to acknowledge helpful assistance received from Dong Kyoon Joo, Jin Hee Choi, Hyung-suk Jang, Chang Hwan Kim, and Kwan Kim.

REFERENCES

1. Boussif, O. et al., A versatile vector for gene and oligonucleotide transfer into cells in culture and *in vivo*: polyethylenimine, *Proc. Nat. Acad. Sci. U.S.A.* 92, 7297, 1995.
2. Kukowska-Latallo, J.F. et al., Efficient transfer of genetic material into mammalian cells using starburst polyamidoamine dendrimers, *Proc. Nat. Acad. Sci. U.S.A.* 93, 4897, 1996.
3. Luo, D. and Saltzman, W.M., Synthetic DNA delivery systems, *Nat. Biotechnol.* 18, 33, 2000.
4. Stiriba, S.-E., Frey, H., and Haag, R., Dendritic polymers in biomedical applications: from potential to clinical use in diagnostics and therapy, *Angew. Chem. Int. Ed.* 41, 1329, 2002.
5. Harris, J.M. (Ed.), *Poly(ethylene glycol) Chemistry: Biotechnical And Biomedical Applications*, Plenum Press, New York, 1992.
6. Wolfert, M.A. et al., Characterization of vectors for gene therapy formed by self-assembly of DNA with synthetic block co-polymers, *Hum. Gene Ther.* 7, 2123, 1996.
7. Kataoka, K. et al., Block copolymer micelles as vehicles for drug delivery, *J. Control. Release* 24, 119, 1993.
8. Katayose, S. and Karaoka, K., Water-soluble polyion complex associates of DNA and poly(ethylene glycol)-poly(L-lysine) block copolymer, *Bioconjugate Chem.* 8, 702, 1997.
9. Kakizawa, Y. and Kataoka, K., Block copolymer micelles for delivery of gene and related compounds, *Adv. Drug Deliv. Rev.* 54, 203, 2002.
10. Nguyen, H.-K. et al., Evaluation of polyether-polyethyleneimine graft copolymers as gene transfer agents, *Gene Ther.* 7, 126, 2000.
11. Tam, P. et al., Stabilized plasmid-lipid particles for systemic gene therapy, *Gene Ther.* 7, 1867, 2000.
12. Luo, D. et al., Poly(ethylene glycol)-conjugated PAMAM dendrimer for biocompatible, high-efficiency DNA delivery, *Macromolecules* 35, 3456, 2002.
13. Choi, J.S., Lee, E.J., Choi, Y.H., Jeong, Y.J., and Park, J.S., Poly(ethylene glycol)-*block*-poly(L-lysine) dendrimer: novel linear polymer/dendrimer block copolymer forming a water-soluble polyionic complex with DNA, *Bioconjugate Chem.* 10, 62, 1999.
14. Choi, J.S., Joo, D.K., Kim, C.H., Kim, K., and Park, J.S., Synthesis of a barbell-like triblock copolymer, poly(L-lysine) dendrimer-*block*-poly(ethylene glycol)-*block*-poly(L-lysine) dendrimer and its self-assembly with plasmid DNA, *J. Am. Chem. Soc.* 122, 474, 2000.

15. Kim, T., Jang, H., Joo, D.K., Choi, J.S., and Park J.-S., Synthesis of diblock copolymer, methoxypoly (ethylene glycol)-*block*-polyamidoamine dendrimer and its generation-dependent self-assembly with plasmid DNA, *Bull. Kor. Chem. Soc.* 24, 123, 2003.
16. Kim, T., Seo, H.J., Jang, H., Choi, J.S., and Park, J.S., manuscript in preparation.
17. Choi, J.H., Choi, J.S., Suh, H., and Park, J.S., Effect of poly(ethylene glycol) grafting on polyethylenimine as a gene transfer vector *in vitro*, *Bull. Kor. Chem. Soc.* 22, 46, 2001.
18. Bayer, E. and Mutter, M., The liquid-phase method for peptide synthesis, in *The Peptides*, Gross, E., and Meienhofer, J. (Eds.), Academic Press, New York, 1979, 285.
19. Chapman, T.M. et al., Hydraamphiphiles: novel linear dendritic block copolymer surfactants, *J. Am. Chem. Soc.* 116, 11195, 1994.
20. Iyer, J., Fleming, K., and Hammond, P.T., Synthesis and solution properties of new linear-dendritic diblock copolymers, *Macromolecules* 31, 8757, 1998.
21. Katayose, S. and Kataoka, K., Remarkable increase in nuclease resistance of plasmid DNA through supramolecular assembly with poly(ethylene glycol)-poly(L-lysine) block copolymer, *J. Pharm. Sci.* 87, 160, 1998.
22. Mosman, T., Rapid colorimetric assay for cellular growth and survival: application to proliferation and cytotoxicity assays, *J. Immunol. Methods* 65, 55, 1983.
23. Choy, J.H. et al., Intercalative nanohybrids of nucleoside monophosphates and DNA in layered metal hydroxide, *J. Am. Chem. Soc.* 121, 1399, 1999.

CHAPTER 11

Water Soluble Lipopolymers for Gene Delivery

Ram I. Mahato and Sung Wan Kim

CONTENTS

11.1 INTRODUCTION

Plasmid-based gene delivery systems utilize synthetic gene carriers to condense and protect plasmid DNA (pDNA) from premature degradation during storage and transportation from the site of administration to the site of gene expression.[1-3] Plasmid DNA is condensed into a highly organized structure through a complex self-assembly process. Commonly utilized synthetic gene carriers are cationic lipids, polymers, and peptides that condense pDNA by virtue of their electrostatic interactions with the anionic phosphate backbone of the nucleic acid chain. Cationic copolymers synthesized by grafting polyethylenimine (PEI) with nonionic polymers, such as poly(ethylene oxide) (PEO) or Pluronic 123 have also been used for gene delivery.[4] However, noncondensing polymers, such as poloxamers and polyvinyl pyrrolidone (PVP), are also being investigated for gene delivery to muscle and tumor tissues[5,6] and will be discussed in a Chapters 19 and 20.

11.2 BASIC COMPONENTS OF CATIONIC LIPIDS

Most cationic lipids used for gene transfer have three parts: a hydrophobic lipid anchor group; a linker group, such as an ester, amide, or carbamate; and a positively charged head-group, which interacts with pDNA, leading to its condensation (Figure 11.1).[7] Hydrophobic group lipid anchors

0-8493-1934-X/05/$0.00+$1.50
© 2005 by CRC Press LLC

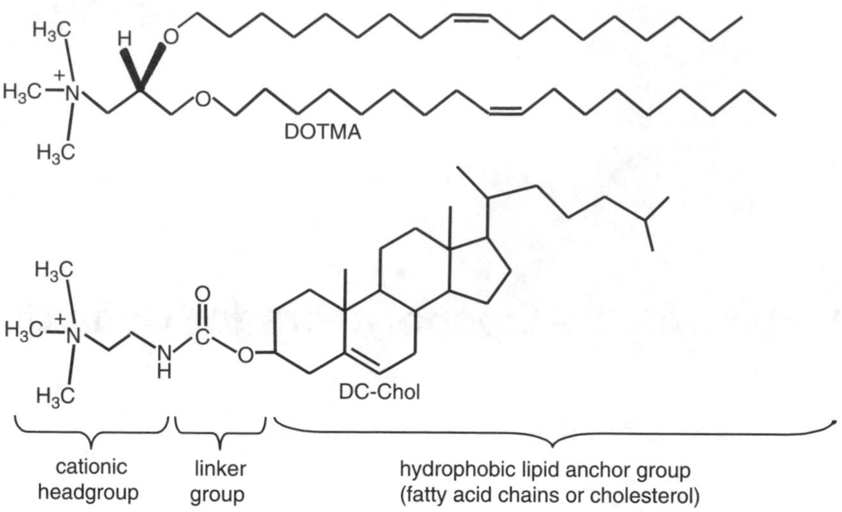

Figure 11.1 Basic components of cationic lipids 3-(N [N′, N′-dimethylaminoethane]–carbamoyl) cholesterol (DC-Chol) and N-(1-[2,3-dioleyloxy]propyl)-N,N,N-trimethylammonium chloride (DOTMA). (a) Hydrophobic lipid group; (b) linker group; (c) cationic head-group.

can be either a cholesterol group or fatty acid chains. The linker group is an important component that determines the chemical stability and biodegradability of the lipid. The linker groups should be biodegradable yet strong enough to survive in a biological environment. Ester linkage between hydrophobic lipid anchors and cationic head-groups is likely to provide biodegradability to cationic lipids.

Among all the basic components of the cationic lipid, the type of head-group has been shown to have a dominant role in transfection efficiency and cytotoxicity. Cholesterol is a naturally occurring lipid and is metabolized in the body. The early success of 3-(N [N′, N′-dimethylamino-ethane]–carbamoyl) cholesterol (DC-Chol) lipid-based gene delivery system spurred interest in the development of novel cholesterol-based cationic lipids.[8] Cationic lipids in T-shape head-groups tend to be more effective than linear counterparts. The levels of gene expression obtained with spermine cholesteryl carbamate and spermidine cholesteryl carbamate in T-shape orientation was 50- to 100-fold higher both *in vitro* and *in vivo* than that observed with DC-Chol, which has only a single protonatable amine.[9]

To combine the advantages of lipids and polycations, Zhou et al.[10] synthesized lipopolysine by mixing poly(L-lysine) of 3300 Da with two molar equivalents of N-hydroxysuccinimide ester of dipalmitoylsuccinylglycerol in dimethyl sulfoxide. Similarly, Choi et al.[11] synthesized 3-(L-lysinea-mide carbamoyl) cholesterol (K-Chol) and 3-(L-ornithinamide carbamoyl) cholesterol (O-Chol) by the solid phase synthesis method. However, liposome preparation with colipid dioleoyl phosphati-dylethanolamine (DOPE) was essential for enhanced gene transfer by these cationic amphiphiles. Yamazaki et al.[12] grafted cetyl groups as hydrophobic lipid anchors onto PEI of 1,800 and 25,000 Da, and prepared polycation liposomes for gene transfer.

11.3 WHY WATER-SOLUBLE LIPOPOLYMERS?

Current cationic lipids are water insoluble and most of them require the formation of liposomes. Complex formation between pDNA and cationic liposomes usually produces large complexes that are poorly extravasated through the capillary endothelia. These liposome–plasmid complexes are efficient at transfecting cells in culture, yet are poor at diffusing within tumor tissues.[13] To prepare liposomes, lipids are dissolved in an organic solvent like chloroform followed by solvent evaporation for lipid film formation and extrusion to prepare small vesicles. The use of organic solvent may be detrimental to health.

PEI of 25 kDa or above has been reported to be effective for gene delivery, because pDNA can be delivered to the cytoplasm after endosomal disruption due to the proton-sponge effect of PEI, in which unprotonated amino moieties on the polymer buffer at the pH inside the endocytic vesicle.[14] However, high molecular weight PEI is toxic to the cells, and PEI–pDNA complexes are prone to aggregation.

To make gene therapy clinically acceptable, delivery systems, or vectors, must deliver DNA to the cells in a transcriptionally active form and fulfill all regulatory agency mandates to be considered safe for use in humans. Therefore, the development of an effective water-soluble nontoxic lipopolymer will be of great importance in gene delivery.

11.4 SYNTHESIS AND CHARACTERIZATION OF WATER-SOLUBLE LIPOPOLYMERS

We sought to develop novel lipopolymeric gene carriers based on a set of predefined design criteria. For efficient transfection and possible clinical application, synthetic gene carriers should meet the following criteria:

- DNA condensation capability
- Endosomolytic property
- Biocompatibility to minimize potential toxic effects on cells and tissue
- Facile synthesis and purification to allow large-scale commercial manufacture[15]

Based on these design criteria, we synthesized novel water-soluble cationic lipopolymers (WSCL) using low molecular weight branched PEI and cholesteryl chloroformate. The cholesterol moiety was used as a lipophilic portion grafted onto the branched PEI, which serves as a hydrophilic head-group due to its ionized primary amino groups in the aqueous environment. The cationic head-group, PEI has particular advantages over conventional cationic DNA condensing agents, such as the following: it can effectively condense pDNA into colloidal particles; it can enhance cellular uptake of pDNA by nonspecific adsorptive mechanisms; and it can enhance endosomal release of DNA due to its proton-sponge effect. As a hydrophobic anchor group, cholesterol can form a stable micellar complex with the hydrophilic head-group in the aqueous environment, and may provide shielding effect to WSCL–pDNA complexes against erythrocytes and plasma proteins at lower charge ratios.

WSCL was synthesized by direct conjugation of cholesteryl chloroformate to branched PEI of 1800 Da through primary amines (Figure 11.2).[16] Following synthesis and purification, we determined the structure and molecular weight of WSCL using ^{1}H NMR and MALDI-TOF mass spectrometry. The mean particle size of WSCL–pDNA complexes was in the range of 40–60 nm and was dependent on N/P ratios. The ζ potential was greatly dependent on the charge ratios and linearly increased with the increase in N/P ratios. It was -41.5 at the N/P ratio of 2.5, but increased to 61.67 mV when formulated at the N/P ratio of 20:1. WSCL–pDNA complexes were nontoxic to CT-26 colon carcinoma cells when formulated at an N/P ratio of 20 and below, whereas PEI25000–pDNA complexes were highly toxic to these cells. WSCL–pDNA complexes did not cause any aggregation of freshly prepared murine erythrocytes, whereas PEI25000–pDNA complexes aggregated the erythrocytes. The result suggests that cholesterol of WSCL may provide some shielding effect to WSCL–pDNA complexes and thus prevent erythrocyte aggregation.[16]

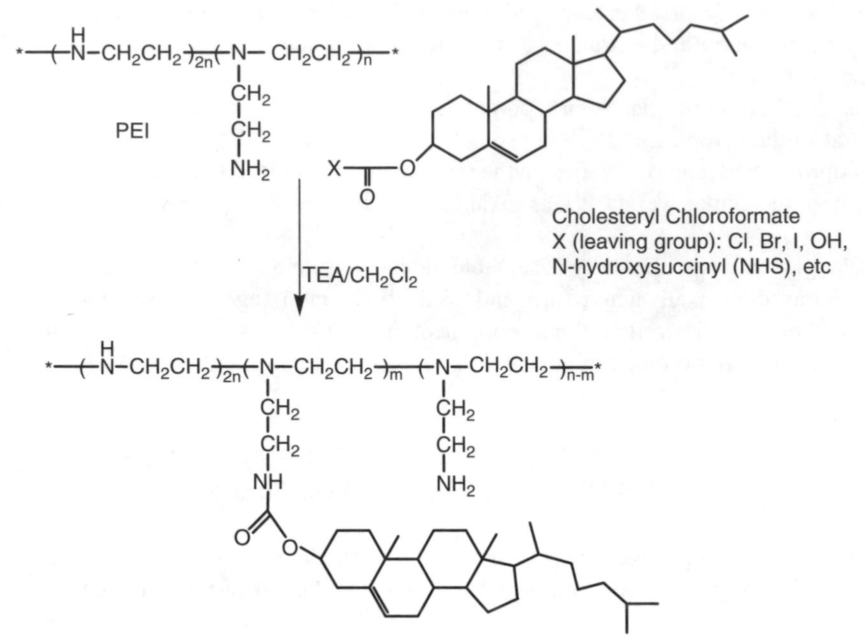

Figure 11.2   Synthesis of water-soluble cationic lipopolymer (WSCL) by direct conjugation of cholesteryl chloroformate to branched polyethylenimine (PEI) of 1800 Da.

11.5 WATER-SOLUBLE LIPOPOLYMERS FOR IL-12 GENE DELIVERY TO TUMORS

Due to high interstitial pressure inside the tumor, cationic liposome–pDNA complexes do not disperse well inside the tumors after direct injection, and gene expression is usually seen near the injection site. Intratumoral injection of DC-Chol cationic liposome–pmIFN-γ complexes into CT-26-tumor-bearing mice have been shown to produce a lower level of transgene expression than those injected with naked DNA.[17] Compared with naked pDNA, a lower level of luciferase expression has also been reported for PEI–pCMV-Luc complexes when injected into subcutaneous-tumor-bearing mice, presumably because of poor diffusion of the complexes within the tumor mass after injection.[18] Therefore, development of an effective water-soluble carrier system is important for gene delivery via local and systemic routes.

We determined the effect of N/P ratio on transfection efficiency of WSCL–p2CMVmIL-12 complexes in CT-26 cells. The levels of mIL-12 increased with an increase in N/P ratios from 5 to 15. No significant difference in the levels of mIL-12 was found when WSCL–p2CMVmIL-12 complexes were compared at the N/P ratios of 15:1, 20:1, 25:1, and 30:1.[19] Therefore, we decided to formulate the complexes at the N/P ratio of 20:1 to avoid any toxic effects that may occur *in vivo*.

We studied mIL-12 gene expression levels after intratumoral injection of WSCL–p2CMVmIL-12 complexes into BALB/c mice bearing CT-26 subcutaneous tumors. Mice injected with naked p2CMVmIL-12, WSCL alone, and 5% w/v glucose were used for comparison. At 48 h after injection, tumors were harvested, chopped into small pieces, and recultured for 24 h. Culture supernatants were analyzed for mIL-12 by ELISA. mIL-12 expression levels were higher than those of naked p2CMVmIL-12. To determine whether complex formation with WSCL can enhance the level and duration of mIL-12 gene expression, tumors were isolated, chopped, and cultured for IL-12 at days 1, 3, and 5 after injection of WSCL–p2CMVmIL-12. The mIL-12 expression level was the highest at day 1 and then decreased slowly with time. At day 3 after injection, IL-12 levels were higher compared with the naked DNA group (Figure 11.3).[19] Upon injection of WSCL alone

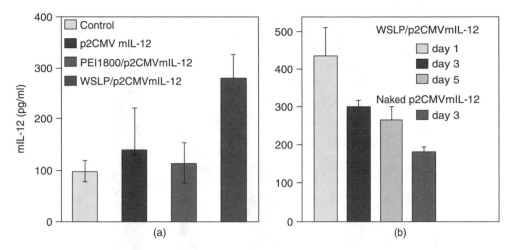

Figure 11.3 mIL-12 expression after intratumoral injection of WSCL–p2CMVmIL-12 complexes into subcutaneous CT-26-tumor-bearing mice. At 48 h postinjection, tumors were isolated, chopped into small pieces, and cultured. Supernatants were analyzed by ELISA. (Reproduced with permission from Mahato, R.I. et al., Intratumoral delivery of p2CMVmIL-12 using water-soluble lipopolymers, *Mol. Ther.* 4, 130–138, 2001.)

and 5% glucose, there was little induction of these cytokines, which confirms that the production was solely due to p2CMVmIL-12 and not due to any change or antigenecity associated with WSCL.

Compared with naked p2CMV-mIL-12, there was only a modest (1.5- to 2-fold) increase in IL-12 expression after intratumoral injection of WSCL–p2CMV-mIL-12 complexes into CT-26 subcutaneous-tumor-bearing BALB/c mice. We believe this is a significant finding, since WSCL is nontoxic and may be used for repeated injections. Although PVP has been used for intratumoral delivery of pmIL-12 into CT-26-tumor-bearing BALB/c mice, no comparison was made between naked pDNA in saline and PVP-based formulations.[5]

There is a strong induction of IFN-γ as a result of rIL-12 administration, which acts synergistically along with mIL-12 to generate a potent antitumor response. Low systemic levels of mIL-12 have been shown to induce mIFN-γ production, and through a feedback control mechanism, mIL-12 activates its own up-regulation through macrophages, dendritic cells, and other stimulatory factors induced by IFN-γ as part of adaptive immune response.[20] Therefore, we determined the induced levels of mIFN-γ after intratumoral injection of WSCL–p2CMV-mIL-12 complexes. The production of induced mIFN-γ was significantly higher compared with control mice injected with naked p2CMV-mIL-12 and 5% glucose (Figure 11.4).[19]

11.6 WHY SECONDARY AMINES FOR CONJUGATION OF CHOLESTEROL TO PEI?

Direct conjugation of cholesterol moiety to the primary amines of PEI is undesirable since the primary amino groups are positively charged at physiological pH and contribute to DNA condensation. Also, grafting of more than one cholesterol molecule onto PEI would result in an enhanced lipophilic surface of the condensing copolymer, leading to reduced aqueous solubility, DNA condensation ability, and endosomal release. Therefore, we recently utilized a novel design whereby only one cholesterol moiety was attached to each PEI molecule and conjugated specifically to the secondary amine. This was achieved by blocking the primary amines using benzenyloxycarbonyl (CBz) before cholesterol conjugation to PEI.

The NMR spectrum of T-shaped PEI-cholesterol (PEI-Chol) is shown in Figure 11.5. The spectrum exhibits only the statistical average degree of product composition. From this result alone we could not confirm that only one cholesterol moiety was grafted onto one PEI chain. Therefore,

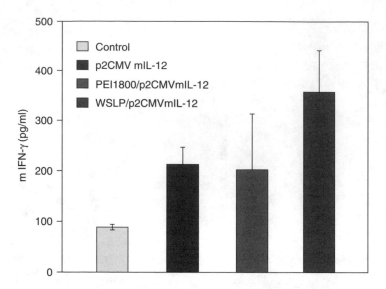

Figure 11.4   Levels of induced mIFN- after intratumoral injection of WSCL–p2CMVmIL-12 complexes into subcutaneous CT-26-tumor-bearing mice. At 48 h postinjection, tumors were isolated, chopped into small pieces, and cultured. Supernatants were analyzed by ELISA. (Reproduced with permission from Mahato, R.I. et al., Intratumoral delivery of p2CMVmIL-12 using water-soluble lipopolymers, *Mol. Ther.* 4, 130–138, 2001.)

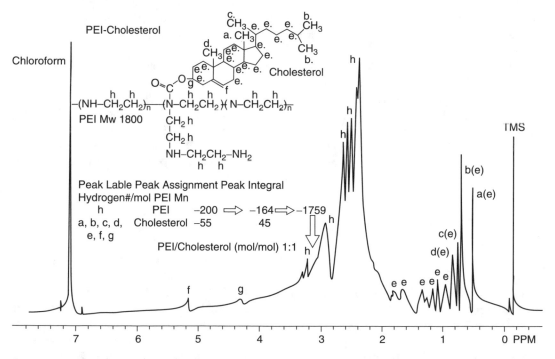

Figure 11.5   ¹H-NMR spectrum of PEI-Chol synthesized by conjugation of cholesteryl chloroformate through the secondary amines of branched PEI of 1800 Da.

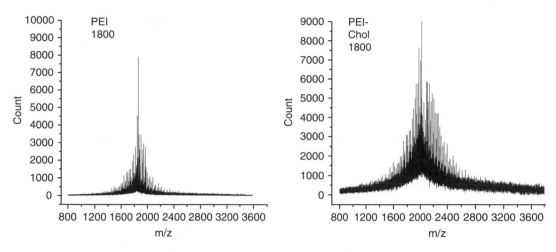

Figure 11.6 MALDI-TOF mass spectrometry of PEI 1800 (A) and PEI-Chol 1800 (B).

we determined the molecular weights of PEI-Chol 1800 and unmodified PEI 1800 using MALDI-TOF mass spectrometry (Figures 11.6A and 11.6B).[21] To study the effect of the molecular weight of the cationic head-group on transfection efficiency, branched PEI of 10,000 Da was also used for synthesis and the product was abbreviated as PEI-Chol 10,000. The mass spectra of PEI-Chol 1800 showed narrow distribution around 2100–2200 m/z, which was expected since the molecular weight of cholesterol is 386.67 Da and thus the molecular weight of PEI-Chol 1800 should be around 2200 Da if only one cholesterol was conjugated per 1800 Da PEI molecule. Furthermore, the mass spectra of PEI 1800 were concentrated around 1800 m/z (Figure 11.6A), suggesting that the conjugation of cholesterol to PEI 1800 was at a 1:1 ratio of PEI:cholesterol. If there were a lot of PEI fractions without any cholesterol moiety or with more than one cholesterol moiety, the overall molecular weight distribution of the products would inevitably be much larger than 2200 Da. However, the conjugation of cholesterol to the secondary amino group in the branch segments of PEI cannot be ruled out. We conclude that the cholesterol moiety grafted onto the PEI secondary amine on the backbone instead of the one on the branch segments is most likely because of the steric hindrance of the CBz group. Furthermore, the grafting location on different secondary amines would have little influence on the physicochemical properties and transfection efficiency of PEI-Chol.

11.7 CHARACTERIZATION OF T-SHAPED WATER-SOLUBLE LIPOPOLYMERS

Conjugation of cholesterol to PEI chains through the secondary amines is expected to result in a T-shaped PEI-Chol lipopolymer, which is likely to have different physicochemical properties and transfection efficiency compared to WSCL synthesized by direct conjugation to the PEI primary amines. Therefore, we determined the physicochemical properties of newly synthesized PEI-Chol by measuring critical micellar concentration (CMC), buffer capacity assay, and DNA condensation by circular dichroism (CD) spectroscopy. Further, we also evaluated the *in vitro* transfection efficiency and cytotoxicity of PEI-Chol–pDNA complexes in murine Jurkat T cell lines at different N/P ratios. In addition, we determined the effect of the orientation and molecular weight of the cationic head-group (PEI 1,800 and PEI 10,000) as well as the effect of serum on transfection efficiency and cytotoxicity.

PEI-Chol lipopolymer is amphiphilic in nature, because PEI is hydrophilic and water soluble, while cholesterol is hydrophobic. With the increase in its concentration, PEI-Chol may form multimolecular micelles or micellar aggregates in water. The micellar property of PEI-Chol depends

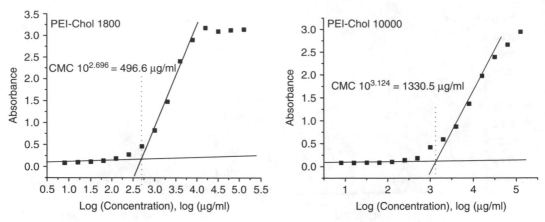

Figure 11.7   Determination of critical micellar concentration (CMC) of PEI-Chol by the dye solubilization method. (A) PEI-Chol 1,800; (B) PEI-Chol 10,000. The CMC values of PEI-Chol 1,800 and PEI-Chol 10,000 are 496.6 and 1330.5 µg/ml, respectively.

on the hydrophilic-hydrophobic balance between the cationic head-group and the lipid tail. 1,6-diphenyl-1,3,5-hexatriene (DPH) is a hydrophobic dye, which is sparingly soluble in water and has a significant UV absorbance at 356 nm. DPH can be solubilized by micelles due to its preferential partitioning into the core of the micelles, resulting in an increase in absorbance in the aqueous media. As shown in Figure 11.7, the CMC of PEI-Chol 1,800 was 496.6 µg/mL and that of PEI-Chol 10,000 was 1330.5 µg/mL. The relatively higher CMC of PEI-Chol 10,000 is due to the larger ratio of the hydrophilic moiety of PEI. Plasmid was condensed at a DNA concentration of 0.1 mg/ml and polymer concentrations of 0.016–0.472 mg/ml, which are below the CMC of these lipopolymers. Therefore, relatively few micellar structures were present in the formulation. At high N/P ratios, it is likely that some lipopolymers are present in the suspension of PEI-Chol–pDNA complexes in the free forms and can affect transfection and cytotoxicity.

The pH at which protonation of PEI takes place is believed to be a function of the pKa values of the primary, secondary, and tertiary amines: 6–7 for the tertiary amine, 8 for the sencondary amines, and 9 for the primary amine. However, in the case of PEI, the local environments for all the amino groups are different, resulting in different pKa values for different amino groups. The transfection properties of PEI rely on its protonation profile, which increases from 20 to 45% between pH 7 and 5, thus making the molecule a virtual "proton sponge." This not only tends to inhibit the action of endosomal nucleases, but also alters the osmolality of the endosome, thus leading to osmotic swelling and eventual rupture of the vesicle.[22] The buffering capacity of the lipopolymer is therefore measured in order to predict the endosomal release of pDNA to the cytoplasm. Both PEI and PEI-Chol showed similar acid-base titration curves, indicating that cholesterol conjugation to PEI did not affect the buffering capacity of PEI.

The CD spectra of pDNA change when pDNA interacts with cationic lipids or polymers. Free PEI-Chol 1,800 and PEI-Chol 10,000 showed little CD signal (Figure 11.8). However, the CD spectrum of naked pDNA presented a typical B-type secondary conformation, which possesses a symmetrical positive peak around 280–290 nm and a negative peak around 240–250 nm. This B geometry has 10 bases per turn, the base plane is perpendicular to the helix axis, and the rotation per residue is 30°. After the addition of PEI-Chol 1800 at an N/P ratio of 2.5:1, the B-type helical conformation began to collapse because of the formation of ion pairs between PEI-Chol and pDNA. Accordingly, the intensity of the positive peak decreased on the CD spectra. This suggests that pDNA within the complex prepared at an N/P of 2.5:1 turned into a C-type geometric conformation, which was similar to the B conformation with about 9.3 bases per turn. When the N/P ratio increased to 5:1, the helical assembly collapsed and a largely negative "tail" came into being on the CD curve at around 220 nm. This distortion implied the construction of -DNA, a highly organized left-handed

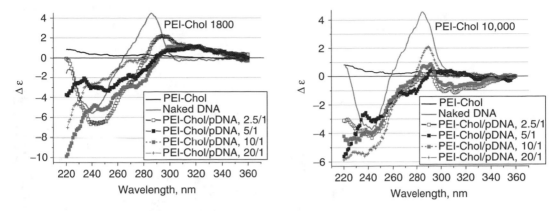

Figure 11.8 CD spectra of naked pDNA, PEI-Chol 1800 and PEI-Chol 1800–pDNA complexes.

chiral assembly (tertiary conformation) of DNA. This liquid-cholesteric-crystal structure of the pDNA assembly "within" the complex was strengthened when the N/P ratio was enhanced to 10:1, which was demonstrated by the dramatic increase of the negative peak and the -DNA "tail" on the spectrum (Figure 11.8). Nonetheless, when the N/P ratio was further enhanced to 20:1, the -tertiary conformation was finally melted, and the DNA assembly was turned into a stack (tertiary) conformation. The corresponding variation on CD spectra was the enhancement of the positive peak and flattening of the negative peak. For PEI-Chol 10,000–pDNA complexes, along with the increase in N/P ratio, the pDNA turned from B-type secondary conformation (naked DNA) to C-type at an N/P ratio of 2.5:1, and then -tertiary conformation came into being at an N/P of 5:1 and higher (Figure 11.8).

We next determined the effect of N/P ratios and the orientation and molecular weight of the cationic head-group on transfection efficiency of PEI-Chol–pCMS-EGFP complexes in murine Jurkat T-cell lines. Flow cytometry results showed the percentage of both GFP positive and viable cells. As shown in Figure 11.9, there was an increase in GFP positive cells with an increase in the N/P ratios of PEI-Chol 1800–pDNA complexes. This could be attributed to the increase in overall positive charges of the complex, which helps in transient cell membrane destabilization, and permeation of the complex into the cells. There was decrease in cell viability with an increase in the N/P ratios. There is, hence, a fine balance between transfection efficiency and cytotoxicity.

The molecular weight of a polymeric gene carrier plays an important role on its transfection efficiency. Therefore, we repeated our transfection experiments using PEI-Chol 10,000–pDNA complexes, while using pDNA complexed with PEI of 1,800, 10,000, and 25,000 Da as positive controls. Unlike PEI-Chol 1,800-based formulations, the transfection efficiency of PEI-Chol 10,000–pDNA complexes was similar to that of PEI 10,000–pDNA complexes (Figure 11.9). There was an increase in transfection efficiency with an increase in the N/P ratios. In case of unmodified PEIs, there was an increase in both transfection efficiency and cytotoxicity with an increase in their molecular weights. PEI 25,000–pDNA complexes showed the highest transfection and cytotoxicity when the complexes were prepared at the N/P ratios of 15:1 and used at a dose of 2 µg pDNA equivalent per 2 million Jurkat cells. The fact that the PEI 25,000–pDNA complexes yielded higher expression implies that the larger PEI molecules afford either better protection to the plasmids or better cellular uptake of the complexes.

Godbey et al. (1999) also compared the transfection efficiency of PEIs of 70, 10, and 1.8 kDa and demonstrated increased transfection with increase in PEI molecular weights.[23] Since one cholesterol molecule was conjugated to PEI of 10,000 Da, we think the influence of cholesterol on hydrophobicity and complex stability would be minimal. This may be the main reason why, unlike PEI-Chol 1,800, there was no increase in transfection when cholesterol was conjugated to PEI of 10,000 Da (Figure 11.9). Increase in the net positive charges promotes the cellular uptake

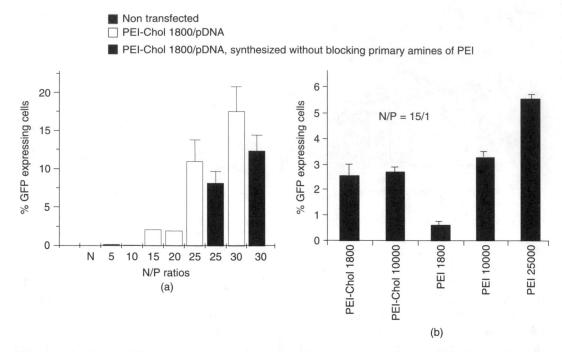

Figure 11.9 Effect of N/P ratio (A) and molecular weight (B) on *in vitro* transfection of PEI-Chol 1800–pDNA complexes into murine Jurkat T cells. To determine the effect of molecular weight on transfection, PEI–pDNA and PEI-Chol–pDNA complexes were prepared at the N/P ratios of 15:1 and used for transfection. The results are expressed as the mean ± SD of the percentage of GFP positive cells determined using flow cytometry. Two micrograms of pDNA were used per two million Jurkat cells. PEI-Chol 1800 was synthesized by conjugating cholesterol to the primary amines of PEI or by first blocking the primary amines with CBz so that cholesterol could be conjugated to the secondary amines of PEI. "Unblocked" means the PEI-Chol, which was synthesized without blocking the primary amines of PEI using CBz.

of the complexes, but also results in destabilization and loss of integrity of the cell membranes, especially at higher concentration of the complexes and high charge ratios, as well as high molecular weight of the PEI moiety — all of which have a common result, namely an increase in the positive charge density received by the cells being transfected.

The purpose of grafting the cholesterol group onto the secondary amine was simultaneously keeping the primary amine for more effective DNA condensation and producing another tertiary amine for more effective endosomal escaping. However, our results showed only modest (1.5-fold) increase in transfection when T-shape grafted PEI-Chol 1800 was used for transfection, as opposed to PEI-Chol 1800 synthesized by direct conjugation of cholesterol to the primary amines of PEIs (Figure 11.9). The transfection efficiency of our lipopolymers is not as high as we would have liked. However, PEI-Chol is a water-soluble lipopolymer and is nontoxic to a variety of cells.

We also studied the effect of fetal bovine serum on the transfection efficiency of PEI-Chol and PEI. As expected, the presence of 10% serum in the transfection media significantly diminished the transfection efficiency of PEI and PEI-Chol, resulting in less than 1% GFP positive cells. However, the cell viability was significantly higher when Jurkat cells were incubated with the PEI–pDNA complexes in presence of 10% serum. Our results are in good agreement with most of the cationic lipids and polymers currently used for gene delivery.[24,25]

11.8 CONCLUSION

In summary, development of an efficient water-soluble lipopolymer will play an important role in gene delivery after local and systemic administrations. We designed a water-soluble lipopolymer using the PEI secondary amines for cholesterol conjugation. While we have achieved significant advantage with respect to making a water-soluble lipopolymer for gene delivery without the use of organic solvents and with enhanced efficiency of transfection, further strides in nonviral gene therapy will come only through the use of a mechanism for cell membrane adhesion of the complexes and enhanced cellular permeability. Effect of lipid tail (cholesterol vs. fatty acids) and fusogenic peptides on the transfection efficiency of these lipopolymers have also been tested by several research groups. For example, Thoma and Klibanov conjugated different fatty acid chains to branched PEI of 2,000 and 25,000 Da and showed enhancement in transfection efficiency.[26] In a similar line, we have also synthesized a water-soluble lipopolymer by linking phosphatidyl ethylene glycol (PhosEG) to the amino groups of branched PEI. In comparison to cholesterol, the molecular weight of PhosEG is almost double (mol wt 387 vs. 767). Moreover, PhosEG has a fatty acid chain that is similar to a cell membrane, which may provide these gene carriers with fusogenic properties with the cell membrane.

ACKNOWLEDGMENTS

We would like to acknowledge the University of Tennessee Health Science Center for financial support to Ram I. Mahato's lab, and Expression Genetics, Inc. for financial support. We would also like to thank Sang-oh Han, Anurag Maheswari, Dong-an Wang, Ajit Narang, Neeraj Kumar, and Zhaoyang Ye for technical assistance.

REFERENCES

1. Mahato, R.I. and Kim, S.W. (Eds.), *Pharmaceutical Perspectives of Nucleic Acid-Based Therapeutics*, Taylor and Francis, London, 2002.
2. Mahato, R.I., Takakura, Y., and Hashida, M., Nonviral vectors for *in vivo* gene delivery: physicochemical and pharmacokinetic considerations, *Crit. Rev. Ther. Drug Carrier Syst.* 14, 133–172, 1997.
3. Mahato, R.I., Smith, L.C., and Rolland, A., Pharmaceutical perspectives of nonviral gene therapy, *Adv. Genet.* 41, 95–156, 1999.
4. Nguyen, H.K. et al., Evaluation of polyether-polyethyleneimine graft copolymers as gene transfer agents, *Gene Ther.* 7, 126–138, 2000.
5. Mendiratta, S.K. et al., Intratumoral delivery of IL-12 gene by polyvinyl polymeric vector system to murine renal and colon carcinoma results in potent antitumor immunity, *Gene Ther.* 6, 833–839, 1999.
6. Lemieux, P. et al., A combination of poloxamers increases gene expression of plasmid DNA in skeletal muscle, *Gene Ther.* 7, 986–991, 2000.
7. Mahato, R.I., Rolland, A., and Tomlinson, E., Cationic lipid-based gene delivery systems: pharmaceutical perspectives, *Pharm. Res.* 14, 853–859, 1997.
8. Gao, X. and Huang, L., A novel cationic liposome reagent for efficient transfection of mammalian cells, *Biochem. Biophys. Res. Commun.* 179, 280–285, 1991.
9. Lee, E.R. et al., Detailed analysis of structures and formulations of cationic lipids for efficient gene transfer to the lung, *Hum. Gene Ther.* 7, 1701–1717, 1996.
10. Zhou, X. and Huang, L., DNA transfection mediated by cationic liposomes containing lipopolylysine: characterization and mechanism of action, *Biochim. Biophys. Acta* 1189, 195–203, 1994.
11. Choi, J.S., Lee, E.J., Jang, H.S., and Park, J.S., New cationic liposomes for gene transfer into mammalian cells with high efficiency and low toxicity, *Bioconjug. Chem.* 12, 108–113, 2001.

12. Yamazaki Y. et al., Polycation liposomes, a novel nonviral gene transfer system, constructed from cetylated polyethylenimine, *Gene Ther.* 7, 1148–1155, 2000.

13. Nomura, T. et al., Intratumoral pharmacokinetics and *in vivo* gene expression of naked plasmid DNA and its cationic liposome complexes after direct gene transfer, *Cancer Res.* 57, 2681–2686, 1997.

14. Boussif, O. et al., A versatile vector for gene and oligonucleotide transfer into cells in culture and *in vivo*: polyethylenimine, *Proc. Nat. Acad. Sci. U.S.A.* 92, 7297–7301, 1995.

15. Pack, D.W., Putnam, D., and Langer, R., Design of imidazole-containing endosomolytic biopolymers for gene delivery, *Biotechnol. Bioeng.* 67, 217–223, 2000.

16. Han, S., Mahato, R.I., and Kim, S.W., Water-soluble lipopolymer for gene delivery, *Bioconjug. Chem.* 12, 337–345, 2001.

17. Nomura, T. et al., Gene expression and antitumor effects following direct interferon (IFN)-gamma gene transfer with naked plasmid DNA and DC-chol liposome complexes in mice, *Gene Ther.* 6, 121–129, 1999.

18. Coll, J.L. et al., *In vivo* delivery to tumors of DNA complexed with linear polyethylenimine, *Hum. Gene Ther.* 10, 1659–1666, 1999.

19. Mahato, R.I. et al., Intratumoral delivery of p2CMVmIL-12 using water-soluble lipopolymers, *Mol. Ther.* 4, 130–138, 2001.

20. Brunda, M.J. et al., Role of interferon-gamma in mediating the antitumor efficacy of interleukin-12, *J. Immunother. Emphasis Tumor Immunol.* 17, 71–77, 1995.

21. Wang, D.A. et al., Novel branched poly(ethylenimine)-cholesterol water-soluble lipopolymers for gene delivery, *Biomacromolecules* 3, 1197–1207, 2002.

22. Ferrari, S. et al., Polyethylenimine shows properties of interest for cystic fibrosis gene therapy, *Biochim. Biophys. Acta* 1447, 219–225, 1999.

23. Godbey, W.T., Wu, K.K., and Mikos, A.G., Size matters: molecular weight affects the efficiency of poly(ethylenimine) as a gene delivery vehicle, *J. Biomed. Mater. Res.* 45, 268–275, 1999.

24. Zelphati, O., Uyechi, L.S., Barron, L.G., and Szoka, F.C., Jr., Effect of serum components on the physico-chemical properties of cationic lipid/oligonucleotide complexes and on their interactions with cells, *Biochim. Biophys. Acta* 1390, 119–133, 1998.

25. Yang, J.P. and Huang, L., Time-dependent maturation of cationic liposome-DNA complex for serum resistance, *Gene Ther.* 5, 380–387, 1998.

26. Thomas, M. and Klibanov, A.M., Enhancing polyethylenimine's delivery of plasmid DNA into mammalian cells, *Proc. Nat. Acad. Sci. U.S.A.* 99, 14640–14645, 2002.

Cyclodextrin-Containing Polymers for Gene Delivery

Suzie Hwang Pun and Mark E. Davis

CONTENTS

12.1 INTRODUCTION

Clinical gene therapy has been largely dominated by viral approaches to gene delivery. As natural gene carriers, viruses have demonstrated success in limited gene delivery applications, most notably in the treatment of severe combined immunodeficiency. However, to date, viral systems are still haunted by difficulties in host immunogenicity, safety, manufacturing, and scale-up. Synthetic delivery systems have the potential to overcome these problems (materials that are not peptide-based are generally not recognized by the adaptive immune system, and large-scale chemical

0-8493-1934-X/05/$0.00+$1.50
© 2005 by CRC Press LLC

synthesis is routine in many industries) but have their own issues to address; e.g., toxicity, efficiency, and *in vivo* stability. It is clear that viruses exploit a variety of methodologies to deliver nucleic acids effectively, and complexity is certainly one key to their high transduction efficiency. Many viral proteins work together to overcome barriers to reaching the cell nucleus, and a mutation in one protein can significantly impair successful cell infection. Similarly, it is becoming obvious that nonviral delivery systems need to incorporate multiple functionalities or components to be efficacious *in vivo*.[1] It is important in the design of nonviral systems that these functionalities are included in a synthetically straightforward manner. Otherwise, the advantages in manufacturing and scalability are lost. The objective of our research is to develop a self-assembling, polymeric delivery system for *in vitro*, *ex vivo*, and *in vivo* gene delivery applications.

The following design strategy was adopted in order to meet our objective. First, a new family of linear, cyclodextrin-based polycations (CDPs) was synthesized. The polycations self-assemble with anionic nucleic acids via electrostatic interactions and condense them into small particles. This self-assembly approach, driven by charge interactions, results in simple formulation procedures that allow the system to be applied to different nucleic acids without alterations in the delivery vectors. These nanoparticles, which are similar in size to most viruses, provide the core of the delivery system. When formulated at net positive charges, the particles are endocytosed by cells via nonspecific pathways that use surface proteoglycan interactions. When modified with a targeting ligand, the particles rely on specific cell surface receptor interactions with the particles to be internalized (receptor-mediated endocytosis). The first section of this chapter describes the design, synthesis, and *in vitro* application of CDPs. Extensive structure-function studies were conducted to determine an optimal polymer structure for gene delivery while maintaining minimal toxicity. In addition, several general design principles for polymeric gene delivery materials were elucidated. Variations of the polymer that include pH-sensitive moieties were also synthesized to enhance intracellular trafficking. The second section describes polycations modified with pendant cyclodextrin, including work from the Uekama Lab (Kumamoto University) on cyclodextrin-modified polyamidoamine dendrimers, and cyclodextrin-modified polyethylenimine polymers (CD-PEIs) synthesized at Insert Therapeutics. These polycations serve the same purpose as the CDPs: self-assembling with and condensing nucleic acids to nanoparticles, thereby protecting against nucleases and facilitating cellular uptake.

The second family of materials, the adamantane (AD) conjugates, provides the potential of introducing multiple functionalities to the nanoparticle surface in a simple, modular approach. The AD conjugates self-assemble onto the particle surfaces by inclusion complex formation between the AD and the cyclodextrin. The formulation process, along with three applications (providing salt stability, tuning surface charge, and targeting to hepatocytes), is discussed in detail in the third section of this chapter. These modifications endow the nanoparticles with properties that make them appropriate for *in vivo* applications. *In vivo* studies using tumor-targeted nanoparticles, including tolerability, biodistribution, and expression data, are presented in the last section of this chapter.

12.2 LINEAR, CYCLODEXTRIN-BASED POLYMERS

12.2.1 Cyclodextrins

Cyclodextrins (CDs) are cyclic oligomers of 6, 7, or 8 glucose units (called α-, β-, and γ-cyclodextrin, respectively). The structure of β-CD is shown in Figure 12.1. Cyclodextrins are cup-shaped molecules that have a hydrophobic cavity and a hydrophilic exterior. Cyclodextrins are water soluble due to their hydrophilicity, and they have the ability to form complexes with hydrophobic guest molecules by inclusion in their core. This ability is exploited by the pharmaceutical industry, which uses CD and CD derivatives as solubilizing agents for lipophilic small molecule drugs. As such, the safety of CDs in humans is well established. CDs are pharmacologically inactive

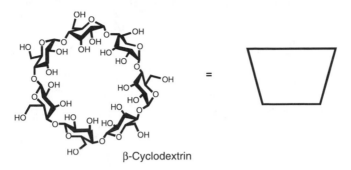

Figure 12.1 Structure of β-cyclodextrin.

and nonimmunogenic, have relatively low toxicity, and are excreted intact from intravenous (IV) administrations. Therefore, CDs were selected as the basis of the new polymeric materials discussed here. Cyclodextrin chemistry, especially with β-CD, has also been well investigated, with significant precedence in the literature for various functionalization procedures.

12.2.2 Polymer Synthesis and Nanoparticle Formulation and Application

Linear, cyclodextrin-based polycations are synthesized by polymerization of difunctionalized cyclodextrin-containing monomers with a second difunctionalized monomer.[2] A schematic for synthesizing CDP6 is shown in Figure 12.2. In this example, 6A, 6D-dideoxy-6A, 6D-di(2-aminoethanethio) β-cyclodextrin is condensed with dimethylsuberimidate. The resulting polycation has several notable properties. First, it contains repeating units of cyclodextrin in the backbone. The relative size of the CDs makes this polycation primarily sugar based. The solubility of β-cyclodextrin is typically limited by crystallization of the cyclodextrins (1.6 mM solubility limit in water). However, CDP6 is extremely water soluble; solutions of at least 0.4 M (by CD monomer) can be prepared, with concentrations limited by viscosity and not solubility. Second, the polycation contains repeating amidine groups. Amidine has a pKa ~12 so the polycation has a fixed charge at physiologic pH for complexation with nucleic acids. Third, the polycations are short and relatively monodisperse with degrees of polymerization (DOP) of 4–8 and polydispersities of around 1.1.[2 6] The low molecular weight is desirable for gene delivery applications for several reasons. Polycations with low molecular weight are more biocompatible in that they exhibit less toxicity and are less prone to elicit a complement response.[7,8] In addition, unpackaging of nucleic acid-bearing nanoparticles has been discussed as a limiting factor in gene expression.[9] The binding of lower molecular weight polycations to nucleic acids should be more reversible than high-molecular-weight polycation analogues.

Nanoparticles are prepared by adding a solution of CDP to a solution of nucleic acid. When mixed at polymer to DNA charge ratios greater than 1, the two components self-assemble and condense into uniform, spherical particles with diameters of 25 nm to 150 nm, depending on formation conditions (Figure 12.3). Nanoparticles containing oligonucleotides tend to be smaller (25–60 nm) than plasmid-loaded particles (60 nm–150 nm). In addition, particle size can be controlled by changing formulation concentrations (average particle size varies directly with DNA and polymer concentrations during formulation).[10] Encapsulation of nucleic acids in the nanoparticle also offers protection from nuclease digestion.[6]

When positively charged nanoparticles are exposed to cultured cells, the particles, like other polyelectrolyte complexes, form aggregates in the physiologic salt concentrations and precipitate on the cells. The particles most likely interact with proteoglycans on the surface of the cells and enter by endocytosis. Nucleic acid delivery efficiency can be monitored by delivering a reporter gene such as luciferase, with gene expression measurements reported in relative light units (RLUs).

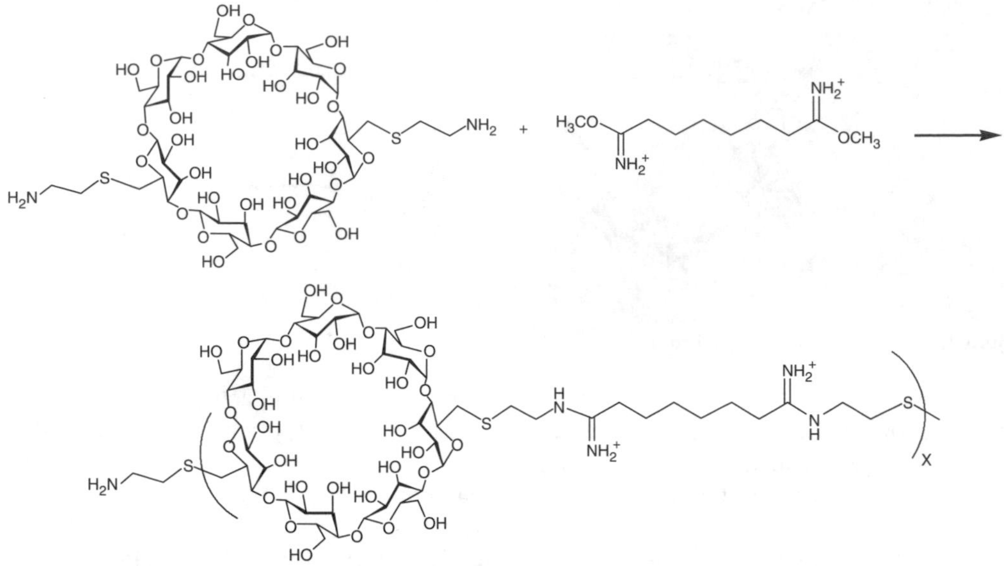

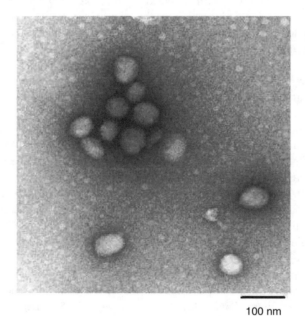

Figure 12.2 Synthesis of CDP6.

Figure 12.3 Transmission electron micrograph of CDP6-based nanoparticles containing plasmid DNA. Bar is 100 nm. (Reproduced with permission from Hwang, S., Bellocq, N., and Davis, M., Effects of structure of beta-cyclodextrin-containing polymers on gene delivery, *Bioconj. Chem.* 12 (2), 280, 2001. Copyright 2001 Am. Chem. Soc.)

Figure 12.4 shows data from a typical CDP6 transfection experiment with PC-3 (human prostatic carcinoma) cells. Luciferase gene expression increases with polymer charge ratio. No significant toxicity is observed by CDP6-mediated delivery of nucleic acid to cultured cells. These experiments demonstrate the potential of CDPs as nucleic acid delivery materials. Next, the effect of polymer structure on delivery efficiency and toxicity is elucidated.

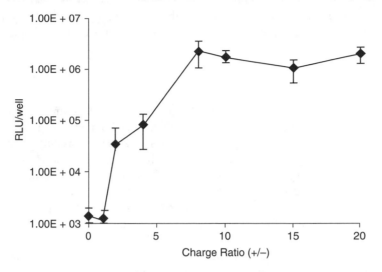

Figure 12.4 Transfection of PC-3 (human prostate carcinoma) cells by CDP6-based nanoparticles formulated at various charge ratios. (Copyright 2003 Insert Therapeutics, Inc. Reproduced with permission.)

12.2.3 Structure-Function Studies

The generalized structure of a CDP is shown below, where A represents comonomer A, B represents comonomer B, C is the cationic charge center, and S is the spacer between comonomer A and the charge center.

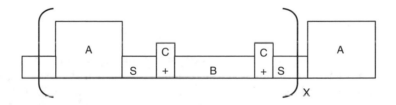

For example, for CDP6:

Systematic changes were made at each site to study the effect of polymer structure on gene delivery properties. The polymers were tested in BHK-21 cells for transfection efficiency and toxicity comparisons.

12.2.3.1 Comonomer A

Reineke and Davis prepared polymers using hexamethylenediamine, trehalose diamine, and β-cyclodextrin diamine as comonomer A, called AP1, AP2, and AP4, respectively, to study variations in comonomer A. The remainder of the polymer structure was kept constant[4] (Figure 12.5A). The polymers had similar DOP (5–10). Studies with other short polycations have indicated that variations in DOP in this range do not have a significant effect on toxicity.[3] Variations in activity and tolerability can therefore be attributed to differences in polymer structure. Hexamethylenediamine is relatively hydrophobic, trehelose is hydrophilic, and β-CD is very hydrophilic, thus providing a controlled study to observe how the degree of polymer hydration affects transfection and toxicity. The transfection efficiency and fraction cell survival of AP1, AP2, and AP4 formulated with pGL3-CV, a luciferase gene-containing plasmid, at 20+/- is shown in Figure 12.5B. The results clearly indicate that toxicity is associated with hydrophobicity of comonomer A. The transfection efficiency of the two sugar-based polycations were similar, while transfection with AP1 was low at all charge ratios tested, most likely due to low cell survival after treatment. Cyclodextrin-based polymers therefore have lower toxicity than analogous polymers synthesized with less hydrophilic monomers. As a follow-up to these studies, Popielarski et al. prepared several β-CD and γ-CD-based polycations.[5] The γ-CD-based polycations consistently revealed no difference in delivery efficiency and slightly lower cellular toxicity than the comparable β-CD-based polycations, further supporting the correlation between polymer hydration and toxicity since the former cyclodextrin is more water-soluble than the latter.

12.2.3.2 Comonomer B

The effect of polymer charge density on transfection and toxicity was studied by polymerization of dicysteamine-substituted β-CD with diimidates containing 4 to 10 methylene units, named CDP4 to CDP10 (Figure 12.6A).[6] All polymers had average DOPs of 4 to 5 and interacted similarly with nucleic acids to form nanoparticles of approximately the same size (120 nm). Exposure of particles

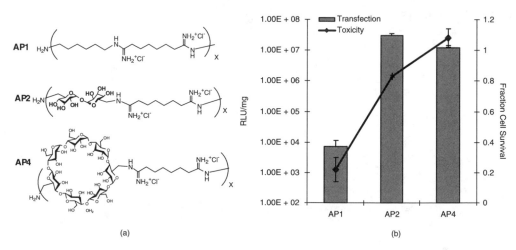

(a)   (b)

Figure 12.5   (a) Variations in comonomer A: chemical structures of AP1, AP2, and AP4. (b) Transfection efficiency and fraction cell survival of AP1, AP2, and AP4-mediated delivery to BHK-21 cells. Polymers were formulated with pGL3-CV (luciferase gene-containing plasmid) at 20 +/–.

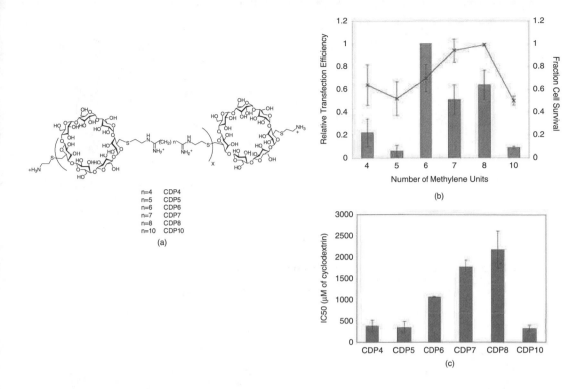

Figure 12.6 (a) Variations in comonomer B: chemical structures of CDP4-CDP10. (b) Transfection efficiency of CDP4-CDP10 mediated delivery to BHK-21 cells. Polymers were formulated with pGL3-CV at 50 +/−. (c) Fraction cell survival of CDP4-CDP10 mediated delivery to BHK-21 cells. Polymers were formulated with pGL3-CV at 50 +/−. (b) and (c) reproduced with permission from Hwang, S., Bellocq, N., and Davis, M., Effects of structure of beta-cyclodextrin-containing polymers on gene delivery, *Bioconj. Chem.* 12 (2), 280, 2001. Copyright 2001 Am. Chem. Soc.

to BHK-21 cells revealed that transfection and toxicity profiles are very sensitive to polymer structure. Optimal transfection was achieved with CDP6 (containing 6 methylene units between amidine charges) (Figure 12.6B). Changes in comonomer B length (and hence distance between charge centers) of only a few Angstroms reduces transgene activity by up to 20-fold. A distinct correlation is also observed between charge density and toxicity (Figure 12.6C): toxicity increases with reduced spacing between charges (the toxicity associated with CDP10 most probably results from its low solubility). Toxicity problems have also been noted with other high-density polycations such as polyethylenimine (PEI) and poly(L-lysine). Based on these studies, the diimidate containing the 6 methylene spacer was selected as the optimal comonomer B and used in the remaining structure-function studies.

12.2.3.3 Spacer (S)

Three sets of polymers were synthesized to test the effect of spacing between CD and charge center on transfection and toxicity. The first set of polymers focused on derivatives of CDP6, prepared from β-CD diamine, β-CD dicysteamine, and β-CD dithiodibutaneamine, called AP4, AP5, and AP6, respectively (Figure 12.7A).[4] The second and third sets were based on β-CD and γ-CD monomers, respectively, that were functionalized at the $3^A,3^B$ hydroxyls with diamines (as opposed to $6^A,6^B$ substitution in CDP6) (Figure 12.8A). Again, all polymers were similar in molecular weight (DOP 4–8). Transfection and toxicity trends were the same for all three sets of polymers (Figure 12.7B, Figures 12.8B and 12.8C). Transfection efficiency and polymer toxicity

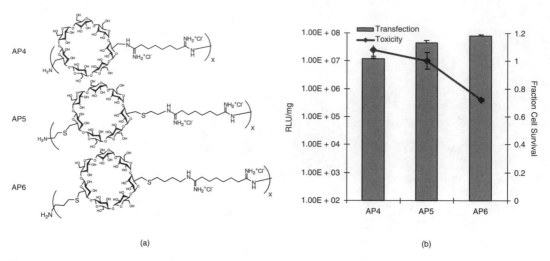

(a)   (b)

Figure 12.7 (a) Variations in spacer: chemical structures of AP4, AP5, and AP6. (b) Transfection efficiency and fraction cell survival of AP4, AP5, and AP6 mediated delivery to BHK-21 cells. Polymers were formulated with pGL3-CV at 20 +/−.

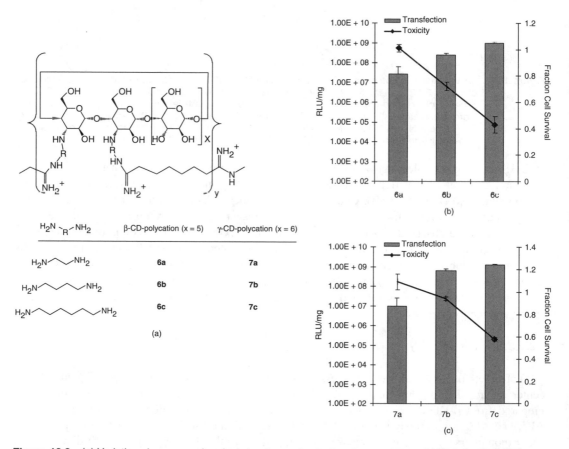

Figure 12.8 (a) Variations in spacer: chemical structures of polymers 6a-c and 7a-c. (b) Transfection efficiency and fraction cell survival of 6a, 6b, and 6c mediated delivery to BHK-21 cells. Polymers were formulated with pGL3-CV at 20 +/−. (c) Transfection efficiency and fraction cell survival of 7a, 7b, and 7c mediated delivery to BHK-21 cells. Polymers were formulated with pGL3-CV at 20 +/−.

increases as the spacer length is increased. (Note that the difference in RLUs/mg between the polymer sets is due to slight variations in transfection conditions; AP5 and polymer 7b have similar transfection efficiencies in direct comparisons.) A likely explanation for these phenomena is that, due to its bulkiness, the CD may be shielding the amidine charge centers and thus effect binding interactions. Indeed, DNA binding studies confirm that polymers 6A and 7A have the lowest plasmid binding affinity.[5] Low polymer-DNA binding may hurt transfection efficiency by reducing nanoparticle stability, thereby impairing the polymer's ability to protect DNA against nucleases. At the same time, the polymer is also less apt to interact nonspecifically with other cellular proteins, reducing toxicity. A trade-off thus exists. For CD polymers, a 2 to 4 methylene spacer is optimal for minimizing toxicity without severely compromising delivery efficiency.

12.2.3.4   Charge (C)

Finally, the effect of charge center was investigated.[3] Reineke and Davis synthesized analogous polymers to four amidine polymers, replacing the amidine charge center with quaternary ammoniums (Figure 12.9A). The quaternary ammonium polymers displayed similar toxicity profiles to the amidine polymers but exhibited consistently lower transgene expression (Figures 12.9B and 12.9C). Based on chloroquine-mediated transfection studies, the authors suggest that the amidine polymers are more efficient in endosomal escape. It is well established that pH-sensitive charge centers are capable of influencing transfection efficiency (see discussion below); this study demonstrates that careful selection of fixed charge centers is also important in polycation design for gene delivery.

12.2.4 pH-Sensitive Polymers

Most polymer–DNA nanoparticles are internalized by cells into vesicles that fuse with endosomes. As the endosomes mature from early endosomes to late endosomes, the pH of the vesicles rapidly drops due to ATP-dependent proton pumps in the membrane. Endosomes eventually fuse with vesicles from the Golgi that contain degradative enzymes. Endosomal material that remains in the fused vesicles, called lysosomes, are degraded by nucleases and proteases.[11] Many viruses use the pH change to elicit protein conformational changes that allow escape of the virus into the cytoplasm. PEI contains secondary amines that protonate as the endosomal pH drops. It is proposed that PEI buffers the pH and protects the material in the endosome from eventual lysosomal degradation.[12] Additionally, concomitant influx of water into the endosome may result in osmotic swelling and vesicular rupture.

Confocal microscopy studies with fluorescein-labeled CDP6 nanoparticles reveal cells with punctuate fluorescence, indicative of endosomal trapping of the delivered material. CDPs show elevating transgene expression with an increasing polymer:DNA charge ratio (Figure 12.4). Formulation experiments where nanoparticles were prepared by mixing CDP6 to plasmid DNA (pDNA) at various charge ratios followed by particle separation and component quantification have revealed that most CDP6 added to DNA at charge ratios greater than 2+/– is not associated with the particle but rather exists free in solution.[10] *In vitro* experiments are generally conducted at high (> 5+/–) charge ratios for optimal gene expression. Thus, free CDP6 may co-accumulate with nanoparticles in cultured cells and assist in intracellular trafficking. For *in vivo* applications, free polymer is quickly diluted from nanoparticles. Techniques for improving the transfection efficiency of nanoparticles formulated at low charge ratios are necessary for efficacious *in vivo* delivery.

The imidazole moiety in histidine also contains a secondary amine that has pKa ~6 and therefore has been studied for its endosomal buffering abilities. Several publications have shown that histidylated or imidazolated polycations offer substantial increases in gene delivery efficiency over their parent polymers.[13-16] In order to avoid interferences with DNA binding, histidine was conjugated to the amine termini of CDP6 (CDP6-Hist) by peptide bond formation chemistry.[17] CDP6-Hist

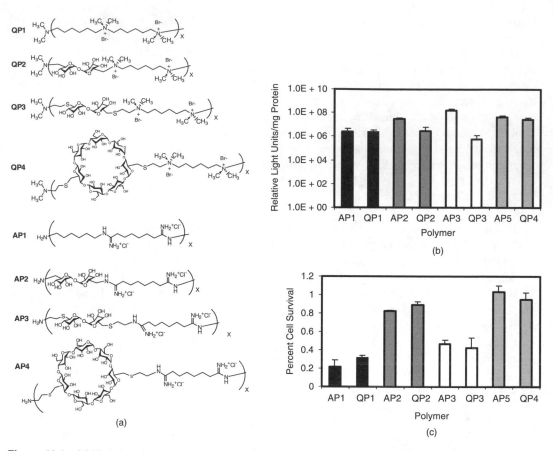

Figure 12.9 (a) Variations in charge centers: chemical structures of polymers QP1-4 and analogous amidine-containing polymers. (b) Comparison of transfection efficiency to BHK-21 cells between quaternary ammonium-containing and amidine-containing polymers. (c) Comparison of toxicity to BHK-21 cells between quaternary ammonium-containing and amidine-containing polymers. (b) and (c) reproduced with permission from Reineke, T.M. and Davis, M.E., Structural effects of carbohydrate-containing polycations on gene delivery. 2. Charge center type, *Bioconj. Chem.* 14 (1), 255, 2003. Copyright 2003 Am. Chem. Soc.

(Figure 12.10A) binds and condenses DNA to give spherical particles with diameters similar to those obtained with CDP6, and is able to transfect cells more efficiently without added toxicity. Delivery of rhodamine-labeled plasmids revealed no difference in cellular uptake between CDP6 and CDP6-Hist. Visualization of fluorescein-labeled nanoparticles by confocal microscopy showed stronger punctate fluorescence in cells exposed to CDP6-Hist nanoparticles (Figure 12.10B). Because there is no difference in DNA uptake between the two polymers, the increase in fluorescence is likely due to the lack of fluorescein quenching at low pH as a result of endosomal buffering.

Synthesis of CDP6-Hist is tedious because the protection chemistry involved requires harsh conditions. A similar polymer, CDP6-Imid, can be synthesized by amidation of the polymer amine termini with 4-imidazoleacetic acid. The reaction is a one-step reaction in aqueous buffer (Figure 12.11A).[18] The polymer can then be purified by dialysis. The CDP6-Imid transfection profile is similar to that of CDP6-Hist; luciferase activity at low charge ratios is over 20-fold higher with CDP6-Imid than CDP6 (Figure 12.11B). No toxicity is observed even at polymer to DNA charge ratios of 50+/-. With its high transfection efficiency at low charge ratios and low toxicity, CDP6-Imid is a suitable polymer for *in vivo* use.

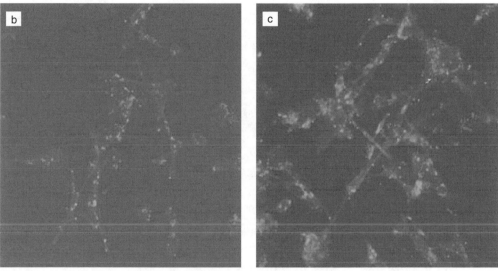

(a)

Figure 12.10   (See color insert following page 336) (a) Chemical structure of CDP-Hist. (b) Confocal images of CDP and CDP-Hist mediated delivery of fluorescein-labeled oligonucleotides to BHK-21 cells. Cell cytoplasms are counter-stained with DiI. (From Davis, M. et al., *Curr. Med. Chem.*, 11 (2), 179, 2004. With permission.)

12.3 POLYCATIONS CONTAINING PENDANT CYCLODEXTRINS

Cyclodextrins possess several desirable properties for pharmaceutical use. When incorporated in the backbone of the linear polycations, cyclodextrins mediate high water solubility and low toxicity. In addition, cyclodextrins can extract cell membrane components such as phospholipids and cholesterol. Can the beneficial properties of cyclodextrins be transferred to existing gene delivery materials by direct conjugation of the cyclodextrin? Monotosylated cyclodextrin can be readily synthesized according to literature protocols[19] and reacted with primary amines of polycations via nucleophilic substitution. For example, Suh et al. synthesized dendrimers of PEI linked to β-cyclodextrins for biomimetic catalysis applications using this approach.[20] Studies involving the synthesis and application of cyclodextrin-conjugated dendrimers and cyclodextrin-conjugated PEI as gene delivery agents are summarized here.

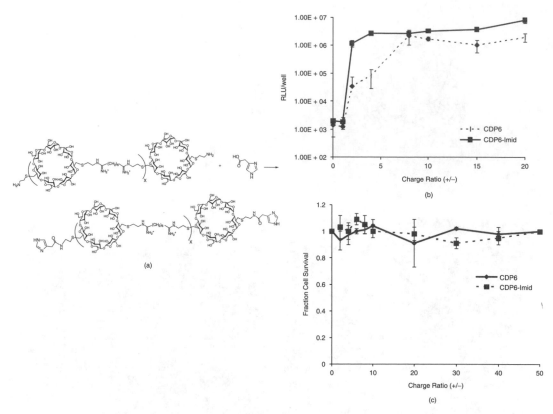

Figure 12.11   (A) Synthesis of CDP-Imid. (B) Comparison of CDP and CDP-Imid transfection efficiency to PC-3 cells. (Copyright 2003 Insert Therapeutics, Inc. Reproduced with permission.) (C) Comparison of CDP and CDP-Imid toxicity to PC-3 cells. (Copyright 2003 Insert Therapeutics, Inc. Reproduced with permission.)

12.3.1   Cyclodextrin-Dendrimer Conjugates

Uekama and colleagues conjugated cyclodextrins to polyamidoamine dendrimers and studied the effect of CD type (α-, β-, or γ-cyclodextrin), dendrimer structure, and cyclodextrin conjugation ratio on gene transfer ability.[21-23] In their first study, tosylated α-, β-, and γ-cyclodextrin were reacted with polyamidoamine dendrimers (G2) at a molar ratio of 1:1.[21] All conjugates formed complexes with DNA, and protected nucleic acids against nuclease degradation. At high charge ratios (200:1 dendrimer to DNA), the three CD-dendrimers had superior transfection efficiency compared to the unmodified dendrimer. However, α-CD-dendrimer showed the highest transfection efficiency at all tested charge ratios, increasing luciferase expression by up to 100-fold higher than physical mixtures of dendrimer–DNA complexes and α-CD. The authors hypothesize that the lipid extraction ability of α-CD assists the particles in translocation from endosome to cytoplasm, as cyclodextrins disrupt phosphatidylcholine liposomes in the order of α-CD $>\beta$-CD $> \gamma$-CD. Confocal microscopy experiments using fluorescently labeled DNA supported this theory; cells transfected with α-CD-dendrimer–DNA complexes had higher fluorescence in the cytoplasm than cells transfected with dendrimer–DNA complexes.

Kihara et al. next studied the effect of dendrimer structure on gene transfer ability by preparing α-CD-dendrimers having different dendrimer generations (G2, G3, and G4).[22] Again, the α-CD-conjugates were more efficient in transfecting cultured cells than the unmodified dendrimers. The G3 α-CD-dendrimer conjugate mediated the highest expression levels of transgene. In their subsequent paper, Kihara et al. synthesized G3 α-CD-dendrimer conjugates with three degrees of substitution of α-CD, 1.1, 2.4, and 5.4 α-CD per dendrimer by mole.[23] Higher substitution levels resulted in increased transfection efficiency both *in vitro* and *in vivo* but also substantial increases in polymer toxicity, most likely because of membrane disruption activity by α-CD. These examples demonstrate that α-CD conjugation to polycations can increase transfection efficiencies, most likely by mediating endosomal release via membrane disruption. This approach may give some advantages over peptide-based transfection enhancing methods due to the ease and low cost of synthesis. However, the increase in efficiency came at the price of increased polycation toxicity.

12.3.2 Cyclodextrin-PEI Conjugates

PEI is a popular polymer used for *in vitro* plasmid transfer due to its relatively high transfection efficiency. Because of its ability to buffer the pH drop in the endosome and the supposedly mediated release of the DNA, PEI is one of the few polymers whose transfection efficiency is not significantly impacted by chloroquine addition. However, PEI's use in systemic applications is still limited by its toxicity, its low critical flocculation concentration, and its tendency to aggregate in physiological salt concentrations. β-cyclodextrin was therefore conjugated to branched $PEI_{25,000}$ (CD-bPEI) (Figure 12.12A) and linear $PEI_{25,000}$ (CD-lPEI) (Figure 12.12B) in an effort to combine the beneficial qualities of β-cyclodextrin (low toxicity, high water solubility, ability to form inclusion complexes) with PEI (chloroquine-independent transfection).[24]

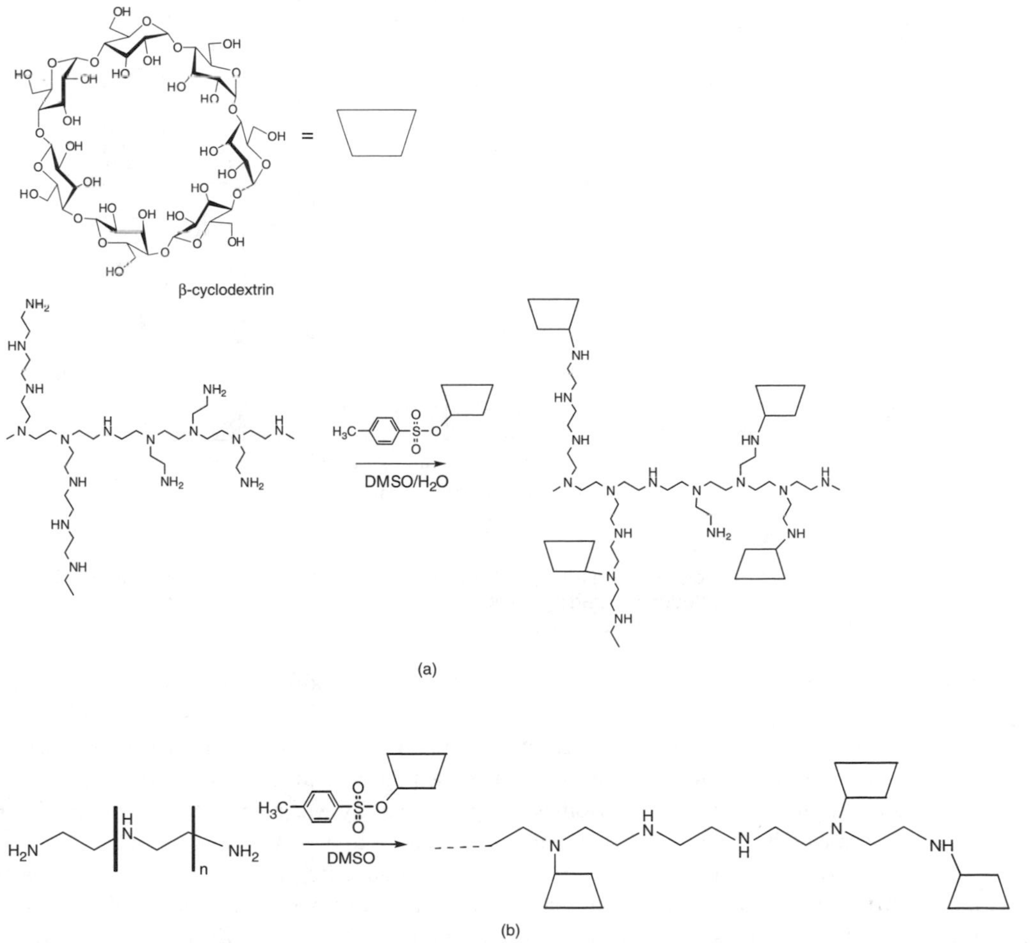

Figure 12.12   (a) Chemical structure of CD-bPEI. (b) Chemical structure of CD-lPEI. (From Pun, S.H. et al., *Bioconj. Chem.* in press, 2004. With permision.)

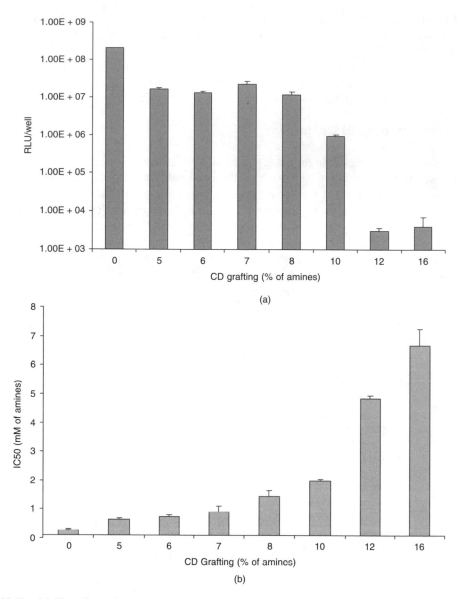

Figure 12.13   (a) The effect of cyclodextrin grafting density on CD-bPEI transfection efficiency to PC-3 cells.
(b) The effect of cyclodextrin grafting density on CD-bPEI toxicity to PC-3 cells. (From Pun, S.H.
et al., *Bioconj. Chem.* in press, 2004. With permision.)

   A series of CD-bPEI with varying degrees of cyclodextrin substitution was synthesized. Transfection efficiency of the polymers was impaired as cyclodextrin grafting increased, with reductions
in luciferase activity of up to 4 orders of magnitude at the highest levels of grafting (Figure 12.13A).
High levels of amine substitution may substantially alter the PEI amine pKa, thereby preventing
protonation and pH buffering in the endosomes. Polymer toxicity was also affected by cyclodextrin
grafting; polymers with higher CD grafting densities also have higher IC_{50}s in cultured cells (Figure
12.13B). As discussed in Section 1, β-cyclodextrins may reduce polycation toxicity by increasing
polymer solubility, by capping primary amines, or by reducing polycation binding affinity. Thus,
the optimal CD-bPEI for plasmid delivery was determined to have 8% PEI amine grafting. This
polymer is used in the remaining discussion in this section.

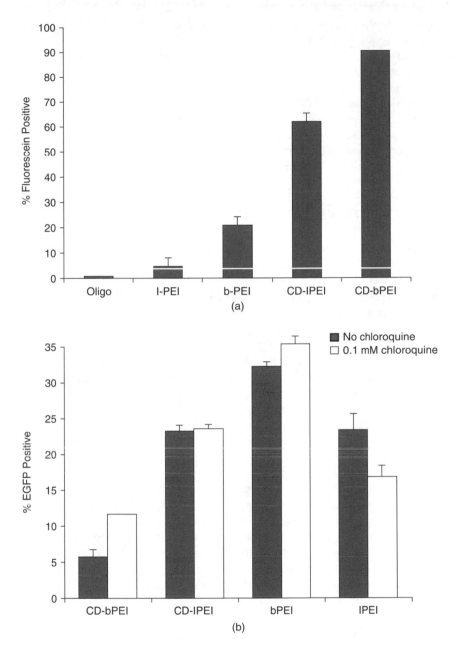

Figure 12.14 (A) Comparison of oligonucleotide delivery efficiencies to PC-3 cells by lPEI, bPEI, CD-bPEI, and CD-lPEI. (B) Comparison of plasmid delivery efficiencies to PC-3 cells by lPEI, bPEI, CD-bPEI, and CD-lPEI in the presence and absence of chloroquine.

The oligonucleotide and plasmid delivery efficiencies of the CD-PEI polymers were then compared with the unmodified PEI polymers. CD-bPEI and CD-lPEI delivered fluorescently labeled oligonucleotides with higher efficiency than the parent polymers (Figure 12.14A). Plasmid transfection efficiency was monitored by delivery of a green fluorescent protein gene (Figure 12.14B). Consistent with the data presented in Figure 12.13A, CD-bPEI transfects cultured cells with lower efficiency than bPEI. The addition of chloroquine partially restores the loss in the transfection efficiency upon conjugation, supporting the hypothesis that some transfection efficiency is lost due to inefficient endosomal escape. In contrast, the CD-lPEI, while having a threefold higher IC_{50}, suffered no loss in transfection efficiency from lPEI levels. CD-lPEI transfection levels are also unaffected by chloroquine addition.

In summary, two examples of using cyclodextrin conjugation to impart additional properties to existing polycations used in gene delivery were described. First, α-CD conjugation to dendrimers increased transfection efficiency by mediating endosomal release. Second, CD-bPEI had lower cellular toxicity than bPEI but at the expense of gene transfer ability. Also, CD-lPEI polymers had reduced toxicity compared with lPEI without a change in gene delivery efficiency. The cyclodextrin polymers described thus far have potential for applications as *in vitro* and *ex vivo* delivery agents. However, in order for systemic administration to be feasible, the issues of nanoparticle stability and cell targeting still need to be addressed.

12.4 FORMULATIONS FOR *IN VIVO* APPLICATIONS

12.4.1 Particle Modification by Adamantane Conjugates

The CDP- and CD-PEI-based nanoparticles, like other polycation–DNA complexes, transfect cells by forming aggregates in physiological salt concentrations and precipitating on cells. For *in vivo* applications, particle aggregation, along with particle interaction with serum components or platelets, can cause animal death, for example by blockage of lung capillaries, and is therefore undesired. For systemic administration, the gene delivery vehicles should avoid nonspecific (charge-mediated) uptake by cells and instead target desired cells in the body. PEGylation by polyethylene glycol (PEG) conjugation is a well-studied method of stabilizing colloidal particles that involves protecting the particles with a hydrophilic polymer layer. Recently, a new method of particle PEGylation was described that takes advantage of cyclodextrin's ability to form inclusion complexes with certain guest molecules with high affinity.[1] Adamantane (AD) conjugates are added to cyclodextrin-based polycations and nucleic acids during formulation. The AD self-assembles with surface cyclodextrins to modify the particle surface properties (Figure 12.15). This approach has several benefits. First, it is modular. The polymer and adamantine modifier can be synthesized and purified separately, simplifying quality control and manufacturing procedures. Second, the particles are modified by self-assembly via inclusion complex formation between cyclodextrin and AD. The formulation procedure is straightforward and reproducible and can be used in traditional "drug in the vial" type preparations. Finally, the delivery system is adaptable and versatile. In addition to being able to deliver oligos, plasmids, or combinations of nucleic acids, the modification technology can be used to impart multiple functionalities to the particles by simply formulating with different AD modifiers. Three examples of the application of this technology are presented in this section: particle stabilization, surface charge tuning, and targeting.

12.4.2 Particle Stabilization

PEGylated nanoparticles were prepared by including AD-PEG conjugates in the formulation with CDP6 and plasmid DNA.[1] AD-PEG can either be added to preformed particles or mixed with the cyclodextrin-based polycation and then added to DNA. With the latter approach, high particle concentrations can be obtained. Because PEGylation occurs via interaction with cyclodextrins (which are not involved in the DNA assembly process) PEGylated particles are obtained without interference with polymer–DNA condensation or changes in particle morphology, as is often reported with PEGylation by direct polymer conjugation.[25] PEGylation introduced salt stability to the particles in a PEG length-dependent manner[1] (Figure 12.16A). The resulting particles maintain small sizes in the presence of 150 m*M* salt for hours. PEGylation also increases the critical flocculation concentration of the particles, an important dose-limiting factor for *in vivo* administrations. The particle size of most polycation–DNA complexes is dependent on formulation

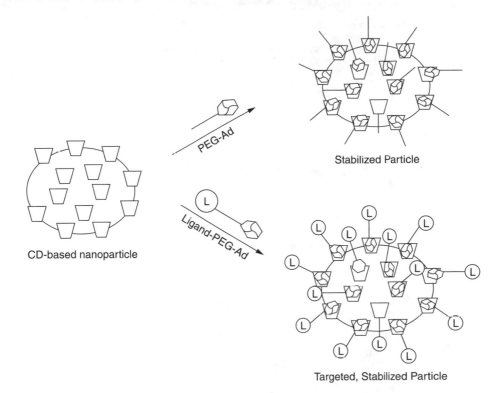

Figure 12.15 Schematic of adamantane-conjugate modification of cyclodextrin-based nanoparticles. (Reproduced in part with permission from Pun, S.H. and Davis, M., Development of a non-viral gene delivery vehicle for systemic application, *Bioconj. Chem.* 13 (3), 630, 2002. Copyright 2002 Am. Chem. Soc.)

concentrations; average particle size increases with polymer and DNA concentration at formulation. The particle size of PEGylated nanoparticles remains independent of formulation concentration (Figure 12.16B); particles of less than 100 nm are obtained even at final DNA concentrations of 10 mg/ml. PEGylation also reduces the amount of DNA lost by particle precipitation (Figure 12.16C). PEGylation technology is also extendable to cyclodextrin-grafted polycations; CD-bPEI- and CD-lPEI-based nanoparticles can also be stabilized against salt-induced aggregation following a similar approach.[24]

12.4.3 Surface Charge Tuning

Although particle PEGylation prevents self-self interactions, the PEGylated particles are still positively charged and are internalized by interaction with proteoglycans on cell surfaces.[26] For many *in vivo* applications, especially systemic administration, this is undesirable. Highly charged particles also tend to aggregate with serum components and blood cells, resulting in particle elimination from the body and potential side effects such as complement activation. AD-PEG modifiers containing a short, anionic region (AD-anionic-PEG) were therefore synthesized. PEGylated particles were prepared by modification with mixtures of AD-anionic-PEG and AD-PEG. The zeta potential of the particles can be tuned by controlling the ratio of AD-anionic-PEG to AD-PEG (Figure 12.17A). Thus, stable particles with near-neutral zeta potentials can be readily obtained. Cellular uptake of several formulations was tested by exposing cultured cells to fluorescently labeled particles for 1 h and analyzing for uptake by flow cytometry. The combination of AD (for anchoring to the particle surface), anionic region (for tuning surface charge to negative), and PEG (for steric inhibition and stabilization) is the most effective in reducing nonspecific cellular uptake (Figure 12.17B).

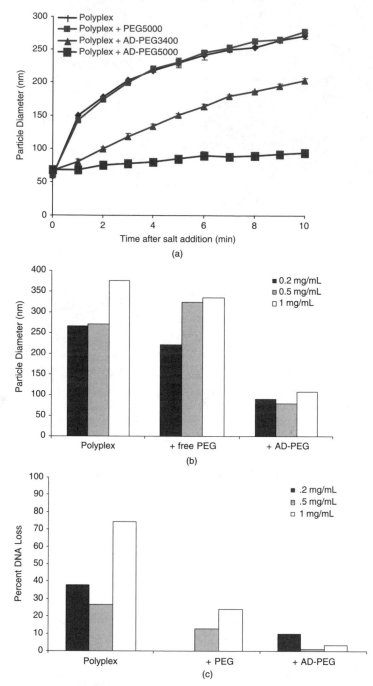

Figure 12.16 (A) Stabilization of CDP-based nanoparticles by PEGylation. AD-PEG stabilizes particles in a length-dependent manner, whereas PEG5k alone does not provide stabilization. (Reproduced in part with permission from Pun, S.H. and Davis, M., Development of a non-viral gene delivery vehicle for systemic application, *Bioconj. Chem.* 13 (3), 630, 2002. Copyright 2002 Am. Chem. Soc.)(B) Particle size of nanoparticles as a function of formulation concentration. NonPEGylated nanoparticles increase in size with higher concentrations. PEGylated nanoparticles retain their small size. (Copyright 2003 Insert Therapeutics, Inc. Reproduced with permission.)(C) Nanoparticle precipitation (measured by DNA loss) as a function for formulation concentration. NonPEGylated nanoparticles aggregate and precipitate at higher concentrations. PEGylation increases the critical flocculation concentration of nanoparticles. (Copyright 2003 Insert Therapeutics, Inc. Reproduced with permission.)

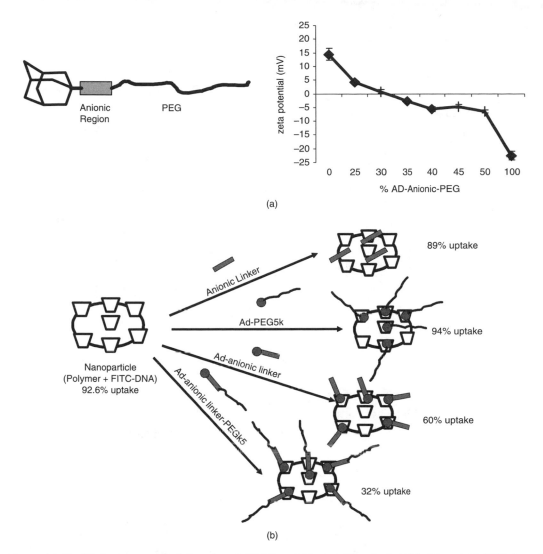

Figure 12.17 (A) Tuning nanoparticle zeta potential by addition of AD-anionic-PEG. (Copyright 2003 Insert Therapeutics, Inc. Reproduced with permission.) (B) Quantification of nanoparticle uptake by flow cytometry. Modification of nanoparticles with AD-anionic-PEG reduces nonspecific, charge-mediated uptake by cultured cells. (Copyright 2003 Insert Therapeutics, Inc. Reproduced with permission.)

12.4.4 Targeting

A targeting ligand was introduced to the AD-anionic-PEG modifer to direct transfection to cells with specified surface receptors. Receptor-mediated endocytosis via the asialoglycoprotein receptor (ASGP-R), which is highly expressed on hepatocytes, was chosen as a model system and two AD modifiers containing galactose (a ligand for ASGP-R) and glucose (as a control) were synthesized (Figure 12.18A).[1] The formulated particles were discrete and uniform with diameters of ~100 nm (Figure 12.18B). Specific nucleic acid delivery was tested by transfection to HepG2 (human hepatoma cells expressing cell surface ASGP-R) and HeLa (human cervical cancer cells that do not express ASGP-R) (Figure 12.19A and 12.19B). The unmodified particles were internalized via nonspecific charge interactions, and were therefore not affected by the addition of excess galactose. The protected polyplex, modified with 40:60 AD-anionic-PEG:AD-PEG, were negatively charged

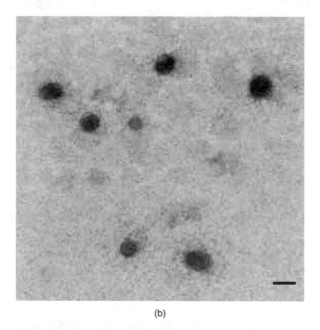

AD-anionic-PEG-galactose

AD-anionic-PEG-glucose

(a)

(b)

Figure 12.18   (A) Chemical structures of AD-anionic-PEG-gal and AD-anionic-PEG-glu. (B) Electron micrograph of polyplexes modified by AD-anionic-PEG-gal. Bar is 100 nm. (Reproduced in part with permission from Pun, S.H. and Davis, M., Development of a non-viral gene delivery vehicle for systemic application, *Bioconj. Chem.* 13 (3), 630, 2002. Copyright 2002 Am. Chem. Soc.)

and therefore did not transfect the cultured cells. The glucose-modified particles (modified with 40:60 AD-anionic-PEG-glucose:AD-PEG) were similarly unable to transfect efficiently, as glucose is not a ligand for ASGP-R. Galactose-modified particles transfected HepG2 cells in a galactose-dependent manner; transgene expression was hampered by the addition of free galactose. Specific, receptor-mediated delivery was further confirmed because the galactose-modified particles do not transfect HeLa cells.

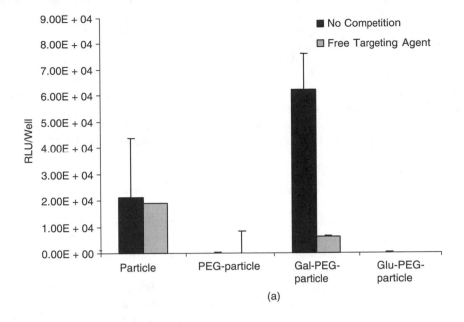

(a)

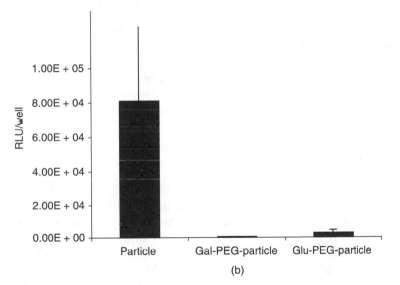

(b)

Figure 12.19 (A) Comparison of plasmid delivery efficiencies to HepG2 cells by unmodified particles, PEGylated particles (modified with AD-anionic-PEG), galactosylated nanoparticles, and glucosylated nanoparticles in the presence and absence of excess galactose. (Copyright 2003 Insert Therapeutics, Inc. Reproduced with permission.) (B) Comparison of plasmid delivery efficiencies to HeLa cells by unmodified particles, PEGylated particles (modified with AD-anionic-PEG), galactosylated nanoparticles, and glucosylated nanoparticles in the presence and absence of excess galactose. (Copyright 2003 Insert Therapeutics, Inc. Reproduced with permission.)

12.5 *IN VIVO* DELIVERY

Treatment of metastatic cancer is one of the holy grails of gene therapy because cancer is the second-highest cause of death in the industrialized world. In addition, metastatic cancer is often unresponsive to traditional treatments. A gene delivery vehicle that is amenable to systemic administration is desirable for accessing disseminated tumor sites. Transferrin receptors are often upregulated in rapidly growing cells like cancer cells. Transferrin-modified, cyclodextrin-based nanoparticles were therefore designed to achieve tumor-specific delivery from intravenous injections.[18] The transferrin-modified nanoparticles (Tf-particles) were prepared by formulation with CDP-Imid for nucleic acid condensation and endosomal release, AD-PEG for particle stabilization, and Tf-PEG-AD for tumor targeting. Transferrin is a large, anionic protein (~80 kD), so interaction of Tf-PEG-AD with the nanoparticles also reduces particle surface charge. The Tf-particles were uniform and spherical in shape, with average diameters of ~50 nm when formulated with oligonucleotides and with average diameters of ~100 nm when formulated with plasmid. At low transferrin modification (< 1% transferrin to cyclodextrin by mole), the particles remained stable in physiologic salt concentrations.

12.5.1 Biocompatibility of Tf-Particles

The LD_{40} of CDP6 in mice is 200 mg/kg.[6] Renal tubular damage, which is typically associated with polycations, was observed at the high dose. In addition, the polymer did not elicit a specific antibody response in rabbit immunogenicity studies. Formulation with nucleic acid reduces polymer toxicity, most probably by neutralizing charge. BALB/c mice were treated with Tf-particles and PEGylated particles (PEG-particles) at 400 μg DNA per mouse (containing 3.3 mg CDP-Imid, 10.6 mg AD-PEG, and 3.4 mg of Tf-PEG-AD) and DOTAP:Chol formulations at 2+/- containing 25 μg DNA per mouse and analyzed after 24 h for liver enzyme levels and platelet counts.[27] DOTAP:Chol at 50 μg DNA per mouse dose was lethal to all mice tested. At 25 μg DNA per mouse, DOTAP:Chol treated mice had significantly elevated liver enzyme levels (10-fold increase in alanine transaminase and 21-fold increase in aspartate transaminase), whereas Tf-particles and PEG-particle-treated mice (at 400 μg DNA per mouse) had no statistically significant increase in liver enzyme levels. In addition, Tf-treated mice had no change in platelet counts, while DOTAP:Chol-treated mice had 60% reduced platelet counts. Thus, the CDP-based delivery system is very well tolerated in mice and is capable of delivering significantly higher nucleic acid payloads than other nonviral delivery agents.

12.5.2 Systemic Plasmid Delivery to Tumors

PEG-particles and Tf-PEG-particles were formulated with a p53 gene-containing plasmid and administered by low-pressure tail vein injection to nude mice bearing PC-3 (prostate carcinoma cells that are p53 null and express the transferrin receptor on the cell surface) tumors.[28] The biodistribution of plasmid DNA in the mice was analyzed 1 h after injection by PCR. Whereas unmodified particles accumulated primarily in the liver (less than 0.2% of the injected dose was found in the tumor), PEGylated particles demonstrated passive targeting to the tumor (10% of injected dose in the tumor). Transferrin modification did not alter the biodistribution of the particles from that of the PEGylated particles. However, expression analysis by RT-PCR revealed that transferrin targeting is required for p53 transgene expression in the tumor tissue. This study demonstrated that Tf-PEG-particle formulations could achieve tumor targeting and expression from low-pressure systemic injections.

12.6 SUMMARY

This chapter described cyclodextrin-based polymeric gene delivery systems that self-assemble with nucleic acid to form small nanoparticles. The properties of the nanoparticles can be easily modified by the addition of AD-based conjugates that also self-assemble with the particles by inclusion complex formation. Thus, this type of gene delivery system is the first to be completely formed via self-assembly. This technology provides for the potential of imparting the functional complexity that is vital to viral delivery efficiency and success with the simplicity of a synthetic, modular system. The components of this system are well tolerated *in vitro* and *in vivo*. Animal studies have confirmed the potential of these materials as systemic delivery vehicles for gene therapy.

REFERENCES

1. Pun, S.H. and Davis, M., Development of a non-viral gene delivery vehicle for systemic application, *Bioconj. Chem.* 13 (3), 630, 2002.
2. Gonzalez, H., Hwang, S., and Davis, M., New class of polymers for the delivery of macromolecular therapeutics, *Bioconj. Chem.* 10, 1068, 1999.
3. Reineke, T.M. and Davis, M.E., Structural effects of carbohydrate-containing polycations on gene delivery. 2. Charge center type, *Bioconj. Chem.* 14 (1), 255, 2003.
4. Reineke, T.M. and Davis, M.E., Structural effects of carbohydrate-containing polycations on gene delivery. 1. Carbohydrate size and its distance from charge centers, *Bioconj. Chem.* 14 (1), 247, 2003.
5. Popielarski, S.R., Mishra, S., and Davis, M.E., Structural effects of carbohydrate-containing polycations on gene delivery. 3. CD type and functionalization, *Bioconj. Chem.* 14, 672, 2003.
6. Hwang, S., Bellocq, N., and Davis, M., Effects of structure of beta-cyclodextrin-containing polymers on gene delivery, *Bioconj. Chem.* 12 (2), 280, 2001.
7. Plank, C., Mechtler, K., Szoka, J., F.C., and Wagner, E., Activation of the complement system by synthetic DNA complexes: a potential barrier for intravenous gene delivery, *Hum. Gene Ther.* 7, 1437, 1996.
8. Fischer, D., Bieber, T., Li, Y., Elasasser, H.-P., and Kissel, T., A novel non-viral vector for DNA delivery based on low molecular weight, branched polyethylenimine: effect of molecular weight on transfection efficiency and cytotoxicity, *Pharm. Res.* 16 (8), 1273, 1999.
9. Schaffer, D., Fidelman, N., Dan, N., and Lauffenburger, D., Vector unpacking as a potential barrier for receptor-mediated polyplex gene delivery, *Biotechnol. Bioeng.* 67, 598, 2000.
10. Davis, M., Pun, S., Bellocq, N., Reineke, T., Popielarski, S., Mishra, S., and Heidel, J., Self-assemblying nucleic acid delivery vehicles via linear, water-soluble, CD-containing polymers, *Curr. Med Chem.* 11 (2), 179, 2004.
11. Smythe, E. and Warren, G., The mechanism of receptor-mediated endocytosis, *Eur. J. Biochem.* 202, 689, 1991.
12. Boussif, O., Lezoualcl'h, F., Zanta, M., Mergny, M., Scherman, D., Demeneix, B., and Behr, J.-P., A versatile vector for gene and oligonucleotide transfer into cells in culture and *in vivo*: polyethylenimine, *Proc. Nat. Acad. Sci. U.S.A.* 92, 7297, 1995.
13. Putnam, D., Gentry, C., Pack, D., and Langer, R., Polymer-based gene delivery with low cytotoxicity by a unique balance of side-chain termini, *Proc. Nat. Acad. Sci. U.S.A.* 98 (3), 1200, 2001.
14. Pack, D., Putnam, D., and Langer, R., Design of imidazole-containing endosomolytic biopolymers for gene delivery, *Biotechnol. Bioeng.* 67, 217, 2000.
15. Midoux, P. and Monsigny, M., Efficient gene transfer by histidylated polylysine/pDNA complexes, *Bioconj. Chem.* 10, 406, 1999.
16. Pichon, C., Roufaï, M., Monsigny, M., and Midoux, P., Histidylated oligolysines increase the transmembrane passage and the biological activity of antisense oligonucleotides, *Nucleic Acids Res.* 28 (2), 504, 2000.
17. Hwang, S., Rational design of a new class of cyclodextrin-containing polymers for gene delivery, in *Chemical Engineering*, California Institute of Technology, Pasadena, 2001, 167.
18. Bellocq, N., Pun, S., Jensen, G., and Davis, M., Synthesis of transferrin-PEG conjugates and preparation of transferrin-modified polyplexes for tumor-targeted gene delivery, *Bioconj. Chem.* 14, 1122, 2003.
19. Brown, S., Oates, J., Coghlan, D., Easton, C., van Eyk, S., Janowski, W., Lepore, A., Lincoln, S., Luo, Y., May, B., Schiesser, D., Wang, P., and Williams, M., Synthesis and properties of 6A-amino-6A-deoxy-α-and β-cyclodextrin, *Aust. J. Chem.* 46, 953, 1993.
20. Suh, J., Hah, S.S., and Lee, S.H., Dendrimer Poly(ethylenimine)s linked to β-cyclodextrin, *Bioorg. Chem.* 25, 63, 1997.

21. Arima, H., Kihara, F., Hirayama, F., and Uekama, K., Enhancement of gene expression by polyami-doamine dendrimer conjugates with α-, β- and γ-cyclodextrins, *Bioconj. Chem.* 12, 476, 2001.

22. Kihara, F., Arima, H., Tsutsumi, T., Hirayama, F., and Uekama, K., Effects of structure of polyami-doamine dendrimer on gene transfer efficiency of the dendrimer conjugate with a-cyclodextrin, *Bioconj. Chem.* 13, 1211, 2002.

23. Kihara, F., Arima, H., Tsutsumi, T., Hirayama, F., and Uekama, K., *In vitro* and *in vivo* gene transfer by an optimized α-cyclodextrin conjugate with polyamidoamine dendrimer, *Bioconj. Chem.* 14(2), 342, 2003.

24. Pun, S., Bellocq, N., Liu, A., Machemer, T., Maneval, D., Quijano, E., Schleup, T., Wen, S., Engler, H., Heidel, J., and Davis, M., Cyclodextrin-modified polyethylenimine polymers for gene delivery, *Bioconj. Chem.* (in press).

25. Kwoh, D., Coffin, C., Lollo, C., Jovenal, J., Banaszczyk, M., Mullen, P., Phillips, A., Amini, A., Fabrycki, J., Bartholomew, R., Brostoff, S., and Carlo, D., Stabilization of poly-L-lysine/DNA poly-plexes for *in vivo* gene delivery to the liver, *BBA* 1444, 171, 1999.

26. Mislick, K. and Baldeschwieler, J., Evidence for the role of proteoglycans in cation-mediated gene transfer, *Proc. Nat. Acad. Sci. U.S.A.* 93, 12349, 1996.

27. Bellocq, N., Davis, M., Engler, H., Jensen, G., Liu, A., Machemer, T., Maneval, D., Quijano, E., Pun, S., Schleup, T., and Wen, S., Transferrin-targeted, cyclodextrin polycation-based gene vector for systemic delivery, American Society of Gene Therapy, Washington, D.C., 2003.

28. Bellocq, N.C., Pun, S.H., Grubbs, B.H., Jensen, G.S., Liu, A., Cheng, J., and Davis, M.E., Development of transferrin-modified, cyclodextrin-based particles for the delivery of RNA-cleaving DNA enzyme (DNAzyme) molecules, 2nd International Symposium on Tumor Targeted Delivery Systems, Rock-ville, MD, 2002.

B. Biodegradable Polymers

CHAPTER 13

Gene Delivery Using Polyimidazoles and Related Polymers

Sharon Wong and David Putnam

CONTENTS

13.1 INTRODUCTION

The potential clinical application of gene delivery is perhaps as diverse as the methods used to deliver DNA to cells. Viral and nonviral DNA delivery vehicles alike are widely reported in the gene therapy literature and several general and critical reviews are available.[1,2] There are distinct advantages and disadvantages to both types of DNA delivery vehicles. For example, arguments that favor viral-based DNA delivery include high efficiency, whereas arguments against it include potential immunogenicity and oncogene activation. The arguments for and against nonviral gene delivery vehicles are essentially the opposite: in general, nonviral vehicles have relatively poor efficiency but superior safety profiles.

Improvements in nonviral vector efficiency are perhaps one of the most active areas of DNA vector research. New delivery vehicles are identified and reported on a monthly basis, some based on existing molecular architectures, others original. A common thread that ties together the development of new delivery vehicles is an underlying hypothesis that defines the rationale for the vector design.

One hypothesis, first reported by Behr and coworkers in 1995, was coined the "proton sponge" hypothesis. This theory postulates that a DNA delivery vector that buffers the endosomal or lysosomal compartments can propagate the rupture of these vesicles by creating an osmotic gradient

0-8493-1934-X/05/$0.00+$1.50
© 2005 by CRC Press LLC

between the vesicle interior and the cytoplasm.[3,4] The material first proposed as a proton-sponge-based delivery system, polyethylenimine (PEI), is now a popular reagent for DNA delivery. However, one of the shortcomings of PEI, particularly branched PEI, is cytotoxicity. Because a cell must remain metabolically viable to synthesize a therapeutic protein following DNA transfer, cytotoxicity of the delivery vehicle is anathema to the eventual utility of the delivery vehicle. Even though PEI can be toxic to cells, it remains an often used material in the field of gene transfer because cells that do remain viable following PEI-mediated DNA delivery express relatively high levels of protein. Derivatives of the PEI molecular scaffold, such as biodegradable PEI[5,6] polyethylene glycol-modified PEI,[7,8] and nonbranched linear forms of PEI[9,10] have been investigated to improve the overall DNA delivery characteristics of the polymer.

Building on the potential utility of a noncytotoxic form of PEI, a number of investigators active in the field of nonviral vector development considered the prospects of engineering new molecular structures based on the proton sponge hypothesis, but with improved cytotoxicity profiles. One result of these efforts is a body of literature that focuses on the application of the imidazole functional group in the molecular architecture of DNA delivery vehicles. The goal of this chapter is to review the literature surrounding the use of imidazole functional groups in polycationic DNA delivery vectors, and to evaluate the future of this line of research.

13.2 THE "PROTON SPONGE" HYPOTHESIS

Cellular internalization of macromolecules is afforded by endocytosis, wherein the cell membrane invaginates and engulfs the macromolecule.[11] Endocytosis of the macromolecule may be triggered by a receptor-ligand interaction (clathrin-dependent endocytosis) or the macromolecule may be nonspecifically absorbed to the cell membrane, for example by ionic interaction, and internalized by constitutive endocytosis (clathrin-independent endocytosis). In either scenario, at early time scales the endocytosed macromolecule is sequestered within an endosome, and at later time scales, the macromolecule is shuttled to the lysosome where it is degraded by one or more classes of enzymes.[12]

For nonviral DNA delivery vehicles, the internalization pathway is the same. However, in the case of therapeutic DNA delivery, the DNA degradation in the lysosome is unacceptable since intact DNA is necessary to direct the synthesis of a therapeutic protein (the goal of gene therapy). Therefore, one goal in the field of nonviral DNA delivery is to identify mechanisms to escape the endosomal/lysosomal pathway while retaining DNA integrity. The proton sponge is functional at this stage of DNA delivery.

Soon after its formation, the endosome is acidified through the activity of membrane-bound V-ATPase proton pumps. Endosome acidification facilitates vesicle fusion with the lysosome and provides the optimum pH for lysosomal enzyme activity. However, if the interior of the endosome is populated with functional groups bearing an apparent pK_a between extracellular pH (~7.2) and acidified endosome pH (~5.5), then endosomal acidification will be counterbalanced by sequestration of the internalized hydronium ion, effectively buffering the endosome to a more neutral pH. Electroneutrality is maintained in the system through the coupling of counterion influx (e.g., chloride) alongside hydronium internalization.[13] The end result is an elevated endosomal osmotic pressure leading to swelling and rupture of the endosome. This process is the mechanistic explanation for the enhanced DNA delivery capability of proton sponge polycations, starting with PEI.[3,4] A schematic of the hypothetical process is shown in Figure 13.1.

The validity of the proton sponge hypothesis remains controversial. Mechanistic reports that both support[14] and challenge[15] the proton sponge mechanism hypothesis are available. There is also evidence that suggests PEI facilitates DNA trafficking to the nucleus.[16] Regardless of mechanism, it is clear that the transfer of plasmid DNA (pDNA) to the cellular transcription and translation

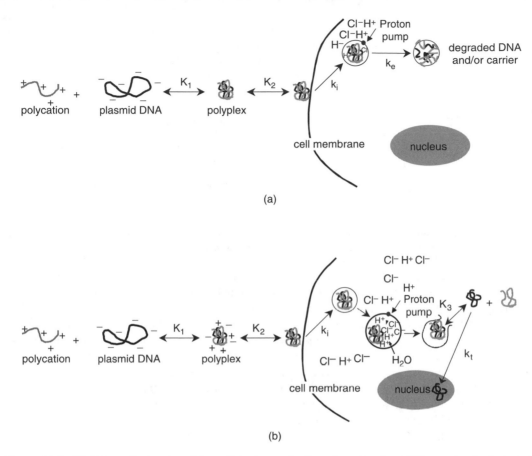

(a)

(b)

Figure 13.1 (A) Schematic drawing of the cellular internalization of a polycation–DNA complex (polyplex) via endocytosis. The electrostatic interaction between a polycation and DNA (K_1) forms a condensed complex that has a certain attraction to the cell surface (K_2). Invagination and eventual pinching off of the cell membrane (k_i) leads to the formation of an early endosome. An endosome membrane-bound ATPase proton pump acidifies the endosome to approximately pH 5.5 and the endosome ultimately fuses with a lysosome (k_e) leading to enzymatic degradation of the internalized polyplex. (B) Schematic drawing of the hypothetical "proton sponge" mechanism and how it can lead to enhanced DNA transfer. Proton-sponge polymers capture the H^+ from the ATPase proton pump to buffer the endosome and eventually increase the vesicle's osmotic pressure. The osmotic pressure differential leads to the influx of water, and ultimately to the rupture of the endosome prior to fusion with the lysosome.

machinery is enhanced by materials that are protonatable in the pH range spanning the endosome and extracellular space.

13.3 HISTIDINE-CONTAINING PEPTIDES AND POLYMERS

The first use of histidine-containing materials for gene transfer that did not involve lipids was reported in 1998 by the research team of Midoux and colleagues.[17] Building upon previous literature that demonstrated the ability of anionic peptide amphiphiles to enhance the transfection efficiency of polylysine-based vectors, these researchers engineered a variant of the HA-2 subunit of influenza virus hemagglutinin, called H5WYG, wherein the natural lysines were substituted with histidyl residues.[18-20] The researchers hypothesized that by substituting lysyl residues (pK_a ~10) with histidyl residues (pK_a ~6), the peptide would not disrupt the cell membrane at extracellular pH, but could disrupt the membranes of acidified intracellular vesicles. In addition, the investigators reported that

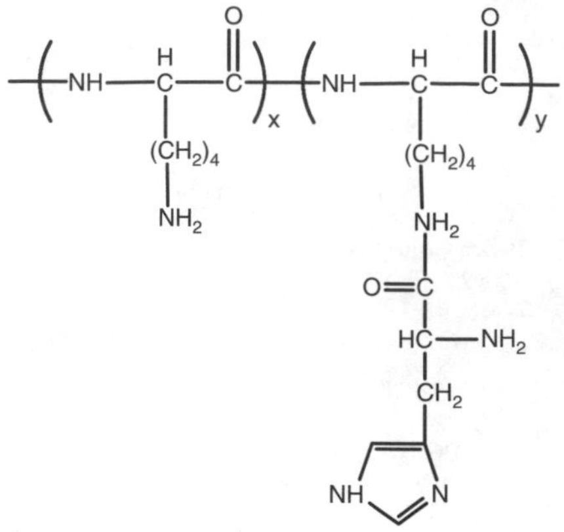

Figure 13.2 The structure of histidylated polylysine. The ε-amines of polylysine were conjugated with histidine at different molar ratios to evaluate the optimal histidine content.

the H5WYG peptide remained more active in the presence of serum than other membrane-permeabilizing amphiphiles. Transfection results were obtained by coadministration of free H5WYG peptide with a lactosylated polylysine–DNA complex.

Unlike previous gene delivery studies where amphiphilic peptides were tethered directly to polylysine scaffolds, these studies built upon the investigators' earlier protonation hypothesis and conjugated histidine directly to the ε-amine of polylysine.[21,22] The polymers (Figure 13.2) were synthesized by the polymer analogous condensation of (Boc)$_2$-protected polyhistidine to polylysine (degree of polymerization [DP] = 190; molecular weight range: 30,000 to 50,000) in the presence of benzotriazol-1-yl-oxy-tris-(dimethylamino) phosphonium hexafluorophosphate (BOP). Polymers with histidyl substitutions ranging from 10 M (20 histidyl residues per polymer chain) to 45 M (83 histidyl residues per polymer chain) were synthesized. Of these polymers, the histidyl:polylysine ratio that afforded the greatest expression of the marker protein luciferase was 38 M (72 histidyl residues per polymer chain). Protein expression plateaued with increased histidyl content. A protein expression plateau was also realized when the DNA:polymer ratio was varied. Protein expression remained constant at DNA:polymer (w/w) ratios of 1:3 and higher. These results corresponded fairly well to the zeta potential of the complexes where the positive zeta potential remained at approximately 25 mV for DNA:polymer ratios of 1:2 (w/w) and greater. The zeta potential of a polyplex can influence its gene transfer characteristics. Because the cell surface's glycoprotein matrix is polyanionic, the cationic nature of a polyplex (measured as zeta potential) can alter its apparent binding affinity to the cell surface. In general, the more postitive the zeta potential, the greater is its affinity to the cell surface.

Perhaps most importantly, the histidylated polymers remained effective in the presence of up to 20% serum. This is quite the opposite result from other polycationic gene transfer reagents where the absorption of serum components adversely affects delivery characteristics.[23,24]

The mechanism by which histidylated polylysines enhance DNA transfer to the cytosol was investigated by performing transfections in the presence and absence of bafilomycin A$_1$, an inhibitor of endosomal acidification.[25] Luciferase expression was reduced approximately 3 orders of magnitude in the presence of bafilomycin A$_1$, suggesting that acidification of internalized vacuoles is necessary for the polymers to more effectively transfer pDNA to the cytosol.

In a follow-up paper these investigators determined the effect that acetylation of the α-amino group of the histidyl residue has on the physicochemical characteristics of the DNA:polymer complex and on the efficiency of DNA transfer.[26] Because acetylation of the α-amino group removes a cationic center on the conjugate, the properties of the acetylated and free amino polymers should be very different. Polyplexes formed between pDNA and the acetylated polymer (Figure 13.3) had

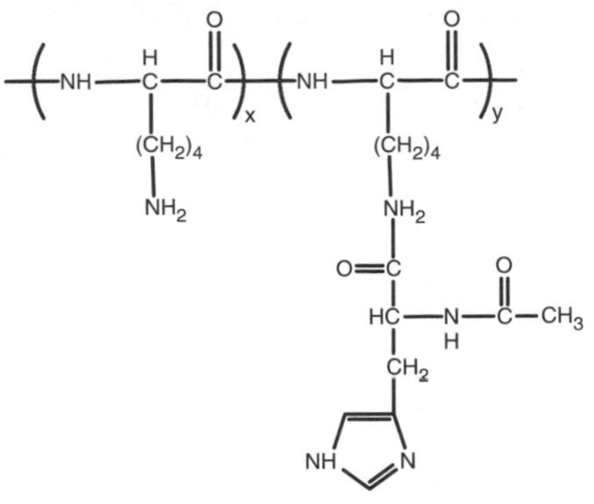

Figure 13.3 The structure of N-acetyl histidylated polylysine. The α-amino group of the conjugated histidine was protected to evaluate its influence upon transfection efficiency and protein absorption to the polyplex.

negative zeta potentials until reaching a DNA:polymer (w/w) ratio of 1:4, whereas polyplexes formed using the free amino polymer had positive zeta potentials starting at a 1:2 ratio. In the presence of serum, the hydrodynamic diameters of polyplexes formed with the acetylated polymer were typically double the size of those formed with the nonacetylated polymer, but in 5% glucose the size difference was reduced roughly 50%. More interestingly, the authors evaluated the type and extent of protein absorption to the complexes and found that acetylation of the α-amino group reduced the binding of serum proteins to the polyplex. Transfection efficiency differences between the two polymers were cell-type dependent and no clear structure-function relationships could be drawn from the data.

Later publications provide initial insights into the structure-function relationships of histidylated polylysine, but not the acetylated form.[27,28] Earlier research focused on polylysine derivatives of a single molecular weight (DP = 190), but the later work investigated a range of molecular weights (DP = 19, 36, 72, 190) along with variable histidyl substitution percentages (22 to 52 M). Interestingly, the DNA delivery efficiency, as determined by the level of marker protein expression (luciferase), increased with increasing histidine substitution for the DP = 190 conjugate, but decreased with increasing histidine content for all other molecular weights tested. Cytotoxicity of the polymers ranged from 4% to 49%, but no clear trend was discernible from the data reported.

A head-to-head comparison among histidylated polylysine (DP = 190, histidine substitution ~37 M) and the most commonly used commercially available transfection reagents, PEI (25 and 800 kDa), Lipofectin®, and LipofectAMINE®, revealed distinct cellular dependencies.[29] Luciferase expression was equal between PEI (25 kDa) and histidylated polylysine (~1 × 10^8 RLU/mg protein) in immortalized human tracheal gland serous cell line, CF-KM4, whereas expression levels mediated by PEI 800 kDa, Lipofectin®, and LipofectAMINE® were all lower (~5 × 10^7 RLU/mg protein). In contrast, luciferase expression levels in the immortalized tracheal epithelial cell line, ΣCFTE29o-, were essentially equal for histidylated polylysine and LipofectAMINE® (~2 × 10^7 RLU/mg protein), but the efficacy of the other reagents was much lower (~3 × 10^6 RLU/mg protein).

Other areas of active research with these materials focus on the delivery of antisense oligonucleotides and understanding the mechanisms by which these materials enhance nucleic acid delivery.[30,31] Mechanistically, these authors describe an increased residence time of histidylated polylysines in endocytic vesicles and propose that the enhanced DNA transfer is attributable to their prolonged endosomal stability. A recent paper reports an unexpected effect of divalent ions, particularly zinc, to further enhance histidylated polylysine-mediated DNA transfer.[32] In their article, the authors suggest that enhanced membrane fusion in the presence of zinc ion leads to an increase in the amount of intracellular plasmid.

13.4 IMIDAZOLE ACETIC ACID-CONTAINING POLYMERS

In an effort to identify the structure-function relationships that optimize pDNA delivery using imidazole-modified polycations, the group of Langer et al. reported the design, synthesis, and *in vitro* evaluation of a series of polylysine-graft-imidazole acetic acid polymers.[33] These authors hypothesized that by optimizing the balance between the polymer's cation and endosomal escape side-chain termini (in this case the ε-amino group of polylysine and imidazole acetic acid, respectively) an optimum balance between the polymer's complexation and condensation of DNA, endosomal escape efficiency, and cytotoxicity could be found.

The polymer conjugates (Figure 13.4) were synthesized by the polymer analogous acylation of the ε-amine of polylysine ($M_r = 34,000$) with imidazole acetic acid using an EDAC/NHS condensation system.[34] Polymers with three different levels of imidazole substitution (73.5 mol %, 82.5 mol %, and 86.5 mol %) were synthesized. Each polymer could electrostatically complex and condense pDNA into nanoscale structures less than 150 nm in diameter (the approximate size limit for endocytic uptake of the complex). The *in vitro* cytotoxicity profile of each polymer was low: at 60 µg/ml both polylysine and PEI killed the majority of P388D1 macrophage, HepG2 hepatoblastoma, and CRL 1476 smooth muscle cell lines, whereas more than 90% of these cells remained viable at equal concentrations of imidazole-substituted polylysine. More interestingly, these imidazole substituted polymers transferred marker pDNA encoding for luciferase as effectively as polyethylenimine, but without the inherent cytotoxicity. These results were not cell-type specific and were consistent among the above-referenced macrophage, smooth muscle and hepatoblastoma cell lines. In contrast to histidylated polylysines, an increase in the side chain substitution of imidazole led to enhanced luciferase expression. It remains unclear why this inverse relationship exists between these two similar polymer classes.

The promising *in vitro* results with these polymers prompted the *in vivo* evaluation of the graft-imidazole polymers. Using a mouse model, the authors assessed the polymer efficacy in delivering a DNA vaccine against HIV.[35] DNA vaccines against HIV in nonhuman models have had limited success and often immunostimulants are necessary to achieve adequate responses.[36] The polymer with the greatest DNA transfer efficacy *in vitro* (86.5 *M* imidazole substitution) was chosen for evaluation in a mouse model. The DNA vaccine construct was based on the sequence of the gp140 envelope from the HIV-2$_{UC2}$ isolate.[37] Using this plasmid construct, nanocomplexes were formulated at a 1:4 DNA:polymer ratio (w/w). Groups of 5 or 10 BALB/c mice were immunized with 20 µg of either naked or nanocomplexed DNA vaccine diluted in a total of 100 µl PBS. Each dose was divided into two equal portions and administered at two sites intradermally at the base of the tail according to the techniques previously developed in the investigators' laboratory. Twice, at

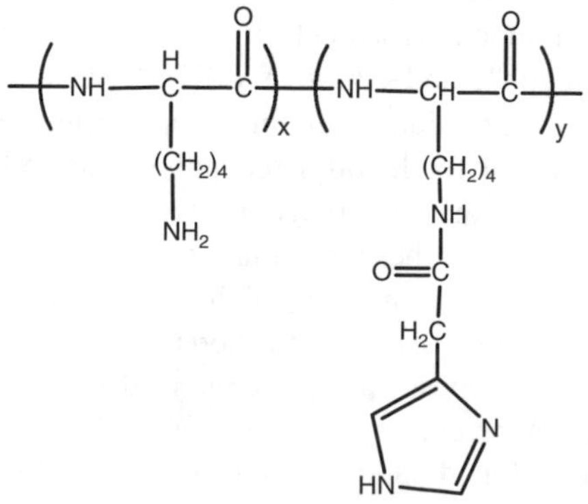

Figure 13.4 The structure of polylysine-graft-imidazole acetic acid. Ratios of the side-chain termini (ε-amino group and the imidazole group) were varied to determine the optimal balance between cationic centers and protonatable groups.

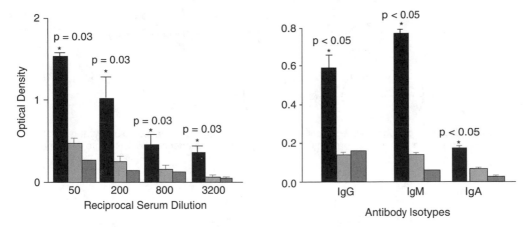

Figure 13.5 Serum antibody isotype titers (A) and IgG subclass titers (B) against the gp140 envelope protein of HIV generated by the intradermal injection of a DNA vaccine formulated with polylysine-graft-imidazole acetic acid (black bars), naked (light gray bars), or formulated with polylysine (dark gray bars).

subsequent 28 day intervals, the mice were boosted with the same amount of DNA by the same route of administration. Control groups were injected with either naked or nanocomplexed sham plasmid. Sero-conversion and antibody titers were determined by ELISA 28 days following the third administration. All antibody isotype titers (IgG, IgM, and IgA) were far greater in mice immunized with the nanocomplexed DNA vaccine, suggesting that the imidazole-conjugated polymers could act as an adjuvant for DNA vaccines (Figure 13.5).

13.5 POLYHISTIDINE-CONTAINING POLYMERS

The pK_a of the imidazole side group of polyhistidine is ~6.5, which limits polyhistidine's aqueous solubility at a pH of greater than 7.0. This low, aqueous solubility limits polyhistidine's utility as a DNA delivery vehicle. To take advantage of polyhistidine as a DNA delivery vehicle, a few investigators have designed and synthesized polyhistidine with various modifications to enhance its aqueous solubility at neutral pH.

The first polyhistidine-based DNA delivery vehicle was modified at the imidazole side chain with gluconic acid.[38] The conjugate (Figure 13.6) was synthesized by the polymer-analogous condensation of gluconic acid's carboxylic acid with one amine group of the imidazole side chain. The final conjugate possessed approximately four gluconic acid residues per polyhistidine chain. This level of substitution was sufficient to enhance the polymer's aqueous solubility at pH 7.0 to at least 100 mg/ml. These authors showed that complexes between the substituted polyhistidine and pDNA were sufficiently strong to retard the electrophoretic mobility of DNA. However, characterization of polyhistidine–DNA complexes using ethidium bromide exclusion and quasi-elastic light scattering revealed that the level of DNA condensation was borderline with respect to the sizes necessary to best deliver genes to cells. To further condense the modified polyhisti-dine–DNA complexes below 150 nm, a ternary complex consisting of pDNA, gluconic acid-modified polyhistidine, and polylysine was formulated. While the transfection efficiency of the polymer system reported was somewhat low relative to other reagent types, the final polymer did have a very good *in vitro* cytotoxicity profile (~95% viability at the highest concentration reported, 20 µg/ml).

Going one step further, the group of Kim et al. synthesized a polymer conjugate consisting of a polylysine backbone (8 kDa) with the ε-amino group substituted with N-acetyl-poly(L-histidine).[39]

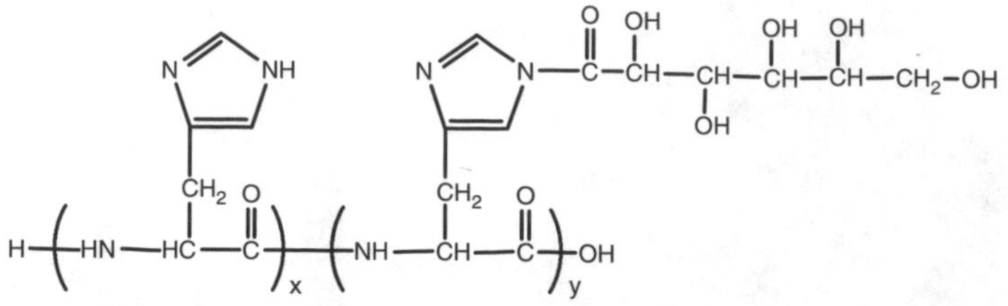

Figure 13.6   Structure of polyhistidine substituted with gluconic acid to enhance solubility at neutral pH. This conjugate required coformulation with polylysine to condense the DNA into sufficiently small polyplexes.

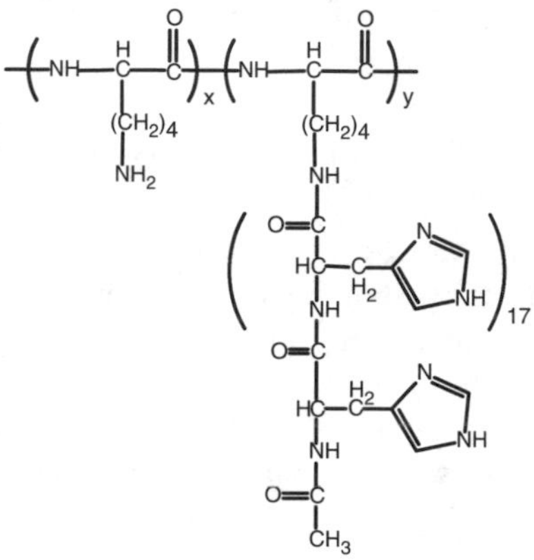

Figure 13.7   Structure of N-Ac-poly(L-histidine)-graft-poly(L-lysine). This conjugate is similar to that shown in Figure 13.2 except with a greater length of the substituted histidine to further enhance endosomal escape.

To prepare the conjugate (Figure 13.7), imidazole-protected polyhistidine (18 residues long) was synthesized by solid-phase peptide synthesis. The C-terminus of the polyhistidine was activated by forming the N-hydroxysuccinimide ester allowing the polymer analogous conjugation to the ε-amino group of polylysine. A total of 25% of the free amino groups on the polylysine were substituted with the histidine 18-mer. Complexation and condensation of pDNA, evidenced by gel retardation, ethidium bromide displacement, and dynamic laser light scattering, showed that the conjugate electrostatically interacted with pDNA in a conjugate concentration-dependent fashion and could form polyplexes on the order of ~100 nm. On a molar basis the transfection efficiency of the reported polyhistidine/polylysine conjugate was greater than polylysine alone, with lower cytotoxicity. However, while the total number of transfected cells was greater with the polyhistidine-containing conjugates, the total level of protein expression was on the same order as polylysine alone.

Lastly, two polyethylene glycol-modified polyhistidines were recently reported.[40] The rationale behind the design of these polymers was to utilize the unique characteristics of polyethylene glycol to minimize protein absorption to the resulting complex with DNA. Both comb (1, 5, and 10 PEG chains per polyhistidine chain) and linear diblock copolymer architectures were synthesized and evaluated with respect to their self-assembly with pDNA in an aqueous environment (Figures 13.8A and 13.8B). Using gel electrophoresis, the authors demonstrated that the comb PEG-polyhistidine polymer, with one PEG chain, bound tightly to pDNA, whereas the multi-PEG-substituted conjugates loosely complexed with the DNA. The diblock copolymer complexed even more tightly with the DNA than the single PEG substituted comb-shaped polymer. The authors concluded that the linear architecture allows a more intimate interaction between the polyhistidine and the DNA. The

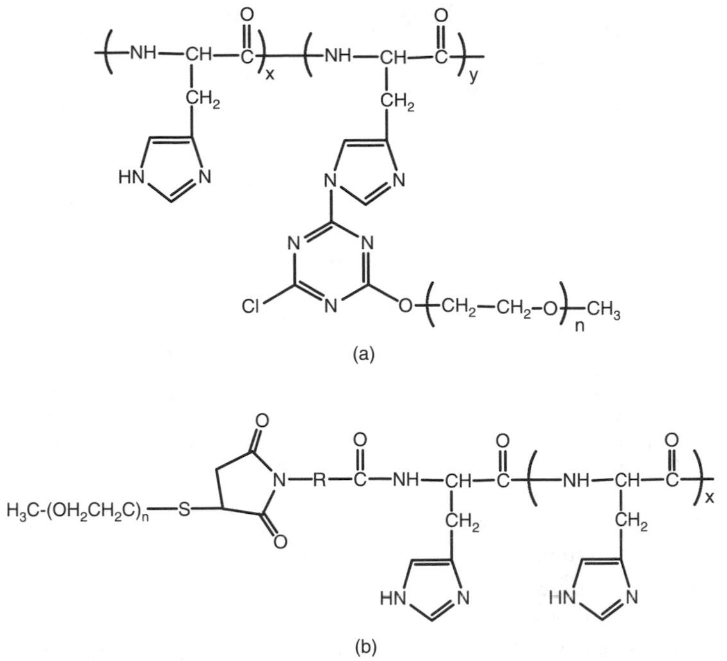

(a)

(b)

Figure 13.8 Structure of polyethylene glycol-modified polyhistidines: (A) comb architecture, (B) block copolymer. Polyethylene glycol was added to enhance solubility at neutral pH and to form a hydrated corona surrounding a polyhistidine–DNA complex.

aqueous behavior of these polymers was determined by laser light scattering. With the exception of the 10 PEG chain comb-shaped conjugate, all the polymers condensed pDNA into nanostructures on the order of 150 nm or less. The stability of these complexes in an aqueous environment was very good, as their hydrodynamic diameters remained essentially constant over 7 days at room temperature. Perhaps the most intriguing of the results is that the zeta potential of these complexes remained negative or only slightly positive. Often, the formation of stable DNA–polycation complexes results in complexes with positive zeta potentials that can lead to nonspecific protein absorption in the bloodstream and the activation of the complement cascade.[23,41] The retention of a negative zeta potential combined with the retention of complexes less than 150 nm is promising for this class of material for DNA delivery. In addition to these characteristics, the cytotoxicity of these polymer conjugates was low (greater than 80% cell viability at 100 μg/ml). While the authors do not report a high level of DNA transfer efficiency in this paper, this is likely due to the lack of adequate binding to the cell surface. Additional research is necessary to fully evaluate the utility of this class of material.

13.6 VINYL IMIDAZOLE-BASED POLYMERS OR POLYMETHACRYLATES

A variety of vinyl copolymers have been synthesized to investigate the structural effect of different side-chain types upon polyplex formation and transfection efficiency.[42,43] One of the materials described in the two referenced papers is a copolymer of 2-(dimethylaminoethyl) methacrylate (DMAEMA) and 4-methyl-5-imidazole methyl methacrylate (HYMIMMA) (Figure 13.9). A polymer with 19 mol % HYMIMMA substitution approached the buffering capacity of polyethylenimine and was able to condense pDNA into polyplexes on the order of ~50 nm in diameter. Interestingly, these authors demonstrate that the incorporation of the imidazole side chain leads to a significantly reduced DNA transfer efficiency. It is unclear whether the reduction in efficiency

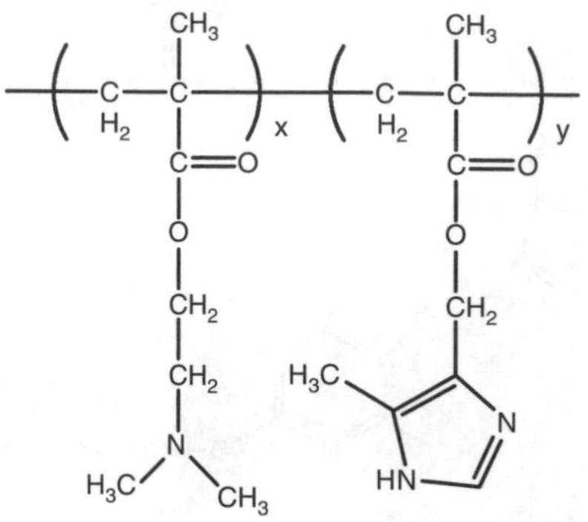

Figure 13.9   Structure of copolymers of 2-(dimethylaminoethyl)methacrylate (DMAEMA) and 4-methyl-5-imida-zoyl methyl methacrylate (HYMIMMA). These polymers were synthesized by free radical polymer-ization. Unlike the polypeptide-based polymers previously described, the addition of 4-methyl imidazole to the polymer reduced transfection efficiency.

caused by the addition of the imidazole functionality is due to the class of polymer backbone (alkyl backbone vs. peptide backbone), the comonomer's amino group type (tertiary vs. primary amine of the lysine side chain), or other factors.

More recently, the ability of poly(4-vinylimidazole) to deliver pDNA to a variety of cells was reported.[44] In this communication, the authors demonstrate that the homopolymer of 4-vinylimi-dazole (Figure 13.10) can effectively condense pDNA into complexes on the order of 175 nm and has a relatively acceptable cytotoxicity profile (> 70% viability) *in vitro* up to a concentration of 20 μg/ml. The *in vitro* DNA transfer efficiency was evaluated in three different cell lines (HeLa, MC3T3E1, and 293). In the cell lines tested, the DNA transfer efficiency (determined by luciferase expression) was ~3 orders of magnitude greater than PEI. However, these transfection experiments were performed at a polymer:DNA ratio that was not the optimum for PEI, so further evaluation is necessary to fully evaluate the transfection efficiency of poly(4-vinylimidazole) relative to established transfection reagents.

13.7 FUTURE TRENDS

With few exceptions, imidazole-containing polymeric gene delivery vectors are statistical dis-tributions of side-chain constituents and molecular weights. Some disadvantages of these random-ized material populations are the inability to fine-tune control over a defined property, and the inherent difficulty of defining clear structure-function relationships. There are at least two novel technologies that could help to define more accurately the molecular architectures that deliver DNA most efficiently.

The synthesis of genetically engineered protein polymers, generated from the biosynthesis of protein structures via a well-defined DNA template, is one way to control the sequence of a material accurately. Borrowing technology from modern molecular biology and biotechnology protocols, investigators are now advancing the field of drug and gene delivery by designing polymer sequences

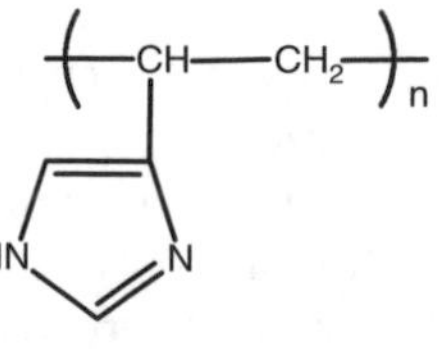

Figure 13.10   Structure of poly(4-vinylimidazole) homopolymer.

with exquisite control over the monomeric sequence as well as molecular weight. Current examples of applications for genetically engineered materials include controlled drug delivery and targeted drug delivery.[45,46] Commercialization of the technology is also under way.[47]

Another approach to understanding the structure-function relationships of statistical polymers for gene delivery is the application of combinatorial approaches to material development. The application of combinatorial chemistry to biomedical polymer development, originally reported by Kohn and colleagues in 1997, is ripe for application in the field of gene delivery.[48] An example of a polymer library approach to identify the structure-function relationships of polymer-based DNA delivery was reported by Langer and coworkers.[49] While the materials reported in these two manuscripts are also statistical populations, the generation of many different structures will help to identify potential "lead" structures for more in-depth investigation. Another factor that suggests a combinatorial approach is necessary to jump-start advancements in nonviral DNA delivery is that not all mechanisms by which vectors deliver DNA to cells are known. Some of the original postulates that directed the rational design of nonviral vectors are now under challenge.[50] It is reasonable to propose that by parsing the parameter space that defines the possible structural compositions of nonviral DNA delivery vectors, investigators could first identify molecular structures that efficiently deliver genes to cells, then deconvolute the delivery mechanisms to allow for a more effective design-based approach in vector development.

The impact of these new approaches to the design and synthesis of DNA delivery vehicles remains to be determined. However, by first understanding the structure-function relationships that govern the effective delivery of DNA, one can imagine real progress toward the design and synthesis of DNA vectors with greater efficacy and potential for clinical use.

ACKNOWLEDGMENTS

The authors would like to thank Dr. Amy Grayson and Dr. Alexander Zelikin for their critical evaluation of the manuscript.

REFERENCES

1. Luo, D. and Saltzman, W.M., Synthetic DNA delivery systems, *Nature Biotechnol.* 18, 33, 2000.
2. St. George, J.A., Gene therapy progress and prospects: adenoviral vectors, *Gene Ther.* 10, 1135, 2003.
3. Boussif, O. et al., A versatile vector for gene and oligonucleotide transfer into cells in culture and *in vivo*: polyethylenimine, *Proc. Nat. Acad. Sci. U.S.A.* 92, 7297, 1995.
4. Behr J-P., The proton sponge: a trick to enter cells the viruses did not exploit, *Chimia* 51, 34, 1997.
5. Forrest, M.L., Koerber, J.T., and Pack, D.W., A degradable polyethylenimine derivative with low toxicity for highly efficient gene delivery, *Bioconj. Chem.* 14, 934–940, 2003.
6. Ahn, C.H. et al., Biodegradable poly(ethylenimine) for plasmid DNA delivery, *J. Control. Release* 80, 273, 2002.
7. Suh, W. et al., An angiogenic, endothelial-cell-targeting polymeric gene carrier, *Mol. Ther.* 6, 664, 2002.
8. Lemieux, P. et al., Block and graft copolymers and NanoGel copolymer networks for DNA delivery into cell, *J. Drug Target.* 8, 91, 2000.
9. Coll, J.L. et al., *In vivo* delivery to tumors of DNA complexed with linear polyethylenimine, *Hum. Gene Ther.* 10, 1659, 1999.
10. Lemkine, G.F. and Demeneix, B.A., Polyethylenimines for *in vivo* gene delivery, *Curr. Opin. Mol. Ther.* 3, 178, 2001.
11. Mishra, S.K. et al., Clathrin-and AP-2-binding sites in HIP1 uncover a general assembly role for endocytic accessory proteins, *J. Biol. Chem.* 276, 46230, 2001.

12. Wattiaux, R. et al., Endosomes, lysosomes: their implication in gene transfer, *Adv. Drug Deliv. Rev.* 41, 201, 2000.
13. Nelson, N., Structure and pharmacology of the proton-ATPases, *Trends Pharmacol. Sci.* 12, 71, 1991.
14. Kichler, A. et al., Polyethylenimine-mediated gene delivery: a mechanistic study, *J. Gene Med.* 3, 135, 2001.
15. Godbey, W.T. et al., Poly(ethylenimine)-mediated transfection: a new paradigm for gene delivery, *J. Biomed. Mater. Res.* 51, 321, 2000.
16. Pollard, H. et al., Polyethylenimine but not cationic lipids promotes transgene delivery to the nucleus in mammalian cells, *J. Biol. Chem.* 273, 7507, 1998.
17. Midoux, P. et al., Membrane permeabilization and efficient gene transfer by a peptide containing several histidines, *Bioconj. Chem.* 9, 260, 1998.
18. Wagner, E. et al., Influenza virus hamagglutinin HA-2 N-terminal fusogenic peptides augment gene transfer by transferrin-polylysine-DNA complexes: toward a synthetic virus-like gene-transfer vehicle, *Proc. Nat. Acad. Sci. U.S.A.* 89, 7934, 1992.
19. Midoux, P. et al., Specific gene transfer mediated by lactosylated poly-L-lysine into hepatoma cells, *Nucleic Acids Res.* 21, 871, 1993.
20. Plank, C. et al., The influence of endosome-disruptive peptides on gene transfer using synthetic virus-like gene transfer systems, *J. Biol. Chem.* 269, 12918, 1994.
21. Wagner, E. et al., Influenza virus hemagglutinin HA-2 N-terminal fusogenic peptides augment gene transfer by transferrin-polylysine-DNA complexes: toward a synthetic virus-like gene-transfer vehicle, *Proc. Nat. Acad. Sci. U.S.A.* 89, 7934, 1992.
22. Midoux, P. and Monsigny, M., Efficient gene transfer by histidylated polylysine/pDNA complexes, *Bioconj. Chem.*10, 406, 1999.
23. Dash, P.R. et al., Factors affecting blood clearance and *in vivo* distribution of polyelectrolyte complexes for gene delivery, *Gene Ther.* 6, 643, 1999.
24. Augouy, S. et al., Serum as a modulator of lipoplex-mediated gene transfection: dependence of amphiphile, cell type and complex stability, *J. Gene Med.* 2, 465, 2000.
25. Bowman, E.J., Siebers, A. and Altendorf, K., Bafilomycins: a class of inhibitors of membrane ATPases from microorganisms, animal cells, and plant cells, *Proc. Nat. Acad. Sci. U.S.A.* 85, 7972, 1988.
26. Bello-Roufai, M. and Midoux, P., Histidylated polylysine as DNA vector: evaluation of the imidazole protonation and reduced cellular uptake without change in the polyfection efficiency of serum stabilized negative polyplexes, *Bioconj. Chem.* 12, 92, 2001.
27. Pichon, C., Goncalves, C., and Midoux, P., Histidine-rich peptides and polymers for nucleic acids delivery, *Adv. Drug Del. Rev.* 53, 75, 2001.
28. Midoux, P., and Pichon, C., Histidylated polycationic molecules for nucleic acids transfer, *Recent Res. Devel. Bioconj. Chem.* 1, 95, 2002.
29. Fajac, I. et al., Histidylated polylysine as a synthetic vector for gene transfer into immortalized cystic fibrosis airway surface and airway gland serous cells, *J. Gene Med.* 2, 368, 2000.
30. Pichon, C. et al., Histidylated oligolysines increase the transmembrane passage and the biological activity of antisense oligonucleotides, *Nucleic Acids Res.* 28, 504, 2000.
31. Goncalves, C. et al., Intracellular processing and stability of DNA complexed with histidylated polylysine conjugates, *J. Gene Med.* 4, 271, 2002.
32. Pichon, C., Zinc improves gene transfer mediated by DNA/cationic polymer complexes, *J. Gene Med.* 4, 548, 2002.
33. Putnam, D. et al., Polymer-based gene delivery with low cytotoxicity by a unique balance of side-chain termini, *Proc. Nat. Acad. Sci. U.S.A.* 98, 1200, 2001.
34. Sehgal, D. and Vijay, I.K., A method for the high efficiency of water-soluble carbodiimide-mediated amidation, *Anal. Biochem.* 218, 87, 1994
35. Locher, C. et al., Enhancement of a human immunodeficiency virus *env* DNA vaccine using a novel polycationic nanoparticle formulation, *Immunol. Lett.* 90, 67–70, 2003.
36. Robinson, H.L., New hope for an AIDS vaccine, *Nature Immunol.* 22, 239, 2002.
37. Barnett, S.W. et al., Molecular cloning of the human immunodeficiency virus subtype 2 strain HIV-2UC2, *Virology* 222, 257, 1996.
38. Pack, D., Putnam, D., and Langer, R., Design of imidazole-containing endosomolytic biopolymers for gene delivery, *Biotechnol. Bioeng.* 67, 217, 2000.

39. Benns, J.M. et al., pH-sensitive cationic polymer gene delivery vehicle: N-Ac-poly(L-histidine-graft-poly(L-lysine) comb shaped polymer, *Bioconj. Chem.* 11, 637, 2000.

40. Putnam, et al., Polyhistidine-PEG:DNA nanocomplexes for gene delivery, *Biomaterials* 24, 4425, 2003.

41. Plank, C. et al., Activation of the complement system by synthetic DNA complexes: a potential barrier for intravenous gene delivery, *Hum. Gene Ther.* 7, 1437, 1996.

42. Dubruel, P., Toncheva, V., and Schacht, E.H., pH sensitive vinyl copolymers as vectors for gene therapy, *J. Bioactive Compat. Polym.* 15, 191, 2000.

43. Dubruel, P. et al., Physicochemical and biological evaluation of cationic polymethacrylates as vectors for gene delivery, *Eur. J. Pharm. Sci.* 18, 211, 2003.

44. Ihm, J-E. et al., High transfection efficiency of poly(4-vinylimidazole) as a new gene carrier, *Bioconj. Chem.* 14, 707, 2003.

45. Megeed, Z., Cappello, J., and Ghandehari H., Genetically engineered silk-elastinlike protein polymers for controlled drug delivery, *Adv. Drug Del. Rev.* 54, 1075, 2002.

46. Chilkoti, A., Dreher, M.R., and Meyer, D.E., Design of thermally responsive, recombinant polypeptide carriers for targeted drug delivery, *Adv. Drug Del. Rev.* 54, 1093, 2002.

47. Protein Polymer Technologies, Inc. San Diego, CA. http://www.ppti.com/.

48. Brocchini, S. et al., A combinatorial approach for polymer design, *J. Am. Chem. Soc.* 119, 4553, 1997.

49. Lynn, D. et al., Accelerated discovery of synthetic transfection vectors: parallel synthesis and screening of a degradable polymer library, *J. Am. Chem. Soc.* 123, 8155, 2001.

50. Forrest, M.L. and Pack, D.W., On the kinetics of polyplex endocytic trafficking: implications for gene delivery vector design, *Mol. Ther.* 6, 1, 2002.

CHAPTER 14

Degradable Poly(β-amino ester)s for Gene Delivery

David M. Lynn, Daniel G. Anderson, Akin Akinc, and Robert Langer

CONTENTS

14.1 INTRODUCTION

The safe and efficient delivery of therapeutic DNA to cells represents a fundamental obstacle to the clinical success of gene therapy. Cationic polymers have been investigated broadly as nonviral gene delivery agents because they can spontaneously self-assemble with and condense plasmid DNA (pDNA) into structures small enough to enter cells via endocytosis.[1] Many different cationic polymers are effective at overcoming these early entry-based barriers to gene delivery. The incorporation of additional functionality into polycationic scaffolds has yielded more sophisticated polymers that further protect DNA and help surmount other important intracellular barriers to efficient delivery and expression.[2-4] Despite recent advances, however, conventional polymeric vectors such as polyethylenimine (PEI)[5,6] have been associated with substantial cytotoxicity and polycations remain far less effective at mediating gene transfer than viral vectors.[7] The efficiency and general safety of synthetic cationic polymers — and an understanding of the structure–property relationships that define and influence them — must ultimately be addressed to support the continued advance of these materials into the clinic.

Recently, several groups have reported the synthesis and evaluation of new biodegradable polycations as potential gene delivery agents.[8-19] Readily degradable cationic polymers are of

0-8493-1934-X/05/$0.00+$1.50
© 2005 by CRC Press LLC

interest both from the standpoint of mitigating the toxicity of conventional materials as well as a potential means through which to affect the timely release of DNA inside transfected cells (an important and often overlooked part of any vector-mediated gene delivery process).[1,20] Examples of biodegradable polycations introduced in the context of gene delivery include poly(4-hydroxy-L-proline ester),[8,13] poly[α-(4-aminobutyl)-L-glycolic acid],[9,10] poly(2-aminoethyl propylene phosphate),[14,15] and degradable cationic hyperbranched[11] or network[12] polymers. As a class of materials, biodegradable polycations are generally less toxic than polymers such as poly(lysine) and PEI and in many cases mediate gene transfer at levels that either approach or exceed those using PEI *in vitro*. Although the development of these materials remains at an early stage, the results demonstrate that degradable polycations represent an attractive approach for the development of safe and effective polymeric gene delivery vectors. A more comprehensive review of the synthesis and gene delivery properties of these and other cationic polyesters can be found in other chapters of this book.

This chapter details the synthesis, discovery, and engineering of gene delivery systems based on a class of polymers known as poly(β-amino ester)s. Poly(β-amino ester)s are hydrolytically degradable, condense plasmid DNA into nanometer-scale structures at physiological pH, and are generally less toxic than polycations such as PEI.[16-19] The synthesis of these materials is straightforward and a variety of different polymer structures can be generated from a diverse pool of commercially available monomers.[16] Related synthetic advantages have enabled the parallel synthesis of combinatorial libraries containing hundreds to thousands of structurally related poly(β-amino ester)s.[17,19] These polymer libraries have been used as discovery-based platforms for the identification of new polymeric vectors through high-throughput cell-based screening assays as well as for the identification of emerging structure–property relationships for this class of materials.[18] Thus far, these methods have led to the identification of at least 46 new poly(β-amino ester)s that mediate gene delivery *in vitro* more efficiently than either PEI or leading lipid-based transfection vector systems.[17,19] Several poly(β-amino ester)s possess pH-dependent solubility characteristics and are suitable for the fabrication of solid microsphere and nanosphere formulations that could be used to trigger or enhance the intracellular release of DNA upon exposure to acidic endosomal vesicles.[21]

14.2 SYNTHESIS OF POLY(β-AMINO ESTER)S

Poly(β-amino ester)s are readily synthesized via the conjugate addition of either primary or bis(secondary) aliphatic amines to diacrylate compounds, as shown in Equations 14.1 and 14.2.[16,22,23] Polymerization occurs by a step-growth mechanism and the resulting linear polymers contain both esters and tertiary amines in their backbones. Side-chain functionalized polymers can be synthesized by incorporation of functionalized amine or diacrylate monomers (e.g., R in Equations 14.1 and 14.2). Poly(β-amino ester)s are structurally related to linear poly(amido amine) materials synthesized via the conjugate addition of amines to bis(acrylamides).[24] A review of poly(amido amine)s and their gene delivery properties can be found in Chapter 17.

$$\hspace{6cm} (14.1)$$

$$\hspace{6cm} (14.2)$$

While poly(amido amine)s are generally synthesized in protic solvents, the synthesis of poly(β-amino ester)s is performed either neat (no solvent) or in anhydrous organic solvents to minimize hydrolytic degradation during synthesis.[16] The average molecular weights of polymers synthesized under these conditions are largely dependent on monomer structure and solvent choice, but generally lie within the range of 2,000 to 50,000 with a modest degree of molecular weight control, depending on experimental conditions.[16,17] Polydispersity indices (PDIs) for these polymers are typically close to 2.0, consistent with the step-growth nature of the polymerization process. The structures of three model poly(β-amino ester)s (1–3) synthesized by the conjugate addition of N,N′-dimethylethylenediamine, piperazine, and 4,4′-trimethylenedipiperidine to 1,4-butanediol diacrylate are shown below.[16]

Poly-1

Poly-2

Poly-3

14.3 DEGRADATION AND CYTOTOXICITY

Poly(β-amino ester)s degrade under physiological conditions via hydrolysis of their backbone esters to yield small molecule bis(β-amino acid) and diol products (Equation 14.3).[16] Although it is possible that degradation could occur via a retro-conjugate addition mechanism rather than by ester hydrolysis, no evidence has been found suggesting that this occurs under physiological conditions.

Poly-2

$$(14.3)$$

Figure 14.1 shows degradation kinetic profiles for polymers 1–3 investigated over the range of pH likely to be encountered by these polymers during transfection. In general, these polymers degrade more rapidly at pH 7.4 ($t_{1/2} < 1$ h) than at pH 5.1 ($t_{1/2} = 10$ h). These degradation profiles are consistent with relative rates of acid- and base-catalyzed ester hydrolysis as well as the pH/degradation profiles observed for other cationic polyesters at physiological pH.[8,10] Although hydrolysis occurs rapidly in aqueous solution, degradation appears to occur much more slowly when these polymers are allowed to self-assemble with plasmid DNA to form electrostatic complexes.[16] In general, solid or neat samples of polymer can be stored dry and used for several months without detectable decreases in molecular weight.

An initial determination of the cytotoxicity of polymers 1–3 and their corresponding degradation products has been made using the MTT/thiazolyl blue dye reduction assay and the NIH 3T3 cell line.[16] The NIH 3T3 cell line is commonly used as a first level screening population for new transfection vectors, and the MTT assay is used to determine the influence of added substances on cell growth and metabolism.[25] As shown in Figure 14.2, cells incubated with polymers 1–3 remained 100% viable relative to controls at concentrations of polymer up to 100 µg/ml. These results

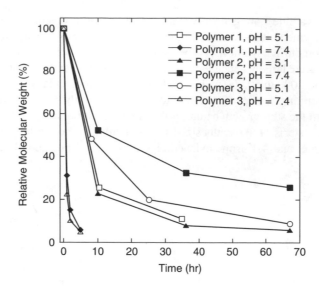

Figure 14.1   Molecular weight vs. time profiles for the hydrolytic degradation of poly(β-amino esters) **1–3** at 37°C at pH 5.1 and pH 7.4. Relative molecular weights were determined by gel permeation chromatography (GPC). (Reprinted with permission from Lynn, D.M. and Langer, R., *J. Am. Chem. Soc.* 122, 10761–10768, 2000.)

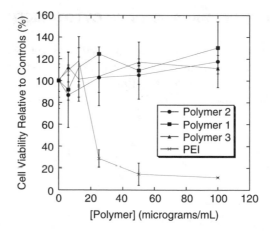

Figure 14.2   Cytotoxicity profiles of poly(β-amino esters) **1–3** relative to PEI. Viability of NIH 3T3 cells is expressed as a function of polymer concentration using the MTT assay. (Reprinted with permission from Lynn, D.M. and Langer, R., *J. Am. Chem. Soc.* 122, 10761–10768, 2000.)

compared impressively to data obtained for cell populations treated with PEI, for which fewer than 30% of cells remained viable at concentrations as low as 25 µg/ml (Figure 14.2). The corresponding hydrolytic degradation products for these three polymers did not affect cell growth or metabolism as determined by these assays.

It should be noted that the MTT assay is only a preliminary indicator of biocompatibility and that additional work must be done to more firmly establish the safety of poly(β-amino ester)s as gene delivery agents *in vivo*. Subsequent work with a broader range of polymers has suggested that cytotoxicity is a function of polymer structure and that self-assembled polymer–DNA complexes formed from certain polymers may be more cytotoxic than the free, uncomplexed polymers themselves.[17,18] This latter observation highlights the importance of evaluating new polymers such as these over the full range of conditions and forms likely to be encountered in the context of gene delivery.

14.4 SELF-ASSEMBLY WITH PLASMID DNA

To deliver DNA to cells, a cationic polymer must be able to self-assemble with DNA through electrostatic interactions and condense it into positively charged, nanometer-scale polymer–DNA conjugates.[1] Self-assembled polymer–DNA complexes can be characterized using a variety of different analytical techniques, including agarose gel electrophoresis, dynamic light scattering, and zeta potential analysis.[2,16,26] Agarose gel electrophoresis separates macromolecules on the basis of both charge and size, and the immobilization of DNA on a gel in the presence of cationic polymer can be used to determine the conditions under which self-assembly and/or charge neutralization occurs. Dynamic light scattering provides information related to the dimensions of the polymer–DNA conjugates formed upon self-assembly, and zeta potential analysis can be used to measure the relative surface charges of these particles. For most cell types, the size requirement for particle uptake via endocytosis is on the order of 200 nm or less, and a net positive charge on the surface of the conjugate has been shown to be important for triggering uptake.[27,28]

Polymers 1–3 are able to interact electrostatically with pDNA and retard the migration of DNA on an agarose gel.[16] As shown in Figure 14.3, for the self-assembly of polymer 1 with pCMV-Luc, DNA migration is completely retarded at DNA:polymer ratios above 1:1.0 (w/w). Polymers 2 and 3 inhibit the migration of plasmid DNA at DNA:polymer ratios (w/w) above 1:10 and 1:1.5, respectively. The higher concentrations of polymer 2 required to achieve complete retardation have been attributed to structural differences and estimated differences in the pK_a values for the amines in these three materials.[16] Subsequent electrophoresis experiments evaluating a broader array of 140 structurally related polymers further underscore the relationship between poly(β-amino ester) structure and the ability to form electrostatic complexes with pDNA under physiological conditions (as discussed below).[17,18]

Dynamic light scattering and zeta potential analyses demonstrate that polymers 1–3 can condense pDNA into small, positively charged polymer–DNA complexes less than 200 nm in diameter under physiological conditions.[16] Figures 14.4 and 14.5 show representative particle size and zeta potential profiles for complexes formed using polymer 3 at different DNA:polymer ratios. These data are consistent with models of DNA condensation observed for other polycations,[13] in which DNA is initially condensed into small, negatively charged particles at low polymer concentrations (Figure 14.4). Observed particle sizes increase dramatically upon the addition of more polymer as charge neutrality is achieved and aggregation occurs. Particle sizes decrease sharply at DNA–polymer concentrations above charge neutrality to yield stable dispersions of positively charged complexes ranging from 100 to 200 nm in diameter (Figure 14.4). This general model is further supported by the analysis of the zeta potentials of complexes formed under each of these conditions.

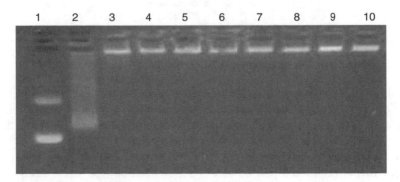

Figure 14.3 Agarose gel electrophoresis retardation of pCMV-Luc DNA by polymer 1. Lane numbers correspond to different DNA/polymer weight ratios as follows: 1) 1:0 (DNA only), 2) 1:0.5, 3) 1:1, 4) 1:2, 5) 1:3, 6) 1:4, 7) 1:5. 8) 1:6, 9) 1:7, 10) 1:8. (Reprinted with permission from Lynn, D.M. and Langer, R., *J. Am. Chem. Soc.* 122, 10761–10768, 2000.)

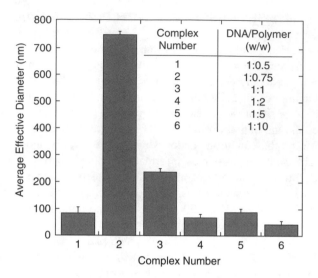

Figure 14.4 Average effective diameters of DNA–polymer complexes formed from pCMV-Luc plasmid and polymer **3** as a function of DNA–polymer weight ratios. (Reprinted with permission from Lynn, D.M. and Langer, R., *J. Am. Chem. Soc.* 122, 10761–10768, 2000.)

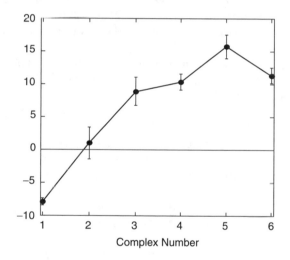

Figure 14.5 Average zeta potentials of DNA–polymer complexes formed from pCMV-Luc plasmid and polymer **3** as a function of DNA–polymer weight ratios. The numbers for each complex correspond to the weight ratios for complexes in Figure 14.4. (Reprinted with permission from Lynn, D.M. and Langer, R., *J. Am. Chem. Soc.* 122, 10761–10768, 2000.)

As shown in Figure 14.5, the zeta potentials of polymer–DNA particles formed from polymer **3** are negative at low polymer concentrations, near neutral at DNA:polymer ratios corresponding to the observation of large aggregates, and approach a limiting value of +10 mV at DNA:polymer ratios above 1:2. An evaluation of the particle sizes and zeta potentials of polymer–DNA complexes formed from a library of 140 different poly(β-amino ester)s suggests that these parameters are affected by polymer structure and molecular weight, and that new polymers need to be evaluated individually to understand, evaluate, and optimize individual polymer performance in subsequent transfection assays.[18]

The optimal biophysical parameters identified above were used in initial transfection assays to evaluate the ability of polymers **1–3** to mediate the expression of firefly luciferase in the NIH 3T3 cell line.[16] Polymers **1** and **2** did not yield observable levels of luciferase expression, but polymer **3** demonstrated transfection efficiencies exceeding those using PEI under certain conditions. These data demonstrated the potential of poly(β-amino ester)s as gene delivery agents and suggested that with further development this general synthetic approach could be used to identify efficacious, noncytotoxic alternatives to conventional cationic polymers used for gene delivery.

14.5 PARALLEL SYNTHESIS AND SCREENING OF POLY(β-AMINO ESTER) LIBRARIES

The process by which new materials are developed for biomedical applications has traditionally been linear and iterative; new polymers are synthesized around a specific set of design criteria and the resulting materials are tested individually for their properties. In recent years, several groups have focused on the development of parallel and combinatorial methods for the synthesis and screening of diverse collections of new biomaterials.[17,29-31] In the context of gene delivery, Murphy et al. recently reported the solid-phase synthesis and subsequent screening of a library of cationic peptoids to identify potential new transfection agents.[30]

The solution-phase synthesis of cationic polyesters generally requires either the design of specialized monomers[10-12,14] or the use of stoichiometric coupling reagents and amine protecting groups.[8,10,13,14] While these requirements have not limited the advent of new degradable polymers for gene delivery, they do place limits on the rates at which different structural analogs can be synthesized to investigate structure–function relationships or further optimize polymer properties. The synthesis of poly(β-amino ester)s is based on the conjugate addition of primary or bis(secondary) amines to diacrylates (Equations 14.1 and 14.2), and is not subject to the same set of limitations. In the context of parallel polymer synthesis, this modular approach provides an attractive framework for the rapid generation of structural analogs, for the following reasons. First, amine and diacrylate monomers are inexpensive, commercially available starting materials. Second, amines participate directly in the bond-forming processes in these reactions, and polymerization can be accomplished without the need for additional protection or deprotection schemes. Third, no byproducts are generated during polymerization, and fourth, the conjugate addition reaction is generally tolerant of additional functionality such as alcohols, ethers, and tertiary amines. This final point significantly expands the range of functionalized cationic polymers that can be synthesized because monomers can be chosen from a diverse pool of functionalized amine and diacrylate monomers.

Lynn et al. reported a parallel approach suitable for the synthesis of hundreds to thousands of structurally unique poly(β-amino ester)s and the application of these libraries to the rapid and high-throughput identification of new gene delivery agents and structure–function trends.[17] Figure 14.6 shows the set of seven diacrylate monomers (**A–G**) and 20 amine-based monomers (**1–20**) used to synthesize a proof-of-concept screening library. Monomers were selected to incorporate varying degrees of hydrophobicity, hydrophilicity, and functional character into the resulting materials. The size of the library constructed from this set of monomers (7 diacrylates x 20 amines = 140 different polymers) was selected to allow a reasonable level of structural and functional diversity to be incorporated without the need for automation in initial experiments. Subsequent work by Anderson et al. has miniaturized and automated certain monomer handling and polymerization procedures and has significantly expanded the number of polymers that can be synthesized and screened using this approach (as discussed below).[19]

The monomers in Figure 14.6 were combined in all possible pairwise amine–diacrylate combinations and all 140 polymerization reactions were conducted simultaneously in individual glass vials.[17] Removal of solvent provided amounts of each material sufficient for routine characterization and all subsequent biophysical and cell-based screening assays. Over 50% of the polymers were characterized by gel permeation chromatography (GPC) and found to have average molecular weights ranging from 2,000 to 50,000 relative to polystyrene standards. Only 70 polymers were

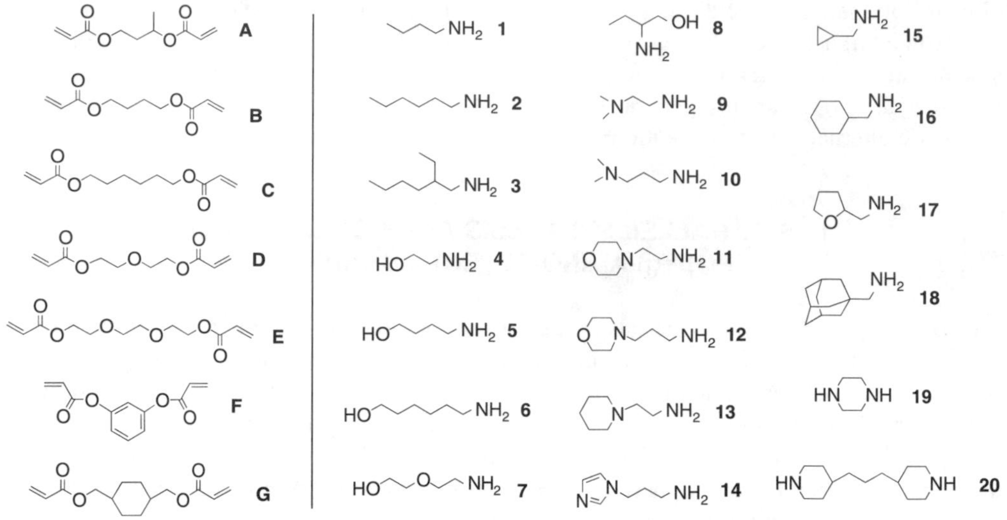

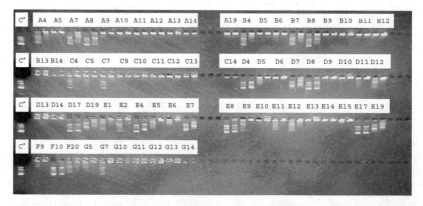

Figure 14.6   Diacrylate (A–G) and amine (1–20) monomers used for the parallel synthesis of a library of 140
structurally unique poly(β-amino esters). (Reprinted with permission from Lynn, D.M., Anderson,
D.G., Putnam, D., and Langer, R., *J. Am. Chem. Soc.* 123, 8155–8156, 2000.)

sufficiently water-soluble to be included in a parallel gel electrophoresis assay to determine the
ability of each polymer to form self-assembled complexes with DNA. As shown in Figure 14.7,
all 70 water-soluble polymers were assayed simultaneously at two different DNA:polymer ratios
(1:5 and 1:20) (w/w). Thirty-nine of these polymers retarded the migration of plasmid DNA through
the gel matrix at a ratio of 1:5 (e.g., **A5**) and 56 polymers were able to retard the migration of
DNA at a ratio of 1:20 (e.g., **A4** and **A5**). Fourteen polymers were unable to bind DNA sufficiently
under either of these conditions (e.g., **A7** and **A8**) and were discarded from further consideration
as gene delivery agents.

The set of 56 polymers identified in the gel electrophoresis assay were evaluated in high-
throughput cell-based transfection screening assays using a firefly luciferase reporter gene (pCMV-
Luc) and the COS-7 cell line.[17] To facilitate the rapid screening of polymers under several different
conditions, stock solutions of the 56 polymers identified above were prepared in 96-well master

Figure 14.7   Parallel gel electrophoresis assay used for the rapid identification of DNA-complexing members
of a 140-member poly(β-amino ester) library. Lane annotations correspond to the 70 water-soluble
members of the screening library. For each polymer, assays were performed at DNA:polymer
ratios of 1:5 (left well) and 1:20 (right well). Lanes marked C˙ contain DNA alone used as a control.
(Reprinted with permission from Lynn, D.M., Anderson, D.G., Putnam, D., and Langer, R., *J. Am.
Chem. Soc.* 123, 8155–8156, 2000.)

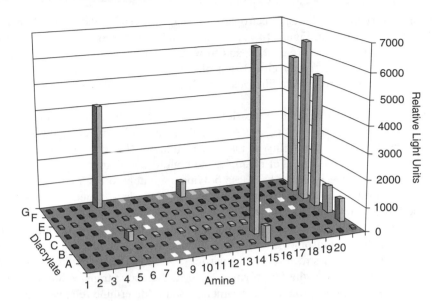

Figure 14.8 Results of a gene expression assay used to identify members of a 140-member poly(β-amino ester) library capable of mediating luciferase expression in COS-7 cells cultured in serum-free media. Light units are arbitrary and not normalized to total cell protein. (Reprinted with permission from Lynn, D.M., Anderson, D.G., Putnam, D., and Langer, R., *J. Am. Chem. Soc.* 123, 8155–8156, 2000.)

plates. Polymer–DNA complexes corresponding to each polymer were formed at desired polymer:DNA weight ratios directly in 96-well daughter plates and transferred directly to 96-well plates containing COS-7 cells in serum-free medium. Relative levels of luciferase expression were determined using a commercially available luciferase assay kit and a 96-well luminescence plate reader. This assay system was easily modified to evaluate transfection as a function of DNA concentration, DNA:polymer ratio, cell seeding densities, and incubation times.

Figure 14.8 shows the results of a screening assay employing pCMV-Luc (600 ng/well) at DNA:polymer ratios of 1:20 (w/w). Most of the polymers screened did not mediate gene expression at levels greater than "naked" DNA (no polymer) controls under these conditions. However, several polymers mediated higher levels of expression and were identified as "hits" in this assay. Two polymers, **B14** and **G5** (see Figure 14.6 for structures), yielded gene expression levels 4 to 8 times higher than control experiments using PEI and levels within or exceeding the range of expressed protein using Lipofectamine 2000, a leading commercially available lipid-based vector system. These results indicate that screening libraries of even modest size can be used to identify rapidly new polymers with interesting gene delivery properties. Additional polymers were identified as "hits" in subsequent screening assays, but levels of gene expression were not significantly higher than PEI. Although the mechanisms behind the efficacy of polymers **B14** and **G5** are not yet completely understood, polymer **B14** may be capable of buffering the pH of endosomal vesicles similar to other imidazole-substituted polymers.[2-4,32]

14.6 STRUCTURE–PROPERTY RELATIONSHIPS AND BIOPHYSICAL ANALYSIS

In addition to providing a screening platform for the identification of new gene delivery vectors, the library of structurally related polymers described above provides an opportunity to investigate structure–property relationships for this class of materials. The relatively small size of the library used and the diverse range of molecular weights obtained have thus far prevented the assignment of definitive structure–function relationships. However, an analysis of the water-soluble members of the polymer library by parallel gel electrophoresis suggests trends related to polymer structure and DNA binding. As shown in Figure 14.7, polymers synthesized from amine monomers bearing an oxygen atom in the β-position (e.g., **4**, **7**, and **8**) were generally unable to bind DNA under the conditions employed.[17,18] The further elucidation of structure–property relationships will require the generation of larger, more diverse libraries as well as greater control over polymer molecular weights.

Although the observation of gene expression in transfected cells is an important tool for the identification of new polymeric vectors, it is essentially an endpoint-driven analysis and provides no specific information about the ability or inability of individual polymers to overcome specific intracellular barriers to transfection. For example, the high levels of gene expression mediated by polymers **B14** and **G5** above indicate that these polymers are able to enter cells and escape from endosomal vesicles prior to fusion with lysosomes. However, the lack of gene expression observed using other polymers does not indicate that these polymers are uptake limited. For example, a polymer may be able to mediate cellular entry but gene expression could ultimately be limited by an inability to escape from endosomes or target the nucleus. Akinc et al. investigated the biophysical relationships between polymer structure and the particle sizes and zeta potentials of polymer–DNA complexes formed from each of the 56 polymers used in the above transfection screens.[18] These results were compared to flow cytometry experiments used to determine relative levels of particle uptake and the ability of internalized polymer–DNA complexes to avoid trafficking to lysosomes.

For polymer–DNA complexes formed from each member of the above library, small particle sizes and positive zeta potentials were found to correlate with higher rates of internalization in NIH 3T3 cells.[18] The majority (70%) of complexes formed at DNA:polymer ratios of 1:20 (w/w) were determined to have negative zeta potentials at pH 7.2. Complexes formed from the family of polymers formed from monomers **10**, **13**, and **14**, having multiple amines per repeat unit, had positive zeta potentials under these conditions and corresponded to the group of polymers mediating the highest levels of uptake. Complexes with effective diameters under 250 nm with negative zeta potentials were not internalized at appreciable levels. One exception to this general trend was complexes formed from polymer **G5**, which yielded complexes about 240 nm in size with a zeta potential of -3.5 mV. Complexes formed from polymer **G5** were internalized efficiently and resulted in high levels of gene expression (Figure 14.8).

The 10 polymers mediating the highest levels of cell uptake were evaluated in additional flow-cytometry assays using fluorescently double-labeled plasmid to determine the ability of their corresponding polymer–DNA complexes to avoid trafficking to lysosomes.[18] The details of these assays have been described in detail.[33,34] Polymers synthesized from monomer **14** were found to experience post-transfection pH environments of around 7.0, suggesting that these polymers were able to escape from endosomal vesicles prior to fusion with lysosomes. These results are consistent with the hypothesis that these polymers are able to act as "proton sponges" similar to other imidazole-functionalized polymers.[2-4,32] This assay also suggested that complexes formed from polymer **G5** were able to avoid trafficking to lysosomes, although the mechanism by which this occurs is less clear because this polymer does not contain an obvious means of promoting endosomal escape. Although polymers **B14** and **G5** both mediate high levels of expression, complexes formed from polymer **G5** are substantially cytotoxic, suggesting that polymer **B14** may be the more promising polymer for gene delivery.[18]

14.7 SYNTHESIS OF LARGER POLYMER LIBRARIES

Anderson et al. extended the parallel synthetic approach above to the synthesis and screening of a library of 2350 different poly(β-amino ester)s from a pool of 94 amines and 25 diacrylates.[19] Two factors limiting library size and screening throughput for the 140 member library described above were monomer handling during synthesis and the manipulation of solid or viscous polymer products during subsequent screening. Key to the extension of this approach, therefore, was the automation of monomer handling steps during synthesis and the development of a synthetic platform allowing product polymers to be stored and dispensed as solutions in DMSO. These modifications allowed polymers to be synthesized in, and later dispensed from, deep-well 96-well plates using either multichannel pipettors or a conventional liquid-handling robot.

All 2350 polymerization reactions were conducted simultaneously, and subsequent transfection screening assays were conducted at a rate of 1000 per day using plasmid encoding firefly luciferase (pCMV-Luc) and the COS-7 cell line.[19] Initial broad-based screening and more refined optimization procedures identified 46 new polymers that transfect as well as or better than PEI in serum-free medium (Figure 14.9); 26 of these polymers were also found to transfect COS-7 cells at levels exceeding those using Lipofectamine 2000, a leading lipid-based transfection vector system. The polymers identified in these assays comprise a fairly diverse set of structures, but many of these polymers have a number of structural features in common. For example, many of the most effective polymers are derived from hydrophobic diacrylates and/or linear bis(secondary) amines. Additionally, 12 of the 26 polymers identified that transfect cells more effectively than Lipofectamine 2000 are derived from hydroxyl-functionalized amines. These polymers contain alcohol-functionalized side chains and are functionally similar to hydroxyl-functionalized polymer **G5** identified in earlier proof-of-concept screening.[17]

14.8 pH-RESPONSIVE POLY(β-AMINO ESTER) MICROSPHERES

Microparticles fabricated from hydrolytically degradable polymers such as poly(lactic-*co*-glycolic acid) (PLGA) have been widely used to sustain the release of encapsulated therapeutic compounds.[35,36] The advent of therapeutics that require intracellular administration and subsequent trafficking to the nucleus has created a demand for related materials that respond to intracellular stimuli such as pH.[37] Lynn et al. reported the fabrication of pH-responsive microspheres using poly(β-amino ester) **3**.[21] Solid samples of polymer **3** are insoluble in water at physiological pH values around 7.2, but the polymer becomes soluble in aqueous media when the pH of the solution is reduced below 6.5. This polymer was suitable for the fabrication of polymer microspheres ranging from 5 to 30 microns in diameter using a double-emulsion fabrication technique (Figure 14.10). Microspheres encapsulating fluorescently labeled dextran were used as a model system to study encapsulation and release as a function of pH.

Suspended microspheres remained intact at pH 7.4 but dissolved immediately when the pH of the suspending medium was lowered to 5.1.[21] Figure 14.11 shows the release profile for encapsulated dextran in response to an environmental change in pH. While very low levels of dextran were released after 24 h at pH 7.4, lowering the pH of the suspending medium to 5.1 resulted in rapid and quantitative release of encapsulated material. As the transition from solid microspheres to dissolved material occurs over the range of extracellular and endosomal pH, these particles may be useful for the delivery of nucleic acid therapeutics that must escape endosomal compartmentalization prior to fusion with lysosomes.

The relatively large (i.e., several micron) size of these microparticles should allow them to be targeted selectively to phagocytic macrophages. Initial uptake experiments demonstrated that these particles could be internalized by macrophages and that encapsulated material was delivered to the cytosol much more rapidly than for experiments employing microspheres fabricated from PLGA.[21] Potineni et al. reported the fabrication of pH-responsive nanoparticles from polymer **3** ranging from 100 to 150 nm in diameter using a solvent displacement technique.[38] Poly(ethylene oxide) modified polymer **3** nanoparticles were suitable for the encapsulation and delivery of either rhodamine-123 or the drug paclitaxel to human breast cancer cells. Experiments employing rhodamine-loaded nanospheres demonstrated the efficient distribution of fluorescent compound to the cytoplasm of these cells within 4 h of administration. The pH-responsive nature of these particles should also

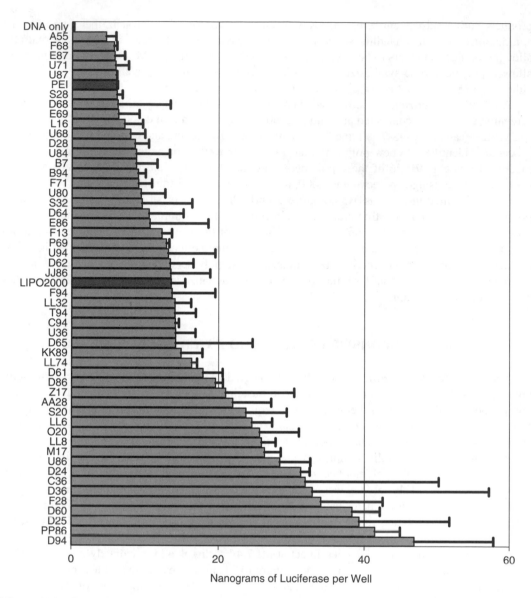

Figure 14.9 Optimized relative transfection efficiencies for the top-performing 50 members of a 2350-member poly(β-amino ester) library relative to PEI and Lipofectamine 2000 (labeled).

be useful for the triggered delivery of therapeutic agents to the low pH environment within solid tumors and could be extended to the intracellular release of nucleic acid materials.

14.9 SUMMARY

Poly(β-amino ester)s comprise a class of degradable cationic polymers with many desirable properties in the context of gene delivery. The synthesis of these materials is straightforward and a variety of structural variants can be generated from a broad array of commercially available amine and diacrylate monomers. This has enabled the parallel synthesis of combinatorial libraries containing hundreds to thousands of structurally related properties that can be screened using high-throughput assays to identify new polymers with enhanced gene delivery properties. Access to

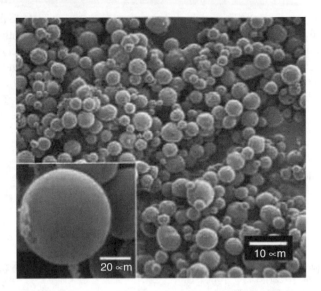

Figure 14.10 SEM image of rhodamine/dextran-loaded microspheres fabricated from polymer **3**. (Reprinted with permission from Lynn, D.M., Amiji, M.M., and Langer, R., *Angew Chem. Int. Ed. Engl.* 40, 1707–1710, 2001.)

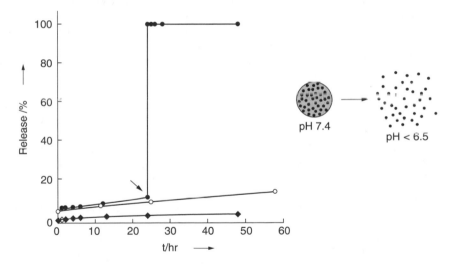

Figure 14.11 Release profiles of rhodamine/dextran from polymer **3** and PLGA microspheres at various pH values: (O) polymer **3** (pH 7.4); (●) polymer **3** (pH 7.5/5.1); (♦) PLGA (pH 7.4/5.1). The arrows indicate the points at which HEPES buffer (pH = 7.4) was exchanged with acetate buffer (pH = 5.1). (Reprinted with permission from Lynn, D.M., Amiji, M.M., and Langer, R., *Angew Chem. Int. Ed. Engl.* 40, 1707–1710, 2001.)

collections of structurally related polymers should also provide new opportunities to investigate structure–function relationships and accelerate the pace at which new polymers are discovered for clinical applications.

The screening of a library of 140 different poly(β-amino ester)s has begun to reveal structure–property relationships that will be useful in understanding and optimizing polymer properties. The cell-based screening of a larger 2350-member library has identified 46 new degradable polymers that mediate higher levels of gene expression than PEI, and 26 polymers that perform better than a leading lipid-based reagent. Important steps for the continued development and optimization of poly(β-amino ester)s for gene delivery applications *in vitro* and *in vivo* include improving the serum stability and demonstrating uniform control over molecular weight during synthesis so that polymer properties can be more directly compared. The use of poly(β-amino ester)s as pH-responsive degradable polymers for the encapsulation of DNA within microspheres and nanospheres could also represent an important approach to the targeted, intracellular delivery of nucleic acid material and other drugs to cells.

REFERENCES

1. Luo, D. and Saltzman, W.M., Synthetic DNA delivery systems. *Nat. Biotechnol.* 18, 33–37, 2000.
2. Putnam, D., Gentry, C.A., Pack, D.W. and Langer, R., Polymer-based gene delivery with low cytotoxicity by a unique balance of side-chain termini. *Proc. Natl. Acad. Sci. U.S.A.* 98, 1200–1205, 2001.
3. Benns, J.M., Choi, J.S., Mahato, R.I., Park, J.S. and Kim, S.W., pH-sensitive cationic polymer gene delivery vehicle: N-Ac-poly(L-histidine)-graft-poly(L-lysine) comb shaped polymer. *Bioconjug. Chem.* 11, 637–645, 2000.
4. Midoux, P. and Monsigny, M., Efficient gene transfer by histidylated polylysine/pDNA complexes. *Bioconjug. Chem.* 10, 406–411, 1999.
5. Godbey, W.T., Wu, K.K. and Mikos, A.G., Poly(ethylenimine) and its role in gene delivery. *J. Control. Release* 60, 149–160, 1999.
6. Kircheis, R., Wightman, L. and Wagner, E., Design and gene delivery activity of modified polyethylenimines. *Adv. Drug Deliv. Rev.* 53, 341–358, 2001.
7. Navarro, J., Oudrhiri, N., Fabrega, S. and Lehn, P.,) Gene delivery systems: Bridging the gap between recombinant viruses and artificial vectors. *Adv. Drug Deliv. Rev.* 30, 5–11, 1998.
8. Lim, Y.B., Choi, Y.H. and Park, J.S., A self-destroying polycationic polymer: Biodegradable poly(4-hydroxy-L-proline ester). *J. Am. Chem. Soc.* 121, 5633–5639, 1999.
9. Lim, Y.B., Han, S.O., Kong, H.U., Lee, Y., Park, J.S., Jeong, B. and Kim, S.W., Biodegradable polyester, poly[alpha-(4-aminobutyl)-L-glycolic acid], as a non-toxic gene carrier. *Pharm. Res.* 17, 811–816, 2000.
10. Lim, Y.B., Kim, C.H., Kim, K., Kim, S.W. and Park, J.S., Development of a safe gene delivery system using biodegradable polymer, poly[alpha-(4-aminobutyl)-L-glycolic acid]. *J. Am. Chem. Soc.* 122, 6524–6525, 2000.
11. Lim, Y.B., Kim, S.M., Lee, Y., Lee, W.K., Yang, T.G., Lee, M.J., Suh, H. and Park, J.S., Cationic hyperbranched poly(amino ester): A novel class of DNA condensing molecule with cationic surface, biodegradable three dimensional structure, and tertiary amine groups in the interior. *J. Am. Chem. Soc.* 123, 2460–2461, 2001.
12. Lim, Y.B., Kim, S.M., Suh, H. and Park, J.S., Biodegradable, endosome disruptive, and cationic network-type polymer as a highly efficient and nontoxic gene delivery carrier. *Bioconjug. Chem.* 13, 952–957, 2002.
13. Putnam, D. and Langer, R., Poly(4-hydroxy-L-proline ester): Low-temperature polycondensation and plasmid DNA complexation. *Macromolecules* 32, 3658–3662, 1999.
14. Wang, J., Mao, H.Q. and Leong, K.W., A novel biodegradable gene carrier based on polyphosphoester. *J. Am. Chem. Soc.* 123, 9480–9481, 2001.
15. Zhao, Z., Wang, J., Mao, H.Q. and Leong, K.W., Polyphosphoesters in drug and gene delivery. *Adv. Drug Deliv. Rev.* 55, 483–499, 2003.
16. Lynn, D.M. and Langer, R., Degradable poly(beta-amino esters): Synthesis, characterization, and self-assembly with plasmid DNA. *J. Am. Chem. Soc.* 122, 10761–10768, 2000.
17. Lynn, D.M., Anderson, D.G., Putnam, D. and Langer, R., Accelerated discovery of synthetic transfection vectors: parallel synthesis and screening of a degradable polymer library. *J. Am. Chem. Soc.* 123, 8155–8156, 2001.
18. Akinc, A., Lynn, D.M., Anderson, D.G. and Langer, R., Parallel synthesis and biophysical characterization of a degradable polymer library for gene delivery. *J. Am. Chem. Soc.* 125, 5316–5323, 2003.

19. Anderson, D. G., Lynn, D. M. and Langer, R., Semi-Automated Synthesis and Screening of a Large Library of Degradable Cationic Polymers for Gene Delivery. *Angew. Chem. Int. Ed. Engl.* 42, 3153–3158, 2003.

20. Schaffer, D.V., Fidelman, N.A., Dan, N. and Lauffenburger, D.A., Vector unpacking as a potential barrier for receptor-mediated polyplex gene delivery. *Biotechnol. Bioeng.* 67, 598–606, 2000.

21. Lynn, D.M., Amiji, M.M. and Langer, R., pH-Responsive Polymer Microspheres: Rapid Release of Encapsulated Material within the Range of Intracellular pH. *Angew. Chem. Int. Ed. Engl.* 40, 1707–1710, 2001.

22. Danusso, F. and Ferruti, P., Synthesis of tertiary amine polymers. *Polymer* 11, 88–113, 1970.

23. Ferruti, P. and Barbucci, R., Linear Amino Polymers - Synthesis, Protonation and Complex-Formation. *Adv. Polym. Sci.* 58, 55–92, 1984.

24. Ferruti, P., Marchisio, M.A. and Duncan, R., Poly(amido-amine)s: Biomedical applications. *Macromol. Rapid Comm.* 23, 332–355, 2002.

25. Hansen, M.B., Nielsen, S.E. and Berg, K., Re-examination and further development of a precise and rapid dye method for measuring cell growth/cell kill. *J. Immunol. Methods* 119, 203–210, 1989.

26. Gonzalez, H., Hwang, S.J. and Davis, M.E., New class of polymers for the delivery of macromolecular therapeutics. *Bioconjugate Chem.* 10, 1068–1074, 1999.

27. Mahato, R.I., Rolland, A. and Tomlinson, E., Cationic lipid-based gene delivery systems: pharmaceutical perspectives. *Pharm. Res.* 14, 853–859, 1997.

28. Wagner, E., Ogris, M. and Zauner, W., Polylysine-based transfection systems utilizing receptor-mediated delivery. *Adv. Drug Deliv. Rev.* 30, 97–113. 1998.

29. Brocchini, S., James, K., Tangpasuthadol, V. and Kohn, J., A combinatorial approach for polymer design. *J. Am. Chem. Soc.* 119, 4553–4554, 1997.

30. Murphy, J.E., Uno, T., Hamer, J.D., Cohen, F.E., Dwarki, V. and Zuckermann, R.N., A combinatorial approach to the discovery of efficient cationic peptoid reagents for gene delivery. *Proc. Natl. Acad. Sci. U.S.A.* 95, 1517–1522, 1998.

31. Brocchini, S., Combinatorial chemistry and biomedical polymer development. *Adv. Drug Deliv. Rev.* 53, 123–130, 2001.

32. Pack, D.W., Putnam, D. and Langer, R. Design of imidazole-containing endosomolytic biopolymers for gene delivery. *Biotechnol. Bioeng.* 67, 217–223, 2000

33. Akinc, A. and Langer, R., Measuring the pH environment of DNA delivered using nonviral vectors: implications for lysosomal trafficking. *Biotechnol. Bioeng.* 78, 503–508, 2002.

34. Forrest, M.L. and Pack, D.W., On the kinetics of polyplex endocytic trafficking: implications for gene delivery vector design. *Mol. Ther.* 6, 57–66, 2002.

35. Okada, H., One- and three-month release injectable microspheres of the LH-RH superagonist leuprorelin acetate. *Adv. Drug Deliv. Rev.* 28, 43–70, 1997.

36. Hanes, J., Cleland, J.L. and Langer, R., New advances in microsphere-based single-dose vaccines. *Adv. Drug Deliv. Rev.* 28, 97–119, 1997.

37. Gerasimov, O.V., Boomer, J.A., Qualls, M.M. and Thompson, D.H., Cytosolic drug delivery using pH- and light-sensitive liposomes. *Adv. Drug Deliv. Rev.* 38, 317–338, 1999.

38. Potineni, A., Lynn, D.M., Langer, R. and Amiji, M.M., Poly(ethylene oxide)-modified poly(beta-amino ester) nanoparticles as a pH-sensitive biodegradable system for paclitaxel delivery. *J. Control. Release* 86, 223–234, 2003.

CHAPTER 15

Cationic Polyesters as Biodegradable Polymeric Gene Delivery Carriers

Yong-Beom Lim, Yan Lee, and Jong-Sang Park

CONTENTS

15.1 INTRODUCTION

Successful gene therapy is largely dependent on the development of a vector that has a high transfection efficiency (TE) and as low a cytotoxicity as possible. Viral vectors include recombinant retroviruses, adenoviruses, and adeno-associated viruses that have demonstrated high TE but whose effectiveness is limited due to adverse effects such as immunogenicity and mutagenesis to generate infectious wild-type viruses.[1] Synthetic nonviral gene carriers, cationic polymers or liposomes, have elicited much attention as alternatives for viral systems.[1-5] They can form soluble complexes with plasmid DNA or antisense oligonucleotide, which can then be carried into cellular

0-8493-1934-X/05/$0.00+$1.50
© 2005 by CRC Press LLC

compartments. With suitable design of synthetic gene carriers, it has become possible to construct highly efficient and targetable nonviral systems. However, backbone linkages of most current gene carriers are composed of amide or carbon-carbon bonds, which hardly spontaneously degrade in aqueous solution. For this reason, the potential for current nonbiodegradable gene carriers to accumulate in an endosomal compartment or cell nucleus and adversely interact with the host gene always exists. These issues present a problem with regard to their use in the gene therapy treatment of human disease.

For this reason, we wished to create spontaneously degradable cationic polymers that could be used in DNA condensation and gene delivery, and eventually in gene therapy, since we hypothesized that such polymers would be minimally toxic to cells. In addition, the degraded oligomers or monomers will be rapidly removed from the cellular compartment followed by metabolism and excretion from the body. Polyesters with pendent amine groups appeared to be well suited for this purpose. After initiation of this project, several cationic polyesters were rationally synthesized and tested in our laboratory as potential gene delivery vectors by producing improvements in each polymer in terms of toxicity and TE.

15.2 POLY(4-HYDROXY-L-PROLINE ESTER) (PHP-ESTER)[6]

15.2.1 Polymer Synthesis

To synthesize a biodegradable, biocompatible, and pendent amine-containing polyester, 4-hydroxy-L-proline was chosen as a monomer. The monomeric unit of PHP-ester is a natural amino acid, 4-hydroxy-L-proline, which is a major constituent of collagen, gelatin, and other proteins. Synthesis of cbz-protected PHP-ester (PHCP-ester) was performed in the melt under vacuum (Scheme 15.1). PHCP-ester had number- and weight-average degrees of polymerization (DPs) of 21 and 36, respectively, which were appropriate for DNA condensation. Amine protecting cbz groups were removed by catalytic transfer hydrogenation-yielding PHP-ester. To investigate the influence of deprotection reaction on the integrity of the polymer, a secondary amine-protecting cbz group was regenerated from PHP-ester. Gel permeation chromatography (GPC) analysis shows that the number- and weight-average DPs of cbz-regenerated polymer were 27 and 34, respectively. This minute change of the DP shows that the integrity of the polymer backbone is largely preserved under this deprotection condition. The structure of PHP-ester was characterized by multiple secondary amine groups that could be protonated near neutral pH and biodegradable ester backbone.

15.2.2 Molecular Weight Distribution Determination and Degradation

As PHP-ester was a basic, water-soluble, and fast-degrading polymer (*vide infra*), it was hard to determine its molecular weight distribution (MWD). The problems associated with traditional

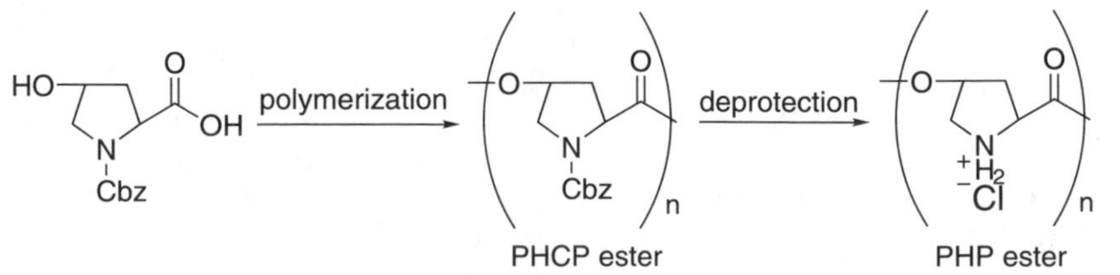

Scheme 15.1   Synthesis of PHP-ester. (From Lim, Y.-b., Choi, Y.H., and Park, J.-s., A self-destroying polycationic polymer: biodegradable poly(4-hydroxy-L-proline ester), *J. Am. Chem. Soc.* 121, 5633, 1999. With permission.)

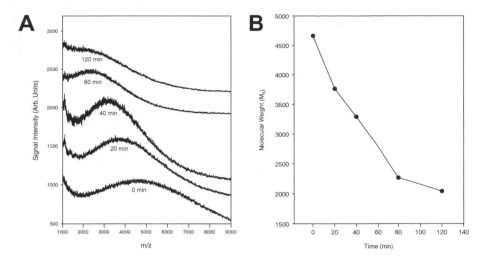

Figure 15.1 (A) MALDI-TOF spectra of PHP-ester degradation. MALDI spectra of PHP-ester at a 37°C buffer condition (pH 7) were obtained by varying incubation time. The times in min represent incubation time. The spectrum is the sum of 32 laser shots and was 7-point Savitzky-Golay smoothed. (B) Plot of M_p change as a measure of PHP-ester degradation. (From Lim, Y.-B., Choi, Y.H., and Park, J.-S., A self-destroying polycationic polymer: biodegradable poly(4-hydroxy-L-proline ester), *J. Am. Chem. Soc.* 121, 5633, 1999. With permission.)

methods, e.g., GPC for accurate measurement of MWD of water-soluble polymers, are well known.[7] As a result, matrix-assisted laser desorption/ionization mass spectrometry (MALDI-TOF MS) was chosen as a method for estimating the MWD of PHP-ester. MALDI is a mild ionization process that has been used to determine the molecular weight of proteins, oligonucleotides, and, in some limited cases, synthetic polymers. Although it was well known that the MWD of synthetic polymers having a polydispersity greater than 1.1 cannot be accurately characterized by MALDI alone because of mass discrimination effects,[8,9] the obtained MWD of PHP-ester was in agreement with the calculated value. As a result, the value of M_p, the most probable peak molecular weight determined from the highest peak intensity in the MALDI spectrum, was chosen to describe the MWD of PHP-ester, rather than M_n or M_w.

By obtaining MALDI spectra of PHP-ester incubated in pH 7 buffer at 37°C at appropriate time intervals, the degradation kinetics of the polymer could be monitored. It was found that PHP-ester was degraded in a very rapid fashion with a half-life of less than 2 h (Figure 15.1). In comparison, a degradation study of biodegradable polyester such as poly(lactide-*co*-glycolide) (PLGA) shows that it needs 2 to 24 months to degrade.[10] Since PLGA is composed of an ester in the backbone with no pendent functional groups, hydrolysis by water is the main reason for PLGA degradation. Polymers degrading at such a rapid rate as that of PHP-ester had never been previously reported, so the degradation mechanism of PHP-ester was regarded as "self-destruction," in which pendent secondary amine groups act as nucleophiles. After 3 months, PHP-ester was degraded nearly completely to its monomeric unit, 4-hydroxy-L-proline (Figure 15.2).

15.2.3 DNA Condensation and Gene Delivery

To study the interaction of PHP-ester with DNA, the shift of DNA electrophoretic migration as a function of the polymer concentration was performed. As the proportion of PHP-ester in the sample increased, there was an increase in the DNA remaining at the well of the gel with the polymer. This result was consistent with the fact that secondary amine groups of PHP-ester with positive charge interact with DNA phosphate groups with negative charge to form neutral self-assembled polyplex. In contrast to the rapid degradation of PHP-ester, the PHP-ester–DNA

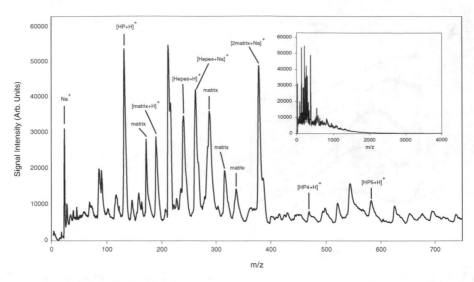

Figure 15.2   Degradation of PHP-ester up to near completion. The PHP-ester solution as described above was incubated at 37°C for 3 months. MALDI spectrum was measured by using an αCHCA as a matrix. HP, hydroxy-proline; HP2, dimer of hydroxy-proline; HP4, tetramer of hydroxy-proline; HP5, pentamer of hydroxy-proline; matrix, peak found in αCHCA background. The full spectrum is shown in the upper-right position. The spectrum is the sum of 128 laser shots and was 7-point Savitzky-Golay smoothed. (From Lim, Y.-B., Choi, Y.H., and Park, J.-S., A self-destroying polycationic polymer: biodegradable poly(4-hydroxy-L-proline ester), *J. Am. Chem. Soc.* 121, 5633, 1999. With permission.)

polyplex was stable over several hours, indicating that PHP-ester retained the ability to form polyplex with DNA. This finding was attributed to the occupation of PHP-ester amine groups by DNA phosphates resulting in the slower degradation of PHP-ester. The size distribution of PHP-ester–DNA measured by dynamic light scattering (DLS) and ζ potential also revealed the formation of polyplex.

In vitro gene delivery ability of the polyplex was assessed by transfecting CPAE cells with PHP-ester–pSV-β-gal polyplex. The TE of PHP-ester was comparable to that of poly(L-lysine), indicating the potency of the polymer in gene delivery.

15.3 POLY(α-[4-AMINOBUTYL]-L-GLYCOLIC ACID) (PAGA)[11,12]

Although PHP-ester demonstrated the possible usefulness of biodegradable cationic polymers as nontoxic gene carriers, the polymer had relatively weak binding ability to DNA and quite low TE. The weakness in the DNA binding ability of PHP-ester should be due to the low pK_a value of secondary amine groups of the polymer. (The pK_a of 4-hydroxy-L-proline: 9.66; L-lysine: 10.54). However, cationic polymers binding strongly to DNA are required to condense DNA into a geometrically compact shape. For this reason, we attempted the synthesis of a polyester with stronger DNA binding ability than PHP-ester in the expectation that this might enhance the TE as well.

15.3.1 Polymer Synthesis and PAGA Structure

Cbz-protected PAGA was synthesized by condensation polymerization of Nε-cbz-L-oxylysine (Scheme 15.2). L-oxylysine is an analogue of L-lysine with a hydroxyl group in the C_α position instead of a primary amine group in L-lysine. Cbz groups were removed by catalytic transfer hydrogenation and the resulting primary amine groups of PAGA were obtained as a hydrochloride

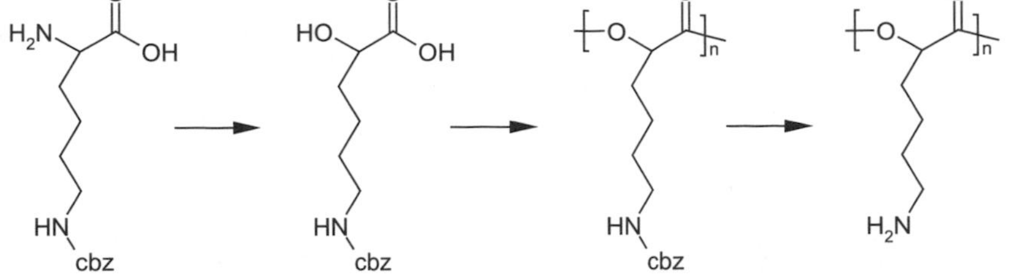

Scheme 15.2 Synthesis of PAGA.

salt. The M_p of PAGA as determined by MALDI was 3300, which was appropriate for DNA condensation. The structure of PAGA is almost identical to poly(L-lysine) (PLL) except for the backbone linkages. The backbone linkages of PAGA and PLL are ester and amide bonds, respectively. As PLL is one of the most intensively studied polymeric gene carriers,[13] comparing PAGA and PLL was thought to be of interest. PLL is known to have cytotoxicity[14] that might arise from its slow degradation *in vivo*. Spontaneously degradable ester linkages in PAGA would make this polymer minimally toxic.

15.3.2 Degradation and DNA Condensation

As revealed by a degradation study of PAGA conducted in buffer solution (pH 7.3) at 37°C, PAGA exhibited rapid degradation kinetics similar to those of PHP-ester. Only 30 min was needed for M_p of the intact polymer to halve (Figure 15.3). Such a fast degradation of PAGA implied that the polymer is another example of a "self-destroying" polymer in which the main chain cleavage occurs by the nucleophilic attack of amine groups of the polymer itself or of the nearby polymer molecules. The final degradation product of the polymer was a degraded monomer (L-oxylysine) and required approximately 5 months.

Gel band shift assay revealed that DNA complexes of PAGA were formed at a charge ratio (+/−) of about 1. Ethidium bromide exclusion assay reconfirmed that the polyplex was formed below a charge ratio of 2. These results show that PAGA produced strong and near stoichiometric DNA condensate.

Since the degradation of PAGA was very fast, as shown in Figure 15.3, we questioned whether the polyplex is stable over the time needed for *in vitro* transfection, which is generally 4 h. Stability studies were carried out by measuring dissociated DNA, in the form of a band, from the polyplex at pH = 7.3 and 37°C. We observed a slower degradation rate for the polyplex than for PAGA alone and there was no indication of DNA dissociation from the polyplex even after 4 h incubation. The polyplex dissociated completely within one day, while PLL–DNA polyplex was stable for 4 days.

For the DNA complexes of polymer to express the encoded gene in the cell nuclei, the polymer should protect DNA for a few hours from nuclease attack in the cellular environment. Fragmentation of DNA was found to be rapid as manifested by a sudden increase in absorbance at 260 nm occurring within 1 min after addition of nuclease into DNA solution. However, the DNA was partially protected from nuclease attack at a charge ratio of 3 and complete protection was achieved at a charge ratio of 5. The protection continued throughout an incubation period of 4 h with nuclease, implying that the degradation of DNA-complexed PAGA is not as fast compared to free PAGA.

15.3.3 Cytotoxicity and Transfection Efficiency

Owing to the biodegradable property of PAGA, the polymer did not display any detectable cytotoxicity even at very high concentrations (300 µg/ml), as expected. There is a notice of greater toxicity of some polymers applied in the absence of DNA.[15] In this regard, PAGA's lack of toxicity

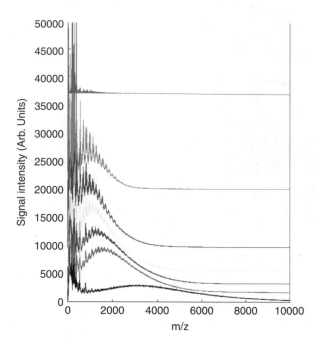

Figure 15.3   Degradation profiles of PAGA incubated at 37°C, pH 7.3. Aliquots were taken at appropriate time intervals and MALDI spectra were measured. From the bottom: M_p at 0 min, 30 min, 1 h, 3 h, 5 h, 3 days, and 6 months. (From Lim, Y.-B. et al., Development of a safe gene delivery system using biodegradable polymer, poly[α-(4-aminobutyl)-L-glycolic acid], *J. Am. Chem. Soc.* 122, 6524, 2000. With permission.)

without DNA complexation is a remarkable finding. To our knowledge, there had been no gene carrier that does not display any noticeable cytotoxicity without DNA complexation. The degradable property of PAGA, with its nontoxic and biocompatible monomers and small oligomers, might be an important contributory factor to the nontoxicity of this polymer.

Under optimized conditions, the TE of PAGA was about three times higher than that of PLL. Although small, the increased TE was ascribed to the near nontoxicity of PAGA and the faster release of DNA from the complexes due to degradation of PAGA after internalization within cells or nuclei. The optimum transfection was observed at a charge ratio of over 60. Such a high polymer:DNA ratio for optimum transfection could be explained by the fact that the gradual degradation of PAGA in the polyplex, which might be faster in the cellular environment than *in vitro* buffer conditions, demands more PAGA molecules for sustained DNA complexation.

15.4 HYPERBRANCHED POLY(AMINO ESTER) (H-PAE)[16]

It has been shown that polyethylenimine (PEI) and starburst PAMAM dendrimers, which have branched structures, are very efficient in gene delivery compared to other polymeric gene carriers. It is generally accepted that their high TE is due to the endosome buffering effect, in which tertiary amine groups in the interior of the polymers protonated at acidic endosomal pH disrupt the endosome by mechanical swelling or osmotic effects.[17] Based on this fact, we hypothesized that a biodegradable cationic polymer with branched structure and interior tertiary amine groups would be both nontoxic and efficient through the endosome buffering effect.

15.4.1 Structure and Synthesis

Dendrimers[18] and hyperbranched polymers[19] are examples of the most widely used branched polymers. Both have merits and drawbacks. Dendrimers have a precisely defined shape and molecular weight, which, however, is achieved at the cost of iterative synthetic steps. In contrast, hyperbranched polymers are synthesized in one-step polymerization but with irregular shape and broad MWD. It has been shown that fractured PAMAM dendrimers are more efficient in transfection than intact PAMAMs due to the increased flexibility of the fractured dendrimers, which enables them to be compact when complexed with DNA and to swell when released from DNA.[20] For this reason, we chose to adopt a hyperbranched structure in designing the branched and endosome disruptive biodegradable polymer.

Hyperbranched-PAE was synthesized by polycondensation of AB_2 type monomers having one hydroxyl and two methyl ester groups in the presence of trifunctional core-forming molecules followed by surface functionalization with primary amino groups. The interesting structural characteristics of the polymer are the cationic surface, biodegradable ester backbone linkages, and multiple tertiary amine groups in the interior.

Hyperbranched poly(amino ester)

There are two different types of amine groups, primary and tertiary, in the polymer. The polymer was designed with the belief that the primary amine groups at the surface will function to condense DNA and tertiary amine groups in the interior to exert an endosome buffering effect.

To calculate degree of branching (DB), model compounds representing terminal and linear units of the polymer were synthesized. The unique ^{13}C-NMR resonances of the model compounds enabled the assignment of different structural subunits (terminal, linear, and dendritic) of the polymer. From the NMR spectrum of the polymer, the peaks occurring at 52.11 and 52.04 ppm were found to be the signals of terminal and linear units, respectively. With this result, the peak at 51.96 ppm could be assigned as the signal of dendritic units. Using a quantitative inverse-gated broadband decoupled technique (INVGATE) and the most common formula reported by Fréchet et al.,[21] we could calculate the DBs of the polymer. Relatively high DBs (0.6 ± 0.02) were obtained for the polymer.

15.4.2 Gene Delivery

With the protonation of the primary amine groups, h-PAE became polycationic at near neutral pH. The formation of interpolyelectrolyte complexes of DNA with the polymer was investigated by measuring the shift of DNA electrophoretic migration as a function of polymer concentration. When the polymer:DNA ratio (w/w) reached 10, no free DNA was in solution indicating the formation of the complexes.

The toxicity of the polymer on the cells was assayed using an MTT assay. After incubating 293 cells for 4 h with the polymer, the cells showed neither a decrease in cell population nor a cell morphology change, indicating that the polymer is not harmful to the cell growth. The cells retained about 85% viability at high polymer concentration (100 μg/ml). In contrast, PLL, PAMAM generation 4 (G4), PAMAM G6, and PEI showed much higher toxicity than the polymer. Although transfection of the cells with the polymer was not as effective as that for PAMAM or PEI, it was increased by a greater than tenfold factor compared with that of PAGA. Since chloroquine was not used in this experiment, this enhanced transfection should be the result of the endosome buffering activity of the polymer.

15.5 NETWORK POLY(AMINO ESTER) (N-PAE)[22]

15.5.1 Structure and Synthesis

Encouraged by the increased transfection of h-PAE, we set out to design another type of branched poly(amino ester). Taking note of the fact that PEIs, one of the most efficient polymeric gene carriers, have an irregular network structure, we hoped that poly(amino ester) with a network-type structure would have a high TE.

The synthesis of n-PAE commenced by polymerizing monomer **1**, which has three hydroxyls, two methyl esters, and one tertiary amine, affording polymer **1** (Scheme 15.3). Coupling of Fmoc-eAhx in hydroxyls of polymer **1**, followed by deblocking of Fmoc groups yields n-PAE. A number of primary amine groups quantified by the fluorescamine method[23] showed that n-PAE had 1.3 μM primary amines per mg. GPC analysis with linear polystyrene standard showed that M_n and M_w of Fmoc-eAhx-coupled polymer **1** (n-PAE precursor) were 3800 and 4500, respectively. It is well known that GPC, based on hydrodynamic radius (R_h) measurement of polymers, underestimates by several times the true MWD of branched polymers.[24] As a result, it is generally accepted that the true MWD of branched polymers should be three to five times higher than the values obtained from GPC. In this respect, the true MWD of n-PAE should be much higher than the values obtained by GPC with the linear polymer standard.

15.5.2 Stability of n-PAE–DNA Polyplex

It has been shown that polyesters having primary or secondary amine groups such as PHP-ester and PAGA degraded very rapidly through the self-destruction mechanism.[6,11,12] For this reason, these polymers had very short half-lives of less than 2 h in aqueous environment. The DNA complexes of these polymers were more stable than the polymers alone with half-lives of less than 8 h *in vitro* buffer condition. However, it might be necessary to extend the speed of degradation to a reasonable rate for the polymeric vector to protect DNA from the harsh extracellular or cytosolic environment until it enters the cell nucleus. It can be expected that the MWD of the linear polyesters would be halved by a single cleavage of the polymer chain, which explains the rapid degradation kinetics of the polymers. We expected that the MWD of polymer is not likely to change by a single cleavage or several chain cleavages in a network-type chain interconnected polymer structure.

The stability of n-PAE polyplex increased with increasing nitrogen/phosphate (N/P) ratio where the polyplex remained intact over 3 days at and above a N/P ratio of 20 (Figure 15.4). The much-increased stability of the polyplex should be the result of the network structure of n-PAE.

15.5.3 Gene Delivery

About 90% of cells treated with n-PAE remain viable even at very high polymer concentration (200 μg/ml). The near nontoxic property of n-PAE should be the result of the biodegradable property

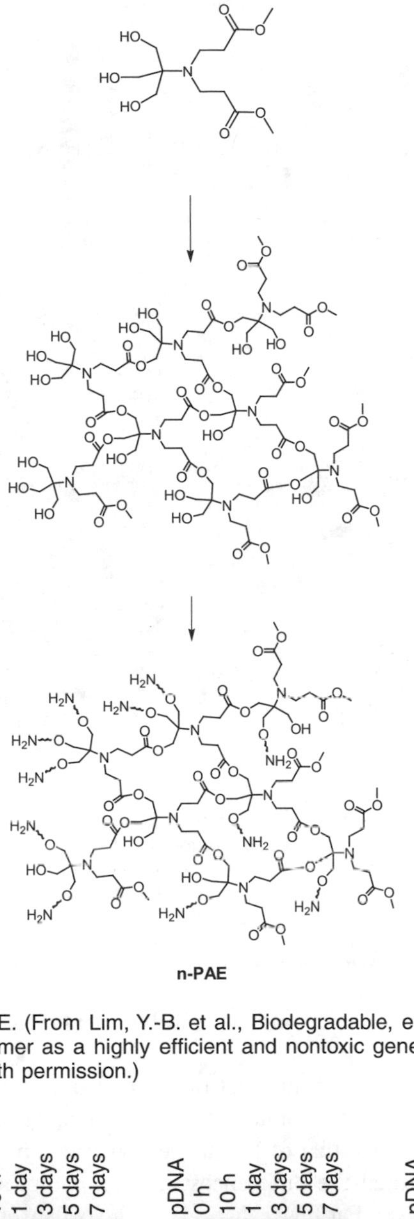

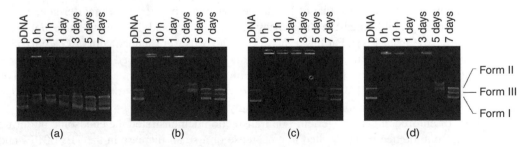

n-PAE

Scheme 15.3 Synthesis of n-PAE. (From Lim, Y.-B. et al., Biodegradable, endosome disruptive, and cationic network-type polymer as a highly efficient and nontoxic gene delivery carrier, *Bioconj. Chem.* 13, 952, 2002. With permission.)

Form II
Form III
Form I

(a)　　　　　　　(b)　　　　　　　(c)　　　　　　　(d)

Figure 15.4 Formation and stability of n-PAE–DNA (pGL3-control) polyplex. The polyplex was formed at an n-PAE:DNA ratio (N/P) of (A) 4, (B) 10, (C) 20, and (D) 40 and incubated at 37°C. Aliquots taken at the indicated times were electrophoresed through 0.7% agarose gel and stained with ethidium bromide to visualize DNA. Form I = supercoiled, form II = nicked circular, and form III = linear DNA. (From Lim, Y.-B. et al., Biodegradable, endosome disruptive, and cationic network-type polymer as a highly efficient and nontoxic gene delivery carrier, *Bioconj. Chem.* 13, 952, 2002. With permission.)

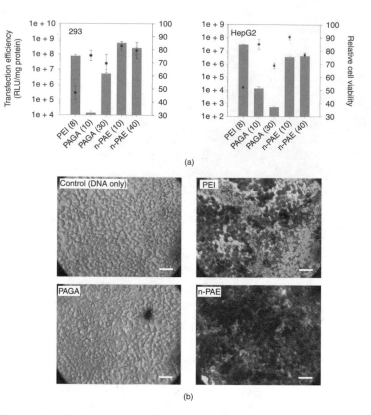

Figure 15.5   (A) Transfection efficiency (bar graph) and cytotoxicity (filled circle) in 293 and HepG2 cells. The cells were transfected with pGL3 control vector encoding firefly luciferase as a reporter gene. Numbers in parentheses indicate N/P ratios. RLU = relative light unit. Mean ± SEM (n = 5). (B) *In situ* X-gal staining of 293 cells transfected with pCN-LacZ vector encoding β-galactosidase gene. The scale bars represent 50 μm. Transfection experiments were performed without serum, chloroquine, and nigericin. (From Lim, Y.-B. et al., Biodegradable, endosome disruptive, and cationic network-type polymer as a highly efficient and nontoxic gene delivery carrier, *Bioconj. Chem.* 13, 952, 2002. With permission.)

of the polymer, which is in accordance with several biodegradable cationic polymers investigated in the previous reports.[6,11,12,16,25-27]

The TEs of n-PAE were very high in most of the cell lines studied. It is remarkable that the TE of n-PAE was not less than that of branched-PEI 25 kDa. In 293 cells, the TE of n-PAE was even about seven times higher than that of PEI as demonstrated by luciferase reporter gene assay. *In situ* X-gal staining highlighted the differences in cytotoxicity and TE among some polymers more demonstrably (Figure 15.5). Far fewer than 1% of the cells were stained with PAGA polyplex, whereas nearly all of the cells transfected with PEI and n-PAE were stained blue. However, there were striking differences among the cells treated with these polymers in terms of viability. The majority of PEI-transfected cells were granulated and dead, whereas most of the cells transfected with PAGA and n-PAE were viable and looked healthy.

The effect of chloroquine and nigericin[28] on transfection demonstrated the endosome buffering of n-PAE. Chloroquine treatment resulted in a decrease or a small increase in the TEs of PEI and n-PAE, respectively, but in a large increase in the TE of PAGA. These results suggest that polyplexes formed with PEI or n-PAE do not need chloroquine assistance for endosome escape, as they are already equipped with endosome buffering. The effect of nigericin on transfection reconfirmed n-PAE's endosome buffering function.

15.6 CONCLUSION

We have developed nontoxic gene carriers based on cationic biodegradable polymers. The results show that the rational design of polymer structure is an appropriate way of making nontoxic and efficient carriers. One of the main drawbacks of current nonviral gene carriers is their low efficiency *in vivo*. We hope that the rational design of polymer structure and conjugation, combined with biodegradable polymeric structure, will ultimately lead to the construction of safe and efficient artificial viruses.

REFERENCES

1. Nishikawa, M. and Huang, L., Nonviral vectors in the new millennium: delivery barriers in gene transfer, *Hum. Gene Ther.* 12, 861, 2001.
2. Han, S.-O. et al., Development of biomaterials for gene therapy, *Mol. Ther.* 2, 302, 2000.
3. Li, S. and Huang, L., Nonviral gene therapy: promises and challenges, *Gene Ther.* 7, 31, 2000.
4. De Smedt, S.C., Demeester, J., and Hennink, W.E., Cationic polymer based gene delivery systems, *Pharm. Res.* 17, 113, 2000.
5. Kabanov, A.V. and Kabanov, V.A., DNA complexes with polycations for the delivery of genetic material into cells, *Bioconj. Chem.* 6, 7, 1995.
6. Lim, Y.-B., Choi, Y.H., and Park, J.-S., A self-destroying polycationic polymer: biodegradable poly(4-hydroxy-L-proline ester), *J. Am. Chem. Soc.* 121, 5633, 1999.
7. Styring, M.G. and Hamielec, A.E., *Determination of Molecular Weight*, Cooper, A.R. (Ed.), John Wiley & Sons, New York, 1989, 263.
8. Schriemer, D.C. and Li, L., Mass discrimination in the analysis of polydisperse polymer by MALDI time-of-flight mass spectrometry. 1. Sample preparation and desorption/ionization issues, *Anal. Chem.* 69, 4169, 1997.
9. Nielen, M.W.F. and Malucha, S., Characterization of polydisperse synthetic polymers by size-exclusion chromatography/matrix-assisted laser desorption/ionization time-of-flight mass spectrometry, *Rapid Commun. Mass Spectrom.* 11, 1194, 1997.
10. Lewis, D.H., Controlled release of bioactive agents from lachidel glycolide polymers, in *Biodegradable Polymers as Drug Delivery Systems*, Chasin, M. and Langer, R. (Eds.), Marcel Dekker, New York, 1990.
11. Lim, Y.-B. et al., Biodegradable polyester, poly[α-(4-aminobutyl)-L-glycolic acid], as a non-toxic gene carrier, *Pharm. Res.* 17, 811, 2000.
12. Lim, Y.-B. et al., Development of a safe gene delivery system using biodegradable polymer, poly[α-(4-aminobutyl)-L-glycolic acid], *J. Am. Chem. Soc.* 122, 6524, 2000.
13. Perales, J.C. et al., Biochemical and functional characterization of DNA complexes capable of targeting genes to hepatocytes via asialoglycoprotein receptor, *J. Biol. Chem.* 272, 7398, 1997.
14. Choi, Y.H. et al., Lactose-poly(ethylene glycol)-grafted poly-L-lysine as hepatoma cell-targeted gene carrier, *Bioconj. Chem.* 9, 708, 1998.
15. Wolfert, M.A. et al., Characterization of vectors for gene therapy formed by self-assembly of DNA with synthetic block co-polymers, *Hum. Gene Ther.* 7, 2123, 1996.
16. Lim, Y.-B. et al., Cationic hyperbranched poly(amino ester): a novel class of DNA condensing molecule with cationic surface, biodegradable three-dimensional structure, and tertiary amine groups in the interior, *J. Am. Chem. Soc.* 123, 2460, 2001.
17. Boussif, O. et al., A versatile vector for gene and oligonucleotide transfer into cells in culture and *in vivo*: polyethylenimine, *Proc. Nat. Acad. Sci. U.S.A.* 92, 7297, 1995.
18. Fischer, M. and Vögtle, F., Dendrimers: from design to application – a progress report, *Angew. Chem. Int. Ed. Engl.* 38, 884, 1999.
19. Fréchet, J.M.J. et al., "Self-condensing" vinyl polymerization: a new approach to dendritic materials, *Science* 269, 1080, 1995.
20. Tang, M.X., Redemann, C.T., and Szoka, F.C., Jr., *In vitro* gene delivery by degraded polyamidoamine dendrimers, *Bioconj. Chem.* 7, 703, 1996.

21. Hawker, C.J., Lee, R., and Fréchet, J.M.J., One-step synthesis of hyperbranched dendritic polyesters, *J. Am. Chem. Soc.* 113, 4583, 1991.

22. Lim, Y.-B. et al., Biodegradable, endosome disruptive, and cationic network-type polymer as a highly efficient and nontoxic gene delivery carrier, *Bioconj. Chem.* 13, 952, 2002.

23. Udenfriend, S. et al., Fluorescamine: a reagent for assay of amino acids, peptides, proteins, and primary amines in the picomole range, *Science* 178, 871, 1972.

24. Feast, W.J. and Stainton, N.M., Synthesis, structure and properties of some hyperbranched polyesters, *J. Mater. Chem.* 5, 405, 1995.

25. Putnam, D. and Langer, R., Poly(4-hydroxy-L-proline ester): low-temperature polycondensation and plasmid DNA complexation, *Macromolecules* 32, 3658, 1999.

26. Lynn, D.M. and Langer, R., Degradable poly(β-amino esters): synthesis, characterization, and self-assembly with plasmid DNA, *J. Am. Chem. Soc.* 122, 10761, 2000.

27. Wang, J., Mao, H., and Leong, K.W., A novel biodegradable gene carrier based on polyphosphoester, *J. Am. Chem. Soc.* 123, 9480, 2001.

28. Uherek, C., Fominaya, J., and Wels, W., A modular DNA carrier protein based on the structure of diphtheria toxin mediates target cell-specific gene delivery, *J. Biol. Chem.* 273, 8835, 1998.

CHAPTER **16**

Poly(amidoamine)s for Gene Delivery

Paolo Ferruti and Jacopo Franchini

CONTENTS

0-8493-1934-X/05/$0.00+$1.50
© 2005 by CRC Press LLC

16.1 INTRODUCTION

One of the most exciting frontiers of macromolecular science is represented by polymer therapeutics, a term used to describe a set of biomedical technologies based on water-soluble polymers. Under this subject, authoritatively reviewed by Duncan,[1] are included for instance polymeric drugs, polymer–protein conjugates, polymeric micelles entrapping drugs by covalent linkage, and bioresponsive polymers used as components of nonviral vectors for intracellular delivery.

Poly(amidoamine)s (PAAs) are a family of synthetic functional polymers endowed with a combination of properties making them suitable for a variety of biomedical applications, mostly related to polymer therapeutics. The aim of this chapter is to provide an updated state-of-the-art report on PAA chemistry as well as some biological data demonstrating their remarkable potential in this field.

16.2 SYNTHETIC ASPECTS

16.2.1 Synthesis and Structure of PAAs

PAAs are synthetic tert-amino polymers obtained by stepwise polyaddition of primary or secondary aliphatic amines to bis(acrylamide)s (Scheme 16.1). The synthetic mechanism is the Michael addition (Scheme 16.2).

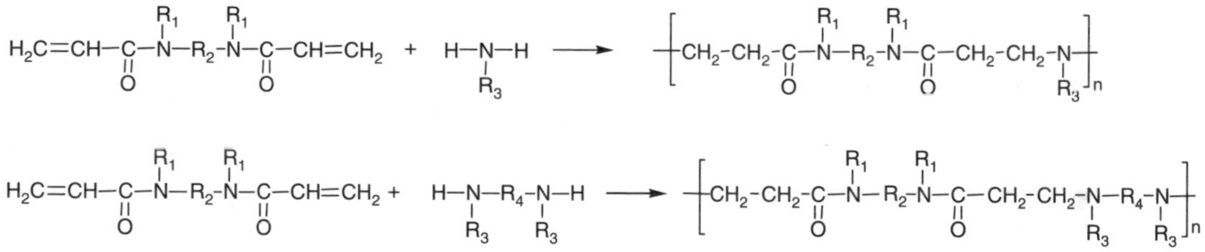

Scheme 16.1 Synthesis of linear PAAs.

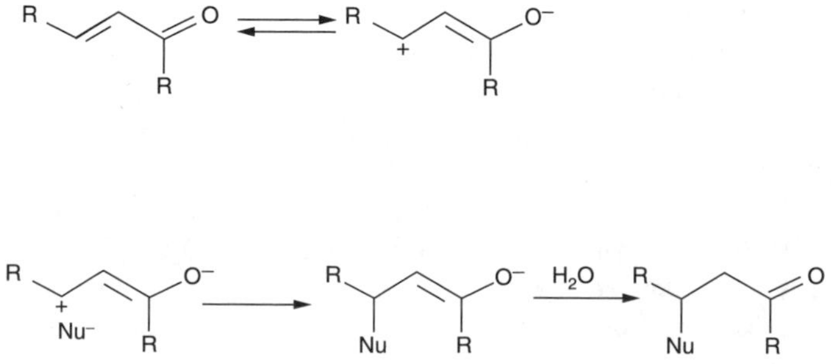

Scheme 16.2 Michael addition mechanism.

The polymerization reaction takes place in solvents carrying mobile protons, such as water or alcohols, at temperatures above about 25°C and without added catalysts.[2-4] High monomer concentrations and relatively low reaction temperatures give the best results. Aprotic solvents, even if highly polar, are unsuitable as reaction media as they yield only low molecular weight products. This is due to the fact that in these systems the only proton source is the amine itself.

The amino groups react only if present as a free base. Both primary mono(amine)s and secondary bis(amine)s lead to linear polymers. The structures of some typical PAAs are reported in Table 16.1.

16.2.2 Polymerization Kinetics

The kinetic constants of the polymerization reactions in water and methanol were determined at 25°C and 0.235 M concentration of both reagents (Malgesini et al., unpublished). The reaction progress was followed by means of UV spectroscopy, by monitoring the decrease of the π - π absorption band due to the disappearing of the acrylic double bond. Typical results obtained in the polymerization of 2-methylpiperazine (2-MePip) and N,N'-bisacryloylpiperazine (BisPip) in both water and methanol at λ = 250 nm are reported in Figure 16.1 for comparison purposes.

The polymerization rate can be expressed through a second order equation with respect to the monomers. Both solvents, due to the presence in their structure of a mobile hydrogen, act as catalysts in the polyaddition reaction and, for this reason, their concentration can be included in the rate constant:

$$-\frac{d[double\,bond]}{dt} = k[2MePip][BisPip][Solvent] = k'[2MePip][BisPip] \tag{16.1}$$

The kinetic constant could, therefore, be determined by plotting the reverse of bisacrylamide concentration versus reaction time. The final calculated values of the kinetic constants are:

$$k'_{H2O} = 0.0024*10^{(-5)}\ [l/(mol)(min)] \tag{16.2}$$

$$k'_{MeOH} = 0.0005*10^{(-5)}\ [l/(mol)(min)] \tag{16.3}$$

These results confirm that the polyaddition rate is more efficiently catalyzed by water than methanol, due to the higher acidic strength of the former.

16.2.3 Functionalization of PAAs

PAAs are inherently highly functional polymers. However, further functionalization of PAAs may be useful for special purposes. Functional groups not capable of undergoing Michael addition under the conditions of PAA synthesis, for instance hydroxy, tert-amino, allyl, amido, and ether groups, do not compete with the polymerization reaction. Therefore, the introduction of these additional functions in PAAs as side substituents can be simply achieved by using as monomers properly functionalized amines or bis(acrylamide)s. Chemical groups capable of reacting with activated double bonds under the conditions of PAA synthesis, such as SH, NH_2, NHR, and PH_2, if unprotected, as a rule cannot be introduced directly as side substituents in PAAs. However, it has been recently ascertained that under proper conditions, PAAs carrying primary amino groups as side substituents can be prepared (see below).

Table 16.1 Some Examples of Poly(amidoamines)*

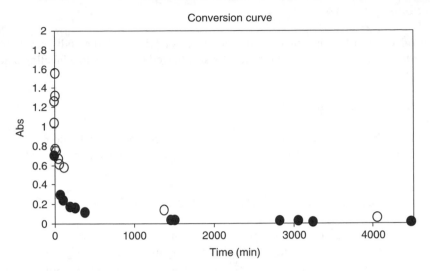

Figure 16.1 Conversion curves of polyaddition reaction in methanol (●) and water (○). Relative absorbance versus reaction time.

16.2.4 Amphoteric PAAs

Amphoteric PAAs are PAAs carrying acidic functions as side substituents, in most cases carboxyl groups (Table 16.2).

They are obtained by using either carboxylated bis(acrylamide)s or amino acids as monomers. In both cases, a stoichiometric amount of strong base must be added to the monomer mixture. As regards amino acids, natural α-amino acids other than glycine yield only oligomeric products, but the polyaddition reaction proceeds reasonably well with all amino acids bearing no substituents on the carbon atom in the α position, as well as with peptides in which the first amino acid residue fulfills the same condition.

Table 16.2 Some Examples of Amphoteric PAAs

16.2.5 Labeling of PAAs

The introduction of labeling agents is important for studying both the body distribution and the intracellular trafficking of PAAs. The former is best followed by radioactive labeling, and the latter by fluorescent labeling.

16.2.5.1 Radioactive Labeling

Radioactive labeling of PAAs can be achieved by substituting a small amount (4% on a molar basis) of tyramine (Figure 16.2) for the same amount of aminic monomers. This allows radioiodination by I^{125}.

16.2.5.2 PAAs with Amino Pendants and Fluorescent Labeling

A wide choice of fluorescent labels for proteins are commercially available. They usually contain an activated carboxyl group that will couple with the ε-amino groups of L-lysine units of proteins. It is obviously of advantage to introduce primary amino groups as side substituents in PAAs, to render them amenable to the same labeling techniques as proteins. However, the potential interest in polymer therapeutics of PAAs carrying pendant primary amino groups is not limited to this. In fact, they may act as general-purpose polymer carriers for carboxylated drugs and will easily yield PAA–protein conjugates by consolidated techniques.

The direct introduction of primary amino groups in PAAs by using primary amino-substituted monomers in the polymerization recipe is not possible under the synthetic conditions normally adopted for PAAs, since all primary amino groups will participate to the polymerization reaction.

Under nonselective conditions, primary bis(amine)s usually give crosslinked products on reaction with bis(acrylamide)s, because they react as tetra-functional monomers.

To prepare PAAs carrying primary NH_2 groups as side substituents (NH_2-PAAs), a general-purpose synthetic strategy is to use as monomers primary diamines, in which one of the aminic functions has been protected by a group capable of preventing its additional reaction with activated double bonds.[5] The protecting group must be stable under the conditions of PAA synthesis, but is easily removable under conditions not affecting PAA stability. We selected 1,2-diaminoethane (EDA) as a primary diamine and tryphenylmethyl as a protecting group. Thus, the starting monomer was 1-triphenylmethylamino-2-aminoethane (TPHMAE) (Scheme 16.3).

TPHMAE was prepared in a pure state by reacting triphenylchloromethane with a large excess of EDA in chloroform solution, extracting the reaction mixture several times with water and then evaporating to dryness *in vacuo*. The choice of reactants made this process particularly easy to

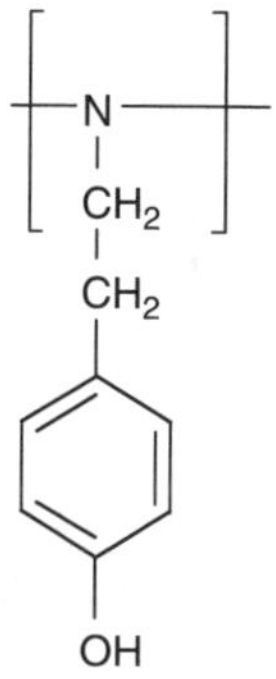

Figure 16.2 Structure of tyramine units in PAA.

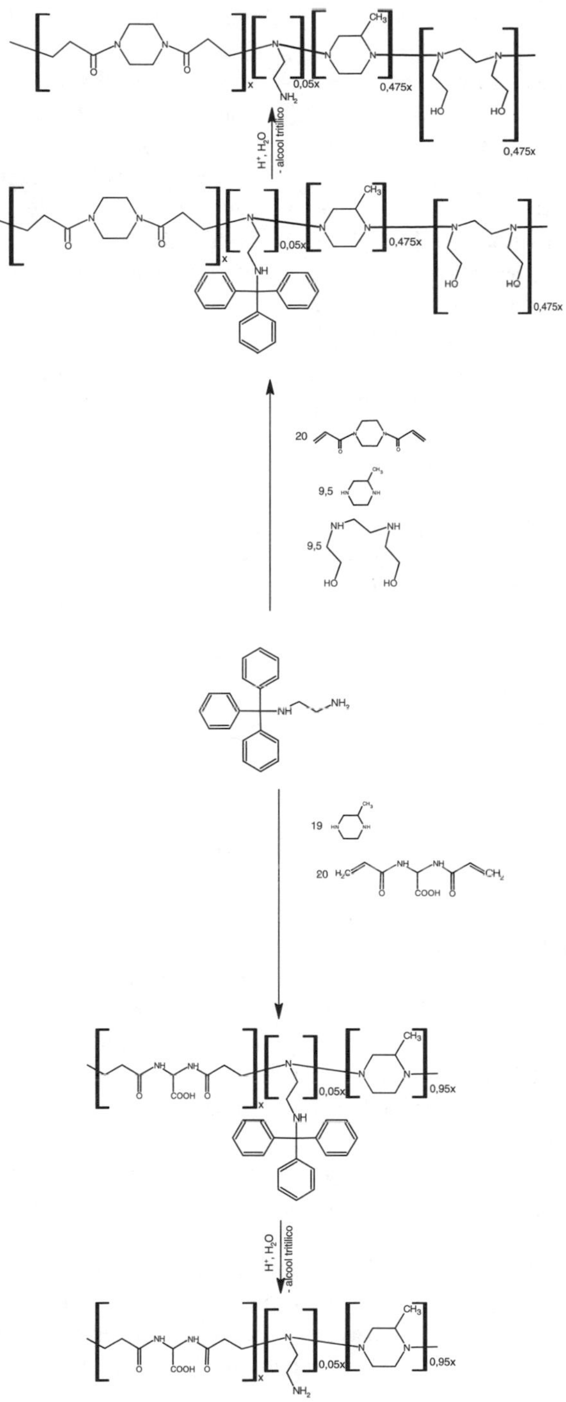

Scheme 16.3 Synthesis of PAAs with amino pendants.

carry out because EDA is very hydrophilic and can be easily and completely extracted by neutral water from its solutions in organic solvents, while the protecting group is very capable of imparting lipophilic properties. The resultant TPHMAE is in fact insoluble in water. Therefore the polymerization reaction must be started in methanol, which as a reaction medium is less effective than water, at least as regards the polyaddition rate. After 1 day reaction in methanol, water-soluble oligomers were obtained. Water was then substituted for methanol and the polymerization was allowed to proceed in the usual way.

An alternative process allows the direct synthesis of NH_2-PAAs from bis(acrylamide)s and EDA. The starting point was a study describing the chemical modification of poly(L-lysine) (PLL) by the Michael-type addition on N,N-dimethylacrylamide (DMAAC) to its primary amino side substituents,[6] showing that even in the presence of excess DMAAC, starting from partially (30%) protonated PLL, the reaction rate became extremely slow and practically stopped at a reaction yield corresponding to the addition of 2 M DMAA per M of PLL unit present as free base, that is, 70% of the total. Since protonated amines do not react with acrylamides under the conditions we used, the overall behavior was consistent with a fixed localization of charges on PLL units. The addition of strong bases induced the addition reaction to start again and go to completion.

This was explained as follows. The addition of acrylamides to a primary or secondary amine invariably results in a very sharp decrease of its basicity, amounting to about 2 pK_a units. Therefore, in the case of PLL, the addition of 1 M DMAA to an unprotonated amino group of a given unit made it very unlikely that the same group would, afterwards, assume a proton by an exchange reaction with an unsubstituted, protonated group belonging to a neighboring unit. Therefore, it was on the already substituted group that a second addition reaction occurred. This further reduced its tendency to assume a proton by deprotonating an unsubstituted amino group. As a consequence, the protons remained nearly fixed on the unsubstituted groups and the reaction stopped after the addition of 2 M DMAA per M of unprotonated units.

The previous result on PLL modification prompted us to investigate if protonation could represent a valid method of protection for one of the amino groups of EDA and, therefore, if monoprotonated EDA would behave in the PAA synthesis as a difunctional monomer leading to linear NH_2-PAAs.[5]

The experiments were performed using bisacrylamido acetic acid (BAC) as the bisacrylamide monomer (Figure 16.3).

BAC is a relatively strong acid, and in PAA synthesis it is usually employed in the form of salt. The two protonation constants of EDA, namely 10.71 and 7.56[7] differ by about three pK_a units. A 1:1 adduct of BAC–EDA contains essentially monoprotonated EDA with little contamination by diprotonated and unprotonated EDA. Therefore, by acting as a difunctional monomer in the polyaddition reaction, it should give rise to linear NH_2-BAC. Soluble NH_2-BAC was in fact obtained by treating a BAC salt of a strong base with monoprotonated EDA (obtained *in situ*) in water and allowing the resultant mixture to react under usual conditions of PAA synthesis. However, the reaction was completely under control only in the presence of some excess acid (5–10% on a molar basis). The NH_2-PAAs so obtained had the same structure as those obtained by the TPHMAE method, but their molecular weight was higher.

Regrettably, this elegant synthesis of NH_2-PAAs is not a general one. For instance, "traditional" PAAs doped with a limited amount of pendant primary amino groups for labeling purposes can hardly be obtained. It is, in fact, difficult to devise reaction conditions that maintain EDA in its monoprotonated form in the presence of other amine-containing monomers with pK_a values similar to those of EDA. It seems to be equally difficult to employ primary amines other than EDA because it is well known that the difference between the two pK_a values of aliphatic diamines gradually vanishes by lengthening the aliphatic chain separating the amino groups.

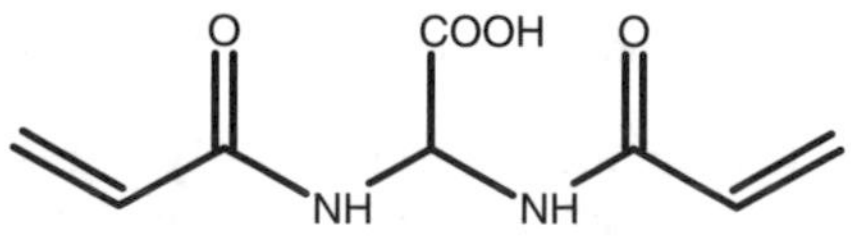

Figure 16.3   Bisacrylamido acetic acid (BAC).

16.2.6 PAA Macromonomers

Consideration of the synthetic process leading to PAAs (Scheme 16.1) shows that if the hydrolytic cleavage of the main backbone during polymerization is minimized, the end groups of the product are either sec-amino or acrylamido groups. Therefore, by performing the polymerization reaction with an excess of one of the two monomers, PAAs prevailingly or totally terminated with either of the two groups can be obtained. End-functionalized PAAs can be employed for preparing block and graft copolymers, as well as crosslinked resins, and therefore can be regarded as macromonomers.

In a sense, PAAs carrying allyl groups as side substituents can be also regarded as macromonomers, since they give hybrid materials by copolymerization with vinyl monomers (see below).

16.2.7 Crosslinked PAAs

In PAA synthesis, aminic monomers carrying more than two mobile hydrogens, such as for instance primary diamines, behave as multifunctional monomers and yield, if no special precautions are taken (see above), crosslinked products (Scheme 16.4). However, it has been reported that under special conditions including low reactant concentrations, low initial temperatures, and a molar excess of bis-amine, it is possible to obtain soluble PAAs carrying secondary instead of tertiary amino groups in their main chain.[8]

Crosslinked PAAs swell but do not dissolve in water, and therefore belong to the vast family of hydrogels.[9] PAA-based hydrogels, as well as PAA-based hybrid materials with reduced ability to swell in water, can be also obtained by radical polymerization of PAA macromonomers either vinyl-terminated or carrying allyl groups as side substituents.

16.2.8 PAA-Related Polymers

Polymers structurally related with PAAs were obtained by substituting either bis(acrylic ester)s or divinylsulfone for bis(acrylamide)s, or hydrazines or phosphines for amines.[2] Other polymers, of poly(amino ketone) structure, were obtained by polycondensation of ketonic bis-Mannich bases with bis(amine)s.[10 12]

The ring-opening addition reaction of secondary bis(amine)s or primary mono(amine)s to ethylene sulfide gave a new family of monomers, namely 2,2′-alkylidene-diiminodiethanethiols.[13] These monomers, besides polymerizing by polyoxidative coupling,[13] could be employed in poly-addition reactions with bis(acrylamide)s, bis(acrylic ester)s, or divinylsulfone, in a way formally

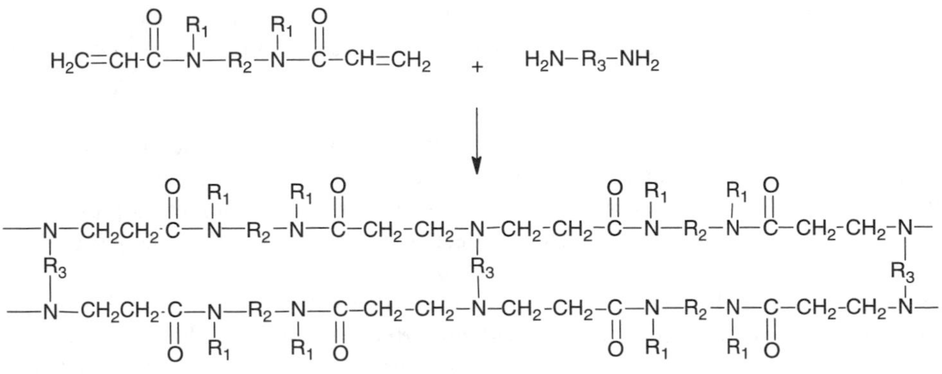

Scheme 16.4 Synthesis of crosslinked PAAs.

similar to that used with secondary bis(amine)s.[14-16] This polyaddition proceeded also with bis(meth-acrylic ester)s[17] and bis(methacrylamide)s,[18] which are extremely sluggish in reacting with bis(amine)s under the usual conditions of PAA synthesis.

16.3 PHYSICO-CHEMICAL PROPERTIES OF PAAs

16.3.1 Molecular Weight, Solubility, and Solution Properties

The number-average and weight-average molecular weights of the PAAs described thus far were in the range 5,000–40,000 (number average) and 10,000–70,000 (weight average), usually with a dispersity index of ~2 depending on the isolation method.[19-22]

Most PAAs are soluble in water as well as chloroform, lower alcohols, dimethyl sulfoxide, and other polar solvents. However, amphoteric PAAs dissolve only in water. The intrinsic viscosities of PAAs in organic solvents or aqueous media usually range from about 0.15 to 1 dl/g. As a rule, PAAs exhibit relatively large hydrodynamic volumes in solution if compared with vinyl polymers of similar molecular weight, indicating a tendency to assume an extended chain conformation in solution.[21,22]

16.3.2 Crystallinity

Owing to their regular structure, many PAAs in the solid state are partially crystalline. In some instances, crystallization can be induced by solvent treatment. Partially crystalline PAAs usually have melting points in the range of 80–120°C, but when cyclic structures are present, melting points as high as 270°C (with decomposition) have been measured.[2]

16.3.3 Thermal Stability

The thermal stability of PAAs is not very high. It may be safely assumed that in most instances decomposition begins at about 140°C in air and 170–180°C under inert atmosphere.[2] Shelf stability in air is not very high for PAAs as free bases, but is good for their salts with strong acids if thoroughly dried and protected from moisture.

16.3.4 Degradation

All PAAs containing amidic bonds in their main chain are degradable in aqueous solution. The degradation of several PAAs has been studied by means of viscometric and chromatographic techniques.[23,24] For instance, Figures 16.4 and 16.5 show the degradation studies of one of the most studied PAAs, ISA 1 (see below). The viscosity and the SEC plots show that degradation occurs in phosphate buffer 0.1 M pH = 7.4 at 37°C.

The structure of both the aminic and amidic moieties has an influence on the degradation rate.[23-25] We have observed in many instances that additional tert-amino groups, if present as side substituents, increase the degradation rate of PAAs. However, we do not have at present any clear evidence that the degradation rate of PAAs is affected by the basicity of the amino groups in the macromolecular backbone.

The effect of side carboxylate groups is less clear. PAAs deriving from glycine or β-alanine apparently degrade faster than similar PAAs with no carboxyl substituents, but PAAs deriving from BAC (Figure 16.3) degrade at a slower rate. For instance, Figure 16.6 shows the results of a SEC degradation study on the polymer ISA 23 (see below) in phosphate buffer 0.1 M pH = 7.4 at 37°C.

The mechanism of PAA degradation seems to be purely hydrolytic, as no vinyl groups, such as those that would have derived from a β-elimination reaction, could be determined by NMR

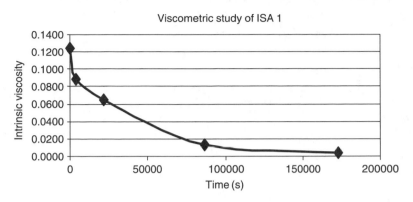

Figure 16.4 Viscometric study of degradation of ISA 1 in phosphate buffer 0.1 *M*, 37°C.

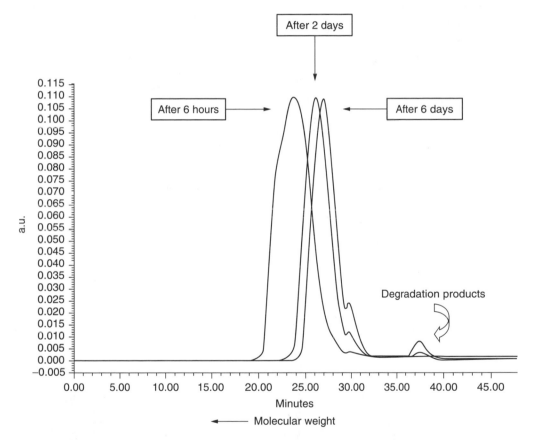

Figure 16.5 SEC study of the degradation of ISA 1 in phosphate buffer 0.1 *M*, 37°C.

spectroscopic analysis.[21] Moreover, the degradation rate was not affected by the presence of a 20-fold excess of 2-mercaptoethanol (on a molar basis). This would have affected degradation if mercaptoethanol was able to act as a scavenger of any activated double bonds resulting from a β-elimination mediated degradation. Finally, it has been demonstrated that degradation of PAAs in aqueous media is strongly influenced by pH but does not seem to be affected by the presence of isolated lysosomal enzymes at pH 5.5.[25]

As a consequence of PAA degradability, when water is used as polymerization medium the average molecular weight of the product increases with reaction time until it reaches a maximum,

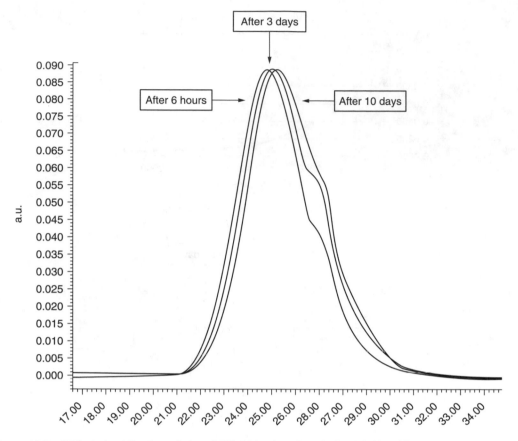

Figure 16.6 SEC study of the degradation of ISA 23 in phosphate buffer 0.1 *M*, 37°C.

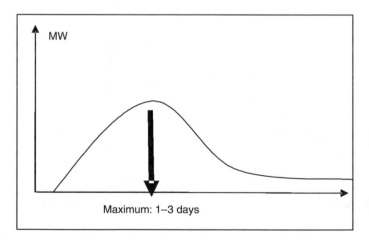

Figure 16.7 Molecular weight growth as a function of time.

after which it steadily decreases. A generic curve describing the variation of molecular weight with time during PAA formation in water is reported in Figure 16.7.

This is explained as follows: in water the polyaddition reaction competes with the hydrolytic cleavage of the amidic bonds. The maximum attainable molecular weight in water depends on the monomer structure and concentration, as well as on the reaction temperature, being higher for higher concentrations and lower temperatures. At relatively high temperatures, for instance 40–60°, the maximum is attained more quickly but is lower.[2]

16.3.5 Acid–Base Properties of PAAs

All PAAs contain tert-amino groups in their main chain and therefore can be classified as polyelectrolytes. Normally, the values of the protonation constants (log K) of polyelectrolytes depend on the degree of protonation of the whole macromolecule. They follow the modified Henderson-Hasselbach equation:

$$\log K_i = \log K_i^0 + (n-1) \log [(1-\alpha)/\alpha] \tag{16.4}$$

where $\log K_i^0$ is the protonation constant of a group present in a completely non-ionized polymer. The protonation constants of polyelecrolytes are usually referred to as "apparent" constants, as opposed to the "real" constants of nonmacromolecular acids and bases.[26] However, in most PAAs the results of the potentiometric titrations are consistent with n values very close to 1. This means that in PAAs the tendency of the aminic nitrogen atoms of each repeating unit to assume a proton in practice does not depend on the degree of protonation of the whole macromolecule. Therefore these groups behave as if belonging to a small molecule.[27]

Consequently, "real" or "quasi-real" basicity constants can be determined. The number of the basicity constants of PAAs is equal to the number of the aminic nitrogen atoms present in their repeating unit and their values are similar to those found for the nonmacromolecular models of PAAs, prepared by hydrogen-transfer addition to 4-acryloylmorpholine of the same amines used in the preparation of the corresponding PAAs. The existence of a linear relationship between the protonation enthalpies and Q_N means that little interaction exists in this respect between the monomeric units. PAAs of similar structure, in which the bis(acryloylpiperazine) moieties had been replaced with noncyclic, hence more flexible bis(acrylamidic) moieties, on the whole, gave very similar results. The unusual behavior of PAAs in the polyelectrolytes domain is probably due to the relatively long distance between the amino groups belonging to different units combined with the high charge-sheltering efficiency of the two amido groups interposed. It is noteworthy that a single amido group bound to a piperazine ring is not sufficient to minimize interactions between neighboring units.[27]

Viscometric titrations in 0.1 M NaCl proved that the conformational freedom is reduced on protonation. The first protonation leads to the formation of a strong hydrogen bond between "onium" ions and carbonyl groups belonging to the same monomeric unit. When the first protonation of all monomeric units is complete, the above effect strongly reduces the conformation freedom of the whole polymer, which tends to assume a rigid structure. This is clearly shown by a pronounced jump of its reduced viscosity at $\alpha = 0.5$ (α = degree of protonation). The second protonation leads to an electrostatic repulsion between the positively charged onium ions belonging to the same monomeric unit and this effect further compels the polymer to adopt a more rigid structure. Consequently a second jump is observed at $\alpha = 1$. The reduction of conformational freedom clearly decreases with the lengthening of the aliphatic chain in each repeating unit.[27]

In the case of PAAs with additional tertiary amino groups in the side chains, thermodynamic values are similar to those of nonmacromolecular models. In these polymers, however, viscometric titrations did not show any jump either after the first or after the second protonation step. This is probably due to the fact that in these PAAs there is only one onium ion (that belonging to the main chain) that can interact with one or another of the two neighboring carbonyl groups. This leads to greater conformational freedom and hence the viscometric titrations do not show any jump even after the second protonation step.

Differently from normal PAAs, amphoteric PAAs deriving from amino acids and therefore carrying both carboxyl and amino groups attached to the same monomer tend to exhibit a typical

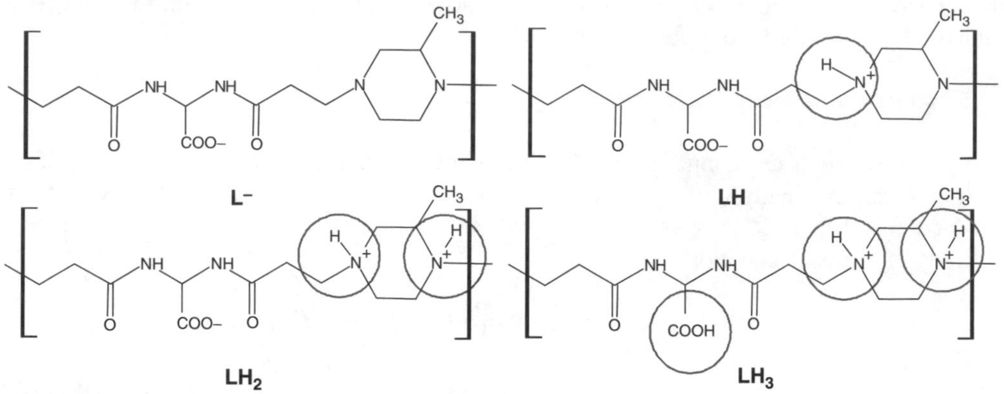

Figure 16.8 Different charged states of the amphoteric PAA ISA 23.

polyelectrolyte behavior, at least as regards the carbyl group.[27] The polyelectrolyte effect is less pronounced in the case of amphoteric PAAs deriving from BAC.[22] These PAAs present a unique interest as bioactive polymers. Besides being generally less toxic than purely cationic PAAs of similar structure, in solution they change their net average charge as a function of pH (Figure 16.8).

By a proper choice of the starting monomers, the acidic and basic strength of the amino and the carboxyl groups can be controlled in such a way that the polymer passes from a prevailingly anionic to a prevailingly cationic state as a consequence of relatively modest pH changes.[22] For instance, this occurs when amphoteric PAAs are internalized by cells via the endocytic pathway. In this process, the pH, which in extracellular fluids is 7.4, drops first to 6.5 (endosomes) and then to 5–5.5 (lysosomes). PAAs that become prevailingly cationic at lysosomal pH can be designed to become, as a consequence, membrane active and capable of displaying endosomolytic properties (see below).

16.4 BIOMEDICAL APPLICATIONS OF PAAs

16.4.1 Applications Mostly Related to Insoluble PAA Derivatives

16.4.1.1 *Neutralization of the Anticoagulant Activity of Heparin, Heparin-Absorbing Resins, and Heparinizable Materials*

Heparin is a well-known anticoagulant agent, widely used in clinical practice. It is a muco-polysaccharide containing carboxyl and sulfonic groups. In aqueous solution it behaves as a polyanion with a high density of negative charge. In many cases, the anticoagulant activity of heparin must be inhibited when no longer needed. It was M. A. Marchisio who first discovered that several linear PAAs are able to neutralize the anticoagulant activity of heparin.[28-30] Different PAAs were evaluated biologically in terms of their anti-heparin activity and hemolytic effect on human red blood cells (HRBC) with very favorable results.[30] Internally buffered amphoteric PAAs deriving from amino acids, though apparently not very strongly basic, were still capable of neutralizing heparin. Heparin neutralization was undoubtedly due to polyelectrolyte complex formation with heparin.

PAA crosslinked resins, based on the same PAAs capable of complexing heparin in solution, behaved in a similar way to their linear counterparts,[30-32] and displayed remarkable heparin absorbing capacity, even when incubated with very dilute solutions of heparin. Hybrid PAA-based crosslinked resins had a heparin-adsorbing capacity proportional to their PAA content.[32,33]

As a further development, heparinizable materials have been obtained. Most nonphysiological materials induce thrombus formation when placed in contact with blood and many have explored the development of nonthrombogenic surfaces by immobilization of heparin on the biomaterial surface. Since the late 1960s, in fact, it has been widely known that heparin ionically adsorbed on the surface of most materials retains its anticoagulant activity locally hindering thrombus formation. This led to the preparation of several families of polyurethane–PAA hybrid materials.[34,35]

A lightly crosslinked polyurethane–PAA (PUPAA) block copolymer containing 5–30 wt% PAA showed approximately the same mechanical properties and hydrolytic stability of the native polyurethane. Although linear segmented PAA–polyurethane copolymers displayed properties similar to PUPAA,[36] their tensile strength was somewhat lower and the elongation at break higher. This was not surprising since PUPAA was slightly crosslinked. Both products were capable of adsorbing relative large amounts of heparin (0.002–0.7 mg/cm^2) and were able to retain it even when subject to multiple extractions, whereas no heparin was retained by native polyurethane. PUPAA provides an excellent coating for poly(vinyl chloride), polyurethane, and other materials.

All PAA-modified materials have been evaluated in terms of their biocompatibility.[34-36] After heparinization the blood compatibility of all the PAA-modified materials was significantly improved, especially as regards clotting formation. These materials have obvious potential for fabrication as thrombo-resistant devices intended for short- to medium-term applications in contact with blood. Recently, a new type of PAA-PMMA copolymer has been also prepared for a use as thrombo-resistant material.[37]

16.4.1.2 Soft PAA Hydrogels as Scaffolds for Tissue Engineering

As previously mentioned, in PAA synthesis, aminic monomers carrying more than two mobile hydrogens (e.g., primary diamines) behave as multifunctional monomers and yield crosslinked products. Crosslinked PAAs belong to the class of hydrogels. Their texture and ability to swell in water depend on both their structure and their degree of crosslinking.

16.4.1.2.1 Water Absorption Capacity

The water content in a hydrogel is a very important property that seems to be connected with biocompatibility. It is defined as

$$\frac{weight\,(swollen) - weight\,(dry)}{weight\,(dry)} * 100 \tag{16.5}$$

If the water content is higher than 90% the hydrogel is defined "super absorbant." Different factors influence the water content of hydrogels. Positive factors are the osmotic potential, strong interactions with water, a high free volume, a high chain flexibility, and a low density of the crosslinking points. Many PAA hydrogels exhibit all these positive factors. As an example, a hydrogel prepared by polyaddition of BAC (Figure 16.3), 2-MePip, and 1,10-diaminodecane in molar proportions 1:0.7:0.15 absorbed water to an extent of nearly 14 times its dry weight (S. Bianchi, F. Chiellini et al., unpublished) — that is, when fully swollen their water content was 92.9%. For this reason PAA hydrogels may be considered super absorbants.

16.4.1.2.2 Degradation in Aqueous Media

Most PAA hydrogels are degradable. The degradation behavior of many PAA hydrogels has been studied. For instance, a hydrogel nicknamed HPA1, prepared from BAC (Figure 16.3),

2-MePip, and 1,10-diaminodecane in molar proportions 1:0.7:0.15, in phosphate buffer 0.1 M at 37°C pH 7.4 (S. Bianchi et al., unpublished), dissolved completely after 10 days.

16.4.1.2.3 Cytotoxicity

Toxicological studies have been performed on many PAA hydrogels in the form of hydrochlorides, sulphates, or lactates. All samples were not significantly cytotoxic.

The assay was performed using fibroblast A31 cell lines. All the data collected have shown that PAA hydrogels are biocompatible. For instance, Figure 16.9 shows the result of a cytotoxicity assay on a sample of ISA 23 hydrogel isolated as lactic acid salt. It may be noticed that the cells are healthy and exhibit a good morphology.

The same hydrogels, modified by including residues of bioactive molecules as side-substituents, have been synthesized and evaluated as polymeric scaffolds for tissue engineering. They have been found to induce excellent cell adhesion and proliferation.[51]

16.4.1.2.4 Other Properties Relevant from a Biological Standpoint

PAA hydrogels are permeable to metabolites and have a soft texture, as well as a low interfacial tension between the gel and the aqueous solution. Their high permeability offers an easy means of purification as small molecules can be washed away from the final product.

All these properties show that PAA hydrogels have a potential as polymeric scaffolds; that is, as three-dimensional supports to aid in cellular adhesion, proliferation, and differentiation.

16.4.2 PAAs in Polymer Therapeutics

16.4.2.1 PAAs as Bioactive and Biocompatible Polymers

Bioactive polymers are synthetic polymers capable either of improving the mode of action of other active substances, as in the case of targetable polymer drug carriers, or of exerting a biological activity of their own, as in the case of membrane-active polymers with a potential to act as transfection promoters. They are in general functional polymers; that is, polymers endowed with reactive chemical functions, as exemplified by the well-known Ringsdorf model of polymer–drug conjugate.[38]

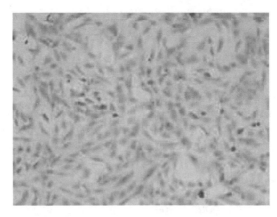

Figure 16.9 (See color insert following page 336) Cytotoxicity assay of ISA 23 lactate hydrogel (F. Chiellini et al., unpublished).

PAAs, and especially amphoteric PAAs, present a unique interest as bioactive polymers. R. Duncan and her colleagues at the School of Pharmacy of the University of London and, more recently, at the Welsh School of Pharmacy of the University of Cardiff have performed the following biological studies.

Most of these studies regard two amphoteric PAAs, ISA 22 and ISA 23, and on two purely cationic polymers labeled ISA 1 and ISA 4. It may be observed that ISA 22 and ISA 4 are the counterparts of ISA 22 and ISA 1, in which a small amount of tyramine (4% on a molar basis) has been substituted for the same amount of aminic monomers for labeling purposes.

16.4.2.2 Toxicity of PAAs

The cytotoxicity of PAAs ISA 1, ISA 4, ISA 22, and ISA 23 toward different cell lines are reported in Table 16.3, in comparison with PLL and dextran.[39]

It may be observed that, with all the cell lines considered, the PAAs were found to be less toxic than PLL by 2 or 3 orders of magnitude. The toxicity of the amphoteric PAAs (ISA 22 and ISA 23) (Figure 16.10) is of the same order as that of dextran.

16.4.2.3 Body Distribution of PAAs

The body distributions of ISA 4 and ISA 22 have been studied in mice. Relevant data are shown in Table 16.4.

It may be noticed that the purely cationic ISA 4 is able to concentrate in a very selective way in the liver and is potentially an excellent liver-targeting carrier. On the contrary, ISA 22 shows a typical "stealth" behavior; that is, when injected in animals it avoids rapid clearance by the reticulo-endothelial system. This property allows ISA 22 to remain in the blood stream until eliminated through the kidneys. In tumor-bearing animals, it may therefore concentrate at the tumor level by passive targeting, owing to the enhanced permeability and retention (EPR) effect.[1] The EPR effect is due to the fact that capillary tubes resulting from the neo-vascularization of tumors have, contrary to those of healthy tissues, a highly permeable endothelial lining (Figure 16.11).

This allows extravasation of circulating macromolecules and particles. As tumors also exhibit limited lymphatic drainage, the final result is a tumor-specific passive accumulation of macromolecules. The basic requirement is that the macromolecules in question are not captured on the way by the body's defense system; that is, they are endowed with "stealth" properties. As a matter of fact, I^{125} labeled ISA 22, after intravenous administration, exhibits remarkable tumor accumulation (Table 16.5).[40]

Table 16.3 Cytotoxicity[a] of Poly(amidoamine)s and Reference Polymers toward Different Cell Lines

Polymer	Molecular Weight (MW)	B16F10	Mewo	Hep G2
ISA	9000	3.05 ± 0.70	2.24 ± 0.36	2.80
ISA	9500	3.45 ± 1.18	1.89 ± 0.51	4.60
ISA	17000	4.00 ± 1.41	4.63 ± 0.53	>5.00
ISA	21500	>5.00	4.23 ± 1.10	>5.00
Poly(L-lysine)	5650	0.05 ± 0.01	0.01 ± 0.01	0.05 ± 0.01
Dextran	70000	>5.00	>5.00	>5.00

[a] Expressed as IC 50 (mg/ml) ± S.D.

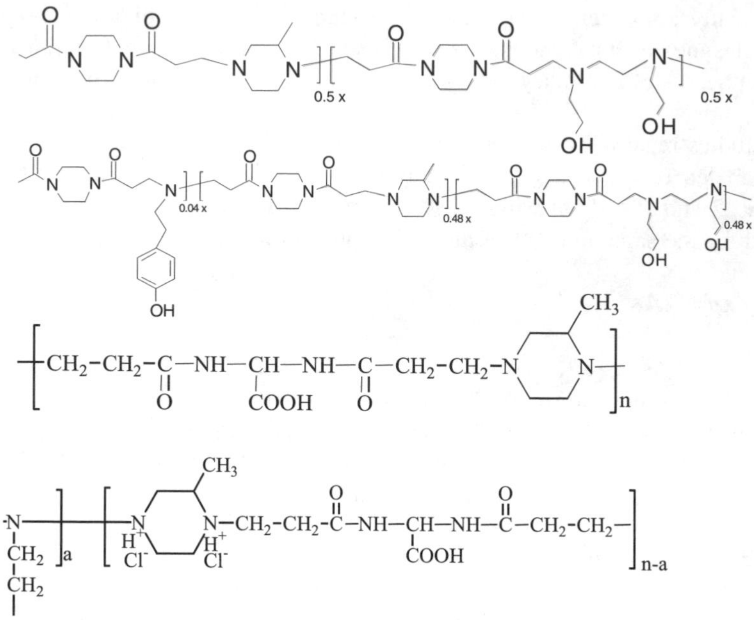

Figure 16.10 Structures of ISA 1, ISA 4, ISA 22, and ISA 23.

Table 16.4 Body Distribution of ISA 4 and ISA 22

Tissue	After 1 h ISA 22	After 1 h ISA 4	After 5 h ISA 22	After 5 h ISA 4
Blood	~70%	~1.5%	~20%	~3%
Kidneys	~2%	~7%	~1%	~4%
Liver	~10%	~83%	~8%	~85%
Lungs	~2%	~4%	~4%	~3%
Urine	~13%	~0%	~65%	~5%

16.4.2.4   PAAs as Polymer Carriers for Anticancer Drugs

Following the rationale adopted for design of other polymer-drug conjugates,[41] PAAs have systematically been developed as water-soluble carriers for known anticancer agents including Mitomycin C (MMC)[42] and Platinates.[43]

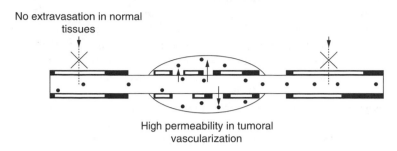

No extravasation in normal
tissues

High permeability in tumoral
vascularization

Figure 16.11 Scheme of the EPR effect.

Table 16.5 Distribution into Tumoral Tissues Due to EPR Effect of ^{125}I-Labeled ISA 22 after Intravenous Administration to Mice with B16F10 Cell Lines

Tissue	After 1h (%)	After 5h (%)
Blood	~10.7	~4.5
Liver	~2.1	~2
Kidneys	~2.2	~1.8
Tumor	~1.8	~2.2

PAA–MMC adducts were synthesized from ISA 4. After *in vitro* studies, preliminary experiments were carried out to investigate the efficacy and toxicity of these PAA–MMC conjugates *in vivo*. The PAA–MMC conjugates were equally active compared to MMC, given the i.p. route, and indeed in this pilot experiment a long-term survivor was observed in each group treated with conjugate. It was found that PAA–MMC conjugates were less toxic than free MMC when administered at an MMC-equivalent dose of 5 mg/kg, the maximum tolerated dose of free MMC that can be given to animals. ISA 23 and two other PAAs containing pendant β-cyclodextrins were subsequently used to synthesise PAA-platinates by reaction with cisplatin.[43] The conjugates had a content of 8–70 wt% platinum, and *in vitro* at pH 5.5 and pH 7.4 they released low-molecular weight platinum species (0–20% in 72 h). *In vitro* the PAA-platinates were generally less toxic than cisplatin towards lung tumor cell lines (IC_{50} for cisplatin was 2–5 µg/ml and 1–130 µg/ml for the PAA-platinates). In *in vivo* experiments the PAA-Pt derivatives were equally active compared to cisplatin in an i.p. L1210 leukaemia model,[43] thus confirming their ability to liberate biologically active platinum species. Whereas ISA 23-platinate was significantly less toxic *in vivo* than cisplatin, the cyclodextrin-containing PAAs did show significant toxicity on repeated dosing and thus are unsuitable for further development.

As regards activity, preliminary experiments have shown that the ISA 23-platinum conjugate displayed similar antitumor activity to that reported for other platinum conjugates of similar molecular weight.[44-46] Further experiments, however, are needed to establish the maximum tolerated dose of ISA 23-platinate and thus its therapeutic index. PAA-platinates have the advantage of good water solubility, which is maintained during storage, in contrast with other polymer-platinates, which often show reduced solubility after very short times (1 to 2 weeks).[47]

16.4.2.5 pH-Mediated Membrane Activity, DNA Complexing, and Transfection Ability of PAAs

A peculiar quality of several PAAs is that they display pH-mediated activity on cell membranes. It had been previously found that PAAs bearing two aminic nitrogens in their repeating unit show a marked conformational change during movement from a neutral pH to acidic one as a result of the modification of their average charge, this effect being more pronounced when the aminic nitrogens were separated by only two methylene groups. This property provides, in principle, the possibility of designing polymer-drug conjugates that are, following intravenous administration, relatively compacted, thus protecting a drug payload in the circulation, where the pH is 7.4, but following pinocytic internalization into acidic (pH 5.5) intracellular compartments unfold, permitting pH-triggered intracellular drug delivery.[48]

Owing to their pH-dependent membrane activity, PAAs have a definite potential to act as fusagenic polymers.[49] Fusagenic polymers are able to form complexes with the DNA chains retarding their degradation. Amphoteric PAAs, even those that at pH 7.4 have a negative net average charge, may form complexes with DNA, probably because positively and negatively charged sites are located in different positions along the polymer chain. For a synthetic fusagenic system to be effective *in vivo*, it must not only be tolerated by the body, but also demonstrate an appropriate pharmacokinetic profile, i.e., a plasma residence time appropriate to its function and the ability to afford tissue tropic targeting.

PAAs, for instance ISA 1 and ISA 23, when studied in this respect did in fact exhibit remarkable efficiency as transfection promoters, the latter being as active as polyethylenimine and LipofectIN, and more active than LipofectACE.[49]

16.4.2.6 *PAAs as Promoters of Intracellular Trafficking of Proteins*

Plant and bacterial toxins have been widely explored as anticancer agents, particularly in the form of immunotoxins. Therefore, we have chosen to investigate the ability of PAAs to mediate intracellular delivery of two ribosome-inactivating toxins, ricin and gelonin.

Ricin, derived from *Ricinus communis* beans, is a highly cytotoxic protein when present in the native dimeric form, consisting of an A-chain (RTA) and a B-chain (RTB) linked by a disulfide bridge. RTB is the cell-binding moiety that acts to promote binding and endocytosis of RTA.

Once inside the cell, RTA is believed to utilize the E-R associated protein degradation pathway (ERAD pathway) and is translocated to the cytosol in a partially unfolded state. Once in the cytosol, RTA refolds and exhibits pharmacological activity by cleavage of the N-glycosidic bond of adenosine nucleoside leading to inhibition of protein synthesis. RTA alone does show some inherent toxicity towards cells but this is a very inefficient process. This effect is greatly reduced compared to the toxicity of the holotoxin. Gelonin toxin does not contain the cell-binding subunit of ricin and is completely nontoxic to intact cells. PAAs shown previously to display pH-dependent haemolysis were chosen to investigate their ability to promote intracytoplasmic delivery.

Two PAAs, ISA 1 and ISA 4, were incubated with B16F10 cells *in vitro* together with two nonpermeant toxins: either ricin A-chain (RTA) or gelonin. The relatively nontoxic PAAs restored activity to the inherently inert toxins.

Ricin and RTA displayed IC_{50} values of 0.3 and 1.4 µg/ml, respectively, and gelonin was nontoxic over the range studied. In all further experiments, nontoxic concentrations of RTA (250 ng/ml) and gelonin (1.4 µg/ml) were used. When incubated with B16F10 cells, ISA 1 and 4 were relatively nontoxic, showing IC_{50} values of more than 2 and 1.8 mg/ml, respectively. When B16F10 cells were incubated with a combination of ISA 1 and RTA (250 ng/ml), the observed IC_{50} fell to 0.65 ± 0.05 mg/ml. Similarly, when a combination of ISA 4 and RTA (250 ng/ml) was added, the IC_{50} value fell to 0.57 ± 0.03 mg/ml. When dextran or ISA 22 was added together with RTA (250 ng/ml), no increase in cytotoxicity was observed. Similarly, when ISA 1 and 4 were incubated in combination with gelonin (1.4 µg/ml), its toxicity toward B16F10 cells remarkably increased. No increased gelonin cytotoxicity was seen in the presence of dextran or ISA 22.

As these experiments show that ISA 1 and 4 but not ISA 22 were able to promote both RTA and gelonin cytotoxicity, subsequent experiments were designed to elucidate the relative potency of the ISA 4–RTA combination in a comparison to ricin. Cytotoxicity assays were again performed using B16F10 cells but in this case a fixed concentration of ISA 4 was combined with increasing concentrations of RTA (0–10 μg/ml). As the ISA 4 concentration increased, less RTA was required to achieve the IC_{50} level.[50]

It can be clearly seen that a mixture of ISA 4 and RTA can be used to promote cytotoxicity to a greater level than seen for the holotoxin at an ISA 4 concentration of 1.5 mg/ml. These observations suggest that specific PAA-toxin combinations warrant further development as novel therapeutics.

16.4.2.7 PAA-Grafted Proteins and Future Trends

As a further development of the above studies, we are presently studying the grafting of PAAs on toxins in order to create a macromolecule with the same chemical and biological properties of both its constituents.

The first step was the determination of the best reacting conditions and of the best the way to isolate our products. A possibly general process has been developed using bovine serum albumine (BSA). The structure of BSA is characterized as a single, carbohydrate-free polypeptide chain consisting of 580–585 amino acid residues stabilized by 17 S-S bridges. It was chosen as a protein model for other proteins and toxins.

BSA contains 60 Lysine units and therefore has 60 free -NH_2 groups. It was modified by grafting PAA chains via the hydrogen-transfer addition reactions of vinyl-terminated PAAs. The PAAs used were ISA 23 and ISA 1 (structures reported in Figure 16.10).

Scheme 16.5 shows the reaction scheme for the grafting of ISA 23 on a protein. The reaction scheme for ISA 1 was similar.

Tests on different ratios of BSA compared to total amount of monomers have shown that a 20% ratio of protein and a 15% dilution give the best results as regards yield and chemical properties. The grafting reaction requires very mild conditions. It takes place in water at room temperature and without an added catalyst. The grafting reactions were controlled on the crude product by SEC analysis (Table 16.6).

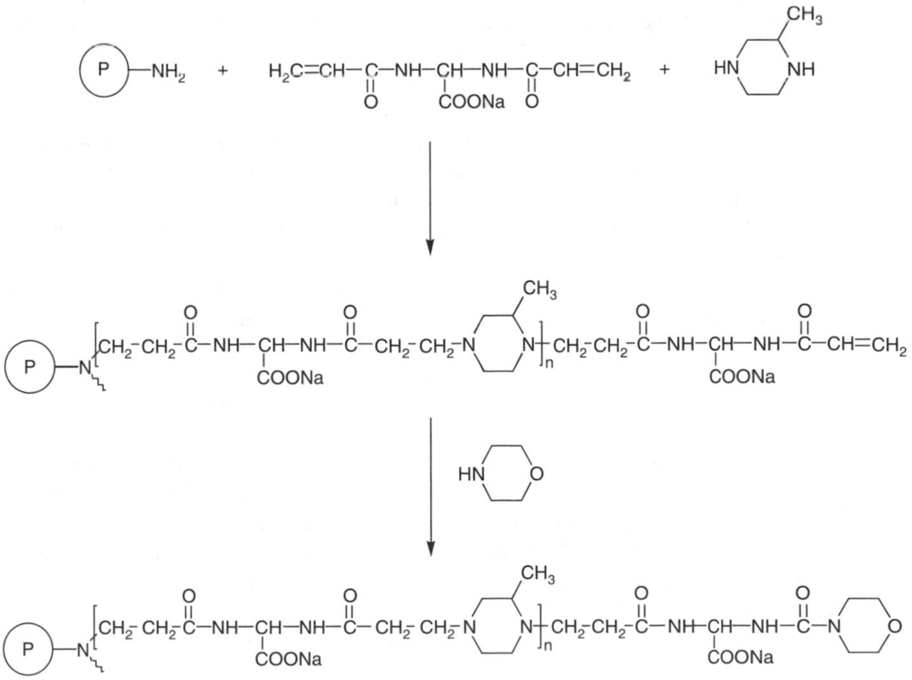

Scheme 16.5 Synthetic mechanism of the grafting reaction.

Table 16.6 SEC[a] Results of the BSA Grafting Reaction

Product	Retention Time (min)	M_n[b]	M_w[b]	M_p[b]
ISA 23 — BSA 20%	16–22	32,000	48,000	43,500
ISA 1 — BSA 20%	18–26	89,000	108,000	82,000
BSA	20–24	25,800	30,500	28,300

[a] Elution conditions: columns TSK Gel G5000 PW, TSK Gel G4000 PW, TSK Gel G3000 PW connected in series purchased from Toso Haas; mobile phase: 0.1 M Tris/hydrochloric acid buffer pH = 8 ± 0.05 + 0.2 M sodium chloride; flow rate: 1 ml/min.

[b] Apparent molecular weight determined against poly(N-vinyl-pyrrolidinone) standards.

The grafted products exhibited lower retention times with respect to BSA, thus indicating an increase of molecular weight, which is consistent with a successful grafting reaction.

REFERENCES

1. Duncan, R., The dawning era of polymer therapeutics, Nature reviews, Drug discovery vol. 2, 347, 2003.
2. Danusso, F. and Ferruti, P., Synthesis of tertiary amine polimers, *Polymer* 11,88, 1970.
3. Ferruti, P., Marchisio, M.A., and Barbucci, R., Synthesis physico-chemical properties and biomedical applications of poly(amido-amine)s, *Polymer* 26, 1336, 1985.
4. Ferruti, P., Ion-chelating polymers (medical applications), in *Polymeric Materials Enciclopedia*, vol. 5, Salamone, J.C. (Ed.), CRC Press, Boca Raton, FL, 1996, 3334–3359.
5. Malgesini, B., Verpilio, I., Duncan, R., and Ferruti, P., Poly(amido-amines) carrying primary amino groups as side substituents, *Macromol. Biosci.* 3, 59–66, 2003.
6. Ferruti, P., Knobloch, S., Ranucci, E., Duncan, R., and Gianasi, E., A novel modification of poly(L-lysine) leading to a soluble cationic polymer with reduced toxicity and with potential as a transfection agent, *Macromol. Chem. Phys.* 199, 2565, 1998.
7. Lide, D.R. (Ed.), *Handbook of Chemistry and Physics*, 76th edition, CRC Press, Boca Raton, FL, 1995, Sect. 8, 47.
8. Caldwell, G., Neuse, E., and Stephanou, A., Synthesis of water-soluble polyamidoamines for biomedical applications. II. Polymers possessing intrachain-type secondary amino groups suitable for side-chain attachment., *J. Appl. Polym. Sci.* 50, 393, 1993.
9. Peppas, N.A., *Biomaterials Science*, Ratner, B.D., Hoffman, A.S., Schoen, F.D., and Lemons, J.E. (Eds), Academic Press, San Diego, 1996, 60.
10. Andreani, F., Angeloni, A.S., Angiolini, L., Costa Bizzarri, P., Della Casa, C., Fini, A., Ghedini, N., Tramontini, M., Ferruti, P., Poly(b-aminoketone)s by polycondensation of bis(b-dialkylaminoketone)s with bisamines, *J. Polym. Sci. Polym. Lett. Ed.* 19, 443, 1981.
11. Angeloni, A.S., Ferruti, P., Tramontini, M., and Casolaro, M., The Mannich bases in polymer synthesis. 3. Reduction of poly(β-amino ketone)s to poly(γ-amino alcohol)s and their N-alkylation to poly(γ-hydroxy quaternary ammonium salt)s., *Polymer* 8, 549, 1982.
12. Angeloni, A.S., Ferruti, P., Laus, M., Tramontini, M., Chiellini, E., and Galli, G., The Mannich bases in polymer synthesis: 4. Further studies of synthesis and characterization of some poly(b-amino-ketone)s containing potentially mesogenic groups, *Polym. Commun.* 24, 87, 1983.
13. Ferruti, P. and Ranucci, E., Poly(1,4-piperazinediylethylenedithioethylene) a first example of a new family of easily biodegradable polymers containing tertiary amino groups, *Makromol. Chem. Rapid Commun.* 8, 549, 1987.
14. Ferruti, P. and Ranucci, E., Poly(esterthioetheramines), a new family of tertiary amino polymers, *J. Polym. Sci. Polym. Lett. Ed.* 26, 357, 1988.
15. Ferruti, P., Ranucci, E., and Depero, L., New basic multifunctional polymers, 4. Polyaddition of 2,2'-alkylenediiminodiethanethiols to divinylsulfone, *Makromol. Chem., Rapid Commun.* 9, 807, 1988.

16. Ferruti, P., Ranucci, E., and Depero, L., New basic multifunctional polymers. III. Synthesis and properties of poly(amidothioetheramine)s, *Polym. Commun.* 30, 157, 1989.

17. Ranucci, E. and Ferruti, P., New basic multifunctional polymers: 5. Poly(esterthioetheramine)s by polyaddition of 2,2'-alkylenediimino diethanethiols to bisacrylic and bismethacrylic esters, *Polymer* 32, 2876, 1991.

18. Ranucci, E., Bigotti, F., Cancelli, E., and Ferruti, P., Tertiary amino polymers by polyaddition of 1,4-piperazinediylethylenedithioethylene to 1,4-bis-methacryloyl piperazine, *Polym. J.* 25, 625, 1993.

19. Tanzi, M.C., Barzaghi, B., Anouchinski, R., Bilenkis, S., Penhasi, A., and Cohn, D., Grafting reactions and heparin adsorption of poly(amidoamine)-grafted poly(urethane amide)s, *Biomaterials* 13, 425, 1992.

20. Tanzi, M.C., Pieghi, G., Botto, P., Barozzi, C., and Cardillo, P., Synthesis and characterization of poly(amido amine)s belonging to two different homologous series., *Biomaterials* 5, 357, 1984.

21. Bignotti, F., Sozzani, P., Ranucci, E., and Ferruti, P., NMR studies, molecular characterization, and degradation behavior of poly(amido amine)s. 1. Poly(amido amine) deriving from the polyaddition of 2-methylpiperazine to 1,4-bis(acryloyl) piperazine, *Macromolecules* 27, 7171, 1994.

22. Ferruti, P., Manzoni, S., Richardson, S.C.W., Duncan, R., Pattrick, N.G., Mendichi, R., and Casolaro, M., Amphoteric linear poly(amidoamine)s as endosomolytic polymers: correlation between physico-chemical and biological properties. *Macromolecules* 33, 7793, 2000.

23. Ferruti, P., Ranucci, E., Sartore, L., Bigotti, F., Marchisio, M.A., Bianciardi, P., and Veronese, F.M., Recent results on functional polymers and macromonomers of interest as biomaterials or for biomaterial modification, *Biomaterials* 15, 1235, 1994.

24. Ferruti, P., Ranucci, E., Bigotti, F., Sartore, L., Bianciardi, P., and Marchisio, M.A., Degradation behaviour of ionic stepwise polyaddition polymers of medical interest, *J. Biomater. Sci. Polym. Ed.* 6, 833, 1994.

25. Ranucci, E., Spagnoli, G., Ferruti, P., Sgouras, D., and Duncan, R., Poly(amidoamine)s with potential as drug carriers: degradation and cellular toxicity, *J. Biomater. Sci. Polym. Ed.* 2, 303, 1991.

26. Mandel, M., in *Encyclopaedia of Polymer Science and Technology,* vol. 11, Mark, H.F., Gaylord, N.G., and Bikales, N.M. (Eds.), Interscience, New York, 1992, 739–829.

27. Ferruti, P., Marchisio, M.A., and Duncan, R., Poly(amido-amine)s: biomedical applications, *Macromol. Rapid Commun.* 23, 332–335, 2002, and references therefrom.

28. Marchisio, M.A., Longo, T., Ferruti, P., and Danusso, F., *Eur. Surg. Res.* 3, 240, 1971.

29. Marchisio, M.A., Ferruti, P., and Longo, T., *Eur. Surg. Res.* 4, 312, 1972.

30. Marchisio, M.A., Longo, T., and Ferruti, P., A selective de-heparinizer filter made of new cross-linked polymers of a poly-amido-amine structure, *Experientia* 29, 93, 1973.

31. Ferruti, P., Casini, G., Tempesti, F., Barbucci, R., Mastacchi, R., and Sarret, M., Heparinizable materials. (III). Heparin retention power of a poly(amido-amine) either as crosslinked resin, or surface-grafted on PVC, *Biomaterials* 5, 234, 1984.

32. Marchisio, M.A., Ferruti, P., Bertoli, S., Barbiano di Belgiojoso, G., Samour, C.M., and Wolter, K.D., A novel approach to the problems of heparin in haemodialysis: the use of a de-heparinizing filter, in *Polymers in Medicine II*, Migliaresi, C. and Chiellini, E. (Eds.), Elsevier Science Publishers B. V., Amsterdam, 1988, 11–118.

33. Marchisio, M.A., Ferruti, P., Longo, T., and Danusso, F., U.S. Patent 3,865,723, 1975; *Chem. Abstr.* 286, 112677j, 1974.

34. Albanese, A., Barbucci, R., Belleville, J., Browry, S., Eloy, R., Lemke, H.D., and Sabatini, L., *Biomaterials* 15, 129, 1994.

35. Barbucci, R., Benvenuti, M., Dal Maso, G., Ferruti, P., Nocentini, M., Russo, R., Tempesti, F., Duncan, R., Bridges, J.F., and McCornick, L.M., A new material for biomedical application, in *Polymers in Medicine II*, Migliaresi, C. and Chiellini, E. (Eds.), Elsevier Science Publishers B. V., Amsterdam, 1988, 3–18.

36. Tanzi, M.C. and Levi, M., Heparinizable segmented polyurethanes containing polyamidoamine blocks, *J. Biomed. Mater. Res.* 23, 863, 1989.

37. Dey, R.K. and Ray, A. R., *Biomaterials* 24 (18), 2985–2993, 2003.

38. Ringsdorf, H., *J. Polym. Sci. Polym. Symp.* 51, 135–153, 1981.

39. Richardson, S.C.W., Ph.D. Thesis at Center of Polymer Therapeutics, School of Pharmacy, Faculty of Medicine, University of London, 1998.

40. Richardson, S., Ferruti, P., and Duncan, R., Poly(amidoamine)s as potential endosomolytic polymers: evaluation *in vitro* and *in vivo* body distribution in normal and tumour-bearing animals, *J. Drug Target.* 6, 391, 1999.

41. Duncan, R., *Anti-Cancer Drugs* 3, 175, 1992.

42. Schacht, E., Ferruti, P., and Duncan, R., Drug delivery agents incorporating mitomycin, W.O. 9505,200, 1994; *Chem. Abstr.* 595, 248301a, 1995.

43. Ferruti, P., Ranucci, E., Trotta, F., Gianasi, E., Evagorou, E.G., Wasil, M., Wilson, G., and Duncan, R., Synthesis, characterisation and antitumor activity of platinum(II) complexes of novel functionalised poly(amido-amine)s, *Macromol. Chem. Phys.* 200, 1644, 1999.

44. Ohya, Y., Masunaga, T., Baba, T., and Ouchi, T., *J. Macromol. Sci. Pure Appl. Chem.* A33(8), 1005, 1996.

45. Sohn, Y.S., Baek, H., Cho, Y.H., Lee, Y., Jung, O., Lee, C.O., and Kim, Y.S., *Int. J. Pharm.* 153, 79, 1997.

46. Gianasi, E., Wasil, M., Evagorou, E.G., Keddle, A., Wilson, G., and Duncan, R., HPMA copolymer platinates as novel antitumor agents: in vitro properties, pharmacokinetics and antitumor activity in vivo, *Eur. J. Cancer* 35, 994, 1999.

47. Neuse, E.W., Caldwell, G., and Perlwitz, A.G., *J. Inorg. Organomet. Polym.* 5, 195, 1995.

48. Duncan, R., Ferruti, P., Sgouras, D., Toboku-Metzger, A., Ranucci, E., and Bignotti, F., A polymer-Triton X-100 conjugate capable of pH-dependent red blood cell lysis: a model system illustrating the possibility of drug delivery acidic intracellular compartments, *J. Drug Target.* 2, 341, 1994.

49. Richardson, S.C., Pattrick, N.G., Man, Y.K., Ferruti, P., and Duncan, R., Poly(amidoamine)s as potential nonviral vectors: ability to form interpolyelectrolyte complexes and to mediate transfection in vitro, *Biomacromolecules* 2, 1023–1028, 2001.

50. Pattrick, N.G., Richardson, S.C.W., Casolaro, M., Ferruti, P., and Duncan, R., Poly(amidoamine)-mediated intracytoplasmic delivery of ricin A-chain and gelonin, *J. Control. Rel.* 77(3), 225–232, 2001.

51. Ferruti, P., Ranucci, E., Chiellini, F., and Cavalli, R., Italian patent application to Universita di Milano, March 24, 2004, MI2004A000435.

CHAPTER 17

Cationic Polysaccharides for Gene Delivery

Tony Azzam and Abraham J. Domb

CONTENTS

17.1 INTRODUCTION

DNA can be delivered into the cell nucleus using physical means or using specific carriers that carry the genes into the cells for gene expression. Of the various methods developed for delivering genes, gene carriers have been extensively investigated as transfecting agents for therapeutic genes in gene therapy. Gene carriers are divided into two main groups: viral carriers, where the DNA to be delivered is inserted into a virus; and cationic molecular carriers that form electrostatic

0-8493-1934-X/05/$0.00+$1.50
© 2005 by CRC Press LLC

interactions with DNA. Successful gene therapy depends on the efficient delivery of genetic materials into the cells nucleus and its effective expression within these cells.[1] Although at present, the *in vivo* expression levels of synthetic molecular gene vectors are lower than for viral vectors and gene expression is transient; these vehicles are likely to present several advantages including safety, low immunogenicity, capacity to deliver large genes, and large-scale production at low cost. The two leading classes of synthetic gene delivery systems that have been mostly investigated involve the use of either cationic lipids or cationic polymers.[2]

17.1.1 Cationic Lipids

Since the introduction of the transfection reagent Lipofectin™, a cationic lipid composed of a 1:1 (w/w) mixture of the quaternary ammonium lipid N(1-[2,3-dioleyloxy]propyl)-N,N,N-trimethylammonium chloride (DOTMA) and a colipid dioleylphosphatidylethanolamine (DOPE),[3] an increasing number of cationic lipids have been developed. Cationic lipids commonly used in gene delivery and transfection can be classified into several categories including quaternary ammonium salt lipids,[4-6] lipopolyamines,[1,7,8] amidinium salt cationic lipids,[9,10] and imidazole-, phosphonium-, and arsonium-based salt lipids.[11] The driving force of such interaction (cationic lipid–DNA) is the release of low-molecular-weight counter-ions associated with the charged lipids into the external media, which is accompanied by a substantial entropy gain.[12,13] The lipid moieties play the role of forming and maintaining a self-assembly system with the DNA and enhancing fusion of the complex through the cell membrane.[14,15]

17.1.2 Cationic Polymers

Cationic polymers, commonly called polycations, are a leading class of molecular gene-delivery systems, in part because their molecular diversity can be modified in order to fine-tune their physicochemical properties.[16,17] Polyelectrolyte complexes (PEC) formed between DNA and polycations have been shown to pack the DNA tightly in the PEC complex, so that the entrapped DNA is shielded from contact with DNase.[18] The polycations commonly used in gene delivery and transfection include polyethyleneimine,[19] poly(L-lysine),[20] dendrimers,[21] polybrene,[22] gelatin,[23] tetraminofullerene,[24] poly(L-histidine)-graft-poly(L-lysine),[25] and cationic polysaccharides. Although PEC systems have some advantages over viral vectors — e.g., low immunogenicity and easy manufacture[26,27] — several problems such as toxicity, lack of biodegradability, low biocompatibility, and, in particular, low transfection efficiency need to be solved prior to practical use.[28] Polycations used in gene delivery are polyamines that become cationic at physiologic conditions. All polymers contain primary, secondary, tertiary, or quaternary amino groups capable of forming electrostatic complexes with DNA under physiological conditions. The highest transfection activity is obtained usually at a 1.1–1.5 charge ratio of polycation to DNA, respectively. Such polycations exhibit a random distribution of cationic sites along the polymer chains. Most polycations are toxic to cells and nonbiodegradable, while the polymers based on amino acids such as poly(L-lysines) are immunogenic.[29] More advanced polymeric gene delivery systems employ macromolecules with high cationic charge density that act as endosomal buffering systems, thus suppressing the endosomal enzymes activity and protecting the DNA from degradation.[19]

Among the various polycations used in gene delivery and transfection, cationic polysaccharides are considered to be the most attractive candidates. They are natural, nontoxic, biodegradable, biocompatible materials that can be simply modified for improved physicochemical properties.[30,31] This review focuses mainly on the chitosan derivatives, diethylaminoethyl dextran (DEAE-dextran), and polysaccharide-oligoamine based conjugates as gene delivery vectors *in vitro* and *in vivo*.

17.2 CHITOSAN

17.2.1 Structure, Chemistry, Physico-Chemical Properties, and Application

The use of chitosan as a gene vector is described in chapter 18 of this book. This section focuses on the chemistry and physicochemical properties of chitosan, condensation with DNA, and applications of chitosan derivatives in gene delivery.

Chitosan is a biodegradable polysaccharide composed of two subunits, D-glucosamine and N-acetyl-D-glucosamine, linked together by -(1,4) glycosidic bonds (Figure 17.1). It is a deacetylated form of chitin, an abundant polysaccharide present in crustacean shells.[32,33] Even though the discovery of chitosan dates from the 19th century, it has only been over the last 2 decades that this polymer has received attention as a material for biomedical and drug delivery applications.[34] The amino groups of chitosan present in the N-deacetylated subunits conferring a highly positive charge density.[35,36] These amino groups exhibit intrinsic pK_a values of 6.5 and thus chitosan behaves as a polycation at acidic and neutral pH.[37] The term chitosan is generally used to describe a series of chitosan polymers with different molecular weights (50 kDa to 2000 kDa), viscosities, and degrees of deacetylation.[38] Chitosan is a linear polyamine with many primary amine groups readily available for chemical modification and salt formation with acids.

Important characteristics of chitosan polymer are its molecular weight, viscosity, degree of deacetylation,[39,40] crystallinity index, number of monomeric units (n), water retention value, pK_a, and energy of hydration.[41] Chitosan is soluble at acidic media and practically insoluble at an alkaline and neutral pH. Upon dissolution, amine groups of the polymer become protonated, positively charged polysaccharides (RNH_3^+). However, chitosan salts (glutamate, chloride, etc.) are usually soluble in water, and their solubility is dependent on the degree of deacetylation. For example, chitosan with a low degree of deacetylation (40%) has been found to be soluble up to pH 9, whereas chitosan with a degree of deacetylation of about 85% is soluble up to pH 6.5. Solubility is also greatly influenced by the addition of salt to the solution. The higher the ionic strength, the lower the solubility of chitosan in aqueous media.[42,43] Chitosans with various molecular weights and degrees of deacetylation are cheap and commercially available.[44,45] The unique physicochemical and biological properties of chitosan led to the recognition of this polymer as a promising candidate for drug delivery, and more specifically, for the delivery of delicate macromolecules such as DNAs and proteins. From a technological point of view, it is extremely important that chitosan be hydrosoluble and positively charged.[33] These unique electric properties enable it to interact with negatively charged polymers, macromolecules, and certain polyanions in an aqueous environment. These interactive forces can be exploited for nanoencapsulation purposes. From a biopharmaceutical point of view, chitosan has the special feature of adhering to mucosal surfaces, a fact that makes it a useful polymer for mucosal drug delivery.[46] Chitosan has the potential to open tight junctions between epithelial cells, thus facilitating the transport of macromolecules through well-organized epithelia.[47] The interesting biopharmaceutical characteristics of this polymer are accompanied by

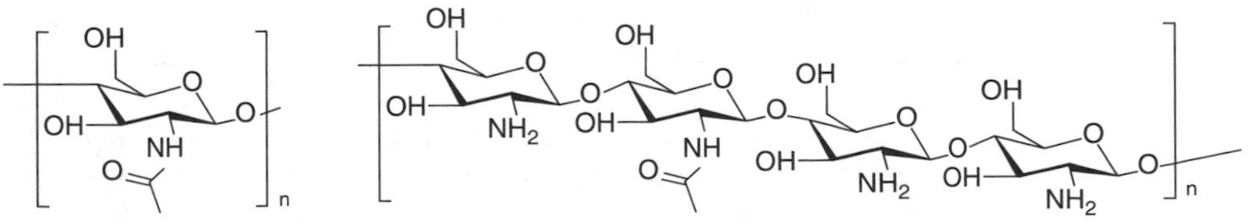

(a)   (b)

Figure 17.1   Structures of the naturally occurring chitin (**A**) and the deacetylated form, chitosan (**B**).

its well-documented biodegradability, biocompatibility, and low toxicity.[30,31,44,48] The electrostatic interactions of chitosan with polyions, including indomethacin,[49] sodium hyaluronate,[50] pectin, and acacia polysaccharide[51] have been well characterized in the literature. Chitosan has been investigated for a number of pharmaceutical applications such as drug delivery,[52,53] controlled drug delivery,[51,54-57] peptide delivery,[58-61] drug delivery for drugs across the intestinal and nasal mucosa,[62–64] colon delivery,[65,66] and so on.

The chemical modification of chitin and chitosan is expected to exploit their full potential. Recent studies on the chemical modification of chitin and chitosan are reviewed by Kurita et al.[67,68] The C-2 groups of chitin are fully acetylated, but usually 5–15% deacetylation takes place by strong alkaline treatment during the production process of chitin. The degree of deacetylation is an important factor in characterizing the properties of both chitin and chitosan. The methods of determining the degree of deacetylation have been reported, including NMR, UV, IR, CD, and colloidal titration.[69-71] For soluble chitin and chitosan in solvents, the NMR method has been successfully adapted for the determination of the degree of deacetylation.[72,73] The N-acetyl groups of chitin can be deacetylated by aqueous sodium hydroxide in heterogeneous conditions to obtain the deacetylated chitin (DAC). However, maximum deacetylation usually stopped at around 70% under these conditions.[74] To obtain further deacetylated chitin, it is necessary to isolate and treat the deacetylated chitin repeatedly with alkali in the same manner.[75,76] Acetylation of chitosan can be easily achieved upon treating the polymer with acetic anhydride or acetyl chloride in homogenous/heterogeneous conditions, with relatively good yields. The degree of acetylation is controllable by the reaction conditions and the acetylation proceeds preferentially at the free amine groups of chitosan and with lower extent at the hydroxyl groups.[77]

17.2.2 Chitosan Derivatives

Mumper et al.[78,79] were the first to describe the potential use of chitosan as a gene carrier. The low toxicity and its nature make chitosan attractive for gene delivery purposes.[80] Chitosan has also been shown to bind mammalian and microbial cells by interacting with surface glycoproteins,[81] and some studies have indicated that chitosan may actually be endocytosed into the cell.[31]

Several chitosan derivatives have been synthesized in the last few years in order to obtain modified carrier with altered physico-chemical characteristics.[82] Such modifications include quaternization of amine groups to increase the net positive charge of the complex, ligand attachment for targeting purposes, conjugation with hydrophilic polymers to increase stability of chitosan–plasmid complex against degrading enzymes, and conjugation with endosomolytic peptides to increase the efficiency of transfection. The following sections summarize the important chitosan derivatives and their potential applications.

17.2.2.1 Deoxycholic Acid-Modified Chitosan

Deoxycholic acid was conjugated to chitosan in methanol and water media using EDC as a coupling agent.[82,83] The degree of substitution (DS) was determined to be 5.1 (5.1 deoxycholic acid groups substituted per 100 anhydroglucose units). Hydrophobically modified chitosan provides colloidally stable self-aggregates in aqueous media having mean diameter of ~160 nm with a unimodal size distribution. Self-aggregate–DNA complexes were formed in aqueous media and found useful in transfecting mammalian cells *in vitro*. The transfection efficiency of this system was relatively higher in comparison to naked DNA but significantly lower than the Lipofectamine®–DNA formulation.

17.2.2.2 Quaternized Chitosan

Despite advantageous properties, chitosan is insoluble at physiological pH values. Free amine groups of chitosan can form ammonium salts with inorganic and organic acids. Reaction of chitosan with excess of methyl iodide in alkaline conditions gives N-trimethyl chitosan derivative. Such quaternary chitosan derivatives are useful for applications using their electronic properties, because the derivatives can keep their cationic character independent of the external conditions, including pH of medium.[84] Trimethyl chitosan oligomers (TMO) of 40% (TMO-40) and 50% (TMO-50) degrees of quaternization were synthesized and examined for their transfection efficiencies in two cell lines: COS-1 and Caco-2.[85,86] Chitosan raises the transfection efficiency 2–4 times compared to the control value (i.e., naked DNA). TMO-50 markedly increases the transfection efficiencies from 5-fold (for complexes with a DNA:oligomer ratio of 1:6) to 52-fold (for a ratio of 1:14). TMO-40 displays even higher transfection efficiencies, ranging from 26-fold (for a ratio of 1:6) to 131-fold (for a ratio of 1:14). However, none of the TMO-based vectors was able to increase the transfection efficiency in differentiated cells like Caco-2. Chitosan and TMO oligomers were found to exhibit significantly lower cytotoxicity than DOTAP, a well-known cationic lipid formulation commonly used in gene transfection.

17.2.2.3 Chitosan Modified with Hydrophilic Polymers

Modification of chitin and chitosan with a hydrophilic polymer such as polyethylene glycol (PEG) would be expected to result in hydrophilic chitin or chitosan while keeping the fundamental skeleton intact. Multiple methods have developed for the grafting of hydrophilic polymers onto chitin or chitosan[87,88] to improve affinity to water or organic solvents. PEG-chitosan derivatives with various molecular weights (mol wt = 550, 2000, 5000) of PEG and DSs were synthesized, and the water solubility of these derivatives was evaluated at pH values of 4, 7.2, and 10.[32] Almost all PEG-chitosan derivatives were soluble in acidic buffer (pH 4). Furthermore, some derivatives dissolved in neutral (pH 7.2) and alkaline buffers (pH 10). The weight ratio of PEG in the derivatives seems to dominate its water solubility. Higher molecular weight PEG was found to enhance water solubility of chitosan with a lower DS of PEG in comparison with the lower molecular weight PEG. PEG modification was found to minimize aggregation and prolong the transfection potency for at least 1 month in storage. Intravenous injection of chitosan-DNA nanoparticles and PEGylated-chitosan-DNA nanoparticles resulted in majority of nanoparticles to localize in kidney and liver within the first 15 min. The clearance of the PEGylated nanoparticles was slightly slower in comparison to nonPEGylated nanoparticles.[89]

17.2.2.4 Galactosylated Chitosan

Although, in many cases, uptake of chitosan-DNA nanoparticles appears to occur in the absence of ligand-receptor interaction, Park et al.[90,91] prepared and examined galactosylated chitosan-graft-dextran–DNA complexes. Galactose groups were chemically bound to chitosan for liver-targeted delivery and dextran was grafted for enhancing the complex stability in aqueous media. This system was found to efficiently transfect Chang liver cells expressing asialoglycoprotein receptor (ASGP-R), which specifically recognize the galactose ligands on modified chitosan. In parallel work, galactosylated chitosan-graft-PEG (CGP)[92] was developed for the same purpose. CGP–DNA complexes were found to be stable due to hydrophilic PEG shielding and increased the protection against DNase. Also, CGP–DNA complexes were found to enhance transfection in HepG2 cells having ASGP-R, indicating galactosylated chitosan will be an effective hepatocyte-targeted gene carrier. A galactosylated chitosan-graft-poly(vinyl pyrolidone) (CGPVP) was also synthesized[93] and showed improved physicochemical properties over the unmodified chitosan. Erbacher et al.[94] synthesized lactosylated-modified chitosan derivatives (having various DSs) and tested their transfection efficiencies in many cell lines. However, the *in vitro* transfection was found to be cell-type dependent. HeLa cells were efficiently transfected by this modified carrier even in the presence of 10% serum, but neither chitosan nor lactosylated chitosans have been able to transfect HepG2 and BNL CL2 cells.

17.2.2.5 *Chitosan Conjugated with Transferrin, KNOB, and Endosomolytic Proteins*

The transferrin receptor responsible for iron import to cells is found in many mammalian cells.[95] As a ligand, transferrin could efficiently transfer low-molecular-weight drugs, macromolecules, and liposomes, through a receptor-mediated endocytosis mechanism.[96] Transferrin has been applied to deliver plasmid DNA (pDNA) and oligonucleotide.[97,98] Mao et al.[89] explored two strategies to bind transferrin onto the surface of chitosan–DNA complex. In the first strategy, aldehyde groups were introduced in transferrin (glycoprotein) after oxidation with periodate, and thereafter allowed to react with chitosan amine groups via the formation of Schiff-base linkages. The transfection efficiencies of transferrin-modified chitosan carriers (at varying degrees of modification) were examined in the HEK293 cell line and found to produce a twofold transgene expression in comparison to unmodified chitosan carrier. In the second strategy, transferrin was introduced to the nanoparticle surface through a disulfide bond. The transferrin-conjugated carrier only resulted in a maximum of fourfold increase in transfection efficiency in HEK293 cells and only 50% increase in HeLa cells. The negligible increase in the transgene efficiencies as a result of ligand-modification (e.g., alactose and transferrin) led the investigators to speculate that chitosan nanospheres may enter the cell via a unique endocytic pathway.[90,93,99-101] To further enhance the transfection efficiency, KNOB (C-terminal globular domain on the fiber protein) was conjugated to the chitosan by the disulfide linkages as well. Conjugation of KNOB to the chitosan-DNA nanoparticles was found to improve gene expression level in HeLa cells approximately 130-fold. Also, the inclusion of pH-sensitive endosomolytic peptide GM227.3 in the formulation enhanced the level of expression *in vitro*. Expression of a plasmid–chitosan–GM225.1 formulation in rabbits after administration in the upper small intestine and colon was observed, in contrast with naked plasmid, which gave no expression. The lipidic formulation DOTMA–DOPE was used as control and was not expressed to as high an extent as the chitosan–lytic peptide formulation.[79]

17.3 DEAE-DEXTRAN MEDIATED TRANSFECTING AGENT

Diethylaminoethyl-dextran (DEAE-Dextran) is a polycationic derivative of dextran, and is obtained by reacting diethylaminoethyl chloride with dextran in basic aqueous medium.[102] Commercial DEAE-Dextran (Sigma-Aldrich, Promega®, Amersham Biosciences®, etc.) is supplied in the hydrochloride form, and the degree of substitution corresponds to approximately one DEAE subunit per three glucose units (Figure 17.2). The average molecular weight (mol wt) of commercial DEAE-Dextran is greater than 500,000 Da, and is carefully purified to remove contaminants arising from the derivatization.

As depicted in Figure 17.2, DEAE-Dextran has two types of subunits: the single tertiary DEAE-group and "tandem" groups with a quaternary amine group. The quaternary group is strongly basic (pK$_b$ 14), whereas the tandem DEAE group has a pK$_b$ of 5.7 and the single DEAE group has a pK$_b$ of 9.5.[103] The polymer is usually supplied as a white, hygroscopic powder, is freely soluble in water, and buffers within a pH range of 4 to 10. The nitrogen content of DEAE-Dextran is in the range of 3–3.5%.

Numerous reports attest to the effect of DEAE-Dextran as an adjuvant in vaccine production. Most reports are related to veterinary applications such as somatropin release inhibiting factor in

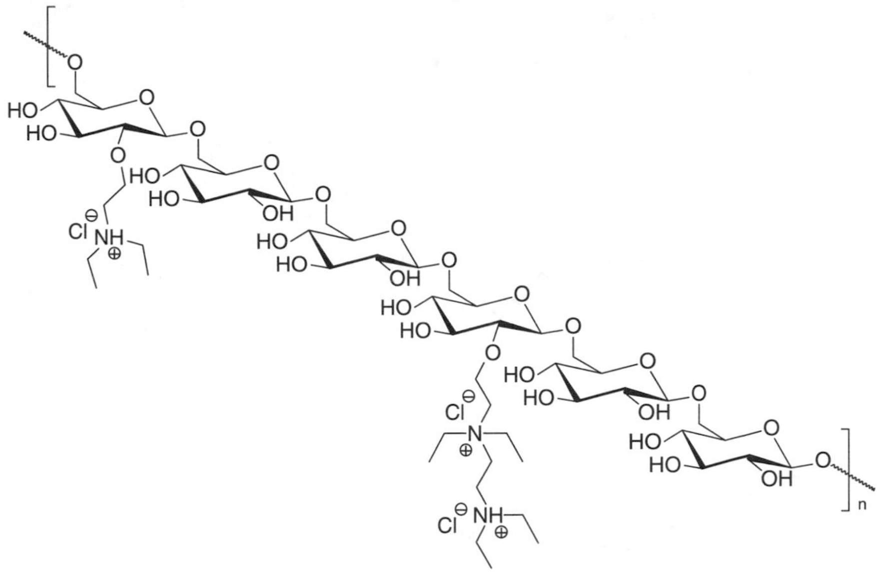

Figure 17.2 Structure of diethylaminoethyl-dextran (DEAE-Dextran).

lambs[104] and effective adjuvant for cholera vaccine in mice.[105] A 10–15% solution of DEAE-Dextran is usually added to an antigen prior to injection. DEAE-Dextran in combination with low molecular weight alcohols was also reported to preserve the activity of vacuum-dried glycerol kinase[106] and oxidase-based biosensors.[107]

DEAE-Dextran was one of the first chemical reagents used for transfer of the foreign genes into cultured mammalian cells.[108] As a positive charged polymer, DEAE-Dextran can associate with negatively charged nucleic acids.[109] An excess of positive charge, contributed by the polymer in the DNA–polymer complex, allows the complex to come into close association with the negatively charged cell membrane.[110] Uptake of the complex presumably takes place by means of an endocytosis process. The DEAE-Dextran mediated gene transfection is successful for delivery of nucleic acids into cells for transient expression. Detailed and modified protocols for transfection techniques using the DEAE-Dextran vector have been published.[108,111,112] Also, DEAE-Dextran was found to allow superior transfection compared to other techniques in the transfer of DNA to human macrophages.[108] However, the efficiency of DEAE-Dextran in transfecting a wide range of cell lines is still very low in comparison to "modern" cationic vectors such as polybrene, PEI, and dendrimers.

In gene therapy, DEAE-Dextran was found to increase adenovirus mediated gene therapy without any additional toxicity.[113] Transfection techniques applying DEAE-Dextran for gene therapy approaches to colon diseases have been reported.[114] Also, complexes of the cystic fibrosis transmembrane conductance regulator with DEAE-Dextran have been described as having potential for gene therapy.[115]

17.4 POLYSACCHARIDE-OLIGOAMINE CONJUGATES FOR GENE DELIVERY

In designing universal polycation systems for gene delivery, one should consider the way in which a plasmid becomes active in the cell and tissue. The plasmid has first to be protected from DNA-degrading enzymes in the extracellular medium, then must penetrate the cell wall, protected from degrading systems such as lysosome and enzymes in the intracellular medium until it is internalized in the nucleus. The plasmid must then allow for the insertion of the genetic material in its active form and, finally, must biodegrade and eliminate from the cell and tissue without causing toxicity.[116]

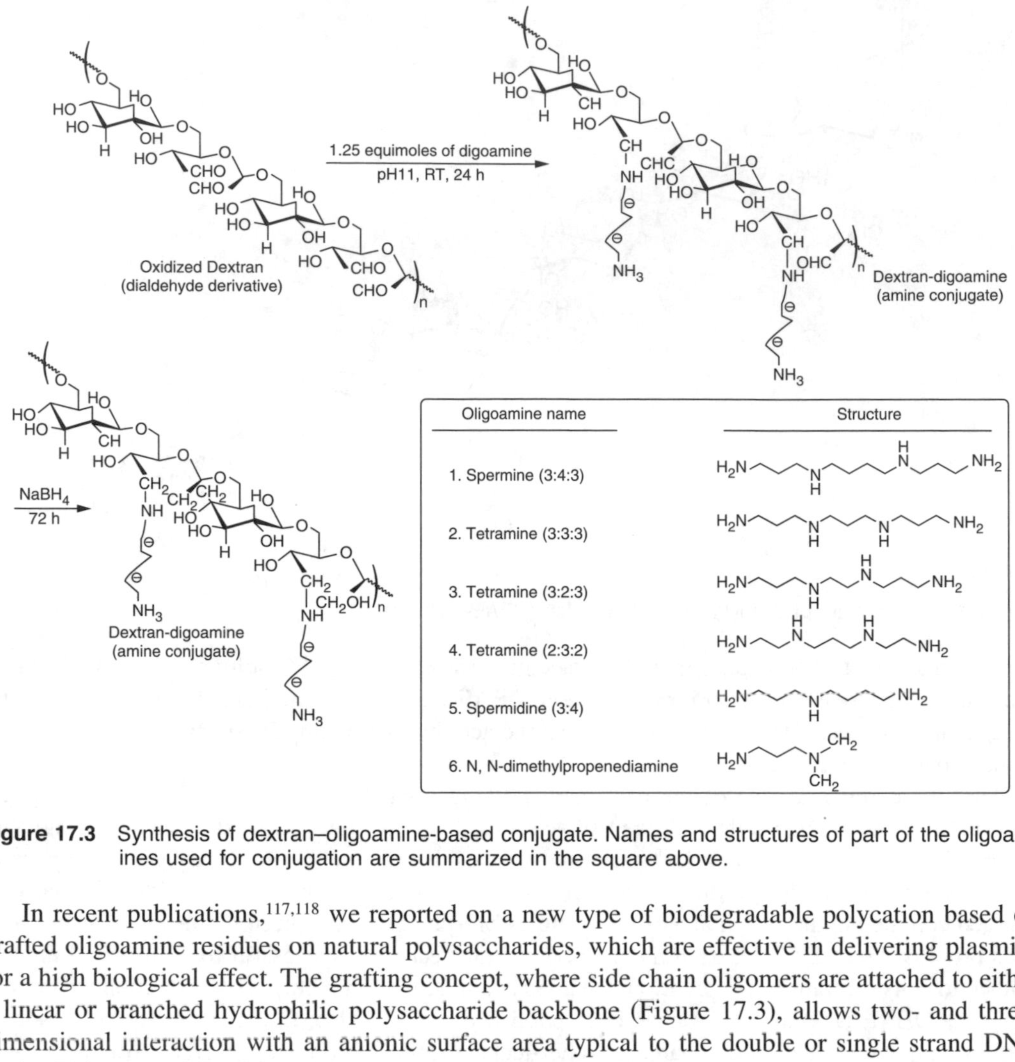

Figure 17.3 Synthesis of dextran–oligoamine-based conjugate. Names and structures of part of the oligoamines used for conjugation are summarized in the square above.

In recent publications,[117,118] we reported on a new type of biodegradable polycation based on grafted oligoamine residues on natural polysaccharides, which are effective in delivering plasmids for a high biological effect. The grafting concept, where side chain oligomers are attached to either a linear or branched hydrophilic polysaccharide backbone (Figure 17.3), allows two- and three-dimensional interaction with an anionic surface area typical to the double or single strand DNA chain. This type of flexible cationic area coverage is not available with nongrafted polycations or low-molecular-weight cations.[14,15] Low-molecular-weight cations and their lipid derivatives, such as the Lipofectin™ and Lipofectamine®, have a localized effect on the DNA, where the degree of complexation is dependent on how these small molecules organize around the anionic DNA.[2] Each molecule has to be synchronized with the other molecules at all times of the transfection process, whereas when the oligoamines are grafted on a polymer they are already synchronized and each side chain helps the other side chain to be arranged to fit the anionic surface of a given DNA.[119] Grafting the functional groups as an average distribution along a polymer chain at a certain distance from each other (for example, grafting an oligoamine chain every one, two, three, or four monomer units) may provide optimal complexation with various DNAs. The use of biodegradable polysaccharide carriers is especially suitable for transfection and biological applications because they are water soluble, can be readily transported to cells *in vivo* by known biological processes, and can act as effective vehicles for transporting agents complexed with them.[120]

17.4.1 Synthesis of Polysaccharide–Oligoamine Conjugates

Polysaccharides used in this study were the highly branched arabinogalactan (AG) (19 kDa) and the linear dextran with an average molecular weight ranging between 9.3 and 500 kDa. The representative polysaccharide was oxidized by reacting it with one equimolar amount of potassium periodate (to saccharide units) in water. Various cationic polysaccharides were prepared by reaction between the desired oligoamine and the corresponding oxidized polysaccharide in basic aqueous medium. The resulting imine-based conjugates were then treated with an excess of sodium borohydride to obtain the stable amine-based conjugates in relatively good yields. The oligoamines used for conjugation were the naturally occurring spermine and spermidine, synthetic spermine analogues (i.e., tetramines at 3:3:3, 3:2:3, and 2:3:2), alkandimines ($NH_2[CH_2]_nNH_2$, where n = 2, 3, 4, 6, and 8), N,N-dimethyl propandiamines, and others (Figure 17.3). The resulting cationic polysaccharides were characterized by nitrogen elemental analysis, primary amine content (TNBS), average molecular weight (GPC), FT-IT, and ^{1}H-NMR.[117]

17.4.2 Condensation Studies of Polysaccharide–Oligoamine Conjugates with DNA

Reduction in ethidium bromide–DNA fluorescence can be used to indicate condensation of the DNA. Excited ethidium bromide fluoresces upon intercalating into DNA, while packing of DNA with a polycation results in ethidium bromide expulsion and fluorescence quenching.[121] A series of polycations were tested for their ability to condense DNA as functions of charge ratios (+/–) and the ionic strength of the medium.[122]

Figure 17.4 shows a typical condensation profile of various cationic polymers complexed with pLuc-DNA at various charge ratios ranging from 0.1 to 2 (+/–) in 20 mM HBS (pH 7.4). The charge ratios were expressed as primary amine content (TNBS) for the positive charge of the polymer, and phosphate groups for the negative charge of the DNA. Dextran-spermine based conjugate (Figure 17.4 [●]) was found to condense DNA efficiently in comparison to other grafted oligoamines.

At 0.1 charge ratio (+/–) nearly 44% condensation of DNA helices was obtained with dextran-spermine, while conjugates grafted with spermidine and spermine analogues (i.e., synthetic tetramines) resulted in a low degree of condensation (~10%). A higher charge ratio (i.e., 0.25 +/–) resulted in maximum condensation in spermine and 3:2:3 tetramine (Figure 17.4, [●] and [▲], respectively). At 0.5 to 2 charge ratios (+/–) maximum condensations (~95%) were obtained in all tested polycations, indicating a complete saturation of the negative charge of the DNA helices. The reason for the complete DNA condensation at low charge ratios (i.e., 0.25–0.5 [+/–]) is probably a result of a contribution of the electrostatic interactions between secondary amine functionalities and negative phosphate groups. In addition to primary amine groups, each oligoamine moiety

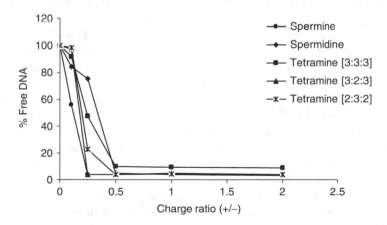

Figure 17.4 Condensation of pLuc-DNA with cationic dextran conjugated with spermine (●), spermidine (♦), N,N-Bis(3-aminopropyl)-1,3-propanediamine (■), N,N-Bis(3-aminopropyl)-ethylenediamine (▲), and N,N-Bis(2-aminoethyl)-1,3-propanediamine (✱). 20 mM HBS (pH 7.4) was used as the medium buffer.

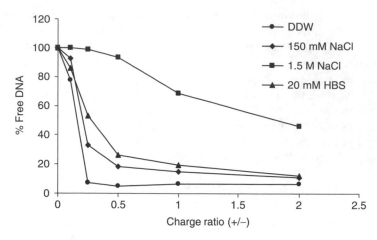

Figure 17.5 Condensation of pLuc-DNA with cationic dextran–spermine based conjugate in DDW (●), 150 mM NaCl (♦), 1.5 M NaCl (■), and 20 mM HBS pH 7.4 (▲).

contains at least two secondary amine groups in the case of spermidine (triamine) and three secondary amine functionalities in the case of spermine and spermine analogues. These secondary amine groups are probably the reason for the ability of these polymers to mask the negative charge of the DNA in relatively low polymer concentrations.

In a similar condensation experiment, dextran-spermine based conjugate was allowed to react with pLuc-DNA in various media and ionic strengths. Figure 17.5 summarizes the condensation profile of dextran-spermine–DNA applying the ethidium bromide quenching assay. When double-distilled water (DDW) was used as the condensation medium, maximum condensation (~90%) was obtained at a 0.25 charge ratio (primary amine groups and phosphate groups, respectively). Further addition of polycation (i.e., 0.5 to 2 [+/–]) did not increase the degree of condensation, probably due to full saturation of the negative charge of the DNA. Media of low ionic strengths (20 mM HBS and 150 mM NaCl, Figure 17.5) resulted in large condensation disruption at low charge ratios (0.1 and 0.25 [+/–]) and minor disruption at 0.5 to 2 charge ratios where nearly ~90% condensation was noticed. High ionic strength medium (1.5 M NaCl) resulted in drastic disruption in the polymer–DNA condensation at all tested charge ratios. At 0.1 to 0.5 charge ratio, slight and negligible condensation was obtained. The highest tested charge ratio (i.e., 2 [+/–]) resulted only in 45% condensation. The dissociation of the DNA–polycation complexes at high salt concentration is a well-known phenomenon and was previously reported.[123,124]

The importance of the primary amine groups for condensation was tested in a cationic dextran lacking primary amine functionalities. N,N-dimethyl 1,3-propane diamine was conjugated to oxidized dextran and this polycation was allowed to react with pLuc-DNA and the condensation profile (ethidium bromide assay) was recorded as a function of charge ratios and ionic strengths of the medium (Figure 17.6). It can be seen clearly from Figure 17.6 that the absence of primary amine groups in this type of polymer strongly reduces the cationic nature of the polymer and hence the low capability of DNA condensation. In DDW, only 40% condensation was obtained with a 10 charge ratio (+/–). Higher charge ratios (up to 20 [+/–]) did not significantly improve condensation (data not shown). On the contrary, when condensation was conducted in low and high ionic strength conditions (i.e., 20 mM HBS, 150 mM and 1.5 M NaCl), low and negligible condensation were recorded at all tested charge ratios. Similar condensation studies were conducted with arabinogalactan- and pullulan-based conjugates. These studies resulted with nearly similar condensation profiles to the dextran-oligoamine based conjugates (data not shown).

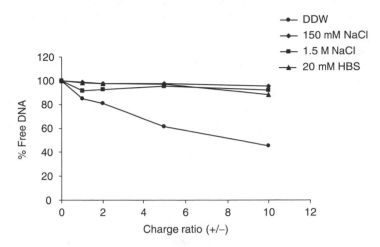

Figure 17.6 Condensation of pLuc-DNA with dextran-(N,N-dimethyl 1,3-propanediamine) based conjugate in DDW (●), 150 mM NaCl (◆), 1.5 M NaCl (■), and 20 mM HBS pH 7.4 (▲). The charge ratio in this case was expressed as nitrogen content (%N, elemental analysis) divided by the calculated amount of phosphate groups.

17.4.3 *In Vitro* Transfection of Polysaccharide–Oligoamine Conjugates

Transfection experiments were performed using three different cell lines: HEK-293, NIH-3T3, and EPC. Plasmids used for these studies were pCMV-GFP encoding to green fluorescent protein, pLNC-luciferase encoding to luciferase protein and pCMV-hGH encoding to the human growth hormone. DOTAP-Chol (Avanti®) and Transfast™ (Promega®) cationic lipids, as well as calcium phosphate (CaPO$_4$) precipitating technique, were used as positive controls. Each single polymer was tested at a wide range of charge ratios (–/+) (phosphate:nitrogen) from 1 to 0.05. Figure 17.7 shows a series of transfection results applying EPC cells and pLNC-luciferase as the marker gene.

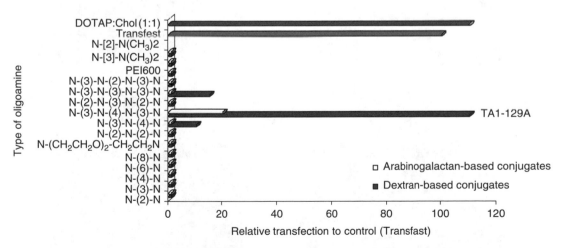

Figure 17.7 Transfection efficiencies of pLNC-luc in EPC cells applying dextran and arabinogalactan grafted oligoamines as vectors. Abbreviations: **N-[2]-N**, Ethanediamine; **N-[3]-N**, Propanediamine; **N-[4]-N**, Butanediamine; **N-[6]-N**, Hexanediamine; **N-[8]-N**, Octanediamine; **N-[CH2CH2O]2-CH2CH2N**, Triethylene glycol diamine; **N-[2]-N-[2]-N**, Diethylene triamine; **N-[3]-N-[4]-N**, Spermidine; **N-[3]-N-[4]-N-[3]-N**, Spermine; **N-[2]-N-[3]-N-[2]-N**, N,N'-bis(2-aminoethyl)-1,3-propanediamine; **N-[3]-N-[3]-N-[3]-N**, N,N'-bis(3-aminopropyl)-1,3-propanediamine; **N-[3]-N-[2]-N-[3]-N**, N,N'-bis(3-aminopropyl) ethylenediamine; **PEI600**, Polyethylenimine with average molecular weight of 600 Da; **N-[3]-N(CH3)2**, N,N'-dimethylpropylene diamine; **N-[2]-N(CH3)2**, N,N'-dimethylethylene diamine. DOTAP-Chol (1:1) and Transfast™ lipid formulations were applied as positive controls.

Transfection results in Figure 17.7 were accounted as luminescence relative light units (LRLUs) and recorded at the best charge ratio relative to Transfast™ control. When simple diamines were applied as the grafting oligoamines (i.e., ethylene, propylene, butane, hexane and octane diamines), no transfection was obtained in both dextran- and arabinigalactan-based conjugates at all tested charge ratios. PEI600, N,N'-dimethylethylene diamine, and N,N'-dimethylpropylene diamine were also grafted on both dextran and arabinogalactan representative polysaccharides. These conjugates were also found to be inactive and showed no transfection at all tested charge ratios. On the contrary, when spermine was used as the grafting oligoamine, a slight protein expression was detected in arabinogalactan-based conjugate (~20% to Transfast™ control). The most active form of these synthetic polymers was dextran–spermine-based conjugate (TA1-129A, Figure 17.7). Applying this polymer as a gene vector resulted in similar transfection to controls (Transfast™ and DOTAP:Chol 1:1). Replacement of spermine with spermidine as the grafted oligoamine resulted in a drastic decrease in transfection in the polysaccharide-based conjugates.

In addition, spermine analogues were grafted in the same manner to both dextran- and arabinoga-lactan-based polysaccharides. When N,N-bis(3-aminopropyl)-1,3-propanediamine (N-3-N-3-N-3-N) was grafted to dextran, a low protein expression was observed (~20% to Transfast™). On the contrary, when N,N-bis(2-aminoethyl)-1,3-propanediamine (N-2-N-3-N-2-N) and N,N-bis(3-aminopropyl)-eth-ylenediamine (N-3-N-2-N-3-N) were used, low and negligible transfection were obtained.

Figure 17.8 shows a typical fluorescence imaging of pCMV-GFP transfected HEK293 and NIH3T3 cells, which shows that Transfast™ (part A) and dextran–spermine-based conjugate (parts B and D) possess a strong expression of GFP (~50% transfection) in HEK293 cells and ~30% transfection in NIH3T3 cells, similar to the calcium phosphate control (part C).

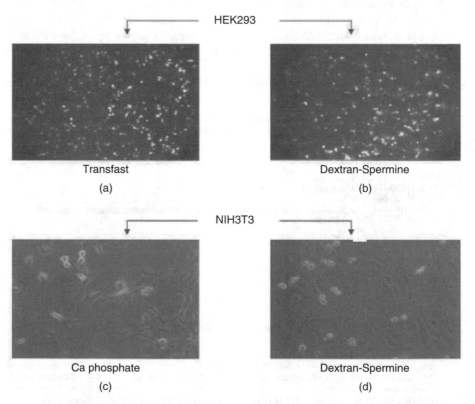

Figure 17.8   (See color insert following page 336) Inverted fluorescent microscope of pCMV-GFP transfected HEK293 and NIH3T3 cells. Parts A and B represent the Transfast™ control and dextran–spermine (respectively) transfected HEK293 cells. Parts C and D represent the calcium phosphate precip-itating technique and dextran–spermine (respectively) transfected NIH3T3 cells.

The *in vitro* transfection of the potential dextran–spermine-based conjugate was also tested in various cell lines (HCT116, CHO, COS-7, C2C12, and HeLa) and found to be as active as the commercial DOTAP-Chol 1:1 transfecting reagent.

In conclusion, more than 300 different polycations were prepared starting from various polysaccharides and oligoamines having two to four amino groups. Although most of these conjugates formed stable complexes with various plasmids as determined by the ethidium–bromide quenching assay, only the dextran–spermine-based polycations were found to be active in transfecting cells *in vitro*. The reason for the transfection efficiency of certain polycations (i.e., dextran–spermine) is probably due to the unique complexation properties formed between the DNA helices and the grafted spermine moieties, which play a crucial role in cell transfection. These results suggest that the structure of the polycation has a significant role in the transfection activity. Current studies focus on understanding the specificity found in the compounds by means of physical characterization of the polycation–DNA complex and trafficking studies.[125]

17.4.4   *In Vivo* Transfection with Dextran–Spermine Vector as a Gene Carrier

Complexation of dextran–spermine with pDNA for the *in vivo* studies was performed by mixing the two components at various weight-mixing ratios in physiological solution. The complex solutions containing varying plasmid contents at increasing weight-mixing ratios (polycation:plasmid) were gently agitated to form cationized dextran–plasmid–DNA complexes.[126]

17.4.4.1   *Intramuscular Injection of pSV-LacZ Gene Complex*

The pSV-LacZ–cationized-dextran complexes containing different amounts of pSV-LacZ were intramuscularly injected at the sutured points of mice muscles. Mice were sacrificed and the muscle samples were taken around the marking points 0.5, 1, 2, 3, 4, and 5 days after administration and analyzed for gene expression. Figure 17.9 shows the average specific β-Gal activities (mU/mg protein) of mice muscles as a function of complex weight-mixing ratio (polycation:pSV-LacZ) 2 days after administration. Intramuscular injection with the control buffer (i.e., PBS) resulted with

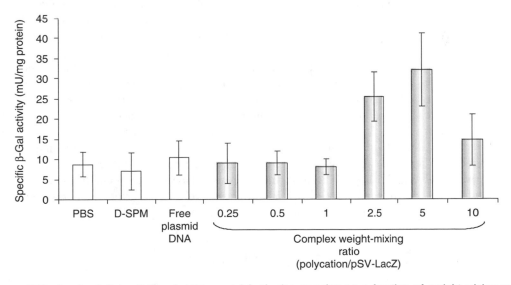

Figure 17.9   *In vivo* β-Gal activities (mU/mg protein) of mice muscles as a function of weight-mixing ratio (polycation:DNA). Mice were intramuscularly injected with dextran–spermine–pSV-LacZ complex at weight-mixing ratios of 0.25 to 10 (polycation:plasmid DNA) and the gene level was evaluated 2 days after injection. PBS, dextran–spermine (D-SPM), and free plasmid DNA solutions were injected as control references.

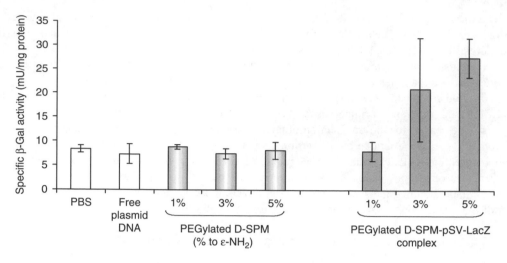

Figure 17.10   *In vivo* β-Gal activities in mice livers as a function of percentage of PEGylation of dextran–spermine polycation. PBS, free plasmid DNA, and PEGylated dextran–spermine derivative (without DNA) solutions were injected as control references.

background β-Gal reading (~7 ± 2 mU/mg protein). Similar background β-Gal activities were obtained with naked pSV-LacZ DNA and the dextran–spermine polycation (coded as D-SPM) control solutions. At complex weight-mixing ratios of 0.25 to 1 (polycation:pSV-LacZ), low to negligible β-Gal activities were obtained. At a 2.5 weight-mixing ratio, high gene expression was detected (25 ± 7 mU/mg protein). The highest β-Gal activity was obtained at a 5 weight-mixing ratio where nearly 34 ± 8 (mU/mg protein) was obtained. Further increase in the polycation ratio (i.e., 10 polycation:pSV-LacZ) resulted in a decrease in the gene expression.

17.4.4.2   *Intravenous Injection of pSV-LacZ Gene Complex*

Preliminary intravenous injection in mice of free pDNA and its complex formulation with dextran–spermine polycation did not induce gene expression in all tested organs. Modification of the dextran–spermine polycations with the nonionic hydrophilic polymer PEG remarkably improved the *in vitro* transfection efficiency of the polycation in serum-rich media.[126] The level of gene expression in the liver after intravenous injection in mice of PEGylated dextran–spermine–pSV-LacZ complex was monitored as a function of percentage of PEGylation and time after treatment. Figure 17.10 shows the calculated β-Gal activities in the liver, 2 days after intravenous injection of PEGylated dextran–spermine–pSV-LacZ complex solutions. The percentages of PEGylated dextran–spermine were 1, 3, and 5% mol/mol (PEG to ε-NH$_2$). PBS, free plasmid DNA, and the PEGylated polycations (without DNA) solutions were similarly injected as controls and resulted as expected with no gene expression. One percent PEGylated polymer showed minor transfection activity, while the 3 and 5% PEGylated dextran–spermine showed high β-Gal transfection yields of 23 and 30 mU/mg protein, respectively. The highest level of gene expression was obtained 2 days after injection, both in intramuscular and intravenous administration.

17.5   CONCLUSIONS

When 10% FCS culture medium was applied instead of SFM   in the *in vitro* transfection experiments applying the dextran–spermine vector, 80% reduction in the gene expression was observed.[126] This phenomenon is well known and is attributed to protein (serum component) adsorption on the surface of the complex (i.e., polycation–DNA), which in part could induce the

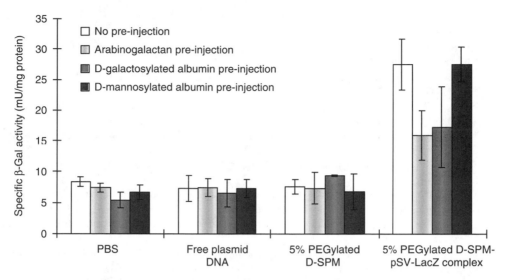

Figure 17.11 Suppression effect of preinjection of blocking agents on the *in vivo* gene expression of liver 2 days after intravenous injection of 5% PEGylated dextrañsperminẽpSV-LacZ complex with or without blocking agents. One hour prior to complex injections, mice were preinjected with a 20 mg/kg dose of arabinogalactan), D-galactosylated bovine albumin (□), and D-mannosylated bovine albumin (■). A direct injection of the complex without preinjection of blocking agents (□) was applied as a positive control. PBS, free plasmid DNA, and 5% PEGylated dextran–spermine polycation (without DNA) solutions were injected as negative controls.

aggregation and deactivation of the complex and finally to reduction in the transfection efficiencies.[127] In accordance with previous studies, it was decided to modify our developed polycations with a hydrophilic polymer such as PEG in order to increase their stability and transfection efficiencies in serum-rich media.[127,128] PEG molecules are considered to be the most attractive materials for this purpose. They are safe, nontoxic, and cheap, and do not interact with plasma components.[129] *In vitro* transfection efficiencies in 10% FCS medium applying the PEGylated dextran–spermine polycations as the gene carriers resulted in a high gene level, indicating a remarkable increase in complex stability.[126]

Intramuscular injection of the dextran–spermine–pSV-LacZ complex solution to mice resulted in high local gene level in comparison to the naked DNA, which resulted with no gene expression. The optimal weight-mixing ratio for proper expression was found to be 5 (polycation:DNA) applying a 50 μg per mouse plasmid injected dose. The nonPEGylated dextran–spermine–pSV-LacZ complex, when applied by intravenous injection to mice, did not show gene expression. On the other hand, when PEGylated dextran–spermine derivatives were applied, a high gene level was obtained in the liver. The 5% PEGylated dextran–spermine was found to be the most active PEGylated derivative for systemic administration.

The gene expression in the liver by the PEGylated dextran–spermine–pSV-LacZ complex was markedly reduced by previous administration of both arabinogalactan and D-galactosylated bovine albumin, whereas the mannosylated albumin had no influence on the liver level of gene expression (Figure 17.11). It is well recognized that both arabinogalactan and D-galactosylated bovine albumin interact with asialoglycoprotein receptor located in the surface of liver parenchymal cells *in vitro* and *in vivo*.[130,131] On the other hand, albumin bovine-α-D-mannopyranosyl phenyl isothiocyanate is bound to the mannose receptor, which is located on the surface of the nonparenchymal cells and is responsible for this uptake.[132] Taken together, the results strongly indicate that the targeting of PEGylated dextran–spermine–pDNA complex to the liver is probably mediated by galactose receptor of the liver parenchymal cells, rather than the mannose receptor of liver nonparenchymal cells.[126]

REFERENCES

1. Gaucheron, J. et al., *In vitro* cationic lipid-mediated gene delivery with fluorinated glycerophospho-ethanolamine helper lipids, *Bioconj. Chem.* 12 (6), 949–963, 2001.

2. Kabanov, A.V., Taking polycation gene delivery systems from *in vitro* to *in vivo*, *Pharm. Sci. Technol. Today* 2 (9), 365–372, 1999.

3. Felgner, P.L. et al., Lipofection: a highly efficient, lipid-mediated DNA-transfection procedure, *Proc. Nat. Acad. Sci. U.S.A.* 84 (21), 7413–7417, 1987.

4. Templeton, N.S. et al., Improved DNA: liposome complexes for increased systemic delivery and gene expression, *Nature Biotechnol.* 15 (7), 647–652, 1997.

5. Zuidam, N.J. et al., Chiral DNA packaging in DNA-cationic liposome assemblies, *FEBS Lett.* 457 (3), 419–422, 1999.

6. Tseng, W.C. et al., Transfection by cationic liposomes using simultaneous single cell measurements of plasmid delivery and transgene expression, *J. Biol. Chem.* 272 (41), 25641–25647, 1997.

7. Vierling, P. et al., Highly fluorinated amphiphiles as drug and gene carrier and delivery systems, *J. Fluor. Chem.* 107 (2), 337–354, 2001.

8. Gaucheron, J. et al., Highly fluorinated lipospermines for gene transfer: synthesis and evaluation of their *in vitro* transfection efficiency, *Bioconj. Chem.* 12 (1), 114–128, 2001.

9. Lewis, J.G. et al., A serum-resistant cytofectin for cellular delivery of antisense oligodeoxynucleotides and plasmid DNA, *Proc. Nat. Acad. Sci. U.S.A.* 93 (8), 3176–3181, 1996.

10. Byk, G. et al., Novel nonviral vectors for gene delivery: synthesis and applications, *Lett. Peptide Sci.* 4 (4–6), 263–267, 1997.

11. Solodin, I. et al., A novel series of amphiphilic imidazolinium compounds for *in vitro* and *in vivo* gene delivery, *Biochemistry* 34 (41), 13537–13544, 1995.

12. Radler, J.O. et al., Structure of DNA-cationic liposome complexes: DNA intercalation in multilamellar membranes in distinct interhelical packing regimes, *Science* 275 (5301), 810–814, 1997.

13. Gershon, H. et al., Mode of formation and structural features of DNA cationic liposome complexes used for transfection, *Biochemistry* 32 (28), 7143–7151, 1993.

14. Koltover, I. et al., An inverted hexagonal phase of cationic liposome-DNA complexes related to DNA release and delivery, *Science* 281 (5373), 78–81, 1998.

15. Koltover, I. et al., DNA condensation in two dimensions, *Proc. Nat. Acad. Sci. U.S.A.* 97 (26), 14046–14051, 2000.

16. Domb, A.J. and Levy, M., *Polymers in Gene Therapy: Frontiers in Biological Polymer Application*, Ottenbrite, R.M. (Ed.), *Technomic*, vol. 2., Lancaster, PA, 1999, 1–16.

17. Putnam, D. et al., Polymer-based gene delivery with low cytotoxicity by a unique balance of side-chain termini, *Proc. Nat. Acad. Sci. U.S.A.* 98 (3), 1200, 2001.

18. Sato, T. et al., Preparation and characterization of DNA-lipoglutamate complexes, *Bull. Chem. Soc. Jap.* 68 (9), 2709–2715, 1995.

19. Boussif, O. et al., A versatile vector for gene and oligonucleotide transfer into cells in culture and *in vivo*: polyethylenimine, *Proc. Nat. Acad. Sci. U.S.A.* 92 (16), 7297–7301, 1995.

20. Oupicky, D. et al., Steric stabilization of poly-L-lysine/DNA complexes by the covalent attachment of semitelechelic poly N-(2- hydroxypropyl)methacrylamide, *Bioconj. Chem.* 11 (4), 492–501, 2000.

21. Bielinska, A.U. et al., Application of membrane-based dendrimer/DNA complexes for solid phase transfection *in vitro* and *in vivo*, *Biomaterials* 21 (9), 877–887, 2000.

22. Mumper, R.J. et al., Polyvinyl derivatives as novel interactive polymers for controlled gene delivery to muscle, *Pharm. Res.* 13 (5), 701–709, 1996.

23. Leong, K.W. et al., DNA-polycation nanospheres as non-viral gene delivery vehicles, *J. Control. Release* 53 (1–3), 183–193, 1998.

24. Isobe, H. et al., Atomic force microscope studies on condensation of plasmid DNA with functionalized fullerenes, *Angew Chem. Int. Ed.* 40 (18), 3364, 2001.

25. Benns, J.M. et al., pH-sensitive cationic polymer gene delivery vehicle: N-Ac- poly(L-histidine)-graft-poly(L-lysine) comb shaped polymer, *Bioconj. Chem.* 11 (5), 637–645, 2000.

26. Deshpande, D. et al., Target specific optimization of cationic lipid-based systems for pulmonary gene therapy, *Pharm. Res.* 15 (9), 1340–1347, 1998.

27. Oupicky, D. et al., Laterally stabilized complexes of DNA with linear reducible polycations: strategy for triggered intracellular activation of DNA delivery vectors, *J. Am. Chem. Soc.* 124 (1), 8–9, 2002.

28. Pouton, C.W. and Seymour, L.W., Key issues in non-viral gene delivery, *Adv. Drug Deliv. Rev.* 46 (1–3), 187–203, 2001.

29. Vanderkerken, S. et al., Synthesis and evaluation of poly(ethylene glycol)-polylysine block copolymers as carriers for gene delivery, *J. Bioact. Compat. Polym.* 15 (2), 115–138, 2000.

30. Berscht, P.C. et al., *In vitro* evaluation of biocompatibility of different wound dressing materials, *J. Mater. Sci. Mater. Med.* 6 (4), 201–205, 1995.

31. CarrenoGomez, B. and Duncan, R., Evaluation of the biological properties of soluble chitosan and chitosan microspheres, *Int. J. Pharm.* 148 (2), 231–240, 1997.

32. Morimoto, M. et al., Control of functions of chitin and chitosan by chemical modification, *Trends Glycosci. Glycotechnol.* 14 (78), 205–222, 2002.

33. Singla, A.K. and Chawla, M., Chitosan: some pharmaceutical and biological aspects: an update, *J. Pharm. Pharmacol.* 53 (8), 1047–167, 2001.

34. Janes, K.A. et al., Polysaccharide colloidal particles as delivery systems for macromolecules, *Adv. Drug Deliv. Rev.* 47 (1), 83–97, 2001.

35. Henriksen, I. et al., Interactions between liposomes and chitosan. 2. Effect of selected parameters on aggregation and leakage, *Int. J. Pharm.* 146 (2), 193–203, 1997.

36. Henriksen, I. et al., Interactions between liposomes and chitosan, *Int. J. Pharm.* 101 (3), 227–236, 1994.

37. Schipper, N.G.M. et al., Chitosans as absorption enhancers for poorly absorbable drugs.1. Influence of molecular weight and degree of acetylation on drug transport across human intestinal epithelial (Caco-2) cells, *Pharm. Res.* 13 (11), 1686–1692, 1996.

38. Illum, L., Chitosan and its use as a pharmaceutical excipient, *Pharm. Res.* 15 (9), 1326–1331, 1998.

39. Bodek, K.H., Potentiometric method for determination of the degree of acetylation of chitosan, in *Chitin World*, Karnicki, Z.S., Pajak, A.W., Breziski, M.M., and Bylowsky, P.J. (Eds.), Springer-Verlag, Bremerharen, Germany, 1994, 456–461.

40. Ferreira, M.C. et al., Optimisation of the measuring of chitin/chitosan degree of acetylation by FT-IR spectroscopy, in *Chitin World*, Karnicki, Z.S., Pajak, A.W., Breziski, M.M., and Bylowsky, P.J. (Eds.), Springer-Verlag, Bremerharen, Germany, 1994, 480–488.

41. Kas, H.S., Chitosan: properties, preparations and application to microparticulate systems, *J. Microencapsul.* 14 (6), 689–711, 1997.

42. Skaugrud, O. et al., Biomedical and pharmaceutical applications of alginate and chitosan, in *Biotechnology and Genetic Engineering Reviews*, vol. 16, 1999, 23–40.

43. Dornish, M. et al., Standards and guidelines for biopolymers in tissue-engineered medical products: ASTM alginate and chitosan standard guides, in *Bioartificial Organs III: Tissue Sourcing, Immunoisolation, and Clinical Trials*, 2001, 388–397.

44. Richardson, S.C.W. et al., Potential of low molecular mass chitosan as a DNA delivery system: biocompatibility, body distribution and ability to complex and protect DNA, *Int. J. Pharm.* 178 (2), 231–243, 1999.

45. Risbud, H. et al., Chitosan-polyvinyl pyrrolidone hydrogels as candidate for islet immunoisolation: *in vitro* biocompatibility evaluation, *Cell Transplant.* 9 (1), 25–31, 2000.

46. Lehr, C.M. et al., *In vitro* evaluation of mucoadhesive properties of chitosan and some other natural polymers, *Int. J. Pharm.* 78 (1), 43–48, 1992.

47. Artursson, P. et al., Effect of chitosan on the permeability of monolayers of intestinal epithelial-cells (caco-2), *Pharm. Res.* 11 (9), 1358–1361, 1994.

48. Mi, F.L. et al., *In vivo* biocompatibility and degradability of a novel injectable-chitosan-based implant, *Biomaterials* 23 (1), 181–191, 2002.

49. Imai, T. et al., Interaction of indomethacin with low-molecular-weight chitosan, and improvements of some pharmaceutical properties of indomethacin by low-molecular-weight chitosans, *Int. J. Pharm.* 67 (1), 11–20, 1991.

50. Takayama, K. et al., Effect of interpolymer complex-formation on bioadhesive property and drug release phenomenon of compressed tablet consisting of chitosan and sodium hyaluronate, *Chem. Pharm. Bull.* 38 (7), 1993–1997, 1990.

51. Meshali, M.M. and Gabr, K.E., Effect of interpolymer complex-formation of chitosan with pectin or acacia on the release behavior of chlorpromazine Hcl, *Int. J. Pharm.* 89 (3), 177–181, 1993.

52. Sabnis, S. and Block, L.H., Improved infrared spectroscopic method for the analysis of degree of N-deacetylation of chitosan, *Polym. Bull.* 39 (1), 67–71, 1997.

53. Sabnis, S. and Block, L.H., Chitosan as an enabling excipient for drug delivery systems. I. Molecular modifications, *Int. J. Biol. Macromol.* 27 (3), 181–186, 2000.

54. Kristmundsdottir, T. et al., Chitosan matrix tablets: the influence of excipients on drug release, *Drug Dev. Ind. Pharm.* 21 (13), 1591–1598, 1995.

55. Luessen, H.L. et al., Bioadhesive polymers for the peroral delivery of peptide drugs, *J. Control. Release* 29 (3), 329–338, 1994.

56. He, P. et al., *In vitro* evaluation of the mucoadhesive properties of chitosan microspheres, *Int. J. Pharm.* 166 (1), 75–88, 1998.

57. He, P. et al., Sustained release chitosan microspheres prepared by novel spray drying methods, *J. Microencapsul.* 16 (3), 343–355, 1999.

58. Bernkop-Schnurch, A., Chitosan and its derivatives: potential excipients for peroral peptide delivery systems, *Int. J. Pharm.* 194 (1), 1–13, 2000.

59. Nishiyama, Y. et al., A conjugate from a laminin-related peptide, Tyr-Ile-Gly-Ser- Arg, and chitosan: efficient and regioselective conjugation and significant inhibitory activity against experimental cancer metastasis, *J. Chem. Soc. Perkin Trans.* 7(1), 1161–1165, 2000.

60. Bernkop-Schnurch, A. and Walker, G., Multifunctional matrices for oral peptide delivery, *Crit. Rev. Ther. Drug Carr. Syst.* 18 (5), 459–501, 2001.

61. Guggi, D. and Bernkop-Schnurch, A., *In vitro* evaluation of polymeric excipients protecting calcitonin against degradation by intestinal serine proteases, *Int. J. Pharm.* 252 (1–2), 187–196, 2003.

62. Fernandez-Urrusuno, R. et al., Enhancement of nasal absorption of insulin using chitosan nanoparticles, *Pharm. Res.* 16 (10), 1576–1581, 1999.

63. Snyman, D. et al., Evaluation of the mucoadhesive properties of N-trimethyl chitosan chloride, *Drug Dev. Ind. Pharm.* 29 (1), 61–69, 2003.

64. Hamman, J.H. et al., N-trimethyl chitosan chloride: optimum degree of quaternization for drug absorption enhancement across epithelial cells, *Drug Dev. Ind. Pharm.* 29 (2), 161–172, 2003.

65. Tozaki, H. et al., Chitosan capsules for colon-specific drug delivery: enhanced localization of 5-aminosalicylic acid in the large intestine accelerates healing of TNBS-induced colitis in rats, *J. Control. Release* 82 (1), 51–61, 2002.

66. Takeuchi, H. et al., Mucoadhesive properties of carbopol or chitosan-coated liposomes and their effectiveness in the oral administration of calcitonin to rats, *J. Control. Release* 86 (2–3), 235–242, 2003.

67. Kurita, K. et al., Synthesis and some properties of nonnatural amino polysaccharides: branched chitin and chitosan, *Macromolecules* 33 (13), 4711–4716, 2000.

68. Kurita, K., Controlled functionalization of the polysaccharide chitin, *Prog. Polym. Sci.* 26 (9), 1921–1971, 2001.

69. Muzzarelli, R.A.A. et al., Methods for the determination of the degree of acetylation of chitan and chitosan, in *Chitin Handbook*, Muzzarelli, R.A.A. and Peters, M.G., (Eds.), European Chitan Society, Ancona, Italy, 1997, p. 109-119.

70. Baxter, A., Dillon, M., Taylor, K.D.A., and Roberts, G.A.F., Improved method for i.r. determination of the degree of N-acetylation of chitosan, *Intl. J. Biol. Macromol.*, 14, 166–169, 1992.

71. Shigemasa, Y. et al., Evaluation of different absorbance ratios from infrared spectroscopy for analyzing the degree of deacetylation in chitin, *Int. J. Biol. Macromol.* 18 (3), 237–242, 1996.

72. Varum, K.M. et al., High-field Nmr-spectroscopy of partially N-deacetylated chitins (chitosans). 1. Determination of the degree of N-acetylation and the distribution of N-acetyl groups in partially N-deacetylated chitins (chitosans) by high-field Nmr-spectroscopy, *Carbohydr. Res.* 211 (1), 17–23, 1991.

73. Hirai, A. et al., Determination of degree of deacetylation of chitosan by H-1-Nmr spectroscopy, *Polym. Bull.* 26 (1), 87–94, 1991.

74. Kurita, K. et al., Studies on chitin. 4. Evidence for formation of block and random copolymers of N-acetyl-D-glucosamine and D-glucosamine by heterogeneous and homogeneous hydrolyses, *Macromol. Chem. Phys.* 178 (12), 3197–3202, 1977.

75. Domard, A. and Rinaudo, M., Preparation and characterization of fully deacetylated chitosan, *Int. J. Biol. Macromol.* 5 (1), 49–52, 1983.

76. Mima, S. et al., Highly deacetylated chitosan and its properties, *J. Appl. Polym. Sci.* 28 (6), 1909–1917, 1983.

77. Kurita, K. et al., Studies on chitin. 14. N-Acetylation behavior of chitosan with acetyl-chloride and acetic-anhydride in a highly swelled state, *Bull. Chem. Soc. Jap.* 61 (3), 927–930, 1988.

78. Mumper, R.J. et al., Novel polymeric condensing carriers for gene delivery, *Proc. Int. Symp. Control. Release Bioact. Mater.* 22, 178–179, 1995.

79. MacLaughlin, F.C. et al., Chitosan and depolymerized chitosan oligomers as condensing carriers for *in vivo* plasmid delivery, *J. Control. Release* 56 (1–3), 259–272, 1998.

80. Rolland, A.P., From genes to gene medicines: recent advances in nonviral gene delivery, *Crit. Rev. Ther. Drug Carr. Syst.* 15 (2), 143–198, 1998.

81. Olsen, R. et al., Biomedical applications of chitin and its derivatives, in *Chitin and Chitosan Sources, Chemistry and Biochemistry, Physical Properties and Application*, Skjak-Braek, G., Antonsen, T., and Sandford, P. (Eds.), Elsevier, New York, 1989, 813–827.

82. Liu, W.G. and De Yao, K., Chitosan and its derivatives: a promising non-viral vector for gene transfection, *J. Control. Release* 83 (1), 1–11, 2002.

83. Lee, K.Y. et al., Preparation of chitosan self-aggregates as a gene delivery system, *J. Control. Release* 51 (2–3), 213–220, 1998.

84. Nudga, L.A. et al., N-alkylation of chitosan, *Zhur. Obshchei Khimii* 43 (12), 2756–2760, 1973.

85. Thanou, M. et al., Quaternized chitosan oligomers as gene delivery vectors *in vitro*, *J. Control. Release* 87 (1–3), 294–295, 2003.

86. Jansma, C.A. et al., Preparation and characterization of 6-O-carboxymethyl-N- trimethyl chitosan derivative as a potential carrier for targeted polymeric gene and drug delivery, *STP Pharm. Sci.* 13 (1), 63–67, 2003.

87. Yalpani, M. et al., Syntheses of poly(3-hydroxyalkanoate) (Pha) conjugates: Pha carbohydrate and Pha synthetic-polymer conjugates, *Macromolecules* 24 (22), 6046–6049, 1991.

88. Aoi, K. et al., Synthesis of novel chitin derivatives having poly(2-alkyl-2-oxazoline) side-chains, *Macromol. Chem. Phys.* 195 (12), 3835–3844, 1994.

89. Mao, H.Q. et al., Chitosan-DNA nanoparticles as gene carriers: synthesis, characterization and transfection efficiency, *J. Control. Release* 70 (3), 399–421, 2001.

90. Park, I.K., et al., Galactosylated chitosan-graft-dextran as hepatocyte-targeting DNA carrier, *J. Control. Release* 69 (1), 97–108, 2000.

91. Park, I.K. et al., Galactosylated chitosan-graft-dextran as hepatocyte-targeting DNA carrier, *J. Control. Release* 69 (1), 97–108, 2000.

92. Park, I.K. et al., Galactosylated chitosan-graft-poly(ethylene glycol) as hepatocyte-targeting DNA carrier, *J. Control. Release* 75 (3), 433, 2001.

93. Park, I.K. et al., Galactosylated chitosan (GC)-graft-poly(vinyl pyrrolidone) (PVP) as hepatocyte-targeting DNA carrier: preparation and physicochemical characterization of GC-graft-PVP/DNA complex (1), *J. Control. Release* 86 (2–3), 349–359, 2003.

94. Erbacher, P. et al., Chitosan-based vector/DNA complexes for gene delivery: biophysical characteristics and transfection ability, *Pharm. Res.* 15 (9), 1332–1339, 1998.

95. Feelders, R.A. et al., Structure, function and clinical significance of transferrin receptors, *Clin. Chem. Lab. Med.* 37 (1), 1–10, 1999.

96. Wagner, E. et al., Delivery of drugs, proteins and genes into cells using transferrin as a ligand for receptor-mediated endocytosis, *Adv. Drug Deliv. Rev.* 14 (1), 113–135, 1994.

97. Cheng, P.W., Receptor ligand-facilitated gene transfer: enhancement of liposome-mediated gene transfer and expression by transferrin, *Hum. Gene Ther.* 7 (3), 275–282, 1996.

98. Wightman, L. et al., Development of transferrin-polycation/DNA based vectors for gene delivery to melanoma cells, *J. Drug Target.* 7 (4), 293–303, 1999.

99. Gao, S.Y. et al., Galactosylated low molecular weight chitosan as DNA carrier for hepatocyte-targeting, *Int. J. Pharm.* 255 (1–2), 57–68, 2003.

100. Saito, A. et al., Characteristics of a Streptomyces coelicolor A3(2) extracellular protein targeting chitin and chitosan, *Appl. Environ. Microbiol.* 67 (3), 1268–1273, 2001.

101. Dufes, C. et al., Niosomes and polymeric chitosan based vesicles bearing transferrin and glucose ligands for drug targeting, *Pharm. Res.* 17 (10), 1250–1258, 2000.

102. Kamidate, T. et al., DEAE-Dextran enhanced firefly bioluminescent assay of ATP, *Chem. Lett.* 3, 237–238, 1996.

103. DEAE-Dextran, Amersham Biosciences[R], http://wwwapczechcz/pdf/DF_DEAE_dextranpdf.

104. Westbrook, S.L. and McDowell, G.H., Immunization of lambs against somatotropin release inhibiting factor to improve productivity: comparison of adjuvants, *Aust. J. Agric. Res.* 45 (8), 1693–1700, 1994.

105. Joo, I. and Emod, J., Adjuvant effect of DEAE-dextran on cholera vaccines, *Vaccine* 6 (3), 233–237, 1988.

106. Gibson, T.D. et al., Stabilization of analytical enzymes using a novel polymer carbohydrate system and the production of a stabilized, single reagent for alcohol analysis, *Analyst* 117 (8), 1293–1297, 1992.

107. Gavalas, V.G. and Chaniotakis, N.A., Polyelectrolyte stabilized oxidase based biosensors: effect of diethylaminoethyl-dextran on the stabilization of glucose and lactate oxidases into porous conductive carbon, *Anal. Chim. Acta* 404 (1), 67–73, 2000.

108. Mack, K.D. et al., A novel method for DEAE-dextran mediated transfection of adherent primary cultured human macrophages, *J. Immunol. Methods* 211 (1–2), 79–86, 1998.

109. Ya-Wun, Y. and Jyh-Chyang, Y., Studies of DEAE-dextran-mediated gene transfer, *Biotechnology and Applied Biochemistry* 25 (1), 47–51, 1997.

110. Calderwood, S.K. et al., The polycation DEAE dextran binds to the CHO cell-surface causing marked modulation of treatment response and surface properties, *Int. J. Radiat. Oncol. Biol. Phys.* 10 (9), 1801–1802, 1984.

111. Pazzagli, M. et al., Use of bacterial and firefly luciferases as reporter genes in DEAE-dextran-mediated transfection of mammalian cells, *Anal. Biochem.* 204 (2), 315–323, 1992.

112. Gonzalez, A.L. and Joly, E., A simple procedure to increase efficiency of DEAE-dextran transfection of COS cells, *Trends Gen.* 11 (6), 216–217, 1995.

113. Kaplan, J.M. et al., Potentiation of gene transfer to the mouse lung by complexes of adenovirus vector and polycations improves therapeutic potential, *Hum. Gene Ther.* 9 (10), 1469–1479, 1998.

114. Liptay, S. et al., Colon epithelium can be transiently transfected with liposomes, calcium phosphate precipitation and DEAE dextran *in vivo*, *Digestion* 59 (2), 142–147, 1998.

115. Welsh, M.J. and Fasbender, A.J., Complexes of adenovirus with cationic molecules for gene therapy, U.S. Patent Appl. 5,962,429, 1999.

116. Huang, L. and Viroonchatapan, E., in *Nonviral Vectors for Gene Therapy*, Huang, L. (Ed.), Academic Press, San Diego, CA, 1999, 3–22.

117. Azzam, T. et al., Polysaccharide-oligoamine based conjugates for gene delivery, *J. Med. Chem.* 45 (9), 1817–1824, 2002.

118. Azzam, T. et al., Cationic polysaccharides for gene delivery, *Macromolecules* 35 (27), 9947–9953, 2002.

119. Tang, M.X. and Szoka, F.C., The influence of polymer structure on the interactions of cationic polymers with DNA and morphology of the resulting complexes, *Gene Ther.* 4 (8), 823–832, 1997.

120. Larsen, C., in *Dextran Prodrugs*, Christesen, V.A. (Ed.), Copenhagen, Denmark, 1990.

121. Toncheva, V. et al., Novel vectors for gene delivery formed by self-assembly of DNA with poly(L-lysine) grafted with hydrophilic polymers, *Biochim. Biophys. Acta* 1380 (3), 354–368, 1998.

122. Azzam, T., Eliyahu, H., Makovitzki, A. and Domb, A.J., Dextran-spermine conjugate: An efficient vector for gene delivery, *Macromolecular Symposia*, 195, 247–261, 2003.

123. Izumrudov, V.A. and Zhiryakova, M.V., Stability of DNA-containing interpolyelectrolyte complexes in water-salt solutions, *Macromol. Chem. Phys.* 200 (11), 2533–2540, 1999.

124. Parker, A.L. et al., Methodologies for monitoring, nanoparticle formation by self-assembly of DNA with poly(L-lysine), *Anal. Biochem.* 302 (1), 75–80, 2002.

125. Eliyahu, H., Makovitzki, A., Azzam, T., Zlotkin, A., Joseph, A., Gazit, D., Barenholz, Y., and Domb, A.J., Novel dextran-spermine conjugates as transfecting agents: comparing water-soluble and micellar polymers, *Gene Therapy*, in press.

126. Hosseinkhani, H., Azzam, T., Tabata, Y., and Domb, A.J., Dextran-spermine polycation: an efficient nonviral vector for *in vitro* and *in vivo* gene transfection, *Gene Therapy*, 11 (2), 194–203, 2004.

127. Ogris, M. et al., PEGylated DNA/transferrin PEI complexes: reduced interaction with blood components, extended circulation in blood and potential for systemic gene delivery, *Gene Ther.* 6 (4), 595–605, 1999.

128. Woodle, M.C. et al., New amphipatic polymer lipid conjugates forming long-circulating reticuloendothelial system-evading liposomes, *Bioconj. Chem.* 5 (6), 493–496, 1994.

129. Pepinsky, R.B. et al., Long-acting forms of sonic hedgehog with improved pharmacokinetic and pharmacodynamic properties are efficacious in a nerve injury model, *J. Pharm. Sci.* 91 (2), 371–387, 2002.

130. Kaneo, Y. et al., Pharmacokinetics and biodisposition of fluorescein-labeled arabinogalactan in rats, *Int. J. Pharm.* 201 (1), 59–69, 2000.

131. Hashida, M. et al., Targeted delivery of plasmid DNA complexed with galactosylated poly(L-lysine), *J. Control. Release* 53 (1–3), 301–310, 1998.

132. Ogawara, K.I. et al., Pharmacokinetic evaluation of mannosylated bovine serum albumin as a liver cell-specific carrier: quantitative comparison with other hepatotropic ligands, *J. Drug Target.* 6 (5), 349–360, 1999.

CHAPTER 18

Gene Delivery Using Chitosan and Chitosan Derivatives

Gerrit Borchard and Maytal Bivas-Benita

CONTENTS

18.1 PROPERTIES OF CHITOSAN AND CHITOSAN DERIVATIVES

Chitosan, a (1→4) 2-amino-2-deoxy-β-D-glucan, is a linear cationic polysaccharide derived by partial alkaline deacetylation of chitin,[1,2] a polymer abundant in nature. Chitin is mostly obtained from exoskeletons of crustacea; main sources are shell wastes of shrimp and crab. The backbone of the copolymer consists of two subunits, D-glucosamine and N-acetyl-D-glucosamine, which are linked by 1→4 glycosidic bonds.

Chitosans, except for high molecular weight chitosans of specific salt forms,[3] are generally shown to be nontoxic,[4] partially due to their biodegradability. Chitosan is degraded by lysozymes into a common amino sugar, which is incorporated into the synthetic pathway of glycoproteins, and is subsequently excreted as carbon dioxide.[5] Due to chitosan's biocompatibility and toxicity profile (LD_{50} in rats of 16g/kg),[6] it has been evaluated as carrier material for drug and gene delivery.[7]

Depending on the molecular weight and degree of deacetylation, which are both determined by the conditions of the deacetylation process, the physicochemical properties of chitosan can differ. Chitosan polymers are available at a molecular weight (mol wt) range between 50 and 2000 kDa, with N-deacetylation degrees between 40 and 98%. At acidic pH, the primary amine functions are protonated, resulting in a cationic polymer of high charge density, which can form stable complexes with plasmid DNA (pDNA), partially protecting DNA from nuclease degradation.[8] Very low-molecular-weight chitosan (VLMWC) of molecular weights of less than 10 kDa show pK_a

0-8493-1934-X/05/$0.00+$1.50
© 2005 by CRC Press LLC

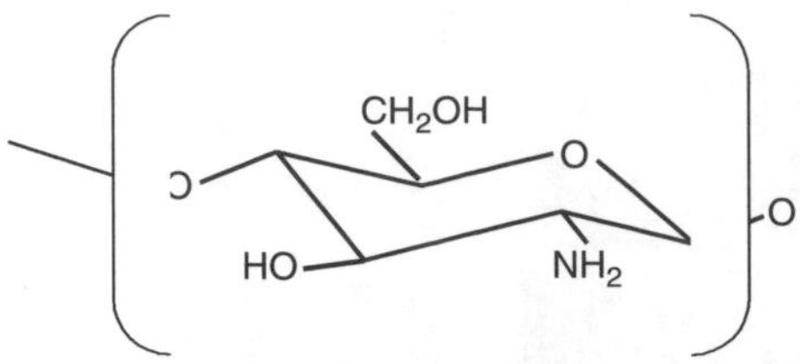

Figure 18.1   Chitosan, a (1->4) 2-amino-2-deoxy-β-D-glucan, is a linear cationic polysaccharide derived by partial alkaline deacetylation of chitin. The amino function may be methylated to obtain more water-soluble *N,N,N*-trimethyl-chitosan (TMC) derivatives. Chitosan can be modified at the amine function or the C_6-position to attach, e.g., peptides or saccharides to the polymer. These may act as targeting moieties to achieve cell-specific targeting and receptor-mediated endocytosis of chitosan–DNA complexes.[52,53]

values with a maximum of 6.8.[9] Low-molecular-weight chitosan (LMWC) (mol wt 22 kDa), however, is highly soluble in water.[10] Chitosan polymers of molecular weights less than 50 kDa show pK_a values of about 5.5, and are therefore insoluble at neutral or basic pH.

To increase charge density of chitosan at physiological pH, and to overcome solubility problems, chitosan derivative *N,N,N*-trimethyl chitosan chloride (TMC) has been synthesized and characterized.[11] Quaternized chitosan shows higher aqueous solubility than chitosan over a broader pH range. The reason for this improved solubility is the substitution of the primary amine with methyl groups and the prevention of hydrogen bond formation between the amine and the hydroxylic groups of the chitosan backbone. *N,N,N*,-trimethyl-chitosan–galactose conjugates[12] and TMCs of low molecular weight (TMOs) have also been tested for their usefulness as nonviral gene delivery materials with regard to transfection efficiency and cytotoxicity.[13] Compared to other nonviral delivery systems, chitosan has the ability to transfect several cell types both *in vitro* and *in vivo*. As will be described in this chapter, these characteristics and properties render chitosan a suitable and safe candidate as carrier material for gene and vaccine delivery.[14]

18.2 INTERACTIONS OF CHITOSAN WITH CELL MEMBRANES

The first step in the transfection process is the mode of interaction of the gene carrier with (parts of) the membrane of the target cell. Recently, several reports on the interaction of chitosan with lipid bilayers, mimicking cell membranes, have appeared in literature.[15,16,17] Examining competitive binding of HeLa cells to chitosan beads in the presence of hypertonic salt solution or 1 *M* methyl-α-D-mannopyranoside, Venkatesh and Smith[15] concluded that cellular attachment prior to internalization occurs through interaction of the chitosan carbohydrate backbone with the cell membrane. Chan et al.[16] employed bilayers of dipamitoyl-*sn*-glycero-3-phosphocholine (DPPC), a major component of cell membranes, to examine the membrane-destabilizing effect of chitosan. Chitosan interacted strongly with DPPC multilamellar vesicles (MLV) by penetration into the acyl chains of the phospholipid bilayer, disrupting membrane organization resulting in reduction of phase transition enthalpy. Disturbances of a lipid bilayer established on a mercaptopropionic acid gold surface by LMWC (4,200 Da, acetylation degree 10%) were measured by Xang et al.,[17] employing cyclic voltammetry, electrochemical impedance spectroscopy, and surface plasmon resonance (SPR). The lipid bilayer consisted of didodecyl dimethylammonium bromide, and interaction was measured using a 0.01 molar chitosan solution in phosphate buffered saline at neutral pH. Incubation with chitosan led to disturbance of the bilayer structure, possibly through repulsion forces between the cationic polymer and the positively charged bilayer. Hydrophobic interactions between chitosan's *N*-acetyl groups and the lipophilic core of the membrane were considered as another driving force of the disturbance measured. These studies suggest that chitosan interacts with cell membranes not only by electrostatic forces, but also by interactions with carbohydrate moieties on cell membranes. These interactions lead to a disturbance of the bilayer structure.

18.3 PREPARATION OF CHITOSAN-BASED GENE DELIVERY SYSTEMS

18.3.1 Polyplexes

Preparation of self-assembling polymeric and oligomeric chitosan–DNA complexes was first described in 1995[18] by mixing a solution of chitosan with pDNA. The complex sizes yielded (150–500 nm) were found to depend upon the molecular weight (108–540 kDa) of the chitosan used, but not on the buffer composition or presence of sugars. Other authors reported comparable complex sizes and the dependence on the molecular weight of chitosan, but also on the chitosan:DNA ratio.[13,19]

The chitosan backbone of glucosamine units shows a high density of amino groups, and requires pH below 6 to be soluble. At physiological pH, not all of the amino groups are protonated. Chitosan:DNA ratios in complexes are therefore better expressed as N/P ratios, the number of polymer nitrogen (N) per DNA phosphate (P). Erbacher et al.[20] found an increase in complex sizes ranging from 1 to 5 μm at an N/P ratio of 2. This coincided with a zeta potential of the complexes close to 0 mV, indicating full retardation of pDNA and aggregation of complexes at neutral complex charge.

Chitosan, hydrophobically modified by binding of deoxycholic acid, yielded spherical self-aggregates of a mean diameter of 160 nm.[21] Other authors noted donut or rod shapes of chitosan–DNA complexes.[20] In a recent study, Koping-Hoggard et al.[22] examined the impact of the physical shape and complex stability on the transfection efficiency of polyplexes prepared with LMWC (< 5 kDa, 6-, 8-, 10-, 12-, 14-, and 24-mers). The number of nonaggregated globular complexes increased with increasing chitosan chain length, increasing N/P ratio, and decrease of pH during preparation. However, studies showed a window of parameters that allowed the preparation of complexes with favorable properties. It was concluded that 24 mer low molecular weight chitosan was preferable to high molecular weight chitosan.

The electrostatic nature of interaction of chitosan with pDNA can be examined by competitive binding studies using ethidium bromide (EtBr) as DNA stain.[8] When chitosan is added to a solution of EtBr-stained DNA, fluorescence decreases as a result of competitive binding of cationic chitosan to DNA.

Extent of DNA complexation or encapsulation into chitosan nanoparticles is determined by various methods. Whereas Mumper et al.[18] indirectly determined loading efficiency by demonstrating the absence of free DNA after formation of complexes by gel electrophoresis, Mao et al.[8] determined unentrapped DNA in the supernatant after sucrose gradient centrifugation by staining with Hoechst 33258 dye and measurement of fluorescence determination. Free chitosan in the supernatant is detected quantitatively by a ninhydrin assay.[23]

Another method for determining DNA loading is based on the Pico Green assay after digestion with chitosanase and lysozyme. Usually, loading efficiencies of more than 95% are reported. A result of encapsulation is the shift of DNA conformation from the supercoiled to the relaxed state, as detected by electrophoresis.[24]

DNA delivered by nonviral carriers is prone to degradation by DNase. A critical parameter of DNA delivery systems therefore is the ability of the carrier material to protect incorporated DNA against degradation by DNase. As physiological concentrations of the enzyme can merely be estimated, protection against DNase is routinely measured by incubation of chitosan–DNA complexes or nanoparticles with DNase I or II as model enzymes at different concentrations,[8] followed by gel electrophoresis. Complexation of DNA with highly purified chitosan fractions (molecular weights of < 5000 Da, 5,000–10,000 Da, and > 10,000 Da) at a charge ratio of 1:1 resulted in almost complete inhibition of degradation by DNase II.[9] Florea et al.[25] demonstrated that N-trimethylated oligomeric chitosan (TMOC) (2–20 chitosan monomers) of different degrees of substitution were able to protect DNA from DNase I degradation. In these studies, oligomeric chitosan insuffiently complexed DNA, and the DNA was released and subsequently cleaved. Correlating with better complexing transfection efficiency of DNA–TMOC complexes in COS-1 cells was two orders of magnitude higher than chitosan–oligomer complexes. In summary, some chitosan poly- and oligomers are able to protect DNA against degradation efficiently, probably by a change in the tertiary DNA structure causing steric hindrance.

18.3.2 Nano- and Microparticles

Chitosan–DNA nanospheres are prepared by a complex coacervation method,[26] using sodium sulfate as a desolvating agent and yielding particle sizes from 200 to 500 nm. A recent study by Mao et al.[8] examined the influence of several parameters on the size of chitosan–DNA nanoparticles. A range of sodium sulphate concentrations (2.5–25 mM) did not have an effect on the particle size, in contrast to a clear dependence on the N/P ratio. Particle sizes between 150 and 250 nm were observed at N/P ratios between 3 and 8, when particles were prepared at a temperature of 55°C and a pH of 5.5. The incorporation of plasmids of various sizes (5.1–11.9 kbp), however, had no effect on the particle sizes. EtBr staining in combination with confocal laser scanning microscopy is also applied to determine the distribution of DNA in chitosan nanoparticles.[24] Studies have shown that cross-linked chitosan–DNA nanoparticles stored in water remained stable for more than 3 months, whereas non-cross-linked nanoparticles stored in PBS remained stable for a few hours only.[24] Lyophilized chitosan–DNA nanoparticles retained their transfection potency for more than 4 weeks.[27]

Two plasmids were coencapsulated by coacervation into chitosan microparticles by Ozbas-Turan et al.,[28] resulting in a sustained release formulation. Release studies were performed in phosphate buffered saline (PBS) (pH 7.4) at 37°C; DNA released was detected photometrically at 260 nm. An initial burst effect was detected followed by a sustained release of DNA from the microparticles over a time period of up to 150 days. Even though the release study conditions were not comparable to physiological conditions, they correlated well with in vivo studies performed in mice. Expression of luciferase (the reporter gene used) activity was significantly higher 12 weeks after intramuscular application of DNA-loaded chitosan microparticles than nonencapsulated DNA.

Okamoto et al.[29] recently described the preparation of dry powders containing chitosan–DNA complexes and mannitol as a carrier, prepared by precipitation using supercritical carbon dioxide. Supercritical fluid technology had been used before to generate DNA-containing powders.[30] In these studies, however, a high degradation rate of DNA during the process was observed, which was due to a drop in pH in the aqueous DNA solution. Formation of chitosan–DNA complexes obviously stabilized DNA against degradation during the supercritical fluid process. Successive in vivo transfection studies showed a 27-fold increase in transfection efficiency of the chitosan–DNA powders in mouse lungs after insufflation compared to results obtained after instillation of DNA solution.

18.4 TRANSFECTION STUDIES

Several transfection studies involving chitosan and its derivatives have been reported over the last few years. Chitosan polymers and depolymerized chitosan have been investigated for their ability to condense and deliver DNA into Cos-1 cells, which are routinely used for screening transfection studies. Chitosans having a molecular weight lower than 10 kDa were forming small complexes of a size range of 100–200 nm.[19] These complexes showed variable stability when challenged with 10% serum, with higher stabilities shown for complexes prepared with higher molecular weight (102 kDa) chitosan. The molecular weight of chitosan, however, had limited influence on the plasmid expression *in vitro*. *In vivo* experiments using complexes of chitosan–DNA and an endosomolytic peptide led to expression of the plasmid in the small intestine of rabbits, exceeding expression levels obtained with DOTMA–DOPE formulations. As shown in other reports as well, *in vitro* transfection results in these studies were not predicitive for *in vivo* transfection efficiency.

Thanou et al.[13] prepared TMOCs of 40 and 50% degrees of quaternization. These oligomers were examined for their potency as DNA carrier systems in two cell lines, Cos-1 and epithelial Caco-2. The latter is a human colon carcinoma cell line representing a model of the intestinal epithelium.[31] It was shown that TMOC was superior to oligomeric chitosan in transfecting Cos-1 cells; however, none of the used chitosan–DNA and lipofectin–DNA complexes was able to increase transfection efficiency in fully differentiated Caco-2 cells.

Chitosan and lactosylated chitosan vectors were prepared and investigated for their transfection efficiencies *in vitro*.[20] In this study, the transfection efficiency of chitosan in HeLa cells in the presence of 19% fetal calf serum (FCS) was found to be comparable to that of another cationic polymer, polyethylenimine (PEI). Lactosylated chitosan was tested as a vector targeted to cells expressing a galactose-specific membrane lectin (HepG2 cells). It was shown that these vectors were poorly efficient in transfecting this specific cell line. This failure probably resulted from aggregation of the complexes due to a decrease of zeta potential after lactosylation, accompanied by lower affinity of the lactosylated polymer to DNA. Presence of chloroquine, a weak base preventing lysosome acidification,[32] did not improve transfection efficiency.

Recently, galactosylated chitosan-graft-dextran–DNA complexes were described.[33] Galactose groups were chemically bound to chitosan for liver specificity and dextran grafted to increase the complex stability in water. This system could effectively transfect Chang liver cells expressing the asialoglycoprotein receptor *in vitro*, indicating a specific interaction of the galactose ligands with this receptor.

Trimethylated chitosan (80% degree of quaternization) polymers, bearing antennary galactose residues through a 6-O-linked carboxymethyl group, were also examined as DNA carrier systems. The complexes were tested for specific targeting to HepG2 cells, which are expressing the galactose receptor, and expression of β-galactosidase activity was monitored. The complexes were found to efficiently transfect HepG2 cells. Transfection was significantly inhibited in the presence of a specific ligand, indicating that the conjugates were internalized via the galactose receptor.[12]

Chitosan–DNA nanospheres were prepared by self-induced complex coacervation method, yielding particles of 200–750 nm in size.[24] These nanospheres were shown to be efficient vectors in the Luciferase-293 cell system. However, their transfection efficiency varied when tested in different cell lines. Interestingly, addition of transferrin for induction of receptor-mediated uptake, or chloroquine (endosomolysis), did not increase transfection efficiency.

In an *in vivo* study, Koping-Hoggard et al.[34] prepared chitosan–DNA complexes using ultrapure chitosan (UPC) (acetylation degree 17%, MW 162 kDa) and pDNA expressing the pLacZ reporter gene. After intratracheal instillation in mice, the chitosan polyplexes transfected epithelial cells in the central airways. Compared to PEI complexes, the onset of gene expression was delayed for chitosan–DNA complexes, which may have been due to retarded endosomal escape and slower breakup of chitosan–DNA complexes. Gene expression of the reporter genes was detected up to 72 hours after administration.

18.5 DNA VACCINATION

The concept of DNA immunization evolved from the observation that intramuscular injection of naked pDNA containing reporter genes resulted in protein expression in the muscle 2 months after injection.[35] Hence, it was suggested that pDNA could be used to express foreign proteins inside cells and potentially induce an immune response against the expressed protein. The major advantage of this vaccination strategy is the induction of cytotoxic T cell responses. The hypothesis of their mechanism of action involves cross priming by antigen transfer from transfected muscle cells and direct transfection of antigen-presenting cells (APCs). However, much is still being learned and the exact mechanism will be important for the development of effective DNA vaccines and carrier systems.[36-38]

Although the intramuscular injection of DNA vaccines has been widely explored, mucosal vaccination is an evolving field of interest, especially in viral and infectious diseases where the mucosal tissue is the portal of entry of the pathogen.[37,39,40] In these diseases, initiating local immune response followed by immune memory induction should be beneficial and result in better protection against the disease. Vaccination in mucosal-associated lymphoid tissues (MALT) like the Peyer's patches in the GI tract (GALT), Waldeyer's ring in the nose (NALT), bronchus-associated lymphoid tissue in the lungs (BALT), and the urogenital lymphoid tissue (UALT) will result in initial immune response in these inductive sites and migration of immune cells via the lymphatics to regional lymph nodes and from there to mucosal effector sites through the blood circulation. This common mucosal immune system initiates immunity at mucosal inductive sites and results in immune responses in distal mucosal effector sites. The response includes both humoral and cellular immunity, but the presence of secretory IgA antibodies at the mucosal surface is the main characteristic for this specific means of vaccination. These antibodies will protect the mucosal surface by inhibition of viral/bacterial adherence, mucus trapping, neutralization of viruses and toxins, inhibition of antigen penetration, and interaction with innate antimicrobial factors.[41] The benefits of mucosal vaccination are also emphasized in health care: noninvasive application methods like tablets for oral delivery, sprays for nasal application, and inhalers for pulmonary administration are always preferred over needle injections and result in better patient compliance.

Chitosan is very appealing as a carrier for mucosally delivered vaccines.[42-44] The most frequently described chitosan–DNA vaccine formulations to date are nanoparticles (from both high and low molecular weight chitosan) that were prepared by the complexation-coacervation method and characterized by Roy et al. for oral DNA vaccine delivery.[8,24,46] The particle size was optimized to be in a narrow submicron range, the zeta-potential was positive at pH values lower than 6.0, and particles were stable without further cross-linking. The particles protected the pDNA from nuclease degradation, their transfection efficiency was cell-type dependent. Transferrin and chloroquine did not alter the transfection efficiency of these particles, suggesting that endocytosis and transfer in the endosomal-lysosomal pathway were not the main mechanisms of transport.[47] Mice that received a pDNA encoding the peanut allergen Arah2 orally generated a Th1 immune response measured by levels of serum IgG2a and were protected against peanut extract induced anaphylaxis.[48] These nanoparticles were also used for intranasal gene transfer of RSV cDNAs.[49] In these studies, mice were immunized nasally with DNA-loaded chitosan nanoparticles and later challenged intranasally with human RSV A2 strain. The loaded nanoparticles were safe as demonstrated by lack of change in methacholine responsiveness in the lungs of the control group and in vaccinated mice. The effectiveness was proven by the reduced damage to lung epithelium of vaccinated mice in comparison to unvaccinated mice and mice vaccinated only with pDNA. Furthermore, a single intranasal dose of RSV DNA incorporated in chitosan nanoparticles induced a 100-fold reduction in viral titers after an acute infection. The RSV pDNA-loaded chitosan nanoparticles significantly induced higher specific neutralizing IgG antibodies, nasal IgA titers, IFN-γ levels in the lung, and stronger CTL response compared with various controls, suggesting a role for chitosan in the increased immunologic potency.

In contrast, in a chicken model for Marek's disease virus (MDV), chitosan nanoparticles gave no significant protection against MDV while the DNA in PBS resulted in protection in four out of seven chickens.[47] In this model, the nanoparticles were given intramuscular and that is probably why the nanoparticles offered no benefit. This was also observed in own studies when vaccinating mice intramuscularly with pDNA encoding Ag85B of *Mycobacterium tuberculosis* (in cooperation with Dr. Kris Huygen, Pasteur Institute of Brussels, Belgium, unpublished data). It is possible that the DNA in buffer is transcribed and translated more efficiently in muscle tissue while the DNA attached to chitosan is not released and degraded before performing its action.

Recently, a nasal vaccine formulation prepared by complexing chitosan with a plasmid containing the CTL epitope of the M2 protein of RSV was described by Iqbal M. et al.[49] Intranasal immunization of BALB/c mice with the chitosan–DNA vaccine complexes induced peptide and virus-specific CTL responses, comparable to the intradermal DNA solution (positive control). Moreover, reduction in the virus loading following challenge was observed in the mice that received three intranasal immunizations with the complexes and also in the mice that received intradermal immunization dose followed by two intranasal doses.

Chitosan and chitosan oligomers formulated with carboxymethylcellulose (CMC) were also investigated as DNA carriers for topical genetic immunization.[50] Gene expression of the reporter gene luciferase was detectable 24 hours after topical application of chitosan–DNA formulations. Furthermore, 28 days after topical application of chitosan oligomer–CMC nanoparticles (300:100) loaded with β-galactosidase gene, β-galactosidase specific IgG titers were comparable to mice immunized by intramuscular injection of naked pDNA.

Another study described the use of chitosan nanoparticles as a nonviral delivery system for GRA1 (a *T. gondii* excreted/secreted dense granule protein 1) encoding pDNA for oral immunization of mice against toxoplasmosis.[51] Results indicated that oral vaccination using chitosan-based formulations as particulate delivery systems can prime the immune response. Boosting with GRA1 pDNA vaccine resulted in high anti-GRA1 antibody levels, but these were biased towards Th2 or mixed Th1/Th2 immune responses, whereas protective immune responses against *T. gondii* are clearly associated with Th1 type responses. Whether mucosal administration of a GRA1 chitosan formulation after GRA1 pDNA priming can enhance the Th1 type elicited immune responses is currently being addressed.

It is yet to be investigated whether chitosan could be the choice carrier for DNA vaccines. The system seems to be safe and efficient in mucosal delivery of vaccines, offering benefits such as low costs and better patient compliance for infectious diseases such as malaria and tuberculosis. It is of great importance to see if chitosan is maintaining the Th1 immune response that is necessary to fight intracellular pathogens and viruses.

Although initiating an immune response with DNA still requires its delivery into the nucleus, there is no proven need for a long-term expression of the antigen. Even so, chitosan is known to be a polymer with controlled release properties and, therefore, can result in long-term protein expression. It is reasonable to assume that chitosan derivatives will also have beneficial mucosal properties, although each derivative will have to be evaluated for immunogenicity, toxicity, and DNA protection separately. It will be of great importance to evaluate whether chitosan initiates a humoral or a cellular immunity and which cytokines will be involved in the immune response to a given vaccine.

REFERENCES

1. Skaugrud, O., Chitosan makes the grade, *Manuf. Chem.* 60, 31–35, 1989.
2. Muzzarelli, R.A., Human enzymatic activities related to the therapeutic administration of chitin derivatives, *Cell Mol. Life Sci.* 53, 131–140, 1997.
3. Carreno-Gomez, B. and Duncan, R., Evaluation of the biological properties of soluble chitosan and chitosan microspheres, *Int. J. Pharm.* 148, 231–240, 1997.
4. Hirano, S., Seino, H., Akiyama, Y., and Nonaka, I., Chitosan: a biocompatible material for oral and intravenous administration, in *Progress in Biomedical Polymers*, Gebelein, C.G. and Dunn, R.L. (Eds.), Plenum Press, New York, NY, 1990, 283–290.

5. Chandy, T. and Sharma, C.P., Chitosan as a biomaterial, *Biomat. Art. Cells Art. Org.* 18, 1–24, 1990.
6. Sandford, P.A., Chitosan and alginate: new forms of commercial interest, *Am. Chem. Soc. Div. Polym. Chem.* 31, 628–629, 1990.
7. Rolland, A.P., From genes to medicines: recent advances in nonviral gene delivery, *Crit. Rev. Ther. Drug Carrier Syst.* 15, 143–198, 1998.
8. Mao, H.Q., Roy, K., Troung-Le, V.L., Janes, K.A., Lin, K.Y., Wang, Y., August, J,T., and Leong, K.W., Chitosan-DNA nanoparticles as gene carriers: synthesis, characterization and transfection efficiency, *J. Control. Release* 70, 399–421, 2001.
9. Richardson, S.C.W., Kolbe, H.V.J., and Duncan, R., Potential of low molecular weight chitosan as a DNA delivery system: biocompatibility, body distribution and ability to complex and protect DNA, *Int. J. Pharm.* 178, 231–243, 1999.
10. Lee, M., Nah, J.-W., Kwon, Y., Koh, J.J., Ko, K.S., and Kim, S.W., Water-soluble and low molecular weight chitosan-based plasmid DNA delivery, *Pharm. Res.* 18, 427–431, 2001.
11. Sieval, A.B., Thanou, M., Kotze, A.F., Verhoef, J.C., Brussee, J., and Junginger, H.E., Preparation and NMR characterization of highly substituted N-trimethyl chitosan chloride, *Carbohydr. Polym.* 36, 157–165, 1998.
12. Murata, J., Ohya, Y., and Ouchi, T., Design of quaternary chitosan conjugate having antennary galactose residues as a gene delivery tool, *Carbohydr. Polym.* 32, 105–109, 1997.
13. Thanou, M., Florea, B.I., Geldof, M., Junginger, H.E., and Borchard, G., Quaternized chitosan oligomers as novel gene delivery vectors in epithelial cell lines, *Biomaterials* 23, 153–159, 2002.
14. Borchard, G., Chitosans for gene delivery, *Adv. Drug Deliv. Rev.* 52, 145–150, 2001.
15. Venkatesh, S. and Smith, T.J., Chitosan-membrane interactions and their probable role in chitosan-mediated transfection, *Biotechnol. Appl. Biochem.* 27, 265–267, 1998.
16. Chan, V., Mao, H.-Q., and Leong, K.W., Chitosan-induced perturbation of dipalmitoyl-*sn*-glycero-3-phosphocholine membrane bilayer, *Langmuir* 17, 3749–3756, 2001.
17. Yang, F., Cui, X., and Yang, X., Interaction of low-molecular-weight chitosan with mimic membrane studies by electrochemical methods and surface plasmon resonance, *Biophys. Chem.* 99, 99–106, 2002.
18. Mumper, R.J., Wang, J., Claspell, J.M., and Rolland, A.P., Novel polymeric carriers for gene delivery, *Proc. Intern. Symp. Control. Rel. Bioact. Mater.* 22, 178–179, 1995.
19. MacLaughlin, F.C., Mumper, R.J., Wang, J., Tagliaferi, J.M., Gill, I., Hinchcliffe, M., and Rolland, A.P., Chitosan and depolymerised chitosan oligomers as condensing carriers for *in vivo* plasmid delivery, *J. Control. Release* 56, 259–272, 1998.
20. Erbacher, P., Zou, S., Bettinger, T., Steffan, A.-M., and Remy, J.-S., Chitosan-based vector/DNA complexes for gene delivery: Biophysical characteristics and transfection ability, *Pharm. Res.* 15, 1332–1339, 1998.
21. Lee, K.Y., Kwon, I.C., Kim, Y.H., Jo, W.H., and Jeong, S.Y., Preparation of chitosan self-aggregates as a gene delivery system, *J. Control. Release* 51, 213–220, 1998.
22. Koping-Hoggard, M., Mel'nikova, Y.S., Vårum, K.M., Lindman, B., and Artursson, P., Relationship between the physical shape and the efficiency of oligomeric chitosan as a gene delivery system *in vitro* and *in vivo*, *J. Gene Med.* 5, 130–141, 2003.
23. Curotto, E. and Aros, F., Quantitative determination of chitosan and the percentage of free amino groups, *Anal. Biochem.* 211, 240–241, 1993.
24. Leong, K.W., Mao, H.Q., Truong-Le, V.L., Roy, K., Walsh, S.M., and August, J.T., DNA-polycation nanospheres as non-viral gene delivery vehicles, *J. Control. Release* 53, 183–193, 1998.
25. Florea, B.I., Ravenstijn, P.G.M., Junginger, H.E., and Borchard, G., *N*-trimethylated oligomeric chitosan (TMO) protects plasmid DNA from DNase I degradation and promotes transfection efficiency *in vitro*, *STP Pharm. Sci.* 12, 243–249, 2002.
26. Mao, H.Q., Roy, K., Truong-Le, V.L., Walsh, S.M., August, J.T., and Leong, K.W., DNA-chitosan nanospheres for gene delivery, *Proc. Int. Symp. Control. Release Bioact. Mater.* 23, 401–402, 1996.
27. Mao, H.Q., Truong-Le, V.L., August, J.T., and Leong, K.W, DNA-chitosan nanospheres: derivatization and storage stability, *Proc. Int. Symp. Control. Release Bioact. Mater.* 24, 671–672, 1997.
28. Özbas-Turan, S., Aral, C., Kabasakal, L., Keyer-Uysal, M., and Akbuga, J., Co-encapsulation of two plasmids in chitosan microspheres as a non-viral gene delivery vehicle, *J. Pharm. Pharmaceut. Sci.* 6, 27–32, 2003.

29. Okamoto, H., Nishida, S., Todo, H., Sakakura, Y., Iida, K., and Danjo, K., Pulmonary gene delivery by chitosan-pDNA complex powder prepared by a supercritical carbon dioxide process, *J. Pharm. Sci.* 92, 371–380, 2003.

30. Tservistas, M., Levy, M.S., Lo-Yim, M.Y., O'Kennedy, R.D., York, P., Humphrey, G.O., and Hoare, M., The formation of plasmid DNA loaded pharmaceutical powders using supercritical fluid technology, *Biotech. Bioeng.* 72, 12–18, 2001.

31. Delie, F. and Rubas, W., A human colonic cell line sharing similarities with enterocytes as a model to examine oral absorption: advantages and limitations of the Caco-2 cell model, *Crit. Rev. Ther. Drug Carrier Sys.* 14, 221–286, 1997.

32. Pouton, C.W. and Seymour, L.W., Key issues in non-viral gene delivery, *Adv. Drug Del. Rev.* 34, 3–19, 1998.

33. Park, K.Y., Park, Y.H., Shin, B.A., Choi, E.S., Park, Y.R., Akaike, T., and Cho, C.S., Galactosylated chitosan-*graft*-dextran as hepatocyte-targeting DNA carrier, *J. Control. Release* 9, 97–108, 2000.

34. Köping-Höggår, M., Tubulekas, I., Guan, H., Edwards, K., Nillson, M., and Vårum, K.M., Chitosan as a nonviral gene delivery system. Structure-property relationships and characteristics compared with polyethylenimine *in vitro* and after lung administration *in vivo*, *Gene Ther.* 8, 1108–1121, 2001.

35. Wolff, J.A., Malone, R.W., Williams, P. et al., Direct gene transfer into mouse muscle *in vivo*, *Science* 247, 1465–1468, 1990.

36. Donnelly, J.J., Ulmer, J.B., Shiver, J.W., and Liu, M.A., DNA vaccines, *Annu. Rev. Immunol.* 15, 617–648, 1997.

37. Donnelly, J.J., and Ulmer, J.B., DNA vaccines for viral diseases, *Braz. J. Med. Biol. Res.* 32, 215–222, 1999.

38. Donnelly, J.J., Liu, M.A., and Ulmer, J.B., Antigen presentation and DNA vaccines, *Am. J. Respir. Crit. Care Med.* 162, S190–193, 2000.

39. Whalen, R.G., DNA vaccines for emerging infectious diseases: what if? *Emerg. Infect. Dis.* 2, 168–175, 1996.

40. McCluskie, M.J., and Davis, H.L., Novel strategies using DNA for the induction of mucosal immunity, *Crit. Rev. Immunol.* 19, 303–329, 1999.

41. Ogra, P.L., Mestecky, J., Lamm, M.E., Strober, W., Bienenstock, J., and McGhee, J.R., *Mucosal Immunology*, Academic Press, 1999.

42. Bacon, A., Makin, J., Sizer, P.J., et al., Carbohydrate biopolymers enhance antibody responses to mucosally delivered vaccine antigens, *Infect. Immun.* 68, 5764–5770, 2000.

43. Illum, L., Jabbal-Gill, I., Hinchcliffe, M., Fisher, A.N., and Davis, S.S., Chitosan as a novel nasal delivery system for vaccines, *Adv. Drug Deliv. Rev.* 51, 81–96, 2001.

44. van der Lubben, I.M., Verhoef, J.C., Borchard, G., and Junginger, H.E., Chitosan for mucosal vaccination, *Adv. Drug Deliv. Rev.* 52, 139–144, 2001.

45. Roy, K., McGrath, J., Kuo, S.C., and Leong, K.W., Chitosan-DNA nanoparticles: transfection mechanism & kinetics, *Int. Symp. Control. Release Bioact. Mater.* 27, 7313, 2000.

46. Kumar, M., Behera, A.K., Lockey, R.F. et al., Intranasal gene transfer by chitosan-DNA nanospheres protects BALB/c mice against acute respiratory syncytial virus infection, *Hum. Gene Ther.* 13, 1415–1425, 2002.

47. Tischer, B.K., Schumacher, D., Beer, M. et al., A DNA vaccine containing an infectious Marek's disease virus genome can confer protection against tumorigenic Marek's disease in chickens, *J. Gen. Virol.* 83, 2367–2376, 2002.

48. Roy, K., Mao, H.Q., Huang, S.K., and Leong, K.W., Oral gene delivery with chitosan—DNA nanoparticles generates immunologic protection in a murine model of peanut allergy, *Nat. Med.* 5, 387–391, 1999.

49. Iqbal, M., Lin, W., Jabbal-Gill, I., Davis, S.S., Steward, M.W., and Illum, L., Nasal delivery of chitosan-DNA plasmid expressing epitopes of respiratory syncytial virus (RSV) induces protective CTL responses in BALB/c mice, *Vaccine* 21, 1478–1485, 2003.

50. Cui, Z., Mumper, R.J., Chitosan-based nanoparticles for topical genetic immunization, *J. Control. Release* 75, 409–419, 2001.

51. Bivas-Benita, M., Laloup, M., Versteyhe, S., Dewit, J., De Braekeleer, J., Jongert, E., and Borchard, G., Generation of *Toxoplasma gondii* GRA1 protein and DNA vaccine loaded chitosan particles: preparation, characterization and preliminary *in vivo* studies, *Int. J. Pharm.* 266, 17–27, 2003.

52. Park, I.K., Ihm, J.E., Park, Y.H., Choi, Y.J., Kim, S.I., Kim, W.J., Akaike, T., and Cho, C.S., Galac-tosylated chitosan (GC)-graft-poly(vinyl pyrrolidone) (PVP) as hepatocyte-targeting DNA carrier. Preparation and physicochemical characterization of GC-graft-PVP/DNA complex (1), *J. Control. Release* 86, 349–359, 2003.

53. Jansma, C.A., Thanou, M., Junginger, H.E., and Borchard, G., Preparation and characterization of 6-O-carboxymethyl-*N*-trimethyl chitosan derivative as potential carrier for targeted polymeric gene and drug delivery, *STP Pharm. Sci.* 13, 63–67, 2003.

PART III

Non-Condensing Polymeric Systems

CHAPTER 19

Pluronic® Block Copolymers for Nonviral Gene Delivery

Alexander V. Kabanov, Srikanth Sriadibhatla, and Valery Yu. Alakhov

CONTENTS

19.1 INTRODUCTION

The recent completion of the human genome project advanced understanding of genetic diseases.[1] It is now theoretically possible to treat diseases of genetic origin by replacing the mutated genes, inhibiting gene expression, or promoting a protective immune response by administering genes encoding specific antigens. Hence DNA and nucleic acid-based drugs have emerged as potential new therapeutic modalities to introduce deficient genes in patients, suppress disease-related genes, or act as genetic vaccines. The field of gene therapy has attracted the attention of many scientists all over the world, which is exemplified by the current volume. Although the majority of gene delivery approaches so far have involved adenoviral or retroviral vectors, more recently nonviral vectors are receiving increasing attention as gene delivery vehicles because of several advantages, such as ease of manipulation, low cost, safety, and high flexibility regarding the size of the transgene delivered. One major approach in nonviral gene therapy is based on polyplexes, complexes formed by mixing DNA with polycations. The polyplexes form spontaneously as a result of electrostatic interactions between the positively charged groups of the polycation

0-8493-1934-X/05/$0.00+$1.50
© 2005 by CRC Press LLC

and the negatively charged phosphate groups of the DNA. This results in DNA condensation, protection from the nuclease digestion, and more efficient delivery within a cell.

A variety of polycation molecules have been proposed for polyplex formation based on homopolymers or random copolymers of linear, branched, or dendrimeric architectures. These systems usually require the use of an excess of polycation with respect to DNA for transgene expression in eukaryotic cells.[2-4] The polyplexes formed are cationic particles, in which DNA is neutralized by the polycation and which display positive surface charge due to excess of polycation groups immobilized at the particle surface. We will collectively refer to such complexes as first generation polyplexes (Figure 19.1(A)). While selected polyplexes of the first generation display high levels of transgene expression *in vitro*, they all have major drawbacks *in vivo*. The primary limiting factor of these polyplexes is the lack of solubility and stability in biological fluids. The particle size is very sensitive to the nature of simple salts present in the buffer as well as the presence of serum components.[5,6] Another drawback is related to high cationic surface charge of these polyplexes, which contributes significantly to their cytotoxicity.[7-9] Consequently, these polyplexes performed unfavorably when tested *in vivo*, due to poor solubility in blood, complement activation, stimulation of leukocytes, and elimination by the cells of reticuloendothelial system.[10-13] These polyplexes allow for little control over the biodistribution of transgene expression — the one organ where substantial levels of transfection are observed is the lung, while all the other organs are refractory to transfection.[14-16] Furthermore, gene expression in the lung and other tissues is often accompanied with severe toxic effects, including embolism. These effects combined provide major obstacles for the development of gene therapies involving the first generation polyplexes.

To overcome issues of dispersion stability, interactions with serum components, and immune activation, block and graft copolymers containing segments from polycations and nonionic water-soluble polymers, such as poly(ethylene oxide) (PEO), were developed.[4,17,18] Binding of these copolymers with DNA results in complexes containing a hydrophobic core formed by the polycation-neutralized DNA, surrounded by hydrophilic PEO chains that form a "brush"-like corona and stabilize the particles in dispersion (Figure 19.1).[4,19-21] We will refer to such complexes as the second generation polyplexes. Overall the PEO-modified polyplexes form stable dispersions and do not interact with serum proteins, which is a major advancement relative to the first generation polyplexes.[21,22] Furthermore, they display extended plasma clearance kinetics, are less toxic, and were shown to transfect liver and tumor cells after systemic administration in the body.[23-25] In addition there is a possibility of targeting of such polyplexes to specific receptors at the surface of the cell, for example, by modifying the free ends of PEO chains with specific targeting ligands.[26-28] At the same time these systems reveal a new major weakness. Namely, they lack the level of gene transfection activity observed with the first generation polyplexes, which is detrimental to their use in gene delivery applications.[5] The decrease in transfection activity has been attributed to the extended structures formed by these complexes and to the steric stabilization effect of PEO hindering the binding of complexes to the membrane and entry of the complexes into cells.[5,29] To address this problem several groups are developing polymers that mimic the membrane disruptive properties of viruses and toxins, enabling endosomal escape of the delivered material.[5,30-33]

Pluronic® block copolymers were proposed as an alternative to PEO chains in the design of graft copolymers for gene delivery.[5,33] Pluronic® block copolymers are amphiphilic molecules having a triblock structure, PEO-*b*-PPO-*b*-PEO (PPO stands for poly(propylene oxide)) that can self-assemble into micelles. This triblock structure confers the characteristic ability of these molecules to interact with cellular membranes. Concurrent work revealed that such interactions have a significant impact upon the properties of the cell membranes.[34-36] For example, Figure 19.2 reveals the time-dependent changes in the membrane microviscosity of bovine brain microvessel endothelial cells (BBMEC) observed following exposure to various block copolymers, each differing in regard to their hydrophilic-lipophilic balance (HLB). This data demonstrates that, by choice of copolymer, either solidification or fluidization of membranes can be achieved. Therefore, the rationale for the substitution of PEO chains with Pluronic® block copolymers in polyplex design is that the

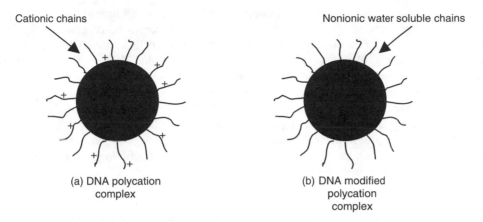

Figure 19.1   Core shell polyplex structure. (a) First generation polyplexes having a core from neutralized DNA and polycation and a corona from polycation chains, displaying net positive charge; (b) second generation polyplexes ("polyion complex micelles" or "block ionomer complexes") with a core from neutralized DNA and polycation and a corona from nonionic water-soluble polymer.

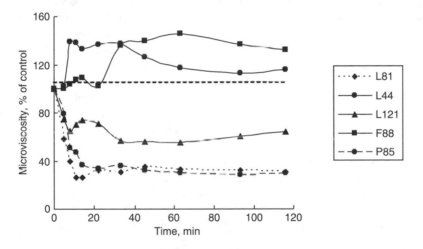

Figure 19.2   Effect of various Pluronic® block copolymers (L81, L44, L121, F88, P85) on membrane microviscosity in BBMEC. Cells were exposed to 0.0001% solution of the block copolymer at 37°C for the indicated times. The microviscosity values were derived from the fluorescence polarization measurements using 1,6-diphenyl-1,3,5-hexatriene as probes to examine the fluidity properties of the hydrocarbon region of cell membranes. (Based on the data presented by Batrakova et al.[35])

hydrophobic PPO chains of the Pluronic® will not only facilitate the formation of more condensed polyplexes, but will also enhance membrane interactions, thus improving transport of the polyplex into the cell interior.[34,37] At the same time, PEO chains of the Pluronic® block copolymers provide for polyplex solubility and masking of the polyplexes from immune recognition. In effect, the incorporation of such block copolymers into the polyplexes results in a structure that more closely mimics viruses, which often display hydrophobic peptide sequences that allow for the fusion of the virus to cell membranes.

On the other hand, Pluronic® block copolymers were shown to enhance gene expression of naked DNA in select tissues, such as skeletal muscle and tumors.[38,39] In these cases, in contrast to polyplexes, the block copolymers enhance gene expression through mechanisms that do not involve DNA condensation. Most likely the block copolymer acts as a polymer adjuvant, sometimes also called a "functional excipient," that alters the response of the cells to the delivered DNA in a way that enhances transgene expression. Although very little is known at this time about the mechanisms involved, these activities of Pluronic® block copolymers appear to be of great importance, in particular, because regular cationic complexes of plasmid DNA (pDNA) are commonly ineffective in local gene transfer applications. Recently, it was disclosed that Pluronic® block copolymer formulations can potentially increase the immune response to DNA vaccines by increasing the tissue infiltration of dendritic cells (DC) following injection of growth factor encoding plasmids.[40]

The current chapter describes the use of Pluronic® block copolymers in gene delivery both as structural and functional elements in polyplex formulations as well as adjuvants for gene expression of naked DNA. Both types of the applications of the Pluronic® block copolymers for gene delivery are considered along with some major aspects of self-assembly and biological properties of these block copolymers, which are essential for their use.

19.2 STRUCTURE AND SYNTHESIS OF PLURONIC® BLOCK COPOLYMERS

Pluronic® block copolymers, also known as "poloxamers," are recognized pharmaceutical excipients listed in the U.S. and British Pharmacopoeia. These molecules contain two hydrophilic ethylene oxide (EO) blocks and a hydrophobic propylene oxide (PO) block arranged in a triblock structure EOx–POy–EOx (Figure 19.3). The lengths of these blocks can be varied and copolymers with various x and y values are characterized by different HLB.[41,42] A particularly attractive feature of the Pluronics® is that they are commercially available in a wide range of molecular weights and block ratios, giving a wide selection of physicochemical properties according to the requirement. Another advantage is that a number of representative block copolymers have been used clinically.

Pluronic® nomenclature includes one letter, "F," "P," or "L," followed by a two- or three-digit numeric code. The letters stand for flakes ("F"), paste ("P") or liquid ("L"). The numeric code defines the structural parameters of the block copolymer. The last digit of this code approximates the weight content of EO block in tens of weight percent (for example, 80% wt if the digit is 8 or 10% wt if the digit is 1). The remaining first one or two digits encode the molecular mass of the central PO block. To decipher the code, one should multiply the corresponding number by 300 to obtain the approximate molecular mass in Daltons (Da). Therefore Pluronic® nomenclature provides a convenient way to estimate the characteristics of the block copolymer in the absence of reference literature. For example, the code "P85" denotes that the block copolymer, which is a

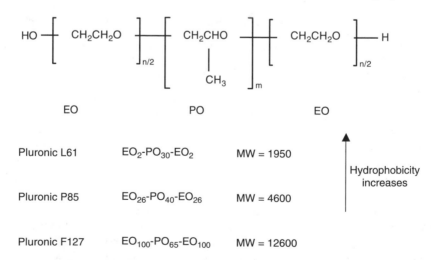

Figure 19.3 Structure of Pluronic® block copolymers.

paste, has a PO block of 2400 Da (8×300) and 50% wt of EO. The precise molecular characteristics of each Pluronic® block copolymer are provided by the manufacturer and can also be located in the literature.[41]

Poloxamer nomenclature is also widely used. From this code the approximate hydrophobe molecular weight is obtained by multiplying the first 2 digits of a poloxamer number by 100. The approximate wt% of the hydrophile is obtained by multiplying the last digit by 10. Thus, poloxamer 407 is derived from a 4000 mol wt hydrophobe with the hydrophile comprising 70% of the final mol wt.[42]

Pluronic® block copolymers are synthesized by sequential polymerization PO and EO monomers in the presence of an alkaline catalyst, such as sodium or potassium hydroxide.[43] The initial stage of the synthesis includes growth of the PO block followed by the growth of EO chains at both ends of the PO block. Anionic polymerization usually produces polymers with a low polydispersity index (Mn/Mw). However, the commercially available Pluronic® preparations may contain admixtures of the PO homopolymer as well as di- and triblock copolymers, exhibiting lower degrees of polymerization than expected. Chromatographic fractionation can be employed in procedures for the manufacture of highly purified block copolymers.[43] This reduces the presence of admixtures, particularly of the PO homopolymer and of the block copolymers containing shorter EO blocks than expected.

19.3 SELF-ASSEMBLY OF PLURONIC® BLOCK COPOLYMERS

Individual block copolymer molecules, termed "unimers," self-assemble into micelles in aqueous solutions. These unimers form a molecular dispersion in water at block copolymer concentrations below the critical micelle concentration (CMC). At concentrations of the block copolymer above the CMC, the unimer molecules aggregate, forming micelles through a process called "micellization." The driving force for the micellization is the hydrophobic interactions of the PO blocks. The PO blocks self-assemble into the inner core of the micelles covered by the hydrophilic corona from EO blocks.[44] Pluronic®–polycation conjugates used in gene delivery retain the ability to self-assemble into micelle-like aggregates. Once reacted with DNA, such conjugates form mixed micelle-like aggregates through hydrophobic interactions of the PO chain segments along with the electrostatic interactions of the polycation and DNA.[5] Hydrophilic EO chains of Pluronic® in such aggregates stabilize particles in aqueous dispersion similar to regular Pluronic® micelles. Self-assembly into micelles is important for stabilization in view of biological adjuvant since it is the unimers that work as modifiers of biological response.[45] Formation of micelles can serve as source for release of these block copolymers into the cell.

19.4 PLURONIC® BLOCK COPOLYMERS AS FUNCTIONAL EXCIPIENTS

Pluronic® block copolymers have been widely used in a variety of pharmaceutical formulations for emulsifying, wetting, dispersing and stabilizing, solubilizing, thickening, coating, and lubricating purposes.[42] Furthermore, Pluronic® block copolymers are used as structural components of micellar drug formulations for the delivery of low molecular mass drugs and polypeptides.[41] In addition, these block copolymer systems also exhibit a variety of useful biological properties. For example, water-in-oil, oil-in-water, and water-in-oil-in-water emulsions formulated with select Pluronic® block copolymers have been used extensively as immunoadjuvants. These studies suggested significant enhancement of both cell-mediated and humoral immune response induced by addition of the block copolymer formulations with respect to a very broad spectrum of antigens. Selected block copolymers, such as Pluronic® F127, have been found to significantly enhance the rate of wound and burn healing, and therefore have been included in cream formulations and skin substitutes for the treatment of burns and for other tissue engineering applications.[41] Several Pluronic®-based formulations were shown to effectively prevent postoperative adhesions or at least to reduce adhesion area after surgery.[46,47] They have been used in artificial blood for their anti-hemolytic properties and also tested as hypolemic agents, for inhibition of intestinal lipid transport.[42] Furthermore, Pluronic® block copolymers can enhance sealing of cell membranes permeabilized by ionizing radiation and electroporation, thus preventing cellular necrosis, which can be helpful for improving drug and gene delivery in skeletal muscle.[48-50] Exposure of the cells to lentivirus in the presence of in Pluronic® F127 (poloxamer 407) at 5%[51] and 15%[52,53] has been shown to enhance viral mediated gene delivery to vascular cells *in vitro* and *in vivo*. The use of Pluronic® block copolymers may be a promising strategy for overcoming challenging drug delivery barriers, including the blood brain barrier (BBB) for drug delivery to the brain and the intestinal epithelium for oral drug delivery, as well as overcoming multidrug resistance (MDR) in cancer.[54-57] Specifically, Pluronic® block copolymers can inhibit Pgp drug efflux protein expressed in brain microvessel endothelial cells and intestinal epithelial cells, which is one way to increase brain and oral absorption of selected drugs.[34,58,59] Furthermore, these block copolymers display a potent chemosensitizing effect in MDR tumors.[59-62] Formulation of anticancer drugs with such block copolymers results in enhancement of the chemotherapy of cancer.[60,63] Currently, one Pluronic®-based formulation of doxorubicin is undergoing clinical trials for treatment of drug-resistant cancers.[64] These examples show that the block copolymers exhibit valuable biological activities that can be of considerable importance for various therapeutic applications.

19.5 PLURONICS® ENHANCE GENE TRANSFER WITH POLYPLEXES

Astafieva et al.[65] have demonstrated that Pluronic® block copolymers can enhance polycation-mediated gene transfer *in vitro*. In this study, a synthetic polycation, poly(N-ethyl-4-vinylpyridinium bromide) (PEVP), was used to prepare complexes with a pDNA, and then these complexes were evaluated for DNA intracellular uptake and transgene expression in cell culture models. PEVP is a relatively low efficiency gene transfer vector compared to some more recently developed poly-cations, such as linear polyethylenimine (PEI), ExGen™ 500, or the polyamidoamine dendrimer, Superfect®.[66] However, when the PEVP and DNA were mixed with 1% Pluronic® P85 and then the cells were exposed to the resulting formulation, both the DNA uptake in the cells and transgene expression were significantly increased compared to the cells treated with the PEVP and DNA complex alone.[65] This study was recently reinforced by a report that another block copolymer, Pluronic® F127, enhances the receptor-mediated gene delivery to hepatic cell line, HepG2, using complexes of a pDNA with an asialo-oroso-mucoid-poly(L-lysine) conjugate.[67] Furthermore, the same authors reported that addition of Pluronic® F127 increased the transgene expression in a cervical cancer cell line transfected with the DNA complex with the papiloma virus proteins-poly(L-lysine) conjugate.[68] It is interesting that in contrast to Pluronic® F127, the endosomolytic compounds, such as fusogenic peptide HA2 and chloroquine, had no effects on the transgene expression.[67] This suggests that the block copolymers might enhance gene transfer through mechanisms other than facilitated DNA release from the endosomes.

Pluronic® block copolymers were also shown to attenuate the transfection activity of the first generation polyplexes in the presence of serum.[69] Six different Pluronic® copolymers with HLB decreasing (hydrophobicity increasing) in the following order were used in this study: F68 > F127 > P105 > P94 > L122 > L61. The copolymers with high HLB (hydrophilic) markedly improved transfection of NIH 3T3 cells in the presence of 10 to 50% fetal bovine serum. Also, higher concentrations (1 and 3%) of these copolymers were more potent than their lower concentrations (0.01%). Overall the authors of this study suggest that Pluronic® block copolymers bind with the polyplexes and prevent their aggregation.[69] However, further studies are needed to demonstrate and characterize self-assembly of Pluronic® block copolymers with these polyplexes.

19.6 USE OF PLURONICS® AS STRUCTURAL COMPONENTS IN POLYPLEXES

Block-graft copolymers synthesized by covalent conjugation of Pluronic® and branched PEI were used as materials for preparation of second generation polyplexes. Such polyplexes usually contain three components: DNA, Pluronic®–PEI conjugate, and free Pluronic® (Figure 19.4). The formulations were prepared with both pDNA and oligonucleotides (ODN), resulting in stable polyplex dispersions with the particle size in the ranges of ~100 to 200 nm. These polyplexes were used successfully for delivery of pDNA and antisense ODN *in vitro* and *in vivo*.[5,70,71]

One such polyplex system, based on a Pluronic® 123 polycation graft, namely P123-*g*-PEI(2K), has demonstrated improved gene delivery characteristics, relative to PEI alone, both *in vitro* and *in vivo*. One major advantage of this system is that the complexes formed by P123-g-PEI(2K) and DNA in the presence of the free Pluronic® are stable in a variety of conditions, particularly in the presence of serum proteins.[5] Confocal microscopy clearly demonstrated that although both systems delivered significant quantity of the pDNA, there were apparent differences in both the quantity and the localization of the internalized DNA; which was likely the result of divergent mechanisms of internalization between the systems.[33] The transfection efficiency of the P123-*g*-PEI (2K)-based polyplexes was evaluated using a luciferase expression reporter construct, demonstrating that transfection with these polyplexes results in high transgene expression.[5,33] In particular, the panel of polycations for gene delivery was examined included ExGen™ 500, branched PEI, PEVP, Superfect®, poly(propyleneimine) dendrimer (Astramol®), and P123-g-PEI(2K) (Figure 19.5). Using a panel of cell lines, ExGen™ 500, Superfect®, branched PEI (25 kDa), and P123-g-PEI(2K) were determined as systems displaying highest transfection activity while exhibiting relatively low cytotoxicity. These systems had activity higher than or comparable to lipid transfection reagents (Lipofectin®, LipofectAMINE®, CeLLFECTIN® and DMRIE-C) but displayed less serum dependence and were less toxic than the lipids.

Most importantly, however, the P123-*g*-PEI(2K)-based system revealed high transfection activity in *in vivo* experiments, exhibiting improved *in vivo* biodistribution compared to unmodified PEI.[5] Furthermore, gene expression was observed in spleen, heart, lungs, and liver 24 h after intravenous injection of this polyplex in mice, revealing a more uniform distribution of gene expression, compared to first generation polyplex and a lipid-based system, between these organs, and allowing for a significant improvement of gene expression in the liver.[5] This was subsequently used for intravenous delivery of the gene encoding the murine ICAM-1 molecule in liver in transgenic ICAM-1 deficient mice.[70] The RT-PCR analysis of ICAM-1 mRNA expression showed that this conjugate induced a dose-dependent expression of ICAM-1 in the liver. Furthermore, this expression of ICAM-1 induced neutrophil invasion in the liver, while no such invasion was observed in mice injected with formulated control plasmid or naked DNA.

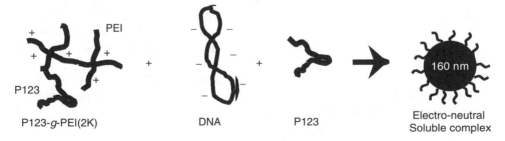

Figure 19.4 Second generation polyplexes containing three components: (1) plasmid DNA; (2) Pluronic® P123–PEI conjugate; and (3) free Pluronic® P123. Without free Pluronic® P123 DNA and Pluronic® P123–PEI conjugate formed large particles of ~600 nm (not presented). Addition of the free Pluronic® P123 resulted in formation of the small stable particles of ~160 nm in diameter. No particle aggregation was observed in the presence of serum and are stable even in presence of serum. Based on the data presented by Gebhart et al.[33]

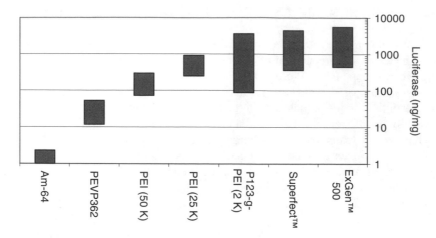

Figure 19.5   Relative transfection activity of various polyplexes in Cos-7 cells as determined using luciferase as a reporter gene. Based on the data presented by Gebhart et al.[66]

Overall, although studies of Pluronic® in gene delivery are incomplete, the results obtained suggest that Pluronic® block copolymers are promising agents for utilization in gene therapy applications. Recently other investigators started developing similar delivery systems, in particular those based on conjugates of Pluronic® block copolymers with polylysine.[72]

19.7 PLURONICS® ENHANCE EXPRESSION IN STABLY TRANSFECTED CELLS

Traditional thinking in polymer-based gene delivery field is that polymers are needed as structural and functional components to enhance delivery of DNA into the cells (the "artificial virus" or polyplex paradigm). At the same time it is entirely possible that some of these polymer components, capable of various nonspecific interactions within the cell, can affect the gene expression machinery, resulting in changes in transgene expression after the DNA is already delivered. To test this hypothesis we recently developed a cell model representing NIH 3T3 cells stably transformed to express luciferase or green fluorescent protein (GFP) genes (both genes under control of the CMV promoter). By exposing these cells to various free Pluronic® block copolymers as well as Pluronic®-PEI for 4 h followed by 24 h incubation in fresh media, we evaluated whether these molecules have an effect on expression of corresponding reporter genes (unpublished data). This study suggested that selected copolymers, such as Pluronic® P85 and P123, as well as polyplexes obtained using P123-*g*-PEI(2K), substantially (by over 10 times) increased gene expression. This result is illustrated in Figure 19.6, using Pluronic® P85 and P123 and cells expressing the luciferase gene. We believe that this effect of enhanced gene expression in the presence of Pluronic® could be important for understanding the effects of the block copolymers on gene expression with polyplexes. It could also contribute to enhancements of gene expression in virus-based formulations supplemented with the block copolymers. Finally, it could be directly related to the effects of Pluronic® on gene expression during administration of naked DNA in the tissues, which are considered in the following section.

19.8 EFFECT OF PLURONICS® ON EXPRESSION OF NAKED DNA *IN VIVO*

pDNA can be injected into skeletal muscle and can generate therapeutically meaningful levels of gene expression.[73] This application of naked DNA was demonstrated using therapeutic genes

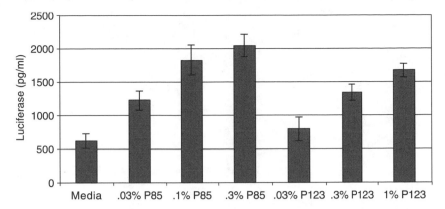

Figure 19.6 Effect of Pluronics on gene expression in NIH3T3 cells stably transformed with gWIZluc vector (Gene Therapy Systems, San Diego, CA). Transformed cells were treated with the block copolymer solutions for 4 h, washed with PBS, incubated in fresh media for 24 h, and then lysed for luciferase analysis.

including systemically acting secreted proteins, such as erythropoietin (EPO) and interleukin-5,[74,75] as well as locally acting proteins, such as basic fibroblast growth factor, vascular endothelial growth factor, and dystrophin.[76-78] However, in many cases the relatively low level of gene expression achieved limits the applicability of naked DNA as a therapeutic agent. Furthermore, commercially available cationic carriers, such as cationic dendrimers and lipids, which are known to improve gene expression in other tissues, significantly inhibit intramuscular gene expression (unpublished data).

Therefore, an alternative approach has been to identify compounds that can enhance gene expression in the muscle. One such compound, poly(N-vinyl pyrrolidone) (PVP), has been widely discussed in the literature as a potential booster of the gene expression using the naked DNA.[79-85] This polymer does not condense the DNA at the concentrations used, a marked distinction compared to the cationic polymers which bind electrostatically to the DNA molecule and cause its condensation.[80]

It was recently reported that certain Pluronic® block copolymers significantly increase expression of pDNA in skeletal muscle in mice.[38,86] In particular, a formulation based on the mixture of the block copolymers Pluronic® L61 and Pluronic® F127 increased expression of both reporter and therapeutic genes by 5- to 20-fold compared to naked DNA. This formulation was termed SP1017. Unlike cationic DNA carriers, SP1017 does not condense DNA. SP1017 exhibited maximal activity at concentrations that are close to the CMC values of the block copolymers present in the SP1017 mixture (Figure 19.7). This suggests Pluronic® unimers, not the micelles, are involved in this activity. With such low doses required for enhanced gene expression, the block copolymer formulation provides for at least a 500-fold safety margin in animals (see next section). Compared to PVP, SP1017 was found to be more efficient and requiring less DNA to produce the same amount of transgene.[38] Furthermore, the histology experiments using β-galactosidase as reporter gene demonstrated that SP1017 significantly enhanced distribution of transgene expression through the tissue, thus increasing the "bioavailability" of the expressed transgene.

Another block copolymer, Pluronic® L64 (PE6400), was also shown to enable gene expression in up to 35% of muscle fibers from mouse tibial cranial muscle and with an efficiency similar to that obtained with electrotransfer in case of skeletal muscle.[87] It was also reported that improvement in gene expression over naked DNA was observed irrespective of the size of the reporter gene, ranging from 0.7 to 3.4 kb. More recently, SP1017 and Pluronic® P123 were shown to enhance and prolong transgene expression following administration of DNA in solid tumors in a manner similar to that observed in the case of DNA administration in the muscles.[39] Finally, Pluronic® F68 (Poloxamer 188) was also used as a carrier for topical gene delivery in eye drops formulation.[88] Following 2 days of administration of the block copolymer-based formulation (three times a day),

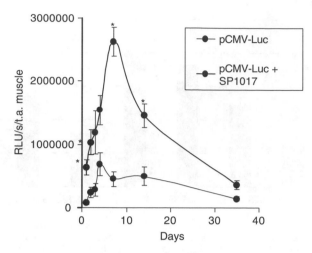

Figure 19.7 Time-course of gene expression in muscles: C57Bl/6 mice (6–8 week old females) were injected with 5μg of pCMV-luc plasmid alone or pCMV-luc formulated with SP1017 in the *tibialis anterior* muscle. The muscles were harvested from day 1 to day 35 after injection and extracted with a lysis buffer to measure the luciferase activity. (Based on the data presented by Lemieux et al.[38])

the intense expression of the β-galactosidase reporter gene was observed in ocular tissue. In contrast, administration of the copolymer-free DNA did not result in any detectable gene expression.

The capacity of SP1017 to enhance plasmid-driven expression of therapeutic genes was illustrated using the murine EPO as an example.[38] The physiological effect that resulted from the expression of EPO was readily quantifiable by measuring the hematocrit levels from the blood of the treated mice. In this study, the hematocrits of mice injected with pCMV-mEPO plasmid were greatly increased when the plasmid was formulated with SP1017. Furthermore, in the case of SP1017 formulated DNA the hematocrit increases were sustained for much longer periods compared to those observed after injection of copolymer-free DNA (50 days versus 21 days after injection). These increases in hematocrits correlated with elevated EPO levels in the blood and muscle, as detected on day 7 after injection. Furthermore, the dose escalation study demonstrated that at high DNA doses (50 μg per mouse) the hematocrits reached the levels comparable to those observed for adeno-associated virus gene delivery.

The mechanism by which the block copolymers enhance gene expression in tissues is still unknown. Pluronic® formulations have no apparent effect of delivery and expression of naked DNA *in vitro* in the murine muscle cell line C2C12,[38] isolated cardiomyocytes, or Cos-7 cell lines.[87] However, this result could indicate that the delivery of the naked DNA from the external solution into the cell, a limiting step for transgene expression, is not affected by Pluronic®. At the same time, when DNA was microinjected into the cell cytoplasm, Pluronic® L64 promoted DNA translocation into the nucleus and induced gene expression.[87] In other words, Pluronic® might have critical effects in the processes beyond just DNA delivery, which is consistent with the observations described in the previous section.

19.9 TOXICOLOGICAL CONSIDERATIONS

Considerable toxicity data are available for many block copolymers. This includes acute and chronic oral toxicity studies, skin and eye sensitivity studies, acute intravenous studies, a three-generation reproduction study, and other toxicological data.[42] For example, the intravenous toxicity studies performed on SP1017 in rodents and non-rodents determined the maximal tolerated doses of this carrier as follows: 1350 mg/kg (rats) and 1080 mg/kg (dogs) for acute toxicity, and 337 mg/kg/day (rats) and 225 mg/kg/day (dogs) for subacute toxicity (14 daily injections). The muscle toxicity caused by injection of both single and multiple doses of various block copolymers was evaluated by morphological examination of the muscle tissue and by monitoring creatine phosphokinase levels. This study concluded that the toxicity of the block copolymers was proportional to their lipophilicity — the more lipophilic the copolymer, the more severe the lesions. Overall, Pluronic® F88 and Pluronic® F127 were considered suitable for incorporation into gel formulations, while Pluronic® P105 and Pluronic® P123 appeared to be more toxic than other accepted vehicles for intramuscular use, such as normal saline, Cremophor EL, and peanut oil.[89] It is noteworthy that the vehicles tested in this study contained a 25% wt Pluronic® gel, which was over a thousand times higher than the concentrations of block copolymers used in the gene delivery applications. Therefore, at least as far as intramuscular administration of block copolymers is concerned, the Pluronic®-based gene delivery systems may have a very substantial safety margin.[41]

19.10 CONCLUSIONS

Although the studies described in this paper are relatively recent, the results obtained already suggest that Pluronic® block copolymers are promising agents for gene therapy applications. Pluronic® block copolymers are versatile molecules displaying potential as adjuvants in naked DNA delivery and also as structural elements in novel self-assembling gene delivery systems that may be superior to currently known vectors. Results from our studies and those of other researchers indicate that Pluronic® block copolymers are potentially useful gene delivery systems capable of delivering a multitude of therapeutic genes. Based on these studies, the use of block copolymers in gene delivery is a very exciting area of research, in which new and important developments are expected in the near future.

ACKNOWLEDGMENTS

The original studies by the authors described in this chapter have been supported by grants and contracts from the National Science Foundation, Supratek Pharma, Inc., and the Nebraska Research Initiative Gene Therapy Program. Alexander V. Kabanov is a consultant and Valery Yu. Alakhov is an employee of Supratek Pharma, Inc.

REFERENCES

1. Gruenert, D.C., Novelli, G., Dallapiccola, B., and Colosimo, A., Genome medicine: gene therapy for the millennium, *Gene Ther.* 9, 653–657, 2002.
2. Kabanov, A.V., Kiselev, V.I., Chikindas, M.L., Astafieva, I.V., Glukhov, A.I., Gordeev, S.A., Izumrudov, V.A., Zezin, A.B., Levashov, A.V., Severin, E.S., and Kabanov, V.A., Increasing of transforming activity of plasmid DNA by incorporating it into an interpolyelectrolyte complex with a carbon chain polycation, *Dokl. Akad. Nauk SSSR* (in Russian) 306, 226–229, 1989.
3. Haensler, J., and Szoka, F.C., Jr., Polyamidoamine cascade polymers mediate efficient transfection of cells in culture, *Bioconj. Chem.* 4, 372–379, 1993.
4. Wolfert, M.A., Schacht, E.H., Toncheva, V., Ulbrich, K., Nazarova, O., and Seymour, L.W., Characterization of vectors for gene therapy formed by self-assembly of DNA with synthetic block copolymers, *Hum. Gene Ther.* 7, 2123–2133, 1996.
5. Nguyen, H.K., Lemieux, P., Vinogradov, S.V., Gebhart, C.L., Guerin, N., Paradis, G., Bronich, T.K., Alakhov, V.Y., and Kabanov, A.V., Evaluation of polyether-polyethyleneimine graft copolymers as gene transfer agents, *Gene Ther.* 7, 126–138, 2000.

6. Ogris, M., Steinlein, P., Kursa, M., Mechtler, K., Kircheis, R., and Wagner, E., The size of DNA/transferrin-PEI complexes is an important factor for gene expression in cultured cells, *Gene Ther.* 5, 1425–1433, 1998.

7. Elferink, J.G., Cytolytic effect of polylysine on rabbit polymorphonuclear leukocytes, *Inflammation* 9, 321–331, 1985.

8. Plank, C., Zatloukal, K., Cotten, M., Mechtler, K., and Wagner, E., Gene transfer into hepatocytes using asialoglycoprotein receptor mediated endocytosis of DNA complexed with an artificial tetra-antennary galactose ligand, *Bioconj. Chem.* 3, 533–539, 1992.

9. Clark, R.A., Olsson, I., and Klebanoff, S.J., Cytotoxicity for tumor cells of cationic proteins from human neutrophil granules, *J. Cell Biol.* 70, 719-23, 1976.

10. Dash, P.R., Read, M.L., Barrett, L.B., Wolfert, M.A., and Seymour, L.W., Factors affecting blood clearance and *in vivo* distribution of polyelectrolyte complexes for gene delivery, *Gene Ther.* 6, 643–650, 1999.

11. Hashida, M., Takemura, S., Nishikawa, M., and Takakura, Y., Targeted delivery of plasmid DNA complexed with galactosylated poly(L-lysine), *J. Control. Release* 53, 301–310, 1998.

12. Oupicky, D., Ogris, M., Howard, K.A., Dash, P.R., Ulbrich, K., and Seymour, L.W., Importance of lateral and steric stabilization of polyelectrolyte gene delivery vectors for extended systemic circulation, *Mol. Ther.* 5, 463–472, 2002.

13. Ward, C.M., Read, M.L., and Seymour, L.W., Systemic circulation of poly(L-lysine)/DNA vectors is influenced by polycation molecular weight and type of DNA: differential circulation in mice and rats and the implications for human gene therapy, *Blood* 97, 2221–2229, 2001.

14. Ferrari, S., Moro, E., Pettenazzo, A., Behr, J.P., Zacchello, F., and Scarpa, M., ExGen 500 is an efficient vector for gene delivery to lung epithelial cells *in vitro* and *in vivo*, *Gene Ther.* 4, 1100–1106, 1997.

15. Goula, D., Benoist, C., Mantero, S., Merlo, G., Levi, G., and Demeneix, B.A., Polyethylenimine-based intravenous delivery of transgenes to mouse lung, *Gene Ther.* 5, 1291–1295, 1998.

16. Zou, S.M., Erbacher, P., Remy, J.S., and Behr, J.P., Systemic linear polyethylenimine (L-PEI)-mediated gene delivery in the mouse, *J. Gene Med.* 2, 128–134, 2000.

17. Kabanov, A.V., and Kabanov, V.A., DNA complexes with polycations for the delivery of genetic material into cells, *Bioconj. Chem.* 6, 7–20, 1995.

18. Katayose, S., and Kataoka, K., Water-soluble polyion complex associates of DNA and poly(ethylene glycol)-poly(L-lysine) block copolymer, *Bioconj. Chem.* 8, 702–707, 1997.

19. Choi, Y.H., Liu, F., Kim, J.S., Choi, Y.K., Park, J.S., and Kim, S.W., Polyethylene glycol-grafted poly-L-lysine as polymeric gene carrier, *J. Control. Release* 54, 39–48, 1998.

20. Kabanov, V.A., and Kabanov, A.V., Interpolyelectrolyte and block ionomer complexes for gene delivery: physico-chemical aspects, *Adv. Drug Deliv. Rev.* 30, 49–60, 1998.

21. Vinogradov, S.V., Bronich, T.K., and Kabanov, A.V., Self-assembly of polyamine-poly(ethylene glycol) copolymers with phosphorothioate oligonucleotides, *Bioconj. Chem.* 9, 805–812, 1998.

22. Itaka, K., Harada, A., Nakamura, K., Kawaguchi, H., and Kataoka, K., Evaluation by fluorescence resonance energy transfer of the stability of nonviral gene delivery vectors under physiological conditions, *Biomacromolecules* 3, 841–845, 2002.

23. Ogris, M., Brunner, S., Schuller, S., Kircheis, R., and Wagner, E., PEGylated DNA/transferrin-PEI complexes: reduced interaction with blood components, extended circulation in blood and potential for systemic gene delivery, *Gene Ther.* 6, 595–605, 1999.

24. Harada-Shiba, M., Yamauchi, K., Harada, A., Takamisawa, I., Shimokado, K., and Kataoka, K., Polyion complex micelles as vectors in gene therapy—pharmacokinetics and *in vivo* gene transfer, *Gene Ther.* 9, 407–414, 2002.

25. Oupicky, D., Ogris, M., and Seymour, L.W., Development of long-circulating polyelectrolyte complexes for systemic delivery of genes, *J. Drug Target.* 10, 93–98, 2002.

26. Choi, Y.H., Liu, F., Park, J.S., and Kim, S.W., Lactose-poly(ethylene glycol)-grafted poly-L-lysine as hepatoma cell-tapgeted gene carrier, *Bioconj. Chem.* 9, 708–718, 1998.

27. Ward, C.M., Pechar, M., Oupicky, D., Ulbrich, K., and Seymour, L.W., Modification of pLL/DNA complexes with a multivalent hydrophilic polymer permits folate-mediated targeting *in vitro* and prolonged plasma circulation *in vivo*, *J. Gene Med.* 4, 536–547, 2002.

28. Vinogradov, S., Batrakova, E., Li, S., and Kabanov, A., Polyion complex micelles with protein-modified corona for receptor-mediated delivery of oligonucleotides into cells, *Bioconj. Chem.* 10, 851–860, 1999.

29. Petersen, H., Fechner, P.M., Martin, A.L., Kunath, K., Stolnik, S., Roberts, C.J., Fischer, D., Davies, M.C., and Kissel, T., Polyethylenimine-graft-poly(ethylene glycol) copolymers: influence of copolymer block structure on DNA complexation and biological activities as gene delivery system, *Bioconj. Chem.* 13, 845–854, 2002.

30. Murthy, N., Campbell, J., Fausto, N., Hoffman, A.S., and Stayton, P.S., Design and synthesis of pH-responsive polymeric carriers that target uptake and enhance the intracellular delivery of oligonucleotides, *J. Control. Release* 89, 365–374, 2003.

31. Murthy, N., Campbell, J., Fausto, N., Hoffman, A.S., and Stayton, P.S., Bioinspired pH-Responsive Polymers for the Intracellular Delivery of Biomolecular Drugs, *Bioconj. Chem.* 14, 412–419, 2003.

32. Stayton, P.S., Hoffman, A.S., Murthy, N., Lackey, C., Cheung, C., Tan, P., Klumb, L. A., Chilkoti, A., Wilbur, F.S., and Press, O.W., Molecular engineering of proteins and polymers for targeting and intracellular delivery of therapeutics, *J. Control. Release* 65, 203–220, 2000.

33. Gebhart, C.L., Sriadibhatla, S., Vinogradov, S., Lemieux, P., Alakhov, V., and Kabanov, A.V., Design and formulation of polyplexes based on pluronic-polyethyleneimine conjugates for gene transfer, *Bioconj. Chem.* 13, 937–944, 2002.

34. Batrakova, E.V., Li, S., Vinogradov, S.V., Alakhov, V.Y., Miller, D.W., and Kabanov, A.V., Mechanism of pluronic effect on P-glycoprotein efflux system in blood-brain barrier: contributions of energy depletion and membrane fluidization, *J. Pharmacol. Exp. Ther.* 299, 483–493, 2001.

35. Batrakova, E.V., Li, S., Alakhov, V.Y., Miller, D.W., and Kabanov, A.V., Optimal structure requirements for pluronic block copolymers in modifying P-glycoprotein drug efflux transporter activity in bovine brain microvessel endothelial cells, *J. Pharmacol. Exp. Ther.* 304, 845–854, 2003.

36. Melik-Nubarov, N.S., Pomaz, O.O., Dorodnych, T., Badun, G.A., Ksenofontov, A.L., Schemchukova, O.B., and Arzhakov, S.A., Interaction of tumor and normal blood cells with ethylene oxide and propylene oxide block copolymers, *FEBS Lett.* 446, 194–198, 1999.

37. Alakhov, V.Y., and Kabanov, A.V., Block copolymeric biotransport carriers as versatile vehicles for drug delivery, *Exp. Op. Invest. Drugs* 7, 1453–1473, 1998.

38. Lemieux, P., Guerin, N., Paradis, G., Proulx, R., Chistyakova, L., Kabanov, A., and Alakhov, V., A combination of poloxamers increases gene expression of plasmid DNA in skeletal muscle, *Gene Ther.* 7, 986–991, 2000.

39. Gebhart, C., Alakhov, V.Y., and Kabanov, A.V., Pluronic block copolymers enhance local transgene expression in skeletal muscle and solid tumor, in *Controlled Release Society*, Glasgow, Scotland, UK, 2003.

40. Sang, H., Pisarev, V.M., Munger, C., Robinson, S., Chavez, J., Hatcher, L., Parajuli, P., Guo, Y., and Talmadge, J.E., Regional, but not systemic recruitment/expansion of dendritic cells by a pluronic-formulated Flt3-ligand plasmid with vaccine adjuvant activity, *Vaccine* 21, 3019–3029, 2003.

41. Kabanov, A.V., and Alakhov, V.Y., Pluronic block copolymers in drug delivery: from micellar nano-containers to biological response modifiers, *Crit. Rev. Ther. Drug Carrier Syst.* 19, 1–72, 2002.

42. Schmolka, I.R., Poloxamers in the pharmaceutical industry, in *Polymers for Controlled Drug Delivery*, Tarcha, P.J. (Ed.), CRC Press, Boca Raton, FL, 1991, 189–214.

43. Schmolka, I.R., A review of block polymer surfactants, *J. Am. Oil Chem. Soc.* 54, 110–116, 1977.

44. Kabanov, A.V., Lemieux, P., Vinogradov, S., and Alakhov, V., Pluronic block copolymers: novel functional molecules for gene therapy, *Adv. Drug Deliv. Rev.* 54, 223–233, 2002.

45. Batrakova, E.V., Lee, S., Li, S., Venne, A., Alakhov, V., and Kabanov, A., Fundamental relationships between the composition of pluronic block copolymers and their hypersensitization effect in MDR cancer cells, *Pharm. Res.* 16, 1373–1379, 1999.

46. Leach, R.E., and Henry, R.L., Reduction of postoperative adhesions in the rat uterine horn model with poloxamer 407, *Am. J. Obstet. Gynecol.* 162, 1317–1319, 1990.

47. Vlahos, A., Yu, P., Lucas, C.E., and Ledgerwood, A.M., Effect of a composite membrane of chitosan and poloxamer gel on postoperative adhesive interactions, *Am. Surg.* 67, 15–21, 2001.

48. Hannig, J., Zhang, D., Canaday, D.J., Beckett, M.A., Astumian, R.D., Weichselbaum, R.R., and Lee, R.C., Surfactant sealing of membranes permeabilized by ionizing radiation, *Radiat. Res.* 154, 171–177, 2000.

49. Lee, R.C., Canaday, D.J., and Hammer, S.M., Transient and stable ionic permeabilization of isolated skeletal muscle cells after electrical shock, *J. Burn. Care Rehabil.* 14, 528–540, 1993.

50. Lee, R.C., Hannig, J., Matthews, K.L., Myerov, A., and Chen, C.T., Pharmaceutical therapies for sealing of permeabilized cell membranes in electrical injuries, *Ann. NY Acad. Sci.* 888, 266–273, 1999.

51. March, K.L., Madison, J.E., and Trapnell, B.C., Pharmacokinetics of adenoviral vector-mediated gene delivery to vascular smooth muscle cells: modulation by poloxamer 407 and implications for cardio-vascular gene therapy, *Hum. Gene Ther.* 6, 41–53, 1995.

52. Feldman, L.J., Pastore, C.J., Aubailly, N., Kearney, M., Chen, D., Perricaudet, M., Steg, P.G., and Isner, J.M., Improved efficiency of arterial gene transfer by use of poloxamer 407 as a vehicle for adenoviral vectors, *Gene Ther.* 4, 189–198, 1997.

53. Dishart, K.L., Denby, L., George, S.J., Nicklin, S.A., Yendluri, S., Tuerk, M.J., Kelley, M.P., Donahue, B.A., Newby, A.C., Harding, T., and Baker, A.H., Third-generation lentivirus vectors efficiently transduce and phenotypically modify vascular cells: implications for gene therapy, *J. Mol. Cell Cardiol.* 35, 739–748, 2003.

54. Kabanov, A.V., Batrakova, E.V., and Alakhov, V.Y., Pluronic block copolymers as novel polymer therapeutics for drug and gene delivery, *J. Control. Release* 82, 189–212, 2002.

55. Kabanov, A.V., Batrakova, E.V., and Yu Alakhov, V., An essential relationship between ATP depletion and chemosensitizing activity of Pluronic((R)) block copolymers, *J. Control. Release* 91, 75–83, 2003.

56. Kabanov, A.V., Batrakova, E.V., and Miller, D.W., Pluronic((R)) block copolymers as modulators of drug efflux transporter activity in the blood-brain barrier, *Adv. Drug Deliv. Rev.* 55, 151–164, 2003.

57. Kabanov, A.V., Batrakova, E.V., and Alakhov, V.Y., Pluronic block copolymers for overcoming drug resistance in cancer, *Adv. Drug Deliv. Rev.* 54, 759–779, 2002.

58. Batrakova, E.V., Han, H.Y., Miller, D.W., and Kabanov, A.V., Effects of pluronic P85 unimers and micelles on drug permeability in polarized BBMEC and Caco-2 cells, *Pharm. Res.* 15, 1525–1532, 1998.

59. Batrakova, E.V., Li, S., Miller, D.W., and Kabanov, A.V., Pluronic P85 increases permeability of a broad spectrum of drugs in polarized BBMEC and Caco-2 cell monolayers, *Pharm. Res.* 16, 1366–1372, 1999.

60. Alakhov, V.Y., Moskaleva, E.Y., Batrakova, E.V., and Kabanov, A.V., Hypersensitization of multidrug resistant human ovarian carcinoma cells by pluronic P85 block copolymer, *Bioconj. Chem.* 7, 209–216, 1996.

61. Batrakova, E.V., Li, S., Elmquist, W.F., Miller, D.W., Alakhov, V.Y., and Kabanov, A.V., Mechanism of sensitization of MDR cancer cells by Pluronic block copolymers: Selective energy depletion, *Br. J. Cancer* 85, 1987–1997, 2001.

62. Venne, A., Li, S., Mandeville, R., Kabanov, A., and Alakhov, V., Hypersensitizing effect of pluronic L61 on cytotoxic activity, transport, and subcellular distribution of doxorubicin in multiple drug-resistant cells, *Cancer Res.* 56, 3626–3629, 1996.

63. Alakhov, V., Klinski, E., Li, S., Pietrzynski, G., Venne, A., Batrakova, E., Bronitch, T., and Kabanov, A.V., Block copolymer-based formulation of doxorubicin. From cell screen to clinical trials, *Colloids Surf. B: Biointerf.* 16, 113–134, 1999.

64. Ranson, M., Ferry, D., Kerr, D., Radford, J., Dickens, D., Alakhov, V., Brampton, M., and Margison, J., Results of a cancer research campaign phase I dose escalation trial of SP1049C in patients with advanced cancer, in *5th International Symposium on Polymer Therapeutics: From Laboratory to Clinical Practice*, The Welsh School of Pharmacy, Cardiff University, Cardiff, UK, 2002, 15.

65. Astafieva, I., Maksimova, I., Lukanidin, E., Alakhov, V., and Kabanov, A., Enhancement of the polycation-mediated DNA uptake and cell transfection with Pluronic P85 block copolymer, *FEBS Lett.* 389, 278–280, 1996.

66. Gebhart, C.L., and Kabanov, A.V., Evaluation of polyplexes as gene transfer agents, *J. Contr. Release* 73, 401–416, 2001.

67. Cho, C.-W., Cho, Y.-S., Lee, H.-K., Yeom, Y.I., Park, S.-N., and Yoon, D.-Y., Improvement of receptor-mediated gene delivery to HepG2 cells using an amphiphilic gelling agent, *Biotechnol. Appl. Biochem.* 32, 21–26, 2000.

68. Cho, C.-W., Cho, Y.-S., Kang, B.T., Hwang, J.S., Park, S.-N., and Yoon, D.-Y., Improvement of gene transfer to cervical cancer cell lines using non-viral agents, *Cancer Lett.* 162, 75–85, 2001.

69. Kuo, J.H., Effect of Pluronic-block copolymers on the reduction of serum-mediated inhibition of gene transfer of polyethyleneimine-DNA complexes, *Biotechnol. Appl. Biochem.* 37, 267–271, 2003.

70. Ochietti, B., Lemieux, P., Kabanov, A.V., Vinogradov, S., St-Pierre, Y., and Alakhov, V., Inducing neutrophil recruitment in the liver of ICAM-1-deficient mice using polyethyleneimine grafted with Pluronic P123 as an organ-specific carrier for transgenic ICAM-1, *Gene Ther.* 9, 939–945, 2002.

71. Ochietti, B., Guerin, N., Vinogradov, S.V., St-Pierre, Y., Lemieux, P., Kabanov, A.V., and Alakhov, V.Y., Altered organ accumulation of oligonucleotides using polyethyleneimine grafted with poly(ethylene oxide) or pluronic as carriers, *J. Drug Target.* 10, 113–121, 2002.

72. Jeon, E., Kim, H.D., and Kim, J.S., Pluronic-grafted poly-(L)-lysine as a new synthetic gene carrier, *J. Biomed. Mater. Res.* 66A, 854–859, 2003.

73. Wolff, J.A., Malone, R.W., Williams, P., Chong, W., Acsadi, G., Jani, A., and Felgner, P.L., Direct gene transfer into mouse muscle *in vivo*, *Science* 247, 1465–1468, 1990.

74. Tokui, M., Takei, I., Tashiro, F., Shimada, A., Kasuga, A., Ishii, M., Ishii, T., Takatsu, K., Saruta, T., and Miyazaki, J., Intramuscular injection of expression plasmid DNA is an effective means of long-term systemic delivery of interleukin-5, *Biochem. Biophys. Res. Commun.* 233, 527–531, 1997.

75. Tripathy, S.K., Svensson, E.C., Black, H.B., Goldwasser, E., Margalith, M., Hobart, P.M., and Leiden, J.M., Long-term expression of erythropoietin in the systemic circulation of mice after intramuscular injection of a plasmid DNA vector, *Proc. Nat. Acad. Sci. U.S.A.* 93, 10876–10880, 1996.

76. Laham, R.J., Chronos, N.A., Pike, M., Leimbach, M.E., Udelson, J.E., Pearlman, J.D., Pettigrew, R.I., Whitehouse, M.J., Yoshizawa, C., and Simons, M., Intracoronary basic fibroblast growth factor (FGF-2) in patients with severe ischemic heart disease: results of a phase I open-label dose escalation study, *J. Am. Coll. Cardiol.* 36, 2132–2139, 2000.

77. Lathi, K.G., Vale, P.R., Losordo, D.W., Cespedes, R.M., Symes, J.F., Esakof, D.D., Maysky, M., and Isner, J.M., Gene therapy with vascular endothelial growth factor for inoperable coronary artery disease: anesthetic management and results, *Anesth. Analg.* 92, 19–25, 2001.

78. Braun, S., Thioudellet, C., Rodriguez, P., Ali-Hadji, D., Perraud, F., Accart, N., Balloul, J.M., Halluard, C., Acres, B., Cavallini, B., and Pavirani, A., Immune rejection of human dystrophin following intramuscular injections of naked DNA in mdx mice, *Gene Ther.* 7, 1447–1457, 2000.

79. Rolland, A.P., and Mumper, R.J., Plasmid delivery to muscle: Recent advances in polymer delivery systems, *Adv. Drug Deliv. Rev.* 30, 151–172, 1998.

80. Mumper, R.J., Wang, J., Klakamp, S.L., Nitta, H., Anwer, K., Tagliaferri, F., and Rolland, A.P., Protective interactive noncondensing (PINC) polymers for enhanced plasmid distribution and expression in rat skeletal muscle, *J. Control. Release* 52, 191–203, 1998.

81. Mumper, R.J., Duguid, J.G., Anwer, K., Barron, M.K., Nitta, H., and Rolland, A.P., Polyvinyl derivatives as novel interactive polymers for controlled gene delivery to muscle, *Pharm. Res.* 13, 701–709, 1996.

82. Fewell, J.G., MacLaughlin, F., Mehta, V., Gondo, M., Nicol, F., Wilson, E., and Smith, L.C., Gene therapy for the treatment of hemophilia B using PINC-formulated plasmid delivered to muscle with electroporation, *Mol. Ther.* 3, 574–583, 2001.

83. Morse, M.A., Technology evaluation: VEGF165 gene therapy, Valentis Inc, *Curr. Opin. Mol. Ther.* 3, 97–101, 2001.

84. Mendiratta, S.K., Quezada, A., Matar, M., Wang, J., Hebel, H.L., Long, S., Nordstrom, J.L., and Pericle, F., Intratumoral delivery of IL-12 gene by polyvinyl polymeric vector system to murine renal and colon carcinoma results in potent antitumor immunity, *Gene Ther.* 6, 833–839, 1999.

85. Mendiratta, S.K., Quezada, A., Matar, M., Thull, N.M., Bishop, J.S., Nordstrom, J.L., and Pericle, F., Combination of interleukin 12 and interferon alpha gene therapy induces a synergistic antitumor response against colon and renal cell carcinoma, *Hum. Gene Ther.* 11, 1851–1862, 2000.

86. Alakhov, V., Klinski, E., Lemieux, P., Pietrzynski, G., and Kabanov, A., Block copolymeric biotransport carriers as versatile vehicles for drug delivery, *Expert Opin. Biol. Ther.* 1, 583–602, 2001.

87. Pitard, B., Pollard, H., Agbulut, O., Lambert, O., Vilquin, J.-T., Cherel, Y., Abadie, J., Samuel, J.-L., Rigaud, J.-L., Menoret, S., Anegon, I., and Escande, D., A nonionic amphiphile agent promotes gene delivery *in vivo* to skeletal and cardiac muscles, *Hum. Gene Ther.* 13, 1767–1775, 2002.

88. Liaw, J., Chang, S.F., and Hsiao, F.C., *In vivo* gene delivery into ocular tissues by eye drops of poly(ethylene oxide)-poly(propylene oxide)-poly(ethylene oxide) (PEO-PPO-PEO) polymeric micelles, *Gene Ther.* 8, 999–1004, 2001.

89. Johnston, T.P., and Miller, S.C., Toxicological evaluation of poloxamer vehicles for intramuscular use, *J. Parenter. Sci. Technol.* 39, 83–89, 1985.

CHAPTER 20

Use of Poly(N-vinyl pyrrolidone) with Noncondensed Plasmid DNA Formulations for Gene Therapy and Vaccines

Michael Nicolaou, Polly Chang, and Mark J. Newman

CONTENTS

20.1 INTRODUCTION

Products and the associated technologies for delivering genes into human tissues for gene therapy can be generally classified into two major categories: viral vectors and plasmid DNA (pDNA). The use of viral vectors and pDNA-based systems to deliver genes that encode vaccine immunogens has evolved, in part, from the field of gene therapy. For vaccination, the primary goal is the same as for gene therapy, to induce *in vivo* protein expression in the host. The most commonly ascribed goal of genetic vaccination is to provide a gene product to the vaccinee in a way that mimics intracellular pathogens and which is, therefore, capable of inducing Major Histocompatibility Complex (MHC) Class I and Class II restricted immune responses, cytotoxic T-lymphocyte (CTL) and helper T-lymphocyte (HTL) responses, respectively.

Viral vectors, which are typically based on attenuated or replication incompetent viruses, have been used for the delivery of multiple experimental gene therapy and vaccine products. Viral vectors were developed prior to pDNA because they have multiple intrinsic attributes that contribute to their utility for delivering genes, including the following:

0-8493-1934-X/05/$0.00+$1.50
© 2005 by CRC Press LLC

1. Their ability to condense DNA
2. Specific interaction with target cells mediated through interactions with cell-surface expressed receptors
3. Their ability to efficiently gain entry to cells, generally through a process that includes membrane fusion
4. Their ability to facilitate the transport of DNA into the cytoplasm and/or across nuclear membranes

Viral vector systems that have been or are currently under study for uses in gene therapy include, but are not limited to, adenovirus, adeno-associated virus, herpes viruses, retroviruses, and alphavirus.[1-3] Vectors based on these viruses are also being developed for vaccine delivery although vaccinia viruses, the related modified vaccinia virus ankara (MVA) and canary pox viruses, and adenoviruses are in the most advanced stages of development.[4,5]

Despite these attributes, a number of disadvantages are associated with the use of viral vectors in gene therapy and vaccine delivery. Limitations in both fields include manufacturing difficulties and related stability issues, pre-existing immunity against the vector and safety in target populations where the immune system may be suppressed to the point that attenuated viruses remain pathogenic, such as cancer chemotherapy patients and HIV-1 infected people. When used in vaccines, immunodominance of viral vector epitopes over the vaccine epitopes has also been a problem. Because of these limitations, nonviral gene delivery systems incorporating pDNA have continued to be developed for use in numerous vaccine fields.

20.2 DEVELOPMENT OF pDNA VACCINES

Initial studies of pDNA vaccines were completed using vaccine without delivery components, commonly referred to as naked DNA vaccines, and were tested in inbred mice, and other small animals, with promising results.[6-10] Based on these studies, several pDNA plasmid vaccines were developed for Phase I clinical trials. One of the first pDNA candidate vaccines tested encoded the *Plasmodium falciparum* circumsporozoite protein (PfCSP) and was developed for use against human malaria. An open label, dose-escalation trial of a PfCSP pDNA vaccine was completed in 20 healthy volunteers. The experimental vaccine was delivered by intramuscular needle injection, in phosphate buffered saline, 3 times at 4-week intervals using a dose escalation format, 20, 100, 500, and 2500 µg per dose, based on pDNA content.

There were no significant adverse events reported in this study, demonstrating the safety of the pDNA, or naked DNA, vaccine format. Unfortunately, the vaccine failed to induce measurable antibody responses and antigen-specific CTL responses were measured only sporadically, even within individuals.[11,12] The CTL responses were observed somewhat more often after the second immunization, compared with the first immunization, and following administration of the higher vaccine doses.

The same PfCSP pDNA malaria vaccine was subsequently evaluated using rabbits and then in a second clinical trial designed to allow for comparison of intramuscular needle injection to the use of a needle-free injection device (NFID), BioJector (BioJect Inc., Portland, OR). The vaccine again proved to be safe and immunogenicity was increased by delivery using the NFID.[13] It was assumed that the NFID increased the efficiency of vaccine uptake by cells, presumably through the increased level of localized destruction of muscle and skin at the injection site, and that this resulted in increased vaccine potency.

The findings of these clinical studies, and several other early studies completed using vaccines for HIV,[14-17] demonstrated that the pDNA vaccine format can safely deliver vaccine immunogens. However, the naked DNA vaccine format has typically failed to produce therapeutic or immunogenic amounts of protein reproducibly, presumably due to pDNA instability and inefficient uptake by cells with minimal nuclear delivery. These limitations mandate the use of additional delivery technologies that impart some, or all, of the properties associated with efficient delivery of genes by viral vectors.

20.3 DELIVERY OPTIMIZATION FOR pDNA VACCINES AND IMMUNE-SYSTEM-BASED THERAPEUTICS

While the need to increase the potency of pDNA vaccines is obvious, the mechanism for accomplishing this goal is not. There are several logical ways to improve the utility of pDNA vaccines, including the following:

1. The use of simple formulation excipients that increase delivery efficiency of pDNA delivery to cells and/or the nucleus[18-21]
2. Approaches to modifying the DNA plasmid, through the use of different promoters, or optimization of the gene encoding the vaccine immunogen so that it is more efficiently produced and/or targeted to the appropriate cellular compartments[22-25]
3. The use of biological response modifiers, such as common adjuvants, cytokines and costimulatory molecules, to augment, modify, or direct the development of particular immune responses following their induction[26-28]

These different approaches are not mutually exclusive and it is conceivable to incorporate multiple technologies into a single pDNA vaccine. However, when bias is applied toward vaccine safety, the use of conceptually simple devices, such as the NFID, or formulation excipients to increase vaccine delivery to cells and tissues is highly logical and the focus of significant effort within the field.

Many commonly used pDNA delivery materials have the ability to associate with pDNA and to condense it, thus providing a form that is more amenable to cellular uptake and less likely to be degraded extracellularly. For example, numerous polycations are known to interact ionically with pDNA and to induce condensation. These types of materials are usually lipids or liposomes and peptides, and they augment pDNA delivery to both the cellular cytoplasm and nucleus.[18-20,29-32] Formulation components to enhance delivery of pDNA to the nucleus, such as DNA-binding nuclear proteins and histone, can also be added as components of lipid-based formulations.[33,34] Thus, there are numerous pDNA-delivery formulations where condensation of the plasmid is central to their function.

Biodegradable polymeric encapsulation, based primarily on the use of poly(lactide-*co*-glycolide) (PLG), represents an alternative approach. The logic behind the use of polymeric microspheres is that noncondensed, encapsulated pDNA is directly targeted to phagocytic cells at the injection site or draining lymph nodes due to the particulate nature of the formulation. These phagocytic cells are presumably macrophages or dendritic cells which represent professional and very efficient antigen-presenting cells (APCs)[35] and as such, this delivery approach obviates expression of the plasmid in muscle or skin cells.

When pDNA is encapsulated, it is protected from extracellular degradation until the polymeric structure of the microsphere breaks down, presumably within a cell. *In vitro* experiments, based on the use of a murine macrophage cell line (P388D1 cells), were used to demonstrate that encapsulated pDNA encoding a reporter gene, luciferase, was released and expressed intracellularly over a period of 3 to 5 days following phagocytosis.[36] Companion immunogenicity studies completed in mice clearly support the value of this type of vaccine delivery. Measurable CTL responses were induced following a single vaccination with encapsulated pDNA vaccine, which were greater than responses induced after two immunizations with nonformulated DNA and protection in an experimental tumor challenge model was demonstrated.[37,38] These data were used to support the development and testing of a therapeutic cervical cancer pDNA vaccine based on human papilloma virus gene. While it is difficult to compare data reported from different clinical trials, this PLG-encapsulated vaccine appeared to be at least as potent for inducing CTL responses as the PfCSP pDNA vaccine for malaria and, more importantly, clinical benefit was reported.[36,39] Thus, the utility of a delivery technology that is not dependent on condensation of pDNA has been established.

20.4 DEVELOPMENT OF POLY(N-VINYL PYRROLIDONE) (PVP) AS A pDNA DELIVERY AGENT

Poly(N-vinyl pyrrolidone) (PVP) is a synthetic homopolymer with mild adhesive properties; it is "sticky" when wet. It is water-soluble, with its maximum concentration being limited only by viscosity, and it does not form micelles or particulate structures in standard formulations. PVP is commonly used in the pharmaceutical industry to formulate drugs that are poorly soluble in aqueous buffers. The most common use is as a tablet binder and tablet coating aid, but it is also used in formulations of antibiotics, chemotherapeutic drugs, hormones, and ophthalmic and topical solutions. It is produced in a wide molecular weight range, 8–1300 kDa, and sold under several trade names including Plasdone (International Specialty Products, Wayne, NJ) and Kollidon (BASF, Ludwigshafen, Germany). PVP used for pDNA delivery and DNA vaccines is typically in the size range of 10–50 kDa.[40]

PVP can bind to the base pairs in the major groove of DNA, through simple hydrogen bonding, at pH 4–6. The PVP–pDNA complex is more hydrophobic, and less negatively charged, than nonformulated pDNA, presumably due to the vinyl backbone of the polymer (Figure 20.1). This interaction between PVP and pDNA does not induce plasmid condensation, which distinguishes it from cationic materials. PVP-formulated pDNA differs from that formulated in PLG because the final product is not particulate nor is the DNA encapsulated. However, the binding of PVP to pDNA decreases its susceptibility to extracellular degradation and thereby increases the likelihood of uptake by cells.[40,41]

We believe the mechanism through which PVP improves the efficiency of DNA uptake by cells is multifactorial. First, the increased hydrophobicity and charge shielding effect contributed by the vinyl backbone of PVP should cause the pDNA complex to become less accessible to the active sites of degradative enzymes, which have evolved to interact with the charged hydrophilic backbone of DNA. This will increase the stability and half-life of the pDNA *in vivo*. Second, shielding of the pDNA negative charge, as the result of PVP presence, should decrease charge-based repulsion between the pDNA and the anionic cellular membrane, which would increase the likelihood of

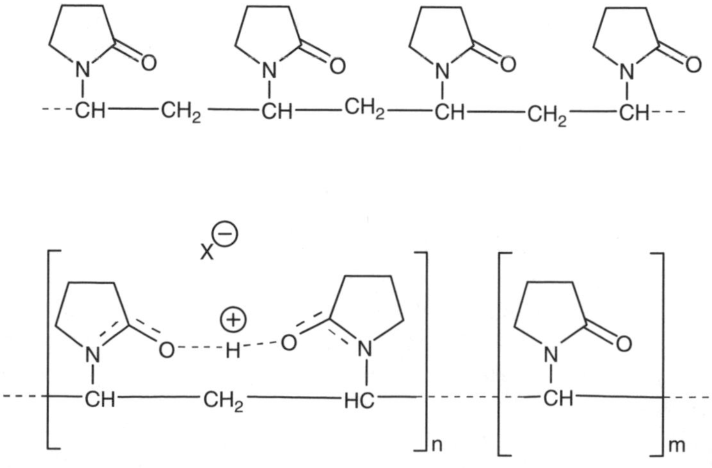

Figure 20.1   The chemical structure of PVP. Under acidic conditions, PVP forms positively charged complexes that pair with anions (X-). A representative example is the disinfectant povidone iodine where X- is I_3^-.

interaction with cellular membranes and subsequent internalization. Third, hydrogen bonding of water with pDNA will be reduced in the presence of PVP. The unfavorable desolvation energy needed, as pDNA partitions into hydrophobic membranes, including the nuclear membrane, will therefore be minimized. Thus the energetics of the pDNA–PVP complex for membrane diffusion will be favored. Finally, the efficiency of intracellular delivery of the pDNA following pinocytosis may be positively affected by PVP because it should inhibit fusion of pinocytotic vesicles and endosomes with lysosomes, which are a known site of degradation within cells. Alternatively, direct destabilization of pinocytotic vesicles and endosomes would promote release of the pDNA into the cytoplasm. Either way, these events would increase the probability of intact pDNA reaching the nucleus.

A series of studies were completed using rats, pDNA encoding the β-galactosidase gene, and formulations that were varied with respect to PVP concentration, pH, and salt concentration. Following intramuscular delivery, β-galactosidase activity was measured and used to determine relative amounts of pDNA uptake. The PVP-formulated pDNA was effectively taken up by muscle cells, approximately 10-fold more efficiently than nonformulated DNA.[40,42] In these studies, formulations at pH 4 and a DNA:PVP ratio of approximately 1:17 proved to be most effective. The use of PVP, and other structurally related materials used as formulation excipients for pDNA delivery, was developed as the Polymer Interaction Non-Condensing (PINC) system by Gene Medicine, Inc. (Houston, TX) and subsequently Valentis, Inc. (Burlingame, CA).[18]

PVP was tested with success in a murine model of cancer immunotherapy using two different tumors: Renca, a spontaneously arising murine renal cell carcinoma, and CT26, a poorly immunogenic adenocarcinoma. Formulations consisting of PVP supplemented pDNA encoding human cytokines, specifically interleukin-12 (IL-12) or alpha-interferon (α-IFN), were administered directly into established tumors.[43-46] Readily detectable levels of these cytokines were produced in the tumors 24 h after injection, and antitumor effects, mediated primarily by CD8+ T-lymphocytes, presumably CTL, were induced by treatments involving the IL-12 encoding plasmid or a combination of both plasmids. Cellular infiltration of tumor sites by activated CD4+ and CD8+ T-lymphocytes, HTL and CTL respectively, as well as natural killer cells and macrophages, or potentially dendritic cells, was also associated with the treatment. Comparison studies using the plasmids encoding the cytokines without PVP were not completed so direct evidence of the effect of the PVP was not demonstrated. However, comparison of protein expression data from the individual studies indicates that the levels of pDNA delivery using PVP were likely comparable to, or exceeded 2- to 20-fold, those obtained using the cationic lipid (DOTMA). Valentis, Inc., evaluated the IL-12 pDNA tumor immunotherapy therapy product in a Phase I clinical trial for the treatment of squamous cell carcinoma of the head and neck; results of the study have not yet been published.

The pDNA encoding β-galactosidase gene was also used in experimental vaccine studies completed using pigs and beagle dogs wherein the β-galactosidase activity in muscle and immune responses, antibody production, and T-lymphocyte proliferation were measured. The use of a NFID was also incorporated into the study to investigate potential synergistic effects of these two delivery methods. The incorporation of the PVP into the formulation increased the potency of the pDNA vaccine; antibody titers were 3- to 4-fold higher than those induced using nonformulated pDNA. When the PVP complexed pDNA was delivered with a NFID, antibody titers were increased more than 15- to 20-fold, compared to nonformulated pDNA, demonstrating a synergistic effect of PVP and NFID.[47] Interestingly, the use of the NFID did not appear to significantly increase the amount of gene delivered to muscle cells, or subsequent expression of the gene product. The increased potency of the vaccines may be due to delivery of the pDNA to other resident cell types such as APC. Thus, PVP may function to deliver the pDNA to muscle, skin, and potentially, directly to APC.

20.5 USE OF PVP IN T-LYMPHOCYTE EPITOPE-BASED pDNA VACCINES

A clear understanding of how T-lymphocytes recognize antigens has emerged over the past decade. It is now well established that small fragments of protein antigens are generated, defined as peptide epitopes, which bind to MHC molecules expressed on the cell surface. These epitope–MHC complexes represent the ligands recognized by T-lymphocytes through the function of T-cell receptors (TCR).[48,49] The CD8+ CTL recognize epitope peptides bound to MHC Class I molecules, whereas the CD4+ HTL recognize epitope peptides bound to MHC Class II antigens. The Epimmune, Inc., approach to vaccine design is based on the use of highly defined epitopes that are used in sets.

There are several vaccine delivery methods amenable for use with epitopes. Synthetic peptides representing CTL or HTL epitopes derived from HIV-1 have been tested in clinical trials delivered in the high quality Incomplete Freund's Adjuvant[50,51] or as lipidated peptides.[52] The numbers of epitopes that can be included in a vaccine is generally limited to only a few and these types of vaccine formats can be difficult to manufacture and maintain as stable formulations. We believe the incorporation of multiple CTL epitopes restricted to numerous HLA types into a vaccine is key to developing this approach for generalized use in human vaccines and therapeutics. Thus, we elected to investigate other vaccine delivery formats.

An alternative approach to multi-epitope vaccine design is to construct "minigenes" and to produce these as pDNA vaccines or viral vectored vaccines. Vaccines of this format were developed in a relatively predictable stepwise manner by several research groups.[53-57] The unique attributes of the Epimmune program include the use of amino acid spacers between T-lymphocyte epitopes, to optimize proteosomal processing,[58,59] and incorporation of the universal HTL epitope, termed PADRE, to augment CTL responses.[57,60] Since the Epimmune vaccines are composed entirely of T-lymphocyte epitopes restricted to human HLA, they cannot be readily tested in common animals models. We therefore utilize HLA-A*0201, -A*1101, and -B*0702 transgenic mice to assess immune responses to CTL epitopes.[57] The most advanced multi-epitope pDNA vaccine in the Epimmune pipeline is the EP HIV-1090 vaccine for HIV therapy and prophylaxis (Figure 20.2).[61] This pDNA vaccine encodes 21 CTL epitopes, 7 each restricted to HLA-A*0201, -A*1101, and -B*0702, and other related HLA allelic products.

The HLA-A transgenic mice will often respond very well to certain CTL epitopes but only poorly to others that are known to be immunogenic in humans. Thus, the successful outcome of testing PVP in multi-epitope pDNA vaccines is augmentation of responses to the less immunogenic epitopes while not impacting responses to others. With this goal in mind, testing was completed using HLA-A*0201 transgenic mice and the clinical candidate HIV vaccine, EP HIV-1090.

Initial studies were completed to assess basic formulation parameters because of our bias to formulate vaccines at pH 7; formulations commonly used for gene therapy applications by Gene Medicine and Valentis were produced at pH 4.[40,41] Acetate or phosphate buffered formulations, ranging from pH 4–8, with a 1:17 ratio of pDNA:PVP (1 mg/ml pDNA), were tested for immunogenicity. Surprisingly, pDNA vaccines formulated at pH 4 were inactive whereas those formulated at higher pH were immunogenic (Figure 20.3), although this is in contrast to results observed for gene therapy formulations. Different ratios of pDNA:PVP were also tested, ranging from 1:10 to 1:20, but no effect on vaccine potency was observed (data not shown). Thus, pH 7. 0 and the 1:17 pDNA:PVP ratio were selected for use in the clinical product.

The clinical product was formulated as 2 mg/ml EP HIV-1090 pDNA vaccine, 3.4% PVP in 100 mM sodium phosphate (pH 7. 0) and 150 mM NaCl. The effect of PVP on the potency of HLA-A*0201 restricted epitopes in HLA transgenic mice is shown in Figure 20.4. The responses to 3 of the 4 epitopes that were weakly to moderately immunogenic in the nonformulated pDNA vaccine, Vpr 62, Gag 386, Pol 448, and Pol 498, were increased approximately tenfold, while responses to other epitopes were not impacted significantly; CTL responses were measured as a function of γ-INF production. These responses were induced after only a single immunization with

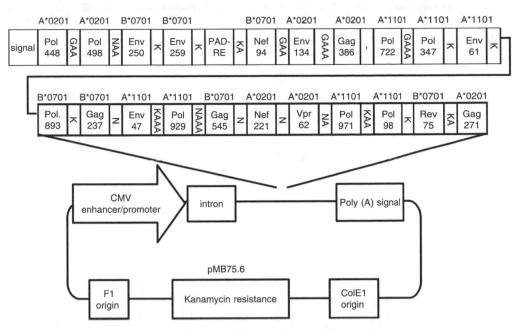

Figure 20.2 Schematic of EP HIV-1090 pDNA vaccine. The epitopes included in the vaccine insert are illustrated. The HLA restriction is shown above each epitope. The flanking amino acids inserted between epitopes are also shown. The vaccine insert is cloned into the multiple cloning sites of the expression backbone pMB75.6. The functional elements of the expression vector are also indicated.

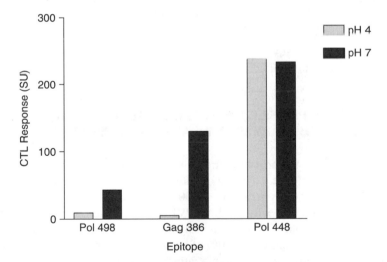

Figure 20.3 PVP augments the potency of CTL epitopes in the EP-HIV-1090 pDNA vaccine at neutral pH. EP-HIV-1090 (100 μg per mouse) formulated with PVP at pH 4 and pH 7 was used to immunize HLA-A*0201 transgenic mice. Eleven days after immunization, splenocytes were harvested and stimulated *in vitro* for 6 days with peptides (1 μg/ml) representing epitopes encoded in the pDNA vaccine and irradiated splenocytes that had been activated for 3 days with LPS and dextran sulfate. After 6 days of *in vitro* culture, CTL were washed once and tested for γ- IFN production in response to peptide stimulation in an *in situ* γ-IFN ELISA.[76]

a suboptimal dose of the vaccine, 20 μg per mouse. As such, the measured CTL responses represent primary responses. Due to the nontoxic nature of the PVP, repeat immunizations can be administered at different sites or repeat administrations can be given within 24 h and response levels can be

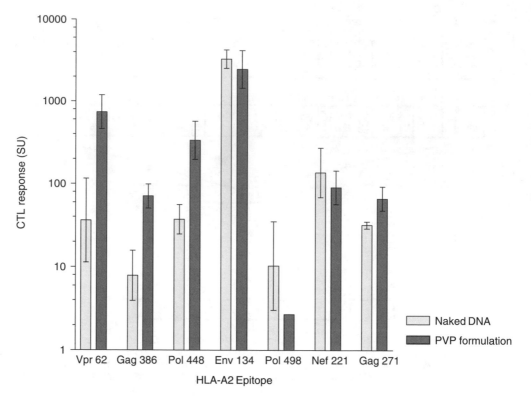

Figure 20.4 PVP augments the potency of weakly to moderately CTL epitopes in the EP-HIV-1090 pDNA vaccine. EP-HIV-1090 (20 µg per mouse) formulated with and without PVP was used to immunize HLA-A*0201 transgenic mice. Eleven days after immunization, splenocytes were harvested and the *in situ* γ-IFN ELISA used to assess CTL responses.

increased. This is a significant advantage when compared to delivery products with less desirable toxicity profiles, such as cationic lipids or viral vectors.

20.6 SAFETY PROFILE OF PVP USED AS A DRUG AND GENE THERAPY FORMULATION EXCIPIENT

As noted, PVP is used in the pharmaceutical industry as a formulation excipient and as such, there exists an extensive database on product safety;[62] PVP has been shown to be essentially nontoxic. The oral LD_{50} for PVP is greater than 100 gm/kg for rats and 40 gm/kg for mice and 12 gm/kg when delivered intraperitoneally to mice. PVP is biocompatible but not biodegradable and it is therefore cleared intact from the body, primarily though the kidneys, at rates that increase with molecular weight. For example, more than 90% of low molecular weight PVP (30 kDa) is typically excreted within 72 h after intravenous administration, whereas higher molecular weight PVP ($\geq$ 100 kDa) requires 7–10 days to be excreted to similar levels.

Studies supporting product development that are probably most relevant for use in vaccines and gene therapy are associated with parenterally administered drugs, such as injectable forms of hormones, antibiotics, and analgesics. Low molecular weight (25 kDa) pyrogen-free PVP has been most widely used in these types of products, without significant toxicity. PVP with molecular weight > 25 kDa, material similar to the Plasdone used in the EP HIV-1090 pDNA vaccine, is used as a formulation excipient for some intramuscularly injected drugs, such as vasopressin, to delay adsorption from the injection site, again without toxicity.[62] It is interesting to note that this effect

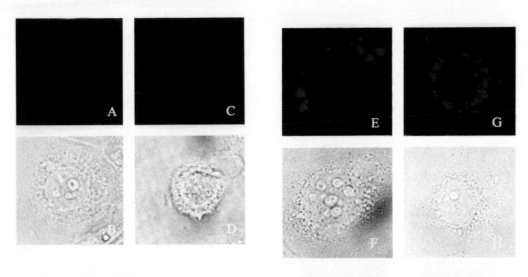

Figure 8.6

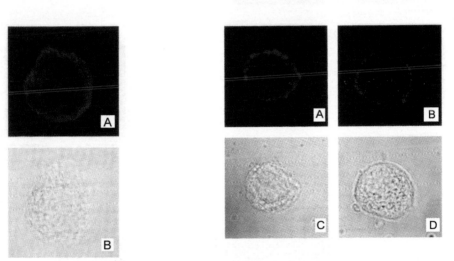

Figure 8.7   **Figure 8.8**

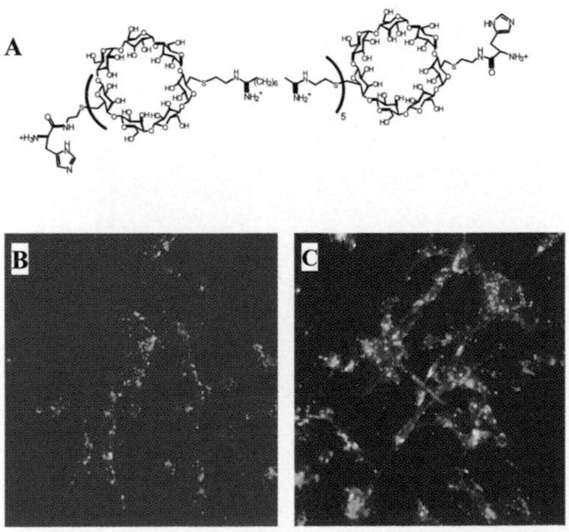

Figure 12.10

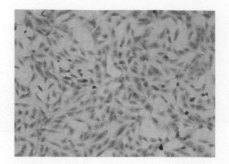

Figure 16.9

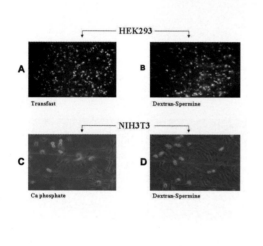

HEK293

A Transfast

B Dextran-Spermine

NIH3T3

C Ca phosphate

D Dextran-Spermine

Figure 17.8

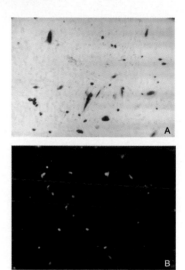

A

B

Figure 30.6

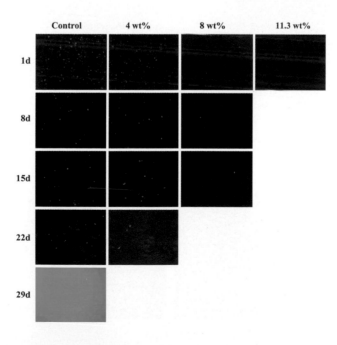

	Control	4 wt%	8 wt%	11.3 wt%
1d				
8d				
15d				
22d				
29d				

Figure 31.9

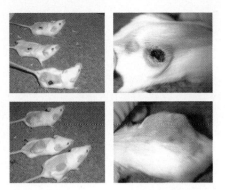

Figure 34.4

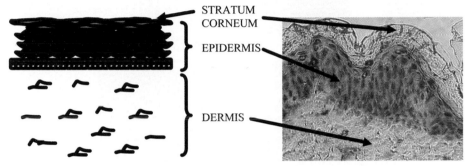

STRATUM CORNEUM

EPIDERMIS

DERMIS

Figure 36.1

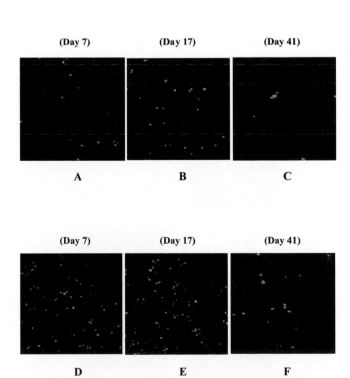

Figure 37.11

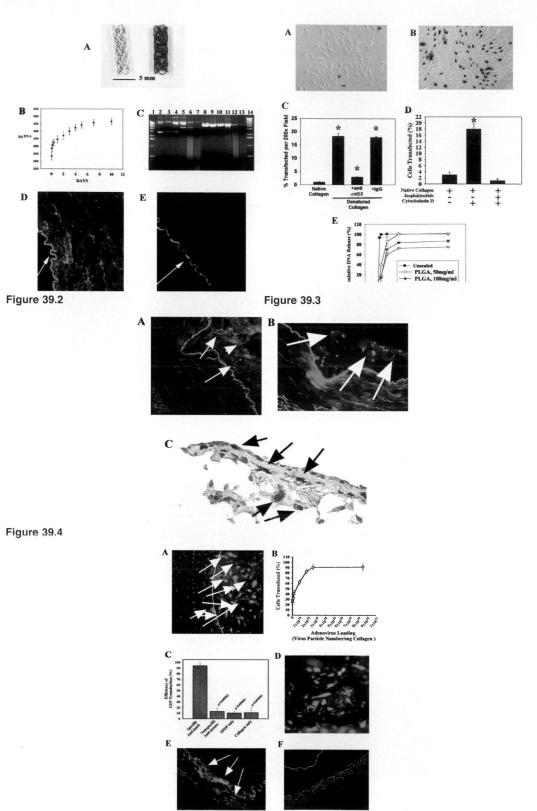

Figure 39.2

Figure 39.3

Figure 39.4

Figure 39.5

of retaining drugs in injected muscle likely contributes to the utility of PVP in pDNA vaccines administered similarly.

As noted previously, Valentis developed a gene therapy pDNA based product designed to produce the cytokine IL-12 *in vivo* in head and neck cancer patients and toxicity studies for the PVP–pDNA formulation were completed in cynomolgus monkeys.[63] Animals received doses of up to 18 mg/kg, administered subcutaneously twice weekly for 4 weeks, and were evaluated at multiple time points, including after a 4-week treatment recovery period. Total PVP dose for these animals over the 4-week treatment was 144 mg/kg. Weights of the male animals ranged from 2.4 to 4.0 kg, which means the cumulative PVP dose could have ranged from 346 to 576 mg. Toxic effects associated with administration of the formulation were not observed; the results of serum chemistry, hematology, urinalysis, and coagulation testing were unaffected. The authors of this report stated that the use of PVP at the level of 5% in the formulation was safe.

For the EP HIV-1090 Phase I clinical studies, the highest monthly dose will be 68 mg, for an individual volunteer, not per kg, and the highest cumulative dose over the study period will be 272 mg, assuming 4 vaccinations are administered. Clearly, the anticipated dose levels of PVP needed for pDNA vaccines are well below those established to be same in animal studies.

20.7 SAFETY TESTING OF PVP FORMULATED EP HIV-1090 pDNA VACCINE FOR HIV THERAPY AND PROPHYLAXIS

Criteria for the types and design of laboratory animal-based preclinical safety studies needed to support clinical testing of pDNA vaccines are now well established in the industry. They typically include the measurements of pDNA biodistribution and clearance after one or more doses followed by genomic integration studies in tissues where total clearance is not observed. Generalized toxicity testing following multiple vaccine doses with assessments completed immediately after treatment, acute phase, and following a rest period, recovery phase, is also commonly employed. Animal species used include mice,[39,64-66] rabbits[67,68] and nonhuman primates.[69] Using these testing approaches, nonformulated pDNA vaccines were shown to be safe and nontoxic, except for some sporadic, but transient, local reactogenicity due to tissue trauma associated with intramuscular injections.

The repeat immunization study is a standard requirement for vaccine testing (Table 20.1). This study was designed to determine the safety of the EP HIV-1090 vaccine formulation at a dose that exceeds the highest dose to be used in the clinical trial; "human" dose equivalents are administered, and the vaccine dose is not adjusted for weight differences. The clinical trial was designed as a dose escalation using doses ranging from 0.5 mg/dose to 4 mg/dose. For the animal studies, dose levels of 1 mg and 6 mg per injection were selected. Five injections were administered over a 44 day period, which is a dosing schedule that is compressed when compared to the proposed clinical dose schedule of 1 injection every 4 weeks. The vaccine was administered at a uniform concentration of 2 mg/ml, with the control and high dose animals receiving a total of 6 injections in volumes of 0.5 ml at each time point. The low-dose animals received a single 0.5 ml injection at each immunization time point. Half the animals (3 per sex per group) were sacrificed 3 days after the last immunization and the remaining animals were allowed to recover for approximately 4 weeks before necropsy.

Measurement of toxicity was completed using physical examination of the injection site, monitoring of body weight, ophthalmologic exam before and after treatment, clinical chemistry and hematology, urinalysis, and necropsy, with detailed histopathological evaluation of all major organs and tissues. Significant, treatment-associated alterations in normal clinical chemistry of hematology parameters or damage to tissues and/or organs were not observed for any of the animals. Thus, the EP HIV-1090 vaccine formulation did not induce systemic acute or chronic toxicity. All

Table 20.1 Safety Studies to Evaluate the EP HIV-1090 pDNA Vaccine

Study Design	Test System	Route[a]	Dose[b]	Dose Multiple of Human Dose[c]	Number of Treatments per Animal	Treatment Period	Assay Time Points
1. Biodistribution	Rabbit	IM			1	1 day	1 week
a. Low dose			1 mg	7			8 weeks
b. High dose			6 mg	42			
2. Integration	Rabbit	IM	6 mg	42	1	1 day	10 weeks
							16 weeks
3 Repeat dose	Rabbit	IM			5	6 weeks	6 weeks
a. Low dose			5 mg	8. 7			(acute)
b. High dose			30 mg	52. 5			10 weeks (recovery)

[a] IM = Intramuscular.

[b] Cumulative dose.

[c] Based on total plasmid DNA and on a mg/kg ratio assuming the average human at 70 kg and the average rabbit at 2.5 kg. For the biodistribution and integration studies, the dose multiple was calculated based on a single human dose of 4 mg, the highest single dose planned for this clinical study. The dose multiple for the repeated-dose safety studies is based on the cumulative highest dose from the multiple injections; four injections will be administered to study volunteers and, therefore, the highest dose will be 4 mg, for a cumulative dose of 16 mg.

animals survived until their scheduled necropsy and no test article-related effects on body temperature, body weights, or food consumption were observed.

The typical biodistribution study is designed to determine the fate of the injected pDNA vaccine, specifically distribution throughout the body and the rate of clearance. The distribution of EP HIV-1090 plasmid vaccine following a single intramuscular immunization in rabbits was evaluated at 2 doses, 1 and 6 mg of EP HIV-1090 in PVP. Tissues were harvested 1, 7, and 56 days after treatment. Control animals were injected with the vehicle, consisting of 3.4% PVP in phosphate buffered saline, and necropsied at the same time points as the test animals. Vaccine-treated and vehicle control groups consisted of five animals per sex for each dose and each observation period. To evaluate the distribution of the pDNA, a PCR-based assay with a detection limit of 10 copies of plasmid per 0.1 μg of genomic DNA was used with multiple tissues at each time point after immunization.[70] The presence of EP HIV-1090 was evaluated in the skin and muscle of the injection site, lymph nodes, spleen, bone marrow, liver, thymus, kidney, heart, brain, lungs, gonads, and plasma. Quantitation of the plasmid was based on SYBR Green double-stranded DNA binding fluorescence techniques.

One day after a single high dose treatment with EP HIV-1090, approximately 10^8 copies of the plasmid were detected in 0.1 μg of genomic DNA obtained from the injection site skin samples and this level was maintained for the first week (Figure 20.5). An average of approximately 10^6 copies of the plasmid were detected in same amount of genomic DNA obtained from the injection site muscle samples and this level dropped to an average of approximately 10^4 copies 1 week later. Levels of pDNA recorded 1 week after administrations are assumed to definitively represent pDNA taken up by cells. Up to 10^6 copies of plasmid were detected in 0.1 μg of genomic DNA obtained draining lymph nodes and lower levels were detected all other distal organs tested one day after administration. The pDNA was cleared from most of these tissues seven days after injection, with less than 10^3 copies per 0.1 μg genomic DNA detected; no pDNA was detected in these tissues on study day 56. A similar pattern of pDNA distribution and clearance was noted for the low dose animals with proportionately lower copy numbers of pDNA measured (Figure 20.6).

Approximately 10^3 to 10^4 copies of EP HIV-1090 pDNA persisted in 0.1 μg of genomic DNA in the site of injection tissues 8 weeks after injection. These tissues were therefore further examined to evaluate the possibility of genomic integration of the EP HIV-1090 pDNA. Genomic DNA was

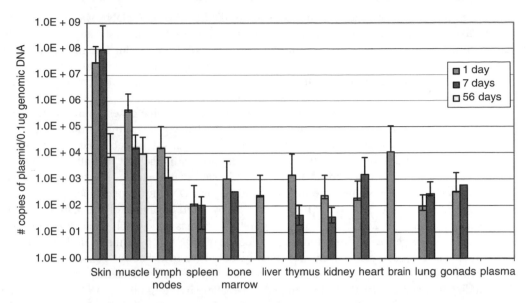

Figure 20.5 Kinetics of distribution and clearance of EP HIV-1090 pDNA vaccine from multiple tissues following high dose administration. Testing was completed following a single high dose (6 mg) treatment. Each data point designates the mean number of copies of pDNA in 0.1μg of genomic DNA for each tissue at the time of necropsy plus the standard error. Each sample was tested at least two times and tissues from 10 animals were tested for each time point.

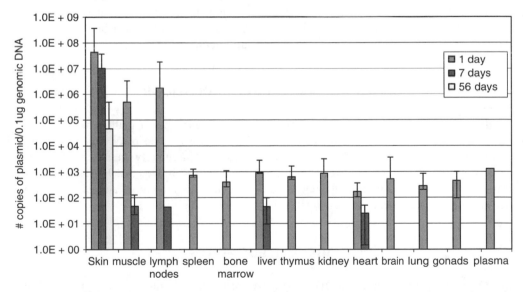

Figure 20.6 Kinetics of distribution and clearance of EP HIV-1090 pDNA vaccine from multiple tissues following low dose administration. Testing was completed following a single low dose (1 mg) treatment. Each data point designates the mean number of copies of pDNA in 0.1 μg of genomic DNA for each tissue at the time of necropsy plus the standard error. Each sample was tested at least two times and tissues from 10 animals were tested for each time point.

extracted from injection site tissues obtained on study days 70 and 112 and assayed for pDNA. Samples were restriction digested with the restriction enzyme Xho I, which is a "rare base cutter" for genomic DNA that cleaves EP HIV-1090 plasmid concatemers into linear monomers. The digested genomic DNA samples were gel purified and PCR analyzed. With the exception of one animal, a specific band at the expected amplicon size for the EP HIV-1090 pDNA was not observed, indicating that all of the detectable plasmid was extra-chromosomal and not integrated into the muscle genome. After 3 rounds of gel purification of the DNA from the final animal, fewer than 20 copies of EP HIV-1090 appeared to be present per 0.1 μg of genomic in the skin genome. Thus, the results of these PCR studies demonstrate that the detected pDNA was only loosely associated with the genomic DNA, and therefore not integrated into the genome.

Differences in study design, such as animal species used and vaccine doses administered, make it difficult to compare data and to determine the potential advantages and limitations of different pDNA vaccine delivery technologies accurately. However, comparison of published data do reveal certain similarities and differences.[66,67,71-74] Of prime importance is the fact that all of the pDNA products evaluated thus far have proved to be generally safe and nontoxic. For both formulated and nonformulated pDNA vaccines, the initial spread of the pDNA to many tissues immediately after injection is commonly observed. This is followed by rapid clearance of the pDNA from all tissues, usually within 1 to 2 weeks, except injection site tissues and lymph nodes that drain the injection sites. With respect to the injection site, the decline of pDNA copy numbers in skin generally occurs at a slower rate, compared to muscle. Finally, definitive evidence for genomic integration has not been observed. Thus, the use of pDNA vaccines appears to be inherently safe.

Our studies indicate that PVP-formulated pDNA behaves similar to other pDNA products after intramuscular injection. The efficiency of pDNA uptake by muscle and skin cells appears to be inceased and the rate of clearance slower than for many of the other products. This may reflect increased levels of pDNA uptake by cells compared to nonformulated pDNA, which is an idea supported by the biodistribution study data, and if this is true, vaccine potency should be augmented. However, delayed clearance may also indicate that the rabbits in the study were incapable of inducing immune responses to the vaccine because it is based totally on epitopes restricted to human HLA; immune system mediated destruction of transfected muscle cells is known to contribute to pDNA vaccine clearance.[75] Data obtained in clinical studies will be the only realistic measure of pDNA vaccine utility and such studies are ongoing.

20.8 SUMMARY AND CONCLUSIONS

Vaccines based on pDNA hold significant promise for use against pathogens where cellular immune responses are needed. Vaccines composed of nonformulated pDNA have proved to be immunogenic in several species of laboratory animals but only marginally immunogenic in humans. The incorporation of PVP as a delivery agent is a logical approach due to its low toxicity and ease of formulation. Safety and immunogenicity studies completed in laboratory animals have provided data to support the continued development of PVP–pDNA vaccines.

ACKNOWLEDGMENTS

Data specific to the Epimmune HIV program were generated with the support of a grant (Dr. Newman, P01-AI48238) and access to contracted services (Dr. Chang) from the Division of AIDS, National Institute of Allergy and Infectious Diseases, National Institutes of Health.

REFERENCES

1. Markovitz, N.S. and Roizman, B., Replication-competent herpes simplex viral vectors for cancer therapy, in *Advances in Virus Research*, Maramorosch, K., Murphy, F.A., and Shatkin, A.J. (Eds.), Academic Press, New York, NY, 2000, 409.

2. Hitt, M. and Graham, F.L., Adenovirus vectors for human gene therapy, in *Advances in Virus Research*, Maramorosch, K., Murphy, F.A., and Shatkin, A.J. (Eds.), Academic Press, New York, NY, 2000, 479.

3. Schlesinger, S., Alphavirus expression vectors, in *Advances in Virus Research*, Maramorosch, K., Murphy, F.A., and Shatkin, A.J. (Eds.), Academic Press, New York, NY, 2000, 565.

4. Cairns, J.S. and Sarver, N., New viral vectors for HIV vaccine delivery, *AIDS Res. Hum. Retrovir.* 14, 1501, 1998.

5. Voltan, R. and Robert-Guroff, M., Live recombinant vectors for AIDS vaccine development, *Curr. Mol. Med.* 3, 273, 2003.

6. Davis, H.L., Michel, M.L., Mancini, M., Schleef, M., and Whalen, R.G., Direct gene transfer in skeletal muscle: plasmid DNA-based immunization against the hepatitis B virus surface antigen, *Vaccine* 12, 1503, 1994.

7. Robinson, H.L. and Pertmer, T.M., DNA vaccines for viral infections: basic studies and applications, in *Advances in Virus Research*, Maramorosch, K., Murphy, F.A., and Shatkin, A.J. (Eds.), Academic Press, New York, NY, 2000, 1.

8. Weiner, D. B. and Kennedy, R. C. Genetic vaccines, *Sci. Am.* 281, 50, 1999.

9. Donnelly, J.J., Ulmer, J.B., and Liu, M.A., Immunization with DNA, *J. Immunol. Methods* 176, 145, 1994.

10. Moelling, K., DNA for genetic vaccination and therapy, *Cytokines Cell Mol. Ther.* 3, 127, 1997.

11. Wang, R., Doolan, D., Le, T., Hedstrom, R., Coonan, K., Charoenvit, Y., Jones, T., Hobart, P., Margalith, M., Ng, J., Weiss, W., Sedegah, M., De Taisne, C., Norman, J., and Hoffman, S., Induction of antigen-specific cytotoxic T lymphocytes in humans by a malaria DNA vaccine, *Science* 282, 476, 1998.

12. Le, T.P., Coonan, K.M., Hedstrom, R.C., Charoenvit, Y., Sedegah, M., Epstein, J.E., Kumar, S., Wang, R., Doolan, D.L., Maguire, J.D., Parker, S.E., Hobart, P., Norman, J., and Hoffman, S.L., Safety, tolerability and humoral immune responses after intramuscular administration of a malaria DNA vaccine to healthy adult volunteers, *Vaccine* 18, 1893, 2000.

13. Wang, R., Epstein, J., Baraceros, F.M., Gorak, E.J., Charoenvit, Y., Carucci, D.J., Hedstrom, R.C., Rahardjo, N., Gay, T., Hobart, P., Stout, R., Jones, T.R., Richie, T.L., Parker, S.E., Doolan, D.L., Norman, J., and Hoffman, S.L., Induction of CD4(+) T cell-dependent CD8(+) type 1 responses in humans by a malaria DNA vaccine, *Proc. Nat. Acad. Sci. U.S.A.* 98, 10817, 2001.

14. Boyer, J.D., Chattergoon, M.A., Ugen, K.E., Shah, A., Bennett, M., Cohen, A., Nyland, S., Lacy, K.E., Bagarazzi, M.L., Higgins, T.J., Baine, Y., Ciccarelli, R.B., Ginsberg, R.S., MacGregor, R.R., and Weiner, D.B., Enhancement of cellular immune response in HIV-1 seropositive individuals: a DNA-based trial, *Clin. Immunol.* 90, 100, 1999.

15. MacGregor, R.R., Boyer, J., Ugen, K., Lacy, K., Gluckman, S., Bagarazzi, M., Chattergoon, M., Baine, Y., Higgins, T., Ciccarelli, R., Coney, L., Ginsberg, R.S., and Weiner, D., First human trial of a DNA-based vaccine for treatment of Human Immunodeficiency Virus Type 1 infection: safety and host response, *J. Infect. Dis.* 178, 92, 1998.

16. Boyer, J.D., Cohen, A.D., Vogt, S., Schumann, K., Nath, B., Ahn, L., Lacy, K., Bagarazzi, M.L., Higgins, T.J., Baine, Y., Ciccarelli, R.B., Ginsberg, R.S., MacGregor, R.R., and Weiner, D.B., Vaccination of seronegative volunteers with a Human Immunodeficiency Virus Type 1 env/rev DNA vaccine induces antigen-specific proliferation and lymphocyte production of β-chemokines, *J. Infect. Dis.* 181, 476, 2000.

17. Calarota, S., Leandersson, A.C., Bratt, G., Hinkula, J., Klinman, D.M., Weinhold, K.J., Sandström, E., and Wahren, B., Immune responses in asymptomatic HIV-1-infected patients after HIV-DNA immunization followed by highly active antiretroviral treatment, *J. Immunol.* 163, 2330, 1999.

18. Rolland, A.P., From genes to gene medicines: recent advances in nonviral gene delivery, *Crit. Rev. Ther. Drug Carrier Syst.* 15, 143, 1998.

19. Mahato, R.I., Rolland, A.P., and Tomlinson, E., Cationic lipid-based gene delivery systems: pharmaceutical perspectives, *Pharm. Res.* 14, 853–859, 1997.

20. Felgner, P.L., Tsai, Y.J., Sukhu, L., Wheeler, C.J., Manthrope, M., Marshall, J.A., and Cheng, S.H., Improved cationic lipid formulations for *in vivo* gene therapy, *Ann. NY Acad. Sci.* 27, 126, 1995.

21. O'Hagan, D., Singh, M., Ugozzoli, M., Wild, C., Barnett, S., Chen, M., Schaefer, M., Doe, B., Otten, G.R., and Ulmer, J.B., Induction of potent immune responses by cationic microparticles with adsorbed Human Immunodeficiency Virus DNA vaccines, *J. Virol.* 75, 9037, 2001.

22. Fu, T.M., Guan, L., Friedman, A., Ulmer, J.B., Liu, M.A., and Donnelly, J.J., Induction of MHC class I-restricted CTL response by DNA immunization with ubiquitin-influenza virus nucleoprotein fusion antigens, *Vaccine* 16, 1711, 1998.

23. Yokoyama, M., Zhang, J., and Whitton, J.L., DNA immunization: Effects of vehicle and route of administration on the induction of protective antiviral immunity, *FEMS Immunol. Med. Microbiol.* 14, 221, 1996.

24. Rodriguez, F., Harkins, S., Redwine, J.M., de Pereda, J.M., and Whitton, J.L., CD4(+) T cells induced by a DNA vaccine: immunological consequences of epitope-specific lysosomal targeting, *J. Virol.* 75, 10421, 2001.

25. Whitton, J.L., Rodriguez, F., Zhang, J., and Hassett, D.E., DNA immunization: mechanistic studies, *Vaccine* 17, 1612, 1999.

26. Kim, J.J., Trivedi, N.N., Nottingham, L.K., Morrison, L., Tsai, A., Hu, Y., Mahalingam, S., Dang, K.S., Ahn, L., Doyle, N.K., Wilson, D.M., Chattergoon, M.A., Chalian, A.A., Boyer, J.D., Agadjanyan, M.G., and Weiner, D.B., Modulation of amplitude and direction of *in vivo* immune responses by co-administration of cytokine gene expression cassettes with DNA immunogens, *Eur. J. Immunol.* 28, 1089, 1998.

27. Sin, J.I. and Weiner, D.B., Improving DNA vaccines targeting viral infection, *Intervirology* 43, 233, 2000.

28. Boyer, J.D., Chattergoon, M., Muthumani, K., Kudchodkar, S., Kim, J., Bagarazzi, M., Pavlakis, G., Sekaly, R., and Weiner, D.B., Next generation DNA vaccines for HIV-1, *J. Liposome Res.* 12, 137, 2002.

29. Legendre, J.Y. and Szoka, F.C., Jr., Delivery of plasmid DNA into mammalian cell lines using pH-sensitive liposomes: comparison with cationic liposomes, *Pharm. Res.* 9, 1235, 1992.

30. Freimark, B.D., Blezinger, H.P., Florack, V.J., Nordstrom, J.L., Long, S.D., Deshpande, D.S., Nochumson, S., and Petrak, K.L., Cationic lipids enhance cytokine and cell influx levels in the lung following administration of plasmid: cationic lipid complexes, *J. Immunol.* 160, 4580, 1998.

31. Mahato, R.I., Monera, O.D., Smith, L.C., and Rolland, A.P., Peptide-based gene delivery, *Curr. Top. Microbiol. Immunol.* 1, 226, 1999.

32. MacLaughlin, F.C., Mumper, R.J., Wang, J., Tagliaferri, J.M., Gill, In., Hinchcliffe, M., and Rolland, A.P., Chitosan and depolymerized chitosan oligomers as condensing carriers for *in vivo* plasmid delivery, *J. Control. Release* 56, 259, 1998.

33. Kaneda, Y., Iwai, K., and Uchida, T., Increased expression of DNA cointroduced with nuclear protein in adult rat liver, *Science* 243, 375, 1989.

34. Kato, K., Nakanishi, M., Kaneda, Y., Uchida, T., and Okada, Y., Expression of hepatitis B virus surface antigen in adult rat liver. Co-introduction of DNA and nuclear protein by a simplified liposome method, *J. Biol. Chem.* 266, 3361, 1991.

35. Lunsford, L., McKeever, U., Eckstein, V., and Hedley, M.L., Tissue distribution and persistence in mice of plasmid DNA encapsulated in a PLGA-based microsphere delivery vehicle, *J. Drug Target.* 8, 39, 2000.

36. Hao, T., McKeever, U., and Hedley, M.L., Biological potency of microsphere encapsulated plasmid DNA, *J. Control. Release* 69, 249, 2000.

37. Hedley, M.L., Curley, J., and Urban, R., Microspheres containing plasmid-encoded antigens elicit cytotoxic T-cell responses. *Nat. Med.* 4, 365, 1998.

38. McKeever, U., Barman, S., Hao, T., Chambers, P., Song, S., Lunsford, L., Hsu, Y.Y., Roy, K., and Hedley, M.L., Protective immune responses elicited in mice by immunization with formulations of poly(lactide-co-glycolide) microparticles. *Vaccine* 20, 1524, 2002.

39. Sheets, E.E., Urban, R.G., Crum, C.P., Hedley, M.L., Politch, J.A., Gold, M.A., Muderspach, L.I., Cole, G.A., and Crowley-Nowick, P.A., Immunotherapy of human cervical high-grade cervical intraepithelial neoplasia with microparticle-delivered human papillomavirus 16 E7 plasmid DNA, *Am. J. Obstet. Gynecol.* 188, 916, 2003.

40. Mumper, R.J., Duguid, J.G., Anwer, K., Barron, M.K., Nitta, H., and Rolland, A.P., Polyvinyl derivatives as novel interactive polymers for controlled gene delivery to muscle, *Pharm. Res.* 13, 701, 1996.

41. Mumper, R.J., Wang, J., Klakamp, S.L., Nitta, H., Anwer, K., Tagliaferri, F., and Rolland, A.P., Protective interactive noncondensing (PINC) polymers for enhanced plasmid distribution and expression in rat skeletal muscle, *J. Control. Release* 52, 191, 1998.

42. Alila, H., Coleman, M.E., Nitta, H., French, M., Anwer, K., Liu, Q., Meyer, T., Wang, J., Mumper, R.J., Oubari, D., Long, S.D., Nordstrom, J.L., and Rolland, A.P., Expression of biologically active human insulin-like growth factor-I following intramuscular injection of a formulated plasmid in rats, *Hum. Gene Ther.* 8, 1785, 1997.

43. Coleman, M., Muller, S., Quezada, A., Mendiratta, S.K., Wang, J., Thull, N.M., Bishop, J., Matar, M., Mester, J., and Pericle, F., Nonviral interferon alpha gene therapy inhibits growth of established tumors by eliciting a systemic immune response, *Hum. Gene Ther.* 9, 2223, 1998.

44. Bishop, J.S., Thull, N.M., Matar, M., Quezada, A., Munger, W.E., Batten, T.L., Muller, S., and Pericle, F., Antitumoral effect of a nonviral interleukin-2 gene therapy is enhanced by combination with 5-fluorouracil, *Cancer Gene Ther.* 7, 1165, 2000.

45. Mendiratta, S.K., Quezada, A., Matar, M., Wang, J., Hebel, H.L., Long, S., Nordstrom, J.L., and Pericle, F., Intratumoral delivery of IL-12 gene by polyvinyl polymeric vector system to murine renal and colon carcinoma results in potent antitumor immunity, *Gene Ther.* 6, 833, 1999.

46. Mendiratta, S.K., Quezada, A., Matar, M., Thull, N.M., Bishop, J.S., Nordstrom, J.L., and Pericle, F., Combination of interleukin 12 and interferon alpha gene therapy induces a synergistic antitumor response against colon and renal cell carcinoma, *Hum. Gene Ther.* 11, 1851, 2000.

47. Anwer, K., Earle, K.A., Shi, M., Wang, J., Mumper, R.J., Proctor, B., Jansa, K., Ledebur, H.C., Davis, S.S., Eaglstein, W., and Rolland, A.P., Synergistic effect of formulated plasmid and needle-free injection for genetic vaccines, *Pharm. Res.* 16, 889, 1999.

48. Townsend, A. and Bodmer, H., Antigen recognition by Class I-restricted T lymphocytes, *Ann. Rev. Immunol.* 7, 601, 1989.

49. Germain, R.N. and Margulies, D.H., The biochemistry and cell biology of antigen processing and presentation, *Ann. Rev. Immunol.* 11, 403, 1993.

50. Bartlett, J.A., Wasserman, S.S., Hicks, C.B., Dodge, R.T., Weinhold, K.J., Tacket, C.O., Ketter, N., Wittek, A.E., Palker, T.J., and Haynes, B.F., Safety and immunogenicity of an HLA-based HIV envelope polyvalent synthetic peptide immunogen, DATRI 010 Study Group, Division of AIDS Treatment Research Initiative, *AIDS* 12, 1291, 1998.

51. Pinto, L.A., Berzofsky, J.A., Fowke, K.R., Little, R.F., Merced-Galindez, F., Humphrey, R., Ahlers, J., Dunlop, N., Cohen, R.B., Steinberg, S.M., Nara, P., Shearer, G.M., and Yarchoan, R., HIV-specific immunity following immunization with HIV synthetic envelope peptides in asymptomatic HIV-infected patients, *AIDS* 13, 2003, 1999.

52. Gahéry-Ségard, H., Pialoux, G., Charmeteau, B., Sermet, S., Poncelet, H., Raux, M., Tartar, A., Lévy, J.P. Gras-Masse, H., and Guillet, J.G., Multiepitopic B- and T-cell responses induced in humans by a Human Immunodeficiency Virus Type 1 lipopeptide vaccine, *J. Virol.* 74, 1694, 2000.

53. An, L.L., Rodriguez, F., Harkins, S., Zhang, J., and Whitton, J.L., Quantitative and qualitative analyses of the immune responses induced by a multivalent minigene DNA vaccine, *Vaccine* 18, 2132, 2000.

54. An, L.L. and Whitton, J.L., Multivalent minigene vaccines against infectious disease, *Curr. Opin. Mol. Ther.* 1, 16, 1999.

55. Hanke, T., Schneider, J., Gilbert, S.C., Hill, A.V.S., and McMichael, A., DNA multi-CTL epitope vaccines for HIV and *Plasmodium falciparum*: immunogenicity in mice, *Vaccine* 16, 426, 1998.

56. Hanke, T., Neumann, V.C., Blanchard, T.K., Sweeney, P., Hill, A.V.S., Smith, G.L., and McMichael, A., Effective induction of HIV-specific CTL by multi-epitope using gene gun in a combined vaccination regime, *Vaccine* 17, 589, 1999.

57. Ishioka, G.Y., Fikes, J., Hermanson, G., Livingston, B., Crimi, C., Qin, M., del Guercio, M.F., Oseroff, C., Dahlberg, C., Alexander, J., Chesnut, R.W., and Sette, A., Utilization of MHC Class I transgenic mice for development of minigene DNA vaccines encoding multiple HLA-restricted CTL epitopes, *J. Immunol.* 162, 3915, 1999.

58. Livingston, B.D., Newman, M., Crimi, C., McKinney, D., Chesnut, R., and Sette, A., Optimization of epitope processing enhances immunogenicity of multi-epitope DNA vaccines, *Vaccine* 19, 4652, 2001.

59. Livingston, B., Crimi, C., Newman, M., Higashimoto, Y., Appella, E., Sidney, J., and Sette, A.A., Rational strategy to design multi-epitope immunogens based on multiple Th-lymphocyte epitopes, *J. Immunol.* 168, 5499, 2002.

60. Alexander, J., Sidney, J., Southwood, S., Ruppert, J., Oseroff, C., Maewal, A., Snoke, K., Serra, H.M., Kubo, R.T., and Sette, A., Development of high potency universal DR-restricted helper epitopes by modification of high affinity DR-blocking peptides, *Immunity* 1, 751, 1994.

61. Wilson, C., McKinney, D.M., Anders, M., MaWhinney, S., Forster, J., Crimi, C., Southwood, S., Sette, A., Chesnut, R., Newman, M., and Livingston, B., Development of a DNA vaccine designed to induce cytotoxic t lymphocyte responses to multiple conserved epitopes in HIV-1, *J. Immunol.*, 171, 5611–5623, 2003.

62. Robinson, B.V., Sullivan, F.M., Borzelleca, J.F., and Schwartz, S.L., *PVP: A Critical Review of the Kinetics and Toxicology of Polyvinylpyrrolidone (Povidone)*, Lewis Publishers, 1990, 1.

63. Quezada, A., Horner, M.J., Loera, D., French, M., Pericle, F., Johnson, R., Perrard, J., Jenkins, M., and Coleman, M., Safety toxicity study of plasmid-based IL-12 therapy in Cynomolgus monkeys, *J. Pharm. Pharmacol.* 54, 241, 2002.

64. Martin, T., Parker, S.E., Hedstrom, R., Le, T., Hoffman, S.L., Norman, J., Hobart, P., and Lew, D., Plasmid DNA malaria vaccine: the potential for genomic integration after intramuscular injection, *Hum. Gene Ther.* 10, 759, 1999.

65. Kawabata, K., Takakura, Y., and Hashida, M., The fate of plasmid DNA after intravenous injection in mice: involvement of scavenger receptors in its hepatic uptake, *Pharm. Res.* 12, 825, 1995.

66. Ledwith, B.J., Manam, S., Troilo, P.J., Barnum, A.B., Pauley, C.J., Griffiths, T.G., Harper, L.B., Beare, C.M., Bagdon, W.J., and Nichols, W.W., Plasmid DNA vaccines: investigation of integration into host cellular DNA following intramuscular injection in mice, *Intervirology* 43, 258, 2000.

67. Parker, S.E., Borellini, F., Wenk, M.L., Hobart, P., Hoffman, S.L., Hedstrom, R., Le, T., and Norman, J.A., Plasmid DNA malaria vaccine: tissue distribution and safety studies in mice and rabbits, *Hum. Gene Ther.* 10, 741, 1999.

68. Winegar, R.A., Monforte, J.A., Suing, K.D., O'Loughlin, K.G., Rudd, C.J., and Macgregor, J.T., Determination of tissue distribution of an intramuscular plasmid vaccine using PCR and *in situ* DNA hybridization, *Hum. Gene Ther.* 7, 2185, 1996.

69. Bagarazzi, M.L., Boyer, J.D., Ugen, K.E., Javadian, M.A., Chattergoon, M., Shah, A., Bennett, M., Ciccarelli, R., Carrano, R., Coney, L., and Weiner, D.B., Safety and immunogenicity of HIV-1 DNA constructs in chimpanzees, *Vaccine* 16, 1836, 1998.

70. Haworth, R. and Pilling, A.M., The PCR assay in the preclinical safety evaluation of nucleic acid medicines, *Hum. Exp. Toxicol.* 19, 267, 2000.

71. Winegar, R.A., Monforte, J.A., Suing, K.D., O'Loughlin, K.G., Rudd, C.J., and Macgregor, J.T., Determination of tissue distribution of an intramuscular plasmid vaccine using PCR and *in situ* DNA hybridization, *Hum. Gene Ther.* 7, 2185, 1996.

72. Martin, T., Parker, S.E., Hedstrom, R.C., Le, T., Hoffman, S.L., Norman, J., Hobart, P., and Lew, D., Plasmid DNA malaria vaccine: the potential for genomic integration after intramuscular injection, *Hum. Gene Ther.* 10, 759, 1999.

73. Ledwith, B.J., Manam, S., Troilo, P.J., Barnum, A.B., Pauley, C.J., Griffiths, T.G., Harper, L.B., Schock, H.B., Zhang, H., Faris, J.E., Way, P.A., Beare, C.M., Bagdon, W.J., and Nichols, W.W., Plasmid DNA vaccines: assay for integration into host genomic DNA, *Dev. Biol.* 104, 33, 2000.

74. Manam, S., Ledwith, B.J., Barnum, A.B., Troilo, P.J., Pauley, C.J., Harper, L.B., Griffiths II, T.G., Niu, Z., Denisova, L., Follmer, T.T., Pacchione, S.J., Wang, Z., Beare, C.M., Bagdon, W.J., and Nichols, W.W., Plasmid DNA vaccines: tissue distribution and effects of DNA sequence, adjuvants and delivery method on integration into host DNA, *Intervirology* 43, 273, 2000.

75. Payette, P.J., Weeratna, R.D., McCluskie, M.J., and Davis, H.L., Immune-mediated destruction of transfected myocytes following DNA vaccination occurs via multiple mechanisms, *Gene Ther.* 8, 1395, 2001.

76. McKinney, D.M., Skvoretz, R., Qin, M.S., Ishioka, G., and Sette, A., Characterization of an *in situ* IFN-γ ELISA assay which is able to detect specific peptide responses from freshly isolated splenocytes induced by DNA minigene immunization, *J. Immunol. Methods* 237, 105, 2000.

CHAPTER 21

Use of HPMA Copolymers in Gene Delivery

David Oupicky

CONTENTS

21.1 INTRODUCTION

   Effective therapeutic application of gene drugs requires noninvasive and cost-effective methods for gene delivery, which should be safe for repetitive use and provide reproducible therapeutic effect. Systematic research efforts in the area of polyelectrolyte complexes of DNA with polycations (polyplexes) resulted in an increased understanding of the principles governing the biological activity of polyplexes and now permit rational design of increasingly more efficient vectors. Despite this tremendous progress, significant constrains limiting the efficiency of the gene delivery process mediated by polyplexes still remain, particularly when they are administered systemically.[1] The growing emphasis on the systemic delivery of genes via intravenous injection reflects the need to gain access to disseminated and widespread disease targets, such as cancer cells, and to expand the available therapeutic modalities. To construct a delivery system suitable for systemic administration, a thorough understanding of the *in vivo* pharmacokinetic and disposition characteristics of the vector is important.[2] The distribution and elimination patterns of systemically administered polyplexes largely follow general behavior observed for other macromolecules and nanoparticles, and depend mainly on the physicochemical properties of the carrier, such as size and molecular weight, electric charge, and hydrophilic/hydrophobic balance.[3,4] Based on the accurate understanding of the relationship between the physicochemical properties of macromolecular carriers and their pharmacokinetics, it is often possible to effectively control their disposition properties.

0-8493-1934-X/05/$0.00+$1.50
© 2005 by CRC Press LLC

Distribution of intravenously injected nanoparticles, such as polyplexes, is usually restricted to the intravascular space due to low capillary permeability in most organs with continuous capillary bed. In such cases, the liver plays the critical function in the disposition of the nanoparticles circulating in the blood stream. The discontinuous endothelial capillaries with about 100 nm fenestrae in the liver permit free contact of circulating nanoparticles with the surface of parenchymal cells and are responsible for the typical wide distribution of nanoparticles in the liver. Similar vascular permeability of many tumor tissues combined with the lack of functional lymphatic drainage often permits accumulation of macromolecules and nanoparticles in solid tumors by the mechanism of passive targeting.[4-6] Macromolecular drugs and nanoparticle delivery systems with adequate molecular weight and size and a slightly anionic nature show prolonged retention in the plasma circulation and relatively large accumulation in the target tumor, which can be often further enhanced by active targeting.

High positive surface charge and poor solubility in physiological conditions, coupled with nonspecific charge-mediated interactions with proteins and cells, are the leading factors responsible for rapid (often first-pass) hepatic clearance of polyplexes from the bloodstream.[7-9] Extension of the plasma circulation times and reduction of the hepatic clearance are the necessary requirements to permit effective systemic delivery of polyplexes into distant targets, such as metastatic tumors. The widely applicable methodology for manipulating *in vivo* disposition profiles of macromolecules and nanoparticles relies on the use of hydrophilic nonionic polymers such as polyethylene glycol (PEG), dextran, and poly(*N*-[2-hydroxypropyl] methacrylamide) (PHPMA). This approach has been used for modification of proteins, nanoparticles, liposomes, and recently also polyplexes and viruses in gene delivery. The presence of the hydrophilic polymers increases solubility and reduces recognition of the macromolecules and nanoparticles by the immune system, and subsequently leads to prolonged plasma circulation times.[10]

HPMA copolymers and their conjugates with drugs are among the most extensively studied in the field of macromolecular therapeutics. PHPMA itself was developed in the late 1960s as a new hydrophilic and biocompatible polymer and extensively tested as a blood plasma expander. The chemical properties of HPMA permit straightforward access to linear polymers with easy control of molecular weight distribution. This enables tailoring the PHPMA properties and ensures easy elimination from the organism. An important observation that oligopeptide sequences attached to HPMA copolymers were degradable *in vivo* and thus had a potential as drug attachment and release sites encouraged development of HPMA-based macromolecular therapeutics. Their biocompatibility was validated in numerous clinical trials, which also established activity of HPMA anticancer drug conjugates against numerous cancer models. HPMA copolymers were also used in a hydrogel form for controlled release of drugs and for preparation of protein conjugates.[11,12]

The use of the HPMA copolymers in the design of viral and nonviral gene delivery vectors focuses on the protective function of the copolymers during systemic gene delivery. This contribution summarizes the findings that resulted from using HPMA-containing copolymers for the design of gene delivery vectors suitable for systemic intravenous application. It focuses on the correlation between physicochemical properties of the delivery vectors and *in vivo* disposition properties, using random copolymers of HPMA with cationic monomers, block and graft copolymers with polycations, and semitelechelic and multivalent HPMA copolymers for surface modifications of polyplexes and adenoviruses.

21.2 RANDOM COPOLYMERS OF HPMA WITH CATIONIC COMONOMERS

Random copolymers of HPMA with cationic comonomers were studied for their potential to reduce high positive surface charge and increase solubility of polyplexes. These copolymers also permit examination of the effect of charge spacing on the properties of polyplexes. For instance, copolymers of HPMA with 2-(trimethylammonio)ethyl methacrylate (P(TMAEM-co-HPMA))

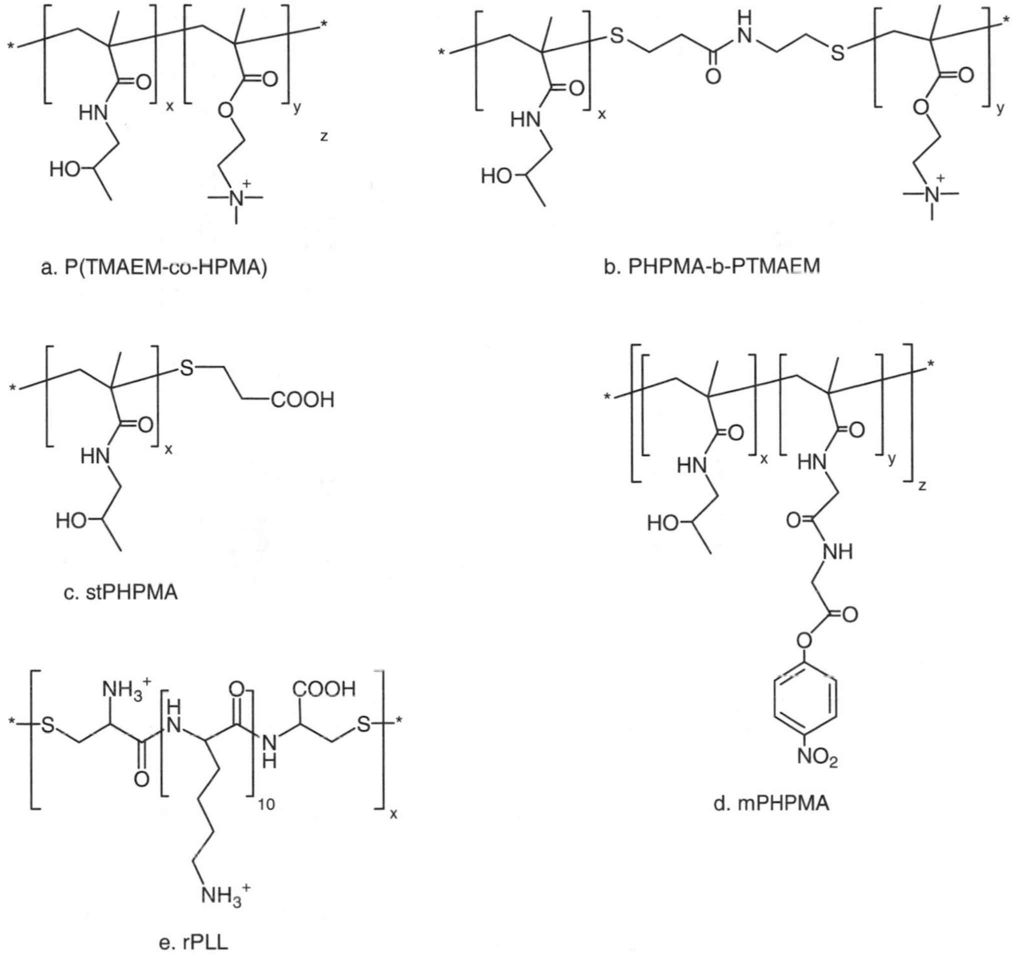

a. P(TMAEM-co-HPMA)

b. PHPMA-b-PTMAEM

c. stPHPMA

d. mPHPMA

e. rPLL

Figure 21.1 Structures of selected HPMA copolymers used in the design of gene delivery vectors (A–D) and the structure of reducible poly(L-lysine) analog (E).

with HPMA content ranging from 25 to 95 mol% were studied (Figure 21.1A).[13,14] Decreasing the charge spacing leads to reduced ability of the P(TMAEM-co-HPMA) copolymers to condense DNA into discrete particles and also negatively affects the ability of the copolymers to protect DNA against enzymatic degradation. In particular, copolymers containing 50% or less cationic units often form extended structures, with strands of DNA protruding from denser complex cores probably representing DNA complexes which are not subject to charge-neutralization-driven self-assembly. In addition, these copolymers are capable of mediating only a very low transfection activity. Increased hydrophilic character and solubility of DNA complexes containing P(TMAEM-co-HPMA) copolymers is probably responsible for the observed decrease of association with isolated leucocytes (especially with granulocytes and monocytes) compared to the parent polycation PTMAEM. This encouraging *in vitro* behavior is however not correlated with improved pharmacokinetics of the polyplexes based on the random copolymers (Figure 21.2). In conclusion, random P(TMAEM-co-HPMA) copolymers cannot drive hydrophobic self-assembly of DNA complexes and instead they form extended structures that are not capable of reducing hepatic clearance and extending plasma circulation time of the polyplexes. As such they are not suitable for systemic intravenous administration.

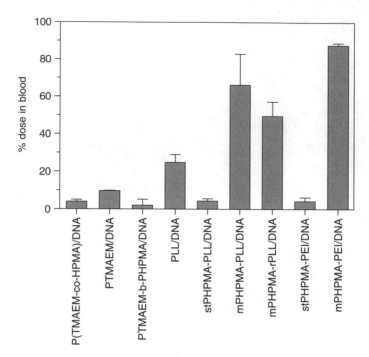

Figure 21.2   Influence of various HPMA copolymers on the levels of polyplexes remaining in the blood circulation 30 min after intravenous injection in mice. (Data adopted from References 14, 17, 22, 23, and 28.)

21.3 BLOCK AND GRAFT COPOLYMERS OF HPMA WITH POLYCATIONS

The block and graft copolymers of hydrophilic nonionic and cationic polymers have been introduced to increase the hydrophilic character of the polyplexes. Separating the DNA-binding domain from the hydrophilic nonionic polymer, the approach relies on the concept of oriented self-assembly to introduce a surface hydrophilic layer protecting the polyplexes from unwanted interactions with components of the immune system (steric stabilization) (Scheme 21.1). Studies were performed using A-B type block copolymers and graft copolymers in which polycations were grafted with hydrophilic nonionic polymers.[15-20] Among the studied copolymers were block copolymers PHPMA-b-PTMAEM (Figure 21.1B) as well as graft copolymers containing more than one PHPMA block linked to PTMAEM or PLL. Both the block and graft copolymers usually exhibit a good ability to bind and condense DNA.[16,19] The presence of the PHPMA blocks in the polyplexes results in an increased solubility in physiological conditions and a reduction of the surface charge (zeta potential) compatible with a level of surface shielding. Despite these improvements of physicochemical properties, *in vivo* biodistribution studies show that these complexes are removed very quickly from the murine bloodstream and often even faster than the polyplexes formed between DNA and simple polycations (Figure 21.2).[17]

The reasons for the apparent discrepancy between the behavior of the polyplexes in the *in vitro* tests and poor performance *in vivo* can be traced to deviations of the actual morphology of the polyplexes from the expected core-shell structure (Scheme 21.1). Considering that the typical size of polyplexes is about 100 nm, it is likely that a considerable amount of hydrophilic PHPMA blocks and grafts becomes entrapped within the core of the polyplexes and will prevent efficient microphase separation which can lead to thermodynamic destabilization of the polyplexes. This possibility gains experimental support from the observation of significantly decreased stability of the polyplexes against polyelectrolyte exchange reactions and from decreasing structural density of the

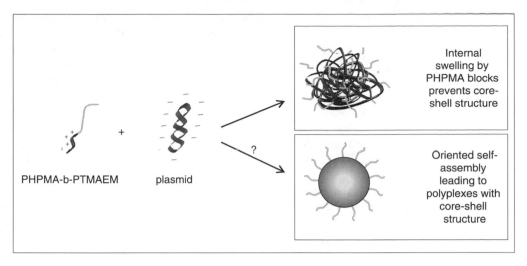

Scheme 21.1 Schematic representation of self-assembly of PHPMA block copolymers with plasmid DNA resulting in polyplexes with core-shell structure (oriented self-assembly) and polyplexes exhibiting internal swelling by PHPMA blocks.

polyplexes with increasing PHPMA content in the block and graft copolymers, suggesting effective internal swelling of the polyplexes with the hydrophilic PHPMA (Figure 21.2).[20]

Better microphase separation of the nucleic acid/polycation core from the PHPMA blocks can be expected when comparably sized polyelectrolytes are used. The HPMA block and graft copolymers were therefore also tested for delivery of antisense oligonucleotides (ASON) in order to improve their biocompatibility and modulate their pharmacokinetic behavior.[21] Examining the ability of various block and graft copolymers to condense the ASON reveals that the copolymer PHPMA-b-PTMAEM with the size of the cationic block matching the size of ASON produces the smallest and least charged complexes with sizes of 36 nm, high stability in physiological conditions, and high solubility that permits achieving DNA concentrations up to 0.5 mg/ml without significant aggregation. Using the ASON complexes with the copolymer PHPMA-b-PTMAEM reduces the urinary clearance of the ASON after intravenous injection into mice but does not prolong blood circulation times, indicating that the polymer-derived complexes are not stable enough in the blood circulation.

21.4 SEMITELECHELIC HPMA COPOLYMERS FOR SURFACE MODIFICATION OF POLYPLEXES

Alternative approaches to oriented self-assembly of block and graft copolymers with DNA that would yield sterically stabilized polyplexes were investigated. Covalent attachment of semitelechelic PHPMA (stPHPMA) (Figure 21.1C) to the surface of the preformed polyplexes was suggested as the possible approach leading to sterically stabilized polyplexes without compromising the stability and compactness of the hydrophobic polycation–DNA core. It was hypothesized that this approach could avoid the incorporation of PHPMA blocks into the hydrophobic core of the polyplex during the self-assembly reaction, resulting in more stable complexes with better microphase separation.[22] PLL–DNA polyplexes were prepared prior to the covalent attachment of stPHPMA to the surface and the effect of the coating on their physicochemical stability, phagocytic uptake *in vitro*, and pharmacokinetics was examined. The stPHPMA-coated polyplexes show good stability against aggregation in physiological conditions as well as reduced binding of albumin, which was chosen as a model for the study of the interactions of the polyplexes with plasma proteins. Further, no effect of the stPHPMA coating on the morphology of the polyplexes can be observed by transmission electron microscopy. To address the possibility that the stPHPMA-coated polyplexes may be able to evade phagocytic capture *in vivo*, an *in vitro* model was employed using mouse peritoneal macrophages. Coated polyplexes showed significantly decreased levels of macrophage uptake compared to the parent PLL–DNA. However, despite all the improvements in resistance to protein binding and reduced phagocytic uptake, no correlation can be observed between these *in vitro* results and the *in vivo* pharmacokinetic behavior (Figure 21.2). No reduction in the rates of hepatic clearance and no increase in the circulation times is observed. The stPHPMA-coated polyplexes are cleared from the mouse bloodstream even more rapidly than unmodified polyplexes. This suggests that the *in vivo* fate of the polyplexes is significantly influenced by factors not reflected in the *in vitro* cell culture model.

Results obtained from the studies of polyelectrolyte exchange reactions with heparin and PLL suggest that the presence of the stPHPMA surface layer has no beneficial effect on the susceptibility to these types of reactions compared to parent PLL–DNA polyplexes. In fact, the results suggest that the stPHPMA layer may be easily lost, notably by displacement of the modified polycations from the polyplex, and this can lead to the observed rapid plasma elimination of stPHPMA-coated polyplexes.

21.5 MULTIVALENT HPMA COPOLYMERS FOR SURFACE MODIFICATION OF POLYPLEXES AND ADENOVIRUS

21.5.1 Surface Modification of Polyplexes

The tendency to undergo polyelectrolyte exchange reactions in the bloodstream is one of the most challenging aspects of the instability of polyplexes. Many cells, surfaces and macromolecules *in vivo* have a polyionic character and can compete with the polyions in the polyplexes for binding, leading to destabilization of the polyplexes and rapid clearance from the plasma. Hypothesizing that not only careful control of surface properties and steric stability but also significantly increased resistance against polyelectrolyte exchange reactions are required to permit increased circulation times of polyplexes, multivalent HPMA copolymers (mPHPMA) were used for surface modification of preformed polyplexes.[23]

The multivalent HPMA copolymers are prepared by copolymerization of HPMA with meth-acryloylglycylglycine 4-nitrophenyl ester and contain 8 mol% of the reactive comonomer (Figure 21.1D). The mPHPMA binds around the surface of the polyplex particles, linking together surface amino groups and providing the polyplexes with a combination of steric and lateral stability (Scheme 21.2). Similar to stPHPMA, the multivalent HPMA copolymers endow polyplex particles with steric stability, which is manifested by increased resistance to salt-induced aggregation and reduced uptake by macrophages *in vitro*. The lateral stabilization is distinctly demonstrated by increased resistance to polyelectrolyte exchange reactions and reduced hepatic clearance *in vivo*. Whereas typical polyplexes or stPHPMA-coated polyplexes release DNA following incubation with polya-nions, polyplexes coated with mPHPMA are laterally stabilized and resistant to polyelectrolyte exchange reactions. The mPHPMA coating also reverses the surface charge of the polyplex particles, leaving them negatively charged. The negative charge arises from a partial hydrolysis of the reactive 4-nitrophenyl groups during the coating, yielding negatively charged carboxylate groups. Notice-ably however, the physicochemical properties of the formed polyplex nanoparticles are predomi-nantly determined by the coating polymer, and are not significantly affected by the used polycation. Greatly improved stability and solubility of the polyplex nanoparticles modified with mPHPMA permit easy cryopreservation and achieve concentrations up to at least 1 mg/ml of DNA without excessive aggregation.

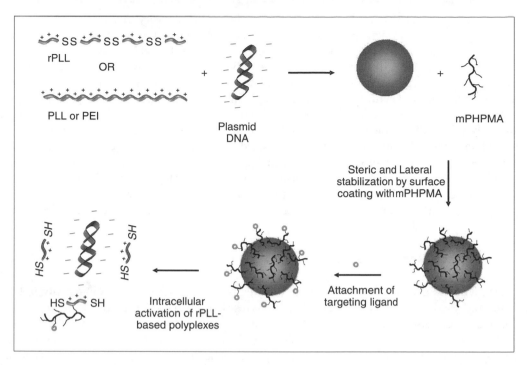

Scheme 21.2 Schematic representation of the approach leading to targeted polyplexes with combined lateral and steric stability achieved by surface modification with multivalent PHPMA and the principle of intracellular specific activation of rPLL-containing polyplexes.

PLL–DNA and PEI–DNA coated with mPHPMA exhibit significantly extended plasma circulation compared with the parent polyplexes.[23] Factors that affect the pharmaco-kinetics of the mPHPMA coated polyplexes include molecular weight of the polycation, molecular weight of the mPHPMA, and the amount of mPHPMA attached to the polyplexes. For example, increasing the concentration and molecular weight of mPHPMA results in increased lateral stability, reduced rate of hepatic clearance, reduced dose-dependence of the pharmacokinetics, and increased circulation times. Of the two polycations tested, PEI-based polyplexes coated with mPHPMA exhibit higher circulation times than PLL-based polyplexes (Figure 21.2). Using the optimized formulation of mPHPMA-coated PEI–DNA polyplexes leads to circulation half-lives of almost 90 min in mice, a significant improvement over the typical 3–5 min observed for typical polycation–DNA polyplexes. The prolonged plasma circulation also enables passive accumulation of the mPHPMA-coated polyplexes in solid subcutaneous tumors at the levels comparable to sterically stabilized liposomes.

Using mPHPMA for coating of the polyplexes permits simultaneous easy linkage of protein targeting ligands to the surface of the polyplexes without the need for heterobifunctional chemistry (Scheme 21.2).[24,25] Such polyplexes then offer an open platform, where a variety of bioactive compounds can be incorporated, allowing an easy control of the tropism of the polyplexes. Vascular endothelial growth factor (VEGF), transferrin, and basic fibroblast growth factor (bFGF) have each been linked to mPHPMA-stabilized polyplexes.[25] Such ligand-targeted polyplexes demonstrate increased uptake into receptor-positive cells that could be antagonized with excess free ligand. Targeted polyplexes also exhibit increased transfection activity and resistance to inhibition by serum when compared to nontargeted polyplexes. Analysis using fluorescence microscopy confirmed enhanced uptake of bFGF-targeted complexes and their efficient delivery into the nucleus, with accumulation of more than 100,000 plasmids per cell within distinct intranuclear compartments. However, relatively low levels of gene expression are typically observed, suggesting that the DNA is not readily available for transcription due to the retention of the polymer coating, restricting enzymatic access, and transcription efficiency. This indicates that the very characteristic that permits to improve the pharmacokinetic profile of the polyplexes is detrimental to their transfection activity.

An intracellularly specific activation mechanism was developed to address the limited transcriptional availability of DNA from laterally stabilized polyplexes.[26] The activation mechanism utilizes the existence of a high redox potential gradient between oxidizing extracellular space and the reducing environment of certain intracellular compartments. The design of the polyplexes relies on the strong dependence of the lateral stabilization induced by mPHPMA on the molecular weight of the polycation. High-molecular-weight reducible PLL analogues containing disulfide bonds in the backbone (rPLL, Figure 21.1E) were synthesized by oxidative polycondensation. These polycations are capable of reductive degradation into low-molecular-weight species in the reducing intracellular environment; they permit reversal of lateral stabilization and facilitate release of the DNA (Scheme 21.2).[26] The design of the rPLL-based vectors hence satisfies the contradictory requirements for high stability in the extracellular environment and easy intracellular availability of DNA.

21.5.2 Surface Modification of Adenovirus

Adenovirus is a very efficient vector for gene delivery because of its high infectivity in both dividing and nondividing cells. The main limitations restricting therapeutic use of adenovirus largely to local or direct administration are neutralization of adenovirus by pre-existing antibodies and wide tissue distribution of the coxsackie and adenovirus receptor (CAR) that precludes specific targeting. Various attempts were made to re-engineer adenovirus in order to improve its systemic properties. Coating the surface of the adenovirus with multivalent HPMA copolymers in the same fashion as polyplexes provides a simple nongenetic approach.[27] Easy incorporation of targeting ligands such as bFGF and VEGF on to the polymer-coated virus produces ligand-mediated, CAR-independent binding and uptake into cells bearing appropriate receptors.

Modification of the surface of adenovirus with mPHPMA suppresses CAR binding, shields the virus from neutralizing antibodies, and permits incorporation of new ligands for virus retargeting. The remarkable finding that adenovirus retargeted to bFGF and VEGF receptors can infect cells efficiently following receptor binding, despite extensive covalent surface modification with a non-biodegradable polymer, has important implications for the development of gene therapy vectors. The possibility to retarget adenoviruses to selected receptors provides the opportunity to refine their tissue tropism and their disposition properties *in vivo*. The avoidance of CAR-mediated infection of nontarget tissues and evasion of neutralization by pre-existing anti-adenovirus antibodies should permit the use of lower viral doses. This can decrease the toxicity associated with adenovirus gene therapy and could significantly improve the efficacy and safety of the approach. PHPMA proves to be an attractive polymer for virus modification, since it is not fusogenic like PEG, hence allowing development of a stealth-like virus without enhanced infection through nonspecific membrane activity. Indeed, studies using PEG to modify the surface of adenovirus have not resulted in such favorable improvements of the adenovirus properties as modification with mPHPMA.

21.6 CONCLUSIONS

Over the last 30 years, HPMA copolymers found use in a variety of biomedical applications. The most recent addition to the utility of HPMA copolymers as components of gene delivery vectors further confirms the biomedical application flexibility of the HPMA copolymers. In particular, surface modification of gene delivery vectors with multivalent HPMA copolymers proves to be an effective way of changing the tropism and interaction with the immune system of both viral and nonviral vectors. The simplicity and generic applicability of this system promises to fulfill several key requirements for successful application of gene therapy. The demonstration of extended plasma circulation of mPHPMA-coated polyplexes is a landmark achievement, and the next major challenge is to combine extended plasma circulation and targeting with efficient transgene expression following arrival at the target site.

REFERENCES

1. Wiethoff, C.M., and Middaugh, C.R., Barriers to nonviral gene delivery, *J. Pharm. Sci.* 92, 203, 2003.
2. Suzuki, H. et al., Design of a drug delivery system for targeting based on pharmacokinetic consideration, *Adv. Drug Deliv. Rev.* 19, 335, 1996.
3. Takakura, Y. et al., Development of gene drug delivery systems based on pharmacokinetic studies, *Eur. J. Pharm. Sci.* 13, 71, 2001.
4. Takakura, Y. et al., Control of pharmacokinetic profiles of drug-macromolecule conjugates, *Adv. Drug Deliv. Rev.* 19, 377, 1996.
5. Nishikawa, M., and Hashida, M., Pharmacokinetics of anticancer drugs, plasmid DNA, and their delivery systems in tissue-isolated perfused tumors, *Adv. Drug Deliv. Rev.* 40, 19, 1999.
6. Jain, R.K., Delivery of molecular and cellular medicine to solid tumors, *Adv. Drug Deliv. Rev.* 46, 149, 2001.
7. Takakura, Y. et al., Influence of physicochemical properties on pharmacokinetics of non-viral vectors for gene delivery, *J. Drug Target.* 10, 99, 2002.
8. Opanasopit, P., Nishikawa, M., and Hashida, M., Factors affecting drug and gene delivery: Effects of interaction with blood components, *Crit. Rev. Ther. Drug Carr. Syst.* 19, 191, 2002.
9. Dash, P.R. et al., Factors affecting blood clearance and *in vivo* distribution of polyelectrolyte complexes for gene delivery, *Gene Ther.* 6, 643, 1999.
10. Lasic, D.D., and Needham, D., The "stealth" liposome: a prototypical biomaterial, *Chem. Rev.* 95, 2601, 1995.
11. Putnam, D., and Kopecek, J., Polymer conjugates with anticancer activity, *Adv. Polym. Sci.* 122, 57, 1995.
12. Kopecek, J. et al., HPMA copolymer-anticancer drug conjugates: design, activity, and mechanism of action, *Eur. J. Pharm. Biopharm.* 50, 61, 2000.
13. Wolfert, M.A. et al., Polyelectrolyte vectors for gene delivery: Influence of cationic polymer on biophysical properties of complexes formed with DNA, *Bioconj. Chem.* 10, 993, 1999.
14. Howard, K.A. et al., Influence of hydrophilicity of cationic polymers on the biophysical properties of polyelectrolyte complexes formed by self-assembly with DNA, *Biochim. Biophys. Acta* 1475, 245, 2000.
15. Toncheva, V. et al., Novel vectors for gene delivery formed by self-assembly of DNA with poly(L-lysine) grafted with hydrophilic polymers, *Biochim. Biophys. Acta* 1380, 354, 1998.
16. Wolfert, M.A. et al., Characterization of vectors for gene therapy formed by self-assembly of DNA with synthetic block copolymers, *Hum. Gene Ther.* 7, 2123, 1996.
17. Oupicky, D. et al., Effect of albumin and polyanion on the structure of DNA complexes with polycation containing hydrophilic nonionic block, *Bioconj. Chem.* 10, 764, 1999.
18. Oupicky, D., Konak, C., and Ulbrich, K., Preparation of DNA complexes with diblock copolymers of poly[N-(2-hydroxypropyl)methacrylamide] and polycations, *Mater. Sci. Eng. C: Biom. Supramol. Syst.* 7, 59, 1999.
19. Oupicky, D., Konak, C., and Ulbrich, K., DNA complexes with block and graft copolymers of N-(2-hydroxypropyl) methacrylamide and 2-(trimethylammonio)ethyl methacrylate, *J. Biomater. Sci. Polym. Ed.* 10, 573, 1999.
20. Oupicky, D., et al., DNA delivery systems based on complexes of DNA with synthetic polycations and their copolymers, *J. Control. Rel.* 65, 149, 2000.
21. Read, M.L. et al., Physicochemical and biological characterisation of an antisense oligonucleotide targeted against the bcl-2 mRNA complexed with cationic-hydrophilic copolymers, *Eur. J. Pharm. Sci.* 10, 169, 2000.
22. Oupicky, D. et al., Steric stabilization of poly-L-lysine/DNA complexes by the covalent attachment of semitelechelic poly [N-(2-hydroxypropyl)methacrylamide], *Bioconj. Chem.* 11, 492, 2000.

23. Oupicky, D. et al., Importance of lateral and steric stabilization of polyelectrolyte gene delivery vectors for extended systemic circulation, *Mol. Ther.* 5, 463, 2002.

24. Dash, P.R. et al., Decreased binding to proteins and cells of polymeric gene delivery vectors surface modified with a multivalent hydrophilic polymer and retargeting through attachment of transferrin, *J. Biol. Chem.* 275, 3793, 2000.

25. Fisher, K.D. et al., A versatile system for receptor-mediated gene delivery permits increased entry of DNA into target cells, enhanced delivery to the nucleus and elevated rates of transgene expression, *Gene Ther.* 7, 1337, 2000.

26. Oupicky, D., Parker, A.L., and Seymour, L.W., Laterally stabilized complexes of DNA with linear reducible polycations: Strategy for triggered intracellular activation of DNA delivery vectors, *J. Am. Chem. Soc.* 124, 8, 2002.

27. Fisher, K.D. et al., Polymer-coated adenovirus permits efficient retargeting and evades neutralising antibodies, *Gene Ther.* 8, 341, 2001.

28. Oupicky, D., and Seymour, L.W., unpublished data, 2002.

PART IV

Polymeric Nanospheres and Microspheres

A. Polymeric Nanospheres

CHAPTER 22

Biodegradable Nanoparticles as a Gene Expression Vector

Swayam Prabha, Wenxue Ma, and Vinod Labhasetwar

CONTENTS

22.1 INTRODUCTION

With advances in polymer science and with the advent of various synthetic biodegradable and biocompatible polymeric materials, the field of polymer-based gene expression vectors has been expanding and gaining significant interest in recent years.[1,2] Some of these polymers are modified or custom synthesized specifically for the delivery of nucleic acids.[3,4] Recent experiences with viral vectors suggest that the safety of these expression vectors is still an unresolved issue[5] and that it cannot be overlooked if the gene therapy approach is to succeed in clinical trials. It is anticipated that the polymer-based nonviral gene delivery systems will be safer, and versatile and efficient in gene expression.

In general, DNA delivery to the nucleus encounters major cellular barriers such as intracellular uptake of a vector, escape of the vector from the endo-lysosomal compartment to the cytoplasmic compartment, dissociation of DNA from the vector inside the cell, and successful localization of DNA into the nucleus.[6] In addition, the protection of DNA from degradation due to nucleases is equally important to achieve efficient gene expression. Among the various nonviral gene delivery systems that have been investigated, cationic lipids, liposomes, polymers, and molecular conjugates

0-8493-1934-X/05/$0.00+$1.50
© 2005 by CRC Press LLC

have shown promising results. However, these vectors still suffer from low transfection efficiencies compared to viral vectors and some of these systems are toxic or not suitable for *in vivo* applications because of their instability in the presence of serum.[7] Current research is focused on investigating various approaches to increase the transfection efficiency while reducing the toxicity associated with these systems. Another limiting factor with the above nonviral systems is the transient gene expression, requiring frequent administration of the vector, which could cause toxicity.

Although various polymeric systems are under investigation, our efforts are focused on investigating biodegradable nanoparticles as a gene delivery system. These nanoparticles are formulated using biodegradable polymers, poly(lactide-*co*-glycolide) (PLG) and polylactide (PLA), with a plasmid DNA (pDNA) entrapped into the nanoparticle polymer matrix.[8] Biodegradable nano- and microparticles have been extensively investigated as carriers for various therapeutic agents, including macromolecules such as proteins and peptides.[8] However, their application as a gene expression vector is recent and has yet to be explored completely. The main advantage of nano- and microparticles is their polymeric nature, which makes the system devoid of the toxicity and immunogenic concerns that are associated with viral vectors. Furthermore, the slow release of the encapsulated DNA from nano- or microparticles is expected to provide sustained gene expression.[9,10]

Significant information is available on the formulation and characterization of nano- and microparticles for drug delivery applications; however, very little is known about their desired formulation characteristics that may be critical to developing nanoparticles as an efficient gene delivery system. Furthermore, the intracellular trafficking of nanoparticles is relatively unknown, and could be important in understanding the mechanism of nanoparticle-mediated gene expression. A strategy based on better understanding of the factors that affect the intracellular uptake of nanoparticles, retention, and their escape from the endo-lysosomal compartment could further enhance the efficacy of nanoparticle-mediated gene expression. The discussion below will review some of these issues and the prospects of nanoparticle-mediated gene therapy.

22.2 NANOPARTICLES VERSUS MICROPARTICLES: THE EFFECT OF SIZE

Nano- and microparticles formulated using biodegradable polymers such as PLA and PLG offer the advantage of being able to control the dosing and drug release kinetics, and are used for sustaining drug levels in tissues and cells.[11] Further, the polymers used for nanoparticle formulation are biocompatible and biodegradable, and are approved by the Food and Drug Administration for human use. Nanoparticles are submicron in size and have been demonstrated to have significantly higher cellular and tissue uptake as compared to microparticles. It has been shown that in some cell lines, only the submicron size particles are taken up inside the cells.[12,13] Thus, there is a size-dependent cutoff for the cellular and tissue uptake of particles, with the exception of macrophages in which larger size particles are also taken up efficiently.[12] Since gene expression requires intracellular delivery of pDNA, nanoparticles could be a better gene delivery vehicle than microparticles.

22.3   DNA INCORPORATION AND THE EFFECT OF FORMULATION CONDITIONS

In general, a multiple water-in-oil-in-water (w/o/w) solvent evaporation technique is used to incorporate DNA in nano- or microparticles. In this procedure, typically a DNA solution in Tris-EDTA buffer is emulsified into a polymer solution in an organic solvent to form a water-in-oil (w/o) emulsion, which is then further emulsified into an aqueous phase containing an emulsifier to form a w/o/w emulsion. Either homogenization or sonication energy is used to form the emulsion. Depending upon the energy input during the emulsification step, particle size could be either in nano or in micron size range. In our studies, nanoparticles used for gene delivery are typically 100 nm in diameter with pDNA encoding for the protein of interest encapsulated inside the polymeric matrix.[14] Unlike viral vectors, PLG–PLA nanoparticles can be used to encapsulate large DNA molecules (> 45 Kb). Since the DNA is encapsulated inside the polymeric matrix, it is protected from the nuclease degradation.

DNA loading in PLG–PLA nanoparticles has been reported from about 0.5% w/w to 2.5% w/w in different studies.[9,15] Our recent study has demonstrated that the efficiency of encapsulation of DNA in nanoparticles is influenced by various factors including polymer composition, molecular weight of polymer, and the emulsifier used for the stabilization of emulsion.[16] One of the limiting factors in enhancing DNA loading further using the above procedure is the increase in viscosity of DNA solution with concentration that influences its emulsification into the polymer solution and entrapment in nanoparticles.

Following encapsulation in nanoparticles, DNA is mostly in supercoiled and open-circular forms. There are reports of partial transformation of supercoiled DNA to open-circular form following its encapsulation in nano- and microparticles. However, the follow-up studies have demonstrated that the transfection of each form of DNA extracted from nanoparticles and then separated on agarose gel is almost similar upon testing with a conventional method of transfection such as by using calcium phosphate.[17] The results thus suggest that there is no or minimal loss in the DNA activity following its incorporation in nanoparticles. DNA released from nanoparticles under *in vitro* conditions over a period of time has also been evaluated for transfectivity. There may be some inactivation of DNA as it is released, which could be either because of the exposure of DNA to releasing medium and conditions or due to the acidic microenvironment that is created inside the nanoparticles with the degradation of PLG–PLA polymer. However, several studies have demonstrated retention in biological activity of the DNA that is released from nanoparticles with time.[9,14,17] The results thus indicate the ability of nanoparticles to provide sustained release of functional DNA.

To enhance DNA entrapment efficiency in nanoparticles, the recent approaches include modifying PLG–PLA with cationic polymers or co-incorporating cationic polymers so that the DNA can be encapsulated or coated onto the nanoparticle surface in a condensed form. Furthermore, it is hypothesized that the condensed DNA would have better transfection efficiency than uncondensed DNA. For example, nanoparticles formulated with PLA grafted with polysaccharide or cationic surfactants have been studied as a DNA carrier system.[18-20] Significant progress has been made in the approach of coating DNA onto cationic PLG particles. DNA-coated particles have been investigated to induce immune response against HIV and also as a cancer vaccine.[20,21]

22.4 MECHANISM OF NANOPARTICLE-MEDIATED SUSTAINED GENE TRANSFECTION

Different strategies and polymer systems have been investigated for the formulation of nanoparticles; however, their intracellular uptake and trafficking is a relatively less investigated aspect. In order to understand the mechanism of nanoparticle-mediated gene transfection, it is necessary to understand how nanoparticles deliver DNA into cells. Previous studies in our laboratory have shown that nanoparticles are internalized efficiently into cells through a concentration- and time-dependent endocytic process.[22] The efficiency of intracellular nanoparticle uptake is reduced at higher doses and with incubation time, suggesting that the uptake pathway is saturable or reaches an equilibrium uptake.[23] Our results demonstrated that the nanoparticle uptake reaches a plateau within 2 h of incubation in most cell lines, suggesting a rapid process of nanoparticle uptake.

Particulate systems could enter the cells through various endocytic processes, such as by phagocytosis, fluid phase pinocytosis, or receptor-mediated endocytosis. We have demonstrated that PLG nanoparticles are internalized, in part, through fluid-phase pinocytosis and in part through clathrin-coated pits in vascular smooth muscle cells. Caveoli and phagocytosis are not involved in nanoparticle uptake in this cell line. Following their cellular uptake, nanoparticles are transported to primary endosomes and then to sorting endosomes. From sorting endosomes, a fraction is sorted back to the outside of the cell through recycling endosomes while the rest is transported to the secondary endosomes and lysosomes. It was further demonstrated that nanoparticles rapidly escape the endo-lysosomes and enter the cytosolic compartment within 10 min following incubation of cells with nanoparticles. Reversal of the nanoparticle surface charge selectively in the acidic pH of the endo-lysosomes has been proposed as the mechanism responsible for the endo-lysosomal escape of nanoparticles. This reversal of the surface charge probably occurs through the transfer of protons from the bulk solution to the nanoparticle surface in the acidic pH of secondary endosomes. Protonation of PLG particles in the acidic pH has been demonstrated by Makino et al.[24] and is attributed to the transfer of protons from the bulk solution to the surface of nanoparticles. We hypothesize that the cationization of nanoparticles results in their interaction with vesicular membranes leading to transient and localized destabilization of the membrane and the escape of nanoparticles into the cytoplasmic compartment.[22] Following their localization into the cytoplasmic compartment, nanoparticles could release the encapsulated DNA intracellularly at a slow rate for nuclear localization. Although we have demonstrated rapid escape of nanoparticles from the endo-lysosomal compartment, the subsequent study demonstrated that a significant fraction of nanoparticles undergoes exocytosis[25] (Figure 22.1). Thus any modification in the nanoparticle properties that would reduce the exocytosis of nanoparticles is expected to increase the intracellular retention of nanoparticles and hence gene expression.

Escape of the expression vector from the endo-lysosomal compartment is a key step in gene expression. Cationic polymers, for example polyethylenimines, have been demonstrated to escape the endo-lysosomal compartment through a "proton sponge" effect, in which the osmotic pressure developed inside the endosomes causes swelling, rupture, and escape of the vector.[26] This mechanism of escape of the vector, in part, is responsible for the toxicity associated with cationic polymeric systems.[27] Furthermore, the interactions of cationic polymers with cell membrane and cytoplasmic

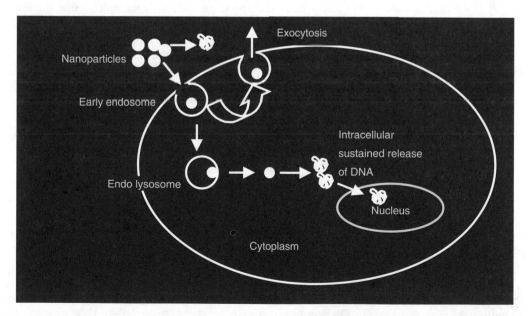

Figure 22.1   Schematic depicting intracellular trafficking of nanoparticles. Nanoparticles are taken up by cells via endocytic process. A fraction of nanoparticles following uptake undergoes exocytosis[25] whereas a fraction enters in the secondary endosomes. Cationization of nanoparticles in the acidic pH of secondary endosomes causes nanoparticles to interact with the endosomal vesicles, resulting in transient destabilization of the endosomal membrane and escape of nanoparticles in the cytoplasmic compartment.[22] It is hypothesized that the DNA is released slowly from the nanoparticles localized in the cytoplasmic compartment, resulting in sustained gene expression.

proteins have been attributed to their toxicity. In our recent studies, we have demonstrated that polyethylenimine causes significant toxicity. With PLG nanoparticles, we did not observe cell toxicity even at a concentration of 1,000 µg/ml *in vitro* in vascular smooth muscle cells, whereas 80% of cell death occurred when they were treated with polyethylenimine (25 kDa) at 2 µg/ml concentration.[28] Efforts are under way to reduce the toxicity of cationic polymers by chemical modification to balance the cationic charge on the polymers so that it can condense the DNA without causing toxicity.[27] However, the cellular toxicity with these polymers still remains an issue despite their better transfection efficiency than other polymeric systems.

22.5 FACTORS AFFECTING NANOPARTICLE-MEDIATED GENE TRANSFECTION

Based on the intracellular trafficking scheme proposed above, nanoparticle uptake and retention, endo-lysosomal escape, and DNA release from nanoparticles are important factors that could influence gene transfection. Various formulation factors such as the type of the polymer and the emulsifier used in the nanoparticle formulation have been shown to influence these parameters, and hence the nanoparticle-mediated gene transfection.

22.5.1 Intracellular Uptake and Endo-Lysosomal Escape

Intracellular uptake of nanoparticles is affected by a number of factors, such as particle size, and surface characteristics, such as hydrophilicity and zeta potential. PLG–PLA nanoparticles prepared by emulsion solvent evaporation technique have a heterogeneous size distribution. In our studies, the nanoparticles fractionated into greater than 100 nm and less than 100 nm sizes demonstrated different levels of gene expression. The smaller size fraction of nanoparticles demonstrated 27-fold higher transfection in COS-7 cells than the larger particle size fraction. However, higher gene transfection with the smaller size fraction was not associated to its higher cellular uptake on weight basis or DNA release rate as compared to the larger size fraction.[14] Although this is yet to be investigated, it could be speculated that the higher gene transfection with smaller size nanoparticles could be due to their greater number as compared to the larger size fraction per unit weight, or it could be that they have different intracellular distribution than the larger size fraction. Effect of particle size on gene transfection has been reported for other systems such as polyplex and DNA-lipid complexes, with smaller size complexes demonstrating better transfection than larger size complexes or aggregates.[29] Thus the smaller size with a uniform particle size distribution is expected to enhance the gene expression of nanoparticles.

Since the nanoparticle surface comes in contact with the cell surface, it is anticipated that the interfacial properties of nanoparticles could influence their cellular uptake properties as well as gene expression. The most ignored aspect with PLG–PLA nanoparticles is the role of surface-associated polyvinyl alcohol (PVA), which is a commonly used emulsifier in the formulation of PLG–PLA nanoparticles, on the interfacial properties of nanoparticles, and its influence on cellular uptake. A fraction of PVA remains associated with the nanoparticle surface and cannot be removed even by multiple washing of nanoparticles.[30] It has been proposed that the hydrophobic region of PVA molecules penetrates into the nanoparticle matrix while the hydrophilic region is present on the surface. PLG–PLA nanoparticle surface can be covered by up to five layers of PVA molecules. In our studies, the surface-associated PVA has been shown to influence the physical as well as the cellular properties of nanoparticles.[31] The amount of PVA associated with nanoparticle surface depends on its concentration used as an emulsifier in the formulation of nanoparticles. Nanoparticles formulated using 2% w/v PVA had about 3% w/w surface-associated PVA whereas those formulated using 5% w/v PVA had about 5% w/w surface associated PVA. In our studies, we observed about 2.8-fold higher nanoparticle uptake in vascular smooth muscle cells for the formulation with lower

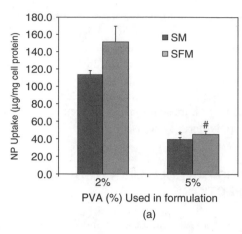 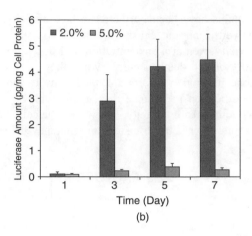

PVA (%) Used in formulation   Time (Day)

(a)   (b)

Figure 22.2   Effect of surface associated polyvinyl alcohol (PVA) on nanoparticle uptake in vascular smooth muscle cells (VSMCs). Nanoparticles were formulated using 2% and 5% w/v PVA as an emulsifier. VSMCs (50,000 cells per plate in 24-well plate) were allowed to attach for 24 h, the medium was changed with a suspension of nanoparticles (100 μg/ml) prepared either in serum (SM) or serum free (SMF) medium. Cells were incubated with particles for 1 h, washed, and the nanoparticle levels in the cell were determined by HPLC.[23] Nanoparticles formulated with 2% PVA demonstrated greater uptake than the uptake of nanoparticles formulated with 5% PVA despite their similar particle size (A). Gene expression of nanoparticles formulated with 2% PVA as an emulsifier was significantly greater than the gene expression with nanoparticles formulated using 5% PVA in MCF-7 cells, despite similar DNA loading and particle size (B). For gene transfection, MCF-7 cells (35,000 per well in 24-well plate) were incubated with nanoparticles (450 μg/ml per well, DNA dose = 13 μg) for 1 day and then the medium in wells was replaced with fresh medium (without nanoparticles). Medium was changed on every alternate day thereafter. (Figure 22.2A is reproduced from Sahoo, S. K. et al., Residual polyvinyl alcohol associated with poly (D,L-lactide-co-glycolide) nanoparticles affects their physical properties and cellular uptake, *J. Control. Release* 82 (1), 105, 2002. with permission from Elsevier.)

amount of surface associated PVA than nanoparticles with a higher amount of surface associated PVA (Figure 22.2A). In our studies, we attributed the reduced cellular uptake of nanoparticles with increase in surface associated PVA to the increase in the hydrophilicity of the nanoparticle surface.[31]

In our recent studies, the surface-associated PVA has been demonstrated to influence the intracellular distribution of nanoparticles. Nanoparticles with a higher amount of surface-associated PVA (formulated using 5% w/v PVA) demonstrated lower nanoparticle levels in the cytoplasmic fraction in MCF-7 cells as compared to the levels for the formulation with lower amount of surface-associated PVA (formulated using 2% PVA).[32] It is suggested that the PVA present on the nanoparticles' surface shields the charge reversal of nanoparticles in the endo-lysosomes, resulting in a lower number of nanoparticles escaping into the cytoplasm. The lower uptake and the reduced cytoplasmic levels for the nanoparticles with higher amount of surface-associated PVA reflected the gene transfection of these nanoparticles. In MCF-7 cells, nanoparticles with higher amount of surface associated PVA demonstrated 12- to 20-fold lower gene expression as compared to that of nanoparticles with a lower amount of surface-associated PVA (Figure 22.2B). We have further demonstrated that the gene expression of nanoparticles depends on the molecular weight of PVA used as well as on its degree of hydrolyzation. Thus the interfacial properties of nanoparticles, influenced by the surface-associated emulsifier, affect their cellular uptake, intracellular distribution, and gene expression.

22.5.2   DNA Release

Release of the encapsulated DNA from nanoparticles inside the cells could be a critical factor in determining the transfection levels, since it would provide the estimate of the amount of DNA that is available in the cytoplasmic compartment for nuclear localization. Release of therapeutic

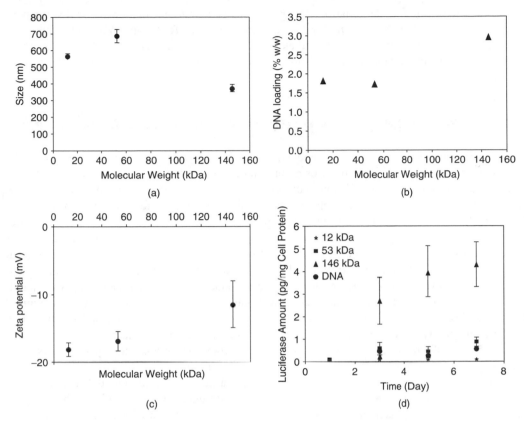

Figure 22.3 Effect of molecular weight of PLG polymer on particle size (A), DNA loading (B), zeta potential (C), and gene expression of nanoparticles in MCF-7 cells (D). The same dose of nanoparticles (450 µg/ml per well) was used in the transfection study.

agents from PLG matrices has been shown to occur via diffusion through the polymer matrix during early stages and then by degradation of the polymeric matrix at later stages. However, for a hydrophilic and high-molecular-weight agent such as pDNA, release could occur through aqueous channels and pores created inside the nanoparticle polymer matrix.[33] Thus, the factors that could influence the DNA loading, diffusion of DNA through polymer matrix, and degradation of the nanoparticle polymer matrix would influence the DNA release from nanoparticles.

DNA loading in nanoparticles is influenced by the factors that affect emulsion stability during nanoparticle formulation (Figure 22.3). In our studies, nanoparticles formulated using higher molecular weight polymer demonstrated relatively smaller particle size, lower polydispersity index, higher DNA loading, and lower zeta potential. The increase in DNA loading with increase in molecular weight of polymer could be because of higher viscosity of the polymer phase and its better emulsification, resulting in greater DNA entrapment in nanoparticles.[15] Thus, due to relatively higher DNA loading and higher DNA release, the nanoparticles formulated using higher molecular weight polymer (143 kDa) demonstrated 10- to 100-fold higher transfection as compared to the transfection with nanoparticles formulated using lower molecular weight polymers (12–53 kDa) (Figure 22.3D). The difference in the transfection levels has been attributed to the DNA loading and its release from nanoparticles. Higher DNA loading resulted in greater DNA release and hence higher gene transfection.

It is known that polymer hydrophobicity affects the release of encapsulated agent from nanoparticles. In our studies, we observed that nanoparticles formulated with 100% lactide (PLA) had slower DNA release as compared to the release from nanoparticles formulated with PLG. Increasing the polymer hydrophobicity could result in reduced diffusion of hydrophilic DNA and also could reduce the rate of polymer degradation. The gene transfection studies demonstrated that the nano-particles formulated from 50/50 glycolide/lactide demonstrate 10-fold higher transfection compared to the transfection of nanoparticles formulated using 100% lactide. Thus, the DNA loading and DNA release rate from nanoparticles seem to be critical determinants in nanoparticle-mediated gene transfection.

22.6 APPLICATIONS OF NANOPARTICLE-MEDIATED GENE TRANSFECTION

In our previous studies, sustained gene expression has been demonstrated both *in vitro* and *in vivo* using nanoparticles. In a rat bone osteotomy model, sustained gene expression was observed in the bone-gap tissue that was filled with marker DNA loaded nanoparticles and retrieved 5 weeks after the initial surgery.[17] This gene delivery strategy could be used to facilitate bone healing using therapeutic genes such as bone morphogenic protein. Also, PLG emulsion containing alkaline phosphatase as a marker gene was used for coating a gut suture. The gene-coated suture was used to close an incision in the rat skeletal muscles. Two weeks after the surgery, the tissue from the incision site demonstrated gene expression.[17] Gene-coated sutures encoding a growth factor such as vascular endothelial growth factor could facilitate wound healing. In another study, Cohen et al. have demonstrated that the gene transfection with nanoparticles was 1 to 2 orders of magnitude greater than that with a liposomal formulation at 7 days following intramuscular injection mice, despite significantly lower transfection observed *in vitro* with nanoparticles as compared to lipo-somes.[9] Further, the nanoparticle-mediated *in vivo* gene transfection was seen up to 28 days in this study, suggesting sustained gene activity of nanoparticles *in vivo*.

Sustained gene expression is a key feature of nanoparticle-mediated gene delivery. When one discusses the relative efficiency of gene expression for different vectors, it is usually the peak gene expression level that is taken into consideration, but not the duration of gene expression. In our studies, we have observed that transfection with a commercially available transfecting agent (FuGENE 6) peaks at 1 to 2 days in COS-7 cells following transfection but then declines rapidly. With nanoparticles, however, the gene expression is relatively lower than the transfecting agent but remains sustained.[16] Thus, the nanoparticle-based strategy is particularly appealing for treating disease conditions which require sustained level of gene transfection. For example, sustained level of growth factor expression in the ischemic tissue to induce angiogenesis is considered better than higher but transient level of gene expression in order to develop functional and matured blood vessels.[34] Unlike polyplexes and lipoplexes, nanoparticles are anionic in physiologic pH and do not aggregate under physiologic conditions or in the presence of serum, and hence nanoparticles are suitable for *in vivo* administration. Analogous to drug therapy, where optimal drug concentration in the target tissue for a desired duration is necessary, the gene therapy approach could similarly be optimized for the level and duration of gene expression based upon the need of the disease condition to achieve a desired therapeutic effect. Biodegradable nanoparticles thus could be one such promising vector that can achieve sustained gene expression in the target tissue without the risk of toxicity concerns.

22.7 SUMMARY

Biodegradable nanoparticles thus offer a nontoxic and nonimmunogenic gene delivery system that is capable of delivering a gene of interest at a slow rate, resulting in sustained gene delivery. Many factors, such as DNA loading, release of the encapsulated DNA from nanoparticles, cellular uptake of nanoparticles, and endo-lysosomal escape of the internalized nanoparticles, could influ-ence nanoparticle-mediated gene transfection. Nanoparticles conjugated to specific ligands could further enhance the potential of the system as a targeted gene delivery vector.

ACKNOWLEDGMENTS

We gratefully acknowledge financial support from the Nebraska Research Initiative, Gene Therapy Program and the National Institutes of Health (HL57234). Swayam Prabha is supported by a predoctoral fellowship from the Department of Defense, U.S. Army Medical Research and Materiel Command (DAMD17-02-1-0506). We would also like to thank Ms. Elaine Payne for providing administrative support.

REFERENCES

1. Maheshwari, A. et al., Biodegradable polymer-based interleukin-12 gene delivery: role of induced cytokines, tumor infiltrating cells and nitric oxide in anti-tumor activity, *Gene Ther.* 9 (16), 1075, 2002.
2. Maheshwari, A. et al., Soluble biodegradable polymer-based cytokine gene delivery for cancer treatment, *Mol. Ther.* 2 (2), 121, 2000.
3. Han, S., Mahato, R.I., and Kim, S.W., Water-soluble lipopolymer for gene delivery, *Bioconjug Chem* 12 (3), 337, 2001.
4. Han, S. et al., Development of biomaterials for gene therapy, *Mol. Ther.* 2 (4), 302, 2000.
5. Thomas, C.E., Ehrhardt, A., and Kay, M.A., Progress and problems with the use of viral vectors for gene therapy, *Nat. Rev. Genet.* 4 (5), 346, 2003.
6. Lechardeur, D. and Lukacs, G.L., Intracellular barriers to non-viral gene transfer, *Curr. Gene Ther.* 2 (2), 183, 2002.
7. Li, S. and Huang, L., Nonviral gene therapy: promises and challenges, *Gene Ther.* 7 (1), 31, 2000.
8. Panyam, J. and Labhasetwar, V., Biodegradable nanoparticles for drug and gene delivery to cells and tissue, *Adv. Drug Deliv. Rev.* 55 (3), 329, 2003.
9. Cohen, H. et al., Sustained delivery and expression of DNA encapsulated in polymeric nanoparticles, *Gene Ther.* 7 (22), 1896, 2000.
10. Wang, J. et al., Enhanced gene expression in mouse muscle by sustained release of plasmid DNA using PPE-EA as a carrier, *Gene Ther.* 9 (18), 1254, 2002.
11. Shive, M.S. and Anderson, J.M., Biodegradation and biocompatibility of PLA and PLGA microspheres, *Adv. Drug Deliv. Rev.* 28 (1), 5, 1997.
12. Ikada, Y. and Tabata, Y., Phagocytosis of polymer microspheres by macrophages, *Adv. Polym. Sci.* 94, 107, 1990.
13. Desai, M. P. et al., The mechanism of uptake of biodegradable microparticles in Caco-2 cells is size dependent, *Pharm. Res.* 14 (11), 1568, 1997.
14. Prabha, S. et al., Size-dependency of nanoparticle-mediated gene transfection: studies with fractionated nanoparticles, *Int. J. Pharm.* 244 (1–2), 105, 2002.
15. Prabha, S. and Labhasetwar, V., Effect of polymer molecular weight and composition on nanoparticle-mediated gene transfection, *Mol. Ther.* 7 (5), S73, 2003.
16. Prabha, S. and Labhasetwar, V., Critical determinants in PLGA/PLA nanoparticle-mediated gene expression, *Pharm. Res.*, 21 (2), 354–364, 2004.
17. Labhasetwar, V. et al., Gene transfection using biodegradable nanospheres: results in tissue culture and a rat osteotomy model, *Coll. Surf. B* 16 (1–4), 281, 1999.
18. Cui, Z. and Mumper, R.J., Plasmid DNA-entrapped nanoparticles engineered from microemulsion precursors: *in vitro* and *in vivo* evaluation, *Bioconj. Chem.* 13 (6), 1319, 2002.
19. Cui, Z. and Mumper, R.J., Genetic immunization using nanoparticles engineered from microemulsion precursors, *Pharm. Res.* 19 (7), 939, 2002.
20. Cui, Z. and Mumper, R.J., Topical immunization using nanoengineered genetic vaccines, *J. Control. Release* 81 (1–2), 173, 2002.
21. Ott, G. et al., A cationic sub-micron emulsion (MF59/DOTAP) is an effective delivery system for DNA vaccines, *J. Control. Release* 79 (1–3), 1, 2002.
22. Panyam, J. et al., Rapid endo-lysosomal escape of poly(DL-lactide-co-glycolide) nanoparticles: implications for drug and gene delivery, *FASEB J.* 16 (10), 1217, 2002.

23. Davda, J. and Labhasetwar, V., Characterization of nanoparticle uptake by endothelial cells, *Int. J. Pharm.* 233 (1–2), 51, 2002.

24. Makino, K., Ohshima, H., and Kondo, T., Transfer of protons from bulk solution to the surface of poly(L-lactide) microcapsules, *J. Microencapsul.* 3 (3), 195, 1986.

25. Panyam, J. and Labhasetwar, V., Dynamics of endocytosis and exocytosis of poly(D,L-lactide-co-glycolide) nanoparticles in vascular smooth muscle cells, *Pharm. Res.* 20 (2), 212, 2003.

26. Boussif, O., Zanta, M.A., and Behr, J.P., Optimized galenics improve *in vitro* gene transfer with cationic molecules up to 1000-fold, *Gene Ther.* 3 (12), 1074, 1996.

27. Putnam, D. et al., Polymer-based gene delivery with low cytotoxicity by a unique balance of side-chain termini, *Proc. Nat. Acad. Sci. U.S.A.* 98 (3), 1200, 2001.

28. Panyam, J. et al., Fluorescence and electron microscopy probes for cell and tissue uptake of poly (D,L-lactide-co-glycolide) nanoparticles., *Int. J. Pharm.*, 262 (1-2) 1–11, 2003.

29. Cherng, J.Y. et al., Effect of size and serum proteins on transfection efficiency of poly ((2-dimethy-lamino)ethyl methacrylate)-plasmid nanoparticles, *Pharm. Res.* 13 (7), 1038, 1996.

30. Murakami, H. et al., Preparation of poly(DL-lactide-co-glycolide) nanoparticles by modified spontaneous emulsification solvent diffusion method, *Int. J. Pharm.* 187 (2), 143, 1999.

31. Sahoo, S. K. et al., Residual polyvinyl alcohol associated with poly (D,L-lactide-co-glycolide) nanoparticles affects their physical properties and cellular uptake, *J. Control. Release* 82 (1), 105, 2002.

32. Prabha, S. and Labhasetwar, V., Effect of residual polyvinyl alcohol on nanoparticle-mediated gene transfection in breast cancer cells, *Mol. Ther.* 7 (5 Suppl.), S67, 2003.

33. Langer, R. et al., Polymers for sustained release of macromolecules: applications and control of release kinetics, in *Controlled Release of Bioactive Materials*, Baker, R. (Ed.), Academic Press, Bend, OR, 1980, 83.

34. Lee, R. J. et al., VEGF gene delivery to myocardium: deleterious effects of unregulated expression, *Circulation* 102 (8), 898, 2000.

CHAPTER 23

Nanoparticles Made of Poly(lactic acid) and Poly(ethylene oxide) as Carriers of Plasmid DNA

Noémi Csaba, Celso Pérez, Alejandro Sánchez, and Maria José Alonso

CONTENTS

23.1 INTRODUCTION

The advantageous properties of biodegradable colloidal systems have already been clearly evidenced during the past few decades. These carriers can be successfully used for the delivery of a wide variety of drugs, including macromolecules that are poorly soluble or that are unstable in the environmental conditions of the living body.[1-3] The association of the drug molecules can be carried out by encapsulation inside the particle matrix or by adsorption onto the particle surface. There exist already numerous techniques for the formation of nanoparticles, the selection of which normally depends on the characteristics of the drug molecule, and also those of the polymer.[4,5] A particularly important feature of these vehicles is that, due to their nanometric size, they are able to overcome biological membranes and even well-organized mucosae, thereby transporting the associated active compound.[6,7] This specific ability of nanoparticles to provide intracellular penetration for the drug molecules has a great importance in gene delivery. Nevertheless, besides particle size, factors such as the surface characteristics and composition also have strong influence on the stability and on the *in vivo* fate of both the drug and the carrier.[8,9]

Poly(lactic acid) (PLA) and poly(lactic-*co*-glycolic acid) (PLGA) are probably the most commonly used biodegradable polymers used for the formation of particulate drug delivery systems. These polymers are well known for their biodegradability, biocompatibility, and ability to control drug release.[10] However, upon degradation, these polymers are converted into highly reactive oligomers, which generate a very acidic microenvironment inside the particles with dramatic consequences on the stability of the incorporated molecules.[11,12] This is especially important in the case of macromolecules, such as proteins and genes, which can lose their biological activity due to these undesired interactions with the surrounding nanoparticle matrix. The incorporation of additional excipients to the polymer matrix, such as poly(ethylene oxide) and derivatives, has been proposed as an efficient approach for overcoming this limitation.[13]

Interactions between PLA- and PLGA-based nanoparticles and the *in vivo* environment can also compromise their efficiency as drug delivery systems. It has been described that, when administered intravenously, these particles are easily recognized by the mononuclear phagocyte system (MPS) and rapidly cleared from the blood stream.[14] This effect can also be attributed to the hydrophobic character of the particle forming polymers that favour the adsorption of blood components, such as opsonins; a process that represents the first step of their elimination from the circulation. On the other hand, it has also been shown that polyester-based nanoparticles can suffer significant aggregation or flocculation in biological media that are rich in enzymes, such as gastrointestinal fluids.[15] The development of surface-modified nanoparticles appears to be an appropriate and efficient solution for these problems.[16]

23.2 POLY(ETHYLENE OXIDE) AND DERIVATIVES: PHYSICOCHEMICAL PROPERTIES AND BIOLOGICAL BEHAVIOR

The terms poly(ethylene oxide) (PEO) and poly(ethylene glycol) (PEG) refer to the same compound. However, the term PEO is normally preferred when referring to high-molecular-weight polymers and also, irrespective of the molecular weight, when referring to block copolymers such as poly(ethylene oxide)–poly(propylene oxide) (PEO–PPO). The usual names of these latter copolymers, as indicated in the USPNF, are poloxamer and poloxamine.

PEO and its derivatives (poloxamers and poloxamines) are neutral and hydrophilic molecules with flexible polymer chains. Additionally, in the case of poloxamine derivatives, apart from the PEO and PPO units, the polymer also comprises an ethylenediamine molecule as a linking element. The chemical structure of these polymers is presented in Figure 23.1. The differences between these molecules rely on their chain lengths, and their PEO and PPO content. These variations in the chemical structure have a consequence on their solubility and surface active properties. Poloxamers and poloxamines can be obtained from BASF Corporation under the commercial names Pluronic® and Tetronic®, respectively. These commercial names are usually followed by a number of 2 or 3 digits in which additional information is given about the composition and the approximate molecular weight of the molecules. In the case of poloxamers, this additional information also

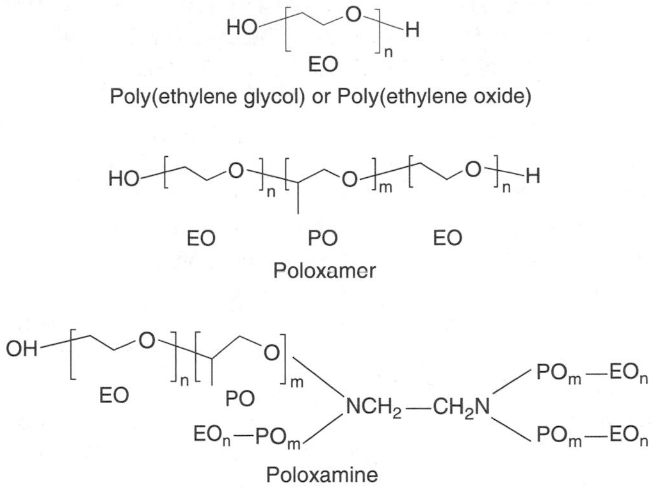

Figure 23.1 Chemical structure of PEG and its derivatives.

includes a letter that indicates the physical form of the compound. According to this, the designation "L," "P," and "F" stand for liquid, paste, and solid (flake) forms.

PEO and its derivatives (poloxamers and poloxamines) have been extensively used as biomaterials. More recently, due to their good biocompatibility and low toxicity, these biomaterials have attracted great attention for the design of advanced drug delivery systems. However, these polymers are not biodegradable and, consequently, their potential application in drug delivery will largely depend on their molecular weight.

Thus, while very high molecular weight polyethylene oxide (up to 7000 kDa) is acceptable for the design of oral drug delivery systems,[17,18] only those with a relative low molecular weight (normally less than 30 kDa and preferably 5–10 kDa) have been proposed for parenteral administration.[19-21] This is obviously due to the difficulties for the glomerular filtration and, hence, elimination of high molecular weight compounds from the blood circulation.

In the particular case of poloxamers, their potential has been continuously increasing due to their versatility and their intrinsic biological behavior, which varies depending on their composition. For example, some specific poloxamers have the ability to evolve reversibly from a sol to a gel state upon physiological conditions, usually upon changes in temperature.[22,23] Due to this interesting behavior these polymers are known as thermosensitive and "*in situ* gelling systems." They have already been used for a wide variety of purposes, their application for ocular drug delivery being of special interest.[24,25]

Another interesting property of poloxamers relies on their ability to increase the permeability of very well-organized epithelia. This specific property and its connotations in drug delivery have been very well illustrated by their successful application as penetration enhancers through the blood-brain barrier.[26-28]

The potential that poloxamers offer as carriers in gene therapy is especially promising. More specifically, there is recent evidence of the ability of these polymers to complex and protect plasmid DNA (pDNA), thereby enhancing gene expression. Furthermore, the efficacy of a combination of two poloxamers, Pluronic® L61 and Pluronic® F 127, as a gene carrier was also confirmed *in vivo*.[29,30]

Besides the intrinsic potential of PEO and its derivatives, which will be extensively reviewed in a different chapter of this book, these polymers are probably the most popular materials for the surface modification of drug delivery vehicles (liposomes, micelles, and nanoparticles). This surface modification can be achieved by adsorbing them onto preformed nanoparticles or by covalent linking to the core forming polymer (i.e., PLA, PLGA) prior to particle formation. More recently, the interest of introducing them into the nanoparticle matrix has also been taken into consideration.

23.3 POLY(ETHYLENE OXIDE) AND DERIVATIVES: NEW APPROACHES FOR IMPROVING DRUG DELIVERY FROM NANOPARTICULATE CARRIERS

Due to the interesting biological properties of PEO and derivatives, there has been a growing interest in developing technological approaches for engineering nanoparticles consisting of a combination of polyesters and PEO. These approaches have been based upon two main principles: (1) physical entrapment or adsorption of PEO and derivatives into or onto PLA and PLGA nanomatrices, and (2) chemical modification of the matrix-forming polymer (PLA and PLGA) by covalent linking with PEO. An illustration of the different structures that can be formed is shown in Figure 23.2. In the first case, the modification will be in the core or on the coating, depending on whether PEO is introduced in the nanoparticles preparation process (Figure 23.2A) or simply adsorbed onto preformed particles (Figure 23.2B). In the second case, the nanostructural organization of the PEG fragments of the copolymer will largely depend on the nanoparticles preparation technique. Accordingly, PEG chains will be oriented toward the surface of the particles (Figure 23.2C) and they can also form small reservoirs in the core of the particles (Figure 23.2D).

As indicated in the introductory section, there are two main purposes for the incorporation of PEO and derivatives to the classical PLA and PLGA nanoparticles. One purpose has been to avoid the interaction of these hydrophobic particles with proteins and enzymes by providing them with a protective hydrophilic stealth. The original idea was to prevent the rapid opsonization of the particles, thus prolonging their circulation time following intravenous administration. At present, it is known that this hydrophilic coating is also beneficial for preventing aggregation of the particles, following mucosal administration (oral, nasal, ocular) and their subsequent transport across mucosal barriers. Moreover, the PEG coating is now considered a very good strategy for introducing ligands on the particle surface and, hence, designing active-targeting delivery vehicles.[31,32] On the other hand, the purpose of the incorporation of PEO and derivatives into the nanoparticle matrix has been the improvement of the stability of nanoencapsulated molecules. As indicated above, the

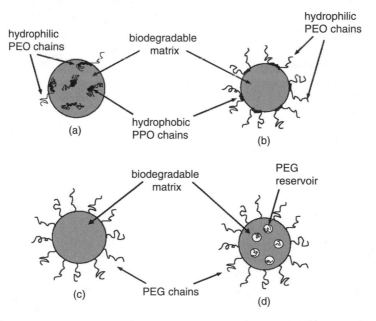

Figure 23.2 Schematic view of the various structures of nanoparticles made of polyesters and PEO and derivatives (poloxamer/poloxamine). (A) Poloxamer/poloxamine-containing nanoparticle. (B) Poloxamer/poloxamine-coated nanoparticle. (C), (D) Nanoparticles made of PLA–PEG copolymers.

dilution of PLA and PLGA with these polymers, forming an intimate blend, will reduce the acidification and, hence, provide a more friendly environment for proteins, oligonucleotides and plasmid DNA. Additionally, it has been hypothesized that the presence of poloxamer and polox-amine, together with plasmid DNA in the nanomatrix, might help improving the transfection efficiency. The next section describes the technological approaches for introducing polyoxyethylene and its derivatives within and/or onto the nanostructures.

23.4 DESIGN OF NANOPARTICLES MADE OF POLYESTERS, PEO, AND DERIVATIVES

23.4.1 Poloxamer/Poloxamine-Coated PLGA Nanoparticles

The easiest approach for surface modification of PLA and PLGA nanoparticles consisted of a simple physical adsorption of poloxamers and poloxamines. It has been reported that this straight-forward surface modification permits the stabilization of polymeric nanoparticles in biological fluids and prolongs their half-life in the blood stream.[21,33] The adsorption of these polymers onto hydro-phobic nanoparticles was possible since the hydrophobic part can easily anchor to the nanoparticles surface whereas the hydrophilic part protrudes toward the aqueous surrounding medium (Figure 23.2B). The protective effect of these polymers can be explained by the steric hindrance that the hydrophilic coating material displays over the hydrophobic matrix polymer. Additionally, from the results of studies performed with radiolabeled particles, it was concluded that these coatings are able to modify the biodistribution of the vehicle. For example, it was observed that some poloxamers and poloxamines enhance the lymphatic drainage and the uptake of particles by the macrophages of the regional lymph nodes.[34,35] This effect could also be advantageous in the development and administration of vaccines and drugs whose target is the lymphatic or the immune system.

Despite the number of articles reporting the biological significance of these coatings, some authors have argued that due to the nonspecific attachment, these polymers can be easily displaced by serum proteins.[36] This unwanted displacement reduces the stability of the vehicles in the circulation and exposes the particles to aggregation. As indicated above, an alternative approach to provide the nanoparticles with a PEG coating would be the synthesis of a copolymer with a PEG fragment (PLA–PEG, PLGA–PEG) and the use of an adequate manufacturing technique that favours the orientation of the PEG fragment toward the particles surface. This will be discussed in the corresponding section.

23.4.2 Poloxamer/Poloxamine-Containing PLGA Nanoparticles

The incorporation of poloxamers into PLA or PLGA micro- and nanoparticles has been a useful tool for the modification of the release profile and for the improvement of the stability of the encapsulated molecules.[13,37-39] The internal organization of poloxamers within the particle structure varies depending on the particles preparation technique. An illustration of a possible structure is presented in Figure 23.2A. For example, we have produced PLGA nanoparticles containing polox-amer reservoirs using the double emulsion–solvent extraction/evaporation technique.[40] In this case, poloxamer 188 was dissolved in the inner water phase of the emulsion, thus forcing its partial entrapment upon precipitation of PLGA. The idea behind this approach was to stabilize the macromolecule that was coencapsulated with the poloxamer. However, this technique is limited to the incorporation of a small amount of a poloxamer with a high HLB.

More recently, we have developed a more versatile technique that allowed us to obtain PLGA–poloxamer/poloxamine blends in the form of nanoparticles. The technique is based upon the principle of the solvent diffusion or solvent displacement. In this case, both PLGA and polox-amer/poloxamine are codissolved in the same organic solvent and, then, coprecipitated upon

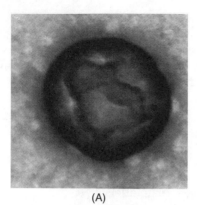
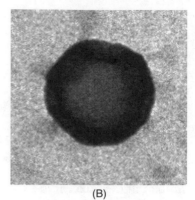

(A)   (B)

Figure 23.3   Nanoparticles prepared from PLGA:poloxamer (Pluronic® F68, HLB = 29) (A) and PLGA:polox-amine (Tetronic® 904, HLB = 14.5) blends (B).

addition to a miscible organic solvent. This technique allowed the effective incorporation of a series of poloxamers and poloxamines with different HLB values into PLGA nanoparticles, as confirmed by ^{1}H-NMR analysis.[41] The formation of these mixed systems is based on the phenomenon that polyesters and poloxamers/poloxamines are readily miscible via hydrogen bonding. Similar to other matrices based on an intimate polymer mixture,[13] the incorporation efficiency of the polox-amer/poloxamine into the particle matrix structure is determined by its thermodynamic compati-bility with the PLGA component and also by the solubility of the poloxamer/poloxamine in the nanoparticles external formation medium. Thus, the composition of the nanomatrix can be adjusted by selecting different polyether types and varying the preparation conditions. The physical appear-ance of some of these nanoparticles is shown in Figure 23.3.

The recognized advantages of the biodegradable PLGA nanoparticles and the previously men-tioned transfection-enhancing effect of poloxamer-based gene delivery systems suggest that these new nanostructures formed by an intimate blend of these two components could be very interesting devices for DNA encapsulation. This will be a specific subject of discussion in the corresponding section.

23.4.3   Design of Nanoparticles from Copolymers of PLA and PEG

Although PEG has also been physically entrapped into PLGA matrices for improving the stability of the encapsulated peptides,[42] the main purpose of using this polymer in the field of colloidal drug carriers has been to modify their surface properties. The only successful approach for forming surface-modified PLA–PLGA nanoparticles with a stable PEG protective coating has been the covalent attachment of PEG to the hydrophobic PLA–PLGA polymers. These copolymers consisting of PLA or PLGA and PEG can be synthesized by a ring opening poly-merization reaction of lactide monomers on the hydroxyl-end of a monomethoxy-poly(ethylene glycol) component in the presence of stannous octoate as catalyst.[19] Depending on the weight ratio of the reaction mixture, a series of PLA–PEG copolymers with different solubilities and hydrophilicities can be obtained. Moreover, besides varying the molecular weights, it is also possible to synthesize multiblock copolymers, for example, consisting of PLGA–PEG–PLGA or PLA–(PEG)$_3$ type block structures.[20,43]

These amphiphilic copolymers have been used to produce surface-modified particles in two different ways: they can be adsorbed onto preformed PLA–PLGA nanoparticles and used as coating materials or, alternatively, they can be processed in order to make nanoparticles with the appropriate surface-core configuration. With respect to the first approach, the adsorption of

PLA–PEG onto the particles is favored, as compared to that of PEG alone, due to the amphiphilic properties of the copolymer. However, in this situation, it is necessary to use low molecular weight PLA–PEG since it must be dissolved in the coating incubation medium. The few reports on the use of this approach indicated that this kind of coating permits a prolongation of the blood circulation time of nanoparticles.[44,45]

The alternative approach, based on the processing of the copolymer to produce PEG-coated PLA nanoparticles, has been the subject of an important number of articles.[19,20,46-48] The results published so far have indicated that not only the exact composition of the polymer, but also the nanoparticles manufacturing technique have a major effect on the polymer organization within the nanostructures and, hence in their surface composition. Most commonly, nanoparticles have been prepared by emulsion-solvent evaporation or by solvent diffusion techniques.[49,50] Because of the different solubility of the PLA and the PEG block, a phase separated structure is formed between the organic solvent and the aqueous phase during particle preparation, leading to the projection of the hydrophilic PEG chains toward the aqueous medium, thus overlaying the solidifying nanoparticle core.[46] The formation of this so-called core-shell structure has been confirmed by various surface analysis techniques[19] and by ¹H-NMR.[20,51]

Classically, the emulsion–solvent extraction/evaporation techniques have required high energy sources (sonication, homogenization) and also the use of surfactants, i.e., cholic acid or polyvinyl alcohol, in order to form submicrometric emulsions. However, very recently, we presented the formation of these PLA–PEG nanoparticles without surfactants and using a mild agitation technique. The key for the formation of these particles relied on the appropriate selection of formulation variables such as polymer molecular weight and concentration.[52]

For an appropriate steric hindrance effect, it is necessary to have a homogeneous distribution of the hydrophilic chains on the nanoparticle surface. We have investigated the influence of some formulation parameters (copolymer molecular weight and preparation technique) on the density of the PEG coating, as determined by ¹H-NMR.[52,53] The results showed that the PEG coating density (amount of PEG per weight of particles) can be modulated by adjusting the formulation conditions. This is particularly important if we take into account that the density of the PEG brush has a remarkable influence on plasma protein adsorption and particle aggregation. Indeed, it has been shown that a high PEG surface density is desirable for preventing protein adsorption[53] and also phagocytic uptake.[47,55,56] In addition, we have observed that the PEG coating has a key role in preventing particle aggregation following oral and nasal administration and facilitating particle permeation across mucosal surfaces.[57,58] More specifically, in a recent work performed in our laboratory we found that nanoparticles prepared by solvent evaporation (single or double emulsion) crossed the nasal mucosa more easily than those prepared by nanoprecipitation, whose PEG surface density was less important.[53]

Depending on the preparation technique, some of the PEG chains can be situated inside the polymeric matrix. For example, using the double emulsion–solvent evaporation technique, some PEG molecules have the chance of being oriented toward the internal aqueous phase of the double emulsion, thus forming, upon solidification, little reservoirs in which the active molecule is entrapped. The presence of these PEG reservoirs increases water uptake, thereby influencing the degradation of the core forming polymer and also the release characteristics of the vehicles.[59] The PEG component present in the interior of the particle can also act upon the stability of the entrapped molecules by blocking unfavorable interactions of the drug with the surrounding environment or with the degradation products of the core polymer. Additionally, the fact that the reactive carboxylic end group of the polyester remains blocked by the attachment of the PEG molecule can also help in preserving the structural integrity of the drug. These protective effects could be confirmed in the case of tetanus toxoid, which maintained its biological activity when encapsulated inside PLGA–PEG–PLGA microspheres[43] and PLA–PEG nanospheres.[60]

23.5 ENCAPSULATION AND DELIVERY OF OLIGONUCLEOTIDES AND PLASMID DNA FROM NANOPARTICLES

Surprisingly, despite the important amount of work in the design of nanostructures consisting of polyesters and PEO and derivatives, there are only a few examples in the literature aimed at exploring the advantages of these novel carriers for gene/oligonucleotide delivery. The main purpose of this section is to describe in detail the advances made so far on the nanoencapsulation and controlled release of these complex macromolecular compounds.

23.5.1 Encapsulation of Oligonucleotides and Plasmid DNA into PLA-PEG Nanoparticles

An interesting approach for the association of oligonucleotides to PLA–PEG nanoparticles was explored by Emile et al.[61] Two different antisense oligonucleotide sequences containing the same bases in reverse orientation, INV-val 13 and AS-val 13, were used for these studies.[62,63] In order to reduce their negative charge, these oligonucleotides were complexed with lysine-containing cationic oligopeptides, prior to their encapsulation. After this charge neutralizing process, the complex was solubilized together with the PLA–PEG copolymer in acetone and then coprecipitated in an aqueous medium. Nanoparticles were prepared of PLA–PEG copolymers with different PEG chain lengths (mol wt 2000 and 5000 Da) and constant PLA content (mol wt 30000 Da). The results showed that the charge neutralization of the oligonucleotide by the positive charges of the oligopeptide is an effective method for improving the encapsulation efficiency (from 5% to about 55%). On the contrary, the type of oligonucleotide and oligopeptide did not have strong influence on the encapsulation. Despite the efficacy of this formulation approach, no information regarding the stability of the oligonucleotide and its release from the nanoparticles has been reported so far. Nevertheless, the previous research on the microencapsulation of oligonucleotides within PLGA micro- and nanospheres (to be covered in Chapters 22 and 28 of this book), suggests that there are formulation options that remain to be explored in order to optimize encapsulation and release of oligonucleotides from nanoparticles.

We have recently investigated the effect of the preparation technique and the presence of protective excipients on the pDNA (pCMV-Luc) encapsulation and delivery from PLA–PEG nanoparticles.[64] For this purpose, we selected an optimized water-in-oil-in-water (w/o/w) solvent evaporation[40] and a newly adapted water-in-oil (w/o) solvent diffusion technique. In both cases, the emulsification energy (sonication) was minimized in order to avoid damage of the pDNA molecules. In addition, we evaluated the potential benefit of the co-encapsulation of protective interacting non-condensing polymers, such as polyvinyl alcohol (PVA) and polyvinyl pirrolidone (PVP). These so-called PINC polymers were reported to be able to protect DNA by complex formation by hydrogen bonds and through this improved stability, they can also enhance the efficiency of transfection.[65,66] Therefore, in the above-mentioned study we incubated the plasmid DNA with PVA and PVP, allowing their complexation in an aqueous medium prior to their encapsulation within PLA–PEG nanoparticles. The scheme for the formation of the pDNA-loaded nanoparticles is presented in Figure 23.4.

Under these conditions, all resulting formulations were of spherical shape (Figure 23.5), in nanometric size range (less than 300 nm), and had surprisingly high DNA loadings.

The physico-chemical characteristics of the formulations were shown to be dependent on the nanoparticle preparation method, however, they were not affected by the co-encapsulation of PVA and PVP. The w/o/w solvent evaporation technique yielded nanoparticles that were larger (270–300 nm) and less negative (~-20 mV) than those prepared by the solvent diffusion technique (125–130 nm, between -22 and -33 mV). These changes were attributed to the different mechanisms of nanoparticle formation and, in particular, to the environment that the different external phases (ethanol or water) provide for particle precipitation.

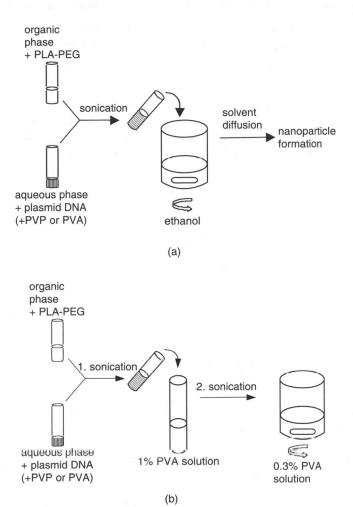

(a)

(b)

Figure 23.4   (A) Encapsulation of plasmid DNA into PLA–PEG nanoparticles by the emulsification–solvent diffusion technique. (B) Encapsulation of plasmid DNA into PLA–PEG nanoparticles by the water/oil/water double emulsion technique.

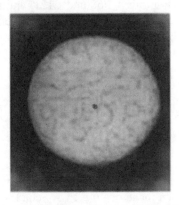

Figure 23.5   TEM micrograph of a plasmid-loaded nanoparticle prepared by the solvent evaporation technique. (From Perez, C., Sanchez, A., Putnam, D., Ting, D., Langer, R., and Alonso, M.J., Poly(lactic acid)-poly(ethylene glycol) nanoparticles as new carriers for the delivery of plasmid DNA, *J. Control. Release* 75, 211–224, 2001. With permission.)

Interestingly, similar tendencies could be observed for the encapsulation efficiencies that were influenced by the particle preparation technique but not by the presence of PVA or PVP. In all cases, the encapsulation efficiencies of the pCMV-Luc were extremely high, around 60–80% for the solvent evaporation and 80–90% for the solvent diffusion method. The higher efficiency provided by the latter technique can be explained by the very quick nanoprecipitation of the polymer upon the diffusion of the polymer dissolving solvent into the polymer precipitating solvent. An additional fact that facilitates high entrapment is the complete insolubility of DNA in the ethanolic external phase of this formulation. However, it was also observed that the addition of the plasmid led to a significant increase in the negative zeta potential of the particles prepared by the solvent diffusion technique. This result suggests that a considerable amount of the plasmid may be entrapped near or associated to the surface of the nanoparticles.

One of the most interesting observations in the above-mentioned work was the positive effect that the PEG in the polymer chain has in the encapsulation efficiency. Indeed, the encapsulation efficiency of pCMV-Luc into PLA nanospheres without PEG was as low as 25%, thus suggesting that PEG favors the interaction and, consequently, the entrapment of the plasmid into the nanoparticles. This interaction could be confirmed by incubating blank PLA–PEG nanospheres with free pDNA. The results showed that 29% of the plasmid was bounded to the surface of the nanoparticles, thus evidencing the interaction between the PEG chains and DNA.

The positive role of PEG in the encapsulation of plasmids was later corroborated for PLA–PEG microspheres[67] where the encapsulation efficiencies of DNA were significantly improved as compared to microspheres prepared from PLGA alone.[68,69] These authors indicated that the hydrophilic PEG chains are able to increase the compatibility between the DNA and the polymer molecules.[70] Other investigations of Liu et al. also revealed the impact of the DNA structure and molecular weight on its encapsulation efficiency in PLA-PEG particles. Their results suggested that small and compact DNA molecules can be more easily incorporated into the particle matrix than large molecules.[71]

23.5.2 Encapsulation of Plasmid DNA into PLGA Nanoparticles Containing Poloxamers and Poloxamines

Very recently, we have investigated the potential of PLGA–poloxamer and PLGA–poloxamine blend nanostructures for DNA delivery.[41,72] Previous results obtained for proteins encapsulated into poloxamer-containing blend matrices suggested that this system might be beneficial for the stability of the DNA molecule. With this objective in mind, we applied a very mild technique that permits the incorporation of poloxamers (Pluronic® F68 and L121) and poloxamines (Tetronic® 908 and 904) within the nanoparticle's structure. This emulsification-solvent diffusion technique differs from the previously applied version[64] in the sense that it does not require any sonication steps (Figure 23.6). The complete elimination of high-energy sources was expected to greatly contribute to the preservation of the initial configuration of the encapsulated plasmid.

The above-described technique yielded plasmid-loaded blend particles in the size range of 150–300 nm, with a spherical shape and a negative surface charge. In general, pDNA encapsulation did not significantly alter particle size. The only exception that could be observed was in the case of the nanoparticles composed of a PLGA–poloxamine 908 blend where nanoparticle size increased significantly (from 170 to 270 nm) upon pDNA incorporation. This phenomenon could probably be explained by a different kind of interaction with DNA that might not have been produced with the other derivatives used. The efficiencies of plasmid encapsulation into all blend formulations were found to be higher (35%) than those of pure PLA vehicles (25%).

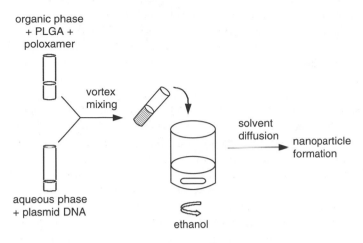

Figure 23.6 Encapsulation of plasmid DNA into PLGA–poloxamer blend nanoparticles by a modified emulsi-fication–solvent diffusion technique.

23.5.3 Controlled Release of Plasmid DNA from Nanoparticles of PLA–PLGA and PEO and Derivatives

The release of macromolecules, i.e., proteins, from PLA–PEG matrices has been studied and documented.[50,59,60] It has been generally accepted that macromolecules are released by a combined mechanism that includes polymer degradation followed by erosion, and also diffusion through the aqueous channels, the dimensions of which increase during polymer degradation and protein release.[59] As compared to the classical PLA–PLGA matrices, those containing PEG are known to have a greater water uptake which might potentially contribute to the release of the encapsulated macromolecule. Nevertheless, despite the work dedicated toward this aim, a clear conclusion of the effect of PEG on protein release has not been established yet. This is due to the additional benefit on the stability of encapsulated macromolecules that has been attributed to PEG, as will be commented on in the following section. Similarly, the co-incorporation of poloxamer within PLGA matrices has been found to affect protein release and stability of macromolecules such as tetanus toxoid and alpha interferon.[13,38]

To our knowledge, the only report on the release of plasmid DNA from PLA–PEG matrices was the one presented by our group for PLA–PEG nanoparticles.[64] The results of this work showed that the nanoparticle's preparation technique had a great impact on the pDNA release, whereas the co-encapsulation of the protective excipients, such as PVP or PVA, did not alter the plasmid release pattern. More specifically, nanoparticles prepared by the solvent diffusion technique led to a very quick plasmid release, whereas those prepared by the double emulsion–solvent evaporation tech-nique provided a slow and continuous release.

In other words, for the nanoparticles prepared by the solvent diffusion technique, the polymer matrix did not appear to have a role in the release, a situation that could in part be attributed to the presence of the previously mentioned surface-associated DNA molecules. A complementary explication was additionally found in the nonaqueous nature of the external phase of this preparation method. The plasmid encapsulated by this anhydrous process can create a considerable osmotic pressure inside the particles that can induce a very quick release by breaking up the particles. The development of this osmotic pressure in the interior of a particle has already been observed for proteins encapsulated similarly in nonaqueous conditions.[73] On the other hand, the nanoparticles prepared by the double emulsion technique displayed a typical biphasic pattern that provides plasmid release in its active form for at least a time period of 28 days. These formulations have an initial fast release that corresponds to the dissociation of the surface-bound DNA, followed by a second, continuous phase that is governed by the degradation of the PLA matrix and by the erosion of the particles.[74,75] On the other hand, in agreement with what was reported for PLGA nanoparticles,[76] we observed that changes in the theoretical plasmid loadings (0.5–1–2%) did not alter significantly the release pattern of PLA–PEG nanoparticles. These results together with the lack of effect of the PINC polymers co-encapsulated with the plasmid suggest that the degradation of the polymer matrix is the most important factor in determining the release characteristics of the nanoparticles prepared by the double emulsion technique.

As indicated in the previous section, we have recently attempted the substitution of a pure polyester matrix by a blend structure comprising a poloxamer-type compound and evaluated its effect on the controlled release of pDNA. A major advantage of these blend systems is that they usually reduce significantly the initial typical burst release of the encapsulated molecules. More specifically, it has been argued that the presence of the poloxamer in the particle can prevent rapid release of the drug by filling the pores of the polymeric core. In addition, poloxamer affects the water uptake of the vehicles and, consequently, the polymer degradation and drug release.[37,77] In agreement with these mechanisms, we have recently observed that the incorporation of poloxamers and poloxamines into PLGA nanoparticles led to an important reduction of the pDNA burst release (normally less than 10% of the encapsulated molecule), as compared to that typically observed for PLGA and PLA–PEG nanoparticles (more than 20%). This reduction of the burst effect was particularly remarkable for the nanostructures that comprise poloxamers and poloxamines with relatively low HLB values (i.e., Pluronic® L121 HLB = 1 and Tetronic® 904 HLB = 14.5). Moreover, the HLB of the poloxamers and poloxamines incorporated into the nanostrucures was observed to affect the pDNA release rate, the rate being slower for those having a lower HLB. Interestingly, for these blend systems, a closely linear pDNA release profile could be observed during a time interval of 2 weeks. Consequently, these results indicated that, as expected, the incorporation of poloxamers and poloxamines into the nanoparticle's structure opens new prospects on the controlled release of pDNA from nanocarriers.

Before making an extrapolation of the *in vitro* release behavior of pDNA from nanoparticles to the *in vivo* situation one must be very cautious. Indeed, a firm conclusion from the important amount of information accumulated in the protein delivery field is that *in vitro* conditions do not reflect the *in vivo* environment to which biodegradable colloidal systems are exposed. Furthermore, it has been clearly shown that *in vitro* release conditions often have a deleterious effect on the stability of the macromolecule being released from the particles.[38]

23.5.4 Stability of Plasmid DNA Released from Nanoparticles of PLA–PLGA and PEO and Derivatives

One of the major challenges in plasmid encapsulation into nanoparticles is the protection of the structural integrity and conformation of the DNA molecules. Indeed, pDNA, as a macromolecule, is very sensitive against the shear and high-energy sources that are normally required for the nanoparticle formation.[78,79] Being conscious of this important limitation, a few years ago Ando et al. proposed the microencapsulation of pDNA within PLGA microspheres using very low temperatures during the emulsification process.[80] The authors indicated that using this cryopreparation approach the exposure of the frozen pDNA to the sonication/homogenization energy was minimized. However, using this technique the authors were able to produce PLGA microparticles rather than nanoparticles. More recently, we investigated the effect of the nanoencapsulation conditions on the stability of pDNA incorporated into PLA–PEG nanoparticles. More specifically, we attempted to minimize the plasmid exposure to shear and cavitation-induced stress by developing a new solvent diffusion technique and using protective polymers (PVA, PVP). The electrophoretic analysis of the

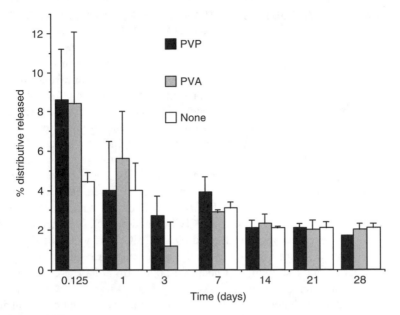

Figure 23.7 *In vitro* release of plasmid DNA from PLA–PEG nanoparticles prepared by double emulsion technique, encapsulated in a free form or associated with PVP or PVA. (From Perez, C., Sanchez, A., Putnam, D., Ting, D., Langer, R., and Alonso, M.J., Poly(lactic acid)-poly(ethylene glycol) nanoparticles as new carriers for the delivery of plasmid DNA, *J. Control. Release* 75, 211–224, 2001. With permission.)

plasmid released from the particles after a 1-day incubation period revealed a conversion of the supercoiled form to the relaxed isoforms. This conversion is not too critical if we take into account that open-circular form still retains an important transfection capacity. However, these results indicate as well that further improvements in particle preparation are necessary in order to maintain intact the original plasmid conformation. An important discovery in this regard is the recently reported approach for producing PLA–PEG nanoparticles without high energy sources and without surfactants.[52] Therefore, it could be assumed that using this single and straightforward procedure the deleterious effect of sonication could be avoided. These new developments hold promise for an optimized DNA nanoencapsulation approach.

Unfortunately, in addition to the shearing and cavitation forces used during encapsulation, the microenvironment generated in the course of the polymer degradation was found to be very critical for the stability of the encapsulated plasmids.[69,81] As indicated in the introductory section, the pH inside PLA–PLGA matrices increases significantly in the course of the polymer degradation due to the accumulation of oligomers.[82] In this regard, it is important to keep in mind that, due to their smaller size and, hence, more important specific surface area, nanoparticles are expected to release more easily the polymer degradation products than microparticles. On the other hand, this very harmful environment, to which proteins are also very sensitive, could be improved thanks to the incorporation of poly(oxyethylene) derivatives. For example, we have verified the positive effect of this strategy for very delicate proteins such as tetanus toxoid and interferon alpha.[13,38] More recently, we have also investigated the stability of pDNA encapsulated in PLGA–poloxamer blend nanostructures and concluded that the incorporation of poloxamer has a very positive effect in the preservation of the pDNA conformation.[83]

Overall, the results published until now indicate that a combination of a mild nanoencapsulation technique with the appropriate stabilizing excipients is required for the adequate formulation of pDNA in nanoparticles. Within this frame, the use of poloxamers and poloxamines as stabilizing excipients appears to be a promising approach.

23.6   *IN VIVO* ADMINISTRATION OF PLASMID-CONTAINING PLA–PEG NANOPARTICLES

Although nanoparticles composed of a PLA–PEG copolymer were originally conceived with the idea of increasing blood circulation times of drugs administered intravenously, the beneficial effects provided by the PEG molecules have proven to be promising for other drug delivery applications. For example, a few years ago we reported the positive effect of a PEG coating around PLA nanoparticles on the nasal/intestinal transport of nanoencapsulated proteins.[58] Moreover, following nasal administration of a model protein (tetanus vaccine), we observed a high and long-lasting immune response.[84] In our attempt to find an explanation for this special behavior we observed that, while PLA nanoparticles aggregated massively upon contact with proteins and enzymes present in mucosal fluids, those coated with PEG remained quite stable. More recently, we were able to visualize the nasal transport of PLA–PEG fluorescent nanoparticles using confocal fluorescent microscopy, and concluded that these PEG-coated particles can cross mucosal surfaces and that the intensity of their transport is affected by the PEG coating.[53] Whether or not this is a direct consequence of their improved stability in the nasal mucosa requires further investigation. Nevertheless, this work clearly supports the potential application of PEG-coated nanoparticles as transmucosal carriers for macromolecules.

Fortunately, the special behavior of PEG-coated particles as transmucosal carriers could be corroborated further for plasmid DNA delivery. In this study PLA–PEG nanoparticles containing the plasmid-encoding beta galactosidase (pCMVβ-Gal) were administered intranasally to mice[51] and the immune response elicited by β-Gal determined by enzyme-linked immunosorbent assays (ELISA). As shown in Figure 23.8, the IgG levels elicited by the encapsulated plasmid were significantly higher (80 ng/ml) than those corresponding to the naked plasmid (20 ng/ml). Moreover, the levels were similar to those elicited by the naked plasmid following intramuscular injection. Consequently, these results clearly show that PLA–PEG nanoparticles act as transmucosal carriers for pDNA. In addition, the fact that the encapsulated plasmid DNA could induce systemic immune response after the intranasal administration indicates that its incorporation into PLA–PEG nanoparticles does preserve the biological function of the plasmid.

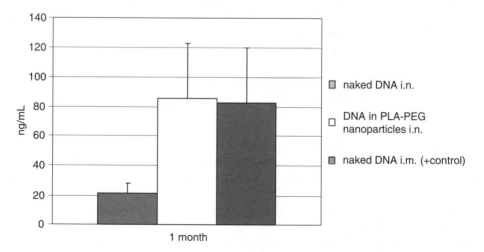

Figure 23.8   IgG antibody levels in mice after intranasal (i.n.) and intramuscular (i.m.) administration of DNA, free and encapsulated in PLA–PEG. (From Vila, A., Sanchez, A., Perez, C., and Alonso, M.J., PLA-PEG nanospheres: New carriers for transmucosal delivery of proteins and plasmid DNA, *Polym. Adv. Technol.* 13, 851–858, 2002. With permission.)

As a main conclusion, these studies reinforced that PLA-PEG nanoparticles hold great potential as carriers through mucosal surfaces. Furthermore, from these studies it has become clear that the use of these vehicles can be extended from protein drugs to pDNA, with the subsequent consequences for the development of new genetic vaccines.

23.7 CONCLUDING REMARKS

Over the last few years, important advances in the design of nanoparticles consisting of biodegradable polyesters (PLA–PLGA) and PEO and derivatives have led to the development of nanostructures with a variety of well-defined core and coat compositions. Even though the number of reports on the utility of these systems for gene delivery is still very limited, a clear conclusion from the results published so far is that these optimized compositions can be beneficially adapted for the adequate encapsulation and controlled release of plasmid DNA. Furthermore, preliminary *in vivo* experiments have evidenced the potential of these new systems as transmucosal DNA carriers. It is expected that additional future information on the intracellular fate of these new carriers will help with their further optimization and, hopefully, their exploitation as gene delivery systems.

ACKNOWLEDGMENTS

The authors are grateful to the Fullbright Commission (Ref. 99112), to the Xunta de Galicia (Ref. PGIDT00BIO20301PR) (Spain) and to the Agencia Española de Cooperación Internacional for the support of some of the work presented in this chapter.

REFERENCES

1. Hirosue, S., Muller, B.G., Mulligan, R.C., and Langer, R., Plasmid DNA encapsulation and release from solvent diffusion nanospheres, *J. Control. Release* 70, 231–242, 2001.
2. Kawashima, Y., Yamamoto, H., Takeuchi, H., Hino, T., and Niwa, T., Properties of a peptide containing DL-lactide/glycolide copolymer nanospheres prepared by novel emulsion diffusion methods. *Eur. J. Pharm. Biopharm.* 45, 41–48, 1998.
3. Sanchez, A., Vila-Jato, J.L., and Alonso, M.J., Development of biodegradable microspheres and nanospheres for the controlled release of cyclosporin A. *Int. J. Pharm.* 99, 263–273, 1993.
4. Alonso, M.J., Nanoparticulate drug carrier technology, in *Microparticulate Systems for the Delivery of Proteins and Vaccines*, Cohen, S. and Bernstein, H. (Eds.), Marcel-Dekker, New York, 1996, 203–242.
5. Quintanar-Guerrero, D., Alleman, E., Fessi, H., and Doelker, E., Preparation techniques and mechanisms of formation of biodegradable nanoparticles from preformed polymers, *Drug. Dev. Ind. Pharm.* 24, 1113–1128, 1998.
6. Florence, A.T., The oral absorption of micro- and nanoparticulates: Neither exceptional nor unusual. *Pharm. Res.* 14, 259-266.
7. Chen, H. and Langer, R. (1998). Oral particulate delivery: status and future trends, *Adv. Drug Del. Rev.* 34, 339–350, 1997.
8. Luck, M., Pistel, K., Li, Y., Blunk, T., Muller, R.H., and Kissel, T., Plasma protein adsorption on biodegradable microspheres consisting of poly(D,L-lactide-co-glycolide), poly(L-lactide) or ABA triblock copolymers containing poly(oxyethylene) Influence of production method and polymer composition, *J. Control. Release* 55, 107–120, 1998.
9. Norris, D.A., Puri, N., and Sinko, P.J., Effect of physical barriers and properties on the oral absorption of particulates, *Adv. Drug Del. Rev.* 34, 135–154, 1998.

10. Johansen, P., Men, Y., Merkle, H.P., and Gander, B., Revisiting PLA/PLGA microspheres: an analysis of their potential in pareteral vaccination, *Eur. J. Pharm. Biopharm.* 50, 129–146, 2000.

11. Fu, K., Pack, D.W., Klibanov, A.M., and Langer, R., Visual evidence of acidic environment within degrading poly(lactic-co-glycolic acid) (PLGA) microspheres, *Pharm. Res.* 17, 100–106, 2000.

12. Zhu, G., Mallery, S.R., and Schwendeman, S.P., Stabilization of proteins encapsulated in injectable poly(lactide-co-glycolide), *Nat. Biotechnol.* 18, 52–57, 2000.

13. Tobio, M., Nolley, J., Guo, Y., McIver, J., and Alonso, M.J., A novel system based on a poloxamer/PLGA blend as a tetanus toxoid delivery vehicle, *Pharm. Res.* 16, 682–688, 1999.

14. Soppimath, K.S., Aminabhavi, T.M., Kulkarni, A.R., and Rudzinski, W.E., Biodegradable polymeric nanoparticles as drug delivery devices, *J. Control. Release* 70, 1–20, 2001.

15. Landry, F.B., Bazile, D.V., Spenlehauer, G., Veillard, M., and Kreuter, J., Influence of coating agents on the degradation of poly(D,L-lactic acid) nanoparticles in model digestive fluids (USP XXII), *S.T.P. Pharm. Sci.* 6, 195–202, 1996.

16. Tobio, M., Sanchez, A., Vila, A., Soriano, I., Evora, C., Vila-Jato, J.L., and Alonso, M.J., The role of PEG on the stability in digestive fluids and *in vivo* fate of PEG-PLA nanoparticles following oral administration, *Coll. Surf. B: Biointerf.* 18, 315–323, 2000.

17. Apicella, A., Cappello, B., Del Nobile, M.A., La Rotonda, M.I., Mensitieri, G., and Nicolais, L., Poly(ethylene oxide) (PEO) and different molecular weight PEO blends monolithic devices for drug release, *Biomaterials* 14, 83–90, 1993.

18. Moroni, A. and Ghebre-Sellassie, I., Application of poly(oxyethylene) homopolymers in sustained release solid formulations, *Drug. Dev. Ind. Pharm.* 21, 1411–1428, 1995.

19. Gref, R., Minamitake, Y., Peracchia, M.T., Trubetskoy, V., Torchilin, V., and Langer, R., Biodegradable long circulating polymeric nanospheres, *Science* 263, 1600–1603, 1994.

20. Peracchia, M.T., Gref, R., Minamitake, Y., Domb, A.J., Lotan, N., and Langer, R., PEG-coated nanospheres from amphiphilicdiblock and multiblock copolymers: Investigation of their drug encapsulation and release characteristics, *J. Control. Release* 46, 223–231, 1997.

21. Redhead, H.M., Davis, S.S., and Illum, L., Drug delivery in poly(lactide-co-glycolid) nanoparticles surface modified with poloxamer 407 and poloxamine 908: *in vitro* characterisation and *in vivo* evaluation, *J. Control. Release* 70, 353–363, 2001.

22. Barichello, J.M., Morishita, M., Takayama, K., and Nagai, T., Absorption of insulin from Pluronic F-127 gels following subcutaneous administration to rats, *Int. J. Pharm.* 184, 189–198, 1999.

23. Moghimi, S.M. and Hunter, A.C., Poloxamers and poloxamines in nanoparticle engineering and experimental medicine, *Tibtech* 18, 412–420, 2003.

24. Bochot, A., Fattal, E., Grossiord, Y., Puisieux, F., and Couvreur, P., Characterisation of a new ocular delivery system based on a dispersion of liposomes in a thermosensitive gel, *Int. J. Pharm.* 162, 119–127, 1998.

25. Felt, O., Baeyens, V., Zignani, M., Bury, P., Gurny, R., Mucosal drug delivery, ocular, in *Encyclopedia of Controlled Drug Delivery*, Vol. 1, Mathiowitz, E. (Ed.), John Wiley & Sons, New York, 1999, 605–626.

26. Batrakova, E.V., Miller, D. W., Li, S., Alakhov, V., Kabanov, A.V., and Elmquist, W.F., Pluronic P85 enhances the delivery of digoxin to the brain: *in vitro* and *in vivo* studies, *J. Pharmacol. Exp. Ther.* 296, 551–557, 2001.

27. Kabanov, A.V., Chekhonin, P., Alakhov, V., Yu, V., Batrakova, E.V., Lebedev, A.S., and Melnik-Nubarov, N.S., The neuroleptic activity of haloperidol increases after its solubilization in surfactant micelles. Micelles as microcontainers for drug targeting, *FEBS Lett.* 258, 343–345, 1989.

28. Kabanov, A.V., Batrakova, E.V., and Miller, D.W., Pluronic block copolymers as modulators of drug efflux transporter activity in the blood-brain barrier, *Adv. Drug Del. Rev.* 55, 151–164, 2003.

29. Lemieux, P., Guérin, N., Paradis, G., Proulx, R., Chistyakova, L., Kabanov, V.A., and Alakhov, V., A combination of poloxamers increases gene expression of plasmid DNA in skeletal muscle, *Gene Ther.* 7, 986–991, 2000.

30. Lemieux, P., Vinogradov, S.V., Gebhart, C.L., Guérin, N., Paradis, G., Nguyen, H.K., Ochietti, B., Suzdaltseva, Y.G., Bartakova, E.V., Bronich, T.K., St-Pierre, Y., Alakhov, V., and Kabanov, V.A., Block and graft copolymers and Nanogel compolymer networks for DNA delivery into cell, *J. Drug Target.* 8, 91–105, 2000.

31. Benns, J.M. and Kim, S.W., Tailoring new gene delivery designs for specific targets, *J. Drug Target.* 8, 1–12, 2000.

32. Otsuka, H., Nagasaki, Y., and Kazunori, K., PEGylated nanoparticles for biological and pharmaceutical applications, *Adv. Drug Del. Rev.* 55, 403–419, 2003.

33. Reich, G., *In vitro* stability of poly(D,L-lactide) and poly(D,L-lactide)/poloxamer nanoparticles in gastrointestinal fluids, *Drug. Dev. Ind. Pharm.* 23, 1191–1200, 1997.

34. Hawley, A.E., Illum, L., and Davis, S.S., Lymph node localisation of biodegradable nanospheres surface modified with poloxamer and poloxamine block copolymers, *FEBS Lett.* 400, 319–323, 1997.

35. Moghimi, S.M., Hawley, A.E., Christy, N.M., Gray, T., Illum, L., and Davis, S.S., Surface engineered nanospheres with enhanced drainage into lymphatics and uptake by macrophages of the regional lymph node, *FEBS Lett.* 344, 25–30, 1994.

36. Neal, J.C., Stolnik, S., Schacht, E., Kenawy, E.R., Garnett, M.C., Davis, S.S., and Illum, L., *In vitro* displacement by rat serum of adsorbed radiolabeled poloxamer and poloxamine copolymers from model biodegradable nanospheres, *J. Pharm. Sci.* 87, 1242–1248, 1998.

37. Park, T.G., Cohen, S., and Langer, R., Controlled protein release from polyethyleneimine-coated poly(L-lactic acid)/Pluronic blend matrices, *Pharm. Res.* 9, 37–39, 1992.

38. Sanchez, A., Tobio, M., Gonzalez, L., Fabra, A., and Alonso, M.J., Biodegradable micro- and nano-particles as long-term delivery vehicles for interferon-alpha, *Eur. J. Pharm. Sci.* 18, 221–229, 2003.

39. Yeh, M., Davis, S.S., and Coombes, A.G.A., Improving protein delivery from microparticles using blends of poly(DL lactide co-glycolide) and poly(ethylene oxide)-poly(propylene oxide) copolymers, *Pharm. Res.* 13, 1693–1698, 1996.

40. Blanco, D. and Alonso, M.J., Development and characterization of protein-loaded poly(lactide-co-glycolide) nanospheres, *Eur. J. Pharm. Biopharm.* 43, 287–294, 1997.

41. Csaba, N., Sanchez, A., and Alonso, M.J., Composite blend nanoparticles for the delivery of plasmid DNA, *1st EUFEPS Conference on Optimising Drug Delivery and Formulation*, Versailles, 2003.

42. Kissel, T., Li, X., Volland, C., Gorich, S., and Koneberg, R., Parenteral protein delivery systems using biodegradable polyesters of ABA block structure, containing hydrophobic poly(lactide-co-glycolide) A blocks and hydrophilic poly(ethylene oxide) B blocks, *J. Control. Release* 39, 315–326, 1996.

43. Pean, J.M., Boury, F., Venier-Julienne, M.C., Menei, P., Proust, J.E., and Benoit, J.P., Why does PEG 400 co-encapsulation imprve NGF stability and release from PLGA biodegradable microspheres? *Pharm. Res.* 8, 1294, 1999.

44. Stolnik, S., Dunn, S.E., Garnett, M.C., Davies, M.C., Coombes, A.G.A., Taylor, D.C., Irving, M.P., Purkiss, S.C., Tadros, T.F., Davis, S.S., and Illum, L., Surface modification of poly(lactide-co-glycolide) nanospheres by biodegradable poly(lactide)-poly(ethylene glycol) copolymers, *Pharm. Res.* 11, 1800–1808, 1994.

45. Hawley, A.E., Illum, L., and Davis, S.S., Preparation of biodegradable, surface engineered PLGA nanospheres with enhanced lymphatic drainage and lymph node uptake, *Pharm. Res.* 14, 657–661, 1997.

46. Gref, R., Domb, A.J., Quellec, P., Blunk, T., Muller, R.H., Verbavatz, J.M., and Langer, R., The controlled intavenous delivery of drugs using PEG-coated sterically stabilized nanospheres, *Adv. Drug Del. Rev.* 16, 215–233, 1995.

47. Gref, R., Lück, M., Quellec, P., Marchand, M., Dellacherie, E., Harnisch, S., Blunk, T., and Müller, R.H., "Stealth" corona-core nanoparticle surface modified by polyethylene glycol (PEG): influence of the corona (PEG chain length and surface density) and of the core composition on phagocytic uptake and plasma protein adsortion, *Coll. Surf. B: Biointerf.* 18, 301–313, 2000.

48. Gref, R., Minamitake, Y., Peracchia, M.T., and Langer, R., Poly(Ethylene Glycol)-coated biodegradable nanospheres for intravenous drug administration, in *Microparticulate Systems for the Delivery of Proteins and Vaccines*, Cohen, S. and Bernstein, H. (Eds.), Marcel-Dekker, New York, 2003, 279–306.

49. Govender, T., Riley, T., Ehtezazi, T., Garnett, M.C., Stolnik, S., Illum, L., and Davis, S.S., Defining the drug incorporation properties of PLA-PEG nanoparticles, *Int. J. Pharm.* 199, 95–110, 2000.

50. Quellec, P., Gref, R., Perrin, D.E., Dellacherie, E., Sommer, F., Verbavatz, J.M., and Alonso, M.J., Protein encapsulation within poly(ethylene glycol)-coated nanospheres I. Physicochemical characterization, *J. Biomed. Mater. Res.* 42, 45–54, 1998.

51. Vila, A., Sanchez, A., Perez, C., and Alonso, M.J., PLA-PEG nanospheres: New carriers for transmucosal delivery of proteins and plasmid DNA, *Polym. Adv. Technol.* 13, 851–858, 2002.

52. Vila, A., Gill, H., McCallion, O., and Alonso, M.J., PLA-PEG nanoparticles as nasal drug carriers: approaches to the modulation of the PEG coating, *Proceedings of the 30th Annual Meeting of the Controlled Release Society*, Glasgow, 2003, 19.

53. Vila, A., Gill, H., McCallion, O., and Alonso, M.J., Towards the optimisation of PLA-PEG nanoparticles production, *Proceedings of the 30th Annual Meeting of the Controlled Release Society*, Glasgow, 2003, 335.

54. Jeon, S.I., Lee, J.H., Andrade, J.D., and De Gennes, P.G., Protein-surface interactions in the presence of polyethylene oxide I. Simplified theory, *J. Colloid Interf. Sci.* 142, 149–158, 1991.

55. Jeon, S.I. and Andrade, J.D., Protein-surface interactions in the presence of polyethylene oxyde II. Effect of protein size, *J. Colloid Interf. Sci.* 142, 159–166, 1991.

56. Mosqueira, V., Legrand, P., Gref, R., Heurtault, B., Appel, M., and Barratt, G., Interactions between a macrophage cell line (J774A1)and surface-modified poly(D,L-lactide) nanocapsules bearing poly(ethylene glycol), *J. Drug Target.* 7, 65–78, 1999.

57. Bazile, D., PrudHomme, C., Bassoullet, M., Marlard, M., Spenlehauer, G., and Veillard, M., Stealth Me.PEG-PLA nanoparticles avoid uptake by the mononuclear phagocytes system, *J. Pharm. Sci.* 84, 493–498, 1995.

58. Vila, A., Sanchez, A., Tobio, M., Calvo, P., and Alonso, M.J., Design of biodegradable particles for protein delivery, *J. Control. Release* 78, 15–24, 2002.

59. Quellec, P., Gref, R., Dellacherie, E., Sommer, F., Tran, M.D., and Alonso, M.J., Protein encapsulation within poly(ethylene glycol)-coated nanospheres II. Controlled release properties, *J. Biomed. Mater. Res.* 47, 388–395, 1999.

60. Tobío, M., Gref, R., Sánchez, A., Langer, R., and Alonso, M.J., Stealth PLA-PEG nanoparticles as protein carriers for nasal administration, *Pharm. Res.* 15, 270–276, 1998.

61. Emile, C., Bazile, D., Herman, F., Helene, C., and Veillard, M., Encapsulation of oligonuclcotidcs in stealth Me.PEG-PLA50 nanoparticles by complexation with structured oligopeptides, *Drug Del.* 3, 187–195, 1996.

62. Saison-Behmoaras, T., Tocque, B., Rey, I., Chassignol, N.T., Thuong, N.T., and Helene, C., Short modified antisense oligonucleotides directed against Ha-ras point mutation induce selective cleavage of the mRNA and inhibit T24 cells proliferation, *EMBO J.* 10, 1111–1118, 1991.

63. Schwab, G., Chavany, C., Duroux, I., Goubin, G., Lebeau, J., and Helene, C., Antisense oligonucleotides adsorbed to polyalkylcyanoacrylate nanoparticles specifically inhibit mutated Ha-ras mediated cell proliferation and tumorigenicity in nude mice, *Proc. Nat. Acad. Sci. U.S.A.* 91, 10460–10464. 1994.

64. Perez, C., Sanchez, A., Putnam, D., Ting, D., Langer, R., and Alonso, M.J., Poly(lactic acid)-poly(cthylene glycol) nanoparticles as new carriers for the delivery of plasmid DNA, *J. Control. Release* 75, 211–224, 2001.

65. Mumper, J.R., Duguid, J.D., Anwer, K., Barron, M.K., Nitta, H., and Rolland, A.P., Polyvinyl derivatives as novel interactive polymers for controlled gene delivery to muscle, *Pharm. Res.* 13, 701–709, 1996.

66. Mumper, J.R., Wang, J., Klakamp, S.L., Nitta, H., Anwer, K., Tagliaferri, F., and Rolland, A.P., Protective interactive noncondensing (PINC) polymers for enhanced plasmid distribution and expression in rat skeletal muscle, *J. Control. Release* 52, 191–203, 1998.

67. Liu, Y. and Deng, X., Influences of preparation conditions on particle size and DNA-loading efficiency for poly(DL-lactic acid-polyethylene glycol) microspheres entrapping free DNA, *J. Control. Release* 83, 147–155, 2002.

68. Jones, D.H., Clegg, J.C.S., and Farrar, G.H., Oral delivery of microencapsulated DNA vaccines, *Dev. Biol. Stand.* 92, 149–155, 1998.

69. Tinsley-Brown, A.M., Fretwell, R., Dowsett, A.B., Davis, S.L., and Farrar, G.H., Formulation of poly(D,L-lactic-co-glycolic acid) microparticles for rapid plasmid DNA delivery, *J. Control. Release* 66, 229–241, 2000.

70. Deng, X., Liu, T., Yuan, M.L., and Li, X.H., Preparation and characterization of poly(DL-lactide)-co-poly(ethylene glycol) microspheres containing lambda DNA, *J. Appl. Polym.*, in press, 2002.

71. Liu, Y., Deng, X., and Li, X.H., Investigation of poly-DL-lactide-co-poly(ethylene glycol)-b-poly(DL-lactide) microspheres containing plasmid DNA, *Chinese J. Polymer Sci.*, 22(3), 205–214.

72. Csaba, N., Sanchez, A., and Alonso, M.J., Design and characterisation of nanoparticulate PLGA-poloxamer blend systems for drug delivery, *30th Annual Meeting of the Controlled Release Society*, Glasgow, CRS, 2003, 231.

73. Schwendeman, S.P., Tobio, M., Joworowitz, M., and Alonso, M.J., New startegies for the microencapsulation of tetanus vaccine, *J. Microencaps.* 15, 299–318, 1998.

74. Luo, D., Woodrow-Mumford, K., Belcheva, N., and Saltzman, W.M., Controlled DNA delivery systems, *Pharm. Res.* 16, 1300–1308, 1999.

75. Wang, D., Robinson, D.R., Kwon, G.S., and Samuel, J., Encapsulation of plasmid DNA in biodegradable poly(D,L-lactic-co-glycolic acid) microspheres as a novel approach for immunogene delivery, *J. Control. Release* 57, 9–18, 1999.

76. Cohen, H., Levy, R.J., Gao, J., Fishbein, I., Kousaev, V., Sosnowski, S., Slomkowski, S., and Golomb, G., Sustained delivery and expression of DNA encapsulated in polymeric nanoparticles, *Gene Ther.* 7, 1896–1905, 2000.

77. Park, T.G., Cohen, S., and Langer, R., Poly(lactic acid)/Pluronic blends: characterization of phase separation behavior, degradation, and morphology and use as protein-releasing matrices, *Macromolecules* 25, 116–122, 1992.

78. Capan, Y., Woo, B.H., Gebredikan, S., Ahmed, S., and DeLuca, P., Influence of formulation parameters on the characteristics of poly(D,L-lactide-co-glycolide) microspheres containing poly(L-lysine) complexed plasmid DNA, *J. Control. Release* 60, 279–286, 1999.

79. Lengsfeld, C.S., Manning, M.C., and Randolph, T.W., Encapsulating DNA within biodegradable polymeric microparticles, *Curr. Pharm. Biotechnol.* 3, 227–235, 2002.

80. Ando, S., Putnam, D., Pack, D.W., and Langer, R., PLGA microspheres containing plasmid DNA: preservation of supercoiled DNA via cryopreparation and carbohydrate stabilization, *J. Pharm. Sci.* 88, 126–130, 1999.

81. Walter, E., Moelling, K., Pavlovic, J., and Merkle, H.P., Microencapsulation of DNA using poly(D,L-lactide-co-glycolide): stability issues and release characteristics, *J. Control. Release* 61, 361–374, 1999.

82. Middaugh, C.R., Evans, R.K., Montgomery, D.L., and Casimiro, D.R., Analysis of plasmid DNA from a pharmaceutical perspective, *J. Pharm. Sci.* 87, 130–146, 1998.

83. Csaba, N., Sanchez, A., and Alonso, M.J., PLGA-poloxamer and PLGA-poloxamine blend nanoparticles: new carriers for gene delivery, submitted, 2003.

84. Vila, A., Sanchez, A., Evora, C., Soriano, I., Vila Jato, J.L., and Alonso, M.J., PEG-PLA nanoparticles as carriers for nasal vaccine delivery, *J. Aerosol Med.*, 17(2), 174–185, 2004.

CHAPTER 24

Poly(alkylcyanoacrylate) Nanoparticles for Nucleic Acid Delivery

Elias Fattal and Patrick Couvreur

CONTENTS

24.1 INTRODUCTION

Nanotechnologies form the basis of support for a number of applications, including biomedical applications such as drug delivery and drug targeting. Among the several submicronic particles developed for drug targeting, biodegradable nanoparticles have been widely used to carry different types of therapeutic agents such as antisense oligonucleotides.[1,2] Once they get inside the cell, these short gene fragments are able to control gene expression and can be potentially used to treat oncogene-related cancers, viral infections, or inflammatory diseases. Because of their low intracellular penetration and poor stability, antisense oligonucleotides need to be transported by a carrier. The choice of nanoparticles was made because of their stability in the systemic circulation and their ability to achieve tissue or cellular targeting. For oligonucleotide delivery, the poly(alkylcyanoacrylate) (PACA) polymer has been the subject of several studies. This review will summarize the achievements that were recently made in the development of this approach.

0-8493-1934-X/05/$0.00+$1.50
© 2005 by CRC Press LLC

24.2 ANTISENSE STRATEGY AND THE NEED FOR PARTICULATE FORMULATIONS

There are two main mechanisms described so far to control gene expression by nucleic acids: antisense oligonucleotides, and the more recently described small interfering RNA (siRNA) methods.

The aim of the antisense strategy is to interfere with gene expression by preventing the translation of proteins from mRNA. Theoretically, an antisense oligonucleotide is a short gene fragment (from 15 to 20 sequence bases) of deoxynucleotides that have a sequence complementary to a portion of the targeted mRNA. The antisense oligonucleotide then hybridizes with the mRNA by Watson-Crick base pairing and blocks sterically the translation of this transcript into a protein.[3] This mechanism is referred to as translational arrest (Figure 24.1). Another mechanism that has been widely described is the destruction of antisense–mRNA hybrids by an enzyme which gets activated: the RNAse H.[4] Inactivation of gene expression by oligonucleotides might also be exerted by triple helix formation between genomic double-stranded DNA and oligonucleotides. This sequence-specific binding is achieved through Hoogsteen hydrogen bonds between thymidine and TA base pairs and between protonated cytosine and CG base pairs.[5] A second motif for triple helix recognition of double-stranded DNA is comprised by a homopurine motif in which a purine-rich oligonucleotide binds to DNA antiparallel to the Watson-Crick purine strand.[6] Pyrimidine unmodified oligodeoxynucleotides or backbone-modified oligonucleotides are able to block gene transcription in a sequence-specific manner.[5,7,8] Another mechanism of RNA inactivation is mediated by ribozymes. The 5' and 3' ends of these ribonucleotides are complementary to the target RNA and contain an intramolecular hairpin loop, which induces the cleavage of the target RNA.[9] However, the limitations of ribozymes as potential antisense therapeutics are similar to those described below for antisense RNA. Finally, it has been recently described that a hairpin structure in the target mRNA can be cleaved by an antisense oligodeoxynucleotide bound to both sides of the hairpin in the presence of Cu^{2+} ions and a reducing agent.[10]

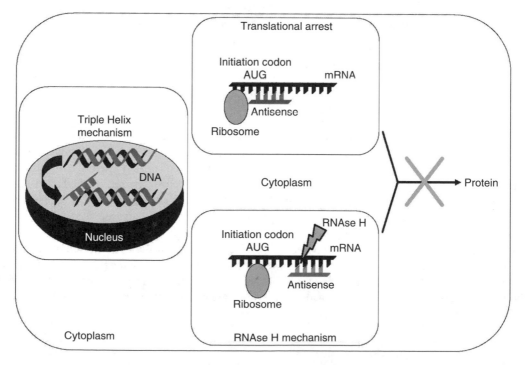

Figure 24.1 Main mechanisms of action of antisense oligonucleotides.

Antisense oligonucleotides consist in natural phosphodiester compounds. However, crucial problems such as the poor stability of these molecules versus nuclease activity *in vitro* and *in vivo* and their low intracellular penetration have limited their use in therapeutics.[11,12] As a result, practical application of antisense oligonucleotides has required modifications with the aim of retaining the hybridization capacity while increasing stability and cellular penetration. The chemical modifications have mainly focused on the phosphodiester backbone and/or the sugar moiety. Replacement of the nonbridging oxygen of the phosphodiester backbone by sulfur results in a phosphorothioate with enhanced stability to enzymatic degradation. Although the duplex formed with the target RNA has a lower melting temperature (T_m) (i.e., a lower affinity) than the phosphodiester parent compound, it remains a substrate for RNase H. Another approach is the replacement of the nonbridging oxygen by a methyl group which results in a greater hydrophobicity due to the loss of the negative charge. However in this case, RNase H activation is also reduced. Alternatively, replacing the hydrogen at the deoxyribose 2' position by a hydroxymethyl group converts the sugar to a modified ribose. The resultant RNA–RNA duplex formed with the target RNA is more stable, as indicated by the elevated T_m. One modification can also be provided by replacement of the deoxyribose phosphate backbone as in peptide nucleic acids.[13] Thus, by these modifications one can synthesize tailored antisense oligonucleotides with a balance of characteristics of hybridization affinity, hydrophobicity, and capacity to recruit RNAse H-mediated hydrolysis of the target RNA.

Although these chemical modifications have provided an improvement of stability and cell penetration, they have also resulted in a variety of non-antisense activities. Thus, when evaluating an oligonucleotide that has been designed as antisense to a target RNA, assaying the pharmacological activity alone is insufficient and may be misleading. The so-called aptameric effect of oligonucleotide is based on the fact that oligonucleotides can fold into three-dimensional structures.[14] Oligonucleotide aptamers are able to bind to receptors, enzymes, and other proteins and affect their function; this can occur in a nonspecific fashion. In this view, oligonucleotides containing four repeated G bases interact specifically with proteins.[15,16] In addition, phosphorothiate oligonucleotides were described to bind to a wide number of proteins in a sequence-independent manner.[17] They can also affect clotting and complement systems.[18] Finally, oligonucleotides containing unmethylated CpG dinucleotides were shown to display immunomodulation properties.[19]

A more recent approach to knocking down the gene is the use of siRNA. In fact, it was shown that introducing long double-stranded RNA (dsRNA) into a variety of hosts can trigger posttranscriptional silencing of all homologous host genes and transgenes. siRNA were discovered after observing that within the intracellular compartment, the long dsRNA molecules are metabolized to small 21–23 nucleotide interfering RNAs by the action of an endogenous ribonuclease, Dicer.[20] The siRNA molecules bind to a protein complex, termed RNA-induced silencing complex (RISC), which contains (1) a helicase activity that unwinds the two strands of RNA molecules, allowing the antisense strand to bind to the targeted RNA molecule,[21] and (2) an endonuclease activity which hydrolyzes the target RNA at the site where the antisense strand is bound. It is unknown whether the antisense RNA molecule is also hydrolyzed or recycles and binds to another RNA molecule. Therefore, as shown in Figure 24.2, RNA interference is an antisense mechanism of action, as ultimately a single-strand RNA molecule binds to the target RNA molecule by Watson-Crick base pairing rules and recruits a ribonuclease that degrades the target RNA. This mechanism makes feasible the use of small double-stranded siRNA in therapeutics instead of oligonucleotides. Indeed, it has been recently shown that unexpectedly small double-stranded RNAs (siRNAs) appear to be very ecient agents to inhibit gene expression in mammalian cells.[22,23] Bertrand et al.[24] have compared the eciency of a nuclease resistant antisense oligonucleotide with siRNA, both being targeted against the green fluorescent protein stably expressed in HeLa cells in cell cultures and in xenografted mice. Using Cytofectin GSV to deliver both inhibitors, the siRNAs appeared to be quantitatively more

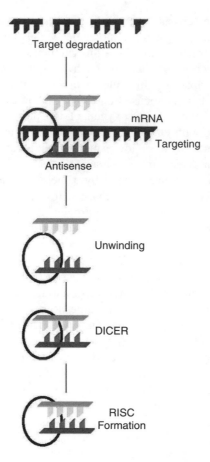

Figure 24.2 Mechanism of action of siRNA. RISC is an RNase III-like enzyme. DICER: RNA-induced silencing complexes.

efficient, and their effects lasted for a longer time in cell culture. In mice, they observed an activity of siRNAs but not of antisense oligonucleotides. The absence of efficiency of antisense oligonucleotides is probably due to their lower resistance to nuclease degradation. However, the need of an appropriate carrier to improve intracellular penetration is still an important parameter to observe any effect. Again, more recently, siRNAs were designed to target the *bcr-abl* oncogene, which causes chronic myeloid leukemia (CML) and *bcr-abl*-positive acute lymphoblastic leukemia.[25] The data demonstrated that siRNA can specifically and efficiently interfere with the expression of the oncogenic fusion gene in hematopoietic cells.[25] This higher efficiency of siRNA seems however to be controversial, since Vickers et al.[26] have recently shown that both siRNA and RNase H-dependent oligonucleotides displayed the same inhibitor activity over the target gene.

Thus, in order to increase nucleic acid stability, to improve cell penetration, and to avoid nonspecific aptameric effect, the use of particulate carriers such as liposomes or nanoparticles may be considered the more realistic approach to deliver short nucleic acids. Indeed, colloidal carriers were found to be able to protect natural unmodified nucleic acids against degradation and, since they are taken up by endocytosis, they could also increase cell penetration of oligodeoxynucleotide (ODN) or siRNA. Among drug carriers, PACA nanoparticles have shown interesting potentialities to bind and deliver ODNs in an efficient manner.[2]

24.3 POLY(ALKYLCYANCRYLATE) NANOPARTICLES: THEIR POTENTIAL FOR ODN DELIVERY

Nanoparticles are defined as being submicronic (< 1 μm) colloidal systems generally made of polymers (biodegradable or not). According to the process used for the preparation it is possible to obtain two types of PACA nanoparticles: nanospheres or nanocapsules. Nanocapsules are vesicular systems in which the drug is confined to a cavity (an oily or aqueous core) surrounded by a unique polymeric membrane; nanospheres are matrix systems in which the drug is dispersed throughout the particles. PACA nanoparticles were first discovered by Couvreur et al.[27] At that time, the research on colloidal carriers was mainly focusing on liposomes, but no one was able to produce stable lipid vesicles suitable for clinical applications. In some cases, nanospheres have been shown to be more efficient drug carriers than liposomes due to their better stability.[28] This is the reason why many drugs were associated to PACA nanospheres in the last decades (e.g., antibiotics, antiviral and antiparasitic drugs, cytostatics, protein, and peptides). The main interest of PACA nanoparticles is their ability to achieve tissue targeting and enhance the intracellular penetration of drugs. The particles are mainly taken up by the cells of the mononuclear phagocyte system (MPS) in the liver, the spleen, lungs, and bone marrow.[29] The uptake of these PACA nanospheres occurs through an endocytosis process after which the particles end up in the lysosomal compartment,[29] where they are degraded, producing low-molecular-weight soluble compounds that are eliminated from the body by renal excretion.[30] As a result of the MPS site-specific targeting, avoidance of some organs was made possible, thus reducing the side effects and toxicity of some active compounds. Due to their lysosomal localization, it may be considered that these PACA nanospheres were not suitable to address ODNs efficiently to the cell cytoplasm. For this reason, and in order to avoid the trapping of ODN-loaded nanospheres within the lysosomal compartment, several compounds able to destabilize the lysosomal membrane were associated with the nanoparticles (e.g., cationic surfactant or cationic hydrophobic peptides)[31,32] as shown in Figure 24.3. Recently, in order to avoid MPS uptake, a strategy consisting of the linkage to PACA nanospheres of polyethylene glycol derivatives was proposed.[33-37] This approach has resulted in a lower uptake of nanoparticles by the MPS and in a longer circulation time.[37] As a consequence, these so-called Stealth ® nanoparticles would be able to extravasate in a selective manner across endothelium that becomes permeable due to the presence of solid tumors[38,39] or inflamed tissues.[40] In addition, it was made possible to attach ligands such as folic acid to the particle surface to make them recognized by cells expressing folate receptors.[41] However, this technology, although available, has not been applied so far for the delivery of ODNs. Since in most cases, the ODNs were adsorbed on the surface of nanospheres containing a cationic surfactant or a hydrophobic peptide, another strategy was recently developed consisting of the design of carriers containing an aqueous core (nanocapsules) (Figure 24.3).[42] The main advantages of nanocapsules are the ability to avoid the use of cationic compounds and the ability to entrap within their aqueous core water-soluble ODN, avoiding their quick desorption or release.

24.4 DELIVERY OF ODN BY POLYALKYCYANOACRYLATE NANOSPHERES

24.4.1 Association of ODN to Nanospheres

Due to their hydrophilicity and polyanionic character, ODNs interact poorly with polymeric materials. Therefore, the association of ODNs to nanospheres is a particularly difficult challenge. Two main strategies have been considered for a successful binding of ODNs to nanospheres: first, ODNs may be covalently linked to a hydrophobic molecule for anchoring at the polymer surface. Second, cationic surfactants may be used to coat the particles giving them a positive charge for successful ionic interactions with the negatively charged ODN molecules.

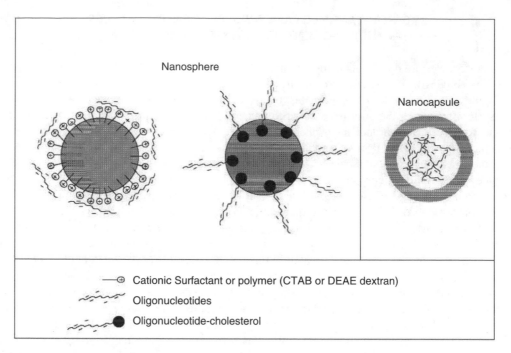

Figure 24.3   Mechanism of association of oligonucleotides with poly(cyanoacrylate) nanospheres and nanocapsules.

Initially, ODNs were efficiently associated onto PACA nanospheres precoated with a hydrophobic cation such as cetyltrimethylammonium bromide (CTAB).[31] Oligonucleotide adsorption
onto nanospheres was therefore mediated by the formation of an ion pair between the negatively
charged phosphate groups of the nucleic acid chain and the positively charged compound pread-
sorbed onto nanospheres (Figure 24.3).[31] The adsorption efficacy of oligonucleotide-cation complexes on PACA nanospheres was found to be highly dependent on several parameters: the oligonucleotide chain length, the nature of the cyanoacrylic polymer, the hydrophobicity of the cation
used as ion-pairing agent, and the ionic strength of the dispersing medium.[31] More recently, Zobel
et al.[43] have replaced CTAB with DEAE–dextran, which was introduced into the polymerisation
medium before nanosphere formation (Figure 24.3). Although the amount of DEAE–dextran
covalently linked to the polymer was determined by the authors, the unloaded nanospheres displayed
a strong positive charge, showing that the DEAE–dextran had been associated with the nanospheres.
With this formulation, the highest loading of ODNs ($35\mu M$ of ODNs per gram of polymer) was
achieved at pH 5.5 using a 10 mM phosphate buffer. Other hydrophobic compounds such as cationic
lipids were also employed as adjuvants for loading ODNs to nanospheres.[44] However, using
compounds such as polyamines (e.g., DOGS), the absorption was lower than with CTAB in similar
conditions of adsorption.[44] Another approach was to use a hydrophobic conjugate of ODNs, such
as ODNs coupled to cholesterol (Figure 24.3).[45] However, because this conjugate was negatively
charged, it was partially repulsed by the negative charges displayed by the polymer surface, which
explains the lower loading efficacy in the absence of CTAB.

24.4.2 *In Vitro* Stability of ODNs Adsorbed onto Nanospheres

When adsorbed onto poly(isohexylcyanoacrylate) (PIHCA) nanospheres through binding to
CTAB, ODNs were efficiently protected against enzymatic degradation even after 5 hours incubation
with phosphodiesterase or in cell culture media.[31] In addition, about 90% of the oligonucleotide

still remained intact after overnight incubation with the enzyme (0.1 mg/ml). Similar results were obtained by Zobel et al.[43] with ODNs adsorbed onto DEAE-containing nanospheres and incubated with DNAse.

ODN–DEAE–nanosphere complexes were found to be more stable in cell culture medium containing 10% fetal calf serum than ODN-associated nanospheres containing CTAB. It was assumed that adsorption of plasma protein could turn to a protective coating in DEAE-containing nanospheres, preventing the action of esterases.[43] On the contrary, the lower stability of the ODNs in CTAB-containing nanospheres was explained by a competition between CTAB and plasma protein inducing a partial desorption of the ODNs from the nanospheres.[32] Thus, the difference between both nanospheres (containing DEAE or CTAB) could arise from a difference in the stability of the coating layer, DEAE being much more strongly attached to the nanospheres than CTAB. The displacement of CTAB–ODN complexes by plasma proteins was confirmed in the presence of more concentrated mice plasma (70%) at 37°C. The half-lives of pdT16 incubated free or bound to the nanospheres were short: 6.0 and 12.5 min, respectively. However, 30 min after incubation in plasma, the percentage of nondegraded 33P-pdT16 was only 2.9% for free ODNs, whereas 28.9% of 33P-pdT 16 associated with nanospheres were still intact.[46] This suggested that at least part of the phosphodiester linkages was not available for nuclease degradation even if in concentrated plasma medium the localization of ODN at the surface of the nanospheres still remains unfavorable.

24.4.3 Cell Interactions with ODN-Loaded Nanospheres

Cell uptake studies of a 15-mer oligothymidilate adsorbed onto PIHCA nanospheres were performed under subtoxic conditions using U937 cells. It was shown that the uptake of the ODNs was dramatically increased when associated with nanospheres. After 24 h incubation, uptake of oligonucleotide was 8 times higher when adsorbed to nanospheres than when incubated as an ODN free solution,[32] and it was markedly reduced (95%) at 4°C as compared to 37°C. These results clearly show that ODNs adsorbed onto nanospheres were internalized by U937 cells through an endocytic/phagocytic process and not simply adsorbed at the membrane surface. This has been confirmed by confocal microscopy with fluorescently labeled nanospheres: after internalization, nanospheres accumulated into phagosomes or lysosomes. Such an intracellular distribution profile is supposed to lead to the rapid degradation of the ODNs after their intracellular release from nanospheres. This is the reason why the intracellular stability of 5′ labeled 15-mer free or adsorbed onto PIHCA nanospheres was investigated:[32] intact ODNs could not be detected in cell lysate after 1.5 h incubation with ODNs free, whereas the 15-mer adsorbed onto PIHCA nanospheres remained intact even after a 6.5 h incubation period. After 24 h, some degradation products appeared, but the fraction of intact ODNs remained considerable with nanospheres. The intracellular distribution of ODNs was also measured after lysis in the presence of Nonidet-P40, a nonionic detergent which protects the nuclear structure. It should be stressed that the cytoplasmic fraction contains also endocellular vesicles like lysosomes or phagosomes. When adsorbed onto nanospheres, intact ODNs were detected in both nuclear and extranuclear fractions. The increased stability of the oligonucleotide associated to the nanospheres which is observed in the extranuclear fraction could be explained by the fact that nanospheres protect them against digestion by lysosomal enzymes. How they escape from the lysosomes still remains unclear, but it is supposed that CTAB, which is a quaternary ammonium, could destabilize the lysosomal membrane after a certain period of time, thus allowing the release of the oligonucleotide into the cell cytoplasm. Escape from the lysosomes was however not observed on Vero cells when ODNs were delivered by nanospheres coated by DEAE instead of CTAB. Indeed, this type of nanosphere remained localized in vesicular structure in the cytoplasm, suggesting that in this case no destabilization of the lysosomal membrane occurred, which might be a strong limitation of these type of nanospheres for ODN delivery.

24.4.4 *In Vitro* Pharmacological Activity of Oligonucleotide-Loaded Nanospheres

The *in vitro* activity of oligonucleotide-loaded PIHCA nanospheres was demonstrated in a few studies. Schwab et al.[47,48] have shown that nanosphere-adsorbed antisense ODNs directed to a point mutation (GU) in codon 12 of the Ha-ras mRNA selectively inhibited the proliferation of cells expressing the point mutated Ha-ras gene. With nanospheres, the efficient concentration was 100 times lower than with ODN free. A sequence-specific inhibition of the proliferation of T24 human bladder carcinoma cells (containing the human Ha-ras oncogene mRNA) was evidenced using the same ODN conjugated to cholesterol and loaded onto PIHCA nanospheres.[45] The efficiency was comparable to that of CTAB containing nanospheres. More recently, Lambert et al.,[49] in an attempt to inhibit PKCα expression in Hep G2 cells have observed that CTAB-coated PIBCA nanospheres in subtoxic concentration were inducing by themselves (without ODNs) a depletion of PKCα. Similar observation was done with lipofectin too, a cationic lipid transfecting agent.[50] These results show that attention should be paid to the fact that the commonly used strategy of ODN targeting with cationic nonviral vectors may display nonspecific effects which can lead to artifactual results.

24.4.5 *In Vivo* Studies with Oligonucleotide Nanospheres

The pharmacokinetic studies carried out with PIBCA nanospheres[46] have shown that although nanospheres did not markedly increase the blood half-life of a ^{33}P 16-mer oligothymidilate, its tissue distribution was significantly modified. Indeed, when transported by PIBCA nanospheres, oligothymidilate was importantly taken up by the liver, whereas a subsequent reduced distribution in the other organs was observed, especially in the kidneys. These data show that with the aid of nanospheres, the oligonucleotide could be delivered to the liver with certain specificity and the urinary excretion decreased. To address the crucial problem of oligonucleotide degradation, the state in which (degraded or not) the oligonucleotide was in the plasma and delivered to the liver was investigated.[46] In this view, an original assay method has been developed that consisted of the use of a TLC radioactivity analyzer allowing, in polyacrylamide gels, to quantify the amount of undegraded 16-mer oligothymidylate extracted from the tissues.[51] This method was able to distinguish between ODNs differing by only one nucleotide in length. Using that methodology, it was found that a significant amount of oligothymidylate was kept intact in the liver and the plasma when administered under the form of nanospheres which was not the case with ODN free.[46]

One study has shown the efficacy of CTAB-coated nanospheres in delivering ODN *in vivo*.[47] In a nude mice model, HBL100ras1 cells (expressing point-mutated Ha-ras genes) were inoculated by subcutaneous administration. These clones were able to induce tumors, since the mutated ras gene is directly involved in cell proliferation and tumorigenicity. Anti-ras oligonucleotide or the sense sequence adsorbed onto PIHCA nanospheres was injected 1 day before and 48 h after implantation of the tumor in the same area where cells were inoculated. Twenty-three days after cell inoculation, animals were sacrificed and the tumors were excised and weighed.[47] It was shown that nanospheres containing the antisense markedly inhibited Ha-ras-dependent tumors in a highly specific manner. Indeed, only the antisense was able to reduce significantly tumors weight and volume as compared to the sense sequence adsorbed onto the same nanospheres.[47]

24.5 DELIVERY OF OLIGONUCLEOTIDES BY POLY(CYANOACRYLATE) NANOCAPSULES

As far as the encapsulation efficiency of oligonucleotide inside the nanocapsules was concerned, it was observed in an early study[42] that when using an aqueous phase with an oligothymidylate (phosphodiester) at 5 μM concentration, an encapsulation efficiency of only 50% was obtained. However, later on, when using a full phosphorothioate oligonucleotide (directed against EWS

Fli-chimeric RNA) at a concentration of 2.5 mM, a yield of 81% was obtained, which is, indeed, an extremely high encapsulation efficiency.[52] This difference was attributed to a possible location of part of the oligonucleotide at the water-oil interface which became more rapidly saturated at a concentration of 2.5 mM. In this condition of concentration, the amount of oligonucleotide located in the aqueous core of the nanocapsules was much more important than that located at the interface. The use of a small preparation volume (2.5 ml instead of 10 ml) might also have played a role by improving the efficacy of stirring, thus leading to a better dispersion of the aqueous phase in the oil.

There are two main issues concerning the encapsulation of oligonucleotide: the location of these molecules (within the aqueous core or adsorbed onto the surface), and the question of whether a heterogenous capsule or a homogenous matrix (spheres) is formed.

The question of the localization of the oligonucleotide has been addressed through fluorescence-quenching experiments using fluorescein-labeled oligonucleotide and potassium iodine as an external quencher. These studies[42] clearly suggested that fluorescent oligonucleotides were located in the aqueous core of the nanocapsules, surrounded by a polymeric wall and thus inaccessible to the quencher. On the contrary, when the fluorescent oligonucleotides were free in solution, the fluorophores were highly accessible and strong quenching occurred. Similar quenching could be obtained with nanoencapsulated oligonucleotides but only after hydrolysis of the polymer wall, releasing the oligonucleotides. The possibility that the obtained particles consisted in porous nanospheres (instead of nanocapsules) with oligonucleotides located in their polymer matrix is not likely because it has been clearly shown that due to their negative charge and to the absence of hydrophobic interactions, oligonucleotides do not interact spontaneously with the poly(alkylcyanoacrylate) polymer.[31] Moreover, in the procedure used for the preparation of the nanocapsules, although washing with water was performed, no significant oligonucleotide release or desorption was observed. In the case of a porous nanosphere structure, an extensive desorption of the oligonucleotide during washings or dilution would have been observed. Finally, when calcein was encapsulated, no increase of fluorescence with time was observed over 24 h, and when the sample was diluted, no reversal of quenching occurred. This again pleads in favor of a core-shell nanocapsule structure rather than a porous nanosphere system which would have released calcein after dilution. Zeta potential experiments have confirmed the localization of oligonucleotide within the aqueous core of the capsule. Indeed, increasing the amount of oligonucleotide in the formulation had no effect on the zeta potential, again suggesting that the main part of the oligonucleotides was located inside the nanocapsules rather than at their surface.

Further studies have also demonstrated that nanoencapsulation was able to protect oligonucleotides against degradation by serum nucleases and that this protection was much more efficient than that obtained with CTAB-coated nanospheres which allowed only simple adsorption of oligonucleotides onto their surface[31,46] rather than encapsulation.

Phosphorothioate oligonucleotides directed against EWS Fli-1 chimeric RNA were encapsulated within PACA nanocapsules and tested *in vivo* for their efficacy against the experimental Ewing sarcoma in mice.[52] Intratumoral injection of antisense-loaded nanocapsules led to a significant inhibition of tumor growth at a cumulative dose of 14.4 nanomoles. No antisense effect could be detected with the free oligonucleotide. In a previous study, using the same antisense sequence as a free drug, Tanaka et al.[53] demonstrated inhibition of a tumor growth in a similar model, but a cumulative dose of 500 nanomoles of oligonucleotide was needed. Thus, with nanocapsules it was possible to obtain an effect comparable to that observed by these authors with a 35-fold lower dose. Therefore, PACA nanocapsule technology would allow lower phosphorothioate doses to be used, thus avoiding toxicity and loss of specificity resulting from phosphorothioates at higher doses.[54,55]

The mechanism by which oligonucleotide in nanocapsules leads to a significant effect on the tumor growth may be explained by both the protection of the oligonucleotide afforded by the nanocapsules and an enhanced intracellular penetration of the ODN mediated by nanocapsules. Thus, the use of phosphorothioates at low doses combined with nanocapsules may represent a new and safe solution for the administration of antisense therapy *in vivo*.

24.6 CONCLUSION

This review shows that polyalkylcyanoacrylate nanoparticle based technologies have great potential for the delivery of oligonucleotides, since they could enhance their biological efficacy. Researchers are now in progress to apply those cyanoacrylate nanotechnologies to the delivery of siRNA. Further strategies should also focus on the use of particles containing recognition systems in order to improve selectivity of such delivery systems.

REFERENCES

1. Fattal, E. and Vauthier, C., Nanoparticles as drug delivery systems, in *Encyclopedia of Pharmaceutical Technology*, Boylan, J.C. (Ed.), Marcel-Dekker, Basel, 2002, 1874–1892.
2. Fattal, E., Vauthier, C., Aynie, I., Nakada, Y., Lambert, G., Malvy, C., and Couvreur, P., Biodegradable polyalkylcyanoacrylate nanoparticles for the delivery of oligonucleotides, *J. Control. Release* 53 (1–3), 137–143, 1998.
3. Stein C.A. and Cheng Y.C., Antisense oligonucleotides as therapeutic agents: is the bullet really magical? *Science* 261, 1004–1012, 1993.
4. Crooke, S.T., Progress toward oligonucleotide therapeutics: pharmacodynamic properties, *FASEB J.* 7 (6), 533–539, 1993.
5. Mergny, J.L., Duval-Valentin, G., Nguyen, C.H., Perrouault, L., Faucon, B., Rougee, M., Montenay-Garestier, T., Bisagni, E., and Helene, C., Triple helix-specific ligands, *Science* 256 (5064), 1681–1684, 1992.
6. Beal, P.A. and Dervan, P.B., Second structural motif for recognition of DNA by oligonucleotide-directed triple-helix formation, *Science* 251 (4999), 1360–1363, 1991.
7. Helene, C., The anti-gene strategy: control of gene expression by triplex-forming oligonucleotides, *Anticancer Drug Des.* 6 (6), 569–584, 1991.
8. Roberts, R.W. and Crothers, D.M., Stability and properties of double and triple helices: dramatic effects of RNA or DNA backbone composition, *Science* 258 (5087), 1463–1466, 1992.
9. Been, M.D., Nucleases that are RNA, in *Nucleases*, Roberts, R.S. (Ed.), Cold Spring Laboratory, Cold Spring Harbor, NY, 1993, 407–437.
10. Francois, J.C., Thuong, N.T., and Helene, C., Recognition and cleavage of hairpin structures in nucleic acids by oligodeoxynucleotides, *Nucleic Acids Res.* 22 (19), 3943–3950, 1994.
11. Loke, S.L., Stein, C.A., Zhang, X.H., Mori, K., Nakanishi, M., Subasinghe, C., Cohen, J.S., and Neckers, L.M., Characterization of oligonucleotide transport into living cells, *Proc. Nat. Acad. Sci. U.S.A.* 86 (10), 3474–3478, 1989.
12. Yakubov, L.A., Deeva, E.A., Zarytova, V.F., Ivanova, E.M., Ryte, A.S., Yurchenko, L.V., and Vlassov, V.V., Mechanism of oligonucleotide uptake by cells: involvement of specific receptors? *Proc. Nat. Acad. Sci. U.S.A.* 86 (17), 6454–6458, 1989.
13. Hanvey, J.C., Peffer, N.J., Bisi, J.E., Thomson, S.A., Cadilla, R., Josey, J.A., Ricca, D.J., Hassman, C.F., Bonham, M.A., Au, K.G., et al., Antisense and antigene properties of peptide nucleic acids, *Science* 258 (5087), 1481–1485, 1992.
14. Ellington, A.D. and Szostak, J.W., Selection *in vitro* of single-stranded DNA molecules that fold into specific ligand-binding structures, *Nature* 355 (6363), 850–852, 1992.
15. Burgess, T.L., Fisher, E.F., Ross, S.L., Bready, J.V., Qian, Y.X., Bayewitch, L.A., Cohen, A.M., Herrera, C.J., Hu, S.S., Kramer, T.B., et al., The antiproliferative activity of c-myb and c-myc antisense oligonucleotides in smooth muscle cells is caused by a nonantisense mechanism, *Proc. Nat. Acad. Sci. U.S.A.* 92 (9), 4051–4055, 1995.
16. Castier, Y., Chemla, E., Nierat, J., Heudes, D., Vasseur, M.A., Rajnoch, C., Bruneval, P., Carpentier, A., and Fabiani, J.N., The activity of c-myb antisense oligonucleotide to prevent intimal hyperplasia is nonspecific, *J. Cardiovasc. Surg. (Torino)* 39 (1), 1–7, 1998.
17. Bennett, C.F., Antisense oligonucleotides: is the glass half full or half empty? *Biochem. Pharmacol.* 55 (1), 9–19, 1998.

18. Galbraith, W.M., Hobson, W.C., Giclas, P.C., Schechter, P.J., and Agrawal, S., Complement activation and hemodynamic changes following intravenous administration of phosphorothioate oligonucleotides in the monkey, *Antisense Res. Dev.* 4 (3), 201–206, 1994.

19. Krieg, A.M., Yi, A.K., Matson, S., Waldschmidt, T.J., Bishop, G.A., Teasdale, R., Koretzky, G.A., and Klinman, D.M., CpG motifs in bacterial DNA trigger direct B-cell activation, *Nature* 374 (6522), 546–549, 1995.

20. Ketting, R.F., Fischer, S.E., Bernstein, E., Sijen, T., Hannon, G.J., and Plasterk, R.H., Dicer functions in RNA interference and in synthesis of small RNA involved in developmental timing in C. elegans, *Genes Dev.* 15 (20), 2654–2659, 2001.

21. Vaucheret, H., Beclin, C., and Fagard, M., Post-transcriptional gene silencing in plants, *J. Cell Sci.* 114 (Pt. 17), 3083–3091, 2001.

22. Elbashir, S.M., Harborth, J., Lendeckel, W., Yalcin, A., Weber, K., and Tuschl, T., Duplexes of 21-nucleotide RNAs mediate RNA interference in cultured mammalian cells, *Nature* 411 (6836), 494–498, 2001.

23. Caplen, N.J., Parrish, S., Imani, F., Fire, A., and Morgan, R.A., Specific inhibition of gene expression by small double stranded RNAs in invertebrate and vertebrate systems, *Proc. Nat. Acad. Sci. U.S.A.* 98, 9742–9747, 2001.

24. Bertrand, J.R., Pottier, M., Vekris, A., Opolon, P., Maksimenko, A., and Malvy, C., Comparison of antisense oligonucleotides and siRNAs in cell culture and *in vivo*, *Biochem. Biophys. Res. Commun.* 296 (4), 1000–1004, 2002.

25. Scherr, M., Battmer, K., Winkler, T., Heidenreich, O., Ganser, A., and Eder, M., Specific inhibition of bcr-abl gene expression by small interfering RNA, *Blood* 101 (4), 1566–1569, 2003.

26. Vickers, T.A., Koo, S., Bennett, C.F., Crooke, S.T., Dean, N.M., and Baker, B.F., Efficient reduction of target RNAs by siRNA and RNase H dependent antisense agents: a comparative analysis, *J. Biol. Chem.* 278, 7108–7118, 2003.

27. Couvreur, P., Kante, B., Roland, M., Guiot, P., Bauduin, P., and Speiser, P., Polycyanoacrylate nanocapsules as potential lysosomotropic carriers: preparation, morphological and sorptive properties, *J. Pharm. Pharmacol.* 31 (5), 331–332, 1979.

28. Fattal, E., Rojas, J., Youssef, M., Couvreur, P., and Andremont, A., Liposome-entrapped ampicillin in the treatment of experimental murine listeriosis and salmonellosis, *Antimicrob. Agents Chemother.* 35 (4), 770–772, 1991.

29. Lenaerts, V., Nagelkerke, J.F., Van Berkel, T.J., Couvreur, P., Grislain, L., Roland, M., and Speiser, P., *In vivo* uptake of polyisobutyl cyanoacrylate nanoparticles by rat liver Kupffer, endothelial, and parenchymal cells, *J. Pharm. Sci.* 73 (7), 980–982, 1984.

30. Lenaerts, V., Couvreur, P., Christiaens-Leyh, D., Joiris, E., Roland, M., Rollman, B., and Speiser, P., Degradation of poly (isobutyl cyanoacrylate) nanoparticles, *Biomaterials* 5 (2), 65–68, 1984.

31. Chavany, C., Ledoan, T., Couvreur, P., Puisieux, F., and Helene, C., Polyalkylcyanoacrylate nanoparticles as polymeric carriers for antisense oligonucleotides, *Pharm. Res.* 9 (4), 441–449, 1992.

32. Chavany, C., Saison-Behmoaras, T., Le Doan, T., Puisieux, F., Couvreur, P., and Helene, C., Adsorption of oligonucleotides onto polyisohexylcyanoacrylate nanoparticles protects them against nucleases and increases their cellular uptake, *Pharm. Res.* 11 (9), 1370–1378, 1994.

33. Peracchia, M.T., Vauthier, C., Passirani, C., Couvreur, P., and Labarre, D., Complement consumption by poly(ethylene glycol) in different conformations chemically coupled to poly(isobutyl 2-cyanoacrylate) nanoparticles, *Life Sci.* 61 (7), 749–761, 1997.

34. Peracchia, M.T., Vauthier, C., Puisieux, F., and Couvreur, P., Development of sterically stabilized poly(isobutyl 2- cyanoacrylate) nanoparticles by chemical coupling of poly(ethylene glycol). *J. Biomed. Mater. Res.* 34 (3), 317–326, 1997.

35. Peracchia, M.T., Vauthier, C., Desmaele, D., Gulik, A., Dedieu, J.C., Demoy, M., d'Angelo, J., and Couvreur, P., Pegylated nanoparticles from a novel methoxypolyethylene glycol cyanoacrylate-hexadecyl cyanoacrylate amphiphilic copolymer, *Pharm. Res.* 15 (4), 550–556, 1998.

36. Peracchia, M.T., Harnisch, S., Pinto-Alphandary, H., Gulik, A., Dedieu, J.C., Desmaele, D., d'Angelo, J., Muller, R.H., and Couvreur, P., Visualization of *in vitro* protein-rejecting properties of PEGylated stealth polycyanoacrylate nanoparticles, *Biomaterials* 20 (14), 1269–1275, 1999.

37. Peracchia, M.T., Fattal, E., Desmaele, D., Besnard, M., Noel, J. P., Gomis, J.M., Appel, M., d'Angelo, J., and Couvreur, P., Stealth PEGylated polycyanoacrylate nanoparticles for intravenous administration and splenic targeting, *J. Control. Release* 60 (1), 121–128, 1999.

38. Papahadjopoulos, D., Allen, T.M., Gabizon, A., Mayhew, E., Matthay, K., Huang, S.K., Lee, K.D., Woodle, M.C., Lasic, D. D., Redemann, C., and Martin, F.J., Sterically stabilized liposomes: improvements in pharmacokinetics and antitumor therapeutic efficacy, *Proc. Nat. Acad. Sci. U.S.A.* 88 (24), 11460–11464, 1991.

39. Gabizon, A. and Papahadjopoulos, D., Liposome formulations with prolonged circulation time in blood and enhanced uptake by tumors, *Proc. Nat. Acad. Sci. U.S.A.* 85 (18), 6949–6953, 1988.

40. Bakker-Woudenberg, I.A., Lokerse, A.F., ten Kate, M.T., Mouton, J.W., Woodle, M.C., and Storm, G., Liposomes with prolonged blood circulation and selective localization in Klebsiella pneumoniae-infected lung tissue, *J. Infect. Dis.* 168 (1), 164–171, 1993.

41. Stella, B., Arpicco, S., Peracchia, M.T., Desmaele, D., Hoebeke, J., Renoir, M., D'Angelo, J., Cattel, L., and Couvreur, P., Design of folic acid-conjugated nanoparticles for drug targeting, *J. Pharm. Sci.* 89 (11), 1452–1464, 2000.

42. Lambert, G., Fattal, E., Pinto-Alphandary, H., Gulik, A., and Couvreur, P., Polyisobutylcyanoacrylate nanocapsules containing an aqueous core as a novel colloidal carrier for the delivery of oligonucleotides, *Pharm. Res.* 17 (6), 707–714, 2000.

43. Zobel, H.P., Kreuter, J., Werner, D., Noe, C. R., Kumel, G., and Zimmer, A., Cationic polyhexylcyanoacrylate nanoparticles as carriers for antisense oligonucleotides, *Antisense Nucleic Acid Drug Dev.* 7 (5), 483–493, 1997.

44. Balland, O., Saison-Behmoaras, T., and Garestier, T., Nanoparticles as carriers for antisense oligonucleotides, in *Targeting of Drugs 5: Strategies for Oligonucleotide and Gene Delivery in Therapy*, McCormack, B. (Ed.), Plenum Press, New York, 1996, 131–142.

45. Godard, G., Boutorine, A.S., Saison-Behmoaras, E., and Helene, C., Antisense effects of cholesterol-oligodeoxynucleotide conjugates associated with poly(alkylcyanoacrylate) nanoparticles, *Eur. J. Biochem.* 232 (2), 404–410, 1995.

46. Nakada, Y., Fattal, E., Foulquier, M., and Couvreur, P., Pharmacokinetics and biodistribution of oligonucleotide adsorbed onto poly(isobutylcyanoacrylate) nanoparticles after intravenous administration in mice, *Pharm. Res.* 13 (1), 38–43, 1996.

47. Schwab, G., Chavany, C., Duroux, I., Goubin, G., Lebeau, J., Helene, C., and Saison-Behmoaras, T., Antisense oligonucleotides adsorbed to polyalkylcyanoacrylate nanoparticles specifically inhibit mutated Ha-ras-mediated cell proliferation and tumorigenicity in nude mice, *Proc. Nat. Acad. Sci. U.S.A.* 91 (22), 10460–10464, 1994.

48. Schwab, G., Duroux, I., Chavany, C., Helene, C., and Saison-Behmoaras, E., An approach for new anticancer drugs: oncogene-targeted antisense DNA, *Ann. Oncol.* 5 (Suppl. 4), 55–58, 1994.

49. Lambert, G., Fattal, E., Brehier, A., Feger, J., and Couvreur, P., Effect of polyisobutylcyanoacrylate nanoparticles and lipofectin loaded with oligonucleotides on cell viability and PKC alpha neosynthesis in HepG2 cells, *Biochimie* 80 (12), 969–976, 1998.

50. Filion, M.C. and Phillips, N.C., Anti-inflammatory activity of cationic lipids, *Br. J. Pharmacol.* 122 (3), 551–557, 1997.

51. Aynie, I., Vauthier, C., Foulquier, M., Malvy, C., Fattal, E., and Couvreur, P., Development of a quantitative polyacrylamide gel electrophoresis analysis using a multichannel radioactivity counter for the evaluation of oligonucleotide-bound drug carrier, *Anal. Biochem.* 240 (2), 202–209, 1996.

52. Lambert, G., Bertrand, J. R., Fattal, E., Subra, F., Pinto-Alphandary, H., Malvy, C., Auclair, C., and Couvreur, P., EWS fli-1 antisense nanocapsules inhibits ewing sarcoma-related tumor in mice, *Biochem. Biophys. Res. Commun.* 279 (2), 401–406, 2000.

53. Tanaka, K., Iwakuma, T., Harimaya, K., Sato, H., and Iwamoto, Y., EWS-Fli1 antisense oligodeoxynucleotide inhibits proliferation of human Ewing's sarcoma and primitive neuroectodermal tumor cells, *J. Clin. Invest.* 99 (2), 239–247, 1997.

54. Monteith, D.K. and Levin, A.A., Synthetic oligonucleotides: the development of antisense therapeutics, *Toxicol. Pathol.* 27 (1), 8–13, 1999.

55. Stein, C.A., Phosphorothioate antisense oligodeoxynucleotides: questions of specificity, *Biotechnology* 14 (5), 47–149, 1996.

CHAPTER 25

Layer-by-Layer Nanoengineering with Polyelectrolytes for Delivery of Bioactive Materials

Dinesh Shenoy, Alexei Antipov, and Gleb Sukhorukov

CONTENTS

25.1 INTRODUCTION

Development of advanced methods for the controlled assembly of multicomponent nanostructures is highly desirable, although it is understood that complex structures as those found in the biological world cannot be fabricated as such. It was attractive to develop a simple approach that would yield nanoarchitectured films with good positioning of individual layers, but whose fabrication would be largely independent of the nature, size and topology of the substrate. For about

0-8493-1934-X/05/$0.00+$1.50
© 2005 by CRC Press LLC

seven decades, controlled fabrication of nanostructured films has been dominated by the conceptually elegant Langmuir-Blodgett (LB) technique, in which monolayers are formed on a water surface and subsequently transferred onto a solid support.[1,2] Although a great variety of substances could be used to prepare LB films and the resultant layers were highly ordered nanostructures having uniform thickness with a possibility of fabricating thick multilayers (a ~3000 multilayer coating was described by Blodgett), this technique could not find an industrial application. There are a few reasons for that: the substrate has to be smooth and homogeneous; any defect in a layer is hard to repair by subsequent layers; the resulting LB multilayers cannot withstand high temperatures or solvent treatment; and, finally, expensive equipment is required, especially to coat relatively large areas.

A possible route to obtain a desired target structure is to use an assembly procedure that prevents equilibrium by trapping every compound kinetically in a predetermined spatial arrangement. The self-organization of polymers has been progressively explored for the preparation of well-defined surfaces and interfaces in recent years.[3-5] The most recent of the self-organization techniques is the alternating physisorption of oppositely charged polyelectrolytes — a step-by-step procedure that allows fine structuring in the third dimension, popularly known as the layer-by-layer (LbL) technique.[6-9] The method is extremely versatile, because not only polyelectrolytes, but also charged nanoobjects, such as molecule aggregates, clusters, or colloids can be used. The method is relatively rapid, as adsorption steps last typically from a few minutes up to an hour. LbL deposition leads to the fabrication of polydisperse supramolecular objects and is a rather general approach for the fabrication of complex surface coatings. Influence of polyelectrolyte concentration on adsorption is relatively weak, except for a few cases reported; especially at very low concentrations the influence may become more sensitive. In a previous work, the overall thickness of the layers differed by 15% while changing the polymer concentration from 1 to 50 mM in the repeat unit.[10] This fact can be attributed to the mechanism of adsorption. It seems to be that during the first adsorption step only some segments of polyelectrolyte bind tightly to the interface — a diffusion-limited step. During the second step polymer loops relax and settle down onto the surface. Probably due to limited diffusion, at lower concentrations, one polyelectrolyte molecule occupies more binding sites than at higher concentrations and, hence, the derived layer becomes thinner. The ability of an incoming oppositely charged species to be adsorbed in a subsequent step on top of the first layer drives forward the process of multilayer build-up. Surplus polymer solution adhering to the support after the first adsorption cycle is removed by simple washing. Under proper conditions, polymeric material with more than the stoichiometric number of charges (relative to the substrate) is adsorbed, so that the sign of the surface charge is reversed.[8,11] As a consequence, when the substrate is exposed to a second solution containing a polyion of opposite charge, an additional polyion layer is adsorbed. This reverses the sign of the surface charge again. Consecutive cycles with alternating adsorption of polyanions and polycations result in the stepwise growth of polymer films (see Figure 25.1 for a schematic representation of the procedure). LbL deposition could be considered as an analogue to a chemical reaction sequence. Like a chemical reaction, the precise structure of each layer depends on a set of control parameters such as concentration, adsorption times, ionic strength, pH, solvent, or temperature, but in general the processing window is rather broad. As in the case of interpolyelectrolyte complex formation, the driving force of the multilayer buildup is the entropy gain due to the ions leaving the polyelectrolyte molecules.[5] Even if the major energetic contribution to the success of the LbL assembly comes from the entropy gain, other associations such as hydrophobic interactions, charge transfer interactions, stacking forces, H-bonding, and other short-range interactions may contribute to the success of LbL nanoengineering for a given system.[12-14]

Materials of construction for multilayer films by the LbL self-assembly could include small organic molecules or inorganic compounds, macromolecules, biomacromolecules such as proteins and DNA, and even colloids (metallic or ceramic or latex particles), as well as certain charged supra-molecular biological assemblies such as viruses or membrane fragments.[5,15-22] The fabrication

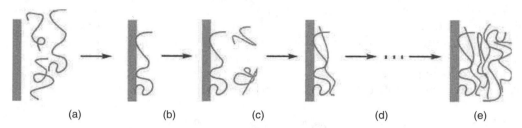

(a)   (b)   (c)   (d)   (e)

Figure 25.1   Scheme of the LbL assembly of two polyelectrolytes on flat substrate: (A) a charged planar surface is being exposed to an oppositely charged polyelectrolyte; (B) physisorption of the polyelectrolyte to form the first layer; (C) exposure to the second polyelectrolyte having complementary charge to that of the first adsorbed layer; (D) planar surface having two adsorbed polyelectrolyte layers; (E) repetition of several cycles of A and C, leading to formation of polyelectrolyte multilayers on planar surface. Note: Intermittent washing steps with water separate the two adsorption cycles. The same concept works in an identical fashion when the planar surface is replaced by a colloidal core, leading to the formation of polyelectrolyte multilayerd shell around the core.

of multicomposite films by the LbL procedure means literally nanoscopic assembly of hundreds of different materials in a single configuration, using an environmentally friendly, low-cost, reproducible, reliable, technologically simple, yet robust procedure. The investigations in this particular field of physical chemistry are mostly motivated by their biological function and biomimicking applications: for imparting biocompatibility to the surfaces, for enzymatic activity, for bio-sensing purposes, or for molecular recognition.[20] The technique can be applied to solvent-accessible surfaces of almost any kind and any shape, the more exotic ones being microspheres, oil droplets, and biological cells.[23-26] Coatings made by the LbL technique are stratified, but do not consist of well-separated, distinguishable alternating layers — instead, adjacent layers interpenetrate each other and have a smeared structure.[6]

LbL engineering has been a welcome addition to the established self-organization techniques. Up to now it has provoked the research interests of physicists, chemists, and scientists in the biomedical field, simply because it is extremely powerful yet simple to use the technique and because it challenges theory at the level of polyelectrolyte adsorption. Advances in the understanding of this particular field of nanoengineering during the past years have rendered discussions on practical applications more realistic.

25.2 LBL ENGINEERING PRINCIPLES IN DESIGN OF COLLOIDS WITH TAILORED ARCHITECTURE

Surface engineering of colloid particles at the nanoscale level is favored for the manufacturing of hierarchically ordered structures. This section concisely explains the extension of the LbL technique from fabrication of simple multilayered functional thin films to complex, ordered nanocomposite colloids with tailored properties.

25.2.1 Extension of LbL Sequential Adsorption Technique from Flat Surfaces to Colloidal Particles

Colloids are routinely employed for scientific investigations and technology-driven applications. Technologically, they have been of importance for centuries, having been exploited in diverse areas such as coatings, catalysis, sensing, separations, drug delivery, electronics, and photonics.[27] A prerequisite for the successful utilization of colloids for many applications is that they remain colloidally stable.[28] The two main components of the intermolecular interactions that influence colloidal stability, namely, the electrostatic repulsion and van der Waals attraction, have been intensively studied. The emerging nanotechnology and biotechnology fields, however, will require tailor-made building blocks for the construction of advanced materials and devices. Since colloid particles are useful components of such systems, the availability of colloids with unique and tailored functional properties (e.g., optical, mechanical, thermal, electrical, magnetic, catalytic, and biological) is highly desired, and routes that allow their preparation have attracted considerable interest.

Several techniques have been used for forming polymer coatings on top of the colloids, although polymerization-based methods are the most commonly employed. These include direct polymerization of monomers adsorbed onto the particle surface,[29,30] heterocoagulation–polymerization,[31-33] and emulsion–polymerization.[34] Inorganic coatings are often formed by precipitation or surface reactions of inorganic precursors with the particle surface[35-37] or by the sonochemical generation of nanoparticles in the presence of larger colloids.[38] The majority of the existing coating methods do not permit nanoscale control of the coating composition and thickness and, often, the particles are not uniformly coated, thus promoting aggregation.

The novel, versatile LbL self-assembly process permits the preparation of coated (and encapsulated) colloids of different shapes and sizes, with uniform layers of diverse composition with a nanoscale control that can be afforded over the wall thickness.[39,40] In the extension of the LbL technology for the synthesis of coated colloids, the first added species usually have an opposite charge to that on the bare colloids, thereby adsorbing through electrostatic interactions. An overcompensation of charge often ends with adsorption of each layer, and at this stage the charge on the surface of the particles is reversed. This facilitates the alternate deposition of subsequent layers of a wide range of charged components. The inter-polyelectrolyte interactions not only act as molecular glue, but also impart enhanced colloidal stability to the coated particles via electrostatic as well as steric contributions (Figures 25.2A to 25.2E illustrate schematic representation of the procedure). Multilayers comprised solely of polymers to low-molecular-weight compounds and inorganic nanoparticle-, protein-, and phospholipid-based coatings on colloids have been established for synthesis of nanocomposite core-shell materials.[41]

25.2.2 Core-Shell Compositions for Drug Delivery Applications

In LbL engineering science, an assembly having a core material (which could be solid or a liquid in nature) that is covered or coated with a multilayered construction is generally referred to as "core-shell" structure. Colloids of different sizes, shapes, and compositions can be employed as core particles in the LbL method for the creation of novel core-shell colloids with various functions (e.g., catalytic, optical, chemical, magnetic, etc.) and can provide a simple means of encapsulating a variety of significant compounds. The core materials could consist of inorganic or organic crystals, polymeric latexes, protein crystals and aggregates, biological cells, and even oil droplets, and

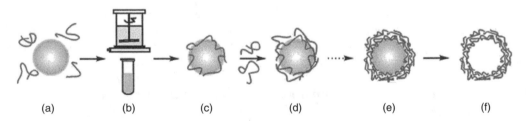

(a)   (b)   (c)   (d)   (e)   (f)

Figure 25.2 Scheme for LbL coating of colloids and formation of hollow capsules: (a) sacrificial template/core exposed to an oppositely charged polyelectrolyte; (b) removal of unadsorbed polyelectrolyte by either filtration through an assembly having a membrane filter (top) or by centrifugation; (c) core-shell assembly having one polyelectrolyte layer; (d) exposure to the second adsorption cycle with second polyelectrolyte having complementary charge to the first one (two adsorption cycles are separated by washing steps with water to remove excess of the preceding polyelectrolyte); (e) core-shell structure having polyelectrolyte multilayered shell after repetitive adsorption-washing cycles; (f) elimination of sacrificial core by suitable means to derive polyelectrolyte hollow capsules. The size of the capsule is a replica of templating colloid.

the shells could be made up of polymers, proteins, lipids, low-molecular-weight compounds, or nanoparticles.[41-43]

Enzyme crystals (e.g., catalase) were successfully encapsulated by coating with polyelectrolytes.[44] The catalase crystals, approximately 10 μm in diameter, exhibit a positive surface charge in water at pH 5 (catalase isoelectric point = 5.8). This positive charge on the surface of the crystals facilitated the adsorption of polyelectrolyte multilayers. These particles contain an extremely high enzyme loading in each polyelectrolyte multilayer coating, and the activity of the encapsulated catalase is preserved after coating. The coatings also act as a protective layer for the enzyme, shielding it from high-molecular-weight inhibitors (like proteases), which cannot penetrate the polyelectrolyte coating and influence the enzyme functionality. Polyelectrolyte multilayer assembly can also be performed on amorphous particles consisting of aggregates of proteins, which are formed prior to multiplayer build-up.[45] The use of micron-sized amorphous protein aggregates as templates for polyelectrolyte multilayer assembly was demonstrated for lactate dehydrogenase and chymotrypsin.[46-48]

A major challenge in the development of advanced drug formulations is the elaboration of delivering systems capable of providing sustained and controlled release of bioactive materials. In this regard, it is advantageous to be able to decrease the layer permeability for low-molecular-weight compounds once they are encapsulated. The formation of a thicker capsule wall (which is a function of number of layers, composition of the multilayers, and the conditions of its assembly) is a straightforward way to decrease shell permeability.[49-51] Figure 25.3 provides a general scheme for bioactive particle coating by LbL deposition and its subsequent dissolution and release through the multilayers. The time of complete dissolution increases by more than 2 orders of magnitude for model drug (fluorescent dye) crystals coated with 18 layers of synthetic polymers.[49,52] Biocompatible polyelectrolytes (of polysaccharide origin, such as chitosan, alginate, dextran, sulfate, etc.) have been used successfully to design controlled release formulations (in oral and sterile dosage forms) by direct surface modification of drug microcrystals obtained by either micronization or by chemical means.[51,53,54] It was shown that employment of a bulky macromolecule (such as a protein like gelatin) as one of the layer constituents leads to a 300-fold increase in the dissolution time of furosemide microcrystals coated with only four polyelectrolyte bilayers.[50] Yet another way of prolonging the dissolution and diffusion profile of the encapsulated and coated drug is the use of lipids as a layer constituent.[55,56]

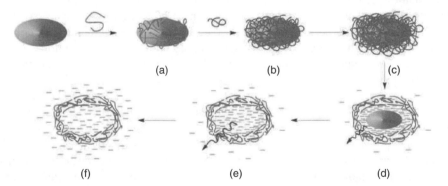

Figure 25.3 Scheme of the polyelectrolyte multilayer deposition process and of the subsequent core dissolution. The initial steps (a–c) involve stepwise shell formation on a fluorescein/biocolloidal core (C represents the core-shell assembly). After the deposition of the desired number of polyelectrolyte layers, the coated particles are exposed to a core dissolution condition (c, d). The release of the core material through the polyelectrolyte multilayered shell wall follows a two-step process: dissolution of the core material within the shell (d) and diffusion-mediated release of the core material (e) resulting finally in fully dissolved cores and core-exhausted shells (f).

Biological cells can also be utilized as cores for encapsulation. Gluteraldehyde fixed echinocytes have been coated with polymer and polymer–nanoparticle multilayers.[57] Such biocolloids have a jagged and highly structured surface as the coating tends to follow the contour with nanoscale precision. Following removal of the core by treatment with a NaOCl solution, it was shown that the structure of the coating left behind mimics the original shape, including spikes of the echinocytes.[58,59] Similar work has been done with living yeast cells.[25] It was shown that the polyelectrolyte shell does not affect the integrity of the cells and their reproductive functions. Permeability of the shell to low-molecular-weight nutrients made possible the cells' division.[25] Coating of biological cells in this manner can provide an attractive means to encase them in ultrathin coatings, thus protecting them from external, undesirable species such as enzymes or pathogens. Such cells can be used in biotechnology and can potentially serve as bioreactors *in vivo* without eliciting significant immune response if surface-modified with appropriate biofriendly species, or they could be designed as circulating reservoirs of bioactive compounds using "stealth" materials (such as polyethylene glycol) as a final envelope. This elegant tuning of the surface properties could be achieved by a simple physico-chemical approach — the electrostatic self-assembly.

25.2.3 Fabrication of Hollow Capsules by the LbL Approach

The assembly of multilayer films onto micrometer and submicrometer sized "sacrificial" template-particles is an extension of the polyion LbL adsorption technology to form micro- and nano-sized three-dimensional ultrathin structures in addition to the well-advanced fabrication of two-dimensional films. This way, novel supramolecular assemblies with controlled nanostructure and composition of the interfacial region could be produced.

Decomposable colloid particles could be coated with polyions by the LbL technique; the core is then decomposed or dissolved in such a manner that the low-molecular-weight core residuals would permeate through the wall composed of polyelectrolyte multilayer assembly (see the complete sequence in Figure 25.2).[39,41,60] Since a wide variety of template-particles can be used, the core removal chemistry would be highly varied.[41] The walls can consist of the whole array of materials that have been used for planar multilayers.[42,43] Upon removal of the templating colloid, what is left behind would be hollow polyelectrolyte structures, that are replicas of the templated structure with respect to size and shape. Because of the great diversity and modularity, permeation of the core material can be controlled by pH, ionic strength, and/or temperature. Small molecules can permeate the walls as the multilayers possess a certain degree of viscoelasticity. Various approaches have been explored for extraction of the sacrificial template depending upon its physicochemical properties; these approaches are summarized in the following section.

When the core materials are made up of weakly cross-linked or polymerized material (e.g., melamine formaldehyde [MF] latex microparticles), effective depolymerization (and hence core decomposition) using mild conditions (e.g., 0.1 M HCl for MF) at which the multilayer shell assembly is stable, leads to formation of hollow capsules.[23] The increase in osmotic pressure during decomposition is currently thought to be responsible for widening the existing pores or for the formation of new pores which then allow passage of the melamine resin degradation products.[61] If the core is made of inorganic crystals (such as calcium, cadmium, or manganese carbonate microparticles), it can be removed by a chemical reaction (e.g., hydrolysis). This can be achieved by changing the pH (e.g., with hydrochloric acid for carbonate cores),[62] or by adding strong chelating agents like ethylenediaminetetraacetic acid.[62,63] The RBCs , when used as sacrificial templates, are decomposed by oxidative degradation (with NaOCl).[57,59]

Polymeric particles such as poly(styrene), poly(DL-lactic acid) (PLA) or poly(DL-lactic-*co*-glycolic acid) (PLGA) can be removed as a whole (core dissolution) using suitable organic solvent(s) following a similar procedure as for core decomposition.[26,64,65] Some of these templates may be of particular interest due to their inherent biocompatibility (e.g., PLA and PLGA) — so that one could derive biocompatible hollow capsules when they are used together with biocompatible coating materials.[65] However, precautions should be taken against rupturing of the capsule wall due to the nascent osmotic pressure.[65]

One could use few other rare techniques for elimination of the templating colloid, such as calcinations or etching, which can be used in the cases when the shell is made up of solely inorganic or composite inorganic and organic materials. This extreme treatment causes decomposition and evaporation of the organic matter (core substance) leaving behind only the inorganic shell.[66] In principle, it should also be possible to destroy cores with light, but this has not yet been attempted successfully.

Special types of cores are precipitates of drug particles, enzymes, or DNA. They are not meant to be destroyed during fabrication, but to carry out a specific function when the shell gets destroyed at a specified target (e.g., after attaching to a cell or its nucleus). It should be remembered that the kinetics of the core dissolution in such cases is limited not only by its solubility at given conditions, but also by the permeability of the shell wall.

25.2.4 Scope and Methods for Loading of Bioactive Materials into Preformed Capsules

Various applications of hollow capsules may be derived from the fact that the inner volume could be loaded via a chemical or a physical process. One may make use of specific physical and chemical processes inside the wall, in particular:

- Precipitation and crystallization (active trapping of substances)[67-69]
- Polymerization and polymer functionalization (stimuli-responsive capsules)[70]
- Enzymatic and heterogeneous catalysis (chemical reactions within restricted volumes)[71]
- Photocatalysis and crosslinking (sensory/catalytic applications)[72]

Various strategies have been successful in achieving a defined chemical environment inside a capsule that is different from the environment outside (see Figure 25.4).

25.2.3.1 Utilization of Donnan Equilibrium

A Donnan equilibrium can be established, making use of the fact that generally the wall is impermeable to polyelectrolytes. Thus by adding a polyacid to the outside, the dissociated protons could penetrate the wall — to create a charge imbalance. Consequently, the proton concentration outside will be larger than inside.[73,74] For example, the negatively charged dye carboxyfluorescein (CF) precipitates at pH lower than 3.0. Hence incorporation of a strong polyanion, such as poly(styrene sulfonate) (PSS), in the capsule interior establishes a Donnan equilibrium with lower pH inside; dye precipitation will commence inside. Thus more monomeric CF will be sucked inside until the capsule is filled with dye precipitates.[67]

25.2.3.2 Capturing of Particulate Material, Polymeric Catalyst, or Enzyme

A particulate material, a polymeric catalyst, or an enzyme may be caged inside the polyelectrolyte capsule. This can be achieved by either adsorbing it onto the template before polyelectrolyte coating or depositing with decomposable components in the inner layers.[68,74-76] Another approach of polymer encapsulation is based on reversible change of the capsule wall permeability. Incubation of capsules that are "open" at, say, acidic pH, with the macromolecular substance of interest, and subsequent "closing" of the wall by increasing to basic pH leads to entrapment of the polymer inside.[63,77,78] Principles to establish chemical nonequilibrium across the polyelectrolyte multilayer coatings have been useful for the fabrication of micro- or nano-containers. An obvious advantage of the system is

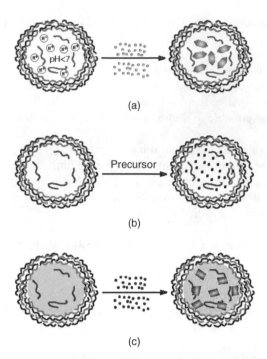

Figure 25.4 Fabrication and utilization of smart capsules; tuning of chemical environment of the capsule interior: (a) employing the Donnan equilibrium (the internal pH of the capsule filled with a polyacid will is more acidic than of the bulk solution); (b) enzyme encapsulated capsule when exposed to a precursor leading to formation of products that are trapped inside the capsule volume — nanoreactors or nanocontainers for enzyme/bioactive compounds; (c) differential interior/exterior solvent composition by utilization of ampiphillic compounds leading to precipitation/accumulation of compounds selectively within the capsule interior.

that large molecules in general do not penetrate the wall but small molecules do (multilayers function as selective barriers). Hence enzymes can be entrapped and protected against proteases and high-molecular-weight inhibitors. Small substrates can penetrate to react inside and products can be released and accumulated[47,77] — a typical requirement in diagnostic applications.

25.2.3.3 Preparing the Double-Walled Capsules

This section deals with preparing the inner wall in a manner that the catalysts or solvents are enriched inside. Verified examples suggest that presence of excess charges at the inner wall would lead to enrichment of an initiator or catalyst of opposite charge, or preparing the inner (or outer) wall with an amphiphilic polymer could be encashed into entrapment of hydrophobic solvents.[72] This leads essentially to a micron-scale system that could act as catalytic or solvent reservoir (or a stabilized emulsion) where many selective reactions could be carried out.

We found that the precipitates formed inside capsules did not show any X-ray diffraction signals unless later annealed, suggesting that they are either amorphous or consist of nanocrystallites.[79] The findings were confirmed with a polyelectrolyte capsule actively loaded with natamycin taking advantage of the pH gradient.[80] This is a highly desirable feature for many medicinal delivery applications, since it facilitates retarded dissolution and hence release.

In the sections above, we have moved from a defined multilayer on a flat surface to a colloidal core-shell assembly to a hollow capsule. The systems were shown to be extremely modular involving inorganic and organic molecules and particles, polymers, proteins, and biomolecules such as DNA.

25.3 EXTENSION OF LBL CONCEPT FOR GENE DELIVERY

Significant advances in the understanding of the genetic abnormalities have made gene therapy one of the most radical notions ever put forward in medicine. Recent years have witnessed an explosion of activity in the field of gene therapy. Following advances in our understanding of the molecular basis of disease, hopes are high that the tremendous potential market for drugs employing antisense nucleotides and genes will one day be fulfilled. However, many obstacles remain in the technology of gene delivery systems. Much of today's research focuses on nonviral approaches to gene delivery. Among these, self-assembling polycation–DNA complexes represent a promising synthetic vector for gene therapy.[81-84] In spite of considerable transfectional activity *in vitro*, such materials are quickly eliminated from the bloodstream following intravenous administration. For targeted systemic delivery, a prolonged circulation of the vector is essential,[85] and this could be governed by further surface modification of the polycation–DNA complex with supramolecular assemblies.[86-95] Of particular importance are supramolecular complexes formed between DNA and various natural and synthetic polymers (polyplexes)[96-98] and lipids (lipoplexes).[99-102] The early studies on DNA condensation induced by polycations provided a model system for better understanding of how DNA is compacted in viral particles.[103] Given that most viruses are negatively charged, the supramolecular electrostatic self-assembly phenomenon has been an additional tool for modeling viral assembly.

DNA is a very good example of a polyelectrolyte of biological relevance. Compared to naked DNA, the viral–polymeric–lipid envelope provides an efficient way for transporting genes into the cells for expression in the nucleus as a recombinant gene. There are two fundamental approaches that are utilized in gene delivery that involve interactions between a charged polymer and DNA. Charge inversion of the DNA double helix by an oppositely charged flexible polyelectrolyte to deduce a condensed form of DNA (as stable submicron-sized particles) is one of them – the post-treatment of which by LbL has been explored to fabricate supramolecular assemblies utilizing the surplus charge. The second approach involves incorporation of the DNA as a polyionic shell component along with a suitable complementary polyelectrolyte by the LbL construction protocol. Parameters like polyelectrolyte molecular weight, charge spacing along the polymer backbone, length of the side chain, and charge type play an important role in the formation of the DNA–polyelectrolyte complex that governs both the approaches.[104-106] Research in this field of biovectorization is still in the budding stages, and this chapter has been an effort to drive various research groups actively engaged in gene delivery toward exploring a potential field of formulation development techniques based in physical chemistry that deserve to be actively explored, and that are supported by certain concrete fundamental facts.

25.3.1 DNA Loading by Direct Surface Modification of the Compacted DNA Complex

In Section 25.2, we have outlined how one could subject micro- or nanoparticulate objects for direct surface modification by the LbL protocol for biomimicking or controlled release purposes. Once we have a stable form of compacted DNA, the LbL procedure can be utilized for further surface stabilization and functionalization.[107,108] This forms a simple and direct means of creating a biovector that could have multiple functionalities — enhanced stability, protection against degradation, targeting, and/or controlled release. If polyelectrolytes have one or more ordered conformations in solution, such as ionic polypeptides or DNA, oppositely charged polyelectrolyte counterparts are able either to stabilize or destabilize these conformations by complexation. It is known that free native DNA in aqueous solution has the B form of the double helix. Complexation with polycations results in transformation of the B-form into the compact C or A forms. The ability of DNA–polycation complexes to function as nonviral gene delivery vectors is influenced by their charge ratios and associated structures. When DNA is in excess (DNA:polycation charge ratio > 1), complexes assemble into daisy-shaped particles with loops of uncondensed DNA that inhibit aggregation of the particles. When the charge ratio is less than 1, then DNA condenses completely within particles that customarily adopt a toroid or rod-like morphology. The applicability of such assemblies as such is limited by poor transfection efficiency due to their tendency to undergo aggregation in physiological salt solution.[109,110] When subjected to surface nanoengineering with the appropriate species or technique, one could derive at a novel nonviral delivery system that meets the requirement for effective trafficking of genetic materials. It is expected that additional charged layers would protect DNA–polycation complex from substitution by the charged plasma components during its voyage within the systemic circulation as well as from nuclease degradation inside the cell.

The DNA payload of the system is generally higher than that of other fabricated systems, as it is presented as a complex (polyplex or lipoplex or lipopolyplex) without encountering any complicated encapsulation procedure, and hence encounters minimal losses. The secondary derivation of the complex is achieved by using charged substances under conditions that offer maximal stability for the primary DNA complex. As stated in the introduction, the LbL self-assembly process can accommodate composite materials that can be decorated with nanometric precision. Along with several other groups, we are engaged in research which could allow inclusion of two categories of substances driven by electrostatic assembly: those that increase the circulation time of the nanoparticles (using block copolymers having a charged polymer linked usually to polyethylene glycol) and those that can incorporate means to maneuver the nanoparticles upon administration (homing systems, such as ligands, magnetic materials, or even elements that can help in tracking the pathway of the nanoparticle itself). In coming years, we anticipate an upsurge in this particular nanotechnology frontier due to the striking simplicity with which one can incorporate several desirable features and elements in one system.

25.3.2 DNA Encapsulation as Shell Component

RNA and DNA are strong polyanions due to the phosphate esters in their backbone, and thus deposit easily.[107,111-114] DNA has been incorporated as one of the charged components in ultrathin films fabricated by the LbL self-assembly technique on noncolloidal substrates.[115-117] Thin films of alternating DNA and poly(L-lysine) (PLL) layers have been fabricated on different surfaces by the LbL self-assembly technique with uniform layers of DNA being fully adsorbed onto each alternative PLL layer. While it was observed that the DNA film thickness increased with the ionic strength of the adsorbing solution, the shape of the DNA molecule was found to undergo a transition from extended linear structure to a more circular or coiled configuration.[17] Besides linear and branched polyelectrolytes, dendronized polymers with protonated amine groups at the periphery and different cylindrically shaped dendronic nanoobjects whose radii and linear charge densities can be varied systematically have been complexed with DNA and subsequently adsorbed on planar surfaces.[118]

Experimental observations for the formation of a smooth film on the colloid particles have been reported, with coatings composed of only one component (DNA), or DNA with an uncharged polymer (e.g., dextran), or DNA as complex with oppositely charged ions or polyelectrolytes.[74,119,120] DNA and dextran were precipitated on the particle surface by the drop-wise addition of ethanol in aqueous solution leading to surface-controlled precipitation (SCP) — a tool for obtaining thick coating on colloidal particles. As revealed by confocal microscopy images, the resulting coating was homogeneous for the investigated systems. The SCP technique is advantageous in a way as it can be applied to compounds that cannot be deposited by means of LbL assembly. The thickness of the polymeric film on colloids can be tuned to the level of a few monomolecular layers by this method.[74,121] In other words, one can increase the DNA payload within the shell if the loading is done this way using appropriate polymer that glues the DNA to the core particle.

DNA has been incorporated as a polyelectrolyte complex along with speramine for coating by the LbL technique onto decomposable PS particles for studying the effect of salt on the complex.[120]

Biodegradable polyester-based particles (such as PLA or PLGA) have provided good antigen delivery systems for proteins and DNA. Such a particle-based approach has been investigated by incorporating the DNA via LbL assembly to form the shell along with a polycation.[122] The polycation formed the first adsorbed layer, which reversed the charge of the PLA particle (originally negative), and onto which plasmid DNA was adsorbed to complete the assembly. Surface immobilization of DNA via complexation with a suitable counter-ion on a colloidal support could be an elegant technique similar to the method outlined in Section 25.3.2, as further surface-stabilization or tailoring could be done with a variety of advanced materials.

25.3.3 Scope for Active Loading of DNA into Preformed Polyelectrolyte Hollow Capsules

The evidence gathered in Section 25.3.2 opens up new avenues for maneuvering active loading of charged/uncharged compounds into preformed micro- or nanocapsules containing a drug-retaining substance via electrostatic or pH- and polarity-based mechanisms. We are engaged in research pertaining to capture of DNA by a matching reaction sequence, which in principle should result in a stable delivery system having a reasonable payload. The LbL engineered capsule (even without a drug capturing material within) could be still loaded, in principle, by utilizing porous diffusion of the compound of interest (when exposed to a highly porous template such as calcium carbonate microparticle) or by taking advantage of the switchable permeability of the polyelectrolyte multilayers constituting the capsule's shell by pH-, solvent-, and ion-mediated mechanisms.[63] The nanocomposite system could be further armed by attachment of ligands onto the surface of the capsule, which would enhance the target specificity of the assembly. All this is still in the theoretical stages or the preliminary stages of research. With the progress in understanding the molecular mechanics of the active loading of the preformed polyelectrolyte capsules, we expect to come out with additional data that would open up new avenues for development of DNA-loaded biomimicking novel vectors.

25.4 CONCLUSIONS AND OUTLOOK

There are several DNA-based therapeutics under development in the biotechnology and pharmaceutical industries. Key delivery challenges must be met before many of these biomacromolecular therapeutic techniques reach the clinic. Two important barriers are the effective targeting of drugs to specific tissues or cells and the subsequent intracellular trafficking into the cellular compartments. The inherent flexibility and unparalleled simplicity of the LbL method has opened new, exciting applications in the physical and life sciences and now could find exciting applications in gene delivery as well. Figure 25.5 gives an idea about the overall applicability and versatility of the LbL technique wherein protocol has been extended to coat, or more specifically to encapsulate, biologically significant materials in the form of bio-colloids. In this chapter, many aspects have only been considered qualitatively since we only have a quantitative understanding in some cases. Most examples in the area of colloid chemistry utilizing the LbL surface nanoengineering technique are from the work of our lab since there are not many groups actively engaged in this emerging field. This chapter is meant to introduce and spread the technique to the rapidly advancing and explosive area of gene delivery systems, because the application perspectives are broad and because much work remains to be done before the basic aspects can be understood. We hope that this chapter stimulates and motivates more medicinal delivery groups to join in this endeavor.

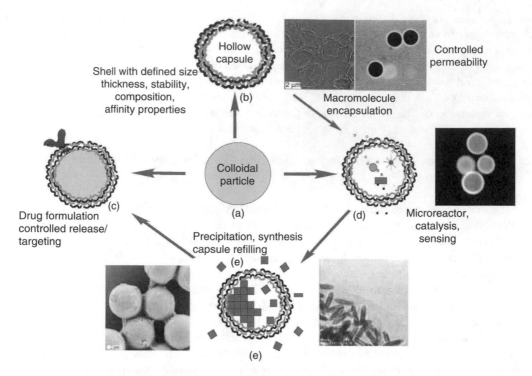

Figure 25.5   Scheme illustrating probable applications of the LbL technique in loading and delivery of bioactive materials: (a) the colloidal particle could be a sacrificial template or a bioactive compound in a suitable condensed form; (b) dissolution of the sacrificial core leading to formation of hollow polyelectrolyte capsule (see SEM image, top left); (c) direct surface modification of the colloidal particle — novel drug formulation; (d) hollow capsules could be loaded with macromolecules or tuned to show controlled permeability (confocal image showing "closed" capsules to a fluorescent high-molecular-weight marker, top right) or could be used to cage enzymes, luminescent nanoparticles; (e) capsules having a polymer prefill leading to special physico-chemical microenvironment within — used to load (see SEM image of a drug-filled capsule) or synthesize metal nanoparticles (see TEM image of microcapsules having iron nanoparticles synthesized within).

ACKNOWLEDGMENTS

We thank our many wonderful colleagues from all over the world, whose names are listed in the bibliography, for their excellent contributions to the research. We thank Prof. H. Möhwald, Director, Department of Interfaces, for support of this work. Generous financial assistance from the Sofja Kovalevskaja program of the Alexander von Humboldt Foundation and the German Ministry of Education and Research, the Volkswagen Foundation, the DFG, the DAAD, and the BMBF is gratefully acknowledged.

REFERENCES

1. Blodgett, K.B., Films built by depositing successive monomolecular layers on a solid surface, *J. Amer. Chem. Soc.* 57 (6), 1007–1022, 1935.
2. Blodgett, K.B. and Langmuir, I., Built-up films of barium stearate and their optical properties, *Phys. Rev.* 51 (11), 964–982, 1937.
3. Ulman, A., Formation and structure of self-assembled monolayers, *Chem. Rev.* 96 (4), 1533–1554, 1996.

4. Sano, M., Lvov, Y., and Kunitake, T., Formation of ultrathin polymer layers on solid substrates by means of polymerization-induced epitaxy and alternate adsorption, *Ann. Rev. Mater. Sci.* 26, 153–187, 1996.

5. Bertrand, P., Jonas, A., Laschewsky, A., and Legras, R., Ultrathin polymer coatings by complexation of polyelectrolytes at interfaces: suitable materials, structure and properties, *Macromol. Rapid Comm.* 21 (7), 319–348, 2000.

6. Decher, G., Polyelectrolyte multilayers, an overview, in *Multilayer Thin Films: Sequential Assembly of Nanocomposite Materials*, Schlenoff, J.B. (Ed.), Wiley/VCH Verlag, Weinheim, 2002, 1–46.

7. Decher, G., Layered nanoarchitectures via directed assembly of anionic and cationic molecules, in *Comprehensive Supramolecular Chemistry (Templating, Self-Assembly and Self-Organization)*, Hosseini, M.W. (Ed.), Pergamon Press, Oxford, 1996, 507–528.

8. Decher, G., Fuzzy nanoassemblies: toward layered polymeric multicomposites, *Science* 277 (5330), 1232–1237, 1997.

9. Arys, X., Jonas, A.M., Laschewsky, A., and Legras, R., Supramolecular polyelectrolyte assemblies, in *Supramolecular Polymers*, Ciferri, A. (Ed.), Marcel-Dekker, New York, 2000, 505–564.

10. Dubas, S.T. and Schlenoff, J.B., Factors controlling the growth of polyelectrolyte multilayers, *Macromolecules* 32 (24), 8153–8160, 1999.

11. Schwarz, S., Eichhorn, K.J., Wischerhoff, E., and Laschewsky, A., Polyelectrolyte adsorption onto planar surfaces: a study by streaming potential and ellipsometry measurements, *Coll. Surf. A: Physicochem. Eng. Aspects* 159 (2–3), 491–501, 1999.

12. Kotov, N.A., Layer-by-layer self-assembly: the contribution of hydrophobic interactions, *Nanostruct. Mater.* 12 (5–8), 789–796, 1999.

13. Sukhishvili, S.A. and Granick, S., Layered, erasable polymer multilayers formed by hydrogen-bonded sequential self-assembly, *Macromolecules* 35 (1), 301–310, 2002.

14. Messina, R., Holm, C., and Kremer, K., Polyelectrolyte multilayering on a charged sphere, *Langmuir* 19 (10), 4473–4482, 2003.

15. Sukhorukov, G.B., Mohwald, H., Decher, G., and Lvov, Y.M., Assembly of polyelectrolyte multilayer films by consecutively alternating adsorption of polynucleotides and polycations, *Thin Solid Films* 285, 220–223, 1996.

16. Sukhorukov, G.B., Montrel, M.M., Petrov, A.I., Shabarchina, L.I., and Sukhorukov, B.I., Multilayer films containing immobilized nucleic acids. Their structure and possibilities in biosensor applications, *Biosen. Bioelec.* 11 (9), 913–922, 1996.

17. Shi, X.Y., Sanedrin, R.J., and Zhou, F.M., Structural characterization of multilayered DNA and polylysine composite films: influence of ionic strength of DNA solutions on the extent of DNA incorporation, *J. Phys. Chem. B* 106 (6), 1173–1180, 2002.

18. He, J.A., Samuelson, L., Li, L., Kumar, J., and Tripathy, S.K., Bacteriorhodopsin thin film assemblies: immobilization, properties, and applications, *Adv. Mater.* 11 (6), 435, 1999.

19. Ariga, K., Sasaki, Y., Horiguchi, H., Horiuchi, N., and Kikuchi, J., Layered nanoarchitectures between cationic and anionic materials: composite assemblies of polyions, lipid bilayers, and proteins, in *Soft Chemistry Leading to Novel Materials*, Scitec, Uetikon-Zuerich, 2001, 35–59.

20. Ai, H., Jones, S.A., and Lvov, Y.M., Biomedical applications of electrostatic layer-by-layer nanoassembly of polymers, enzymes, and nanoparticles, *Cell Biochem. Biophys.* 39 (1), 23–43, 2003.

21. Lvov Yu, M. and Sukhorukov, G.B., Protein architecture: assembly of ordered films by means of alternated adsorption of oppositely charged macromolecules, *Membr. Cell Biol.* 11 (3), 277–303, 1997.

22. Lvov, Y., Haas, H., Decher, G., Mohwald, H., Mikhailov, A., Mtchedlishvily, B., Morgunova, E., and Vainshtein, B., Successive deposition of alternate layers of polyelectrolytes and a charged virus, *Langmuir* 10 (11), 4232–4236, 1994.

23. Donath, E., Sukhorukov, G.B., Caruso, F., Davis, S.A., and Mohwald, H., Novel hollow polymer shells by colloid-templated assembly of polyelectrolytes, *Angew. Chem. Int. Ed.* 37 (16), 2202–2205, 1998.

24. Voigt, A., Donath, E., and Mohwald, H., Preparation of microcapsules of strong polyelectrolyte couples by one-step complex surface precipitation, *Macromol. Mater. Eng.* 282 (9), 13–16, 2000.

25. Diaspro, A., Silvano, D., Krol, S., Cavalleri, O., and Gliozzi, A., Single living cell encapsulation in nano-organized polyelectrolyte shells, *Langmuir* 18 (13), 5047–5050, 2002.

26. Donath, E., Soukhorukov, G.B., Moya, S., Voigt, A., Lichtenfeld, H., Dahne, L., Gao, C.Y., and Mohwald, H., Polyelectrolyte-multilayer capsules templated on latex particles and biological cells: fabrication and properties, *Abstr. Pap. Am. Chem. Soc.* 219, 464–465, 2000.

27. Caruso, F., Nanoengineering of particle surfaces, *Adv. Mater.* 13 (1), 11, 2001.

28. Fuchs, A. and Killmann, E., Adsorption of polyelectrolytes on colloidal latex particles, electrostatic interactions and stability behaviour, *Coll. Polym. Sci.* 279 (1), 53–60, 2001.

29. Solodovnik, V.D., *Microcapsulation*, Chimiya, Moscow, 1980.

30. Spange, S., Silica surface modification by cationic polymerization and carbenium intermediates, *Prog. Polym. Sci.* 25 (6), 781–849, 2000.

31. Li, H., Han, J., Panioukhine, A., and Kumacheva, E., From heterocoagulated colloids to core-shell particles, *J. Coll. Interf. Sci.* 255 (1), 119–128, 2002.

32. Pham, H.H. and Kumacheva, E., Core-shell particles: building blocks for advanced polymer materials, *Macromol. Symp.* 192, 191–205, 2003.

33. Okubo, M., He, Y., and Ichikawa, K., Studies on suspension and emulsion : analysis of stepwise heterocoagulation process of small Cationic polymer particles onto large anionic polymer particles using dynamic light-scattering, *Coll. Polym. Sci.* 269 (2), 125–130, 1991.

34. Antonietti, M. and Landfester, K., Polyreactions in miniemulsions, *Prog. Polym. Sci.* 27 (4), 689–757, 2002.

35. Bourgeat-Lami, E., Organic-inorganic nanostructured colloids, *J. Nanosci. Nanotechnol.* 2 (1), 1–24, 2002.

36. Sharifi-Sanjani, N., Naderi, N., and Soltan-Dehghan, M., Organic-inorganic nanocomposites with core-shell structure via sol-gel process, in *Advanced Materials Processing II*, Lai, M.O. and Lu, L., Eds., Trans Tech Publications, Zurich-Uetikon, 2003, 419–422.

37. Antipov, A.A., Sukhorukov, G.B., Fedutik, Y.A., Hartmann, J., Giersig, M., and Mohwald, H., Fabrication of a novel type of metallized colloids and hollow capsules, *Langmuir* 18 (17), 6687–6693, 2002.

38. Breen, M.L., Dinsmore, A.D., Pink, R.H., Qadri, S.B., and Ratna, B.R., Sonochemically produced ZnS-coated polystyrene core-shell particles for use in photonic crystals, *Langmuir* 17 (3), 903–907, 2001.

39. Sukhorukov, G.B., Donath, E., Lichtenfeld, H., Knippel, E., Knippel, M., Budde, A., and Mohwald, H., Layer-by-layer self assembly of polyelectrolytes on colloidal particles, *Coll. Surf. A: Physicochem. Eng. Aspects* 137 (1–3), 253–266, 1998.

40. Sukhorukov, G.B., Donath, E., Davis, S., Lichtenfeld, H., Caruso, F., Popov, V.I., and Mohwald, H., Stepwise polyelectrolyte assembly on particle surfaces: a novel approach to colloid design, *Polym. Adv. Technol.* 9 (10–11), 759–767, 1998.

41. Caruso, F. and Sukhorukov, G.B., Coated colloids: preparation, characterization, assembly and utilization, in *Multilayer Thin Films: Sequential Assembly of Nanocomposite Materials*, Decher, G. and Schlenoff, J.B., (Ed.), Wiley/VCH Verlag, Weinheim, 2002, 331–362.

42. Sukhorukov, G.B., Multilayer hollow microspheres, in *Dendrimers, Assemblies, Nanocomposites*, Arshady, R. and Guyot, A. (Ed.), Citus Books, London, 2002, 111–147.

43. Sukhorukov, G.B., Designed nano-engineered polymer films on colloidal particles and capsules, in *Novel Methods to Study Interfacial Layers*, Moebius, D. and Miller, R. (Eds.), Elsevier, 2001, 384–415.

44. Caruso, F., Trau, D., Mohwald, H., and Renneberg, R., Enzyme encapsulation in layer-by-layer engineered polymer multilayer capsules, *Langmuir* 16 (4), 1485–1488, 2000.

45. Balabushevitch, N.G., Sukhorukov, G.B., Moroz, N.A., Volodkin, D.V., Larionova, N.I., Donath, E., and Mohwald, H., Encapsulation of proteins by layer-by-layer adsorption of polyelectrolytes onto protein aggregates: factors regulating the protein release, *Biotechnol. Bioeng.* 76 (3), 207–213, 2001.

46. Bobreshova, M.E., Sukhorukov, G.B., Saburova, E.A., Elfimova, L.I., Shabarchina, L.I., and Sukhorukov, B.I., Lactate dehydrogenase in interpolyelectrolyte complex. Function and stability, *Biofizika* 44 (5), 813–820, 1999.

47. Tiourina, O.P., Antipov, A.A., Sukhorukov, G.B., Larionova, N.L., Lvov, Y., and Mohwald, H., Entrapment of alpha-chymotrypsin into hollow polyelectrolyte microcapsules, *Macromol. Biosci.* 1 (5), 209–214, 2001.

48. Tiourina, O.P. and Sukhorukov, G.B., Multilayer alginate/protamine microsized capsules: encapsulation of [alpha]-chymotrypsin and controlled release study, *Int. J. Pharm.* 242 (1–2), 155–161, 2002.

49. Tieke, B., van Ackern, F., Krasemann, L., and Toutianoush, A., Ultrathin self-assembled polyelectrolyte multilayer membranes, *Eur. Phys. J. E* 5 (1), 29–39, 2001.

50. Ai, H., Jones, S.A., de Villiers, M.M., and Lvov, Y.M., Nano-encapsulation of furosemide microcrystals for controlled drug release, *J. Control. Release* 86 (1), 59–68, 2003.

51. Qiu, X.P., Leporatti, S., Donath, E., and Mohwald, H., Studies on the drug release properties of polysaccharide multilayers encapsulated ibuprofen microparticles, *Langmuir* 17 (17), 5375–5380, 2001.

52. Antipov, A.A., Sukhorukov, G.B., Donath, E., and Mohwald, H., Sustained release properties of polyelectrolyte multilayer capsules, *J. Phys. Chem. B* 105 (12), 2281–2284, 2001.

53. Qiu, X.P., Donath, E., and Mohwald, H., Permeability of ibuprofen in various polyelectrolyte multi-layers, *Macromol. Mater. Eng.* 286 (10), 591–597, 2001.

54. Shenoy, D.B. and Sukhorukov, G.B., Unpublished results describing the synthesis of surface stabilized microcrystals of naproxen and subsequent direct surface modification with biocompatible polyelec-trolytes for sterile controlled release applications.

55. Sukhorukov, G.B., Donath, E., Moya, S., Susha, A.S., Voigt, A., Hartmann, J., and Mohwald, H., Microencapsulation by means of step-wise adsorption of polyelectrolytes, *J. Microencap.* 17 (2), 177–185, 2000.

56. Moya, S., Donath, E., Sukhorukov, G.B., Auch, M., Baumler, H., Lichtenfeld, H., and Mohwald, H., Lipid coating on polyelectrolyte surface modified colloidal particles and polyelectrolyte capsules, *Macromolecules* 33 (12), 4538–4544, 2000.

57. Neu, B., Voigt, A., Mitlohner, R., Leporatti, S., Gao, C.Y., Donath, E., Kiesewetter, H., Mohwald, H., Meiselman, H.J., and Baumler, H., Biological cells as templates for hollow microcapsules, *J. Microen-cap.* 18 (3), 385–395, 2001.

58. Leporatti, S., Voigt, A., Mitlohner, R., Sukhorukov, G., Donath, E., and Mohwald, H., Scanning force microscopy investigation of polyelectrolyte nano- and microcapsule wall texture, *Langmuir* 16 (9), 4059–4063, 2000.

59. Moya, S., Dahne, L., Voigt, A., Leporatti, S., Donath, E., and Mohwald, H., Polyelectrolyte multilayer capsules templated on biological cells: core oxidation influences layer chemistry, *Coll. Surf. A: Physicochem. Eng. Aspects* 183, 27–40, 2001.

60. Caruso, F., Hollow capsule processing through colloidal templating and self-assembly, *Chem. Eur. J.* 6 (3), 413–419, 2000.

61. Ibarz, G., Dahne, L., Donath, E., and Mohwald, H., Resealing of polyelectrolyte capsules after core removal, *Macromol. Rapid Comm.* 23 (8), 474–478, 2002.

62. Antipov, A.A., Shchukin, D., Fedutik, Y., Petrov, A.I., Sukhorukov, G.B., and Mohwald, H., Carbonate microparticles for hollow polyelectrolyte capsules fabrication, *Coll. Surf. A: Physicochem. Eng. Aspects* 224 (1–3), 175–183, 2003.

63. Antipov, A.A., Sukhorukov, G.B., Leporatti, S., Radtchenko, I.L., Donath, E., and Mohwald, H., Polyelectrolyte multilayer capsule permeability control, *Coll. Surf. A: Physicochem. Eng. Aspects* 198, 535–541, 2002.

64. Donath, E., Sukhorukov, G.B., and Mohwald, H., Submicrometric and micrometric polyelectrolyte capsules, *Nachricht. Chem. Tech. Labor.* 47 (4), 400, 1999.

65. Shenoy, D.B., Antipov, A.A., Sukhorukov, G.B., and Mohwald, H., Layer-by-layer engineering of biocompatible, decomposable core-shell structures, *Biomacromolecules* 4 (2), 265–272, 2003.

66. Caruso, F., Hollow inorganic capsules via colloid-templated layer-by-layer electrostatic assembly, in *Current Topics in Chemistry – Colloid Chemistry II*, Antonietti, M. (Ed.), Springer-Verlag, Berlin, 2003, 145–168.

67. Radtchenko, I., Sukhorukov, G., and Mohwald, H., A novel method for encapsulation of poorly water-soluble drugs: precipitation in polyelectrolyte multilayer shells, *Int. J. Pharm.* 242 (1–2), 219–223, 2002.

68. Shchukin, D.G., Radtchenko, I.L., and Sukhorukov, G.B., Micron-scale hollow polyelectrolyte cap-sules with nanosized magnetic Fe_3O_4 inside, *Mater. Lett.*, 57, 1743–1747, 2003.

69. Sukhorukov, G., Dahne, L., Hartmann, J., Donath, E., and Mohwald, H., Controlled precipitation of dyes into hollow polyelectrolyte capsules based on colloids and biocolloids, *Adv. Mater.* 12 (2), 112–115, 2000.

70. Dahne, L., Leporatti, S., Donath, E., and Mohwald, H., Fabrication of micro reaction cages with tailored properties, *J. Amer. Chem. Soc.* 123 (23), 5431–5436, 2001.

71. Antipov, A., Shchukin, D., Fedutik, Y., Zanaveskina, I., Klechkovskaya, V., Sukhorukov, G., and Mohwald, H., Urease-catalyzed carbonate precipitation inside the restricted volume of polyelectrolyte capsules, *Macromol. Rapid Comm.* 24 (3), 274–277, 2003.

72. Moehwald, H., Donath, E., and Sukhorukov, G.B., Smart capsules, in *Multilayer Thin Films: Sequential Assembly of Nanocomposite Materials*, Decher, G. and Schlenoff, J.B. (Eds.), Wiley/VCH Verlag, Weinheim, 2002, 363–392.

73. Sukhorukov, G.B., Brumen, M., Donath, E., and Mohwald, H., Hollow polyelectrolyte shells: exclusion of polymers and donnan equilibrium, *J. Phys. Chem. B* 103 (31), 6434–6440, 1999.

74. Radtchenko, I.L., Sukhorukov, G.B., and Mohwald, H., Incorporation of macromolecules into polyelectrolyte micro- and nanocapsules via surface controlled precipitation on colloidal particles, *Coll. Surf. A: Physicochem. Eng. Aspects* 202 (2–3), 127–133, 2002.

75. Radtchenko, I.L., Giersig, M., and Sukhorukov, G.B., Inorganic particle synthesis in confined micronsized polyelectrolyte capsules, *Langmuir* 18 (21), 8204–8208, 2002.

76. Radtchenko, I.L., Sukhorukov, G.B., Leporatti, S., Khomutov, G.B., Donath, E., and Mohwald, H., Assembly of alternated multivalent ion/polyelectrolyte layers on colloidal particles. Stability of the multilayers and encapsulation of macromolecules into polyelectrolyte capsules, *J. Coll. Interf. Sci.* 230 (2), 272–280, 2000.

77. Lvov, Y., Antipov, A.A., Mamedov, A., Mohwald, H., and Sukhorukov, G.B., Urease encapsulation in nanoorganized microshells, *Nano Lett.* 1 (3), 125–128, 2001.

78. Sukhorukov, G.B., Antipov, A.A., Voigt, A., Donath, E., and Mohwald, H., pH-controlled macromolecule encapsulation in and release from polyelectrolyte multilayer nanocapsules, *Macromol. Rapid Comm.* 22 (1), 44–46, 2001.

79. Saphiannikova, M., Radtchenko, I., Sukhorukov, G., Shchukin, D., Yakimansky, A., and Ilnytskyi, J., Molecular-dynamics simulations and x-ray analysis of dye precipitates in the polyelectrolyte microcapsules, *J. Chem. Phys.* 118 (19), 9007–9014, 2003.

80. Shenoy, D.B. and Sukhorukov, G.B., Unpublished results, PSS filled polyelectrolyte capsules were fabricated using a method as described in reference 74 and exposed to natamycin, which led to active loading of the drug within the capsules.

81. Reschel, T., Konak, C., Oupicky, D., Seymour, L.W., and Ulbrich, K., Physical properties and *in vitro* transfection efficiency of gene delivery vectors based on complexes of DNA with synthetic polycations, *J. Control. Release* 81 (1–2), 201–217, 2002.

82. Wolfert, M.A., Schacht, E.H., Toncheva, V., Ulbrich, K., Nazarova, O., and Seymour, L.W., Characterization of vectors for gene therapy formed by self-assembly of DNA with synthetic block copolymers, *Hum. Gene Ther.* 7 (17), 2123–2133, 1996.

83. Thomas, M. and Klibanov, A.M., Non-viral gene therapy: polycation-mediated DNA delivery, *Appl. Microbiol. Biotechnol.* 62 (1), 27–34, 2003.

84. Merdan, T., Kopecek, J., and Kissel, T., Prospects for cationic polymers in gene and oligonucleotide therapy against cancer, *Adv. Drug Deliv. Rev.* 54 (5), 715–758, 2002.

85. Oupicky, D., Ogris, M., Howard, K.A., Dash, P.R., Ulbrich, K., and Seymour, L.W., Importance of lateral and steric stabilization of polyelectrolyte gene delivery vectors for extended systemic circulation, *Mol. Ther.* 5 (4), 463–472, 2002.

86. Zhang, S., Fabrication of novel biomaterials through molecular self-assembly, *Nat. Biotechnol.* 21 (10), 1171–1178, 2003.

87. Safinya, C.R., Structures of lipid-DNA complexes: supramolecular assembly and gene delivery, *Curr. Op. Struct. Biol.* 11 (4), 440–448, 2001.

88. Pouton, C.W., Nuclear import of polypeptides, polynucleotides and supramolecular complexes, *Adv. Drug Del. Rev.* 34 (1), 51–64, 1998.

89. Pouton, C.W. and Seymour, L.W., Key issues in non-viral gene delivery, *Adv. Drug Del. Rev.* 34 (1), 3–19, 1998.

90. Niemeyer, C.M. and Blohm, D., DNA microarrays, *Angew. Chem. Int. Ed.* 38 (19), 2865–2869, 1999.

91. Niemeyer, C.M., Boldt, L., Ceyhan, B., and Blohm, D., Evaluation of single-stranded nucleic acids as carriers in the DNA-directed assembly of macromolecules, *J. Biomol. Struct. Dyn.* 17 (3), 527–538, 1999.

92. Zuidam, N.J., Barenholz, Y., and Minsky, A., Chiral DNA packaging in DNA-cationic liposome assemblies, *FEBS Lett.* 457 (3), 419–422, 1999.

93. Trubetskoy, V.S., Budker, V.G., Hanson, L.J., Slattum, P.M., Wolff, J.A., and Hagstrom, J.E., Self-assembly of DNA-polymer complexes using template polymerization, *Nucleic Acids Res.* 26 (18), 4178–4185, 1998.

94. Katayose, S. and Kataoka, K., Remarkable increase in nuclease resistance of plasmid DNA through supramolecular assembly with poly(ethylene glycol) poly(L-lysine) block copolymer, *J. Pharm. Sci.* 87 (2), 160–163, 1998.

95. Liu, F. and Huang, L., Development of non-viral vectors for systemic gene delivery, *J. Control. Release* 78 (1–3), 259–266, 2002.

96. Oupicky, D., Ogris, M., and Seymour, L.W., Development of long-circulating polyelectrolyte complexes for systemic delivery of genes, *J. Drug Target.* 10 (2), 93–98, 2002.

97. Howard, K.A. and Alpar, H.O., The development of polyplex-based DNA vaccines, *J. Drug Target.* 10 (2), 143–151, 2002.

98. Benns, J.M. and Kim, S.W., Tailoring new gene delivery designs for specific targets, *J. Drug Target.* 8 (1), 1–12, 2000.

99. Kirby, A.J., Camilleri, P., Engberts, J.B., Feiters, M.C., Nolte, R.J., Soderman, O., Bergsma, M., Bell, P.C., Fielden, M.L., Garcia Rodriguez, C.L., Guedat, P., Kremer, A., McGregor, C., Perrin, C., Ronsin, G., and van Eijk, M.C., Gemini surfactants: new synthetic vectors for gene transfection, *Angew. Chem. Int. Ed.* 42 (13), 1448–1457, 2003.

100. Duzgunes, N., De Ilarduya, C.T., Simoes, S., Zhdanov, R.I., Konopka, K., and Pedroso de Lima, M.C., Cationic liposomes for gene delivery: novel cationic lipids and enhancement by proteins and peptides, *Curr. Med. Chem.* 10 (14), 1213–1220, 2003.

101. Zhdanov, R.I., Podobed, O.V., and Vlassov, V.V., Cationic lipid-DNA complexes-lipoplexes-for gene transfer and therapy, *Bioelectrochemistry* 58 (1), 53–64, 2002.

102. Dass, C.R. and Burton, M.A., Lipoplexes and tumours: a review, *J. Pharm. Pharmacol.* 51 (7), 755–770, 1999.

103. Mahato, R.I., Takakura, Y., and Hashida, M., Nonviral vectors for *in vivo* gene delivery: physicochemical and pharmacokinetic considerations, *Crit. Rev. Ther. Drug Carrier Syst.* 14 (2), 133–172, 1997.

104. Wolfert, M.A., Schacht, E.H., Toncheva, V., Ulbrich, K., Nazarova, O., and Seymour, L.W., Characterization of vectors for gene therapy formed by self-assembly of DNA with synthetic block copolymers, *Hum. Gene Ther.* 7 (17), 2123–2133, 1996.

105. Wolfert, M.A., Dash, P.R., Nazarova, O., Oupicky, D., Seymour, L.W., Smart, S., Strohalm, J., and Ulbrich, K., Polyelectrolyte vectors for gene delivery: influence of cationic polymer on biophysical properties of complexes formed with DNA, *Bioconj. Chem.* 10 (6), 993–1004, 1999.

106. Toncheva, V., Wolfert, M.A., Dash, P.R., Oupicky, D., Ulbrich, K., Seymour, L.W., and Schacht, E.H., Novel vectors for gene delivery formed by self-assembly of DNA with poly(L-lysine) grafted with hydrophilic polymers, *Biochim. Biophys. Acta* 1380 (3), 354–368, 1998.

107. Trubetskoy, V.S., Loomis, A., Hagstrom, J.E., Budker, V.G., and Wolff, J.A., Layer-by-layer deposition of oppositely charged polyelectrolytes on the surface of condensed DNA particles, *Nucleic Acids Res.* 27 (15), 3090–3095, 1999.

108. Trubetskoy, V.S., Wong, S.C., Subbotin, V., Budker, V.G., Loomis, A., Hagstrom, J.E., and Wolff, J.A., Recharging cationic DNA complexes with highly charged polyanions for *in vitro* and *in vivo* gene delivery, *Gene Ther.* 10 (3), 261–271, 2003.

109. Ogris, M., Steinlein, P., Kursa, M., Mechtler, K., Kircheis, R., and Wagner, E., The size of DNA/transferrin-PEI complexes is an important factor for gene expression in cultured cells, *Gene Ther.* 5 (10), 1425–1433, 1998.

110. Ross, P.C. and Hui, S.W., Lipoplex size is a major determinant of *in vitro* lipofection efficiency, *Gene Ther.* 6 (4), 651–659, 1999.

111. Montrel, M.M., Sukhorukov, G.B., Shabarchina, L.I., Apolonnik, N.V., and Sukhorukov, B.I., Proton-induced conformation changes and sensoric properties of polycytidylic acid immobilized in Langmuir and polyelectrolyte multilayer films, *Mater. Sci. Eng. C: Biomimet. Mater. Sens. Sys.* 5 (3–4), 275–279, 1998.

112. Montrel, M.M., Sukhorukov, G.B., Petrov, A.I., Shabarchina, L.I., and Sukhorukov, B.I., Spectroscopic study of thin multilayer films of the complexes of nucleic acids with cationic amphiphiles and polycations: their possible use as sensor elements, *Sens. Actuat. B: Chem.* 42 (3), 225–231, 1997.

113. Zu, X.L., Lu, Z.Q., Zhang, Z., Schenkman, J.B., and Rusling, J.F., Electroenzyme-catalyzed oxidation of styrene and cis-beta-methylstyrene using thin films of cytochrome P450cam and myoglobin, *Langmuir* 15 (21), 7372–7377, 1999.

114. Lvov, Y., Decher, G., and Sukhorukov, G., Assembly of thin-films by means of successive deposition of alternate layers of DNA and poly(allylamine), *Macromolecules* 26 (20), 5396–5399, 1993.

115. Lvov, Y.M. and Sukhorukov, G.B., Protein architecture: assembly of ordered films by means alternated adsorption of opposite charged macromolecules, *Biolog. Membr.* 14 (3), 229–250, 1997.

116. Pei, R., Cui, X., Yang, X., and Wang, E., Assembly of alternating polycation and DNA multilayer films by electrostatic layer-by-layer adsorption, *Biomacromolecules* 2 (2), 463–468, 2001.

117. Dennany, L., Forster, R.J., and Rusling, J.F., Simultaneous direct electrochemiluminescence and catalytic voltammetry detection of DNA in ultrathin films, *J. Am. Chem. Soc.* 125 (17), 5213–5218, 2003.

118. Gossl, I., Shu, L., Schluter, A.D., and Rabe, J.P., Molecular structure of single DNA complexes with positively charged dendronized polymers, *J. Am. Chem. Soc.* 124 (24), 6860–6865, 2002.

119. Mumper, R.J. and Cui, Z., Genetic immunization by jet injection of targeted pDNA-coated nanoparticles, *Methods* 31 (3), 255–262, 2003.

120. Schuler, C. and Caruso, F., Decomposable hollow biopolymer-based capsules, *Biomacromolecules* 2 (3), 921–926, 2001.

121. Radtchenko, I.L., Sukhorukov, G.B., Gaponik, N., Kornowski, A., Rogach, A.L., and Mohwald, H., Core-shell structures formed by the solvent-controlled precipitation of luminescent CdTe nanocrystals on latex spheres, *Adv. Mater.* 13 (22), 1684–1687, 2001.

122. Trimaille, T., Messai, I., Pichot, C., and Delair, T., Elaboration of PLA-based DNA particulate vectors via the layer-by-layer approach, in *30th Annual meeting of the Controlled Release Society*, Glasgow, Scottland, 2003.

CHAPTER 26

Ex Vivo and *In Vivo* Adenovirus-Mediated Gene Delivery into Refractory Cells via Nanoparticle Hydrogel Formulation

Ales Prokop, Gianluca Carlesso, and Jeffrey M. Davidson

CONTENTS

26.1 POLYMER-DRIVEN ADENOVIRAL GENE DELIVERY

The field of nonviral gene delivery was created as a result of breakthrough work involving the utilization of cationic lipids for gene transfer.[1] In general, the ability to vary and control process parameters in liposome dispersions is relatively limited. In addition, the lipoplex (DNA–lipid complex) systems are poorly water soluble and their macroscopic characteristics are unstable over time, limiting their pharmaceutical applications. The lipoplex instability is due to the fact that such complexes may dissociate because of the charge screening effect of the electrolytes (and polyelectrolytes) in biological fluids. In contrast, self-assembly of polyplexes (DNA–polymer complexes) offer greater flexibility by varying the composition, the polymer molecular mass, and the polycation architecture. However, all such systems are inherently less efficient as compared to viral gene systems. The estimated gap between the efficiency of viral and nonviral systems can be as much as 10,000-fold.[2] Lately, it has been reported that adenoviral transduction of cells that are normally refractory to infection can be enhanced by complexing virus particles with cationic lipids or cationic polymers. The mechanism of cellular uptake of adenovirus has recently been clarified and shown to consist of two distinct steps. Initial attachment to the cell occurs via an interaction between the fiber knob and several cellular receptors that are still being identified but include the Coxsackie's virus and adenovirus receptor (CAR)[3,4] and the $\alpha2$ domain of major histocompatibility complex (MHC) class I molecules.[5] This initial binding allows close proximity with the cell and subsequent binding of the viral penton base protein to cellular $\alpha_v \beta_3$ and $\alpha_v \beta_5$ integrins, promoting receptor-mediated endocytosis of the viral particle.[6] The abundance of the CAR receptor, MHC class I

0-8493-1934-X/05/$0.00+$1.50
© 2005 by CRC Press LLC

receptors, and $\alpha_v \beta_3/\alpha_v \beta_5$ integrins on the surface of different cell types greatly influences the level of infection by adenoviral vectors.[7,8] Thus, the cells lacking such receptors are refractory to AdV gene transfer. The cationic lipid component ensures efficient cellular uptake, but endosomal escape and nuclear uptake are significant barriers to non–virus-mediated gene transfer.[9] Adenovirus possesses an inherent capacity to destabilize the endosome and a nuclear targeting function,[10] and a combination of polycations and adenovirus should permit efficient delivery to the nucleus. Precomplexing adenovirus with polycations enables a smaller dose of adenovirus to be administered in comparison with adenovirus alone, to achieve efficient transduction. In addition, a partial protection of adenovirus particles against the neutralizing effect of adenovirus antiserum when complexed with polycations has been observed.[11]

As to the selection of viral vectors, retroviral vectors have been used with limited success. The limitations include the lower concentration attainable (number of infectious units) and the necessity of cell division for effective gene transfer.[12,13] Adenoviral vectors provide advantages in that they can be produced in high concentrations (infectious units), exhibit substantially higher stability, and can easily infect both dividing and nondividing cells. The nonintegration into the host chromosome obviates the fear of insertional oncogenesis, in addition to the benefit of transient expression that doesn't have to consider lingering transgenic proteins staying beyond their welcome. The adenovirus vectors account for nearly 27% of current gene therapy trials.[14] These vectors have been used successfully for *in vivo* gene transfer in several therapeutic fields.[15-29]

Polymer- and cationic-lipid-enhanced adenoviral-mediated delivery has been reported extensively.[23-39] When the AdV vector is complexed with cationic lipids or polymers, gene transfer is improved. It appears that the binding of cationic substances is dependent on an electrostatic interaction with the viral and cell surfaces. The entry does not require an interaction of adenovirus fiber protein with the cell surface and the complexes then enter cells via a pathway different from that utilized by adenovirus alone.[40] The enhanced adenoviral infection via polymer components offers the greater advantage for cells that are not otherwise easily infected.

Among the polymers, poly(ethylene) glycol (PEG) occupies a special prominence. A special coating technique has been invented to shield viral proteins by a simple coat (via a simple physical adsorption) or by coupling the surface of the adenovirus using bifunctional PEG molecules.[40,41] The hydrophilic coat of PEG decreases an access of antibody to virus and produces a two- to threefold increase in transduction in the lung when administered by intratracheal injection and a fivefold increase in transduction in the liver when administered intravenously. In addition, the storage stability of such conjugated adenovirus particles is improved at higher temperature. At $-20°C$, even glycerol could be eliminated from the formulation.[42] Romanczuk[41] also employed a homing (targeting) peptide, obtained by a phage display library screen, to target the PEG-modified AdV to airway epithelium NHBE cells. A re-targeted delivery of AdV vectors through FGF receptors involves another "biopolymer"-based retargeting.[43] In fact, a methacrylamide-based linker has been also used to attach FGF-2 and $VEGF_{165}$ as targeting sequences, allowing for adenovirus type 5 (AdV5) 10-fold enhancement of gene transfer into tumor cells.[44]

Some research has been reported on the effect of polymers to facilitate retroviral gene transfer. Thus RV–Polybrene–chondroitin sulfate complex increased 10- to 20-fold the *in vitro* gene transfer in NIH 3T3 murine fibroblasts and primary human fibroblasts.[45] Again, as with AdV-polymer complexes, charged polymers are thought to act on an early receptor-independent step of infection and increase the electrostatic repulsion between viruses and cells.[46,47] By capturing a large percentage of viral particles (> 80%), the polymer complex may be able to deliver virus more effectively to cells than by simple diffusion. The mechanism of gene enhancement is not understood. Polybrene-enhanced RV delivery was reported as well as a simple mathematical model to describe essential steps of gene transfer, allowing a quantification of transduction rate constant.[48] Such a model could be applied for polymer-mediated AdV delivery as well.

Another modality for enhancing the AdV gene transfer is the formulation of nanoparticulate (or microparticulate) delivery vehicles.[49-53] In the case of nanoparticles, the mechanism of enhancement involves intracellular uptake of nanoparticles and rapid escape from the endo-lysosomal compartment into cytosol. DNA entrapment is enhanced by prior condensing with polymers prior to their entrapment in nanoparticles or by synthesizing novel polymers with cationic groups that can condense DNA into the nanoparticles.[49,50] The sustained gene effect is another benefit. That has been shown for microparticles.[52-54] For certain microparticle categories, however, the intracellular uptake is also possible, particularly for macrophages and dendritic cells.[55] The employment of polymers or micro- and nanoparticles provides for a possibility by targeting of AdV vectors via changing tropism of AdV particles. A more specific molecular targeting to required sites and reduction in the unwanted systemic toxicity (e.g., in liver) has been discussed in several papers.[56,57]

26.2 GENE TRANSFER INTO DENDRITIC CELLS

Developments of *in vitro* methods for generation of large numbers of dendritic cells (DCs) from CD34+ hematopoietic stem cells or peripheral blood mononuclear cells (PBMCs) make DC a suitable candidate for immunotherapy (vaccine development) against infectious and malignant diseases.[58-61] Regarding viral gene transfer, previous reports demonstrate that human DCs can be successfully transduced by RV or AdV vectors *ex vivo*, resulting in 40–90% transgene expression,[62-64] and that DCs are relatively refractory to adenoviral-mediated gene transfer at low multiplicity of infection (MOI), but the transduction rate increases substantially at MOI of 100–1000.[65-67]

Dendritic cells are a heterogeneous population of antigen-presenting cells (APCs) identified in various tissues, including the skin (Langerhans cells), lymph nodes (interdigitating and follicular DC), spleen, and thymus. They are described as professional APCs because of their superior T-cell stimulatory capacity. For this reason, attention is being focused on using DCs for clinical applications to treat cancer patients. *Ex vivo* strategies include gene delivery into tumor cells and into cellular components of the immune system, including cytotoxic T-cells, NK, macrophages, and dendritic cells. Although preclinical studies are promising, the majority of clinical studies with DC have not fulfilled expectations.[68] The field of DC biology has progressed rapidly over the past years, leading to several options for the improvement of vaccination. The parameter efficiency and biological and functional consequences of different gene transfer methods into different subsets of human DC should be studied. Another important consideration for DC-based vaccination is the elucidation of the role of maturation and apoptosis during DC differentiation.

Properties of DC include the following:

- The ability to capture, process, and present foreign antigens (especially in early, immature state)
- The ability to migrate to lymphoid-rich tissue
- The ability to stimulate innate and adaptive antigen-specific immune responses

Until recently, the ability to study DC has been limited by the lack of adequate culture systems.[68] It is now known that specific cytokines can be used to expand DC to numbers sufficient for their *in vitro* evaluation and for their use in human immunotherapy trials. Human DC can be derived from hematopoietic progenitors (CD34+-derived DC) or from adherent peripheral blood monocytes (monocyte-derived DC). Cultured DC can be recognized by a typical veiled morphologic appearance and expression of surface markers that include major histocompatibility complex class II, CD86/B7.2, CD80/B7.1, CD83, and CD1a. DCs are susceptible to a variety of gene transfer protocols, which can be used to enhance biological function *in vivo*. Transduction of DC with genes for defined tumor (or pathogen) antigens results in sustained protein expression and presentation of multiple tumors peptide antigens to host T-cells. Alternatively, DC may be transduced with genes for chemokines or immunostimulatory cytokines. Although the combination of *ex vivo* DC expansion and gene transfer is relatively new, preliminary studies suggest that injection of genet-ically modified autologous DC may be capable of generating antitumor immune responses in patients with cancer or may be able to function as a vaccine.[68]

The working hypothesis for utilization of DC-based cancer vaccines is that lack of efficient tumor antigen presentation on mature DC, which is frequently observed in tumor-bearing individ-uals, can be bypassed by direct loading of DC with oncoproteins *in vitro*, thus ensuring the transfer of immunostimulatory peptides on the respective antigen-presenting molecules.[69] To enhance load-ing of DC with oncoproteins *in vitro* and to increase the efficacy of the vaccines, a variety of genetic manipulations have been proposed and shown to be efficient in experimental tumor models. DC were transfected with oligonucleotides, DNA or RNA coding for tumor-associated antigens, or with DNA encoding immunostimulatory cytokines and costimulatory molecules. The delivery of genes coding for antigenic epitopes or other molecules with a recombinant retrovirus, adenovirus, or poxvirus into dendritic cells has also been used for transduction and therapy.[70]

DC can be pulsed *ex vivo* with antigens and then reinfused to autologous or syngeneic recipi-ents.[71] A key obstacle to research is the preparation of large number of DCs for purposes of vaccination and their subsequent maturation. Granulocyte-macrophage colony-stimulating factor (GM-CSF) and IL-4 are critical cytokines being used to expand DCs from proliferating precursors and for differentiating DCs from nonproliferating monocyte precursors.[72-75] The selection of a specific stage of development of DCs is also critical. More mature DCs are more immunogenic in mice and in humans;[76,77] on the other hand, only immature DCs are capable of uptake of particulate antigen. To induce DCs maturation and to prolong their life span, a treatment with CD40L (a T-cell product) is required,[78] since most DCs in injected inoculum do not survive more than 2 days. The development of methods for loading DCs with antigen is also needed. One of them is gene transfer.

To facilitate gene transfer, both nonviral and viral-vector-mediated methods have been used for DC-based immunotherapy in animal models and human clinical trials, with the majority of viral-mediated transductions employing recombinant adenovirus or retroviral vectors.[79-81] DC can be transduced directly with retroviral vectors or in the presence of cationic polymers, such as polybrene and protamine sulfate. Results show a low efficiency of retroviral infection of DC in the presence of polybrene. This cationic polymer was found to be directly cytotoxic to murine DC and thus favored the growth of contaminating macrophages. This effect was not observed using protamine sulfate.[82] Mature DCs isolated from mouse spleen have been reported to be refractory to AdV infection.[83] More recent reports indicate modest transfection efficiencies for DC generated from mouse spleen or bone marrow[84] and human peripheral blood.[85] Very often, AdV5 and the pAdEasy-1 vector are used. The pAdEasy-1 vector is a 33.4 kb plasmid, containing most of the human adenovirus serotype 5 (ad5) genome, with deletions in the E1 and E3 genes (thus it is replication-defective). This vector has been developed to infect but not replicate in nonpermissive target cells. The AdEasy adenoviral vector system simplifies the production of recombinant adenoviruses for versatile gene delivery and expression. The construction of a recombinant adenoviral vector is a two-step system in which the desired expression cassette is first subcloned into a shuttle vector, and subsequently transferred into the adenoviral genome by means of homologous recombination in *E. coli*. This method allows one to save weeks of time and easily select the recombinant adenovirus vectors compared to traditional methods.

AdV5 derived vectors offer significant promise in gene, cancer, and vaccine gene therapy. AdV5 is a DNA virus that primarily causes asymptomatic or mild respiratory infections in young children and induces life long immunity. Death due to AdV5 is extremely rare. AdV5 is easily grown and very well studied. Its viral genome can be manipulated by mutagenesis and insertion of foreign sequences.

An example of application of nanoparticulate- (and AdV-) mediated AdV gene transfer into DC cells is presented in Figure 26.1.

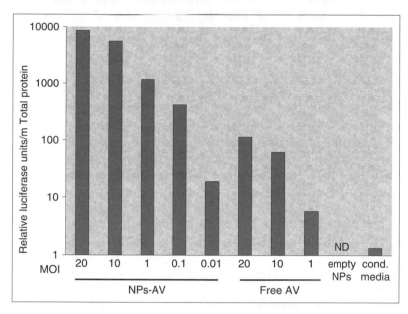

Figure 26.1   Gene transfer into dendritic cells, 3 days post-infection; relative units, as compared to free AdV. NPs-AV: nanoparticles loaded with adenovirus (AV); AV: free adenovirus; empty NPs: no adenovirus; cond. media: only conditioned medium used. The nanoparticles were assembled using a proprietary technology[86] and employing a protocol already published.[87] Particles were generated using a droplet-forming polyanionic solution composed of 0.05 wt% HV sodium alginate (SA-HV), 0.05 wt% CS in water, required amount of adenoviral gene construct (titer tested by plaque assay using 293 cells) and also containing 2 wt% NaCl (Sigma), and a corona-forming polycationic solution composed of 0.05 wt% SH, 0.05 wt% PMCG hydrochloride, 0.05 wt% calcium chloride, and 1 wt% F-68 in water. Poly(methylene-*co*-guanidine) hydrochloride (PMCG) is from Scientific Polymer Products, Inc. (Ontario, NY), average molecular weight 5000, a peptide mimetic. F-68 is a Pluronic polymer from Sigma, a water-soluble nonionic block polymer composed of polyoxyethylene and polyoxypropylene segments. The particles were formed instantaneously via mixing 2 ml of the core solution with 20 mls of the corona solution that were allowed to react for 1 h with stirring. The encapsulation efficiency was 40% based on DNA measurements. The nanoparticle size and charge was evaluated in the reaction mixture by centrifugation at 15,000 g. The average size was 330 nm and the average charge +15.2 mV. Dendritic cell generation: peripheral blood mononuclear cells were separated from buffy coats of healthy donors through Ficoll-Hypaque (Pharmacia). To generate immature dendritic cells (DCs), CD14+ monocytes were purified using the MACS system (Milteni Biotech) and cultured in the presence of IL-4 (100 ng/ml) and GM-CSF (50 ng/ml) (both from R&D) for 4–6 days. DCs were then cultured with varying concentration of adenovirus–nanoparticle complexes. After two days, DCs were stained with PE-conjugated anti-CD86 (which is upregulated in mature DCs) and analyzed using flow cytometry (FACS calibur) and FlowJo software.

CD40 ligand is a 33-kDa type II membrane protein, a member of the tumor necrosis factor (TNF) gene family that is expressed on activated CD4+ T-cells.[88] The receptor for CD40L is CD40, a member of the TNF receptor family.[89] CD40 is expressed on antigen presenting cells, including DC. The CD40–CD40L interaction, besides other functions, is essential in activating the Th1 cells, leading to production of cytokines that favor development of Th1 type immune response.[90] The inefficiency of AdV-mediated gene transfer into DC (due to deficiencies in CAR) can be overcome by surface modification of AdV particles (or of AdV-gene polymeric delivery vehicles), employing the CD40L molecule.[91] In addition, vaccination with DC engineered to express CD40L, via transfection of DC with recombinant AdV–CD40L, can promote DC maturation.[92,93] We propose that the incorporation of CD40L into the AdV delivery system would result in enhanced DC targeting and maturation strategy, thus increasing the efficacy of DC-based vaccination.

26.3 GENE TRANSFER INTO PANCREATIC CELLS

Pancreatic cells represent another category of cells, which are refractory to gene delivery. The transplantation of pancreatic islets has great potential as an effective means of treating Type-1 diabetes. Despite recent success, islet transplantation still lags behind other organ transplants primarily because a large number of transplanted islets often fail to function. This results in the need for multiple islet infusions to cure a single patient, which necessitates the use of more than one donor. This loss of function affects about 50% of all islet grafts and is ascribed to technical failures during isolation and purification, or to nonspecific inflammation that results in islet destruction following transplantation. Moreover, the small proportion of functioning islet grafts is gradually destroyed presumably due to alloimmune rejection or autoimmune destruction.[94-96] To achieve successful islet transplantation, rejection has to be eliminated by addressing all of the above factors simultaneously. Previous work has resulted in improved recovery and viability of human islets by optimization of islet culture conditions.[97] A nonobese diabetic–severe combined immunodeficient (NOD-SCID) mouse model, which allows more accurate assessment of islet *in vivo* function, has been developed.[97]

Islet destruction following transplantation can be prevented by genetically engineering β-cells to express (1) growth factor genes to promote revascularization, and (2) anti-apoptotic and anti-inflammatory genes to prevent apoptosis of the islets.[98-101] Gene therapy strategies can provide high local concentrations of therapeutic proteins in and around the islets without resulting in high systemic toxicity.[102] Adenoviral and adeno-associated viral vectors have shown some promise in abrogation of primary islet nonfunction.[103,104] However, high MOI is often required, which leads to excess fibrosis and malignant matrix formation. Therefore, there is a growing need to develop nonviral gene carriers and plasmid-based expression systems which would not elicit counterproductive immune response or cytotoxicity.

Besides the fact that islets are terminally differentiated and do not proliferate, the low transfection efficiency is due to the islets being a cluster of approximately 1000 (nondividing) cells. Some polymer-mediated (not AdV) gene transfer has been reported.[105] Generally, polymer-based formulations can transfect genes into human islets, but the efficiency is low and requires prolonged incubation.[105] Our data on AdV- (and nanoparticle-) mediated gene transfer into mouse islets are presented in Figure 26.2. The nanoparticulate chemistry is identical to that of Figure 26.1.

26.4 *IN VIVO* GENE DELIVERY

As an example of our successful application of nanoparticulate delivery vehicles for *in vivo* targeted gene delivery, nanoparticles were traced by incorporation of an adenoviral luciferase vector (Ad-luc) into the nanoparticulate core and corona. To allow retention of the targeting peptide (a peptide fragment of thrombospondin-1 protein, in this case TSP-517[106]) within nanoparticles, the sequence was conjugated to polyethylene glycol (M_R 50,000). Conjugate was separated from free peptide by dialysis and then purified by affinity chromatography on heparin–Sepharose. Interestingly, the highest yields of conjugate were obtained with a 2:1 ratio of PEG to peptide, shown by a gradient elution and mobility by SDS-PAGE. The conjugate was incorporated into the nanoparticle corona during their fabrication. The chemistry was similar to that of Figure 26.1 and Figure 26.2. To evaluate the biodistribution of targeted nanoparticles mice were administered either free Ad-luc or conjugated TSP–PEG nanoparticles containing the same amount of adenovirus by tail vein injection. Luciferase activity was evaluated 4 days after injection. Free virus localized predominantly to liver with minor distribution to lung and spleen. It also partitioned heavily into neovasculature in an animal model (polymeric sponge[107]). In contrast, luciferase expression was more widely distributed in TSP–PEG nanoparticles. The lung was a significant reservoir, and significant luciferase activity was detected in sponge homogenates. The targeted nanoparticles were much less

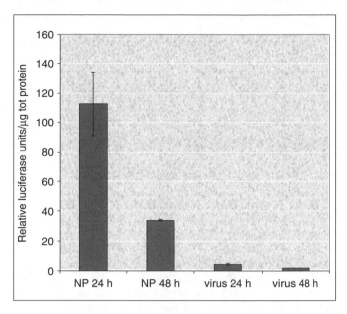

Figure 26.2   Gene transfer into mouse islets, 1 to 2 days post-infection, normalized per protein, as compared to free AdV. NP: nanoparticulate formulation of AdV; virus: free virus. The NP chemistry was identical to that in Figure 26.1.

partitioned into the RES system (to the same degree, in fact, as for nontargeted nanoparticles) and more into proliferating endothelial cells and pericytes of neovascular models. It is likely that the optimization of conjugate loading will further modify the distribution of nanoparticles. In addition, the reporter system used in the pilot studies uses a virus that may contribute to nanoparticulate distribution. Original data will be reported in an upcoming paper.

REFERENCES

1. Lee, R.J. and Huang, L., Lipidic vector systems for gene transfer, *Crit. Rev. Ther. Drug Carrier Syst.* 14, 173, 1997.
2. Kabanov, A.V., Taking polycation gene delivery systems from *in vitro* to *in vivo*, *Pharm. Sci. Technol. Today* 2, 365, 1999.
3. Tomko, R.P., Xu, R., and Phipson, L., HCAR and MCAR: the human and mouse cellular receptors for subgroup C adenovirus and group B coxsackievirus, *Proc. Nat. Acad. Sci. U.S.A.* 94, 3352, 1997.
4. Bergelson, J.M. et al., The murine CAR homologue is a receptor for coxsackie B viruses and adenovirus, *J. Virol.* 72, 415, 1998.
5. Hong, S.S. et al., Adenovirus type 5 fiber knob binds to MHC class I alpha2 domain at the surface of human epithelial and B lymphoblastoid cells, *EMBO J.* 16, 2294, 1997.
6. Wickham, T.J. et al., Integrin $\alpha_v \beta_5$ selectively promotes adenovirus mediated cell membrane permeabilisation, *J. Cell Biol.* 127, 257, 1994.
7. Croyle, M.A. et al., Role of integrin expression in adenovirus mediated gene delivery to the intestinal epithelium, *Hum. Gene Ther.* 9, 561, 1998.
8. Takayama, K. et al., The levels of integrin $\alpha_v \beta_5$ may predict the susceptibility to adenovirus-mediated gene transfer in human lung cancer cells, *Gene Ther.* 5, 361, 1998.
9. Zabner, J. et al., Cellular and molecular barriers to gene transfer by cationic lipid, *J. Biol. Chem.*, 270 18997, 1985.
10. Seth P., Mechanism of adenovirus-mediated endosome lysis: role of the intact adenovirus capsid, *Biochem. Biophys. Res. Commun.* 205, 1318, 1994.

11. Dodds, E. et al., Cationic lipids and polymers are able to enhance adenoviral infection of cultured mouse myotubes, *J. Neurochem.* 72, 2105, 1999.
12. Kotani, H. et al., Improved methods of retroviral vector transduction and production for gene therapy, *Hum. Gene Ther.* 5, 19, 1994.
13. Miller, D.G., Adamand, M.A., and Miller, A.D., Gene transfer by retrovirus vectors occurs only in cells that are actively replicating at the time of infection, *Mol. Cell Biol.* 10, 4239, 1990.
14. Anonymous, Gene therapy clinical trials (charts and statistics), http://www.wiley.co.uk/genmed/clinical/.
15. Makower, D. et al., Phase II clinical trial of intralesional administration of the oncolytic adenovirus ONYX-015 in patients with hepatobiliary tumors with correlative p53 studies, *Clin. Cancer Res.* 9, 693, 2003.
16. Reid, T., Warren, R., and Kirn, D., Intravascular adenoviral agents in cancer patients: lessons from clinical trials, *Cancer Gene Ther.* 9, 979, 2002.
17. Rajagopalan, S. et al., Phase I study of direct administration of a replication deficient adenovirus vector containing the vascular endothelial growth factor cDNA (CI-1023) to patients with claudication, *Am. J. Cardiol.* 90, 512, 2002.
18. Buller, R.E. et al., A phase I/II trial of rAd/p53 (SCH 58500) gene replacement in recurrent ovarian cancer, *Cancer Gene Ther.* 9, 553, 2002.
19. Fuster, V., Charlton, P., and Boyd, A., Clinical protocol: a phase IIb, randomized, multicenter, double-blind study of the efficacy and safety of Trinam (EG004) in stenosis prevention at the graft-vein anastomosis site in dialysis patients, *Hum. Gene Ther.* 12, 2025, 2001.
20. Merritt, J.A., Roth, J.A., and Logothetis, C.J., Clinical evaluation of adenoviral-mediated p53 gene transfer: review of INGN 201 studies, *Semin. Oncol.* 28 (5 Suppl. 16), 105, 2001.
21. Miles, B.J. et al., Prostate-specific antigen response and systemic T cell activation after *in situ* gene therapy in prostate cancer patients failing radiotherapy, *Hum. Gene Ther.* 12, 1955, 2001.
22. Hillenkamp, J. et al., Topical treatment of acute adenoviral keratoconjunctivitis with 0.2% cidofovir and 1% cyclosporine: a controlled clinical pilot study, *Arch. Ophthalmol.* 119, 1487, 2001.
23. Chung-Faye, G. et al., Virus-directed, enzyme prodrug therapy with nitroimidazole reductase: a phase I and pharmacokinetic study of its prodrug, CB1954, *Clin. Cancer Res.* 7, 2662, 2001.
24. Sung, M.W. et al., Intratumoral adenovirus-mediated suicide gene transfer for hepatic metastases from colorectal adenocarcinoma: results of a phase I clinical trial, *Mol. Ther.* 4, 182, 2001.
25. Nanda, D., Driesse, M.J., and Sillevis Smitt, P.A., Clinical trials of adenoviral-mediated suicide gene therapy of malignant gliomas, *Prog. Brain Res.* 132, 699, 2001.
26. Hoffman, J.A. et al., Adenoviral infections and a prospective trial of cidofovir in pediatric hemato-poietic stem cell transplantation, *Biol. Blood Marrow Transplant.* 7, 388, 2001.
27. Benjamin, R. et al., A phase I/II dose escalation and activity study of intravenous injections of OCaP1 for subjects with refractory osteosarcoma metastatic to lung, *Hum. Gene Ther.* 12, 1591, 2001.
28. Runde, V. et al., Adenoviral infection after allogeneic stem cell transplantation (SCT): report on 130 patients from a single SCT unit involved in a prospective multi center surveillance study, *Bone Marrow Transplant.* 28, 51, 2001.
29. Perricone, M.A., et al., Aerosol and lobar administration of a recombinant adenovirus to individuals with cystic fibrosis. II. Transfection efficiency in airway epithelium, *Hum. Gene Ther.* 12, 1383, 2001.
30. Toyoda, K. et al., Cationic polymer and lipids enhance adenovirus-mediated gene transfer to rabbit carotid artery, *Stroke*, 29, 2181, 1998.
31. Goldman, C.K. et al., *In vitro* and *in vivo* gene delivery mediated by a synthetic polycationic amino polymer, *Nat. Biotechnol.* 15, 462, 1997.
32. Feldman, L.J. et al., Improved efficiency of arterial gene transfer by use of poloxamer 407 as a vehicle for adenoviral vectors, *Gene Ther.* 4, 189, 1997.
33. Steg, P.G. et al., Arterial gene transfer to rabbit endothelial and smooth muscle cells using percutaneous delivery of an adenoviral vector, *Circulation*, 90, 1648, 1994.
34. Doebis, C. et al., Efficient *in vitro* transduction of epithelial cells and keratinocytes with improved adenoviral gene transfer for the application in skin tissue engineering, *Transpl. Immunol.* 9, 323, 2002.
35. Rosser, C.J., Benedict, W.F., and Dinner, C.P., Gene therapy for superficial bladder cancer, *Expert Rev. Anticancer Ther.* 1, 531, 2001.

36. Kobayashi, N. et al., Rapidly functional immobilization of immortalized human hepatocytes using cell adhesive GRGDS peptide-carrying cellulose microspheres, *Cell Transpl.* 10, 387, 2001.

37. Marini, F.C. et al., Advances in gene transfer into haematopoietic stem cells by adenoviral vectors, *Expert Opin. Biol. Ther.* 2, 847, 2002.

38. O'Ricordan, C.R., Song, A., and Manciotti, J., Strategies to adapt adenoviral vectors for targeted delivery, *Methods Mol. Med.* 76, 89, 2003.

39. Meurnir-Durmont, C. et al., Adenovirus enhancement of polyethyleneimine-mediated transfer of regulated genes in differentiated cells, *Gene Ther.* 4, 808, 1997.

40. Chillon, M. et al., Adenovirus complexed with polyethylene glycol and cationic lipid is shielded from neutralizing antibodies *in vitro*, *Gene Ther.* 5, 995, 1998.

41. Romanczuk, H. et al., Modification of an adenoviral vector with biologically selected peptides: a novel strategy for gene delivery to cells of choice, *Hum. Gene Ther.* 10, 2615, 1999.

42. Croyle, M.A., Qian-Chun, Y.U., and Wilson, J.M., Development of a rapid method for the PEGylation of adenovirus with enhanced transduction and improved stability under harsh storage conditions, *Hum. Gene Ther.* 11, 1713, 2000.

43. Doukas, J. et al., Retargeted delivery of adenoviral vectors through fiborblast growth factor receptors involves unique cellular pathways, *FASEB J.* 13, 1459, 1999.

44. Fisher, K.D. et al., Polymer-coated adenovirus permits efficient retargeting and evades netralizing antobodies, *Gene Ther.* 8, 341, 2001.

45. Le Doux, J.M. et al., Complexation of retrovirus with cationic and anionic polymers increases the efficiency of gene transfer, *Hum. Gene Ther.* 12, 1611, 2001.

46. Toyoshima, K. and Vogt, P.K., Enhancement and inhibition of avian sacroma viruses by polycations and polyanions, *Virology*, 38, 414, 1969.

47. Manning, J.S., Hackett, A.J., and Darby, N.B., Jr., Effect of polycations on sensitivity of BALB/3T3 cells to morine leukemia and sacoma virus infectivity, *Appl. Microbiol.* 22, 1162, 1971.

48. Kwon, Y.J. and Peng, C.-A., Transduction rate constant as more realiable index quantifying efficiency of retroviral gene delivery, *Biotechnol. Bioeng.* 77, 668, 2002.

49. Panyam, J. and Labhasetwar, V., Biodegradable nanoparticle for drug and gene delivery to cells and tissues, *Adv. Drug Del. Rev.* 55, 329, 2003.

50. Leong, K.W. et al., DNA-polycation nanospheres as non-viral gene delivery vehicles, *J. Control. Release* 53, 183, 1998.

51. Beer, S.J. et al., Poly (lactic-glycolic) acid copolymer encapsulation of recombinant adenovirus reduces immunogenicity *in vivo*, *Gene Ther.* 5, 740, 1998.

52. Sailaja, G. et al., Encapsulation of recombinant adenovirus into alginate microspheres circumvent vector-specific immune response, *Gene Ther.* 9, 1722, 2002.

53. Cavanagh, H.M.A. et al., Cell contact dependent extended release of adenovirus by microparticles *in vitro*, *J. Virol. Meth.* 95, 57, 2001.

54. Beer, S.J., Hilfinger, J.M., and Davidson, B.L., Extended release of adenovirus from polymer microspheres: potential use in gene therapy for brain tumors, *Adv. Drug Del. Rev.* 27, 59, 1997.

55. Lomotan, E.A. et al., Aqueous-based microcapsules are detected primarily in gut-associated dendritic cells after oral inoculation of mice, *Vaccine* 15, 1959, 1997.

56. Einfeld, D.A., and Roelvink, P.W., Advances towards targetable adenovirus vectors for gene therapy, *Curr. Opin. Mol. Ther.* 4, 444, 2002.

57. Palmer, D.H., Mautner, V., and Kerr, D.J., Clinical experience with adneovirus in cancer therapy, *Curr. Opin. Mol. Ther.* 4, 423, 2002.

58. Gilboa, E., Immunotherapy of cancer with DC-based vaccines, *Cancer Immunol. Immunother.* 46, 82, 1998.

59. Sato, K., Nagayama, H., and Takahashi, T.A., Generation of dendritic cells from fresh and frozen cord blood CD34+ cells, *Cryobiology*, 37, 362, 1998.

60. Thurner, B. et al., Generation of large numbers of fully mature and stable dendritic cells from leukapheresis products for clinical application, *J. Immunol. Methods* 223, 1, 1999.

61. Timares, L., Quantitative analysis of the immunopotency of genetically transfected DC, *Proc. Nat. Acad. Sci. U.S.A.* 95, 13147, 1998.

62. Song, W. et al., Dendritic cells genetically modified with an adenovirus vector encoding the cDNA for a model antigen induce protective and therapeutic antitumor immunity, *J. Exp. Med.* 186, 1247, 1997.

63. Butterfield, L.H. et al., Generation of melanoma-specific cytotoxic T lymphocytes by dendritic cells transduced with a MART-1 adenovirus, *J. Immunol.* 161, 5607, 1998.

64. Reeves, M.E. et al., Retroviral transduction of human dendritic cells with a tumor-associated antigen gene, *Cancer Res.* 56, 5672, 1996.

65. Bregni, M. et al., Adenovirus vectors for gene transduction into mobilized blood CD34+ cells, *Gene Ther.* 5, 465, 1998.

66. Dietz, A.B., and Vuk-Pavlovic, S., High efficiency adenovirus-mediated gene transfer to human dendritic cells, *Blood* 91, 392, 1998.

67. Tillman, B.W. et al., Maturation of dendritic cells accompanies high-efficiency gene transfer by a CD-40-targeted adenoviral vector, *J. Immunol.* 162, 6378, 1999.

68. Foley, R., Tozer, R., and Wan, Y., Genetically modified dendritic cells in cancer therapy: implications for transfusion medicine, *Transf. Med. Rev.* 15, 292, 2001.

69. Lundqvist, A. and Pisa, P., Gene-modified dendritic cells for immunotherapy against cancer, *Med. Oncol.* 19, 197, 2002.

70. Bubenik, J., Genetically engineered dendritic cell-based cancer vaccines (review), *Int. J. Oncol.* 18, 475, 2001.

71. Inaba, K. et al., Dendritic cells pulsed with protein antigens *in vitro* can prime antigen-specific, MHC-restricted T-cells in situ, *J. Exp. Med.* 172, 631, 1990.

72. Inaba, K. et al., Generation of large numbers of dendritic cells from mouse bone marrow cultures supplemented with granulocyte/macrophage colony-stimulating factor, *J. Exp. Med.* 176, 1693, 1992.

73. Romani, N. et al., Generation of mature dendritic cells from human blood: An improved method with special regard to clinical applicability, *J. Immunol. Meth.* 196, 137, 1996.

74. Araki, H. et al., Efficient *ex vivo* generation of dendritic cell from CD14+ blood monocytes in the presence of human serum albumin for use in clinical vaccine trials, *Brit. J. Haematol.* 114, 681, 1991.

75. Lundqvist, A. et al., Recombinant adenovirus vector activates and protects human monocyte-derived dendritic cells from apoptosis, *Hum. Gene Ther.* 13, 1541, 1992.

76. Inaba, K. et al., The formation of immunogenic MHC class II-peptide ligands in lysosomal compartments of dendritic cells is regulated by inflammatory stimuli, *J. Exp. Med.* 191, 927, 2002.

77. Dhodapkar, M.V. et al., Antigen specific inhibition of effector T-cell function in humans after injection of immature dendritic cells, *J. Exp. Med.* 193, 233, 2001.

78. Korst, R.J. et al., Effect of adenovirus gene transfer vectors on the immunologic functions of mouse dendritic cells, *Molec. Ther.* 5, 307, 2002.

79. Arthur, J.F. et al., A comparison of gene transfer methods in human dendritic cells, *Cancer Gene Ther.* 4, 17, 1997.

80. De Veerman, M. et al., Retrovirally transduced bone marrow-derived dendritic cells require CD4+ T cell help to elicit protective and therapeutic antitumor immunity, *J. Immunol.* 162, 144, 1999.

81. Timmerman, J.M., and Levy, R., Dendritic cell vaccines for cancer immunotherapy, *Annu. Rev. Med.* 50, 507, 1999.

82. Fresnay, S. et al., Polybrene and interleukin-4: two opposing factors for retroviral transduction of bone-marrow-derived dendritic cells, *J. Gene Med.* 4, 601, 2002.

83. Jooss, K. et al., Transduction of dendritic cells by DNA viral vectors directs the immune response to transgene products in muscle fibers, *J. Virol.* 72, 4212, 1998.

84. Zhang, Y. et al., CD40 ligand-dependent activation of cytotoxic T lymphocytes by adeno-associated virus vectors *in vivo*: Role of immature dendritic cell, *J. Virol.* 74, 8003, 2000.

85. Liu, Y. et al., Transduction and utility of the granulocyte-macrophage colony stimulating factor gene into monocytes and dendritic cells by adeno-associated virus, *J. Interferon Cytokine Res.* 20, 21, 2000.

86. Prokop, A., Micro- and nano-particulate polymeric delivery system, 1997, U.S. patent pending, International patent appl. WO 99/18934; Prokop, A. et al., Polymeric encapsulation system promoting angiogenesis, U.S. patent 6,383,478, May 7, 2002; also PCT application WO 0064954; Prokop, A., Drug delivery system exhibiting permeability control, U.S. patent 6,482,439, November 19, 2002.

87. Prokop, A. et al., Hydrogel-based colloidal polymeric system for protein and drug delivery: Physical and chemical characterization, permeability control and applications, *Advan. Polymer Sci.* 160, 119, 2002.

88. Nair, S. et al., Regression of tumors in mice vaccinated with professional antigen-presenting cells pulsed with tumor extracts, *Int. J. Cancer* 70, 706, 2000.

89. Smith, C., Farrah, T., and Goodwin, R., The TNF receptors superfamily of cellular and viral proteins: Activation, costimulation and death, *Cell* 76, 959, 1994.

90. Cella, M. et al., Ligation of CD40 on dendritic cells triggers production of high levels of IL-12 and enhances T cell stimulatory capacity: T-T help via APC activation, *J. Exp. Med.* 184, 747, 1996.

91. Tillman, B.W. et al., Adenoviral vectors targeted to CD40 enhance the efficacy of dendritic cell-based vaccination against human popillomavirus 16-induced tumor cells in a murine model, *Cancer Res.* 60, 5456, 2000.

92. Liu, Y. et al., Adenovirus-mediated CD40 ligand gene-engineered dendritic cells elicit enhanced CD8+ cytotoxic T-cell activation and antitumor immunity, *Cancer Gene Ther.* 9, 202, 2002.

93. Saudemont, A. et al., Gene transfer of CD154 and IL12 cDNA induces an anti-leukemic immunity in a murine model of acute leukemia, *Leukemia* 16, 1637, 2002.

94. Hering, B., and Ricordi, C., Results, research priorities, and reason for optimism: islet transplantation for patients with type I diabetes, *Graft* 2, 12, 2000.

95. Giannoukakis, N., Thomson, A., and Robbins, P., Gene therapy in transplantation, *Gene Ther.* 6, 1499, 1999.

96. Gaber, A.O. et al., Improved *in vivo* pancreatic islet function after prolonged *in vitro* islet culture, *Transplant.* 72, 1730, 2001.

97. Fraga, D.W. et al., A comparison of media supplement methods for the extended culture of human islet tissue, *Transplantation*, 65, 1060, 1998.

98. Gorden, D.L. et al., Vascular endothelial growth factor is increased in devascularized rat islets of Langerhans *in vitro*, *Transplantation*, 63, 436, 1997.

99. Giannoukakis, N. et al., Adenoviral gene transfer of the interleukin-1 receptor antagonist protein to human islets prevents IL-1B-induced β-cells impairment and activation of islet cell apoptosis *in vitro*, *Diabetes* 48, 1730, 1999.

100. Contreras, J.L. et al., Cryoprotection of pancreatic islets before and soon after transplantation by gene transfer of the anti-apoptotic Bcl-2 gene, *Transplantation*, 7, 1015, 2001.

101. Duville, B. et al., Increased cell proliferation, decreased apoptosis, and greater vascularization leading to beta cell hyperplasia in mutant mice lacking insulin, *Endocrinology*, 143, 1530, 2002.

102. Levine, F., and Leibowitz, G., Towards gene therapy of diabetes mellitus, *Mol. Med. Today* **5**, 165, 1999.

103. Efrat, S. et al., Prolonged survival of pancreatic islet allografts mediated adenovirus immunoregulatory transgenes, *Proc. Nat. Acad. Sci. U.S.A.* 92, 6947, 1995.

104. Flotte, T. et al., Efficient *ex vivo* transduction of pancreatic islet cells with recombinant adeno-associated virus vectors, *Diabetes* 50, 515, 2001.

105. Mahato, R.I. et al., Cationic lipid and polymer-based gene delivery to human pancreatic islets, *Molec. Ther.* 7, 89, 2003.

106. Taraboletti, G. et al., Platelet thrombospondin modulates endothelial cell adhesion, motility, and growth: a potential angiogenesis regulatory factor, *J. Cell Biol.* 111, 765, 1990.

107. Buckley, A. et al., Sustained release of epidermal growth factor accelerates wound repair, *Proc. Nat. Acad. Sci. U.S.A.* 82, 7340, 1985.

CHAPTER **27**

Protein Nanoparticles for Gene Delivery

Goldie Kaul and Mansoor Amiji

CONTENTS

0-8493-1934-X/05/$0.00+$1.50
© 2005 by CRC Press LLC

27.1 INTRODUCTION

27.1.1 Description of Gene Therapy

Gene therapy represents a novel approach to treatment using DNA as a drug. It is designed either to alleviate the genetic defect in cells or to provide additional protective effect. Gene therapy is defined as an approach to treating, curing, or preventing disease by changing the expression of a gene. Given that genes regulate all the basic physiological processes in the body, there is tremendous potential for genes to be employed as therapeutic agents.[1] The two broad divisions of gene therapy are the somatic and the germ line approaches. In the somatic gene therapy, the transformations are not conserved or passed on to the future generations.[2,3] Exogenously administered DNA rarely integrates with the chromosomes of the host genome and persists as an extra-chromosomal element (episome) capable of expressing gene products for a period of time before it is eliminated from the host cell by nuclease degradation.[3]

In vivo gene therapy necessitates the formulation and evaluation of vector systems to deliver therapeutic genes to a specific cell population either locally or systemically in order to efficiently express encoded proteins at the target site. The principle underlying gene delivery is that DNA, or more commonly the plasmid, is a particulate material with a net negative charge, a negatively charged surface and a hydrodynamic diameter of >100 nm. Target cells in the body usually have many barriers that would preclude penetration by the DNA. For instance, DNA cannot cross barriers like the endothelium, the gastrointestinal tract, the keratinized epithelium, and the blood brain barrier. In addition, because of its particulate nature and a net negative charge, DNA is likely to be opsonized from the body by the cells of the reticulo-endothelial system (RES).[3]

Current gene medicine, therefore, comprises optimizing a multicomponent based system that includes a gene encoding a specific therapeutic protein, a plasmid-based gene expression system that is aimed at controlling the functioning of a gene within a target cell, and a gene delivery system that controls the delivery of the gene expression plasmid to specific locations in the body.[4] The vectors for gene delivery must have qualities of safety for repeated use and must provide reproducible levels of the gene product. Gene delivery systems should ideally control the location of a gene in the body by influencing the distribution and access of a gene expression system to the target cell and/or recognition by a cell-surface receptor followed by intracellular trafficking and nuclear translocation. They should also protect a gene expression system from premature degradation in the extracellular milieu and to effect nonspecific or cell-specific delivery to the target cell.[4]

27.1.2 Viral versus Non-Viral Vectors

Viral vectors, like the adenoviruses, retroviruses, and adeno-associated viruses, are very efficient transfection agents. However, viruses are plagued by issues of integration with the host genome to alter permanently their genetic structure, self-replication capability, recombination potential, and the possible complement activation (immunogenecity). The safety issue with viral vectors was particularly highlighted by the tragic death of Jesse Gelsinger, an 18 year old, in a clinical trial of gene therapy experiment to correct for ornithine transcarbamylase deficiency.[5,6] Another child given the pioneering treatment for treatment of SCID (severe combined immunodeficiency disorder) succumbed to leukemia-like syndrome as a result of insertional mutagenesis with viral vectors.[7] Viral vectors are unable to bypass the immune defense mechanism of the host and be limited in the amount of genetic information that they can carry. They are also relatively expensive to manufacture and difficult to scale up.

In the last decade, there has been greater focus on the development of nonviral gene delivery vectors. An ideal vector, in order to transfect cells successfully at a remote site, needs to be inert and stable while in circulation, yet once it reaches the target site, it should release its payload and promote an efficient and specific transfection of the target cells.[3] Specific virus-like characteristics that must be included in nonviral vectors are small size; stability against aggregation in blood, serum or extracellular fluid; the ability to be efficiently internalized by the target cells; and the ability to disassemble and release the DNA in the cell once internalized. The design of an optimal synthetic gene carrier is still the most important limiting step for effective nonviral gene therapy.[8]

27.1.3 Nanoparticle Systems for Gene Delivery

Polymeric nanoparticulate systems are attractive methods of DNA delivery because of their versatility, ease of preparation, and protection of the encapsulated plasmid DNA (pDNA). These carrier systems can efficiently encapsulate the DNA and protect it during transit in the systemic circulation. They can also be targeted to reach specific tissues and cells in the body and avoid uptake by the mononuclear phagocytic system after systemic administration through the use of cell-specific ligands and attachment of poly(ethylene glycol) (PEG) chains on the nanoparticle surface. Nanoparticles usually have a high surface area to volume ratio and thus are able to efficiently encapsulate DNA even without a precondensing step. Nanoparticles can also be made to reach a target site by virtue of their size and charge. Microspheres can be used to direct DNA to specific cells in the body such as for the delivery of DNA vaccines to professional antigen presenting cells like macrophages.[9] Lastly, for industrial production, these systems are amenable to being scaled up and manufactured under the GMP guidelines. DNA-containing microspheres and nanoparticles have been prepared with natural polymers like gelatin,[10-14] chitosan[15,16] and alginates[17,18] as well as synthetic polymers like PLGA[19,20] and poly(β-amino ester)s.[21,22]

27.1.4 Protein Nanoparticles

For the development of polymeric nanoparticulate carriers with improved encapsulation and efficient intracellular delivery of DNA, researchers have outlined several design criteria. First, the matrix polymer must be biocompatible, and preferably biodegradable. Second, the carrier system should efficiently encapsulate the DNA and protect it during transit in the systemic circulation. Third, the carrier should be able to reach specific tissues and cells in the body and avoid uptake by the mononuclear phagocytic system after systemic administration. This is achieved through the use of cell-specific ligands and attachment of PEG chains on the nanoparticle surface. Fourth, the nanoparticulate system should be able to deliver the DNA inside the cell through nonspecific or receptor-mediated endocytosis and protect the DNA during cellular transit. Lastly, for industrial production, the system should be amenable to being scaled up and manufactured under the GMP guidelines. Most of the proteins employed in polymeric nanoparticulate systems are biocompatible and biodegradable in the body. They are also available as "Generally Regarded as Safe" (GRAS) substances — sterile, pyrogen-free forms approved by the U.S. Food and Drug Administration (FDA). Proteins are also importantly capable of forming systems that can effectively encapsulate the payload and protect it from degradation and breakdown under *in vivo* conditions. The surface of most of the proteins is also amenable to cross-linking and derivation, which can help in attachment of surface ligands capable of specific cell type targeting and afford protection under systemic conditions in the body and also cellular transit (Figure 27.1). Protein nanoparticles additionally afford ease of manufacture and scale-up, as they either employ salt-mediated phase separation or coacervation, solvent evaporation, or emulsion techniques, which are relatively easy to scale up.

The main disadvantages of using products containing animal- or plasma-derived proteins, such as gelatin and albumin, respectively, is the potential to induce harmful inflammatory or immune responses in humans and pose the risk of contamination with potentially life-threatening pathogens, such as viruses or prions. In addition, the current process of making gelatin by denaturing collagen (a protein extracted from the skin, bones, and tissues of animal carcasses) yields a highly variable material that is not easily traced to its source once incorporated into consumer products. Several companies are exploring the potential of preparing gelatin and similar proteins by recombinant

Ideal Vector Design Criteria for Gene Therapy

Vector Requirements	Protein Nanoparticles for Gene Delivery
Biocompatible and biodegradable	Versatile, easy to prepare, biocompatible and biodegradable
Efficient pDNA encapsulation and offer protection during cellular transit	Efficient pDNA encapsulation because of high surface to volume ratio and offer protection during cellular transit
Deliver pDNA by non-specific or receptor-mediated endocytosis	
Avoid uptake by the reticuloendothelial system	Deliver pDNA by non-specific (charge mediated or size) or receptor-mediated endocytosis
Ability to reach specific areas in the body by targeting	Avoid uptake by the reticuloendothelial system by PEGylation, etc.
Amenable to scale-up under GMP	Ability to reach specific areas in the body by targeting by virtue of derivatization or attachment of targeting moieties
	Amenable to scale-up under GMP

Figure 27.1   Protein nanoparticles as gene delivery vectors.

technology. Gelatin derived by recombinant technology has been shown to yield a consistent product that is free from extraneous contamination.

27.2 COLLAGEN

27.2.1 Collagen as a Biomaterial

Collagen is the structural building material of vertebrates and the most abundant mammalian protein that accounts for 20–30% of total body proteins. It is mainly present in tissues that carry out mechanical functions. Collagen in the body is mainly synthesized by the fibroblasts that originate from the pluripotent cells of the adventitial matrix or the reticulum cells. In the body, collagen is mainly associated with the tissues such as the tendons and the ligaments. The innate structure of collagen is a triple helical structure that forms a fiber and is predictive of its ability to bear loads. It has a unique structure, size, and amino acid sequence. The collagen molecule comprises three polypeptide chains intertwined around one another, with each chain having an individual twist in the opposite direction. Helix formation in the molecule is facilitated by the presence of a high number of glycine amino acid residues. The three strands are primarily held by hydrogen bonds between the adjacent -CO and the -NH groups along with some covalent bonds. The orderly arrangement of the triple helix tropocollagen molecules results in formation of fibrils with typically distinct periodicity. Nonhelical telopeptides are attached to both ends of the molecule and serve as the major source of antigenicity.[23-30]

Atelocollagen, which is produced by the elimination of the telopeptide moieties using pepsin, has demonstrated its potential as a drug carrier, especially for gene delivery.[31-32] Collagen is regarded as one of the most useful biomaterials due to the fact that it has excellent biocompatibility, biodegradation, and ease of availability.[33] Improvement of the physicochemical and biological properties of the material includes adjusting the structure of the collagen matrix by the addition of other proteins like elastin, fibronectin, and glycosaminoglycans.[34-36] The biodegradability and subsequent release of payload from collagen matrices can be additionally controlled by use of cross-linking agents such as glutaraldehyde, chromium tanning, formaldehyde, acyl azide,[37-40] carbodiimides,[28] and ultraviolet and gamma irradiation.[23] Collagen can additionally form fibers with sufficient strength and stability through its self-aggregation and crosslinking properties. In addition to formulating nanoparticles for gene delivery, collagen is useful in other drug delivery applications in the form of shields in ophthalmology,[41,42] sponge films for burns and wounds,[30] mini-pellets and tablets for protein delivery,[43] gel formulation in combination with liposomes,[44] as a controlling material for transdermal delivery,[45] and as microparticles for gene delivery.[46] Collagen has also been used as scaffold for tissue engineering including skin replacement, bone substitutes, and artificial blood vessels and valves, surgical sutures, and hemostatic agents.[47,48] Thus because of its excellent biocompatibility characteristics and safety aspects, the use of collagen in biomedical application has been growing rapidly. However, there are some inherent disadvantages of collagen based systems that include poor mechanical strength, ineffectiveness in management of infected sites, the difficulty of assuring adequate supplies, and relatively long degradation times in the biological environment. Adding polar functionalities to the polymers may rectify the problem associated with longer degradation times. Collagen has also been combined with other systems to form collagen–liposome and collagen–silicone matrices. Collagen can be solubilized in aqueous media, especially a slightly acidic environment. Although collagen is a stable protein in the body, it can still degrade specifically by means of enzymes such as collagenase, which binds tightly to the triple helices and degrades collagen starting from the surface and telopeptide-cleaving enzymes.[49,50] Reports of adverse reactions to collagen include localized redness and swelling following plastic surgery with implants. Clinical reactions to collagen, although rare, mainly include Ig-E mediated reactions to bovine collagen.[51] Table 27.1 summarizes the various advantages and disadvantages of collagen as a biomaterial for delivery.

27.2.2 Collagen Nanoparticles for Gene Delivery

27.2.2.1 Preparation of Nanoparticles

The crystallites in the gel aggregates appear as multiple chain segments in the collagen-fold configuration. This property has been utilized to prepare aggregates as colloidal systems for drug and gene delivery. The formation of the nanospheres is driven by a combination of electrostatic and electropic forces with sodium sulfate mainly employed as a dissolving reagent to facilitate greater charge interactions between pDNA and collagen.[52] The molecular weight of collagen/gelatin has been found to have influence on the stability of the manufactured gelatin nanoparticles,[53] and the molecular weight profile of collagen was found to be affected by pH and temperature, both of which further influenced the noncovalent interactions responsible for the molecular structure of

Table 27.1 Summary of the Advantages and Disadvantages of Collagen as a Biomaterial

Advantages	Disadvantages
Available in abundance and can be purified from living organisms	Expensive nature of pure Type-I collagen
Nonantigenic	Variability of isolated collagen (cross-linking density, fiber size, impurities)
Bioreabsorbable and biodegradable	Increased hydrophilicity that leads to swelling and a more rapid release
Nontoxic and biocompatible	Variability in the enzymatic degradation rate as compared to hydrolytic degradation
High tensile strength, synergic with bioactive components	
Hemostatic	Complex handling properties
Biodegradability may be regulated by crosslinking	Side effects, e.g., bovine spongioform encephalopathy (BSE) and mineralization
Easily modified/derivatized using functional moieties	
Capable of conjugation with synthetic polymers	

collagen.[54] On examining the relationship between electropic forces and gene factors, it was found that the main driving force between the acidic gelatin and the basic fibroblast growth factor was one of electrostatic interaction. The biodegradable collagen based nanoparticles are stable to heat, and thus could be sterilized readily.[46] Moreover, the nanoparticles could be taken up by the reticulo-endothelial system (RES) and enable enhanced uptake of exogenous compounds, such as anti-HIV drugs, into macrophages.[55] Thus collagen nanoparticles have been used as systemic carriers for cytotoxic agents and other therapeutic compounds, such as campthocin and hydrocortisone.[56]

Collagen can also be formulated into nanoparticles or nanospheres for drug or gene delivery system. A new two-step desolvation method[53] for manufacturing gelatin nanoparticles has been developed that enables the production of gelatin nanoparticles with a reduced tendency for aggregation. The molecular weight of gelatin has a decisive influence on the stability of the manufactured gelatin nanoparticles. In addition, two fluorescent dyes, Texas red and fluoresceinamine, were coupled to the nanoparticles for cell uptake studies. The fluorescent nanoparticles showed a high uptake into monocytes and macrophages. The molecular weight profile of the collagen solution was affected by pH and temperature, both of which further influenced the noncovalent interactions responsible for the molecular structure of collagen.[53]

27.2.3 Other Collagen-Based Gene Delivery Systems

27.2.3.1 Collagen Films and Matrices

Collagen films and matrices have been used as gene delivery carriers for promoting bone formation. A composite of recombinant human bone morphogenetic protein 2 (rhBMP-2) and collagen was used to monitor bone development and absorbent change of carrier collagen. The rh-BMP-2/collagen implant resulted in active bone formation, whereas the collagen alone did not.[57]

Collagen additionally provides anchorage for cell differentiation and remains as an artificial matrix in woven bone.[58] In another study, collagen matrix loaded with BMP and placed in close contact with osteogenic cells achieved direct osteoinduction without causing cartilage formation.[58] Collagen discs and films have some advantages — for example, systems that isolate transplanted cells from the host immune system may be useful and economically attractive, as they would allow the use of allogeneic or even xenogeneic cells in many patients.[59,60] The use of genetically modified cells for a long-term delivery of a therapeutic transgene product has been an attractive option for the treatment of monogenetic hereditary as well as many multifactorial nongenetic diseases.[61,62] Biodegradable collagen films also serve as scaffolds for survival of transfected fibroblasts.[63] In most of the animal models tested, long-term gene expression of a foreign gene after implantation of transfected cells on collagen scaffold has not been achieved. Combination of collagen with polymers like the atelocollagen matrix added onto the surface of polyurethane films enhanced the attachment and proliferation of fibroblasts and supported growth of cells.[64,65] Transplantation of cells embedded in polytetrafluoroethylene (PTFE) fibers coated with rat collagen followed by mixing with matrigel and basic fibroblast growth factor showed a long term expression of human β-glucuronidase by retroviral-transduced murine or canine fibroblasts was achieved using a collagen-based system.[66,67]

27.2.3.2 Collagen Shields for Gene Delivery

The collagen shield was originally designed for the bandage contact lenses that gradually dissolve in the cornea.[68] The mechanical properties of the shield protect the healing corneal epithelium from the blinking action of the eyelids in corneal transplantation and radial keratomy.[69,70] Clinical studies have verified the fact that healing is faster and more complete with the use of collagen shield than the conventional formulations, with less stromal edema at the wound sites in collagen-treated corneas.[41,42] Additionally, the collagen shield afforded protection to the keratocytes adjacent to the wound site. Full-thickness filtering procedures were performed on eyes of New Zealand albino rabbits. Naked pDNA in saline was either injected beneath Tenon's capsule at the filtration site or absorbed into a collagen shield that was then placed external to the sclera and under the Tenon's capsule. Levels of the reporter gene chloramphenicol acetyltransferase (CAT) were measured 48 h after surgery in samples of ocular tissues. Additionally, the β-galactosidase (β-Gal) reporter gene expression was localized histologically.[71]

Injection of pDNA in saline vehicle into the filtration bleb produced readily detectable CAT activity in bleb tissue (conjunctiva, Tenon's capsule, and sclera) whereas CAT activity was nearly undetectable in samples of the cornea, iris–ciliary body, and tissues located opposite the bleb site. Delivery of the pDNA into the bleb through a collagen shield increased CAT activity 30-fold over injection of plasmid in saline (2711 ± 567 mU/mg versus 92 ± 38 mU/mg). β-Gal activity was imaged only in the region of the bleb, and microscopic examination showed that the β-Gal activity was localized to Tenon's capsule fibroblasts, with minimal β-Gal activity observed in inflammatory cells or scleral fibroblasts. It could thus be concluded that the transfection of filtration tissues was enhanced by absorption of naked DNA into a collagen shield. Furthermore, transfection was localized to the fibroblasts and inflammatory cells of the filtration bleb site. Gene therapy using naked pDNA and a simple collagen shield delivery vehicle may be useful for regulating wound healing after glaucoma surgery.[71]

27.2.3.3 Collagen Sponges

Collagen sponges have been used with human collagen membrane as a major source of collagen. Collagen sponges have the ability to absorb large amounts of exudates easily and adhere well to the wet wound bed while preserving the low moist climate, as well as shielding against mechanical harm and secondary bacterial infection.[72] Collagen sponges have been combined with other materials like elastin, fibronectin, or glycosaminoglycans[34-36] to achieve better fluid building capacity and resilience. An absorbable collagen sponge containing the bone morphogenetic protein rhBMP-2 was tested in a rat model for the evaluation of the efficacy of rhBMP-2 produced in *E. coli* on promoting bone healing.[73] Bacterially expressed rhBMP-2 loaded in collagen sponge was osteogenic *in vivo*. Also, an absorbable collagen sponge containing rhBMP-2 was found to stabilize endosseous dental implants in bony areas and normal bone formation restored without any complications.[74]

27.2.3.4 Collagen Hydrogels

Hydrogels of collagen have also been used as gene delivery vehicles. Plasmids bound noncovalently to the Fab portions of the antibody can be efficiently introduced into cells that express the polymeric immunoglobulin receptor (pIgR). Human tracheal epithelial cells grown on plastic, a condition that downregulates the expression of the receptor failed to express the reporter gene, whereas cells from the same trachea maintained on collagen gels were transfected.[75] Clonal murine fibroblast lines transfected with the G-CSF were found to survive with the grafts and express the transgene as well as collagen *in vivo*.[63] Atellocollagen, which is produced from the peptic digestion of the telopeptide moieties, has also been used as a carrier for chondrocytes to repair cartilage defects.[76-78]

27.2.3.5 Collagen Pellets

Collagen-based pellets as a gene delivery vector have been employed by Kohmura et al., who studied the effect of atelocollagen-based minipellets on the mRNA expression and functional status of facial nerve in rats. The facial nerve transmission and immediate repair was accelerated by this system and regeneration achieved.[31] Also studied are minipellets with dimensions of 0.6 mm diameter and 10 mm long cylindrical pellets containing 50 μg of pDNA and human HST-1/FGF-4 cDNA.[32]

27.3 GELATIN

27.3.1 Gelatin as a Biomaterial

Gelatin is a proteinaceous biopolymer, obtained by hydrolysis of collagen, with proven safe use in pharmaceuticals, cosmetics, and food products for a long time. It is a colorless, odorless, non-irritating, biocompatible, and biodegradable material. Gelatin is a relatively inexpensive material that is available in sterile and pyrogen-free form. Gelatin is also considered a GRAS excipient by the FDA. Type-A gelatin is obtained by acid treatment of collagen and has an isoelectric point (pI) between 7.0 and 9.0. Type-B gelatin, on the other hand, is obtained by alkaline hydrolysis of collagen and has a pI between 4.8 and 5.0. The large number of pendant functional groups in the gelatin structure aid in chemical cross-linking and derivatization. It is commercially available by hydrolytic degradation of naturally occurring porcine or bovine collagen. It is most importantly nonirritating, biocompatible and biodegradable.[79] It is a natural polyampholyte that gels below 35–40°C. It can be made into nanoparticulate dosage forms encapsulating DNA. Surface modification of gelatin is also possible with attaching PEG or PEO chains to avoid RES uptake. At pH below 5, positively charged gelatin can form complexes with DNA.[79-80]

27.3.2 Gelatin Nanoparticles for Gene Delivery

27.3.2.1 Gelatin-DNA Coacervates

Truong-Le et al.[81] prepared and characterized nanoparticles made by a salt-induced complex coacervation of DNA and gelatin. Injection of this preparation in mice footpad and the tibialis muscle resulted in enhanced expression of *LacZ* gene product compared to equal doses of naked DNA and Lipofectamine®-DNA. Truong-Le et al.[82] also prepared DNA-gelatin nanoparticulate coacervate with chloroquine as an endosomolytic agent, calcium, and transferrin bound to the gelatin molecule. These nanoparticles offered limited protection against DNAse I enzyme *in vitro*. Transfection of human tracheal epithelial cells (9HTEo) with the gelatin nanoparticle system containing the cystic fibrosis transport regulator (CFTR) gene resulted in 50% transfection of the cells *in vitro*.

Leong et al.[83] prepared DNA–polycation nanospheres as a potential gene delivery vehicle by a salt-induced complex coacervation of cDNA and polycations such as gelatin and chitosan, in the size range of 200–750 nm. Although the transfection efficiency of these nanospheres was found to be typically lower than that of Lipofectamine® and calcium phosphate controls in cell culture, the beta-gal expression in muscle of BALB/c mice was higher and more sustained than that achieved by naked DNA and Lipofectamine® complexes. This gene delivery system was found to be a good system as it combined the potential for surface modification for targeting or stimulating receptor-mediated endocytosis with the possibility to incorporate lysosomolytic agents. Additionally, other boactive agents and multiple plasmids could be coencapsulated and the bioavailability of pDNA was considerably improved because of protection from serum nucleases. Finally, the nanoparticles could be lyophilized for storage without loss of bioactivity.

Coester et al.[84] prepared uniform nanoparticles of gelatin covalently linked to avidin by a two-step desolvation process as carriers for biotinylated peptide nucleic acid (PNA). The surface of the nanoparticles was thiolated and avidin covalently attached via a bifunctional spacer at high levels. Biotinylated PNA was effectively complexed to these avidin gelatin nanoparticles.

27.3.2.2 PEG-Modified Gelatin Nanoparticles

Kaul and Amiji[85-87] have developed and characterized long-circulating nanoparticulate formulation of gelatin by PEG modification of the biopolymer (Figure 27.2 and Figure 27.3). PEG–gelatin nanoparticles had a decreased degradation in the presence of protease enzyme due to steric repulsion. They prepared PEG-modified gelatin nanoparticles as long-circulating intracellular delivery systems. The control gelatin and PEGylated gelatin nanoparticles, prepared by the solvent displacement technique, had a mean particle size of 300 nm and could efficiently encapsulate hydrophilic macromolecules including pDNA. Cellular uptake and trafficking studies showed that the nanoparticles were internalized by tumor cells and were found near the nucleus after 12 h. The nanoparticles containing rhodamine–dextran (Rho–Dex) were found to be localized mainly in the perinuclear region of the BT-20 cells after 12 h of incubation. Also interesting was that the cells remained viable during the course of the study, and as such, these nanoparticles did not confer any overt cytotoxicity. At initial time points, they observed that the nanoparticles

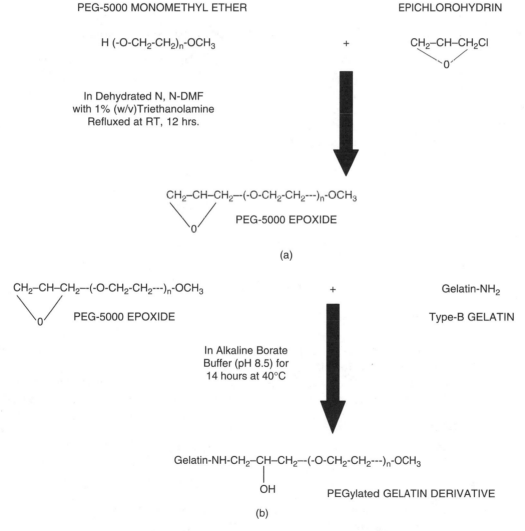

Figure 27.2 Synthesis of methoxypoly(ethylene glycol)-epoxide (A) and poly(ethylene glycol)-modified gelatin derivative (B).

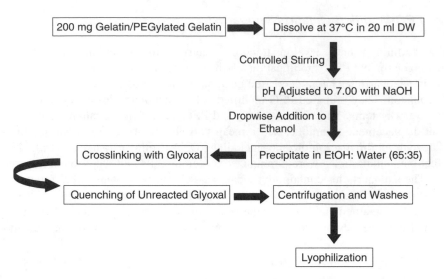

Figure 27.3   Preparation of gelatin and poly(ethylene glycol)-modified gelatin (PEGylated gelatin) nanoparticles.

were mainly present on the cells' surface with subsequent uptake through the vesicular transport system. Once the nanoparticles were endocytosed, they were able to escape the endosome and were found primarily in the cytoplasm around the nuclear membrane. In the case of gelatin nanoparticles, the fluorescence confocal image shows that some of the fluorophore was released and stained the nucleus. PEGylated gelatin nanoparticles, on the other hand, remained intact as the fluorescence image shows discrete particles around the nucleus. When Rho–Dex was added to the cells in solution and incubated for 12 h, the fluorescence was completely diffused throughout the cell. Kinetic analysis of GFP expression showed that the protein was expressed 24 h after administration of the DNA-containing control and PEGylated gelatin nanoparticles. GFP expression also remained stable for up to 96 h. The PEGylated gelatin nanoparticles were also very efficient in expressing GFP. The results proved that this system could be used for intracellular delivery of hydrophilic macromolecules such as DNA.

27.4 ALBUMIN

27.4.1 Albumin as a Biomaterial

Human serum albumin (HSA) or bovine serum albumin (BSA) are abundant proteins widely used as a material for microsphere and nanosphere preparation.[88] Albumin is a biodegradable and biocompatible protein. HSA is also nonantigenic. As a major component of serum, HSA circumvents the problems often encountered with positively charged complexes *in vivo*, such as the rapid opsonization by serum proteins. Furthermore, the surface of the HSA nanoparticles possesses many amino and carboxylic groups that are available for covalent modification and drug or protein attachment. These functional groups can be employed to couple targeting ligands such as antibodies, viral proteins, sugars, and tissue- or cell-specific targeting ligands, as well as to bind covalently to some drugs. Furthermore, albumin may alleviate some of the undesired interactions between transfection complexes and serum components *in vivo* because albumin happens to be the most abundant serum protein. The low surface charge of the most prepared albumin nanoparticles also favorably reduces nonspecific binding to cells.[88]

27.4.2 Albumin Nanoparticles for Gene Delivery

27.4.2.1 Preparation of Albumin Nanoparticles

Rhaese et al.[89] prepared nanoparticles in the range of 300–700 nm of pDNA, HSA and polyethyleneimine by a complex coacervation technique and tested the transfection efficiency *in vitro*. These HSA–PEI–DNA nanoparticles contained the pGL3 vector coding for luciferase as the reporter gene and were formed by charge neutralization. These particles were characterized by various assays such as the gel retardation assay, dynamic light scattering for size, and electrophoretic measurements for charge. The stability was determined using spectrophotometric analysis and transfection efficiency evaluated in cell culture using human embryonic epithelial kidney 293 cells. The nanoparticles formed were prepared by coencapsulation of polyethylenimine (PEI) as a lysosomotropic agent at various nitrogen to phosphate ratios. The optimum transfection efficiency was achieved with N/P ratios between 4.8 and 8.4. Also these nanoparticles were relatively less cytotoxic.

The HSA–PEI–DNA nanoparticles were found to completely retain DNA during gel electrophoresis even with the lowest amounts of PEI tested, demonstrated by the lack of DNA migration after binding to the particles. The dye ethidium bromide was also completely excluded from the staining of the particles at a minimal N/P ratio of 4.8. The higher stability of the pDNA to degradation against serum nucleases with the increasing PEI was attributed to the stronger condensation of DNA. The aggregation effect of the nanoparticles with time on storage was prevented by lyophilization of these nanoparticles.[89]

Shohet and coworkers[90,91] have also addressed the hypothesis that albumin-coated microbubbles could be used to deliver an adenoviral transgene effectively to the rat myocardium by ultrasound-mediated microbubble destruction. They formulated a system with a recombinant adenovirus containing β-galactosidase and attached it to the surface of albumin-coated, perfluoropropane microbubbles. These bubbles were infused into the jugular vein of rats with or without simultaneous echocardiography. Additional controls included ultrasound of microbubbles without the virus, virus alone, and virus with ultrasound. Another control was a group with ultrasound mediated microbubble destruction, followed by an adenoviral infusion. The livers of all the rats receiving the virus showed extensive β-galactosidase activity, whereas there was no expression in the skeletal muscle. The hearts of all the rats in the experimental group showed a blue stain with 5-bromo-4-chloro-3-indolyl-β-D-galactopyranoside (X-gal), indicating the expression of β-galactosidase. None of the control rats showed any myocardial staining, which confirmed that the destruction of the microbubbles containing the virus was responsible for he observed expression in the rat myocardium.

Orson et al.[92,93] developed a novel intravenous method to deliver small quantities of plasmid to lung tissue using nontoxic quantities of polyethyleneimine in combination with albumin. The results of intravenous administration in mice with 1 μg quantities of plasmid resulted in a highly efficient gene expression in the lung interstitial and endothelial tissues (0.5–1 ng per μg of pDNA), while larger amounts of plasmid reduced gene expression. This enhancement in gene expression of the protein-coupled PEI–pDNA system was not protein specific, as many other soluble proteins were found to be equally efficacious. Also in contrast to widespread gene expression after administration of cationic lipid-condensed pDNA, the distribution of gene expression with an intravenous dose of protein–PEI–plasmid showed that the major transfection occurred in the lung, the first capillary bed encountered after intravenous administration. Single injections of these systems resulted in maximal gene expression from 24 to 48 h, with a rapid decline to 1% of the maximum by 96 h. Repeated dosing did not increase the maximal detected activity, but did result in constant gene expression, demonstrating that the response to repeated injection was not inhibited. This lack of refractory period may reflect the low quantities of pDNA required for gene expression since the pDNA CpG induced cytokines that mediate inhibition are elicited with high doses of pDNA. All these results taken together suggest that the protein–PEI method of delivering expression vectors may be particularly useful for repeated direct infusions via a catheter to specific tissues where localized gene expression is desirable as in the case of gene therapy of individual tumors.

Weber et al.[94] prepared pDNA coding for either green fluorescent protein (GFP), luciferase, or beta-galactosidase bound HSA nanoparticles by a modified complex coacervation process with varying amounts of PEI and sodium sulfate. Transferrin was additionally conjugated to the nanoparticles by cross-linking the particles with transferrin solution and dialyzing against saline for 24 h. This process led to the formation of particles in the range of 200–800 nm with a slight negative surface charge.

Roser et al.[95] measured the surface charges on biodegradable albumin nanoparticles that were introduced by covalent coupling different primary amines to examine their influence on phagocytosis by macrophages under *in vitro* conditions. Albumin particles with a zeta potential close to zero showed a reduced phagocytic uptake in comparison with charged particles, especially nanoparticles with a positive zeta potential. The phagocytic uptake in the present study was examined using an established cell culture model based on primary mouse peritoneal macrophages and a human hematopoietic monocytic cell line (U-937) treated with phorbol-12-myristic-13-acetate to induce cell differentiation. The influence of opsonins on *in vitro* phagocytosis experiments was characterized using carriers pretreated with human serum. In the presence of human serum the phagocytic activity of U-937 cells was found to be similar to primary mouse macrophages without serum. In contrast to peritoneal macrophages, U-937 cells showed no phagocytic activity in the absence of serum. In particular, only the C3b- complement deposition on the particle surface seems to promote the phagocytic process. The *in vivo* distribution of albumin carriers in rats was investigated using magnetic resonance imaging (MRI). No differences in blood circulation times and organ accumulation between different nanoparticle preparations with positive, neutral, and negative surface charges could be observed in rats, suggesting that the *in vivo* fate of albumin nanoparticles is significantly influenced by factors not reflected in the *in vitro* cell culture models.

27.5 ADDITIONAL EXAMPLES OF PROTEIN NANOPARTICLES

27.5.1 Lipoproteins

Pan et al.[96] reported the development and evaluation of a novel artificial lipoprotein deliver system that could carry DNA for effective *in vitro* gene transfections in tumor cells. This system is comprised of nanoemulsion cores made of natural lipids and surface lipidized poly(L-lysine) that replaces the surface proteins in natural lipoproteins. Additionally the cytotoxicity of this system was lower than the commercial gene transfection systems using cationic lipids as the lipids used in this system were all natural substances. It could also be formulated using commercially available phospholipids, cholesterol and poly(L-lysine) and the chemical composition, particle size, and type of surface polypeptide or surface protein could be controlled and optimized allowing widely diversified gene or drug delivery applications.

Pan et al.[96] developed and evaluated a novel artificial lipoprotein delivery system for *in vitro* gene transfection in human SF-767 glioma cells. The nanoemulsion was formulated with similar lipid compositions present in natural lipoproteins. The oil phase of nanoemulsion was composed of triolein (70.0%), egg phosphatidylcholine (22.7%), lysophosphatidylcholine (2.3%), cholesterol oleate (3.0%), and cholesterol (2.0%). To replace the surface protein as in natural lipoprotein, poly(L-lysine) was modified to add palmitoyl chains and incorporated onto the nanoemulsion particles through hydrophobic interactions. pDNA, pSV-beta-gal coding for beta-galactosidase, was carried by the nanoemulsion/poly(L-lysine) particles. The charge variation of so formed complex was examined by agarose gel electrophoresis and zeta potential measurements. After standard X-Gal staining, transfected cells were observed under light microscope. The effect of chloroquine on the transfection was examined and, finally, the cytotoxicity of this new system was evaluated in comparison with commercial Lipofectamine® gene transfection system. The pDNA was effectively carried by this artificial lipoprotein delivery system and the reporter gene was expressed in the glioma cells. Transfection efficiency was significantly increased by the treatment of chloroquine, indicating that endocytosis possibly was the major cellular uptake pathway. Compared to the Lipofectamine® system, this new delivery system demonstrated similar transfection efficiency but a much lower cytotoxicity. In the experiment, the cell viability showed up to 75% using this system compared to only 24% using Lipofectamine® system. Thus a new artificial lipoprotein delivery system was developed for *in vitro* gene transfection in tumor cells. The new system showed similar transfection efficiency but a much lower cytotoxicity compared with the commercial Lipofectamine® system.

27.5.2 Epidermal Growth Factor

An efficient receptor-mediated nonviral gene delivery formulation based on mono-PEGylated recombinant human epidermal growth factor (EGF) was developed using a streptavidin–biotin system by Lee et al.[97] Biotin-derivatized and mono-PEGylated EGF was prepared by conjugating a biotin-PEG-NHS derivative to EGF and purified through a chromatographic method. Luciferase pDNA and PEI were complexed to form positively charged nanoparticles on which negatively charged streptavidin was first coated and then biotin-PEG-EGF conjugate was immobilized via streptavidin–biotin interaction. The EGF-PEG-biotin–streptavidin–PEI–DNA complexes were characterized in terms of their effective diameter and surface zeta potential value under various formulation conditions. The formulated complexes exhibited high transfection efficiency (~10^8 in luciferase activity) with no inter-particle aggregation. This was attributed to enhanced cellular uptake of the resultant complexes via receptor-mediated endocytosis. Furthermore, in the presence of serum proteins, a slight decrease in transfection efficiency was observed due to the presence of PEG chains on the surface.

In their study, Lee et al.[97] proposed an efficient EGF receptor-mediated gene delivery system based on biotin–streptavidin interaction. This system utilized specific noncovalent immobilization of PEG-conjugated EGF on the surface of the complexes. Biotin-derivatized and multi- or mono-PEGylated EGF was prepared for formulating the PEI–DNA complexes. Negatively charged streptavidin was bound to positively charged PEI–DNA complexes by ionic interaction, and biotin-PEG-EGF conjugates were introduced to the streptavidin-coated complexes by specific streptavidin–biotin affinity interaction. The EGF-PEG-biotin–streptavidin–PEI–DNA complexes prepared by using multi-PEGylated or mono-PEGylated EGF were analyzed in terms of their size and surface charge value under various formulation conditions. The formulated complexes containing luciferase pDNA as a reporter gene were transfected to A431 cells to examine whether or not the transfection efficiency was enhanced due to the presence of PEG-tethered EGF on their surface. Additionally, the effect of serum proteins on the transfection efficiency was studied. The formulation was based on the concept that mono-PEGylated EGF could be immobilized onto the surface of the positively charged PEI–DNA complexes by the specific biotin–streptavidin interaction. The complexes were tolerant to the nuclease attack. The surface exposed EGF ligands on the complexes facilitated the process of receptor-mediated endocytosis by recognizing EGF receptors on cell membrane, while densely seeded PEG spacers could serve as a protective layer against non-specific protein adsorption. Thus, the surface immobilized EGF, especially in the form of mono-PEGylated EGF, is a versatile approach to be applied for any positively charged cationic polymer–DNA complexes. Furthermore, delivering foreign genes to certain cancer cells over-expressing EGF receptors could be readily achieved in a target-specific manner.

27.5.3 Terplex System

Kim et al.[98,99] modified poly(L-lysine) (PLL) to improve the delivery and expression of transfected pSV-β-gal DNA by developing a novel system consisting of stearyl-poly(L-lysine) (stearyl-PLL) and low-density lipoprotein (LDL) in weight ratios of 3:5:3. The results of this system indicated that pDNA, when formulated with stearyl-PLL and LDL, formed a stable hydrophobicity and charge balanced terplex system of optimal size for efficient cellular uptake and that the pDNA was still intact after the terplex formation. Modification of the PLL by the stearyl chains was chosen for two reasons:

1. A hydrophobic interaction between the stearyl groups in PLL and fatty acyl chains in LDL for stable complex formation
2. An increased hydrophobicity for PLL for facilitating cellular uptake

LDL was used for receptor-mediated endocytosis by the smooth muscle cells as well as hydrophobic interaction. Apart from the electrostatic interaction between the positively charged stearyl-PLL and the negatively charged pDNA, the hydrophobic interaction between stearyl group and the fatty acyl chain in the LDL was found to be the main driving force for the formation of the soluble DNA terplex system. A soluble DNA complex was formed at a wide range of concentration of pDNA, stearyl-PLL, and LDL. Gel retardation revealed that complete complexation was achieved at a 3:3:5 weight ratio of DNA:LDL:stearyl-PLL, although the transfection efficiency of the 1:1:1 derivative was found to be better. The DNA condensation by the polymeric system was confirmed by atomic force microscopy (AFM) that revealed that the size of pDNA shrank from 600 nm to 100 nm in the long axis.

Yu et al. tested the stability of the DNA terplex system *in vitro* with a serum incubation assay. The DNA terplex PK/PD studies were conducted by quantitation of terplex and radiolabeled DNA (CTP alpha-32P) complexes after rat-tail vein injection. The effect of the DNA terplex system on gene expression in mouse major organs was analyzed by measuring luciferase activities after systemic administration. The DNA terplex gene carrier was found to show significantly longer retention in the vascular space than naked pDNA alone. At early time points (1 h after venous injection), the lung was the major organ of the DNA terplex distribution, followed by the liver as a major distribution organ at later time points (24 h postinjection). The major organs of transgene expression after intravenous injection were the liver and heart. Thus it could be concluded that the DNA terplex system has the potential for *in vivo* applications due to its higher bioavailability of pDNA in the tissues, and due to its organ-specific distribution.

Affleck et al.[101] improved delivery of luciferase and beta-galactosidase genes into the hearts of rabbits and rats by complexing pDNA with stearyl-PLL LDL. It is well known that gene therapy for the treatment of cardiovascular disease is limited by inefficient methods of gene transfer. Affleck et al. demonstrated in a study utilizing DNA terplex systems for gene delivery that the system could be used to augment significantly the myocardial transfection rate in a live rabbit and rat model. The rate of transfection in the rabbit myocardium was found to be 20- to 100-fold higher than that achieved using naked pDNA. The terplex system was found to produce a more widespread uniform transfection at the injection site as demonstrated for the beta-gal histologic staining.

To improve efficiency, Affleck et al. complexed reporter genes, luciferase and beta-galactosidase, with stearyl-PLL LDL (DNA terplex) and injected either the complexed pDNA or naked pDNA as controls into the hearts of New Zealand white rabbits.

Affleck et al. examined left heart myocardial cell lysates to quantitate luciferase expression at days 3 (D3) and 30 (D30) after injection. On D3, lysates produced 44571 ± 8730 relative light units (RLU) (RLU = total light units/mg protein) in the rabbits injected with DNA terplex. Those injected with naked DNA produced 1638 ± 567 RLU ($p = 0.002$) on D3. On D30, lysates from hearts injected with DNA terplex produced 677 ± 52 RLU and those from naked-DNA treated hearts produced 18 ± 3 RLU. Histologic examination for beta-galactosidase expression revealed that DNA terplex improved the area and depth of transfection over naked DNA. Affleck and coworkers continued the experiments in Dawley rats and assessed reporter gene expression on D1, D3, D5, D10, D15, D25, and D30 after injection. They reported that luciferase production increased up to D5 and then dropped off in rats injected with naked DNA. In contrast, the DNA-terplex-induced luciferase production increased up to D30.

27.6 CONCLUSION

It is evident from the discussion of various examples in the preceding sections that the use of protein biomaterials, both synthetic and naturally occurring, for gene therapy is an area that needs more work. Protein biomaterials can offer ideal properties of a gene vector according to the design criteria mentioned earlier, and they are also amenable to modifications and derivatization that make them suitable as targeting vectors in the body.

With the advent of recombinant forms of the proteins, the inherent immunogenecity issue of protein vectors can be laid to rest. Also, the humanized or the human-derived forms of protein may alternatively be used for gene therapy. It is an exciting area that definitely warrants more interest.

REFERENCES

1. Ledley, F.D., Non-viral gene therapy. *Curr. Opin. Biotechnol.* 5, 626–636, 1994.
2. www.haverford.edu/biology/HHMI/Definition.html.
3. Fenske, B.D., MacLachlan, I., and Cullis, P.R., Long-circulating vectors for the systemic delivery of genes. *Curr. Op. Mol. Ther.* 3, 153–158, 2001.
4. Orkin, S.H. and Motulsky, A.G., Report and recommendation of the panel to assess the NIH investment in research on gene therapy, December 1995. Available from NIH at http://www.nih.gov/news/panel-rep.html.
5. Lehrman, S., Virus treatment questioned after gene therapy death, *Nature* 401, 517–518, 1999.
6. Marshall, E., Gene therapy on trial, *Science* 288, 951–956, 2000.
7. Check, E., Gene therapy: a tragic setback, *Nature* 420, 116–118, 14 November 2002.
8. Blessing, T., Remy, J.S., and Behr, J.P., Monomolecular collapse of plasmid DNA into stable virus-like particles, *Proc. Nat. Acad. Sci. U.S.A.* 95, 1427–1431, 1998.
9. Schatzlein A.G., Non-viral vectors in cancer gene therapy: principles and progress, *Anti-Cancer Drugs* 12, 275–304, 2001.
10. Truong, V.L., August, J.T., and Leong, K., Gene delivery by DNA-gelatin nanospheres, *Hum. Gene Ther.* 9, 1709–1717, 1998.
11. Truong, V.L., Walsh, S.M., Schweibert, E., Mao, H.Q., Guggino, J.T., and Leong, K.W., Gene transfer by DNA-gelatin nanospheres, *Arc. Biochem. Biophys.* 361, 47–56, 1999.
12. Leong, K.W., Mao, H.Q., Truong-Le, V.L., Roy, K., Walsh, S.M., and August, J.T., DNA-polycation nanospheres as non-viral gene delivery vehicles, *J. Control. Release* 53, 183–193, 1998.
13. Kaul, G. and Amiji, M., Long-circulating Poly(ethylene glycol)-modified gelatin nanoparticles for intracellular delivery, *Pharm. Res.* 19 (7), 1061–1067, 2002.
14. Kaul, G., Potineni, A., Lynn, D.M., Langer, R., and Amiji, M., Surface-modified polymeric nanoparticles for tumor-targeted delivery, *SurFacts Biomater.* 7, 1–6, 2002.
15. Roy, K., Mao, H.Q., Huang, S.K., and Leong, K.W., Oral gene delivery with chitosan-DNA nanoparticles generates immunologic protection in a murine model of peanut allergy, *Nat. Med.* 5, 387–391, 1999.
16. Richardson, S.C., Kolbe, H.V., and Duncan, R., Potential of low molecular mass chitosan as a DNA delivery system: biocompatibility, body distribution and ability to complex and protect DNA, *Int. J. Pharm.* 178, 231–243, 1999.

17. Aggarwal, N., HogenEsch, H., Guo, P.,. Suckow M.A. and Mittal S.K., Biodegradable alginate microspheres as a delivery system for naked DNA, *Can. J. Veter. Res.* 63, 148–152, 1999.

18. Mittal, S.K., Aggarwal, N., Sailaja, G., van Olphen, A., HogenEsch, H., North, A., Hays, J., and Moffatt, S., Immunization with DNA, adenovirus or both in biodegradable alginate microspheres: effect of route of inoculation on immune response, *Vaccine* 19, 253–263, 2000.

19. Rafati, H., Coombes, A.G.A, Adler, J., Holland, J., and Davis, S.S., Protein loaded poly(dl-lactide-co-glycolide) microparticles for oral administration: Formulation, structural and release characteristics, *J. Control. Release* 43, 89–102, 1997.

20. Hsu, S., Hao, T., and Hedley, M.L., Parameters effecting DNA encapsulation in PLGA based microspheres, *J. Drug Target.* 7, 313–323, 1999.

21. Lynn, D.M. and Langer, R., Degradable poly(β-amino esters): synthesis, characterization, and self-assembly with plasmid DNA, *J. Am. Chem. Soc.* 122, 10761–10768, 2000.

22. Lynn, D.M., Amiji, M.M., and Langer, R., pH-responsive biodegradable polymer microspheres: rapid release of encapsulated material within the range of intracellular pH, *Angew. Chem. Int. Ed.* 40, 1707–1710, 2001.

23. Harkness, R.D., Biological functions of collagen, *Biol. Rev. Camb. Philos. Soc.* 36, 399–463, 1961.

24. Bergeon, M.T., Collagen: a review, *J. Okla. State Med. Assoc.* 60 (6), 330–332, 1967.

25. Piez, K.A., History of extracellular matrix: a personal view, *Matrix Biol.* 16 (3), 85–92, 1997.

26. Harkness, R.D., Collagen, *Sci. Prog.* 54 (214), 257–274, 1966.

27. Traub, W. and Piez, K.A., The chemistry and structure of collagen, *Adv. Protein Chem.* 25, 243–352, 1971.

28. Harkness, R.D. and Nimni, M.E., Chemical and mechanical changes in the collagenous framework of skin induced by thiol compounds, *Acta Physiol. Acad. Sci. Hung.* 33 (3), 325–343, 1968.

29. Miyata, T., Taira, T., and Noishiki, Y., Collagen engineering for biomaterial use, *Clin. Mater.* 9 (3–4), 139–148, 1992.

30. Rao, K.P., Recent developments of collagen-based materials for medical applications and drug delivery systems, *J. Biomater. Sci. Polym. Ed.* 7 (7), 623–645, 1995.

31. Kohmura, E., Yuguchi, T., Yoshimine, T., Fujinaka, T., Koseki, N., Sano, A., Kishino, A., Nakayama, C., Sakaki, T., Nonaka, M., Takemoto, O., and Hayakawa, T., BDNF atelocollagen mini-pellet accelerates facial nerve regeneration, *Brain Res.* 849 (1–2), 235–238, 1999.

32. Ochiya, T., Takahama, Y., Nagahara, S., Sumita, Y., Hisada, A., Itoh, H., Nagai, Y., and Terada, M., New delivery system for plasmid DNA *in vivo* using atelocollagen as a carrier material: the Minipellet, *Nat. Med.* 5, 707–710, 1999.

33. DeLustro, F., Condell, R.A., Nguyen, M.A., and McPherson, J.M., A comparative study of the biologic and immunologic response to medical devices derived from dermal collagen, *J. Biomed. Mater. Res.* 20, 109–120, 1986.

34. Lefebvre, F., Gorecki, S, Bareille, R., Amedee, J., Bordenave, L., and Rabaud, M., New artificial connective matrix-like structure made of elastin solubilized peptides and collagens: elaboration, biochemical and structural properties, *Biomaterials* 13 (1), 28–33, 1992.

35. Rabaud, M., Lefebvre, F., and Ducassou, D., *In vitro* association of type III collagen with elastin and with its solubilized peptides, *Biomaterials* 12 (3), 313–319, 1991.

36. Doillon, C.J. and Silver, F.H., Collagen-based wound dressing: effects of hyaluronic acid and fibronectin on wound healing, *Biomaterials* 7 (1), 3–8, 1986.

37. Barbani, N., Giusti, P., Lazzeri, L., Polacco, G., and Pizzirani, G., Bioartificial materials based on collagen: 1. Collagen cross-linking with gaseous glutaraldehyde, *J. Biomater. Sci. Polym. Ed.* 7 (6), 461–469, 1995.

38. Bradley, W.G. and Wilkes, G.L., Some mechanical property considerations of reconstituted collagen for drug release supports, *Biomater. Med. Devices Artif. Organs* 5 (2), 159–175, 1977.

39. Ruderman, R.J, Bernstein, E., Kairinen, E., and Hegyeli, A.F., Scanning electron microscopic study of surface changes on biodegradable sutures, *J. Biomed. Mater. Res.* 7 (2), 215–229, 1973.

40. Petite, H., Rault, I., Huc, A., Menasche, P., and Herbage, D., Use of the acyl azide method for cross-linking collagen-rich tissues such as pericardium, *J. Biomed. Mater. Res.* 24 (2), 179–87, 1990.

41. Kanai, A., Waltman, S., Polack, F.M., and Kaufman, H.E., Electron microscopic study of hereditary corneal edema, *Invest Ophthalmol.* 10 (2), 89–99, 1971.

42. Unterman, S.R., Rootman, D.S., Hill, J.M., Parelman, J.J., Thompson, H.W., and Kaufman, H.E., Collagen shield drug delivery: therapeutic concentrations of tobramycin in the rabbit cornea and aqueous humor, *J. Cataract Refract. Surg.* 14 (5), 500–504, 1988.

43. Lucas, P.A, Syftestad, G.T., Goldberg, V.M., and Caplan, A.I., Ectopic induction of cartilage and bone by water-soluble proteins from bovine bone using a collagenous delivery vehicle, *J. Biomed. Mater. Res.* 23 (A1 Suppl.), 23–39, 1989.

44. Fonseca, M.J., Alsina, M.A., and Reig, F., Coating liposomes with collagen (Mr 50,000) increases uptake into liver, *Biochim. Biophys. Acta* 13;1279, 259–265, 1996.

45. Thacharodi, D. and Rao, K.P., Rate-controlling biopolymer membranes as transdermal delivery systems for nifedipine: development and *in vitro* evaluations, *Biomaterials* 17 (13), 1307–1311, 1996.

46. Rossler, B., Kreuter, J., and Scherer, D., Collagen microparticles: preparation and properties, *J. Microencapsul.* 12 (1), 49–57, 1995.

47. Kemp, P.D., Tissue engineering and cell-populated collagen matrices, *Methods Mol. Biol.* 139, 287–293, 2000.

48. Chvapil, M., Speer, D.P., Holubec, H., Chvapil, T.A., and King, D.H., Collagen fibers as a temporary scaffold for replacement of ACL in goats, *J. Biomed. Mater. Res.* 27 (3), 313–325, 1993.

49. Eyre, D.R., Wu, J.J., and Woolley, D.E., All three chains of 1 alpha 2 alpha 3 alpha collagen from hyaline cartilage resist human collagenase, *Biochem. Biophys. Res. Commun.* 14; 118, 724–729, 1984.

50. Woolley, D.E., Collagenolytic mechanisms in tumor cell invasion, *Cancer Metastasis Rev.* 3 (4), 361–372, 1984.

51. Mullins, R.J., Richards, C., Walker, T., and Aust, N.Z., Allergic reactions to oral, surgical and topical bovine collagen: anaphylactic risk for surgeons, *J. Ophthalmol.* 24 (3), 257–260, 1996.

52. Marty, J.J., Oppenheim, R.C., and Speiser, P., Nanoparticles: a new colloidal drug delivery system, *Pharm. Acta Helv.* 53 (1), 17–23, 1978.

53. Coester, C.J., Langer, K., Van Briesen, H., and Kreuter, J., Gelatin nanoparticles by two step desolvation—a new preparation method, surface modifications and cell uptake, *Journal of Microencapsulation,* 17, 187–193, 2000.

54. Farrugia, C.A. and Groves, M.J., Gelatin behaviour in dilute aqueous solution: designing a nanoparticulate formulation, *J. Pharm. Pharmacol.* 51 (6), 643–649, 1999.

55. Bender, A.R., von Briesen, H., Kreuter, J., Duncan, I.B., and Rubsamen-Waigmann, H., Efficiency of nanoparticles as a carrier system for antiviral agents in human immunodeficiency virus-infected human monocytes/macrophages *in vitro*, *Antimicrob. Agents Chemother.* 40 (6), 1467–1471, 1996.

56. Berthold, A., Cremer, K., and Kreuter, J., Collagen microparticles: carriers for glucocorticosteroids, *Eur. J. Pharm. Biopharm.* 45 (1), 23–29, 1998.

57. Murata, M., Maki, F., Sato, D., Shibata, T., and Arisue, M., Bone augmentation by onlay implant using recombinant human BMP-2 and collagen on adult rat skull without periosteum, *Clin. Oral Implants Res.* 11 (4), 289–295, 2000.

58. Nakagawa, T. and Tagawa, T., Ultrastructural study of direct bone formation induced by BMPs-collagen complex implanted into an ectopic site, *Oral Dis.* 6 (3), 172–179, 2000.

59. Hortelano, G., Al-Hendy, A., Ofosu, F.A., and Chang, P.L., Delivery of human factor IX in mice by encapsulated recombinant myoblasts: a novel approach towards allogeneic gene therapy of hemophilia B, *Blood* 15, 5095–5103, 1996.

60. Al-Hendy, A., Hortelano, G., and Tannenbaum, G.S., Growth retardation: an unexpected outcome from growth hormone gene therapy in normal mice with microencapsulated myoblasts, *Chan. Hum. Gene Ther.* 7 (1), 61–70, 1996.

61. Barr, E. and Leiden, J.M., Systemic delivery of recombinant proteins by genetically modified myoblasts, *Science* 6;254 (5037), 1507–1509, 1991.

62. Heartlein, M.W., Roman, V.A., Jiang, J.L., Sellers, J.W., Zuliani, A.M., Treco, D.A., and Selden, R.F., Long-term production and delivery of human growth hormone *in vivo*, *Proc. Nat. Acad. Sci. U.S.A.* 91 (23), 10967–10971, 1994.

63. Rosenthal, F.M. and Kohler, G., Collagen as matrix for neo-organ formation by gene-transfected fibroblasts, *Anticancer Res.* 17 (2A), 1179–1186, 1997.

64. Park, J.C., Hwang, Y.S., Lee, J.E., Park, K.D., Matsumura, K., Hyon, S.H., and Suh, H., Type I atelocollagen grafting onto ozone-treated polyurethane films: cell attachment, proliferation, and collagen synthesis, *J. Biomed. Mater. Res.* 52, 669–677, 2000.

65. Park, J.C., Han, D.W., and Suh, H.A., Bone replaceable artificial bone substitute: morphological and physiochemical characterizations, *Yonsei Med. J.* 41 (4), 468–476, 2000.
66. Moullier, P., Marechal, V., Danos, O., and Heard, J.M., Continuous systemic secretion of a lysosomal enzyme by genetically modified mouse skin fibroblasts, *Transplantation* 56 (2), 427–432, 1993.
67. Moullier, P., Bohl, D., Heard, J.M., and Danos, O., Correction of lysosomal storage in the liver and spleen of MPS VII mice by implantation of genetically modified skin fibroblasts, *Nat. Genet.* 4 (2), 154–159, 1993.
68. Wedge, C.I. and Rootman, D.S., Collagen shields: efficacy, safety and comfort in the treatment of human traumatic corneal abrasion and effect on vision in healthy eyes, *Can. J. Ophthalmol.* 27 (6), 295–298, 1992.
69. Shaker, G.J., Ueda, S., LoCascio, J.A., and Aquavella, J.V., Effect of a collagen shield on cat corneal epithelial wound healing, *Invest. Ophthalmol. Vis. Sci.* 30 (7), 1565–1568, 1989.
70. Marmer, R.H., Therapeutic and protective properties of the corneal collagen shield, *J. Cataract Refract. Surg.* 14 (5), 496–499, 1988.
71. Angella, G.J., Sherwood, M.B., Balasubramanian, L., Doyle, J.W., Smith, M.F., van Setten, G., Goldstein, M., and Schultz, G.S., Enhanced short-term plasmid transfection of filtration surgery tissues, *Invest. Ophthalmol. Vis. Sci.* 41 (13), 4158–4162, 2000.
72. Lee, C.H., Singla, A., and Lee, Y., Biomedical applications of collagen, *Int. J. Pharm.* 221 (1–2), 1–22, 2001.
73. Kimura, M., Zhao, M., Zellin, G., and Linde, A., Bone-inductive efficacy of recombinant human bone morphogenetic protein-2 expressed in Escherichia coli: an experimental study in rat mandibular defects, *Scand. J. Plast. Reconstr. Surg. Hand Surg.* 34 (4), 289–299, 2000.
74. Cochran, D.L., Jones, A.A., Lilly, L.C., Fiorellini, J.P., and Howell, H.J., Evaluation of recombinant human bone morphogenetic protein-2 in oral applications including the use of endosseous implants: 3-year results of a pilot study in humans, *Journal of Periodontology*, 71 (8), 1241–1257, 2000.
75. Ferkol, T., Kaetzel, C.S., and Davis, P.B., Gene transfer into respiratory epithelial cells by targeting the polymeric immunoglobulinreceptor, *J. Clin. Invest.* 92 (5), 2394–2400, 1993.
76. Katsube, K., Ochi, M., Uchio, Y., Maniwa, S., Matsusaki, M., Tobita, M., and Iwasa, J., Repair of articular cartilage defects with cultured chondrocytes in Atelocollagen gel. Comparison with cultured chondrocytes in suspension, *Arch. Orthop. Trauma Surg.* 120 (3–4), 121–127, 2000.
77. Kuriwaka, M., Ochi, M., Uchio, Y., Maniwa, S., Adachi, N., Mori, R., Kawasaki, K., and Kataoka, H., Optimum combination of monolayer and three-dimensional cultures for cartilage-like tissue engineering, *Tissue Eng.* 9 (1), 41–49, 2003.
78. Uchio, Y., Ochi, M., Matsusaki, M., Kurioka, H., and Katsube, K., Human chondrocyte proliferation and matrix synthesis cultured in Atelocollagen gel, *J. Biomed. Mater. Res.* 50 (2), 138–143, 2000.
79. Ward, A.G. and Courts, A., *The Science and Technology of Gelatin*, Academic Press, New York, 1977.
80. Veis, A., *The Macromolecular Chemistry of Gelatin*, Academic Press, New York, 1964.
81. Truong, V.L., August, J.T., and Leong, K., Gene delivery by DNA-gelatin nanospheres, *Hum. Gene Ther.* 9, 1709–1717, 1998.
82. Truong, V.L., Walsh, S.M., Schweibert, E., Mao, H.Q., Guggino, J.T., and Leong, K.W., Gene transfer by DNA-gelatin nanospheres, *Arc. Biochem. Biophys.* 361, 47–56, 1999.
83. Leong, K.W., Mao, H.Q., Truong-Le, V.L., Roy, K., Walsh, S.M., and August, J.T., DNA-polycation nanospheres as non-viral gene delivery vehicles, *J. Control. Release* 53, 183–193, 1998.
84. Coester, C., Kreuter, J., von Briesen, H., and Langer, K., Preparation of avidin-labelled gelatin nanoparticles as carriers for biotinylated peptide nucleic acid (PNA), *Int. J. Pharm.* 196, 147–149, 2000.
85. Kaul, G. and Amiji, M., Long-circulating Poly(ethylene glycol)-modified gelatin nanoparticles for intracellular delivery, *Pharm. Res.* 19 (7), 1061–1067, 2002.
86. Kaul, G., Potineni, A., Lynn, D.M., Langer, R., and Amiji, M., Surface-modified polymeric nanoparticles for tumor-targeted delivery, *SurFacts Biomater.* 7, 1–6, 2002.
87. Kaul, G., Lee-Parsons, C., and Amiji, M., Poly(ethylene glycol)-modified gelatin nanoparticles for intracellular delivery, *Pharm. Eng.* 111–116, Sept.-Oct. 2003.
88. Peters, T., *All about Albumin: Biochemistry, Genetics and Medical Applications*, Academic Press, San Diego, 1996.

89. Rhaese, S., von Briesen, H., Rubsamen-Waigmann, H., Kreuter, J., and Langer, K., Human serum albumin-polyethylenimine nanoparticles for gene delivery, *J. Control. Release* 92, 199–208, 2003.

90. Shohet, R.V., Chen, S., Zhou, Y.T., Wang, Z., Meidell, R.S., Unger, R.H., and Grayburn, P.A., Echocardiographic destruction of albumin microbubbles directs gene delivery to the myocardium, *Circulation* 6;101 (22), 2554–2556, 2000.

91. Chen, S., Kroll, M.H., Shohet, R.V., Frenkel, P., Mayer S.A., and Grayburn, P.A., Bioeffects of myocardial contrast microbubble destruction by echocardiography, *Echocardiography* 19 (6), 495–500, 2002.

92. Orson, F.M., Song, L., Gautam, A., Densmore, C.L., Bhogal, B.S., and Kinsey, B.M., Gene delivery to the lung using protein/polyethylenimine/plasmid complexes, *Gene Ther.* 9 (7), 463–471, 2002.

93. Orson, F.M., Kinsey, B.M., Bhogal, B.S., Song, L., Densmore, C.L., and Barry, M.A., Targeted delivery of expression plasmids to the lung via macroaggregated polyethylenimine-albumin conjugates, *Methods Mol. Med.* 75, 575–590, 2003.

94. Weber, C., Reiss, S., and Langer, K, Preparation of surface modified protein nanoparticles by introduction of sulfhydryl groups, *Int. J. Pharm.* 211, 67–78, 2000.

95. Roser, M., Fischer, D., and Kissel, T., Surface-modified biodegradable albumin nano- and microspheres. II: effect of surface charges on *in vitro* phagocytosis and biodistribution in rats, *Eur. J. Pharm. Biopharm.* 46 (3), 255–263, 1998.

96. Pan, G., Shawer, M., Oie, S., and Lu, D.R., *In vitro* gene transfection in human glioma cells using a novel and less cytotoxic artificial lipoprotein delivery system, *Pharm. Res.* 20 (5), 738–744, 2003.

97. Lee, H., Kim, T.H., and Park, T.G., A receptor-mediated gene delivery system using streptavidin and biotin-derivatized, pegylated epidermal growth factor, *J. Control. Release* 83, 109–119, 2002.

98. Kim, J.S., Maruyama, A., Akaike, T., and Kim, S.W., Terplex DNA delivery system as a gene carrier, *Pharm. Res.* 15 (1), 116–121, 1998.

99. Kim, J.S., Kim, B.I., Maruyama, A., Akaike, T., and Kim, S.W., A new non-viral DNA delivery vector: the terplex system, *J. Control. Release* 53 (1–3), 175–182, 1998.

100. Yu, L., Suh, H., Koh, J.J., and Kim, S.W., Systemic administration of TerplexDNA system: pharmacokinetics and gene expression, *Pharm. Res.* 18 (9), 1277–1283, 2001.

101. Affleck, D.G., Yu, L., Bull, D.A., Bailey, S.H., and Kim S.W., Augmentation of myocardial transfection using TerplexDNA: a novel gene delivery system, *Gene Ther.* 8 (5), 349–353, 2001.

B. Polymeric Microspheres

CHAPTER 28

Gene Delivery Using Poly(lactide-*co*-glycolide) Microspheres

Mary Lynne Hedley

CONTENTS

28.1 INTRODUCTION

Efforts to create effective DNA-based therapeutics have been hampered by well-known factors, including the body's defense system, which serves to inactivate them. Poly(lactide-*co*-glycolide) (PLG)-based delivery systems have inherent characteristics that offer promise in this field of research. This chapter will focus on information related to DNA expression vectors formulated

0-8493-1934-X/05/$0.00+$1.50
© 2005 by CRC Press LLC

with PLG microparticles. These formulations can protect DNA from nuclease-based attack, facilitate DNA localization to cells and tissues of interest, and promote gene expression; furthermore, they have the potential to provide for controlled release. Safety and the application of PLG formulations to clinical medicine have been proven by the commercial success of PLG depot systems that allow for slow release of peptides (LHRH agonists, Lupron Depot®, Zoladex®) and proteins (human growth hormone, Nutropin Depot®). The following sections provide an overview of key characteristics of PLG–DNA formulations, processes for their manufacture, and their biological activity.

28.2 ADVANTAGES OF PLG MICROPARTICLES FOR GENE DELIVERY

28.2.1 Protection of DNA

The interstitial and intracellular barriers to nonviral DNA delivery are recognized as significant hurdles that must be overcome in order for DNA-based medicines to achieve clinical success.[1] These barriers include the natural response of a mammalian host to destroy foreign, naked DNA (not associated with a delivery system), and the need for therapeutic DNA to be internalized by cells. Interstitial nucleases are remarkably effective and result in the degradation of 99% of exogenous, naked DNA within approximately 90 min of delivery.[2] Encapsulation of DNA in PLG microparticles can protect nucleic acid from nuclease-based degradation *in vitro* (Figure 28.1); and it is likely that protection from interstitial nucleases is also afforded to DNA *in vivo* as long as it remains associated with microparticles. Rapid cellular uptake of nucleic acid formulated in small microparticles, less than 10 µm in diameter, also aids in the escape from interstitial nuclease-mediated destruction. The characteristics of small PLG particles have focused the attention of researchers on formulation of the particles with genetic expression vectors; and experimentation with the particles has led to significant progress in overcoming some of the barriers to clinical success so far observed with DNA-based medicine.

28.2.2 Biocompatibility and Biodegradability

PLG degrades by a process of hydrolysis in which water molecules promote a nucleophilic attack on the chemical bond between glycolide and lactide monomers. This results in bond breakage and the subsequent shortening of the polymer strands.[3] Over time, degradation compromises the integrity of the PLG microparticle and individual monomers of lactic and glycolic acid are released until the polymer strand is destroyed. Information collected from several animal studies suggests

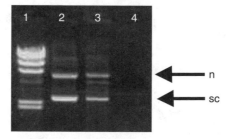

Figure 28.1 Encapsulated DNA is protected from nuclease damage. Particles containing encapsulated DNA, or particles mixed with naked DNA, were incubated in the presence of DNAse for 1 h at 37° C. DNAse was then removed from the system and the treated DNA was analyzed by gel electrophoresis. Lanes: (1) markers; (2) input DNA not exposed to DNAse; (3) DNA extracted from particles that had been treated with DNAse; (4) DNA that had been mixed with particles and incubated with DNAse. Arrows indicate supercoiled (sc) and nicked DNA (n).

that both the polymer and its degradation products are well-tolerated.[4-8] The tissue response following an intramuscular injection of PLG microparticles (> 10 μm) proceeds through a well-defined continuum, which includes an acute inflammatory response followed by a foreign body reaction. The local response resolves as PLG begins to degrade — a process that is dependent on many factors, including drug loading, drug characteristics, PLG molecular weight, L:G ratio, and the tissue type in which the formulation resides.[8,9] Particulates that are produced as a consequence of degradation are phagocytosed by macrophages, wherein breakdown of the polymer into lactic and glycolic acid monomers continues.

28.2.3 Cell Targeting

Targeting of the PLG microparticles to certain cell populations can easily be directed by the administration of formulations within defined size ranges that include particles from hundreds of nanometers to microns in diameter. Larger particles (> 10 μm) remain localized at the site of injection, as they cannot be engulfed by phagocytic cells, and release the drug in a time-dependent fashion. In contrast, particles less than 10 μm in diameter localize to tissues in a manner dictated by injection route, size, and surface characteristics (Figure 28.2).[10-12] Particles ranging from 1 to 10 μm in diameter are phagocytosed by specialized cells, whereas nanoparticles are internalized by many cell types through a fluid-based process known as pinocytosis.

Depending on the route of injection, cell-based internalization can occur at the site of administration, in organs of the reticuloendothelial system, or in the lymphatic tissue.[13-17] The release of entrapped material occurs within phagocytic cells as the PLG degrades, the rate of which is dependent on the PLG formulation. For example, macrophage uptake of microparticles, composed of a 50:50 L:G PLG polymer, with a molecular weight of 6,000 or 60,000 kD, results in drug release to the cytoplasm for at least 7 days, and the rate of drug release is demonstrably faster from the 6,000 kD polymer.[18] In another example, particles composed of 50:50 L:G with a molecular weight of 3,000 kD degrade entirely within 7 days of internalization by macrophages,[11,19] thereby releasing all entrapped drug by the end of this time period.

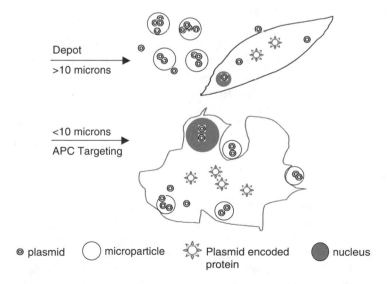

Figure 28.2 Depot versus cell targeting. DNA encapsulated in particles greater than10 μm in diameter will act as a depot system, releasing plasmid over time. DNA is taken up by cells, traffics to the nucleus, and is expressed. Smaller particles (< 10 μm) are taken up by cells (e.g., antigen presenting cells, APC). DNA is released from the microparticle, enters the nucleus, and is expressed.

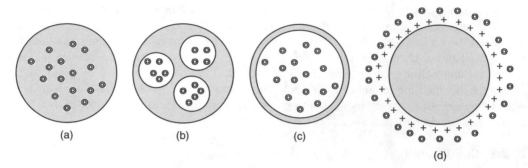

Figure 28.3 PLG–plasmid DNA formulations. DNA is encapsulated in PLG particles and can be embedded within the PLG polymeric matrix (A), or can be located in hollow cavities (B) or hollow cores (C) that are surrounded by a polymeric matrix composed of PLG. DNA can also be adsorbed to the surface of PLG microparticles that are coated with a cationic substance (D).

The facts that particles are readily phagocytosed by cells and that the drug can be released inside cells support the usefulness of this approach for DNA delivery.

28.3 GENERATION OF DNA/PLG MICROPARTICLE FORMULATIONS

At least two approaches for formulating DNA in the context of PLG microparticles have been reported. The first requires encapsulation of DNA inside microparticles so that DNA is embedded in the polymer matrix of a solid particle (Figure 28.3A) or is contained within hollow cavities (Figure 28.3B) or the core of microparticles (Figure 28.3C). The second approach requires formation of empty particles and the subsequent coating of the external particle surface with DNA (Figure 28.3D). Both approaches have been successfully utilized for creating DNA–PLG particle formulations that enable gene expression and result in a biological response *in vivo*.

28.3.1 Encapsulation of Gene Expression Vectors

The most common methods for the encapsulation of water-soluble molecules in PLG particles include, but are not limited to, spray drying and the water-in-oil-in-water double emulsion technique. Both procedures require the use of volatile organic solvents and significant shear forces. DNA is susceptible to the damage that can result from these processing parameters, and for this reason it can be a challenge to retain biological activity and preserve the supercoiling characteristic that defines the most stable isoform of DNA. Despite this, careful attention to processing parameters has resulted in effective processes for the encapsulation of biologically active DNAs.

28.3.1.1 Double Emulsion Technique

Double emulsions are formed by mixing an aqueous solution of nucleic acid with a solution of PLG dissolved in an organic solvent. Mixing is most often performed by sonication or homogenization and results in the formation of an emulsion in which aqueous droplets containing DNA are suspended in the oily organic phase (Figure 28.4A). A balance must be maintained between the intensity of the shear force required to create droplets of the desired size in the primary emulsion, and the minimization of shear, which is necessary to retain DNA supercoiling.[20,21] Several mechanisms can be utilized to maintain this balance. DNA can be condensed with poly(L-lysine) to reduce its Stoke's radius and, therefore, make it less susceptible to shear forces.[22-24] In addition, solutions can be used with reduced temperatures to create a more viscous emulsion that is less affected by the shear forces during creation of the DNA–PLG emulsion.[25] The primary emulsion

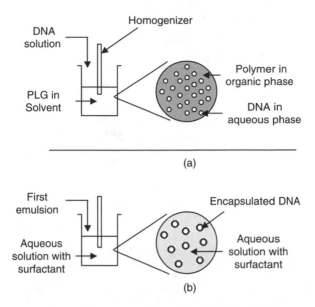

Figure 28.4 Double emulsion method for encapsulating plasmid DNA. (A) An aqueous solution of DNA is prepared and emulsified with a solution of PLG dissolved in an organic solvent. (B) The first emulsion is then homogenized with an aqueous solution containing a surfactant to create a second emulsion.

is mixed with a second aqueous solution that contains a surfactant to prevent the aqueous droplets from coalescing. This creates a secondary emulsion wherein polymer precipitates around the aqueous droplets to create nascent particles (Figure 28.4B). Organic solvent is removed from the particles by extraction or evaporation. Since the polymer degrades through hydrolysis, the resulting particles are often lyophilized or freeze-dried to generate final particles with very low moisture content for long-term storage. Cryoprotective sugars can be added to the formulation to maintain DNA supercoiling during drying or lyophilization.[25] The double emulsion technique can be used to encapsulate single or multiple plasmids in particles less than 10 μm in diameter in a manner that preserves up to 50–90% of the plasmid in a supercoiled form.[20,25-30]

Low encapsulation efficiencies of water-soluble drugs, such as DNA, are often observed as a result of encapsulation with the double emulsion process. This is due, in part, to the escape of DNA from primary droplets during the creation of the second emulsion, and during removal of the organic solvent. Mechanisms by which this loss can be reduced include initiating the process with DNA that is condensed,[23,24,31] reducing the water solubility of DNA by neutralizing negative charge, reducing temperature,[25] and modifying the volume ratio of the oily to the aqueous phases. Such modifications to the basic double emulsion procedure can encapsulate DNA with efficiencies that range from 50–70% and result in DNA loadings that range from 0.5–1% on a weight/weight basis. Encapsulated DNA produced by a water-in-oil-in-water double emulsion process remains stable in a supercoiled isoform for at least 2.5 years when stored as a lyophilized powder at either 4°C or –20°C.

28.3.1.2 Spray Drying Technique

The spray drying technique can combine emulsification and spraying steps to form final particles with entrapped DNA. In an example of this process, an aqueous solution of DNA is emulsified with an organic solvent containing PLG, while controlling shear forces to maintain supercoiling. Alternatively, a powdered form of DNA is suspended in the organic solvent–PLG solution. The DNA–PLG mixture is sprayed through an atomizer in the presence of heated air. The heat evaporates

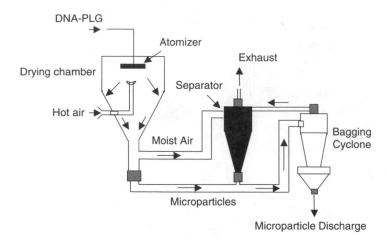

Figure 28.5 Spray drying method for encapsulating plasmid DNA. A mixture of DNA and PLG dissolved in organic solvent is introduced into the spray dryer apparatus. The mixture is atomized to produce fine droplets that are exposed to heated air. The particles dry in the heated air to create a powder. The powder and moist air are transferred from the drying chamber to a cyclone separator where the two are separated using centrifugal action that is caused by increasing the air speed when the particles and air enter the cyclone unit. The dense particles settle to the bottom of the cyclone and are transferred to a closed container for subsequent fill and finish procedures. Alternatively, the particles can be conveyed from the first cyclone unit to a bagging cyclone by a stream of cold air. From this unit the particles are discharged into a closed container for subsequent fill and finish procedures.

the organic solvent and any water in the system to result in the formation of PLG particles and the encapsulation of DNA (Figure 28.5).[32] The flow rate into the atomizer and the air temperature in part determine the size of the particles that form during the process. Emulsion characteristics, such as the size of the aqueous droplets, can also contribute to particle size. The particles are separated from any moist air, by a centrifugal force created by the increased air speed present in the cyclone. The dense microparticles settle to the bottom of the cyclone where they can be collected. If desired, the particles can be cooled and transported to a bagging cyclone by a stream of cool air (Figure 28.5). Once produced, the particles are transferred to a container and the free-flowing powder is filled into vials. This process avoids the loss of DNA into the second aqueous phase that can be problematic in the double emulsion technique, thereby potentially leading to higher encapsulation efficiencies. The method is also advantageous from a manufacturing perspective as it is easily amenable to scaleup and the production of kilogram quantities of material.

28.3.2 DNA-Coated PLG Particles

PLG particles that are externally coated with DNA can be made by standard methods including the double emulsion technique. However, unlike the previously described processes, particles are produced in the absence of DNA to create empty microspheres. When the double emulsion technique is employed, the second aqueous phase of the double emulsion contains a cationic agent that actively coats the empty particles as they are formed. During incubation with a solution of nucleic acid, the positively charged surface of the empty particles promotes DNA binding to the external surface via electrostatic interactions. This process typically creates particles with 1% DNA (weight/weight) loadings, and can be used to load multiple plasmids onto a population of particles.[33]

Each of the methods described above produces DNA–PLG formulations that have demonstrated activity in research-based experiments. In addition, the biological activity of particles produced using the double emulsion technique has been demonstrated in preclinical and clinical studies.

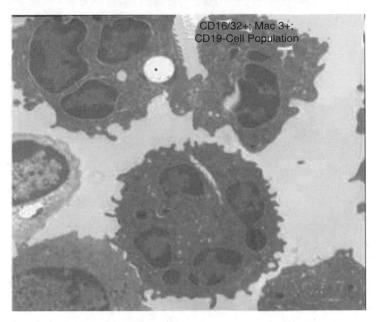

Figure 28.6   APC ingested microparticles containing DNA. PLG microparticles containing DNA (mean diameter 2 μm) were injected into the peritoneum of C57/Bl6 mice. The cell infiltrate was removed and examined by electron microscopy for the presence of internalized particles, which appear as dense white circles. The cells were phenotyped for cell surface markers by analysis on a fluorescence activated cell sorter, and the resulting phenotype (CD16/32+; Mac3+; CD19-) indicates that the cells are macrophages.

28.4 PARTICLE INTERNALIZATION AND GENE EXPRESSION

PLG–plasmid microparticle formulations are internalized by phagocytic cells such as macrophages (Figure 28.6)[26,31,34,35] and dendritic cells.[16,34,35] Several types of nonphagocytic cells also internalize PLG–DNA microparticles at least under certain tissue culture conditions and at high particle to cell ratios.[36] As a result of cellular uptake, particles are transported to endosomal compartments possessing low pH environments that are detrimental to DNA integrity. The particle, or the encapsulated DNA, must escape from these vesicles to allow for gene expression. Some studies have suggested that PLG becomes protonated in the acidic environment of the phagolysosome. Protonation leads to interaction with and disruption of the phagolysosomal membrane, and the subsequent expulsion of the particle into the cytoplasm.[37-40] As the PLG degrades over time, DNA is released into the cytoplasm[35,41] and the free DNA migrates into the nucleus to initiate gene expression.[41] It is unclear whether DNA-coated particles release DNA with a similar type mechanism, or if the cationic coating on the particle promotes interaction with the endosomal membrane to result in particle or DNA release into the cytoplasm.

The length of time particles reside in cells and continue to release active DNA is a function of the particle's degradation profile; however, perhaps more critical are the effects of the low pH environment in the phagolysosome and the lactic and glycolic acid degradation products of PLG, all of which are damaging to DNA. Encapsulated or coated DNA must survive each of these effects to remain active. The integrity of the DNA will be compromised in cases where the particles remain in the phagolysosome for too long, or if degradation of the particle is prolonged. However, DNA release is likely to continue even under these circumstances, although DNA damage may be so great that biological activity is ultimately lost.

Data from several reports suggest that *in vitro* uptake of particles containing plasmid DNA results in expression of exogenous DNA.[26,31,34-36,39,41-44] DNA expression is also detected following

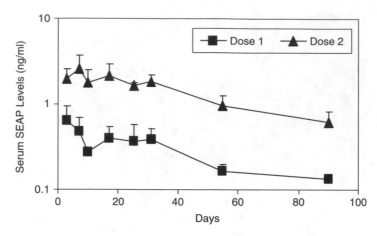

Figure 28.7 Expression of DNA following IM injection of DNA encapsulated in PLG microparticles. A mammalian expression vector encoding secreted embryonic alkaline phosphatase (SEAP) was encapsulated in PLG microparticles. The particles were suspended in saline and injected into the tibialis
and hamstring muscles of C57/Bl6 mice at two different DNA doses (dose 1:50 μg ■; dose 2:100
μg ▲). Serum was collected at different days after injection (x-axis) and assayed for secreted
bioactive SEAP (y-axis).

in vivo administration of PLG particles formulated with DNA. Intramuscular (IM) injection of
particles coated with plasmid DNA resulted in gene expression at the site of injection for up to 14
days post administration. The levels of expression were higher than those seen with naked DNA
administration.[45] This is in contrast to observations made following intramuscular injection of rats
with encapsulated DNA, where in some cases lower levels of protein expression were observed in
comparison to naked DNA.[30,41] However, in these studies, expression resulting from a bolus administration of naked DNA is compared to that following administration of an identical amount of
encapsulated DNA. The majority of encapsulated DNA is not immediately biologically available,
but rather becomes available over time as it is released from particles. Thus, particle delivery of
encapsulated DNA may prolong the bioavailability of DNA and result in expression for a longer
time period.[30,41] Following a single IM injection of mice with encapsulated DNA encoding the
secreted protein, secreted embryonic alkaline phosphatase (SEAP), SEAP protein was detected in
the sera from 3 to 90 days after injection (Figure 28.7), whereas injection of naked DNA is reported
to typically result in a loss of gene expression within 20 days.

The localization of gene expression following injection of particle-associated DNA may differ
from that of naked DNA. A single IM injection of 30 μg encapsulated DNA resulted in transgene
expression in the liver, lymph node, spleen, and muscle at 15 days after injection as detected by
RT-PCR, whereas three IM injections of 100 μg of the same naked DNA led to expression that
was restricted to liver and muscle tissues.[46]

Data supporting the expression of DNA delivered in PLG formulations by routes other than
IM injection have also been reported. Oral administration of encapsulated plasmid resulted in
intestinal targeting and gene expression in the small and large intestines of mice up to 7 days after
the last administration.[47] Protein expression in cells from the draining lymph nodes and spleens of
mice could be detected as early as day 1 and continued for at least 7 days following intranasal
delivery of DNA coated particles.[45] The demonstration that PLG-associated DNA is internalized
and expressed by phagocytic cells provided a rationale for studying the ability of such formulations
to promote immune responses to DNA encoded antigens.

28.5 IMMUNE ACTIVATION

Certain phagocytic cells, such as macrophages and dendritic cells, have the capacity to be professional antigen-presenting cells (APCs). These cell types internalize and digest foreign material, and present the resultant antigenic peptides on the cell surface in the context of an HLA molecule. A naïve T-cell recognizes this complex and an ensuing cross-talk between activated APCs and T-cells occurs via interaction of membrane-associated costimulatory molecules. This interaction results in the production of cytokines that further promote T-cell differentiation and activity. In the absence of costimulatory signals, antigen presentation can result in T-cell inactivation, tolerance, or depletion.[48] Once activated, the T-cells can provide help to activate B-cell and antibody responses and can act as effectors themselves (Figure 28.8). T-effectors will home to relevant sites within the body where an infectious organism or transformed cell population resides. Here, they will have the expected effect of eliminating tumor cells or those cells harboring disease-causing agents.

Work in the early 1990s successfully showed the usefulness of PLG microparticle-based protein and peptide antigen formulations for effective stimulation of immune responses. In mouse studies, Eldridge et al. demonstrated that PLG particles of less than 10 μm target Peyer's patches, gut-associated lymphatic tissues rich in lymphocytes and APCs, following oral administration,[14,15] and that immunization of mice with PLG-encapsulated polypeptides activates the immune response to produce antibodies specific to the encapsulated antigen.[49,50] Other routes of administration also target PLG particles to APCs. Intraperitoneal (IP) injection of PLG particles promotes uptake by CD14+ macrophages, whereas intradermal injections result in uptake by CD86+ dendritic cells.[17] Data from *in vivo* studies performed with 0.5–1.0 μm particles demonstrated that the particles are phagocytosed by immature monocytes, which then differentiate into mature dendritic cells, the most potent APC, and traffic to draining lymph nodes.[51]

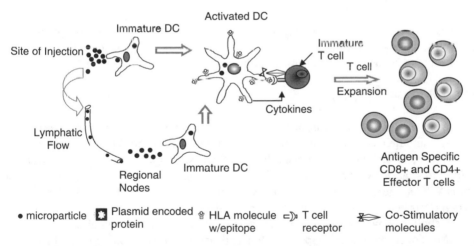

Figure 28.8 Antigen presentation and T-cell activation. Particles or other antigens are internalized by immature APC (e.g., dendritic cells [DC]) at the site of injection. Particles, or cells with internalized particles, travel through the lymphatics to the draining lymph nodes where free particles are phagocytosed by immature APC. The ingested particles release DNA, which is ultimately expressed. The exogenously expressed protein is digested into peptides by the proteasome and these peptides are presented on the surface of the now activated APC in the context of an HLA molecule. Immature T-cells recognize this complex by virtue of the T-cell receptor. Costimulatory signals for T-cells include nonspecific cytokines and the interactions between cell surface molecules. Such signals drive the differentiation and proliferation of naïve T-cells. The activated effector cells include (1) CD4+ T-helper cells, which promote the activity of CD8+ T-cells and B-cells; and (2) CD8+ T-cells, which eliminate cells expressing the antigen encoded by the transgene, including transformed or virally infected cells.

28.5.1 Immune Responses Elicited By PLG–Plasmid Formulations

The ability to target antigen expression plasmids directly to APCs formed the foundation for continued work with microparticle based PLG–DNA formulations as potential immuno-stimulating agents. Although less data have been reported regarding the specifics of PLG–DNA cell-mediated uptake *in vivo*, findings are not expected to vary significantly from what is known for formulations containing protein or peptide antigens. Several groups have successfully shown the internalization of PLG–DNA particles by various APCs. Intravenous (IV) administration results in uptake of particles by splenic dendritic cells[16] and IP injection promotes uptake by macrophages (Figure 28.6). In *in vitro* experiments, dendritic cells and bovine macrophages internalize PLG–DNA formulations[31,34,35] and express the exogeneous DNA.[31,34] Protein expression in CD11b+ and CD11c+ dendritic cells, known to be potent APCs, was detected following intranasal administration of DNA-coated particles.[45] In an elegant series of *in vitro* experiments, it was demonstrated that APCs ingest particles coated with DNA, express the DNA, and present the encoded protein to antigen-specific T-cells.[34] That the immune response generated by DNA-coated particles is likely to be dependent on cellular uptake of PLG formulated DNA is based on the finding that particles too large for phagocytosis (> 30 μm) were not effective in eliciting antibody responses.[52] Interestingly, the same study showed that 300 nm particles, which can be internalized by phagocytic as well as nonphagocytic cell types, were more effective than 1 μm particles at eliciting antibody responses to DNA-encoded antigen. Together, these data support the hypothesis the PLG formulation enhances cellular uptake of DNA, promotes gene expression, and enables antigen presentation by APCs, factors that could improve the potency of DNA vaccines.

28.5.2 Delivery Systems to Elicit Immune Responses with PLG–Plasmid Particle Formulations

The successes of *in vitro* experiments with PLG–DNA formulations have led to increased testing of these formulations in animal studies. Administration of PLG-associated DNA via oral[53-55] or parenteral routes[26,46] has now been shown to elicit both humoral and cellular immune responses.

28.5.2.1 Mucosal Delivery

Oral immunization with encapsulated DNA induces significant levels of serum antibodies, including IgA and IgG isotypes, which are specific for the DNA encoded protein. Oral immunization of mice with encapsulated plasmid also elicits antigen specific mucosal IgA.[53,55] High levels of serum and intestinal IgA were observed following administration of encapsulated DNA encoding a rotavirus VP6 antigen, or coencapsulated DNAs encoding rotavirus VP4 and VP7 antigens. These studies demonstrated for the first time that oral delivery of DNA encoded antigen could protect mice from challenge with rotavirus, as determined by a reduction in viral shedding after challenge.[27,54] Moreover, these data support the concept that plasmids encoding more than one antigen can be delivered in a single formulation, with the expectation that immunity to both proteins will be generated.[27]

In addition to mucosal IgA, oral vaccination with encapsulated DNA encoding HIV gp160 induced cytotoxic T-cells in spleen, Peyer's patches, and the lamina propia of immunized mice. The immune response was robust and able to overcome a rectal challenge with a recombinant vaccinia virus expressing gp160.[47] Priming the immune response to gp160 by oral immunization with encapsulated DNA, followed by boosting via a mucosal route with vaccinia virus recombinants expressing the same antigen, further enhanced the immune response as determined by increased IgA and splenic CTL responses.[56] Particles coated with DNA are also effective immunogens when administered by the mucosal route. Activated B- and T-cells specific for the DNA-encoded antigen have been found in the lymph nodes and spleen following nasal administration of such particles.[45] In summary, data collected to date on the use of particles to stimulate mucosal immunity demonstrate that oral, nasal, and rectal routes are suitable for administration of particle–DNA formulations.

28.5.2.2 Parenteral Delivery

Parenteral administration of encapsulated DNA[26,46,57] or DNA-coated PLG microparticles[45,52,58,59] is effective at stimulating systemic B-cell and T-cell responses. Mice immunized with minigenes encoding individual T-cell epitopes by either IM, subcutaneous, or IP routes, generate cytotoxic T-cell responses specific for the encoded epitopes.[26] In this study, as little as 2 μg of encapsulated DNA was as effective as 200 μg of naked DNA in eliciting T-cell responses. Administration of encapsulated DNA formulations encoding reporter antigens by IM or IV routes was shown to promote antibody responses, T-cell proliferation and T-cell-mediated release of gamma interferon.[57] Data from this report also demonstrated that a single immunization with encapsulated DNA encoding a tumor antigen was able to protect mice from a subsequent tumor challenge, whereas all control mice developed tumor nodules (Figure 28.9).[57]

At low doses (1 μg), immunization of mice with PLG microparticles, coated with DNA encoding HIV antigen, enhances the immune response approximately 250-fold over that achieved with naked DNA. At higher doses, the degree of enhancement is not sustained, but the PLG formulation continues to produce better responses than the injection of naked DNA.[52] Immunization of guinea pigs with populations of particles containing two adsorbed plasmids, encoding HIV gp140 and HIV gag, resulted in high antibody titers for both antigens.[33] Although markedly more DNA (1 mg HPV gp140 and 0.5 mg HIV gag) than was necessary to elicit responses to either DNA encoded antigen individually (50 μg) was required to see the effect, the data certainly support the utility of multiple plasmid formulations.

28.5.2.3 Adjuvants for Microparticle-Based Immunization

Adjuvants, cytokines, and lipids can be added to PLG particle formulations or delivered with them to further enhance immune responses. Trehalose dimicolate (TDM) promotes the proliferation of NK cells, release of gamma interferon, and the activation of macrophages.[60] The inclusion of TDM in DNA-particle formulations boosts immunity as measured by interferon gamma release and IgG2a antibody responses. Interferon levels produced by splenic T-cells are significantly higher when the TDM adjuvant is incorporated in the particles than when it is not present. Particles without TDM (30 μg DNA) elicit responses that are similar to those observed following injection of 300

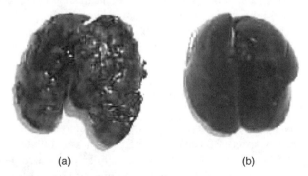

(a)   (b)

Figure 28.9 Tumor protection model. BALB/c mice were immunized one time with β-galactosidase DNA encapsulated in microparticles. Mice were then challenged with a tumor cell line expressing the β-galactosidase protein. These cells promote the development of lung nodules in control mice (A), whereas mice receiving the encapsulated DNA are protected from the tumor challenge and do not develop lung tumor nodules (B).

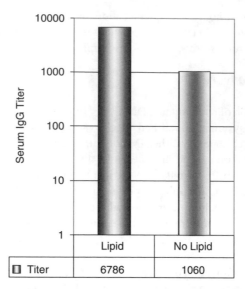

Figure 28.10 Serum titers increase when particles contain lipid. BALB/c mice were immunized one time with β-galactosidase DNA encapsulated in microparticles with or without lipid. β-galactosidase-specific serum antibody titers were tested by ELISA at 6 weeks after immunization.

μg of naked DNA, whereas administration of 30 μg DNA in particles containing TDM results in stronger immune responses than 300 μg of naked DNA. Both particle formulations (30 μg DNA, with or without TDM), elicit enhanced production of gamma IFN by lung cells following *in vivo* challenge with hsp65 antigen in comparison to 10 times as much naked DNA, but the TDM containing particles are more effective in controlling mycobacterium replication in the lungs.[46]

The immune response elicited by immunization with DNA-coated particles can be enhanced by a prime boost regimen using recombinant vaccinia virus,[33] or by coadministration of known adjuvants such as alum[33] and plasmid encoding granulocyte-macrophage colony-stimulating factor (GM-CSF).[59] Administration of particles coated with DNA encoding both CEA and GM-CSF protected 50% of mice from challenge with a tumor expressing human CEA antigen.[59] Microparticles containing an anionic lipid in the DNA formulation are more effective in stimulating immune responses than particles without this lipid.[57] The responses were observed in more animals and were higher than those seen in mice immunized with particles that did not contain lipid. An example of this can be seen in Figure 28.10; the titer of serum antibody specific for β-galactosidase following a single IM injection of DNA encoding β-galactosidase encapsulated in PLG particles with lipid is approximately six times higher than those of animals receiving the particles without lipid. The reason for the observed enhancement is not clear but could include mechanisms that (1) promote increased uptake of DNA and (2) cause increased release of particles and/or DNA from the endosomal compartments.

28.5.2.4 Nonhuman Primates and Clinical Studies

The utility of the DNA–particle approach has also been demonstrated in nonhuman primates. Antibody and CTL responses specific for the HIV gag protein were elicited by immunizing rhesus macaques with DNA-coated particles.[33] A prime boost regimen performed by priming intrarectally with PLG particles containing encapsulated DNA encoding an HIV multi-CTL epitope plasmid, and then boosting intradermally with recombinant vaccinia virus, was effective at eliciting SIV-specific CTL responses in rhesus macaques that persisted for at least 12 weeks after the final boost.[61] The CTL response was detected in the peripheral and regional lymph nodes as expected, but not in the spleen of these mucosally primed animals.

Even though several animal models have demonstrated the usefulness of the PLG–DNA formulations, the true test of these systems is through the demonstration of clinical benefit in patients. To date, data highlighting the clinical effectiveness of PLG-based DNA formulations have been collected in three clinical trials. Three to four IM injections of a formulation containing DNA encoding an HPV antigen encapsulated in PLG particles, in patients diagnosed with high-grade cervical or anal dysplasia, elicited T-cell responses in nearly all of the individuals.[62,63] The T-cells were specific for antigens encoded by the DNA and were present in many patients at levels that permitted their detection in the absence of *in vitro* expansion.[63] Although these studies were not placebo controlled, it is of interest to note that several patients experienced resolution of their dysplasia. Data from a placebo-controlled trial performed with a similar formulation demonstrated that three IM administrations of plasmid, encoding several immunodominant regions from HPV proteins, encapsulated in PLG particles, resulted in resolution of precancerous cervical lesions in significantly more patients receiving the formulation than placebo.[64] In all of these trials, the formulation was found to be well tolerated, the only signs of drug-related side effects being limited to the expected injection site reactions.

28.6 SUMMARY

Researchers in the field of DNA-based medicine have tried many methods to deliver DNA *in vivo*. Effective approaches to gene therapy must protect DNA, target gene expression, and sustain biologically relevant levels of expression for extended periods of time, often for the life of a patient. DNA vaccines require relatively lower levels of expression for short periods of time and are likely to benefit from DNA protection and targeted gene expression. PLG is a versatile, biocompatible polymer that can be an attractive delivery system for DNA. Through the selection of PLG with different characteristics, such as L:G ratio and molecular weight, delivery systems can be created to provide formulations that target different uptake pathways, demonstrate controlled degradation rates and, in turn, control the release of DNA.

Gene therapy based approaches which require long-term, high-level gene expression are in the exploratory phases of research, and it is in this application of DNA based medicine that the most significant difficulties in developing PLG delivery systems will be encountered. Sustained expression of biologically active DNA will require that DNA be protected by its delivery system. Although many studies have demonstrated the ability of PLG encapsulation to afford DNA protection, this protection is time limited since PLG degradation leads to the production of acidic components that can subsequently promote DNA degradation. The combination of PLG with other active agents that condense and protect DNA, or that minimize the activity of acidic byproducts created by degradation of the polymer, may enable researchers to overcome this difficulty. Perhaps more importantly, released DNA will likely be subject to the same degradation processes that destroy naked DNA so soon after it is administered, necessitating the use of other protective agents in these formulations.

In contrast to gene therapy applications, optimization of PLG–DNA particle delivery systems for the activation of effective immune responses to DNA-encoded antigens has progressed rapidly in the last few years. In these systems, expression is required for a shorter duration and at lower levels than in the field of gene therapy. Clinically, use of these particle formulations has shown significant improvement over what has been observed following administration of naked DNA. If there is continued clinical success in the utilization of PLG-based formulations designed for activation of the immune response, regulatory approval of such a novel immunotherapeutic could occur in the foreseeable future.

REFERENCES

1. Lechardeur, D. and Lukacs, G.L., Intracellular barriers to non-viral gene transfer, *Curr. Gene Ther.* 2, 183, 2002.
2. Barry, M.E. et al., Role of endogenous endonucleases and tissue site in transfection and CpG-mediated immune activation after naked DNA injection, *Hum. Gene Ther.,* 10, 2461, 1999.
3. Okada, H. and Toguchi, H., Biodegradable microspheres in drug delivery, *Crit. Rev. Ther. Carrier Syst.* 12, 1, 1995.
4. Visscher, G. et al., Biodegradation of and tissue reaction to 50:50 poly (DL-lactide-co-glycolide) microcapsules, *J. Biomed. Mater. Res.* 19, 349, 1985.
5. Visscher, G. et al., Note: biodegradation of and tissue reaction to 50:50 poly (DL-lactide-co-glycolide) microcapsules, *J. Biomed. Mater. Res.* 20, 667, 1986.
6. Visscher, G., Robison R., and Argentieri, G., Tissue response to biodegradable injectable microcapsules, *J. Biomater. Appl.* 2, 118, 1987.
7. Kamei, S. et al., New method for analysis of biodegradable polyesters by high-performance liquid chromatography after alkali hydrolysis, *Biomaterials* 13, 953, 1992.
8. Anderson, J.M. and Shive, M.S., Biodegradation and biocompatibility of PLA and PLG microspheres, *Adv. Drug Deliv. Rev.* 28, 5, 1997.
9. Tracy, M.A. et al., Factors affecting the degradation rate of poly(lactide-co-glycolide) microspheres *in vivo* and *in vitro*, *Biomaterials* 20, 1057, 1999.
10. Artursson, P. et al., Biodegradable microspheres V: stimulation of macrophages with microparticles made of various polysaccharides, *J. Phar. Sci.* 76, 127, 1987.
11. Tabata, Y. and Ikada, Y., Macrophage phagocytosis of biodegradable micropheres composed of L-lactic acid/glycolic acid homo-and copolymers, *J. Biomed. Mater. Res.* 22, 837, 1988.
12. Tabata, Y. and Ikada, Y., Phagocytosis of polymeric microspheres, in *High Performance Biomaterials,* (M. Szycher, Ed.), Technomic Publishing, Lancaster, UK, pp 621–646 (1991).
13. Davis, S.S. and Illum, L., Polyceric microspheres as drug carriers, *Biomaterials* 9, 111, 1988.
14. Eldridge, J.H. et al., Biodegradable microspheres: vaccine delivery system for oral immunization, *Curr. Topics Microbiol. Immunol.,*146, 59, 1989.
15. Eldridge, J. et al., Controlled vaccine release in the gut-associated lymphoid tissues. I. Orally administered biodegradable microspheres target the Peyer's patches, *J. Control. Release* 11, 205, 1990.
16. Lunsford, L. et al., Tissue distribution and persistence in mice of plasmid DNA encapsulated in a PLG-based microsphere delivery vehicle, *J. Drug Target.* 8, 39, 2000.
17. Newman, K.D. et al., Uptake of poly(d,l-lactic-co-glycolic acid) microspheres by antigen-presenting cells *in vivo*, *J. Biomed. Mater. Res.* 60, 480, 2002.
18. Newman, K.D. et al., Cytoplasmic delivery of a macromolecular fluorescent probe by poly(d,l-lactic-co-glycolic acid) microspheres, *J. Biomed. Mater. Res.* 50, 591, 2000.
19. Tabata, Y. and Ikada, Y., Phagocytosis of polymer microspheres by macrophages, *Adv. Polym. Sci.* 94, 107, 1990.
20. Hsu, Y.Y., Hao, T., and Hedley, M.L., Comparisons of process parameters for microencapsulation of plasmid DNA in poly (D, L- lactic-co-glycolic) acid microspheres, *J. Drug Target.* 7, 313, 1999.
21. Tinsley-Brown, A.M. et al., Formulation of poly(D, L-lactic-co-glycolic acid) microparticles for rapid plasmid DNA delivery, *J. Control. Release* 66, 229, 2000.
22. Capan, Y. et al., Stability of poly(L-lysine)-complexed plasmid DNA during mechanical stress and DNase I treatment, *Pharmacol. Develop. Tech.* 4, 4, 1999.
23. Capan, Y. et al., Influence of formulation parameters on the characteristics of poly (D, L- lactide -co-glycolide) microspheres containing poly (L-lysine) complexed plasmid DNA, *J. Control. Release* 60, 279, 1999.
24. Capan, Y. et al., Preparation and characterization of poly (D, L-lactide-co-glycolide) microspheres for controlled release of poly (L-lysine) complexed plasmid DNA, *Pharm. Res.* 16, 509, 1999.
25. Ando, S. et al., PLG microspheres containing plasmid DNA: preservation of supercoiled DNA via cryopreparation and carbohydrate stabilization, *J. Pharm. Sci.* 88, 126, 1999.
26. Hedley, M.L., Curley, J., and Urban, R., Microspheres containing plasmid-encoded antigens elicit cytotoxic T-cell responses, *Nat. Med.* 4, 365, 1998.

27. Herrmann, J.E. et al., Immune response and protection obtained by oral immunization with rotavirus VP4 and VP7 DNA vaccines encapsulated in microparticles, *Virology* 259, 148, 1999.

28. Wang, D. et al., Encapsulation of plasmid DNA in biodegradable poly(D,L-lactic-co-glycolic acid) microspheres as a novel approach for immunogene delivery, *J. Control. Release* 57, 9, 1999.

29. Barman, S.P. et al., Two methods for quantifying DNA extracted from poly(lactide-co-glycolide) microspheres, *J. Control. Release* 69, 337, 2000.

30. del Barrio, G.G., Novo, F.J., and Irache, J.M., Loading of plasmid DNA into PLG microparticles using TROMS (total recirculation one-machine system): evaluation of its integrity and controlled release properties, *J. Control. Release* 86, 123, 2003.

31. Benoit, M.A. et al., Studies on the potential of microparticles entrapping pDNA - poly (aminoacids) complexes as vaccine delivery systems, *J. Drug Target.* 9, 253, 2001.

32. Walter, E. et al., Microencapsulation of DNA using poly(DL-lactide-co-glycolide) stability issues and release characteristics, *J. Control. Release* 61, 361, 1999.

33. O'Hagan, D. et al., Induction of potent immune responses by cationic microparticles with adsorbed human immunodeficiency virus DNA vaccines, *J. Virol.* 75, 9037, 2001.

34. Denis-Mize, K.S. et al., Plasmid DNA absorbed onto cationic microparticles mediates target gene expression and antigen presentation by dendritic cells, *Gene Ther.* 7, 2105, 2000.

35. Walter, E. et al., Hydrophilic poly (DL-lactide-co-glycolide) microspheres for the delivery of DNA to human-derived macrophages and dendritic cells, *J. Control. Release* 76, 149, 2001.

36. Walter, E. and Merkle, H.P., Microparticle-mediated transfection of non-phagocytic cells *in vitro*, *J. Drug Target.* 10, 11, 2002.

37. Makino, K., Ohshima, H. and Kondo, T., Transfer of protons from bulk solution to the surface of poly (L-lactide) microcapsules, *J. Microencaps.* 3, 195, 1986.

38. Stolnik, S. et al., The colloidal properties of surfactant-free biodegradable nanospheres from poly(-malic acid-co-benzyl malate) poly(lactic acid-co-glycolide), *Coll. Surf. A: Physiochem. Eng. Aspects* 97, 235, 1995.

39. Panyam, J. et al., Rapid endo-lysosomal escape of poly (DL-lactide-co-glycolide) nanoparticles: implications for drug and gene delivery, *FASEB J.* 16, 1217, 2002.

40. Sahoo, S.K. et al., Residual polyvinyl alcohol associated with poly (D,L-lactide-co-glycolide) nano-particles affects their physical properties and cellular uptake, *J. Control. Release* 82, 105, 2002.

41. Cohen, H. et al., Sustained delivery and expression of DNA encapsulated in polymeric nanoparticles, *Gene Ther.* 7, 1896, 2000.

42. Jeong, J.H. and Park, T.G., Poly (L-lysine)-g-poly (D, L-lactic-co-glycolic acid) micelles for low cytotoxic biodegradable gene delivery carriers, *J. Control. Release* 82, 15, 2002.

43. Prabha, S. et al., Size-dependency of nanoparticle-mediated gene transfection: studies with fractionated nanoparticles, *Inter. J. Pharm.* 244, 105, 2002.

44. Sandor, M. et al., Transfection of HEK cells via DNA-loaded PLG and P(FASA) nanospheres, *J. Drug Target.* 10, 497, 2002.

45. Singh, M. et al., Mucosal immunization with HIV-1 gag DNA on cationic microparticles prolongs gene expression and enhances local and systemic immunity, *Vaccine* 20, 594, 2002.

46. Lima, K.M. et al., Single dose of a vaccine based on DNA encoding mycobacterial hsp65 protein plus TDM-loaded PLG microspheres protects mice against a virulent strain of mycobacterium tuber-culosis, *Gene Ther.* 10, 678, 2003.

47. Kaneko, H. et al., Oral DNA vaccination promotes mucosal and systemic immune responses to HIV envelope glycoprotein, *Virology* 267, 8, 2000.

48. Lu, P., Wang, Y.L., and Linsley, P.S., Regulation of self-tolerance by CD80/CD86 interactions, *Curr. Opin. Immunol.* 9, 858, 1997.

49. Eldridge, J. et al., Biodegradable and biocompatible poly (DL-Lactide-Co-Glycolide) microspheres as an adjuvant for staphylococcal enterotoxin B toxoid which enhances the level of toxin-neutralizing antibodies, *Infect. Immunol.* 59, 2978, 1991.

50. Eldridge, J. et al., Biodegradable microspheres as a vaccine delivery system, *Mol. Immunol.* 28, 287, 1991.

51. Randolph, G.J. et al., Differention of phagocytic monocytes into lymph node dendritic cells *in vivo*, *Immunity,* 11, 753, 1999.

52. Briones, M. et al., The preparation, characterization, and evaluation of cationic microparticles for DNA vaccine delivery, *Pharm. Res.* 118, 709, 2001.

53. Jones, D.H. et al., Poly (DL-lactide-co-glycolide)-encapsulated plasmid DNA elicits systemic and mucosal antibody responses to encoded protein after oral administration, *Vaccine* 15, 814, 1997.

54. Chen, S.C. et al., Protective immunity induced by oral immunization with a rotavirus DNA vaccine encapsulated in microparticles, *J. Virol.* 72, 5757, 1998.

55. Jones, D.H., Clegg, J., and Farrar, G.H., Oral delivery of micro-encapsulated DNA vaccines, in *Modulation of the Immune Response to Vaccine Antigens: Developmental Biological Standards*, Brown, F. and Haaheim, L.R. (Eds.), Karger, Basel, 1998, 149.

56. Wierzbicki, A. et al., Immunization strategies to augment oral vaccination with DNA and viral vectors expressing HIV envelope glycoprotein, *Vaccine* 20, 1295, 2002.

57. McKeever, U. et al., Protective immune responses elicited in mice by immunization with formulations of poly(lactide-co-glycolide) microparticles, *Vaccine* 20, 1524, 2002.

58. Singh, M. et al., Cationic microparticles: a potent delivery system for DNA vaccines, *Proc. Nat. Acad. Sci. U.S.A.* 97, 811, 2000.

59. microparticles induces protective immunity against colon cancer in CEA-transgenic mice, *Vaccine* 21, 1938, 2003.

60. Lima, V.M.F. et al., Role of trehalose dimycolate in recruitment of cells and modulation of production of cytokines and NO in tuberculosis, *Infect. Immunol.* 69, 5305, 2001.

61. Sharpe, S.A. et al., Mucosal immunization with PLG-microencapsulated DNA primes a SIV-specific CTL response revealed by boosting with cognate recombinant modified vaccinia virus Ankara, *Virology,* 313, 13–21, 2003.

62. Klencke, B. et al., Encapsulated plasmid DNA treatment for human papillomavirus 16-associated anal dysplasia: a phase I study of ZYC101, *Clin. Cancer Res.* 8 (5), 1028, 2002.

63. Sheets, E.E., Immunotherapy of human cervical high-grade cervical intraepithelial neoplasia with microparticle-delivered human papillomavirus 16 E7 plasmid DNA, *Amer. J. Obstet. Gynecol.* 188, 916, 2003.

64. Garcia, F. et al., A randomized, multicenter, placebo controlled trial of ZYC101a for the treatment of high grade intraepithelial cervical neoplasia, *Obstet. Gynecol.* 103(2), 317–326, 2004.

CHAPTER 29

Polyanhydride Microspheres for Gene Delivery

Yong S. Jong, Camilla A. Santos, and Edith Mathiowitz

CONTENTS

29.1 INTRODUCTION

Polyanhydrides have been investigated in drug delivery research since the early 1980s. Gliadel®, a polyanhydride-based delivery system, is currently FDA approved for delivery of BCNU in the treatment of brain cancer. However, compared to some of the biodegradable polyesters, polyanhydrides are not being widely used in current drug delivery applications. Polyanhydrides can exhibit rapid degradation rates due to the extremely reactive nature of the anhydride bond. They also require special storage conditions at low temperatures. Moreover, polyanhydrides cannot be purchased commercially and require custom synthesis, thereby limiting accessibility. However, the distinct physicochemical properties of polyanhydrides, such as surface erosion, rapid degradation, and bioadhesion, make them particularly attractive for certain applications.

29.2 POLYANHYDRIDES

Polyanhydrides refer to a family of thermoplastic polymers that contain anhydride linkages within a hydrophobic polymer backbone.[1] Polyanhydrides used for drug delivery have been extensively reviewed.[2,3,4] Monomers used for polymer and copolymer synthesis include carboxyphenoxypropane,

0-8493-1934-X/05/$0.00+$1.50
© 2005 by CRC Press LLC

sebacic acid, fatty acid dimer, fumaric acid, and adipic acid.[5,6,7] One anhydride, poly(fumaric-*co*-sebacic anhydride) (P[FA:SA]) is commonly synthesized by melt condensation.[2,8] P(FA:SA) is bio-erodible via hydrolytic degradation, ultimately resulting in monomers of fumaric and sebacic acids that are metabolized by the Krebs/citric acid cycle or the β-oxidation pathway.[9] These polymers are hydrophobic and display a high degree of crystallinity. Depending on the ratio of fumaric to sebacic acid, P(FA:SA) exhibits melting temperatures in the range of ~86–246°C and glass transition temperatures of ~45°C.[10] Typically, the anhydrides degrade on the order of days to weeks, compared to poly(lactide-*co*-glycolide) (PLGA), which degrades on the order of months to years, and are beneficial where "rapid" degradation is desired.[11,12] Another distinguishing feature of polyanhydride matrices is the potential for surface front erosion, because the hydrophobic regions of the polymer backbone prevent penetration of water into the polymer bulk.[13] Theoretically, surface front erosion prevents degradation of the encapsulated drug still entrapped within the polymer matrix. In contrast, bulk hydration and degradation exhibited by some polyesters can lead to a localized decrease in pH and reduced stability of encapsulated drug.[14,15,16]

29.3 BIOADHESION WITH POLY(FUMARIC-*CO*-SEBACIC ANHYDRIDE)

Most work in bioadhesive material development has focused on the use of hydrogels such as polyacrylic acid, polymethacrylic acid (Eurdragit®), and hydroxypropyl-methylcellulose. In the early 1990s, a group at Brown University identified P(FA:SA) and other anhydride polymers as strong bioadhesive materials through a series of *in vitro* and *in vivo* assays.[17-21] Mechanistically, it was hypothesized that the carboxylic acid groups that become exposed on the surface of these microspheres during degradation interact with the mucus glycoproteins to form secondary bonds between the polymer and the biological substrate.[17] A number of different anhydrides were screened for bioadhesion using a variety of *in vitro* testing methods, including everted sac binding assays, tensile stress measurements, and contact angle measurements with mucus. One polyanhydride in particular, polyfumaric-*co*-sebacic acid (FA:SA), showed greater adhesive forces (> 50 nM/cm^2) than any other material (polylactic acid [PLA], PLGA, polystyrene [PS], polycaprolactone [PCL]) tested. Bioadhesion was not strictly limited to anhydrides; the bioadhesive properties of otherwise inert polymers were also increased by addition of anhydride-based oligomers.[22] Subsequently, it was demonstrated that oral feeding of P(FA:SA) microspheres containing radiopaque markers resulted in a significant delay of GI transit time compared to unencapsulated barium sulfate, illustrating the efficacy of this bioadhesive polymer material with respect to increasing residence time.[17]

29.4 PHASE INVERSION NANOENCAPSULATION

A number of microencapsulation techniques have been used with polyanhydrides, including solvent evaporation, solvent removal, spray drying, hot melt, compression molding, and phase inversion nanoencapsulation.[13,21,23,24,25] Due to the hydrolytic instability, encapsulation techniques which involve lengthy exposure to aqueous phases are not particularly suited for polyanhydrides. One microencapsulation technique that was first developed with P(FA:SA) is phase inversion nanoencapsulation (PIN). Conceptually, PIN is similar to precipitation and results in the spontaneous fragmentation of the polymer solution, without the need for droplet formation such as emulsification or atomizing sprays, into discrete microspheres. The particle size is controlled by a variety of factors including: polymer solution concentration, solvent selection, presence or absence of surfactants in the nonsolvent, and polymer material. For example, PIN microspheres of P(FA:SA)

Figure 29.1   P(FA:SA) microspheres produced by PIN. Left: 500 nm particles. Right: 2 μm particles.

can be fabricated by making a 3% (w/v) solution of P(FA:SA) in methylene chloride and pouring the solution into pentane. The microspheres are subsequently recovered by simple filtration. The primary advantage of this technique for a water labile polymer such as P(FA:SA) is that it occurs entirely in a nonaqueous environment, which provides for retention of polymer integrity and superior loading efficiencies for water soluble drugs. No heat is required, and the process occurs instantaneously at room temperature, without mechanical stirring or homogenization, promoting retention of drug stability during encapsulation. However, PIN is limited to producing small microspheres (100 nm to 5 μm) and cannot be used with all polymers (Figure 29.1). PIN has been used to encapsulate a variety of small molecule drugs, proteins, and nucleic acids with a number of different polymers.[19,25,26,27,28,29,48]

29.5 IMPROVING PARTICLE UPTAKE FOR GENE DELIVERY

For effective gene delivery, biologic, and synthetic vectors all face the same hurdles: extracellular degradation, penetration of the cell membrane, intracellular degradation, cytoplasmic diffusion, and penetration of the nuclear membrane. The rationale for developing a vector with P(FA:SA) was to increase the efficacy of gastrointestinal particle uptake by coupling the PIN process with a bioadhesive polymer. Particle uptake along the gastrointestinal tract is an established biological phenomenon that has been adapted for delivery of DNA.[25,30,31,32,33] It is thought that the uptake phenomenon is primarily a function of size, in that small particles (<10 μm) can cross the GI barrier. Microspheres delivered orally are able to cross the intestinal epithelium through the Peyer's patch or the absorptive epithelium.[34,35] While both transcellular and paracellular routes have been associated with uptake, the exact mechanism governing the transepithelial migration of microspheres remains unclear. However, the current view suggests minimizing particle size is critical for maximizing uptake. Following uptake, microspheres have been localized primarily in the mesenteric lymph vessels, lymph nodes, liver, and spleen.[30,34,35] The selection of polymer material also plays a role in mediating the uptake of particles.[36,37] Numerous studies have shown that the efficiency of particle uptake in the gastrointestinal tract is higher with PS than with PLGA. This observation has generally been attributed to the more hydrophobic nature of PS as compared to PLGA.[35,38] However, due to its nondegrading nature, PS has not been used for functional drug delivery. Although quantification of particle uptake is technically difficult, studies in rats and rabbits using light, confocal, and tunneling electron microscopy have shown that P(FA:SA) PIN microspheres exhibit uptake comparable to PS.[25,27] It should be noted that other mucoadhesive polymers such as chitosan have been developed for similar applications.[33,39]

29.6 PROPERTIES OF PLASMID DNA

Plasmid DNA presents some unique challenges for microencapsulation. Plasmid DNA is an extremely high-molecular-weight substance; a typical plasmid is on the order of millions of Daltons, with hydrodynamic diameters in the range of hundreds of nanometers. It is extremely sensitive to physical shearing (homogenization, sonication), which can nick or linearize circular DNA.[40,41] Moreover, it is difficult to solubilize plasmid DNA at high concentrations. These characteristics directly affect the micronization of plasmid DNA for subsequent encapsulation. For example, solvent evaporation for encapsulation of water-soluble drugs generally requires emulsification of solubilized drugs within a polymer solvent phase, through a combination of vortex, homogenization, and sonication. The efficiency of emulsification dictates the eventual particle size of a drug that can be encapsulated in the polymer matrix. The sensitivity of plasmid DNA to physical agitation has been addressed by various strategies such as addition of excipients and modifying process conditions.[40,42,43] An alternative approach is to encapsulate plasmid DNA in the dry state similar to the process used for Prolease® as most biopharmaceuticals are more stable in lyophilized form.[44] However, obtaining a dispersable dry powder of small plasmid DNA particles is challenging, particularly for encapsulation within very small particles. In contrast to lower molecular weight RNA, lyophilized or alcohol precipitated DNA form fibrillar aggregates (Figure 29.2) that cannot be physically micronized without inducing DNA shearing.

29.7 *IN VITRO* TRANSFECTION WITH P(FA:SA) PIN

The application of naked plasmid DNA *in vitro* does not result in transfection of mammalian cells. Surprisingly, in 1990 Jon Wolff et al. showed that direct *in vivo* injection of plasmid DNA resulted in significant transfection in muscle tissue.[45] This apparent dichotomy makes it difficult to evaluate *in vitro* bioactivity using traditional release studies. In addition, *in vitro* release of plasmid DNA from polymer matrices can result in conformational changes (supercoiled to open-circular) that can occur from polymer degradation products and/or the incubation environment, which that may not accurately reflect the *in vivo* condition.[41,46,47] *In vitro* transfection with P(FA:SA) PIN microspheres has been reported in conjuction with chloroquine, a lysosomal enzyme inhibitor.[48] PIN microspheres (~1.5 µm) containing 0.5% (w/w) or 2% (w/w) DNA loadings were assayed for transfection efficiencies. Release studies with P(FA:SA) PIN demonstrated 16% plasmid DNA cumulative release at low loading, which increased to 62% at the higher loading within 10 min of hydration. Both formulations continued to release plasmid DNA up to 83 days. Transfection studies with human embryo kidney (HEK) cells showed that in the presence of chloroquine, both PLGA

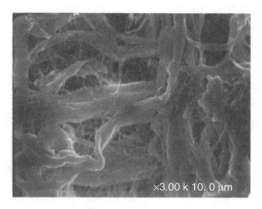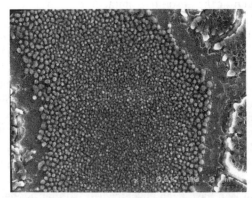

Figure 29.2 Scanning electron micrograph. Left: alcohol-precipitated calf thymus DNA. Right: alcohol-precipitated yeast RNA.

and P(FA:SA) PIN microspheres encapsulating a reporter plasmid pCMV–GFP were effective in delivering the plasmid to the nuclei for functional expression. Control studies with naked pCMV–GFP, even in the presence of chloroquine, did not result in GFP expression, suggesting that the encapsulation was critical in promoting plasmid DNA uptake across the cell membrane. While the PIN microspheres were not as effective compared to Lipofectamine®, the results illustrated that the PIN microspheres can be used to deliver intact plasmid DNA for gene expression.

29.8 *IN VIVO* GENE DELIVERY WITH P(FA:SA) PIN

PIN gene transfer into Peyer's patch tissue using P(FA:SA) PIN microspheres was demonstrated by Mathiowitz et al.[25] A single bioadhesive polymer delivery system composed of P(FA:SA) in a 20:80 molar ratio was used to encapsulate pCMV–β-gal (0.1% w/w), a commercially available reporter plasmid with bacterial β-galactosidase gene, using the PIN method. P(FA:SA) (20:80) was synthesized using a melt condensation protocol. *In vitro* release studies with microspheres (10 mg) were conducted by incubation in 0.5 ml of Tris-EDTA (TE) buffer at room temperature. Supernatants were analyzed by agarose gel electrophoresis at 24 h, 73 h, 1 week, and 2 weeks following hydration. The migration pattern indicated that plasmid DNA remained in supercoiled and open-circular conformation at all time points. Five days following either a single oral dose of these microspheres (50 mg containing 50 μg DNA) or unencapsulated pCMV–β-gal (500 μg), β-galactosidase activity was quantified in the stomach, small intestine, and liver of rats using a chemiluminescent based assay. Animals that were fed encapsulated pCMV–β-gal showed increased levels of β-galactosidase activity in both the small intestine and the liver, compared to animals fed unencapsulated pCMV–β-gal or unfed control animals. The reporter gene activity measured in animals which received the encapsulated pCMV/β-gal was highest in intestinal tissue (> 54 mU versus 24 mU for the unencapsulated plasmid, and 18 mU for the background levels of activity found in untreated control animals; nsd for the three groups). These same animals averaged 11 mU of activity in the liver versus less than 1 mU for plain CMV-fed (ss, p < 0.045) or untreated control animals (ss, p < 0.025). Reporter gene expression in stomach homogenates was not significantly different and generally low (< 11 mU) in all groups. Visual localization of transfected cells following oral administration was performed using X-gal histochemical techniques on whole tissue and frozen sections. Whole tissue X-gal staining showed that the serosal surface of small intestine from encapsulated pCMV–β-gal-fed rats showed individual cells that stained intensely in localized areas containing Peyer's patches. Similar histochemical X-gal staining of frozen sections of Peyer's patches revealed that although there were a few β-galactosidase positive cells within the central lymphoid tissue mass, the majority of transfected cells were located in the muscularis mucosae and adventitia below the Peyer's patches. Neither groups of control animals (unencapsulated pCMV–β-gal or unfed normal rats) showed any false-positive β-galactosidase staining in the Peyer's patches region. Histological examination of the tissue revealed near normal histology in all experimental groups with no evidence of mucosal damage or inflammation. The results of the study confirmed that plasmid DNA can be delivered by the oral route using PIN nanoparticle formulations.

29.9 SUMMARY

Gene vectors are a formidable challenge due to the heavy burden on the technology, which requires delivery of DNA all the way into the nucleus. The bioadhesive nature of polyanhydride-based microspheres provides an alternative for DNA delivery. In addition to protection of DNA from extracellular degradation by virtue of encapsulation, polyanhydrides can offer increased cell membrane penetration by virtue of their bioadhesive properties. Clearly, work with polyanhydrides for nucleic acids is in its early stages and additional progress is required. The addition of compounds to enhance lysosomal escape or increase nuclear translocation to the existing system may provide dramatic improvements in transfection efficiency. Oral vaccine applications, where the target cell type (M-cell) is inherently more suited for particle uptake and where the therapeutic efficacy (immune response) does not require high, long-term systemic expression, is particularly attractive.

REFERENCES

1. Bucher, J.E. and Slade, W.C., The anhydrides of isophthalic and terephthalic acids, *J. Am. Chem. Soc.* 31, 1319, 1909.
2. Domb, A.J. et al., Polyanhydrides: synthesis and characterization, *Adv. Pol. Sci.* 107, 93, 1993.
3. Gopferich, A., Biodegradable polymers: polyanhydrides, in *Encyclopedia of Controlled Drug Delivery*, Mathiowitz, E. (Ed.), John Wiley & Sons, New York, 1999, 61.
4. Kumar, N., Langer, R.S., and Domb, A.J., Polyanhydrides: an overview, *Adv. Drug. Del. Rev.* 54, 889, 2002.
5. Domb, A.J. and Langer, R., Polyanhydrides. I. Preparation of high molecular weight polyanhydrides, *J. Pol. Sci.* 25, 3373, 1987.
6. Tabata, Y. and Langer, R., Polyanhydride microspheres that display near-constant release of water-soluble model drug compounds, *Pharm. Res.* 10 (3), 391, 1993.
7. Albertsson, A-C. and Lundmark, S., Melt polymerization of adipic anhydride (oxepane-2, 7-dione), *J. Macromol. Sci. Chem.* A27 (4), 397, 1990.
8. Domb, A.J. et al., Polyanhydrides. IV. Unsaturated and crosslinked polyanhydrides, *J. Pol. Sci.* 29, 571, 1991.
9. Katti, D.S. et al., Toxicity, biodegradation and elimination of polyanhydrides, *Adv. Drug Del. Rev.* 54, 993, 2002.
10. Mathiowitz, E. et al., Morphological characterization of bioerodible polymers. 1. Crystallinity of polyanhydride copolymers, *Macromolecules* 23 (13), 3212, 1990.
11. Santos, C.A., et al., Poly(fumaric-*co*-sebacic anhydride) A degradation study as evaluated by FTIR, DSC, GPC and X-ray diffraction, *J. Control. Release* 60, 11, 1999.
12. Sandor, M., Harris, J., Mathiowitz, E., A novel polyethylene depot device for the study of PLGA and P(FA:SA) microspheres *in vitro* and *in vivo*, *Biomaterials* 23, 4413, 2002.
13. Leong, K.W., Brott, B.C., Langer, R., Bioerodible polyanhydrides as drug-carrier matrices. I: Characterization, degradation, and release characteristics, *J. Bio. Mater. Res.* 19, 941, 1985.
14. Sanchez, A., Formulation strategies for the stabilization of tetanus toxoid in poly(lactide-*co*-glycolide) microspheres, *Int. J. Pharm.* 185 (2), 255, 1999.
15. Brunner, A., Maeder, K., Goepferich, A., pH and osmotic pressure inside biodegradable microspheres during erosion, *Pharm. Res.* 16, 847, 1999.
16. Fu, K. et al., Visual evidence of acidic environment within degrading poly(lactic-*co*-glycolic acid) (PLGA) microspheres, *Pharm. Res.* 17, 100, 2000.
17. Chickering, D.E., Harris, W.P., Mathiowitz, E., A microtensiometer for the analysis of bioadhesive microspheres, *Biomed. Instrum. Technol.* 29 (6), 501, 1995.
18. Chickering, D.E., Jacob, J.S., Mathiowitz, E., Poly(fumaric-co-sebacic) microspheres as oral drug delivery systems, *Biotech. Bioeng.* 52, 96. 1996.
19. Carino, G.P. et al., Bioadhesive, bioerodible polymers, in *Bioadhesive Drug Delivery Systems: Fundamentals, Novel Approaches and Development*, Mathiowitz, E., Chickering, D.E., and Lehr, C-M. (Eds.), Marcel-Dekker, New York, 1999, 459.
20. Santos, C.A. et al., Correlation of two bioadhesion assays: the everted sac technique and the CAHN microbalance, *J. Control. Release* 61 (1–2), 113, 1999.
21. Vasir, J.K. et al., Bioadhesive microspheres as a controlled drug delivery system, *Int. J. Pharm.* 255, 13, 2003.
22. Santos, C.A. et al., Evaluation of anhydride oligomers within polymer microsphere blends and their impact on bioadhesion and drug delivery *in vitro*, *Biomaterials* 24 (20), 3571, 2003.
21. Pekarek, K.J. and Mathiowitz, E., Double-walled polymer microspheres for controlled drug release, *Nature* 367, 258, 1994.

22. Pekarek, K.J. et al., *In vitro* and *in vivo* degradation of double-walled polymer microspheres, *J. Control. Release* 40, 169, 1996.

23. Mathiowitz, E. and Langer, R., Polyanhydride microspheres as drug carriers I. Hot-melt microencapsulation, *J. Control. Release* 5, 13, 1987.

24. Mathiowitz, E. et al., Polyanhydride microspheres. IV. Morphology and characterization of systems made by spray-drying, *J. App. Pol. Sci.* 45,125, 1992.

25. Mathiowitz, E. et al., Biologically erodible microspheres as potential drug delivery systems, *Nature* 386 (6623), 410, 1997.

26. Egilmez, N.K. et al., Cytokine immunotherapy of cancer with controlled release biodegradable microspheres in a human tumor xenograft/SCID mouse model, *Cancer Immunol. Immunother.* 46 (1), 21, 1998.

27. Carino, G.P. and Mathiowitz, E., Oral insulin delivery, *Adv. Drug Deliv. Rev.* 35 (2–3), 249, 1999.

28. Hill, H.C. et al., Cancer immunotherapy with interleukin-12 and granulocyte-macrophage colony-stimulating factor-encapsulated microspheres: co-induction of innate and adaptive antitumor immunity and cure of disseminated disease, *Cancer Res.* 62 (24), 7254, 2002.

29. Thanos, C.G. et al., Enhancing the oral bioavailability of the poorly soluble drug dicumarol with a bioadhesive polymer, *J. Pharm. Sci.* 92 (8), 1677, 2003.

30. Sanders, E. and Ashworth, C.T., A study of particulate intestinal absorption and hepatocellular uptake, *Exp. Cell. Res.* 22, 137, 1961.

31. Florence, A.T., The oral absorption of micro- and nano-particulates: Neither exceptional nor unusual, *Pharm. Res.* 14 (3), 259, 1997.

32. Jones, D.H. et al., Poly(DL-lactide-co-glycolide)-encapsulated plasmid DNA elicits systemic and mucosal antibody responses to encoded protein after oral administration, *Vaccine* 15 (8), 814, 1997.

33. Roy, K. et al., Oral gene delivery with chitosan DNA nanoparticles elicits immunological protection in a murine model of peanut allergy, *Nat. Med.* 5 (4), 387, 1999.

34. Ermak, T.H. et al., Uptake and transport of copolymer biodegradable microspheres by rabbit Peyer's patch M cells, *Cell Tissue Res.* 279, 433, 1995.

35. Jani P.U. et al., The uptake and translocation of latex nanospheres and microspheres after oral administration to rats. *J. Pharm. Pharmacol.* 41 (12), 809, 1989.

36. Simon, L., Shine, G., Dayan, A.D., Translocation of particulates across the gut wall—a quantitative approach. *J. Drug Target.* 3 (3), 217, 1995.

37. Gibaud, S. et al., Cells involved in the capture of nanoparticles in hematopoietic organs. *J. Pharm. Sci.* 95, 994, 1996.

38. Eldridge, J.H. et al., Controlled release vaccine release in the gut-associated lymphoid tissues. I. Orally administered biodegradable microspheres target the Peyer's patches, *J. Control. Release* 11, 205, 1990.

39. Chew, J.L. et al., Chitosan nanoparticles containing plasmid DNA encoding house dust mite allergen, Derp 1 for oral vaccination in mice, *Vaccine* 21, 2720, 2003.

40. Ando, S. et al., PLGA microspheres containing plasmid DNA: Preservation of supercoiled DNA via cytopreparation and carbohydrate stabilization, *J. Pharm. Sci.* 88 (1), 126, 1998).

41. Walter, E. et al., Microencapsulation of DNA using poly(DL-lactide-co-glycolide): stability issues and release characteristics, *J. Control. Release* 61, 361, 1999.

42. Hirosue, S. et al., Plasmid DNA encapsulation and release from solvent diffusion nanospheres, *J. Control. Release* 70, 231, 2001.

43. Perez, C. et al., Poly(lactic acid)-poly(ethylene glycol) nanoparticles as new carriers for the delivery of plasmid DNA, *J. Control. Release* 75, 211, 2001.

44. Putney, S.D. and Burke, P.A., Improving protein therapeutics with sustained-release formulations, *Nat. Biotech.* 16, 153, 1998.

45. Wolff, J.A. et al., Direct gene transfer into mouse muscle tissue *in vivo*, *Science* 247,1465,1990.

46. Jong, Y.S. et al., Controlled release of plasmid DNA, *J. Control. Release* 47, 123, 1997.

47. Berstrom, D.E., Zhang, P., Paul, N., dsDNA stability dependence on pH and salt concentration, *Biotechniques* 244, 992, 1998.

48. Sandor, M. et al., Transfection of HEK Cells via DNA-loaded PLGA and P(FA:SA) Nanospheres, *J. Drug Target.* 10 (6), 497, 2002.

CHAPTER 30

Microspheres Formulated from Native Hyaluronan for Applications in Gene Therapy

Yang H. Yun and Weiliam Chen

CONTENTS

30.1 INTRODUCTION

30.1.1 Microspheres as Gene Delivery Vectors

Microspheres formulated from biodegradable polymer have many inherent properties that are advantageous for applications in gene therapy. For example, microspheres can be lyophilized for long-term storage and "off the shelf" usage. Also, the DNA is protected from enzymatic degradation as a consequence of the entrapment.[1,2] The most significant feature of microspheres, which is generally not available for traditional gene delivery vehicles such as viruses, naked DNA, or liposomes, is the controlled delivery of DNA for prolonged durations. The extended release of DNA from microspheres could alleviate the disadvantage of transient expression inherent in all nonviral vectors.[3]

Another key advantage of using biodegradable microspheres as gene delivery vehicles is that they do not require surgical implantation or retrieval. Microspheres can be administered locally into a tissue by bolus injections or intravenously for systemic applications. Once injected, the

0-8493-1934-X/05/$0.00+$1.50
© 2005 by CRC Press LLC

microspheres will degrade, the entrapped DNA will be released, and the released DNA will be available for transfection. The rate of polymeric degradation and thus the rate of DNA release is dependent upon the composition and formulation of the microspheres. Since a wide variety of polymers is available, microspheres offer flexibility in dosage and release kinetics.

A meaningful therapeutic intervention will likely involve the delivery of a gene "cocktail," with each gene delivered at the correct dosage and in the proper spatial and temporal sequence, because all cellular mechanisms involve the expression and regulation of multiple growth, differentiation, and transcription factors. Microspheres can easily accommodate these needs since the contents within the microsphere matrix are discrete and the physicochemical properties of the microspheres can be tuned for optimal delivery. The dosage can be easily adjusted by weight once the microspheres have been lyophilized or by volume when they have been suspended in liquid. Once the microspheres of various formulations and DNA contents have been aliquoted, they can be mixed and numerous genes can be delivered simultaneously.[4-6] Since polymeric vehicles offer multiple degrees of freedom for optimization, the dosage and the release of DNA can be tailored to the pathology of a particular disease as well as the minimization of the host response.

30.1.2 Microspheres Prepared from Hyaluronan

We have chosen to investigate hyaluronan (HA) microspheres because HA is a naturally occurring glycosaminoglycan found in the extracellular matrix, connective tissues, and organs of all higher animals.[7,8] Since HA is native to the body, it is nonimmunogenic and could be an ideal biomaterial for tissue engineering[9,10] and gene delivery.[11,12] The structure of HA consists of repeating disaccharide units of D-glucuronic acid and (1-β-3) N-acetyl-D-glucosamine (see Figure 30.1).[13-15] The molecular weight typically ranges from 1 to 5×10^6 Daltons.[15,16] To date, HA has been successfully utilized for biomedical applications as hydrogels for viscosurgery and viscosupplementation,[17,18] scaffolds for wound healing applications,[19] and hydrophilic coatings for medical devices.[20]

However, few options are available for preparing HA into microspheres. For these formulations, HA has been esterified, blended with other polymers, or exposed to elevated temperature for stabilization. Fidia Advanced Biopolymers has prepared microspheres from HA wherein the carboxyl groups of the glucuronic acid moieties have been esterified.[21] Lim et al. have prepared microcapsules by complex coacervation using pure HA or HA blended with gelatin or acacia.[22] Microsphere preparations using water-in-oil emulsions with native HA (without chemical crosslinking) and HA blended with chitosan also have been described by Lim et al.[22,23]

In this chapter, we describe a method for preparing HA microspheres for gene delivery applications by adapting adipic dihydrazide chemistry (see Figure 30.1). This method has been traditionally used to construct HA hydrogels.[7,20] Formulating HA–DNA microspheres with adipic dihydrazide has several advantages. HA derivatives are well defined and nontoxic.[24,25] The DNA can be easily incorporated into the HA matrix before derivatization. The structural integrity and molecular size of HA and DNA are unaffected by the mild reaction conditions.[7] Since the crosslinking of HA occurs in an aqueous solution, the byproducts, such as urea and unreacted reagents, can be easily removed by conventional methods such as dialysis, precipitation, and ultra-filtration. Our study has shown microspheres formulated with this method released intact DNA at a controlled rate for several weeks. Most significantly, the released DNA is bioactive *in vitro* and *in vivo*.

30.2 PREPARATION AND CHARACTERIZATION OF HYALURONAN MICROSPHERES

An in emulsion cross-linking technique has been modified and developed to prepare HA–DNA microspheres from native HA.[6] The initial water-in-oil emulsion is formed by homogenization of 0.5% HA (20 ml, 1.6 to 3.3×10^6 Daltons) (Kraeber GMBH & Co., Waldhofstr, Germany) solution with

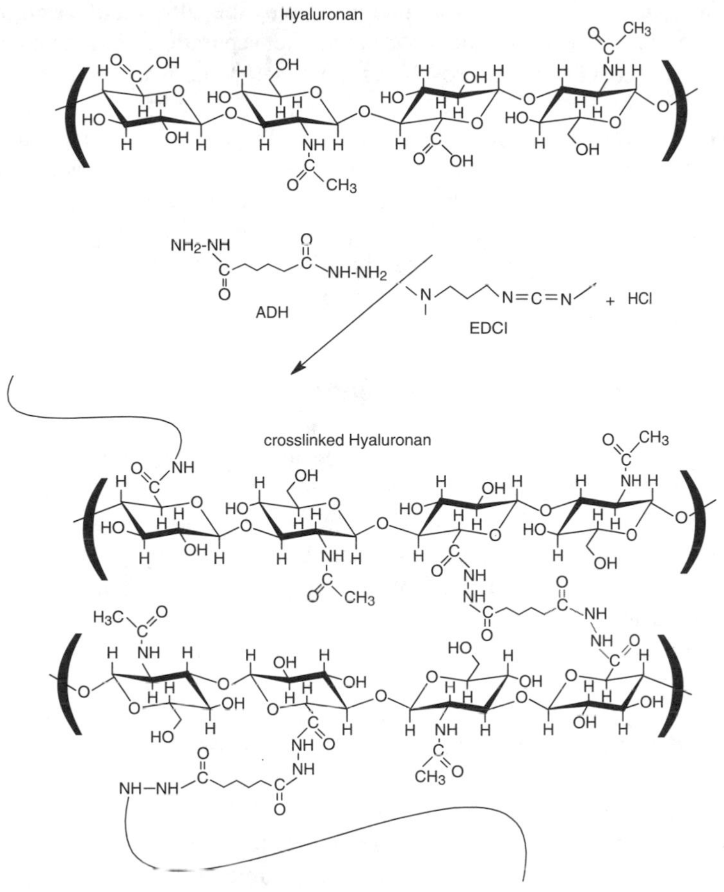

Figure 30.1 A schematic of chemical reaction for the cross-linking of HA polymer with adipic dihydrazide. (Adapted from Prestwich, G.D. et al., Chemical modification of hyaluronic acid for drug delivery, biomaterials, and biochemical probes, in *The Chemistry, Biology, and Medical Applications of Hyaluronan and Its Derivatives*, Laurent, T.C. [Ed.], Portland Press, Miami, 1998, 43.)

100 mg of dissolved adipic dihydrazide (ADH) (Sigma-Aldrich, St. Louis, MO), mineral oil (80 ml) (Sigma-Aldrich, St. Louis, MO), and Span 80 (1 ml) (ICI Chemicals, Wilmington, DE). This solution is mixed for 30 min at 1000 RPM using a mechanical stirrer (LR400D Lab Stirrer, Fisher Scientific, Pittsburgh, PA) fitted with a 0.75 inch diameter impeller (Cole-Parmer, Vernon Hills, IL).

DNA can be incorporated into the water phase of the emulsion before cross-linking the microspheres. For this study, plasmid DNA (pDNA) regulated by the cytomegalovirus immediate early promoter and encoding β-galactosidase (β-gal) (BD Biosciences Clonetech, Palo Alto, CA) or green fluorescent protein (GFP) (BD Biosciences Clonetech, Palo Alto, CA) has been utilized. The pDNA loading is 1 or 5 mg (1 and 5% respectively). For plain HA microspheres, this step is omitted. To ensure even distribution of pDNA into the aqueous phase, the emulsion is stirred for an additional 20 min after the addition of pDNA. Afterwards, 120 mg of ethyl-3-(3-dimethyl amino) propyl carbodiimide (EDCI) (Sigma-Aldrich, St. Louis, MO), dissolved in 2 ml of distilled and deionized water (DH$_2$O), is slowly instilled and mixed for 30 min.

30.2.1 Cross-linking of HA Microspheres

To initiate the cross-linking, hydrochloric acid (HCl) (0.1 N, 0.3 ml) is slowly added to the emulsion. The chemical reaction, occurring at room temperature, is allowed to proceed for 24 h.

Afterwards, the HA–DNA microspheres are precipitated by the addition of isopropyl alcohol (IPA) (150 ml) under vigorous agitation. The aqueous layer, after separation from the mineral oil, contains the HA–DNA microspheres. These microspheres are isolated and washed 3 times by centrifuge at 1500 RPM for 5 min with IPA.

After the final wash, the microspheres are resuspended in a 90% IPA reagent mixture containing 100 mg of ADH and 120 mg of EDCI (90% dimethylformamide also has been used instead of IPA and produced microspheres with similar characteristics). This mixture is stirred gently and the addition of HCl (0.1 N, 0.3 ml) initiates the second cross-linking reaction. After 24 h, the microspheres are collected by centrifugation and washed 3 times with 90% isopropyl alcohol. The HA microspheres are then resuspended in DH_2O, frozen on dry ice, and lyophilized (Virtis Freezemobile 12EL, Gardener, NY).

30.2.2   HA Microspheres and Size Distribution

The resulting microspheres have been suspended in DH_2O, qualitatively evaluated with light microscopy, and quantitatively analyzed with a laser scattering particle size analyzer. The HA microspheres cross-linked with adipic dihydrazide are stable in an aqueous solution, as shown in Figure 30.2. Their structural conformation is spherical with heterogeneous size distribution. The median and mean diameters of the HA–DNA microspheres are 6 μm and 20 μm, respectively, as indicated by a laser scattering particle size analyzer (LA-190, Horiba Ltd., Irvine, CA). The size distribution, seen in Figure 30.3, indicates that more than 60% of the HA–DNA microspheres are between 3 and 10 μm in diameter.

30.2.3   Release Kinetics of DNA from HA–DNA Microspheres

To analyze the release profile and the integrity of the DNA incorporated into HA microspheres, a release study has been completed.[6,26-28] The release profiles of HA–DNA microspheres, as shown in Figures 30.4A and 30.4B, demonstrate an initial burst release of pDNA, but the rate of release moderates thereafter. This initial release is independent of the enzyme activity in the buffer and the amount of pDNA loading. Afterwards, the HA–DNA microspheres steadily release pDNA until the termination of the experiment.

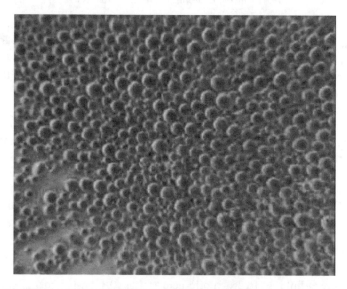

Figure 30.2   Light microscopy of HA microspheres dispersed in DH_2O.

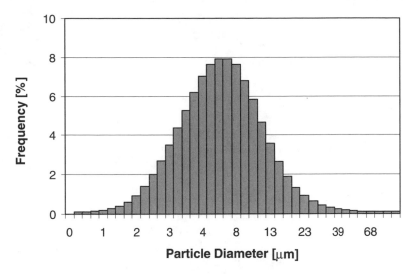

Figure 30.3 Size distribution of HA–DNA microspheres prepared from adipic dihydrazide chemistry. (From Yun, Y.H., Goetz, D.J., Yellen, P., and Chen, W., Hyaluronan microspheres for sustained gene delivery and site specific targeting, *Biomaterials*, 25, 147–157, 2004. With permission.)

For a period of 63 days, the HA–DNA microspheres cumulatively release approximately 610 ng of DNA per 100 µg of microspheres (1% loading; see Figure 30.4A) in hyaluronidase (HAse) buffer. Thus, approximately 60% of the microsphere's DNA contents have been cumulatively released at the termination of this experiment. As seen in Figure 30.4B, microspheres with 5% DNA loading release approximately 5 times more pDNA than HA microspheres with 1% DNA loading (approximately 3000 ng of pDNA per 100 µg of microsphere), which also correspond to 60% of the microsphere's contents. Overall, HA–DNA microspheres incubated in PBS release approximately one third of the DNA as compared to HA–DNA microspheres incubated in HAse solutions, regardless of the loading conditions.

The release profiles for HA–DNA microspheres at the first two time points show equal amounts of pDNA release for microspheres dispersed in HAse and in PBS buffer. This initial release is likely the consequence of free pDNA bound to the surfaces of the HA microspheres but not fully incorporated into the structure. Since the initial release is mediated by diffusion instead of degradation, the amount of pDNA released is independent of DNA loading and the presence of HAse in the release medium.

At the end of the release study, the HA–DNA microspheres in HAse solution release three times more DNA than microspheres in PBS alone. The microspheres incubated in PBS are degraded only by hydrolysis; thus, both the rate and amount of pDNA release are significantly less than HA–DNA microspheres in HAse solution. When the DNA loadings are compared, the microspheres with 5% loading release approximately five times more pDNA than the microspheres with 1% loading.

Native HA cross-linked with adipic dihydrazide has extended the release profile of HA–DNA microspheres from days to months when compared to HA microspheres prepared by heat precipitation without any chemically mediated cross-linking.[22] The release profiles and the release capabilities for HA-DNA microspheres (cross-linked with adipic dihydrazide chemistry) in PBS are similar to PLGA–DNA microspheres (mol. wt. 6000, 50:50 lactic to glycolic acid ratio, 2% DNA loading), which have been prepared by a solvent evaporation in emulsion method and degraded in PBS.[27]

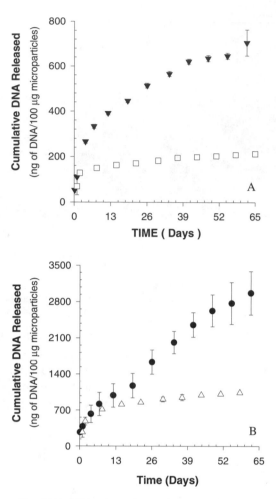

Figure 30.4   DNA release profile of HA–DNA microspheres. (From Yun, Y.H., Goetz, D.J., Yellen, P., and Chen, W., Hyaluronan microspheres for sustained gene delivery and site specific targeting, *Biomaterials*, 25, 147–157, 2004. With permission.) (A) HA–DNA microspheres with 1% DNA loading (▼) in 10 units/ml of HAse and (■) in PBS. (B) HA–DNA microspheres with 5% DNA loading (●) in 10 units/ml of HAse and (△) in PBS.

30.2.4   Gel Electrophoresis Analysis of DNA Released from HA–DNA Microspheres

The structural conformation of the released pDNA has been analyzed by gel electrophoresis of the DNA samples collected during the course of the release study. As shown in Figure 30.5A, lanes 1 and 2 have been loaded respectively with the 1 kb DNA marker and the stock pDNA used to prepare the HA–DNA microspheres. As expected, the stock pDNA has two distinct bands corresponding to relaxed (upper band) and supercoiled (lower band) conformations. Lanes 3 to 6 have been loaded with representative pDNA samples collected during the release kinetics experiment (days 1, 4, 7, and 14 respectively). As indicated by the double bands on the gel, the pDNA released from the HA–DNA microspheres is intact (although faint, both the relaxed and supercoil bands are present for lane 3). However, an increased proportion of the relaxed conformation is present for the pDNA collected from the release studies as compared to the stock pDNA. The faint fluorescence observed in the areas below the supercoiled DNA (the lower bands) is likely the released pDNA that was partially degraded. These results could be attributable to the combination of a brief exposure to high shear stresses during their preparation, the agitation required to suspend the microspheres during the release studies, or the process of lyophilization.

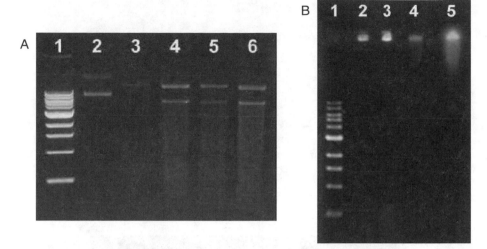

Figure 30.5   Gel electrophoresis. (A) Released DNA samples. (1) DNA molecular weight marker; (2) stock DNA used for preparing HA–DNA microspheres; and (3, 4, 5, and 6) DNA samples collected after 1, 4, 7, and 14 days, respectively, during the release study. (B) HA and pDNA mixtures. (From Yun, Y.H., Goetz, D.J., Yellen, P., and Chen, W., Hyaluronan microspheres for sustained gene delivery and site specific targeting, *Biomaterials*, 25, 147–157, 2004. With permission.) (1) DNA molecular weight marker; (2) HA–DNA microspheres with 1% DNA loading; (3) HA–DNA microspheres with 5% DNA loading; (4) HA solution containing 1% DNA; and (5) HA solution containing 5% DNA.

Figure 30.5B depicts the results obtained by performing a gel electrophoresis on HA–DNA microspheres and non-cross-linked HA–DNA mixtures. Lane 1 has been loaded with the DNA molecular weight marker. Lanes 2 and 3 have been loaded with HA–DNA microspheres with 1% and 5% DNA loading, respectively. The intense fluorescence present in the well indicates that pDNA is trapped inside the microsphere. Although leaching has been detected in the past,[6] it is not detectable for this study. Lanes 3 and 6 have been loaded with HA solutions (non-cross-linked) containing 1% and 5% DNA, and fluorescent smears are observed along lanes in conjunction with the fluorescence of pDNA inside the wells. These results could be due to an association with HA (possibly through extensive hydrogen bonding). Only the pDNA molecules that bind to the HA of lower molecular weight range (HA is a mixture with heterogeneous chain length, and thus is composed of various molecular weight HA) could migrate into the gel. This association could serve as a mechanism that protects pDNA from degradation, and could account for the minimal DNA degradation observed in lanes 3 to 6 in Figure 30.5A. Due to the higher DNA loading, the fluorescent intensity in lane 5 (5% DNA loading) is much greater than the fluorescent intensity in lane 4 (1% DNA loading).

The results from the electrophoresis studies show that the dihydrazide cross-linking chemistry is not destructive to the incorporated pDNA. Furthermore, the HA microspheres have released structurally intact pDNA over a period of weeks. Although some fraction of the released pDNA shows degradation as the result of microsphere preparation or physical perturbations inherent in the release studies, the majority of the released pDNA are structurally intact and available for transfection.

30.3 *IN VITRO* CELLULAR TRANSFECTION

The bioactivity of the pDNA recovered during the course of the release kinetics experiments has been assessed initially with transfection studies using Chinese hamster ovarian (CHO) cells

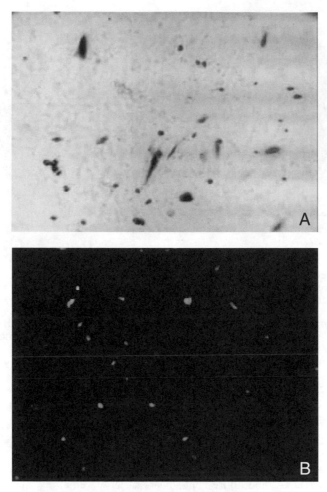

Figure 30.6 **(See color insert following page 336)** Transfection of CHO cells in culture with Lipofectamine™ transfection medium contained the (A) pDNA encoding for β-gal eluded from HA–DNA microspheres (from released studies) and (B) pDNA encoding for GFP eluded from HA–DNA microspheres.

(ATCC, Rockville, MD).[6] As seen in Figures 30.6A and 30.6B, the pDNA encoding for β-gal and GFP released from HA microspheres transfects cells grown in culture. These results indicate that the pDNA released from HA–DNA microspheres is bioactive. The transfection results of the pDNA released from HA microspheres are similar to the published reports of the pDNA associated with a transfection reagent,[29-32] but lower than viruses.[33] When the transfection reagent has been excluded from the DNA medium, transfected cells could not be detected with X-gal staining.

The relative level of transfection over a time course of two months, depicted in Figure 30.7, has been determined with pDNA encoding for β-gal released from a representative HA–DNA microsphere (1% DNA loading). Initially, a relatively high number of CHO cells are transfected but gradually decrease with time. The level of cellular transfection is generally in agreement with the DNA release profiles depicted in Figure 30.5A. The lower levels of transfection observed toward the end of the study are attributed to the decreased amounts of pDNA released from the HA–DNA microspheres and/or the loss of activity of the released pDNA due to degradation.

The transfection study of the released DNA illustrates that the sustained release of DNA could result in sustained transfection. Transfection of cells in culture shows good correlation with the release of DNA from HA–DNA microspheres (see Figure 30.5). As expected, the initial burst

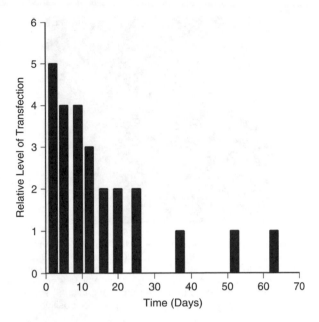

Figure 30.7 Relative levels of transfection of cells in culture from DNA samples collected from the release study. The transfection decreased with time and corresponded to the amount of DNA released from the HA–DNA microspheres. (From Yun, Y.H., Goetz, D.J., Yellen, P., and Chen, W., Hyaluronan microspheres for sustained gene delivery and site specific targeting, *Biomaterials*, 25, 147–157, 2004. With permission.)

release of pDNA results in a high degree of initial transfection. Although the level of transfection decreases with respect to time, the bioactivity of the released pDNA is still measurable.

30.4 *IN VIVO* TRANSFECTION OF RAT HIND LIMB MUSCLES

The *in vivo* efficacy of HA–DNA microspheres has been determined with a rat hind limb model. To determine if the pDNA delivered from the HA microspheres is being transcribed, the RNA extracted from the muscles, 3 weeks after injection, has been purified (DNA digested) and analyzed with reverse transcriptase polymerase chain reaction (RT-PCR).[6] The forward primer sequence is 5′–ACCCGCATTGACCCTAAC–3′, and the reverse primer sequence is 5′–TGTATCGCTCGC-CACTTC–3′. A one step RT-PCR (Qiagen Inc., Valencia, CA) has been performed according to the manufacturer's instructions. Thirty cycles are used for the denaturing (94°C for 1 min), annealing (53°C for 30 sec), and extension (68°C for 1 min) steps with a final 10 min extension at 75°C. The RT-PCR products have been analyzed with gel electrophoresis, and the amplicons (234 base pairs) have been sequenced to ensure they are of β-gal origin.

The RT-PCR analysis (see Figure 30.8) show positive signals for rat hind limb muscles injected with HA–DNA and PLGA–DNA microspheres. Lanes 3 and 4 have been loaded respectively with amplicons from the hind limb tissues of two animals injected with HA–DNA microspheres. The amplicon in lane 5 (a positive signal was present though it was faint) has been generated from the RNA of a muscle injected with PLGA microspheres loaded with pDNA encoding for β-gal. The hind limb muscle injected with saline shows no signal (lane 2), and the stock pDNA encoding β-gal, used as a template for RT-PCR, shows a positive signal (lane 6). As expected, the amplicon from the stock pDNA is the same size as the hind limb muscles transfected from HA–DNA and PLGA microspheres.

The RT-PCR analysis shows that the hind limb muscles injected with HA–DNA microspheres are transcribing a messenger RNA for β-gal, which is foreign to the rat's genome. This result

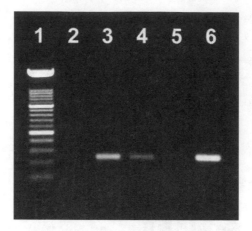

Figure 30.8   RT-PCR from rat hind limbs that have been injected with HA–DNA microspheres, saline, or PLGA. Tissues were harvested 3 weeks after injection. (From Yun, Y.H., Goetz, D.J., Yellen, P., and Chen, W., Hyaluronan microspheres for sustained gene delivery and site specific targeting, *Biomaterials*, 25, 147–157, 2004. With permission.) (1) 100 base pair ladder; (2) saline; (3 and 4) HA–DNA microspheres with pDNA encoding for β-gal from 2 animals; (5) PLGA microspheres loaded with DNA encoding for β-gal; (6) stock β-gal DNA template for positive control.

confirms that the skeletal tissues have been transfected by the pDNA released from HA or PLGA microspheres since the muscle injected with saline was negative. Furthermore, the detection of the messenger RNA, 3 weeks after the injection, could confirm the controlled released characteristics exhibited by HA–DNA microspheres *in vitro* are also apparent *in vivo*.

30.5 CONCLUSIONS

In conclusion, microspheres prepared from native HA by adipic dihydrazide cross-linking chemistry, a method that was originally used to prepare HA hydrogels, are biodegradable yet robust in an aqueous environment. The advantages for formulating HA–DNA microspheres with dihydrazide cross-linking chemistry are the following:

1. Native HA is utilized (instead of modified HA as used by Fidia Advanced Biopolymers[21])
2. The chemical reaction occurs at room temperature (in contrast to the elevated temperature used by Lim et al.[22])
3. The use of chemical reagents such as hexane, chloroform, methylene chloride, or glutaraldehyde (as described in Lim et al.[21,22]) is not required
4. Complicated mechanical equipment is not needed

As the HA–DNA microspheres degrade, the DNA is released for approximately 2 months. The released pDNA is structurally intact and able to transfect cells in culture. More significantly, the HA–DNA microspheres are able to transfect skeletal muscle *in vivo* with a DNA sequence encoding for an enzyme that is not native to the rat genome.

As in any nonviral vehicles, the initial transfection efficiency of HA–DNA microspheres is lower than viral vectors as observed in studies using cultured cells. This limitation could be alleviated by multiple administrations or by increasing the temporal availability of the desired gene. Other delivery vehicles such as naked DNA (without encapsulation into a matrix such as microspheres) and polyelectrolytes have short persistency *in vivo*, and the principal option for increasing their effectiveness is readministration. In contrast, viral vectors trigger an aggressive immune response when they are readministered; however, they initially have a high transfection rate.

An effective gene therapy will likely depend upon the optimal release[5,34] instead of maximal gene transfer for a short duration. For HA–DNA microspheres, the dose and the duration of release could be easily controlled by adjusting the DNA loading and the extent of cross-linking. Although the duration of cross-linking for HA–DNA microspheres has not yet been thoroughly investigated, our recent study using HA matrix cross-linked with adipic dihydrazide has shown that the DNA release profile can be modulated by altering this parameter.[12] In addition, the HA–DNA microspheres can be administered repeatedly, since HA is not immunogenic and high-molecular-weight HA does not elicit an inflammatory response.[35] These properties offer numerous options for the treatment of various diseases using gene therapy. For example, HA microspheres with two distinct release profiles and respectively loaded with genes encoding for vascular endothelial growth factor and platelet-derived growth factor could be used to stimulate angiogenesis in an ischemic tissue.[4,5] The flexibility of DNA loading and HA cross-linking, the extended release of DNA, and the bioactivity of the released DNA *in vivo* make HA microspheres an attractive vehicle for gene therapy.

ACKNOWLEDGMENTS

This study was supported by a grant from the National Heart, Lung, and Blood Institute (R01 HL65175; WC). The authors would like to thank Dr. Rajinder Bhasin of Corner Stone Pharmaceutical, Inc., for his assistance in performing the particle size analysis.

REFERENCES

1. Langer, R., Drug delivery and targeting, *Nature* 392, 5, 1998.
2. Koping-Hoggard, M. et al., Chitosan as a nonviral gene delivery system. Structure-property relationships and characteristics compared with polyethylenimine *in vitro* and after lung administration *in vivo*, *Gene Ther.* 8, 1108, 2001.
3. Leong, K.W., Biopolymer-DNA nanospheres, in *Nonviral Vectors for Gene Therapy*, Huang, L., Hung, M., and Wagner, F. (Eds.), Academic Press, San Diego, 1999, 267.
4. Richardson, T.P. et al., Polymeric system for dual growth factor delivery, *Nat. Biotechnol.* 19, 1029, 2001.
5. Carmeliet, P. and Conway, E.M., Growing better blood vessels, *Nat. Biotechnol.* 19, 1019, 2001.
6. Yun, Y.H. et al., Hyaluronan microspheres for sustained gene delivery and site specific targeting, *Biomaterials*, 25, 147–157, 2004.
7. Prestwich, G.D. et al., Chemical modification of hyaluronic acid for drug delivery, biomaterials, and biochemical probes, in *The Chemistry, Biology, and Medical Applications of Hyaluronan and Its Derivatives*, Laurent, T.C. (Ed.), Portland Press, Miami, 1998, 43.
8. Fraser, J.R., Brown, T.J., and Laurent, T.C., Catabolism of hyaluronan, in *The Chemistry, Biology, and Medical Applications of Hyaluronan and Its Derivatives*, Laurent, T.C. (Ed.), Portland Press, Miami, 1998, 85.
9. Peppas, N.A. and Langer, R., New challenges in biomaterials, *Science* 263, 1715, 1994.
10. Freed, L.E. et al., Biodegradable polymer scaffolds for tissue engineering, *Biotechnology (N.Y.)* 12, 689, 1994.
11. Langer, R., Biomaterials in drug delivery and tissue engineering: one laboratory's experience, *Acc. Chem. Res.* 33, 94, 2000.
12. Kim, A., Checkla, D.M., and Chen, W., Characterization of DNA-hyaluronan matrix for sustained gene transfer, *J. Control. Release* 90, 2003.
13. Goa, K.L. and Benfield, P., Hyaluronic acid. A review of its pharmacology and use as a surgical aid in ophthalmology, and its therapeutic potential in joint disease and wound healing, *Drugs* 47, 536, 1994.
14. Laurent, T.C. and Fraser, J.R., Hyaluronan, *FASEB J.* 6, 2397, 1992.

15. Toole, B.P., Hyaluronan and its binding proteins, the hyaladherins, *Curr. Opin. Cell Biol.* 2, 839, 1990.

16. Bertolami, C.N., Berg, S., and Messadi, D.V., Binding and internalization of hyaluronate by human cutaneous fibroblasts, *Matrix* 12, 11, 1992.

17. Balazs, E.A. and Denlinger, J.L., Clinical uses of hyaluronan, *Ciba Found. Symp.* 143, 265, 1989.

18. Davidson, J.M. et al., Hyaluronate derivatives and their application to wound healing: preliminary observations, *Clin. Mater.* 8, 171, 1991.

19. Juhlin, L., Hyaluronan in skin, *J. Intern. Med.* 242, 61, 1997.

20. Luo, Y., Kirker, K.R., and Prestwich, G.D., Modifications of natural polymers: hyaluronic acid, in *Methods of Tissue Engineering*, Atala, A. and Lanza, R.P. (Eds.), Academic Press, San Diego, 2001, 539.

21. Benedetti, L.M., Topp, E.M., and Stella, V.J., Microspheres of hyaluronic acid esters — fabrications methods and *in vitro* hydrocortisone release, *J. Control. Release* 13, 33, 1990.

22. Lim, S.T. et al., Preparation and evaluation of the *in vitro* drug release properties and mucoadhesion of novel microspheres of hyaluronic acid and chitosan, *J. Control. Release* 66, 281, 2000.

23. Deasy, P.B., *Microencapsulation and Related Drug Processes*, Marcel-Dekker, New York, 1984, 195.

24. Pouyani, T., Harbison, G.S., and Preswich, G.D., Novel hydrogel of hyaluronic acid: synthesis, surface morphology, and solid state NMR, *J. Am. Chem. Soc.* 116, 7515, 1994.

25. Pouyani, T. and Prestwich, G.D., Functionalized derivatives of hyaluronic acid oligosaccharides: drug carriers and novel biomaterials, *Bioconj. Chem.* 5, 339, 1994.

26. Wang, J. et al., Enhanced gene expression in mouse muscle by sustained release of plasmid DNA using PPE-EA as a carrier, *Gene Ther.* 9, 1254, 2002.

27. Wang, D. et al., Encapsulation of plasmid DNA in biodegradable poly(D, L-lactic-co- glycolic acid) microspheres as a novel approach for immunogene delivery, *J. Control. Release* 57, 9, 1999.

28. Tinsley-Bown, A.M. et al., Formulation of poly(D,L-lactic-co-glycolic acid) microparticles for rapid plasmid DNA delivery, *J. Control. Release* 66, 229, 2000.

29. Ohki, E.C. et al., Improving the transfection efficiency of post-mitotic neurons, *J. Neurosci. Methods* 112, 95, 2001.

30. Tang, M.X., Redemann, C.T., and Szoka, F.C., Jr., *In vitro* gene delivery by degraded polyamidoamine dendrimers, *Bioconj. Chem.* 7, 703, 1996.

31. Gonzalez, H., Hwang, S.J., and Davis, M.E., New class of polymers for the delivery of macromolecular therapeutics, *Bioconj. Chem.* 10, 1068, 1999.

32. Yamamoto, M. et al., High efficiency gene transfer by multiple transfection protocol, *Histochem. J.* 31, 241, 1999.

33. Forsayeth, J.R. and Garcia, P.D., Adenovirus-mediated transfection of cultured cells, *Biotechniques* 17, 354, 1994.

34. Back, S. and March, K.L., Gene therapy for restenosis: getting nearer the heart of the matter, *Circ. Res.* 82, 295, 1998.

35. Noble, P.W., McKee, C.M., and Horton, M.R., Induction of inflammatory gene expression by low-molecular-weight hyaluronan fragments in macrophages, in *The Chemistry, Biology, and Medical Applications of Hyaluronan and Its Derivatives*, Laurent, T.C. (Ed.), Portland Press, Miami, 1998, 219.

PART V

Specialized Delivery Systems

CHAPTER 31

Genetically Engineered Protein-Based Polymers: Potential in Gene Delivery

Zaki Megeed and Hamidreza Ghandehari

CONTENTS

31.1 INTRODUCTION

The relatively recent explosion of information on protein structure and function has led to a shift in how proteins are perceived. Detailed structural information, obtained by techniques such as x-ray crystallography, have provided an intricate picture of the conformational properties that underlie protein function. Coupling of this structural information with molecular modeling and physical manipulation techniques, such as steered molecular dynamics simulations and atomic force spectroscopy, has enabled the detailed characterization of proteins' structural and biochemical responses to stimuli such as applied stress.[1,2] In parallel with these molecular characterizations of protein structure and function, an interest has developed in the synthesis and characterization of protein-based materials, or *materials biotechnology*. While scientists have long studied the material properties of structural proteins, such as silk, elastin, collagen, and keratin, it is only more recently that the interesting material properties of other proteins, and their subunits, are being recognized.

0-8493-1934-X/05/$0.00+$1.50
© 2005 by CRC Press LLC

The extensive mechanical characterization of the giant muscle protein titin, and its immunoglobulin and fibronectin subunits, is one example.[3]

As the material properties of protein subunits are further characterized, applications for protein-based materials are emerging in several fields, including the biomedical arena. The purpose of this chapter is to explain the rationale behind the utilization of protein-based biomaterials, their synthesis and characterization, and biomedical applications including gene delivery. Though applications specifically intended for gene delivery have just begun to emerge, the perceptive reader will be able to extrapolate the potential for gene delivery applications rather easily.

31.2 GENETICALLY ENGINEERED PROTEIN-BASED POLYMERS

Genetically engineered protein-based polymers (hereafter referred to simply as *genetically engineered polymers* or *protein-based polymers*) are a class of polymers that can potentially include poly(amino acid)s, natural proteins or protein domains, and hybrids of these to produce artificial proteins. The key distinction that separates this class of materials from chemically synthesized polymers is that the composition, sequence, and molecular weight of genetically engineered polymers are programmed by recombinant techniques at the DNA level.[4] By contrast, sequential polypeptides are synthesized by chemical oligomerization of short amino acid sequences, often synthesized by solid-phase techniques. Chemically synthesized poly(amino acid) homopolymers or copolymers are generally synthesized by random chemical reactions of a homogeneous or heterogeneous pool of amino acids. The synthesis of both sequential polypeptides and poly(amino acid)s results in a mixture of products with a distribution of molecular weights. Though sequential polypeptides offer a limited degree of control over amino acid sequence and composition, poly(amino acid) random copolymers have a random sequence, and their composition depends on the reactivity of the monomers. Recombinant methods allow the construction of DNA templates encoding protein-based polymers that can be utilized in a number of expression systems, including bacteria, fungi, and plants.[5-8]

31.2.1 Synthesis and Characterization of Genetically Engineered Polymers

Although there are several synthetic strategies for the production of genetically engineered polymers, all rely on the basic principle of self-ligation (concatamerization) of DNA monomers to form concatameric DNA sequences.[9,10] These polymeric genes are then transcribed and translated by the cellular machinery to produce a protein-based polymer. The synthesis of genetically engineered polymers begins with the conceptual design of an oligonucleotide sequence encoding the desired monomer. In designing this sequence, there are several biological constraints that must be balanced with the desire to make a given material. First, the codon usage preference of the organism in which the polymers are to be synthesized (usually *Escherichia coli*) must be considered. Repetitive usage of rare codons can lead to transfer RNA (tRNA) depletion, completely inhibiting the synthesis of some polymers and causing truncation of others. Second, while polymeric gene sequences require repetition by nature, the repetition of identical codons should be minimized. The redundancy of the genetic code can be exploited to achieve this objective while simultaneously maintaining a repetitive amino acid sequence. The consequences of highly repetitive codon usage include tRNA depletion and genetic instability. Some organisms, including *E. coli*, truncate repetitive genes through recombination or deletion.[11-13] Genetic instability arising from recombination or deletion may make it difficult or impossible to produce polymers from some DNA sequences. Third, the sequence should be designed to minimize complementarity, which can cause secondary structure formation at the messenger RNA (mRNA) level. Formation of mRNA secondary structures can lead to pausing and disengagement of the ribosome during translation of the mRNA, leading to truncated polymers.

After it has been designed, the monomer oligonucleotide sequence is synthesized by automated chemical synthesizers. The oligonucleotide size limit allowed by the current technology is approximately 100 bases. This limit sometimes requires the synthesis of multiple oligonucleotides, which can be enzymatically ligated to form monomers with a length greater than 100 nucleotides. After synthesis, the monomers are purified and annealed with their complementary strands to make double-stranded DNA that is suitable for cloning.

During synthesis, two types of plasmids (or vectors) are frequently used. These are the *cloning* vector and the *expression* vector. The cloning vector is a plasmid that lacks the DNA sequences necessary for transcription of an inserted gene. Cloning vectors are normally used to "store" genes in a stable form until they can be placed in expression vectors. Expression vectors contain the DNA sequences necessary for gene transcription, and in combination with a complementary cell line can be used to express a protein (polymer) of interest. To minimize genetic instability early in the process, a cloning vector is often used in the first stage of polymer synthesis. The monomer gene is inserted into the vector and the vector is transformed into *E. coli*, which can be grown to produce large quantities of the vector and hence the monomer gene that it contains. The presence of the monomer gene in the vector is confirmed by restriction digestion or the polymerase chain reaction (PCR) and its sequence is confirmed by DNA sequencing.[14]

After the presence and identity of the monomer gene have been confirmed, DNA concatamers are synthesized. The first step in the synthesis of concatamers is the production of relatively large amounts of monomer. This can be done either by performing large-scale plasmid preparations or by PCR. The classic technique for concatamerization of DNA monomers involves the incubation of monomers with ligation enzymes such as T4 DNA ligase. This strategy produces an assortment of DNA concatamers in which the size can be roughly controlled by the reaction time and the concentration of the monomers. However, this technique suffers from three primary limitations. First, because of the random nature of the ligation reaction, there is no guarantee that a concatamer of a desired size will be obtained. Second, as concatamer size increases, circularization can occur, preventing ligation with a vector. Third, the method requires the inclusion of unique restriction enzyme recognition sites that may require the insertion of extraneous codons (and hence amino acids) between monomers. These three limitations have been addressed by the development of slightly altered synthetic strategies. Recursive directional ligation is a method that can be used to synthesize concatameric DNA sequences of precisely defined length while avoiding circularization problems.[15] Another technique, relying on the isolation and ligation of previously ligated concatamers of defined length, has also been utilized to control length and limit circularization.[16] Finally, a technique, termed *seamless cloning*, takes advantage of the fact that type IIs restriction endonucleases remove their own recognition site, to produce monomers and concatamers without any extraneous codons between monomers.[17,18]

After synthesis, the DNA concatamers are ligated into either another cloning vector or an expression vector. The choice of the vector is partially based upon the availability of the necessary restriction sites and desired purification tag, and will vary depending on the cloning strategy. When the polymer gene is cloned into an expression vector, it is usually transformed first into a strain of bacteria that is incapable of expressing it. As with the use of a cloning vector, this approach avoids potential problems related to genetic stability or cytotoxicity due to basal levels of polymer expression. However, the production of polymers eventually requires that an expression plasmid containing the polymer gene be transformed into an expression host.

Most *E. coli* expression systems in use today contain an inducible promoter that can be used to turn on polymer production once the cells have grown to an optimal density. This strategy is intended to maximize the efficiency of polymer production. Production of recombinant proteins redirects metabolic resources that are normally used for growth, maintenance, and division, often leading to a slowing or arrest of growth. By allowing the cells to reach a relatively high density prior to inducing polymer expression, the need for further division can be somewhat circumvented and the yield of the polymer can be increased.

Purification of polymers is performed by standard techniques that have previously been established for recombinant proteins. Frequently these include the use of affinity tags (e.g., poly[histidine], glutathione-s-transferase) that can be used to chromatographically purify recombinant proteins. If necessary, the tag can subsequently be removed by enzymatic cleavage. Some polymers have been purified on the basis of their physicochemical properties. For example, silk-like polymers have been purified by taking advantage of their low solubility in aqueous medium.[19] Elastin-like polymers (ELPs) have been purified by temperature cycling above and below their inverse temperature transition (T_t).[20] This technique has been extended to produce an ELP tag that can be used to purify a number of recombinant proteins by temperature cycling, which is faster and less expensive than column chromatography.[21]

After purification, polymers are typically characterized by a standard set of techniques that can include amino acid content analysis, mass spectrometry, sodium dodecyl sulfate polyacrylamide gel electrophoresis (SDS-PAGE), and immunoblotting. These methods are intended to verify the identity of the polymer. Depending on the type of polymer being synthesized, a series of polymer-specific characterizations will then be performed. For example, ELPs are typically characterized in terms of their T_t.

While biological synthesis of polymers confers many advantages, it also imposes some limitations. As described previously, there are several issues that must be considered when designing the monomer and concatamer gene sequences. Furthermore, the potential toxicity of the genetically engineered polymer to the expression host, and the resultant effect on the product, must be considered. As discussed later, highly charged proteins may be particularly problematic.

While the 20 natural amino acids do provide significant structural diversity, the natural amino acid pool can also be viewed as a limitation. Kiick et al. have begun to address this limitation by incorporating artificial amino acid analogs into genetically engineered polymers.[22] Incorporation of the amino acid azidohomoalanine into the medium of methionine-depleted bacterial cultures resulted in the replacement of natural methionine with azidohomoalanine. The significance of this substitution is that it allows chemical modification of the azide group by the Stuadinger ligation.

31.3 BIOMEDICAL APPLICATIONS OF GENETICALLY ENGINEERED POLYMERS

31.3.1 Biocompatibility and Biodegradation

The fact that genetically engineered polymers are composed of naturally occurring amino acids, and often naturally occurring protein motifs, raises the possibilities that these polymers could be designed to be enzymatically biodegradable and responsive to physiological stimuli. The insertion of proteolytic enzyme recognition sequences into the polymer backbone allows for fine control over the rate of biodegradation of protein-based polymers, which are thought to eventually degrade to their amino acid constituents.[23,24] However, these desirable attributes must be tempered by the possible immunogenicity of some protein sequences and their degradation products.

The silk-elastin-like polymers (SELPs) are one class of genetically engineered polymers for which biocompatibility and biodegradation have been extensively characterized. SELPs are a family of copolymers with the general structure $[(S)_x(E)_y]_z$, where (S) is a motif from *Bombyx mori* (silkworm) silk (Gly-Ala-Gly-Ala-Gly-Ser) and (E) is a motif from mammalian elastin (Gly-Val-Gly-Val-Pro).[5] Studies of implanted SELP films have shown a mild, localized immune response with some macrophage surveillance up to approximately 1 week.[23] After 1 week, the implanted SELP films were observed to be surrounded by fibroblasts and collagen, indicating that healing was occurring, while the number of macrophages around the implant decreased. Implanted SELP sponges were infiltrated by fibroblasts, indicating a high level of biocompatibility with living cells. Similar results were observed in a model of wound healing. Rabbits injected with SELPs were found to develop antibodies to the silk-like blocks of the polymer, but not to the elastin-like blocks.

Biodegradation of SELPs was found to occur in a manner influenced by their composition *and* sequence.[23]

31.3.2 Applications of Genetically Engineered Polymers

Genetically engineered polymers have been used for early-stage studies in several applications of biomedical interest. Perhaps the class of polymer most widely studied has been the ELPs. Chemically synthesized sequential ELP polypeptides and their genetically engineered counterparts have the general sequence (Gly-Xaa-Gly-Val-Pro), where (Xaa) is a substitutable residue that can be used to tailor the properties of the ELP. ELPs exhibit a phase transition that can be triggered by changes in ionic strength, pH, and polymer concentration. Hence, by varying the molecular weight and (Xaa), ELPs can be designed to precipitate under a variety of environmental conditions.[25]

Cappello et al. have synthesized silk-like polymers that contain periodically spaced fibronectin sequences.[26,27] These polymers, called SLP-F, have been shown to support the attachment of a variety of cell types, including epithelial cells, fibroblasts, and cancer cells.[26] In addition, silk-like polymers containing regions of human laminan have been synthesized (SLP-L).[23] The SLP-L polymers promoted greater attachment and spreading of fibrosarcoma and rhabdomyosarcoma cell lines than polylysine or laminan coatings. The inclusion of these attachment sequences in between silk-like blocks allows them to withstand autoclaving without loss of activity.[23] Similarly, Panitch et al. have incorporated periodically spaced fibronectin CS5 domains into ELPs.[28] These polymers enhanced the attachment of endothelial cells to a glass substrate. Genetically engineered polymers containing cellular attachment motifs could be promising materials for tissue engineering and applications requiring the delivery of genes from surfaces.

Halstenberg et al. have synthesized hybrid protein-*graft*-poly(ethylene glycol) polymers for tissue repair.[24] The engineering of this biomaterial was particularly elegant, as it included motifs for rationally controlling both cellular attachment and biodegradation. Hydrogels prepared by cross-linking the polymer acted as scaffolds for three-dimensional outgrowth of fibroblasts, which attached to the polymer via specific interactions between cellular integrins and an RGD sequence on the polymer. Outgrowth involved serine protease degradation of the polymer at sites engineered for this purpose.

Localized hyperthermia has been used to target systemically administered ELPs.[29] ELP sequences were chosen so that the T_t of the polymer (40° C) was above body temperature (37° C) but below a temperature induced by localized heating of tissue (42° C). Using this technique, rhodamine-ELP conjugates were observed to accumulate in ovarian tumors by *in vivo* fluorescence microscopy.

ELPs have also been investigated as biomaterials for cartilaginous tissue repair.[30] Chondrocytes cultured in ELPs retained their phenotype and synthesized significant amounts of extracellular matrix molecules that are important for cartilage function. This highlights the potential use of genetically engineered polymers as biocompatible extracellular matrices for tissue repair and engineering. It may also be possible to incorporate drugs or genes into these systems to guide the development of new tissue.

Recently, ELPs have been used for the patterning of proteins on surfaces. ELP fusion proteins self-assemble onto hydrophobic surfaces above T_t and desorb below T_t.[31] In combination with hydrophilic surfaces, which do not adsorb ELP, this technique can be used to create micropatterned surfaces for cell growth and protein arrays.

Hydrogels are cross-linked, water-swollen polymer networks that have numerous biomedical applications, including the controlled delivery of genes.[32,33] Several genetically engineered polymers have been synthesized that can form hydrogels through either physical or chemical cross-linking. Those forming hydrogels by physical cross-linking include polymers based on the leucine zipper motif[34] and some SELPs.[35,36] Polymers containing the leucine zipper motif reversibly self-assemble in response to changes in temperature and pH.[34] Some SELPs (e.g., SELP-47K) (Figure 31.1)

MDPVVLQ[RR]DWENPGVTQLN[R]LAAHPPFASDPM
GAGSGAGAGS[(GVGVP)₄G[K]GVP(GVGVP)₃(GAGAGS)₄]₁₂
(GVGVP)₄G[K]GVP(GVGVP)₃ (GAGAGS)₂GAGA
MDPG[R]YQDL[R]SHHHHHH

Figure 31.1 The 884 amino acid SELP-47K sequence has a molecular weight of 69,814 Daltons. It is composed of a head and tail sequence, and a series of silk-like (GAGAGS) and elastin-like (GVGVP) repeats. Residues that are predominantly positively charged at pH 7.4 are boxed.

irreversibly self-assemble via hydrogen bonding between their silk-like blocks.[35,36] The release of solutes such as drugs, proteins, DNA, and macromolecular probes from SELP hydrogels has been investigated.[33,35,37] ELP-based hydrogels have also been fabricated. These include chemically cross-linked gels[38] and thermally reversible micellar aggregates that act as virtual cross-links formed between elastin domains.[39] ELP-based hydrogels have been observed to phase or volume transitions in response to changes in temperature.[40,41]

Thermoreversible nanoparticles have been produced from ELPs with hydrophilic and hydrophobic blocks.[42] These nanoparticles were found to form a micellar core-shell structure, with the formation of the nanoparticles regulated by the dehydration of the hydrophobic block above the T_t. Dynamic light scattering studies indicated that the size of the nanoparticles was dependent on the temperature of the medium. There is significant interest in the synthesis of polymeric micelles for gene delivery.[43] Genetic synthesis of micelle-forming polymers that are well defined and responsive to stimuli may enhance their effectiveness for some gene delivery applications.

Another class of related biomaterials combines genetically engineered polymers with chemically synthesized polymers (i.e., hybrid polymers). Combining these two synthetic techniques allows the synthesis of a broad range of materials that is not accessible by either means alone. Some examples of materials synthesized by these techniques include *N*-(2-hydroxypropyl)methacrylamide (HPMA) copolymers cross-linked by coiled coils and immunoglobulin domains.[44,45] Both materials form hydrogels that are responsive to temperature, with the former collapsing with increased temperature, while the latter expands. These hydrogels could potentially be useful for thermally controlled gene delivery applications. The previously discussed protein-*graft*-poly(ethylene glycol) polymers are another example of a hybrid polymer.[24]

31.4 GENE DELIVERY APPLICATIONS FOR GENETICALLY ENGINEERED POLYMERS

In order for gene therapy to be successful as a treatment, gene delivery must first reliably overcome a series of biological barriers. These barriers are discussed in detail in other chapters of this book. The research in our laboratory is currently focused on overcoming biological barriers on two levels: (1) site-specific controlled delivery of naked DNA and viral vectors, and (2) condensation of DNA and endosomal escape at the intracellular level. Our results from the first branch of research will be discussed and the rationale and strategies for approaching the second problem will be presented.

One significant problem with current gene delivery systems is targeting them to the desired site of action. Targeting can improve therapeutic benefit and reduce toxicity, by increasing the concentration of the vector at the site of action while simultaneously minimizing the delivery of the vector to other parts of the body. The targeting strategy is dependent on the route of administration: *systemic* or *site specific*.

Targeting after systemic administration can be broadly categorized as *active* or *passive*. Active systemic targeting refers to targeting of a gene delivery system via a specific molecular interaction (e.g., between a motif on the vector and a receptor in the targeted tissue). Ligands used for active targeting include antibodies,[46] transferrin,[47] folate,[48] and peptides,[49] among others. Ideally, actively targeted vectors accumulate only in the targeted tissue. In reality, active targeting is complicated by the fact that cellular receptors and ligands are often broadly expressed on many different cell types. This can lead to significant delivery of the vector outside of the intended tissue.

Passive systemic targeting relies on the intrinsic, nonspecific relationship between the physicochemical properties of a molecule (or vector) and the systemic physiology. Targeting ligands and receptors are not used in passive targeting. The enhanced permeability and retention effect, causing accumulation of macromolecules in solid tumors, is one example of passive targeting.[50]

In contrast to systemic targeting, site-specific controlled delivery targets a specific location by implantation of and release from a matrix or device. An example of this approach is controlled gene delivery from polymeric matrices.[51-55] The use of polymeric matrices to deliver genes confers three primary advantages over bolus administration. First, the release profile of the gene can be manipulated by altering the physicochemical properties of the polymer and fabrication of the matrix. This allows gene delivery to occur in a prolonged, sustained, and controlled manner, possibly increasing its effectiveness. Second, encapsulation of DNA or a vector in a polymeric matrix may offer some protection against degradative enzymes such as nucleases and proteases. Third, polymeric matrices allow precise spatial localization, obviating the need for active targeting ligands that may be required for systemic administration. Our research has focused on investigating the potential of genetically engineered SELPs as *in situ* gel-forming matrices for the controlled intratumoral delivery of plasmid DNA and adenoviral vectors.[33,56]

Given the current limitations of actively targeted, systemic cancer gene delivery systems, intratumoral delivery offers a logical alternative. The primary advantages of intratumoral gene delivery stem from its very precise spatial localization. Direct injection into the tumor obviates the need for active or passive targeting on the systemic level. The efficiency of the therapy may be further increased by attaching tumor-specific ligands to a vector.[47] The primary disadvantage of intratumoral gene therapy is that each tumor must be individually identified and injected. Hence, by virtue of the difficulty in their detection and distributed nature, small metastases could be impossible to treat by this method. Intratumoral cancer gene therapy may thus be best suited for the treatment of cancers that grow slowly (e.g., prostate cancer) or where potential complications or disability make tumor resection less desirable (e.g., prostate cancer, head and neck cancer).

31.4.1 Controlled Gene Delivery from Silk-Elastinlike Hydrogels

One member of the SELP family, SELP-47K (Figure 31.1) forms hydrogels spontaneously through hydrogen bonding between the silk-like blocks. Formation of this physically cross-linked hydrogel does not require the use of organic solvents or toxic cross-linking reagents that may damage DNA or viruses, or cause cytotoxicity. While hydrogel formation is a kinetic process at any temperature, it is substantially accelerated at 37°C versus room temperature, allowing polymer–DNA or polymer–virus solutions to be prepared at room temperature and to form matrices *in situ* within a few minutes of injection.

As a step toward gene delivery applications, the degree of swelling of SELP-47K hydrogels was characterized as a function of environmental conditions. The degree of swelling describes the amount of water imbibed by a hydrogel network. This parameter provides insight into the relative cross-linking density of the network and hence its porosity and solute transport characteristics. Swelling of SELP-47K hydrogels was insensitive to changes in ionic strength, temperature, and pH.[36] The insensitivity of SELP-47K hydrogels to changes in environmental conditions was explained by the immobilization of the elastin-like blocks through the irreversible self-assembly of the silk-like blocks. However, the degree of swelling of the hydrogel decreased with increasing polymer concentration and cure time.[36] These results are consistent with the increased network density expected from increasing the amount of polymer in the matrix or increasing the time allowed for matrix formation, since it is a kinetic process. SELP copolymers containing one silk-like block have been synthesized that exhibit responsiveness to pH, ionic strength, temperature, and concentration.[57,58] These polymers do not form physically cross-linked hydrogels under conditions studied thus far but can form aggregated particles depending on the above-mentioned conditions.

In order to assess its potential as a matrix for controlled gene delivery, the release of plasmid DNA from SELP-47K hydrogels was evaluated over 28 days.[33,56] DNA-containing hydrogels were fabricated by mixing aqueous DNA solution with aqueous SELP-47K solution. DNA–polymer solutions were incubated at 37°C to induce gelation.

DNA release was initially evaluated as a function of polymer concentration, DNA concentration, ionic strength, and cure time.[33] Increasing the polymer concentration and/or cure time decreased the rate of DNA release from the hydrogels, an effect consistent with an increased network density. Over the range studied, the DNA concentration did not influence the rate of release, with the fraction released at each time point being identical for hydrogels containing 50 or 250 μg/ml DNA. The ionic strength of the medium strongly influenced the rate of DNA release, with virtually no release observed at ionic strengths below $\mu = 0.17$ M, while identical release profiles were obtained above this ionic strength, up to $\mu = 0.50$ M (Figure 31.2A). This ionic strength dependence was thought to arise from an ionic interaction between the positively charged lysine and arginine residues on SELP-47K (Figure 31.1, boxed) and the negatively charged DNA phosphates.

To verify that DNA release was modulated by the effect of buffer ionic strength on the polymer-DNA interaction, a turbidity study was performed that evaluated the interaction between DNA and SELP-47K in buffers with various ionic strengths. Polymer and DNA were mixed at various molar charge ratios and the turbidity of each solution was determined using a spectrophotometer, at 400 nm. This method permitted the detection of insoluble interpolyelectrolyte complexes formed by electrostatic interaction between SELP-47K and DNA. As shown in Figure 2B, the relative turbidity of the mixtures showed a substantial increase when complexes were prepared in PBS with low

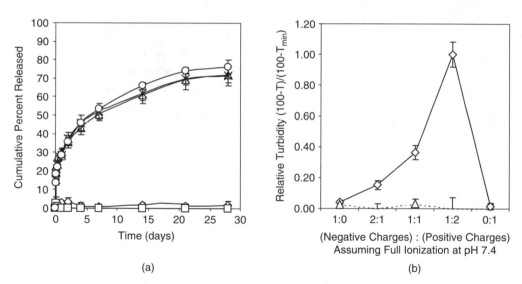

(a)   (b)

Figure 31.2   (A) Cumulative release of pRL-CMV from 12 wt% SELP-47K hydrogels, in PBS with $\mu = 0.03$ M ($\diamond$), 0.10 M ($\square$), 0.17 M ($\triangle$), 0.25 M (X), and 0.50 M ($\circ$). Hydrogels were cured for 1 h at 37° C before placement in the appropriate buffer. Each point represents average ± standard deviation (n = 3). (B) Effect of ionic strength on the formation of insoluble complexes between SELP-47K and pRL-CMV, in PBS with $\mu = 0.03$ M ($\diamond$, solid line) and $\mu = 0.17$ M ($\square$, dashed line). Ratios on the x-axis indicate the molar ratio of negative (DNA) charges to positive (polymer) charges, assuming 100% ionization of each at pH 7.4. The y-axis represents relative turbidity (100 T)/(100 T_{min}). Data points represent average ± standard deviation (n = 3). (From Megeed, Z., Cappello, J., and Ghandehari, H., Controlled release of plasmid DNA from a genetically engineered silk-elastinlike hydrogel, *Pharm. Res.* 19, 954–959, 2002. With permission.)

ionic strength ($\mu = 0.03$), indicating that at this ionic strength insoluble complexes form between SELP-47K and DNA. No increase in turbidity was observed in the buffer with an ionic strength of 0.17 M, indicating that DNA and polymer do not interact at this ionic strength. These results are consistent with the observed release data (Figure 31.2A).

DNA release from the hydrogels was observed for at least 28 days, with apparent diffusion coefficients on the order of 10^{-9} to 10^{-10} cm²/sec, depending on the polymer concentration and cure time.[33] Transfection studies were performed on COS-7 cells to evaluate the bioactivity of plasmid DNA containing the *Renilla* luciferase gene, after encapsulation in SELP-47K hydrogels for up to 28 days.[56] Figure 31.3 shows the results of these transfection assays, which indicate that polymer-encapsulated DNA retained *in vitro* bioactivity equivalent to stock DNA for at least 28 days.

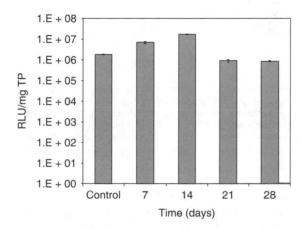

Figure 31.3   *In vitro* bioactivity of *Renilla* luciferase plasmid DNA after encapsulation in SELP-47K hydrogels for various periods of time. DNA encapsulated in the hydrogels retained bioactivity for at least 28 days. Control DNA was used to prepare the hydrogels prior to incubation at 37°C. Each data point represents the mean ± standard deviation for n = 3 samples. (From Megeed, Z., Haider, M., Li, D., O'Malley, Jr. B. W., Cappello, J., and Ghandehari, H., *In vitro* and *in vivo* evaluation of recombinant silk-elastinlike hydrogels for cancer gene therapy, *J. Control. Release*, 94, 433–445, 2004. With permission.)

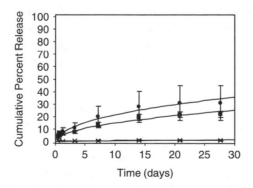

Figure 31.4   Effect of pRL-CMV conformation on release from SELP-47K hydrogels: (•) linear; (■) supercoiled; (×) open-circular; (—) theoretical release based on an equation for two-dimensional diffusion from a cylinder.[61] Each data point represents the mean ± standard deviation for n = 3 samples. (From Megeed, Z., Haider, M., Li, D., O'Malley, Jr. B. W., Cappello, J., and Ghandehari, H., *In vitro* and *in vivo* evaluation of recombinant silk-elastinlike hydrogels for cancer gene therapy, *J. Control. Release*, 94, 433–445, 2004. With permission.)

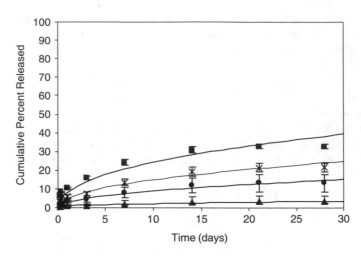

Figure 31.5   Effect of plasmid DNA size on release from SELP-47K hydrogels: (■) pUC 18 (2.6 Kbp), (×) pRL-CMV (4.08 Kbp), (•) pCFB-EGSH-Luc (8.5 Kbp), (p) pFB-ERV (11 Kbp); (—) theoretical release based on an equation for two-dimensional diffusion from a cylinder.[61] Each data point represents the mean ± standard deviation for n = 3 samples. (From Megeed, Z., Haider, M., Li, D., O'Malley, Jr. B. W., Cappello, J., and Ghandehari, H., *In vitro* and *in vivo* evaluation of recombinant silk-elastinlike hydrogels for cancer gene therapy, *J. Control. Release*, 94, 433–445, 2004. With permission.)

The conformation of plasmid DNA, namely supercoiled, open-circular, or linear, has been hypothesized to play a role in its transfection efficiency. While conventional thinking has generally indicated a preference for supercoiled DNA, at least one study has shown that the delivery of linear DNA results in prolonged transgene expression *in vivo*.[59] Other work has shown that the conformation of DNA is largely irrelevant to the transfection efficiency *in vitro* and *in vivo*.[60] It is possible that the conformational requirements of plasmid DNA may depend on the delivery system, whether the delivery is taking place *in vitro* or *in vivo*, and the concentration of nucleases at the site of delivery.

In order to evaluate the influence of conformation on DNA release from SELP-47K hydrogels, plasmid DNA, predominantly in the supercoiled, open-circular, and linear conformations, was produced and its release from SELP-47K hydrogels was evaluated.[56] The linear form of plasmid DNA was released most rapidly from the hydrogels, followed by the supercoiled form (Figure 31.4). The open-circular form was practically not released, probably due to its impalement on the polymer chains. The influence of plasmid size on release was also investigated (Figure 31.5), with size-dependent release observed for plasmids from 2.6 to 11 kilobases (kb) in size, from 10 wt% SELP hydrogels.[56] Since hydrogels form from SELPs with concentrations as low as 4 wt%, it is possible that plasmids larger than 11 kb could be delivered from SELP hydrogels. The ability to deliver larger plasmids would be advantageous for large genes. One limitation of viral vectors is the size of the transgene.

The effect of hydrogel geometry on DNA release was also studied by fabricating hydrogels in the form of cylinders and flat discs.[56] Disc-like hydrogels released DNA faster than their cylindrical counterparts (Figure 31.6). This was attributed to their larger surface to volume ratio and was accurately described by fitting the release data to an equation that describes two-dimensional diffusion from a cylinder with geometric considerations.[61,62]

While the applications of SELP-mediated controlled gene delivery are numerous, we have focused our efforts on controlled delivery to solid tumors. Our initial studies in this arena involved the intratumoral delivery of a reporter plasmid (*Renilla* luciferase) to solid tumors in a murine (athymic *nu/nu*) model of human breast cancer (MDA-MB-435 cell line).[56] The injected SELP-47K

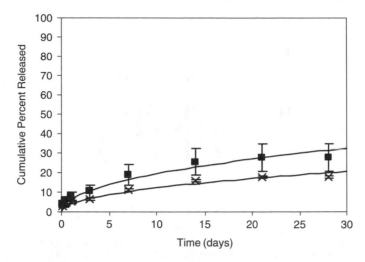

Figure 31.6 Effect of SELP-47K hydrogel dimensions on release of pRL-CMV: (■) disc; (×) cylindrical; (—) theoretical release based on an equation for two-dimensional diffusion from a cylinder.[61] Each data point represents the mean ± standard deviation for n = 3 samples. (From Megeed, Z., Haider, M., Li, D., O'Malley, Jr. B. W., Cappello, J., and Ghandehari, H., *In vitro* and *in vivo* evaluation of recombinant silk-elastinlike hydrogels for cancer gene therapy, *J. Control. Release*, 94, 433–445, 2004. With permission.)

solutions contained either 4, 8, or 12 wt% polymer and 70 µg DNA per 100 µl injection. At predetermined time points, animals were euthanized, tumors were homogenized, and luciferase expression was assayed. The Mann-Whitney test was used to compare treatment groups.

Delivery of the *Renilla* luciferase plasmid from SELP-47K matrices resulted in significantly enhanced tumor transfection for up to 21 days when compared to naked DNA (Figure 31.7). In particular, delivery of the plasmid from matrices containing 4 or 8 wt% polymer resulted in enhanced transfection up to 21 days, while 12 wt% matrices enhanced transfection up to 3 days (Figure 31.7). These results are consistent with sustained delivery from the 4 and 8 wt% matrices and entrapment of the DNA within the 12 wt% matrix.

The levels of tumor transfection mediated by the three concentrations of polymer were statistically equivalent until 7 days, when the 4 and 8 wt% matrices were both more effective than 12 wt% (Figure 31.7). The greater transfection persisted until 21 days for 4 wt% polymer and 14 days for 8 wt% polymer. Overall, the delivery of DNA from 4, 8, and 12 wt% hydrogels resulted in a mean 142.4-fold, 28.7-fold, and 3.5-fold increase in tumor transfection, respectively, compared with naked DNA over the entire 28-day period.

In addition to the delivery of DNA that occurs within the tumor, it can be expected that some DNA will diffuse from the matrix into the surrounding tissue. In order to evaluate this, transfection of the skin approximately 1 cm around the tumor was measured. The enhancement of delivery to the tumor was compared to the levels of transfection in the tumors (Figure 31.7) and skin (Figure 31.8) at each time point and polymer concentration. While statistically significant differences were not detected between all compositions, the mean tumor transfection was 42.0, 27.2, and 4.6 times greater than skin transfection for 4, 8, and 12 wt% hydrogels, respectively, over the entire 28-day period. This is in contrast to a 1.3-fold difference between tumor and skin transfection for naked DNA.

31.4.2 Controlled Release of Adenovirus from Silk-Elastinlike Hydrogels

Despite the promising results obtained from delivering plasmid DNA from SELP hydrogels, the delivery of naked DNA results in relatively low transfection efficiency and thus has limited

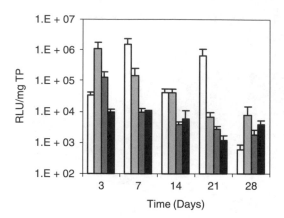

Figure 31.7   Expression of *Renilla* luciferase in MDA-MD-435 tumors grown subcutaneously in athymic *nu/nu* mice, after intratumoral injection. Bars represent 4 wt% polymer (white), 8 wt% polymer (light gray), 12 wt% polymer (dark gray), and naked DNA without polymer (black). Each bar represents the mean ± standard error of the mean for n = 4 or n = 5 samples. (From Megeed, Z., Haider, M., Li, D., O'Malley, Jr. B. W., Cappello, J., and Ghandehari, H., *In vitro* and *in vivo* evaluation of recombinant silk-elastinlike hydrogels for cancer gene therapy, *J. Control. Release*, 94, 433–445, 2004. With permission.)

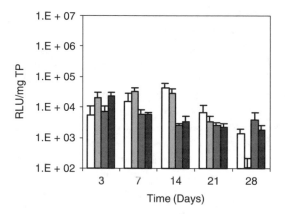

Figure 31.8   Expression of *Renilla* luciferase in the skin directly surrounding (~1 cm) MDA-MD-435 tumors grown subcutaneously in athymic *nu/nu* mice. Bars represent 4 wt% polymer (white), 8 wt% polymer (light gray), 12 wt% polymer (dark gray), and naked DNA without polymer (black). Each bar represents the mean ± standard error of the mean for n = 4 or n = 5 samples. (From Megeed, Z., Haider, M., Li, D., O'Malley, Jr. B. W., Cappello, J., and Ghandehari, H., *In vitro* and *in vivo* evaluation of recombinant silk-elastinlike hydrogels for cancer gene therapy, *J. Control. Release*, 94, 433–445, 2004. With permission.)

applications. For this reason, we sought to explore the potential of SELP-47K to act as a matrix for the controlled delivery of viral vectors. The use of viruses that integrate into the host genome (e.g., retrovirus) can ensure long-term expression of a transgene. However, as was recently observed in a study of cross-linked severe combined immunodeficiency, insertion of the viral DNA at or near an oncogenic regulation site can lead to the development of cancer.[63] Thus, there is also an intense interest in nonintegrating viral vectors, such as adenovirus, which may have a higher margin of safety than integrating viruses. However, without integration into the host genome the duration of transgene expression is limited. A controlled release approach may extend the duration of transgene expression by continuously delivering adenovirus at the site of action. In addition, it is conceivable that the immune response to the virus may be modulated by encapsulation in a polymeric matrix and only allowing the release of very small quantities of virus in a given time. Current state of the art polymers used in drug delivery do not provide the capability to deliver viable adenoviral vectors effectively over a prolonged periods of time. Coacervate microspheres of gelatin and alginate have shown poor encapsulation efficiency, low virus bioactivity, and poor virus release kinetics.[64] In the case of natural polymers such as collagen, control over the rate of release is complicated by the limited control over polymeric structure and cross-linking density.[65,66] Genetic engineering techniques may allow the design and synthesis of new polymers with precisely defined architecture, where biodegradation and biorecognition can be controlled to release viable viral vectors at specific sites and rates, in response to local stimuli.

As a first step toward that goal, we evaluated the potential of genetically engineered SELP-47K hydrogels to act as matrices for the controlled delivery of an adenovirus containing the green fluorescent protein (GFP) gene (AdGFP).[56] SELP-47K–AdGFP solutions were prepared at 4, 8, and 11.3 wt% polymer. The mixtures were allowed to gel and hydrogel discs were placed in a PBS release medium. At predetermined time points, release medium was collected and used to transfect HEK-293 cells. These cells contain a relatively high density of adenovirus receptors and are thus a good screening tool for the presence and bioactivity of adenovirus in the release medium. Transfection was observed up to 22 days with the viruses released from the 4 wt% hydrogel (Column 2, Figure 31.9). The number of transfected cells obtained with the viruses released from the 8 wt% hydrogel (Column 3, Figure 31.9) was less than that obtained from the 4 wt% hydrogel. The 11.3 wt% hydrogel did not release any detectable adenovirus after the first day (Column 4, Figure 31.9). This demonstrates that adenoviral release can be controlled over a continuum by controlling polymer composition from no release (11.3 wt% gel) to greater release (4 wt% gel). Control samples (viral particles without hydrogels in release media; Column 1, Figure 31.9) were bioactive until day 29 (with few GFP+ cells on day 29). However, as anticipated, bioactivity decreased over time. We are now focused on quantifying the amount of adenovirus released, the proportion that is bioactive, and the *in vivo* biodistribution, efficacy, and toxicity of adenoviral particles delivered from SELP-47K. The long-term goal is to engineer polymers tailor-made for specific needs.

31.4.3   *Gene Delivery from Derivatized Recombinant Polymers*

A derivatized genetically engineered polymer, Pronectin F, has recently been studied as a nonviral vector for gene delivery.[67] Pronectin F is a silk-like polymer containing one fibronectin segment between every nine silk-like repeats. This polymer was originally synthesized by Cappello et al. and has been explored as a substrate to enhance cellular attachment to hydrophobic materials.[27]

Hosseinkhani and Tabata cationized Pronectin F by reaction of ethylenediamine, spermidine, and spermine with the hydroxyl groups of the serine in the silk-like blocks. For comparison, similarly cationized derivatives of gelatin were prepared. When complexed with plasmid DNA at a weight ratio of 50 protein:1DNA, all forms of cationized Pronectin F (Pronectin F+) induced the formation of particles with a slightly positive (~10 mV) zeta potential and a particle size of approximately 200 nm. Both the zeta potential and the particle size were a function of the amount of Pronectin F+ added to the DNA.

All three Pronectin F+ derivatives increased the transfection of rat gastric mucosal cells by a reporter (luciferase) plasmid in comparison to naked plasmid. The spermine derivative was found to be significantly more effective than the ethylenediamine and spermidine derivatives. This was attributed to the higher buffering capacity of the spermine derivative, which was comparable to polyethylenimine. Furthermore, cellular attachment mediated by the Pronectin F+ derivatives, containing 13 RGD motifs, was found to be significantly greater than the cationized gelatin, which contained only one RGD motif. The uptake of plasmid DNA into the cells, as mediated by the

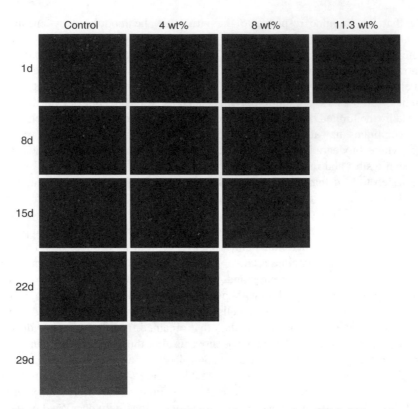

Figure 31.9 **(See color insert following page 336)** Adenovirus release from SELP-47K and the corresponding bioactivity results. The percentage of polymer increases from left to right. The time of release of virus from gels used to transfect cells or control increases from top to bottom. The images are from fluorescent microscopy at 40X magnification. Bright spots represent individual cells transfected with AdGFP. (From Megeed, Z., Haider, M., Li, D., O'Malley, Jr. B. W., Cappello, J., and Ghandehari, H., *In vitro* and *in vivo* evaluation of recombinant silk-elastinlike hydrogels for cancer gene therapy, *J. Control. Release*, 94, 433–445, 2004. With permission.)

three Pronectin F+ derivatives, was quantified using fluorescent labels. In all cases, the amount of plasmid uptake mediated by the Pronectin F+ derivatives was found to be greater than that mediated by cationized gelatin. This study shows the potential of biorecognizable genetically engineered polymers as soluble nonviral vectors and surface coatings for gene delivery.

31.5 FUTURE POTENTIAL OF GENETICALLY ENGINEERED POLYMERS IN GENE DELIVERY

In addition to the localized delivery of naked DNA and adenoviral particles, genetic engineering techniques may have potential to produce well-defined polymers for systemic gene delivery. Chemically synthesized cationic polymers, with a distribution of compositions and molecular weights have been widely investigated in nonviral gene delivery. However, the influence of polymer structure on the physicochemical properties and hence transfection efficiency of the polymer–DNA complexes is poorly understood. Some studies have sought to systematically define these relationships, but are limited by the inherent limitations of chemical polymer synthesis and derivatization, such as random sequences, polydispersity, and fractured structures.[68,69]

The biological synthesis of cationic polymers is attractive in that it could provide a series of macromolecules with well-defined sequence, composition, and molecular weights for DNA complexation and transfection experiments. However, recombinant synthesis of cationic polymers presents a special problem, as many cationic proteins are inherently toxic to bacteria.[70] In fact, some proteins secreted by cells in the immune system (e.g., eosinophil cationic protein) serve this very purpose.[71] To decrease the toxicity of genetically engineered cationic polymers to the expression system, several approaches may be evaluated, including the inclusion of fusion tags that decrease the toxicity of the polymer, and the use of tightly regulated expression systems such as the pLysS-containing expression hosts.[72] Another approach, which may circumvent some of the problems associated with the expression of high-molecular-weight cationic polymers, would be to attach a short cationic tail to a targeting moiety. Multiple cationic tails could then theoretically interact with and condense DNA, while the targeting moiety would enhance its delivery to cells. The inclusion of endosomolytic peptides at various points in the linear polymer may also enhance the efficiency of the vector by promoting endosomal escape. This research is in its embryonic stages. The long-term objective is to synthesize polymers for which structure-transfection relationships can be defined *in vitro* and *in vivo*. Some initial attempts, in our laboratory, to clone multimer genes encoding such polymers have been described elsewhere.[73]

31.6 CONCLUSION AND FUTURE DIRECTION

The potential of genetically engineered polymers in gene delivery has been demonstrated by the delivery of plasmid DNA and adenoviral vectors from SELP-47K.[33,56] The ability to form hydrogel depots *in situ* after injection through a needle and the lack of exposure to organic solvents are key advantages that SELP hydrogels hold over several other matrix-mediated controlled gene delivery systems. Genetic engineering techniques enable precise control over polymer structure and therefore phase transitions, biodegradation, and biorecognition.

Genetically engineered polymers also hold promise for the synthesis of nonviral vectors for systemic and targeted administration. It may be possible to use recombinant methods to synthesize modular gene delivery components that contain well-defined sequences for condensation, targeting, endosomal escape, and nuclear localization. The synthesis of polymeric gene carriers by this technique would also offer a way to examine structure-activity relationships of cationic polymers with unprecedented fidelity.

As our knowledge about natural proteins continues to grow, new genetically engineered polymers will be synthesized. Some of these molecules will have no precedent in the chemically synthesized polymer arena and may be designed to interact favorably with the complex physiological environment encountered by a gene delivery system on its way from administration to transcription.

ACKNOWLEDGMENTS

The authors would like to acknowledge financial support from DOD (Grant # DMAD 17-03-0237) (Hamidreza Ghandehari) and a National Cancer Center Predoctoral Fellowship (Zaki Megeed).

REFERENCES

1. Krammer, A., Lu, H., Isralewitz, B., Schulten, K., and Vogel, V., The fibronectin type III10 module: A tensile molecular recognition switch, *Biophys. J.* 76, A9, 1999.
2. Tinoco, I. and Bustamante, C., The effect of force on thermodynamics and kinetics of single molecule reactions, *Biophys. Chem.* 101, 513–533, 2002.

3. Li, H.B., Linke, W.A., Oberhauser, A.F., Carrion-Vazquez, M., Kerkviliet, J.G., Lu, H., Marszalek, P.E., and Fernandez, J.M., Reverse engineering of the giant muscle protein titin, *Nature* 418, 998–1002, 2002.

4. Cappello, J., Synthetically designed protein-polymer biomaterials, in *Controlled Drug Delivery: Challenges and Strategies*, Park, K. (Ed.), American Chemical Society, Washington, D.C., 1997, 439–453.

5. Cappello, J., Crissman, J., Dorman, M., Mikolajczak, M., Textor, G., Marquet, M., and Ferrari, F., Genetic engineering of structural protein polymers, *Biotechnol. Prog.* 6, 198–202, 1990.

6. Herzog, R.W., Singh, N.K., Urry, D.W., and Daniell, H., Expression of a synthetic protein-based polymer (elastomer) gene in Aspergillus nidulans, *Appl. Microbiol. Biotechnol.* 47, 368–372, 1997.

7. Zhang, X.R., Urry, D.W., and Daniell, H., Expression of an environmentally friendly synthetic protein-based polymer gene in transgenic tobacco plants, *Plant Cell Rep.* 16, 174–179, 1996.

8. Teule, F., Aube, C., Abbott, A.G., and Ellison, M.S. Production of customized novel fiber proteins in yeast (Pichia pastoris) for specialized applications, *Proceedings of the 3rd International Silk Conference*, Montreal, Quebec, Canada, 2003.

9. Ferrari, F., Richardson, C., Chambers, J., Causey, S.C., Pollock, T.J., Cappello, J., and Crissman, J.W., Protein Polymer Technologies, Inc., U.S. Patent 5,243,038, 1993.

10. Ferrari, F. and Cappello, J., Biosynthesis of protein polymers, in *Protein-Based Materials*, Kaplan, D.L., (Ed.), Birkhauser, Boston, 1997, 37–60.

11. Lohe, A.R. and Brutlag, D.L., Multiplicity of satellite DNA sequences in *Drosophila melanogaster*, *Proc. Nat. Acad. Sci. U.S.A.* 83, 696–700, 1986.

12. Sadler, J.R., Tecklenburg, M., and Betz, J.L., Plasmids containing many tandem copies of a synthetic lactose operator, *Gene* 8, 279–300, 1980.

13. Carlson, M. and Brutlag, D., Cloning and characterization of a complex satellite DNA from *Drosophila melanogaster*, *Cell* 11, 371–381, 1977.

14. Sambrook, J. and Russell, S.J., *Molecular Cloning: A Laboratory Manual*, 3rd edition, Cold Spring Harbor Laboratory, Cold Spring Harbor, NY, 2001.

15. Meyer, D.E. and Chilkoti, A., Genetically encoded synthesis of protein-based polymers with precisely specified molecular weight and sequence by recursive directional ligation: examples from the elastin-like polypeptide system, *Biomacromolecules* 3, 357–367, 2002.

16. Won, J.I. and Barron, A.E., A new cloning method for the preparation of long repetitive polypeptides without a sequence requirement, *Macromolecules* 35, 8281–8287, 2002.

17. McMillan, R.A., Lee, T.A.T., and Conticello, V.P., Rapid assembly of synthetic genes encoding protein polymers, *Macromolecules* 32, 3643–3648, 1999.

18. Goeden-Wood, N.L., Conticello, V.P., Muller, S.J., and Keasling, J.D., Improved assembly of multimeric genes for the biosynthetic production of protein polymers, *Biomacromolecules* 3, 874–879, 2002.

19. Cappello, J., Ferrari, F.A., Buerkle, T.L., and Textor, G., Protein Polymer Technologies, Inc., U.S. Patent 5,235,041, 1993.

20. McPherson, D.T., Xu, J., and Urry, D.W., Product purification by reversible phase transition following *Escherichia coli* expression of genes encoding up to 251 repeats of the elastomeric pentapeptide GVGVP, *Protein Expr. Purif.* 7, 51–57, 1996.

21. Meyer, D.E. and Chilkoti, A., Purification of recombinant proteins by fusion with thermally-responsive polypeptides, *Nat. Biotechnol.* 17, 1112–1115, 1999.

22. Kiick, K.L., Saxon, E., Tirrell, D.A., and Bertozzi, C.R., Incorporation of azides into recombinant proteins for chemoselective modification by the Staudinger ligation, *Proc. Nat. Acad. Sci. U.S.A.* 99, 19–24, 2002.

23. Cappello, J., Genetically engineered protein polymers, in *Handbook of Biodegradable Polymers*, Wiseman, D.M., (Ed.), Harwood Academic Publishers, Amsterdam, 1997, 387–416.

24. Halstenberg, S., Panitch, A., Rizzi, S., Hall, H., and Hubbell, J.A., Biologically engineered protein-graft-poly(ethylene glycol) hydrogels: A cell adhesive and plasmin-degradable biosynthetic material for tissue repair, *Biomacromolecules* 3, 710–723, 2002.

25. Urry, D.W., Physical chemistry of biological free energy transduction as demonstrated by elastic protein-based polymers, *J. Phys. Chem. B* 101, 11007–11028, 1997.

26. Cappello, J., Genetic production of synthetic protein polymers, *MRS Bull.* 17, 48–53, 1992.

27. Cappello, J. and Ferrari, F., Microbial production of structural protein polymers, in *Plastics from Microbes*, Mobley, D.P., (Ed.) Hanser Publishers, Munich, 1994, 35–92.

28. Panitch, A., Yamaoka, T., Fournier, M.J., Mason, T.L., and Tirrell, D.A., Design and biosynthesis of elastin-like artificial extracellular matrix proteins containing periodically spaced fibronectin CS5 domains, *Macromolecules* 32, 1701–1703, 1999.

29. Meyer, D.E., Shin, B.C., Kong, G.A., Dewhirst, M.W., and Chilkoti, A., Drug targeting using thermally responsive polymers and local hyperthermia, *J. Control. Release* 74, 213–224, 2001.

30. Betre, H., Setton, L.A., Meyer, D.E., and Chilkoti, A., Characterization of a genetically engineered elastin-like polypeptide for cartilaginous tissue repair, *Biomacromolecules* 3, 910–916, 2002.

31. Frey, W.G., Meyer, D.E., and Chilkoti, A., Thermodynamically reversible addressing of a stimuli responsive fusion protein onto a patterned surface template, *Langmuir* 19, 1641–1653, 2003.

32. Hoffman, A.S., Hydrogels for biomedical applications, *Adv. Drug Del. Rev.* 54, 3–12, 2002.

33. Megeed, Z., Cappello, J., and Ghandehari, H., Controlled release of plasmid DNA from a genetically engineered silk-elastinlike hydrogel, *Pharm. Res.* 19, 954–959, 2002.

34. Petka, W.A., Harden, J.L., McGrath, K.P., Wirtz, D., and Tirrell, D.A., Reversible hydrogels from self-assembling artificial proteins, *Science* 281, 389–392, 1998.

35. Cappello, J., Crissman, J.W., Crissman, M., Ferrari, F.A., Textor, G., Wallis, O., Whitledge, J.R., Zhou, X., Burman, D., Aukerman, L., and Stedronsky, E.R., *In situ* self-assembling protein polymer gel systems for administration, delivery, and release of drugs, *J. Control. Release* 53, 105–117, 1998.

36. Dinerman, A.A., Cappello, J., Ghandehari, H., and Hoag, S.W., Swelling behavior of a genetically engineered silk-elastinlike protein polymer hydrogel, *Biomaterials* 23, 4203–4210, 2002.

37. Dinerman, A.A., Cappello, J., Ghandehari, H., and Hoag, S.W., Solute diffusion in genetically engineered silk-elastinlike protein polymer hydrogels, *J. Control. Release* 82, 277–287, 2002.

38. Urry, D.W., Harris, C.M., Luan, C.X., Luan, C.-H., Channe Gowda, D., Parker, T.M., Peng, S.Q., and Xu, J., Transductional protein-based polymers as new controlled release vehicles, in *Controlled Drug Delivery: Challenges and Strategies*, Park, K., (Ed.) American Chemical Society, Washington, D.C., 1997, 405–438.

39. Wright, E.R. and Conticello, V.P., Self assembly of block copolymers derived from elastin-mimetic polypeptide sequences, *Adv. Drug Del. Rev.* 54, 1057–1073, 2002.

40. Wright, E.R., McMillan, R.A., Cooper, A., Apkarian, R.P., and Conticello, V.P., Thermoplastic elastomer hydrogels via self-assembly of an elastin-mimetic triblock polypeptide, *Adv. Funct. Mater.* 12, 149–154, 2002.

41. McMillan, R.A., Caran, K.L., Apkarian, R.P., and Conticello, V.P., High-resolution topographic imaging of environmentally responsive, elastin-mimetic hydrogels, *Macromolecules* 32, 9067–9070, 1999.

42. Lee, T.A.T., Cooper, A., Apkarian, R.P., and Conticello, V.P., Thermo-reversible self-assembly of nanoparticles derived from elastin-mimetic polypeptides, *Adv. Mater.* 12, 1105–1110, 2000.

43. Harada-Shiba, M., Yamauchi, K., Harada, A., Takamisawa, I., Shimokado, K., and Kataoka, K., Polyion complex micelles as vectors in gene therapy — pharmacokinetics and *in vivo* gene transfer, *Gene Ther.* 9, 407–414, 2002.

44. Wang, C., Stewart, R.J., and Kopecek, J., Hybrid hydrogels assembled from synthetic polymers and coiled-coil protein domains, *Nature* 397, 417–420, 1999.

45. Chen, L., Kopecek, J., and Stewart, R.J., Responsive hybrid hydrogels with volume transitions modulated by a titin immunoglobulin module, *Bioconj. Chem.* 11, 734–740, 2000.

46. Lee, C.H., Hsiao, M., Tseng, Y.L., and Chang, F.H., Enhanced gene delivery to HER-2-overexpressing breast cancer cells by modified immunolipoplexes conjugated with the anti-HER-2 antibody, *J. Biomed. Sci.* 10, 337–344, 2003.

47. Xu, L.N., Pirollo, K.F., and Chang, E.H., Transferrin-liposome-mediated p53 sensitization of squamous cell carcinoma of the head and neck to radiation *in vitro*, *Hum. Gene Ther.* 8, 467–475, 1997.

48. Hofland, H.E.J., Masson, C., Iginla, S., Osetinsky, I., Reddy, J.A., Leamon, C.P., Scherman, D., Bessodes, M., and Wils, P., Folate-targeted gene transfer *in vivo*, *Mol. Ther.* 5, 739–744, 2002.

49. Shadidi, M. and Sioud, M., Identification of novel carrier peptides for the specific delivery of therapeutics into cancer cells, *FASEB J.* 16, U478–U494, 2002.

50. Maeda, H., Fang, J., Inutsuka, T., and Kitamoto, Y., Vascular permeability enhancement in solid tumor: various factors, mechanisms involved and its implications, *Int. Immunopharmacol.* 3, 319–328, 2003.

51. Bonadio, J., Smiley, E., Patil, P., and Goldstein, S., Localized, direct plasmid gene delivery *in vivo*: prolonged therapy results in reproducible tissue regeneration, *Nat. Med.* 5, 753–759, 1999.

52. Luo, D., Woodrow-Mumford, K., Belcheva, N., and Saltzman, W.M., Controlled DNA delivery systems, *Pharm. Res.* 16, 1300–1308, 1999.

53. Shea, L.D., Smiley, E., Bonadio, J., and Mooney, D.J., DNA delivery from polymer matrices for tissue engineering, *Nat. Biotechnol.* 17, 551–554, 1999.

54. Jong, Y.S., Jacob, J.S., Yip, K.-P., Gardner, G., Seitelman, E., Whitney, M., Montgomery, S., and Mathiowitz, E., Controlled release of plasmid DNA, *J. Control. Release* 47, 123–134, 1997.

55. Gebrekidan, S., Woo, B.H., and Deluca, P.P., Formulation and *in vitro* transfection efficiency of poly (D, L-lactide-co-glycolide) microspheres containing plasmid DNA for gene delivery, *AAPS Pharm-SciTech.* 1, Article 28, 2000.

56. Megeed, Z., Haider, M., Li, D., O'Malley, Jr. B. W., Cappello, J., and Ghandehari, H., *In vitro* and *in vivo* evaluation of recombinant silk-elastinlike hydrogels for cancer gene therapy, *J. Control. Release*, 94, 433–445, 2004.

57. Nagarsekar, A., Crissman, J., Crissman, M., Ferrari, F., Cappello, J., and Ghandehari, H., Genetic synthesis and characterization of pH- and temperature-sensitive silk-elastinlike protein block copolymers, *J. Biomed. Mater. Res.* 62, 195–203, 2002.

58. Nagarsekar, A., Crissman, J., Crissman, M., Ferrari, F., Cappello, J., and Ghandehari, H., Genetic engineering of stimuli-sensitive silk-elastinlike protein block copolymers, *Biomacromolecules* 4, 602–607, 2003.

59. Chen, Z.Y., Yant, S.R., He, C.Y., Meuse, L., Shen, S., and Kay, M.A., Linear DNAs concatemerize *in vivo* and result in sustained transgene expression in mouse liver, *Mol. Ther.* 3, 403–410, 2001.

60. Bergan, D., Galbraith, T., and Sloane, D.L., Gene transfer *in vitro* and *in vivo* by cationic lipids is not significantly affected by levels of supercoiling of a reporter plasmid, *Pharm. Res.* 17, 967–973, 2000.

61. Fu, J.C., Hagemeir, C., and Moyer, D.L., A unified mathematical model for diffusion from drug-polymer composite tablets, *J. Biomed. Mater. Res.* 10, 743–758, 1976.

62. Siepmann, J., Ainaoui, A., Vergnaud, J.M., and Bodmeier, R., Calculation of the dimensions of drug-polymer devices based on diffusion parameters, *J. Pharm. Sci.* 87, 827–832, 1998.

63. Hacein-Bey-Abina, S., von Kalle, C., Schmidt, M., Le Deist, F., Wulffraat, N., McIntyre, E., Radford, I., Villeval, J.L., Fraser, C.C., Cavazzana-Calvo, M., and Fischer, A., A serious adverse event after successful gene therapy for X-linked severe combined immunodeficiency, *New Engl. J. Med.* 348, 255–256, 2003.

64. Kalyanasundaram, S., Feinstein, S., Nicholson, J.P., Leong, K.W., and Garver, R.I., Jr., Coacervate microspheres as carriers of recombinant adenoviruses, *Cancer Gene Ther.* 6, 107–112, 1999.

65. Chandler, L.A., Doukas, J., Gonzalez, A.M., Hoganson, D.K., Gu, D.L., Ma, C., Nesbit, M., Crombleholme, T.M., Herlyn, M., Sosnowski, B.A., and Pierce, G.F., FGF2-targeted adenovirus encoding platelet-derived growth factor-β enhances de novo tissue formation, *Mol. Ther.* 2, 153–160, 2000.

66. Doukas, J., Chandler, L.A., Gonzalez, A.M., Gu, D., Hoganson, D.K., Ma, C., Nguyen, T., Printz, M.A., Nesbit, M., Herlyn, M., Crombleholme, T.M., Aukerman, S.L., Sosnowski, B.A., and Pierce, G.F., Matrix immobilization enhances the tissue repair activity of growth factor gene therapy vectors, *Hum. Gene Ther.* 12, 783–798, 2001.

67. Hosseinkhani, H. and Tabata, Y., *In vitro* gene expression by cationized derivatives of an artificial protein with repeated RGD sequences, Pronectin, *J. Control. Release* 86, 169–182, 2003.

68. Tang, M.X. and Szoka, F.C., The influence of polymer structure on the interactions of cationic polymers with DNA and morphology of the resulting complexes, *Gene Ther.* 4, 823–832, 1997.

69. Haider, M. and Ghandehari, H., Influence of poly (amino acid) composition on the complexation of plasmid DNA and transfection efficiency, *J. Bioact. Compat. Polym.* 18, 93–111, 2003.

70. Hancock, R.E.W., Host defense (cationic) peptides: what is their future clinical potential? *Drugs* 57, 469–473, 1999.

71. Rosenberg, H.F., Recombinant human eosinophil cationic protein — ribonuclease-activity is not essential for cytotoxicity, *J. Biol. Chem.* 270, 7876–7881, 1995.
72. Piers, K.L., Brown, M.H., and Hancock, R.E.W., Recombinant-DNA procedures for producing small antimicrobial cationic peptides in bacteria, *Gene* 134, 7–13, 1993.
73. Megeed, Z., Genetically engineered polymers for cancer gene therapy, PhD thesis, Department of Pharmaceutical Sciences, University of Maryland, Baltimore, 2003.

CHAPTER 32

Glycopolymer Tools for Studying Targeted Nonviral Gene Delivery

Kevin G. Rice, Ji-Seon Kim, and Dijie Liu

CONTENTS

32.1 INTRODUCTION

Some of the earliest vectors developed for non-viral gene delivery were glycoconjugates.[1–4] Pioneering studies by Wu and coworkers in the late 1980s recognized the potential to target DNA to gain cell entry via cell surface lectins such as the asialoglycoprotein receptor. The earliest of these glycoconjugates was composed of an asialoglycoprotein attached covalently to a polylysine. These constructs offered the advantage of simplicity in chemical synthesis from commercially available molecules. Typically, a glycoprotein such as human orosomucoid was enzymatically desialylated to expose terminal galactose residues on the bi-, tri-, and tetraantennary N-glycans. The glycoprotein was then chemically conjugated in a random derivatization to the polylysine ε amines to form a covalent linkage.[3] These glycoconjugates were mixtures composed of polydisperse polylysines derivatized with differing numbers of glycoproteins at different locations throughout the polymer. Despite this heterogeneity, these glycoconjugates maintained their ability to bind to DNA through ionic binding of the phosphate backbone with the remaining lysine side chains on the glycoconjugate. Likewise, exposed asialo N-glycans on the glycoproteins maintained their binding affinity to the asialoglycoprotein receptor. These early gene delivery formulations were reportedly able to bind to the asialoglycoprotein receptor and mediate gene delivery and expression in hepatocytes both *in vitro* and *in vivo*.[1,4-8]

During the subsequent decade, gene therapists have developed glycoconjugates in an attempt to repeat and improve upon these early studies.[9-13] The lessons learned during the last decade are

0-8493-1934-X/05/$0.00+$1.50
© 2005 by CRC Press LLC

numerous and some of the conclusions derived from these studies could not have been foreseen by the early pioneers.

Several major advances have changed the approach used to design glycoconjugates for gene delivery. First, the heterogeneity of glycoconjugates has decreased significantly such that even homogeneous glycopeptides may now be prepared and used for nonviral gene delivery *in vivo*.[14,15] This important advance allows systematic analysis of structure-activity relationships between the carrier molecule, the gene formulation, and the resulting gene expression. However, those that have taken this approach now realize that low-molecular-weight homogenous glycoconjugates that mediate gene expression *in vitro* often fail *in vivo* due to the low DNA binding affinity afforded by short polycations.[16] This realization has led to revised schemes that either permanently or transiently cross-link carriers bound to DNA in an effort to create sufficient DNA binding affinity.[17-21]

Second, a great deal more is known about the fate of DNA delivered to animals.[13] Nonspecific uptake of naked plasmid DNA by the Kupffer cell scavenger receptor, and the rapid metabolism of naked DNA in the blood by endogenous DNAse,[16,22] both contribute to the removal of DNA from the blood. Even when DNA is protected from metabolism by condensation with a polycation, the nonspecific biodistribution of cationic colloidal DNA particles to the lung prevents targeted delivery to the liver.[23-27] This artifact arises from the cationic nature of DNA condensates which leads to rapid protein binding,[28,29] significantly increasing their size and propensity for entrapment in the capillary beds of the lung. Most gene therapists have since adopted the use of polyethylene glycol (PEG) to block the surface of electropositive cationic DNA condensates in an effort to block both protein binding in the blood and Kupffer cell recognition.[14-16,30-36] But this apparent advance leads to other complications. The size of PEGs (3–5 kDa) simultaneously conceals the recognition of small carbohydrate ligands that are attached directly to cationic polymers bound to DNA. Thus, newer more complicated glycoconjugate carriers have been developed using heterobifunctionalized PEGs that allow small carbohydrate ligands to be attached to the end of the PEG.[33,37,38]

Third, although nonviral gene delivery remains a major challenge, the advent of hydrodynamic dosing amply proved that a small quantity of DNA (5–10 μg) is sufficient to produce prolonged therapeutic levels of gene expression in animals.[39-41] These studies, along with the discovery of polyethylene amine (PEI) as a potent gene transfer agent[12] that mediates endosomal escape, clearly established that a pharmaceutically elegant nonviral gene delivery system is feasible provided that both extracellular and intracellular barriers to delivery can be overcome.

The following review highlights some of the more innovative experimental glycoconjugates that have been developed and tested as nonviral gene delivery carriers. As with most sciences, looking back with more than a decade of hindsight provides many insights as to why a glycoconjugate either worked or didn't. This type of reflection, at perhaps the midpoint in the 20-year development of targeted nonviral gene delivery systems, is both instructive and insightful.

32.2 GENE TARGETING TO C-TYPE LECTINS

There are many excellent detailed reviews that focus on C-type lectins.[42-44] The following discussion differs from prior reviews in its attempt to convey only the essential information necessary for designing glycoconjugates to delivery DNA to bind to cell surface C-type lectins. The family of C-type lectins was discovered in the early 1970s by observing the rapid clearance of glycoproteins that were desialylated.[45-47] The C-type lectin family includes more than 69 carbohydrate-binding proteins. These individually unique carbohydrate recognition domains (CRD) are related by sequence homology, including 14 invariant and 23 conserved residues along their 115 to 130 amino acid length.[48] Although there is similarity in the topography of their binding sites, differences arise from their precise sugar binding specificity and from the valency of multiple CRDs.[49-54]

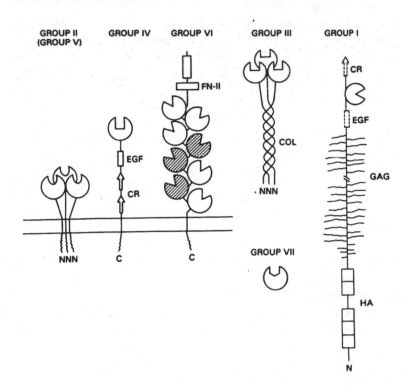

Figure 32.1 A representative structure of the different groups (I–VII) of C-type lectins illustrating the location and valency of CRDs.

Group I C-type lectins are monomeric proteoglycans with domains flanking the C-terminal CRD. Group II members are integral membrane proteins with an intracellular N-terminus and an extracellular C-terminus, the most prominent being the asialoglycoprotein receptor. Group III collectins and V lymphocyte antigens are trimeric structures.[52] Some members of the Group III collectins are soluble proteins that can oligomerize into higher-ordered structures, such as pulmonary surfactant protein, which forms 12-mer cruciform structures. Group IV members include E-, P-, and L-selectins which are transmembrane lectins that regulate leukocyte rolling and extravagation into sites of inflammation.[53] Group VI receptors contain multiple lectin domains in a single polypeptide and illustrated by the macrophage mannose receptor, phospholipase A$_2$ receptor, DEC 205 receptor, and S4GGnM receptors.[54] Consistent with the observation that group VI lectins serve to endocytose ligands, they are referred to as the endocytic C-type lectins. Although it may be noted that a number of group II C-type lectins, such as asialoglycoprotein receptors, are also endocytic. Group VII C-type lectins are the simplest structures consisting of a CRD without flanking domains[49]. There are a number of C-type lectins that have been identified and are not currently categorized into groups I–VII, including a number of homologous proteins which are unlikely to display binding to carbohydrates (Figure 32.1).

The prototypic type II C-type lectin, the asialoglycoprotein receptor, is a cell surface endocytosing receptor located on hepatocytes.[55] The receptor is composed of three noncovalently associated subunits, each of which possesses a CRD. Each subunit is capable of recognizing a non-reducing end galactose or N-acetyl galactosamine (GalNAc) by calcium chelation of the sugar hydroxyl groups in a shallow binding pocket.[44,56] A single galactose binds to the receptor with negligible (Kd = mM) affinity, whereas a significant "cluster effect" occurs that amplifies the affinity to μM or nM when two or three galactose residues are positioned to simultaneously occupy the remaining CRDs that make up the receptor.[57] A great deal is known regarding the fine specificity of the receptor recognition for galactose analogues. Most notably, the receptor binds GalNAc with nearly tenfold higher affinity.[57] The natural ligands for the asialoglycoprotein receptor appear to be N-glycans.[55,58-62] These branched oligosaccharides found on glycoproteins possess the proper valency to project terminal galactose residues and regulate the binding affinity. Numerous artificial ligands have been prepared to study the binding specificity of the asialoglycoprotein receptor and to target drugs and DNA.

The type VI mannose receptor resides on the surface of Kupffer cells in the liver as well as lung macrophages. Even though the mannose receptor possesses eight CRDs, only one of these is active in binding mannose residues.[63] Thus, high-mannose N-glycans are poor ligands for the mannose receptor, which requires mannose residues spaced over longer distances to cross-link between the single active CRD on two adjacent mannose receptors. Thereby, glycoconjugates designed to target Kupffer cells via the mannose receptor must attempt to orient and space the mannose residues properly on a DNA condensate for high affinity recognition rather than create a tight cluster, such as is necessary for galactose residues recognized by the asialoglycoprotein receptor.

Other well-known cell surface mammalian lectins, such as the selectins, could also serve as potential tissue specific targets for gene delivery.[64] However, targeting DNA to these receptors requires ligands of sufficient affinity to allow binding in the presence of endogenous ligands. Often the number of these receptors per cell is insufficient to support appreciable glycotargeting *in vivo*. Likewise, these receptors are located on cells within tissues that are perfused with a small percentage of the blood volume relative to the liver or that are protected by the blood brain barrier. Consequently, the more facile nonspecific biodistribution of DNA condensates to the liver and spleen is still the major barrier to targeting DNA to minor tissues.

32.3 GLYCOSYLATED POLYLYSINE, PEI, AND PEPTIDES

Polylysine was selected as an early scaffold on which to attach glycoproteins, glycopeptides, oligosaccharides, and simple sugars.[2,3,10,65-68] The side chains of polylysine are protonated at physiological pH allowing it to bind ionically to the phosphate backbone of DNA. Upon neutralization with polylysine, a 5 to 6 Kbp plasmid transforms from an open random coil of approximately 200 nm in diameter to a condensed particle of peptide and DNA with dimensions of approximately 50 to 100 nm in diameter as measured by quasielastic light scattering (QELS).[69,70] The stoichiometry of binding and the charge neutralization point are most often expressed as a nitrogen to phosphate (N/P) ratio based on a rough calculation starting from a weighed sample of polylysine. This measurement is inherently inaccurate since it relies upon an approximated average molecular weight for the polydisperse polylysine and does not take into account the mass of the counter ion on the side chain, which is typically bromide.

The length of polylysine influences the formation and efficacy of the resulting DNA condensates.[71] Generally speaking, short polylysine of less than 20 residues has been found to form DNA condensates unreliably, whereas polylysines of 100 to 200 residues are most often used, with longer polylysines generally considered to be more toxic to cells.[71] In contrast to these generalizations, short homogenous polylysines peptides (less than 20 residues) prepared by solid phase peptide synthesis were found to be effective *in vitro* gene transfer agents.[72] An explanation for this discrepancy was revealed by a detailed investigation that determined the actual length of heterogeneous polylysines (dp 20) was incorrectly determined by the vendor.[73]

Early glycosylated gene transfer agents were assembled by functionalizing the side chains of polylysine by attaching asialoorosomucoid (ASOR).[74] ASOR is a human glycoprotein that is known to possess primarily terminal triantennary and tetraantennary N-glycans that have high affinity for the asialoglycoprotein receptor following removal of sialic acid residues.[75] The derivatization chemistry used involved active ester coupling of N-sucinimidyl 3-(2-pyridyldithio) propionate to

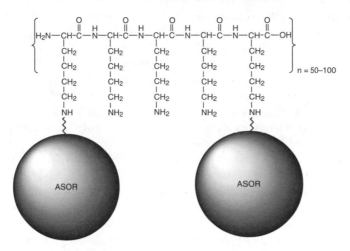

Figure 32.2 A representative chemical structure of an ASOR–polylysine carrier used for gene delivery.[3]

the side chain of polylysine followed by reaction of an ASOR thiol group(s) to displace the pyridyl to form a new disulfide bond with polylysine.[2,3] These conjugates were purified by gel filtration chromatography and characterized by performing a compositional analysis for lysine relative to the incorporation of an ^{125}I-ASOR. The earliest of these conjugates were reportedly 5:1, ASOR:polylysine. The overall heterogeneity of the conjugates must have been great since the coupling of N-sucinimidyl 3-(2-pyridyldithio) propionate proceeded randomly and incompletely on dp 500 polylysine. Likewise, the conjugation of ASOR was most likely incomplete, resulting in a random distribution of glycoproteins dispersed across the polylysine backbone. The difficulty in gene formulation with such a delivery system is that some of the polylysine chains remain underivatized while others contain one or more glycoproteins. The underivatized chains are more likely to combine with DNA since they are not blocked by steric hindrance. Depending upon the precise chemical conditions and reagents, each new batch of ASOR-polylysine would be different. As a consequence, this would lead to DNA condensates with unpredictable charge ratios, resulting in difficulty in reproducing gene transfer (Figure 32.2).

In an attempt to simplify these conjugates, several researchers investigated the use of glycosylated polylysines.[37,68,76-78] The synthesis of these delivery carriers is simplified over that used to prepare ASOR-polylysine in that the reaction of polylysine primary amines with 4-isothiocyanatophenyl-α-glycosides results in the covalent attachment of phenyl-sugars to the side of polylysine in a single step.[76,77] These glycosylated polymers were purified by removal of unreacted sugars using either precipitation or dialysis and then characterized for the degree of derivatization by colorimetrically determining the sugar per weight of polymer. Unlike the derivatization of polylysine with ASOR, the stoichiometry of the reaction was designed to substitute a significant number (> 30%) of the lysine side chain amines with sugar.[68,79] Following condensation, a sufficient density of sugar would be presented on the surface of the DNA condensate to allow artificial clustering and high affinity binding to the asialoglycoprotein receptor or mannose receptor. However, this derivatization is also random such that the precise spacing of sugar moieties cannot be known. Likewise, the derivatization masks amine groups, decreasing the binding affinity of glycosylated polylysine for DNA. As with ASOR–polylysine, the charge ratio resulting in complete DNA condensation can only be estimated by band retardation assay on gel electrophoresis or by intercalator dye fluorescence based measurements.[38] Likewise, it is impossible to determine which of the glycosylated polylysines from the mixture binds to DNA, since those that are most heavily glycosylated also possess the weakest affinity. Even with this limitation, the direct glycosylation of polylysine provides a clear advantage of being able to prepare polylysine decorated with many different types of sugar. This advantage has led to the discovery of a variety of structure activity relationships, correlating the sugar structure with either increased or decreased gene transfer efficiency on mammalian cells, some of which may result from lectin binding[80-83] (Figure 32.3).

The extensive work of Monsigny and coworkers provides numerous illustrations of the use of glycosylated polylysines to transfect hepatocytes, macrophages, and epithelial and many other cell types in culture.[76,77,80,84-86] In most cases, one or more of the sugars galactose, glucose, N-acetyl-glucosamine (GlcNAc), GalNAc, fucose, mannose, or lactose in the α and β configuration were found to confer specificity. Galactosylated and mannosylated polylysine have also been reported to mediate gene transfer *in vivo* but with relatively low efficiency.[87,88]

The approach described above has also been used to derivatize 25-kDa branched PEI resulting in glycosylated polymers with many of the same properties as glycosylated polylysine, but with improved gene transfer efficiency due to the endosomal buffering capacity of the polymer.[89] In an effort to improve on this design, tetrasaccharides possessing terminal galactose residues were incorporated into PEI using reductive amination.[90] The glycosylated polymers reportedly formed smaller DNA condensates intended for mediating *in vivo* gene transfer via the asialoglycoprotein receptor.

Gene formulation scientists now recognize that cationic colloidal DNA condensates bind to albumin in the blood, resulting in rapid growth in particle size and their entrapment in the lung capillary beds.[27] This, along with the well-recognized role of the scavenger receptor on Kupffer cells to mediate binding and phagocytosis of circulating foreign particles[13] provides a significant physiological barrier to achieving targeted gene delivery.

Early investigators avoided the deleterious effect of electropositive charge on nonspecific bio-distribution of DNA condensates to the lung.[1] To circumvent the RES, they chose to manipulate the polylysine to DNA charge ratios to form condensates that were close to neutrality or even slightly electronegative. However, without the use of more modern zeta potential measurements it was difficult to establish the charge of DNA condensates. If the charge ratio was too low, exposed DNA within electronegative DNA condensates was susceptible to metabolism by endogenous DNAses in the blood.[91] Likewise, electronegative DNA condensates are also taken up by the scavenger receptor on Kupffer cells.[13]

More recent attempts to mask the surface of DNA particles utilized polyethylene glycol (PEG),[30,32,33,36,38] to avoid protein binding and to block recognition by the reticuloendothelial system

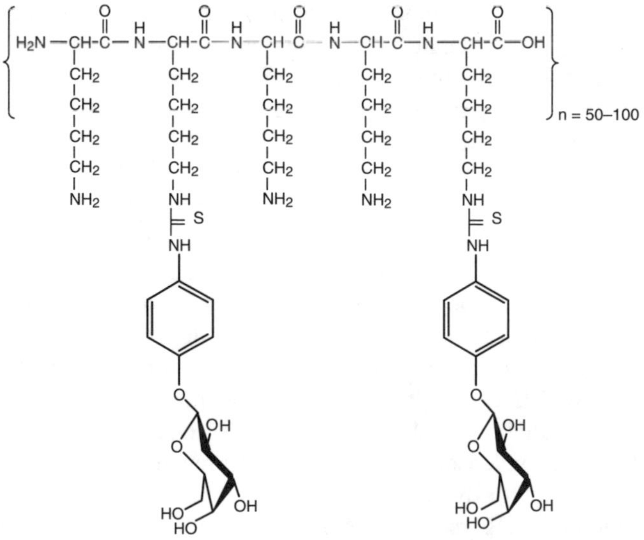

Figure 32.3 The chemical structure of a glycosylated polylysine carrier used for gene delivery.[68]

(RES). Unfortunately, PEG chains can also mask the ligand recognition by steric repulsion of receptor proteins attempting to bind to small ligands attached to the surface of the DNA condensate.[92] An alternative means to circumvent the RES is to incorporate small carbohydrate ligands onto the ends of PEG chains, thereby avoiding the masking effect. This approach utilized heterobifunctional PEG derivatives to attach ligand to one end with subsequent attachment of polylysine to the other.[37,38] However, the inherent heterogeneity of PEG, coupled with a random derivatization of polylysine, complicates the synthesis of these complex glycoconjugates and limits the ability to project a sufficient sugar density to mimic a clustered glycan. Likewise, the random coil conformation of PEG offers no certain guarantee that the ligand will be oriented properly for receptor recognition.

Another way to circumvent the RES is to incorporate larger ligands that could easily protrude above a stealth layer of PEG coating the cationic DNA condensate surface. The advantage of this approach for targeting the asialoglycoprotein receptor was that a clustered high-affinity ligand could be used, such that fewer copies of the ligand would be necessary to mediate binding to the receptor. One of the first gene delivery ligands of this type was a neoglycopeptide composed of a branched tripeptide derivatized with GalNAc residues.[93]

More recently, several gene delivery systems were developed using purified N-glycans.[15,94-96] An early attempt attached a triantennary N-glycan to a heterogeneous dp 20 polylysine and demonstrated *in vitro* gene transfer via the asialoglycoprotein receptor in HepG2 cells.[94] A refinement of this concept has led to the synthesis of homogenous glycopeptides prepared by attaching a purified N-glycan to a single cysteine residue on a synthetic polylysine peptide.[97] This well-characterized triantennary glycopeptide possessing a natural cluster of terminal galactose has been incorporated in low copy number into gene formulations and shown to target hepatocytes selectively in mice and mediate transient gene expression.[15,95] In this case, PEG-peptides were simultaneously incorporated into the gene formulation to mask protein binding and block scavenger receptor recognition. A similar gene formulation composed of a high-mannose glycopeptide and a PEG-peptide was engineered to target the mannose receptor of Kupffer cells.[96] To optimize targeting to Kupffer cells, the copy number of glycopeptide per DNA condensate was significantly higher than when targeting the asialoglycoprotein receptor, which reflects the different valency and spacing of the CRDs of these two receptors (Figure 32.4).

An inevitable consequence of using low-molecular-weight synthetic glycoconjugates is the need to stabilize the resulting DNA condensates in the circulation.[19,98] Several types of reversible and irreversible cross-linking strategies have been invoked to control the stability of these gene

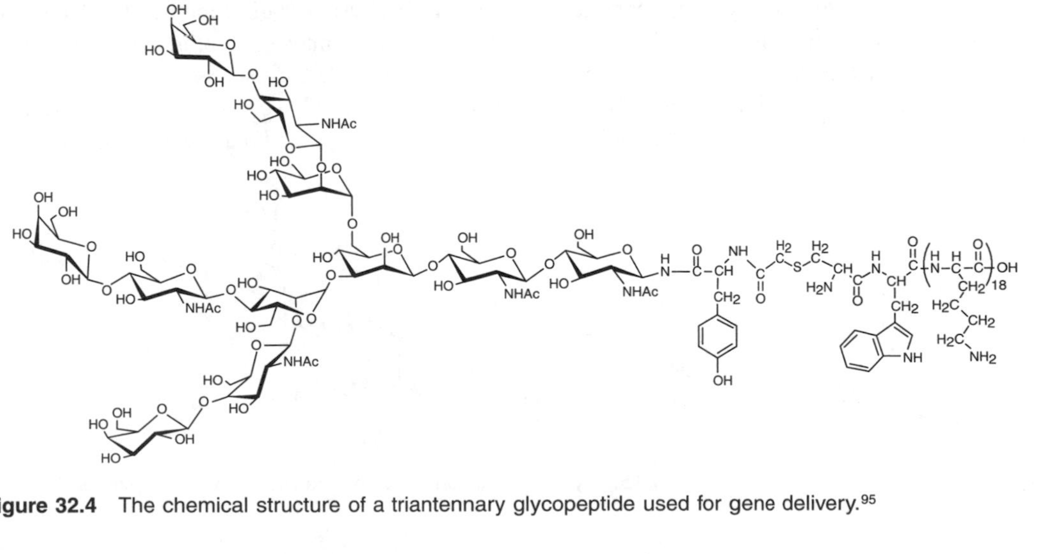

Figure 32.4 The chemical structure of a triantennary glycopeptide used for gene delivery.[95]

formulations.[30] The major disadvantage with these strategies is the relative difficulty in preparing N-glycan ligands through a multiple-step biochemical process, leaving it accessible only to groups capable of specializing in carbohydrate chemistry.

32.4 GLYCOSYLATED LIPOPLEXES

The glycosylation and PEGylation of liposomes was used as one of the earliest means to target these to liver hepatocytes via the asialoglycoprotein receptor.[99-102] An extensive literature on the use of cationic lipids to mediate gene expression and other chapters within this volume are devoted to this subject. Hashida and coworkers proposed the use of a lipid formulation composed of diolexyoxy-propyl-triethylammonium chloride (DOTMA) and cholesterol to form DNA lipoplexes of 100 to 400 nm, which mediated luciferase expression in the lung, liver, and heart when dosed intravenously in mice.[103-106]

To target DNA lipoplexes, Hashida and coworkers synthesized galactosyl-C4-cholesterol and substituted this into the DNA formulation with DOTMA. Galactosylated polyplexes were demonstrated to transfect both HepG2 cells and mouse hepatocytes *in vivo* with specificity for the asialoglycoprotein receptor (Figure 32.5).[105]

This concept was further extended by synthesizing mannosylated-C4-cholesterol.[104] Depending on the precise stoichiometry of the formulation with either dioleoyl phosphatidylethanolamine (DOPE) or DOTMA, these lipoplexes produced luciferase expression in either the liver or the lung following intravenous administration in mice. The targeting specificity to the liver was reportedly via the mannose receptor as demonstrated through competition experiments with coadministration with Man-BSA.

32.5 CONCLUSIONS

The studies outlined in this chapter demonstrate a steady evolution of glycosylated carrier molecules toward smaller and more chemically defined structures. A variety of attempted *in vitro/in vivo* correlations have revealed that it is the preprogrammed physiological-chemical barriers encountered *in vivo*, and not receptor recognition, that defeat experimental gene formulations in the *in vivo* experiment. Attempts to circumvent these physiological-chemical barriers using well-established pharmaceutical principles often require the careful design of increasing complex carrier molecules involving PEG. The majority of glycotargeted gene delivery systems are aimed at the liver, with ligands that target either the asialoglycoprotein receptor or the mannose receptor. This is not likely to change given the limited number of C-type lectins now revealed by sequencing of the human genome. Continued investigation and advancement in the design of glycotargeted gene delivery to the liver will likely lead to innovative solutions to physiological and cellular barriers that will have

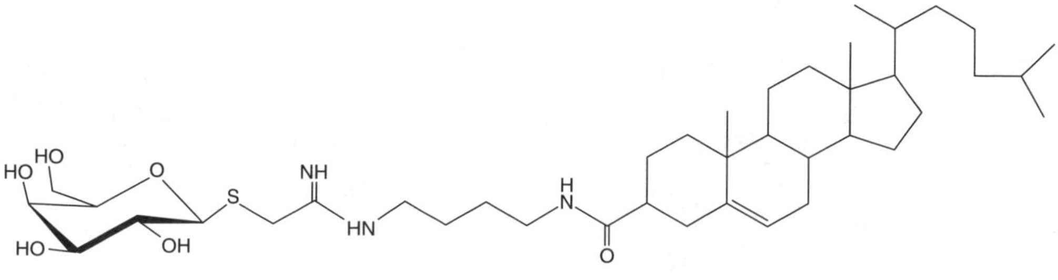

Figure 32.5   The chemical structure of galactosylated cholesterol incorporated into cationic liposomes that were used for gene delivery.[105]

broad reaching impact on the design of nonviral gene delivery systems. Given the ubiquitous nature of the asialoglycoprotein receptor in mammals, and the importance of the liver as a factory for the production of host derived proteins, it is easy to envision that the fulfillment of efficient glycotargeting to hepatocytes may someday lead to a family of gene therapeutics to treat human disease.

ACKNOWLEDGMENT

The authors gratefully acknowledge financial support from the National Institutes of Health (DK063196).

REFERENCES

1. Wu, G.Y. and Wu, C.H., Receptor-mediated gene delivery and expression *in vivo*, *J. Biol. Chem.* 263, 14621–14624, 1988.
2. Wu, G.Y. and Wu, C.H., Evidence for targeted gene delivery to Hep G2 hepatoma cells *in vitro*, *Biochemistry* 27, 887–892, 1988.
3. Wu, G.Y. and Wu, C.H., Receptor-mediated *in vitro* gene transformation by a soluble DNA carrier system, *J. Biol. Chem.* 262, 4429–4432, 1987. (Published erratum appears in *J. Biol. Chem.* 263 (1), 588, 1988.)
4. Wu, C.H., Wilson, J.M., and Wu, G.Y., Targeting genes: delivery and persistent expression of a foreign gene driven by mammalian regulatory elements *in vivo*, *J. Biol. Chem.* 264, 16985–16987, 1989.
5. Wu, G.Y. and Wu, C.H., Delivery systems for gene therapy, *Biotherapy* 3, 87–95, 1991.
6. Wu, G.Y. and Wu, C.H., Targeted delivery and expression of foreign genes in hepatocytes, *Target. Diagn. Ther.* 4, 127–149, 1991.
7. Wu, G.Y. et al., Receptor-mediated gene delivery *in vivo*: partial correction of genetic analbuminemia in nagase rats, *J. Biol. Chem.* 266, 14338–14342, 1992.
8. Wu, G.Y. and Wu, C.H., Carrier system and method for the introduction of genes into mammalian cells (to University of Connecticut, Storrs, CT). Application: U.S. 504,064, 1990. U.S. Patent 5,166,320, 1992.
9. Wilson, J.M. et al., Hepatocyte-directed gene transfer *in vivo* leads to transient improvement of hypercholesterolemia in low density lipoprotein receptor-deficient rabbits, *J. Biol. Chem.* 267, 963–967, 1992.
10. Midoux, P. et al., Specific gene transfer mediated by lactosylated poly-L-lysine into hepatoma cells, *Nucleic Acids Res.* 21, 871–878, 1993.
11. Haensler, J. and Szoka, J.F.C., Synthesis and characterization of a trigalactosylated bisacridine compound to target DNA to hepatocytes, *Bioconj. Chem.* 4, 85–93, 1993.
12. Boussif, O. et al., A versatile vector for gene and oligonucleotide transfer into cells in culture and *in vivo*: polyethylenimine, *Proc. Nat. Acad. Sci. U.S.A.* 92, 7297–7301, 1995.
13. Kawabata, K., Takakura, Y., and Hashida, M., The fate of plasmid DNA after intravenous injection in mice: involvement of scavenger receptors in its hepatic uptake, *Pharm. Res.* 12, 825–830, 1995.
14. Park, Y. et al., Synthesis of sulfhydryl crosslinking poly (ethylene glycol) peptides and glycopeptides as carriers for gene delivery, *Bioconj. Chem.* 13, 232–239, 2002.
15. Kwok, K.Y., Park, Y., Yongsheng, Y., McKenzie, D.L., Rice, K.G., *In vivo* gene transfer using sulfhydryl crosslinked PEG-peptide/glycopeptide DNA co-condensates, *J. Pharm. Sci.* 92, 1174–1185, 2003.
16. Collard, W.T. et al., Biodistribution, metabolism, and *in vivo* gene expression of low molecular weight glycopeptide polyethylene glycol peptide DNA co-condensates, *J. Pharm. Sci.* 89, 499–512, 2000.
17. McKenzie, D.L., Kwok, K.Y., and Rice, K.G., A potent new class of reductively activated peptide gene delivery agents, *J. Biol. Chem.* 275, 9970–9977, 2000.
18. McKenzie, D. et al., Low molecular weight disulfide cross-linking peptides as nonviral gene delivery carriers, *Bioconj. Chem.* 11, 901–911, 2000.

19. Trubetskoy, V.S. et al., Self-assembly of DNA-polymer complexes using template polymerization, *Nucleic Acids Res.* 26, 4178–4185, 1998.

20. Trubetskoy, V.S. et al., Caged DNA does not aggregate in high ionic strength solutions, *Bioconj. Chem.* 10, 624–628, 1999.

21. Oupicky, D., Carlisle, R.C., and Seymour, L.W., Triggered intracellular activation of disulfide crosslinked polyelectrolyte gene delivery complexes with extended systemic circulation *in vivo*, *Gene Ther.* 8, 713–724, 2001.

22. Adami, R.C. and Rice, K.G., Metabolic stability of glutaraldehyde cross-linked peptide DNA condensates, *J. Pharm. Sci.* 88, 739–746, 1999.

23. Nishida, K. et al., Hepatic disposition characteristics of electrically charged macromolecules in rat *in vivo* and in the perfused liver, *Pharm. Res.* 8, 437–444, 1991.

24. Mahato, R.I., Takakura, Y., and Hashida, M., Nonviral vectors for *in vivo* gene delivery: physicochemical and pharmacokinetic considerations, *Crit. Rev. Ther. Drug Carrier Syst.* 14, 133–172, 1997.

25. Mahato, R.I. et al., Physicochemical and disposition characteristics of antisense oligonucleotides complexed with glycosylated poly(L-lysine), *Biochem. Pharmacol.* 53, 887–895, 1997.

26. Mahato, R.I. et al., *In vivo* disposition characteristics of plasmid DNA complexed with cationic liposomes, *J. Drug Target.* 3, 149–157, 1995.

27. Oupicky, D. et al., Effect of albumin and polyanion on the structure of DNA complexes with polycation containing hydrophilic nonionic block, *Bioconj. Chem.* 10, 764–772, 1999.

28. Takakura, Y. and Hashida, M., Pharmacokinetic of macromolecules and synthetic gene delivery systems, in *Self-Assembly Complexes for Gene Delivery: From Laboratory to Clinical Trial*, Kabanov, A.V., Felgner, P.L., and Seymour, L.W. (Eds.), John Wiley & Sons, Chichester, 1998, 295–306.

29. Wolfert, M.A. et al., Polyelectrolyte vectors for gene delivery: influence of cationic polymer on biophysical properties of complexes formed with DNA, *Bioconj. Chem.* 10, 993–1004, 1999.

30. Kwok, K.Y. et al., Formulation of highly soluble poly(ethylene glycol)-peptide DNA condensates, *J. Pharm. Sci.* 88, 996–1003, 1999.

31. Kwok, K.Y. et al., Strategies for maintaining the particle size of peptide DNA condensates following freeze-drying, *Int. J. Pharm.* 203, 81–88, 2000.

32. Lee, H., Jeong, J.H., and Park, T.G., A new gene delivery formulation of polyethylenimine/DNA complexes coated with PEG conjugated fusogenic peptide, *J. Control. Release* 76, 183–192, 2001.

33. Lee, H., Jeong, J.H., and Park, T.G., PEG grafted polylysine with fusogenic peptide for gene delivery: high transfection efficiency with low cytotoxicity, *J. Control. Release* 79, 283–291, 2002.

34. Lim, D.W., Yeom, Y.I., and Park, T.G., Poly(DMAEMA-NVP)-b-PEG-galactose as gene delivery vector for hepatocytes, *Bioconj. Chem.* 11, 688–695, 2001.

35. Mannisto, M. et al., Structure-activity relationships of poly(L-lysines): effects of pegylation and molecular shape on physicochemical and biological properties in gene delivery, *J. Control. Release* 83, 169–182, 2002.

36. Ogris, M. et al., PEGylated DNA/transferrin-PEI complexes: reduced interaction with blood components, extended circulation in blood and potential for systemic gene delivery, *Gene Ther.* 6, 595–605, 1999.

37. Choi, Y.H. et al., Lactose-poly(ethylene glycol)-grafted poly-L-lysine as hepatoma cell-tapgeted gene carrier, *Bioconj. Chem.* 9, 708–718, 1998.

38. Choi, Y.H. et al., Characterization of a targeted gene carrier, lactose-polyethylene glycol-grafted poly-L-lysine and its complex with plasmid DNA, *Hum. Gene Ther.* 10, 2657–2665, 1999.

39. Liu, F., Song, Y., and Liu, D., Hydrodynamics-based transfection in animals by systemic administration of plasmid DNA, *Gene Ther.* 6, 1258–1266, 1999.

40. Liu, D. and Knapp, J.E., Hydrodynamics-based gene delivery, *Curr. Opin. Mol. Ther.* 3, 192–197, 2001.

41. Zhang, G., Song, Y., and Liu, D., Long-term expression of human alpha 1-antitrypsin gene in mouse liver achieved by intravenous administration of plasmid DNA using a hydrodynamics-based procedure, *Gene Ther.* 7, 1344–1349, 2000.

42. Drickamer, K., Multiple subfamilies of carbohydrate recognition domains in animal lectins, *Ciba Found. Symp.* 145, 45–58, 1989.

43. Drickamer, K., Ca2+-dependent carbohydrate-recognition domains in animal proteins, *Curr. Opin. Struct. Biol.* 3, 393–400, 1993.

44. Weis, W.I. and Drickamer, K., Structural basis of lectin-carbohydrate recognition, *Ann. Rev. Biochem.* 65, 441–473, 1996.

45. Ashwell, G. and Morell, A.G., The role of surface carbohydrates in the hepatic recognition and transport of circulating glycoproteins, *Adv. Enzymol. Relat. Areas Mol. Biol.* 41, 99–128, 1974.

46. Pricer, W.E., Jr. and Ashwell, G., The binding of desialylated glycoproteins by plasma membranes of rat liver, *J. Biol. Chem.* 246, 4825–4833, 1971.

47. Hudgin, R.L. et al., The isolation and properties of a rabbit liver binding protein specific for asialoglycoproteins, *J. Biol. Chem.* 249, 5536–5543, 1974.

48. Drickamer, K., Ca+2 dependent carbohydrate-recognition domains in animal proteins, *Curr. Opin. Struct. Biol.* 3, 393–400, 1993.

49. Drickamer, K., Recognition of complex carbohydrates by Ca(2+)-dependent animal lectins, *Biochem. Soc. Trans.* 21, 456–459, 1993.

50. Day, A.J., The C-type carbohydrate recognition domain (CRD) super family, *Biochem. Soc. Trans.* 22, 83–94, 1994.

51. Bezouska, K. et al., Evolutionary conservation of intron position in a subfamily of genes encoding carbohydrate-recognition domains, *J. Biol. Chem.* 266, 11604–11609, 1991.

52. Weis, W.I. and Drickamer, K., Trimeric structure of a C-type mannose-binding protein, *Structure* 2, 1227–1240, 1994.

53. McEver, R.P., Moore, K.L., and Cummings, R.D., Leukocyte trafficking mediated by selectin-carbohydrate interactions, *J. Biol. Chem.* 270, 11025–11028, 1995.

54. Wu, K., Yuan, J., and Lasky, L.A., Characterization of a novel member of the macrophage mannose receptor type C lectin family, *J. Biol. Chem.* 271, 21323–21330, 1996.

55. Lee, Y.C. et al., Binding of synthetic oligosaccharides to the hepatic Gal/GalNAc lectin. Dependence on fine structural features, *J. Biol. Chem.* 258, 199–202, 1983.

56. Ng, K.K.S. et al., Structural analysis of monosaccharide recognition by rat liver mannose-binding protein, *J. Biol. Chem.* 271, 663–674, 1996.

57. Lee, Y.C., Binding modes of mammalian hepatic Gal/GalNAc receptors, *Ciba Found. Symp.* 145, 80–95, 1989.

58. Baenziger, J.U. and Fiete, D., Galactose and N-acetylgalactosamine-specific endocytosis of glycopeptides by isolated rat hepatocytes, *Cell* 22, 611–620, 1980.

59. Lee, Y.C., et al., Binding of synthetic oligosaccharides to the hepatic Gal/GalNAc lectin. Dependence on fine structural features, *J. Biol. Chem.* 258, 199–202, 1983.

60. Chiu, M.H. et al., *In vivo* targeting function of N-linked oligosaccharides with terminating galactose and N-acetylgalactosamine residues, *J. Biol. Chem.* 269, 16195–16202, 1994.

61. Chiu, M.H. et al., Tissue targeting of multivalent Le(x)-terminated N-linked oligosaccharides in mice, *J. Biol. Chem.* 270, 24024–24031, 1995.

62. Rice, K.G. et al., *In vivo* targeting function of N-linked oligosaccharides: pharmacokinetic and biodistribution of N-linked oligosaccharides, *Glycoimmun. Adv. Exper. Med. Biol.* 376, 271–282, 1995.

63. Taylor, M.E. and Drickamer, K., Structural requirements for high affinity binding of complex ligands by the macrophage mannose receptor, *J. Biol. Chem.* 268, 399–404, 1993.

64. Thomas, V.H., Yang, Y., and Rice, K.G., *In vivo* ligand specificity of E-selectin binding to multivalent sialyl Lewisx N-linked oligosaccharides, *J. Biol. Chem.* 274, 19035–19040, 1999.

65. Wagner, E. et al., Transferrin-polycation conjugates as carriers for DNA uptake into cells, *Proc. Nat. Acad. Sci. U.S.A.* 87, 3410–3414, 1990.

66. Cotten, M. et al., Transferrin-polycation-mediated introduction of DNA into human leukemic cells: stimulation by agents that affect the survival of transfected DNA or modulate transferrin receptor levels, *Proc. Nat. Acad. Sci. U.S.A.* 87, 4033–4037, 1990.

67. Zenke, M. et al., Receptor-mediated endocytosis of transferrin-polycation conjugates: an efficient way to introduce DNA into hematopoietic cells, *Proc. Nat. Acad. Sci. U.S.A.* 87, 3655–3659, 1990.

68. Erbacher, P. et al., Glycosylated polylysine/DNA complexes: gene transfer efficiency in relation with the size and the sugar substitution level of glycosylated polylysines and with the plasmid size, *Bioconj. Chem.* 6, 401–410, 1995.

69. Kabanov, A.V. and Kabanov, V.A., DNA complexes with polycations for the delivery of genetic material into cells, *Bioconj. Chem.* 6, 7–20, 1995.

70. Kabanov, A.V., Felgner, P.L., and Seymour, L.W. (Eds.), *Self-Assembling Complexes for Gene Delivery*, John Wiley & Sons, New York, 1998, 422.
71. Ziady, A.G. et al., Chain length of the polylysine in receptor-targeted gene transfer complexes affects duration of reporter gene expression both *in vitro* and *in vivo*, *J. Biol. Chem.* 274, 4908–4916, 1999.
72. Wadhwa, M.S. et al., Peptide-mediated gene delivery: influence of peptide structure on gene expression, *Bioconj. Chem.* 8, 81–88, 1997.
73. McKenzie, D.L., Collard, W.T., and Rice, K.G., Comparative gene transfer efficiency of low molecular weight polylysine DNA-condensing peptides, *J. Pept. Res.* 54, 311–318, 1999.
74. McKee, T.D. et al., Preparation of asialoorosomucoid-polylysine conjugates, *Bioconj. Chem.* 5, 306–311, 1994.
75. Stubbs, H.J., Shia, M.A., and Rice, K.G., Preparative purification of tetraantennary oligosaccharides from human asialyorosomucoid, *Anal. Biochem.* 247, 357–365, 1997.
76. Erbacher, P. et al., Gene transfer by DNA/Glycosylated polylysine complexes into human blood monocyte-derived macrophages, *Hum. Gene Ther.* 7, 721–729, 1996.
77. Kollen, W.J.W. et al., Gluconoylated and glycosylated polylysines as vectors for gene transfer into cystic fibrosis airway epithelial cells, *Hum. Gene Ther.* 7, 1577–1586, 1996.
78. Nishikawa, M. et al., Pharmacokinetics and *in vivo* gene transfer of plasmid DNA complexed with mannosylated poly(L-lysine) in mice, *J. Drug Target.* 8, 29–38, 2000.
79. Erbacher, P. et al., The reduction of the positive charges of polylysine by partial gluconoylation increases the transfection efficiency of polylysine/DNA complexes, *Biochim. Biophys. Acta* 1324, 27–36, 1997.
80. Roche, A.C. et al., Glycofection: facilitated gene transfer by cationic glycopolymers, *Cell. Mol. Life Sci.* 60, 288–297, 2003.
81. Monsigny, M. et al., Glycotargeting: influence of the sugar moiety on both the uptake and the intracellular trafficking of nucleic acid carried by glycosylated polymers, *Biosci. Rep.* 19, 125–132, 1999.
82. Fajac, I. et al., Sugar-mediated uptake of glycosylated polylysines and gene transfer into normal and cystic fibrosis airway epithelial cells, *Hum. Gene Ther.* 10, 395–406, 1999.
83. Fajac, I. et al., Uptake of plasmid/glycosylated polymer complexes and gene transfer efficiency in differentiated airway epithelial cells, *J. Gene Med.* 5, 38–48, 2003.
84. Midoux, P. et al., Specific gene transfer mediated by lactosylated poly-L-lysine into hepatoma cells, *Nucleic Acids Res.* 21, 871–878, 1993.
85. Duverger, E. et al., Sugar-dependent nuclear import of glycoconjugates from the cytosol, *Exp. Cell Res.* 207, 197–201, 1993.
86. Midoux, P. et al., Membrane permeabilization and efficient gene transfer by a peptide containing several histidines, *Bioconj. Chem.* 9, 260–267, 1998.
87. Perales, J.C. et al., Gene transfer *in vivo*: sustained expression and regulation of genes introduced into the liver by receptor-targeted uptake, *Proc. Nat. Acad. Sci. U.S.A.* 91, 4086–4090, 1994.
88. Ferkol, T. et al., Receptor-mediated gene transfer into macrophages, *Proc. Nat. Acad. Sci. U.S.A.* 93, 101–105, 1996.
89. Zanta, M.A. et al., *In vitro* gene delivery to hepatocytes with galactosylated polyethylenimine, *Bioconj. Chem.* 8, 839–844, 1997.
90. Godbey, W.T. et al., Improved packing of poly(ethylenimine)/DNA complexes increases transfection efficiency, *Gene Ther.* 6, 1380–1388, 1999.
91. Adami, R.C. et al., Stability of peptide-condensed plasmid DNA formulations, *J. Pharm. Sci.* 87, 678–683, 1998.
92. Leamon, C.P., Weigl, D., and Hendren, R.W., Folate copolymer-mediated transfection of cultured cells, *Bioconj. Chem.* 10, 947–957, 1999.
93. Merwin, J.R. et al., Targeted delivery of DNA using YEE (GalNAcAH)3, a synthetic glycopeptide ligand for the asialoglycoprotein receptor, *Bioconj. Chem.* 5, 612–620, 1994.
94. Wadhwa, M.S. et al., Targeted gene delivery with a low molecular weight glycopeptide carrier, *Bioconj. Chem.* 6, 283–291, 1995.
95. Collard, W.T. et al., Biodistribution, metabolism, and *in vivo* gene expression of low molecular weight glycopeptide polyethylene glycol peptide DNA co-condensates, *J. Pharm. Sci.* 89, 499–512, 2000.

96. Yang, Y. et al., Cross-linked low molecular weight glycopeptide mediated gene delivery: relationship between DNA metabolic stability and the level of transient gene expression *in vivo*, *J. Pharm. Sci.* 90, 2010–2022, 2001.

97. Collard, W.T., Evers, D.L., and McKenzie, D.L., Synthesis of homogenous glycopeptides and their utility as DNA condensing agents, *Carb. Res.* 323, 176–184, 2000.

98. Kwok, K.Y., Yang, Y., and Rice, K.G., Evolution of cross-linked non-viral gene delivery systems, *Curr. Opin. Mol. Ther.* 3, 142–146, 2001.

99. Ghosh, P. and Bachhawat, B.K., Grafting of different glycosides on the surface of liposomes and its effect on the tissue distribution of 125I-labelled gamma-globulin encapsulated in liposomes, *Biochim. Biophys. Acta* 632, 562–572, 1980.

100. Allen, T.M. et al., Liposomes containing synthetic lipid derivatives of poly(ethylene glycol) show prolonged circulation half-lives *in vivo*, *Biochim. Biophys. Acta* 1066, 29–36, 1991.

101. Hara, T. et al., Specific uptake by asialofetuin-labeled liposomes by isolated hepatocytes, *Int. J. Pharmaceut.* 42, 69–75, 1988.

102. Lasic, D.D. et al., Sterically stabilized liposomes: a hypothesis on the molecular origin of the extended circulation times, *Biochim. Biophys. Acta* 1070, 187–192, 1991.

103. Kawakami, S. et al., Asialoglycoprotein receptor-mediated gene transfer using novel galactosylated cationic liposomes, *Biochem. Biophys. Res. Comm.* 252, 78–83, 1998.

104. Kawakami, S. et al., Mannose receptor-mediated gene transfer into macrophages using novel mannosylated cationic liposomes, *Gene Ther.* 7, 292–299, 2000.

105. Kawakami, S. et al., *In vivo* gene delivery to the liver using novel galactosylated cationic liposomes, *Pharm. Res.* 17, 306–313, 2000.

106. Hashida, M. et al., Cell-specific delivery of genes with glycosylated carriers, *Adv. Drug Del. Rev.* 52, 187–196, 2001.

CHAPTER 33

Targeted Gene Delivery via the Folate Receptor

Shih-Jiuan Chiu and Robert J. Lee

CONTENTS

33.1 INTRODUCTION

Gene transfer is an emerging therapeutic modality for the treatment of cancer. Tumor-targeted gene delivery is a promising strategy for enhancing the efficacy of cancer gene therapy. The folate receptor (FR) is a 38 kDa membrane glycoprotein that is generally absent in normal tissues and is frequently amplified in human cancers. Folate-conjugated drug carriers, including gene transfer vectors, retain high affinity for the FR. As a low-molecular-weight ligand for tumor targeting, folate has many unique advantages, such as nonimmunogenicity and convenient availability, in addition to its high tumor specificity. A variety of gene transfer vectors have been conjugated to folate for folate-receptor-mediated tumor cell-specific delivery, including adenoviral particles, polyplexes, lipoplexes, and lipopolyplexes. In this review, recent reports on FR-targeted gene delivery are summarized. The potential applications of these vectors in cancer gene therapy are discussed.

0-8493-1934-X/05/$0.00+$1.50
© 2005 by CRC Press LLC

33.2 GENE THERAPY FOR CANCER

Therapeutic gene transfer is an emerging modality for the treatment of tumors that are refractory to existing therapy.[1–4] Clinical application of gene therapy has, however, been limited by the unavailability of safe, efficient, and well-characterized gene transfer vectors. Early efforts in vector design have focused primarily on genetically engineered viruses, such as retrovirus, adenovirus, and adeno-associated virus (see recent reviews[5–7]). However, inherent obstacles associated with these vectors, such as immunogenicity and safety concerns, have limited their clinical adoption.[8] Synthetic vectors are, therefore, being developed as alternatives to viral vectors.[8–11] These vectors provide formulation design flexibility and can be tailored to the size and topology of the DNA cargo as well as a specific route of vector administration. Compared to viral vectors, synthetic vectors are potentially less immunogenic, are relatively easy to produce in clinical quantities, and are associated with fewer safety concerns.[8,10,11] Synthetic vectors designed for parenteral administration encompass a wide range of formulations, including unmodified (naked) DNA,[12] cationic polymer–DNA complexes (polyplexes),[13] cationic lipid–DNA complexes (lipoplexes),[14] and polymer–lipid–DNA ternary complexes (lipopolyplexes).[14–16] While unmodified DNA is designed for direct intra-tissue injection, most other types of synthetic vectors are designed for systemic or airway administration. (See recent reviews on nonviral gene transfer vectors[17–24].)

33.3 THE FOLATE RECEPTOR AS A TUMOR MARKER

The FR is a 38 kDa glycosyl phosphatidylinositol (GPI)-anchored glycoprotein that exists in 3 isoforms, 2 membrane bound (α and β)[25,26] and 1 truncated (γ).[27,28] FR expression in normal tissues is highly restricted and its physiological role is clear only in specific instances.[29] Transmembrane transport of folate coenzymes in most normal tissues is primarily mediated by a low affinity reduced folate carrier with K_t in the μM range.[30]

FR expression is frequently amplified in a variety of human cancers and in tumor cell lines.[31–33] Specifically, type α FR is often overexpressed among epithelial-lineage tumors,[32] whereas both acute and chronic myelogenous leukemias express type β FR.[33] Garin-Chesa et al. demonstrated, by histochemical staining using an anti-FR-α antibody, elevated FR expression in ovarian, endometrial, colorectal, breast, lung, and renal cell carcinomas, as well as in brain metastases.[32] Increased FR expression has also been identified by reverse transcriptase polymerase chain reaction (RT-PCR) analyses in choriocarcinomas, meningiomas, uterine sarcomas, osteosarcomas, non-Hodgkins lymphomas, and choroid plexus tumors.[33] FR expression in nonmucinous ovarian carcinomas has been shown to approach 100%.[34] Furthermore, high FR expression in ovarian cancer has been associated with an increase in the percentage of S-phase cells indicating aggressive tumor growth.[34] For recent reviews on the biology of FR, please see.[35–37]

33.4 TARGETED DRUG DELIVERY VIA THE FR

The consistent overexpression of FR in ovarian cancers suggests its potential utility as a cellular target for drug delivery. Radiolabeled monoclonal antibodies against FR-α, MOv18 and MOv19, have been evaluated as potential agents for radioimaging and radiotherapy in ovarian cancer patients.[38,39] In addition, chimeric and bispecific constructs of anti-FR antibodies have been evaluated as potential immunotherapy agents against ovarian cancer.[40,41]

Covalent coupling of folic acid to a drug or drug carriers has been evaluated as an alternative strategy for targeting the FR. Folic acid (Figure 33.1) is a high-affinity ligand of FR ($K_d \sim 10^{-10}$ M). Covalent derivatization of the γ-carboxyl of folic acid has been shown to result in only a moderate reduction in receptor binding affinity.[42] Following receptor binding, folate conjugates

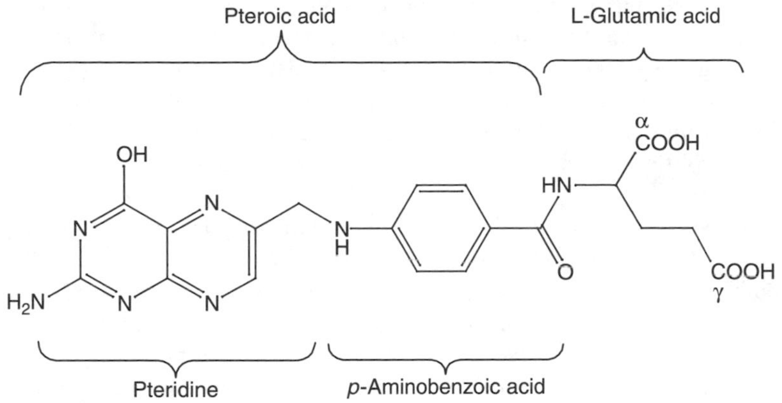

Pteroic acid
L-Glutamic acid
α
COOH
OH
HN
N
N
N
H
O
COOH
γ
H_2N
N
N
Pteridine
p-Aminobenzoic acid

Figure 33.1 Structure of folic acid.

are internalized by cells via receptor-mediated endocytosis which appears to follow a nondegradative pathway.[43] A schematic diagram of the FR-mediated pathway is shown in Figure 33.2.

Folate conjugation, therefore, presents a potential strategy for targeting diagnostic and therapeutic agents non-destructively to FR+ tumor cells. This approach has been explored in the targeted delivery of protein toxins,[44–46] prodrug converting enzyme,[47] cytokine,[48] anti-T-cell antibody,[40] radionuclide chelates,[49,50] chemotherapeutics,[51,52] starburst dendrimers,[53] liposomes,[51,54,55] and gene transfer vectors.[15,56,57] Lee et al. was the first to show FR targeting of liposomes.[54] They discovered that, although folate can be directly linked to low-molecular-weight agents or macromolecules while retaining its high affinity for the FR, targeting of relatively large particles such as liposomes requires the introduction of a long PEG linker.[54] FR-targeted drug delivery has recently been reviewed in.[58–63]

H+
H+ H+

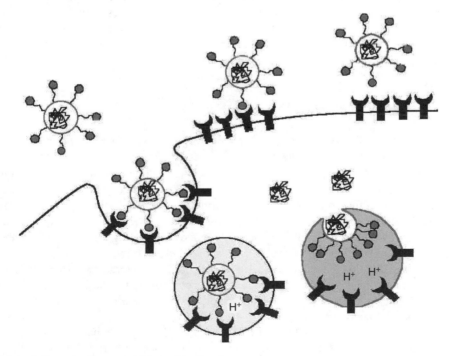

Figure 33.2 Schematic diagram of FR-mediated pathway.

Folic acid (M_w 441) has several important advantages over macromolecular ligand:[58-63] (1) unlimited availability; (2) superior functional stability; (3) relatively simple and defined conjugation chemistry, providing easy incorporation into a gene transfer vector; and (4) presumed lack of immunogenicity. In addition, the FR endocytosis pathway is reportedly nondestructive, allowing the receptor to recycle and to continuously accumulate folate conjugates. Moreover, the high frequency of FR amplification in a variety of human tumors makes this delivery strategy widely applicable.

Potential obstacles for the use of folic acid in tumor targeting are (1) binding interference from circulating folates in the plasma and (2) FR expression in the apical membrane of kidney proximal tubules. Competition from the endogenous form of folate, N-5-methyltetrahydrofolate (at 1–50 nM in human plasma), should not significantly impede FR binding of multivalent folate conjugates such as folate-derivatized gene transfer vectors. Although low-molecular-weight folate conjugates have been shown to accumulate at high levels in the kidneys,[50] the size of high-molecular-weight drug carriers, such as folate-derivatized gene transfer vectors, prevents glomerular filtration, thus eliminating the potential for FR-mediated kidney accumulation.[64]

A phase I/II clinical trial has been completed at the M.D. Anderson Cancer Center evaluating the use of ^{111}In-DTPA-folate for the diagnostic imaging of ovarian carcinomas. The γ-scintigraphy images of patients showed high radionuclide localization in malignant tumors but not in benign tumors. For detection of malignant ovarian carcinomas, a sensitivity of 92% and a specificity of 83% were achieved in this study. No significant accumulation of the folate conjugate has been found in any major organs beside the tumor and the kidneys. These findings show that using folate conjugation to target therapeutics is feasible in humans.

33.5 TARGETED GENE DELIVERY VIA THE FR

Recently, a variety of gene transfer vectors have been conjugated to folate for FR-mediated tumor cell-specific delivery, including adenoviral particles, polyplexes, lipoplexes, and lipopolyplexes. Schematic illustrations of these formulations with or without a PEG linker are shown in Figure 33.3. FR-targeted gene delivery is the subject of a number of recent review articles.[58,61-63,65,66] In the present article, recent literatures on FR-targeted gene delivery are summarized (Table 33.1). The potential applications of these vectors in cancer gene therapy are briefly discussed.

33.5.1 FR-Targeted Adenoviral Vectors

Direct covalent coupling of folate to viral vectors has been evaluated as a method for targeting FR[+] cells. Reddy et al.[67] directly conjugated folate to a sarcoma retrovirus and a murine adenovirus containing the β-galactosidase (β-gal) reporter gene. Without folate modification, the amphotropic adenovirus efficiently transfected KB cells while the ecotropic retrovirus did not. Although folate derivatization enhanced binding of both types of viruses to the FR[+] KB cells, it failed to promote retroviral transfection in these cells and actually reduced the transfection activity of the adenoviral vector. These results suggested that the FR-mediated endocytic pathway was incapable of promoting retroviral transfection and was incompatible with adenoviral infection.

In an alternative strategy, the tropism of adenoviral vectors was altered by masking the specificity of the viral particle using an anti-fiber antibody that was conjugated to a targeting ligand.[68] Douglas et al.[57] reported that a recombinant adenoviral vector carrying a luciferase reporter gene could be retargeted to the FR using an anti-fiber F_{ab}–folate conjugate. The F_{ab}–folate conjugate coated adenoviral particles efficiently transfected FR[+] KB oral carcinoma cells but not FR[-] cells.[57]

The ineffectiveness of the direct folate conjugation approach and effectiveness of the indirect coupling method illustrated the importance of both receptor binding and intracellular trafficking pathways for effective targeted viral infection.

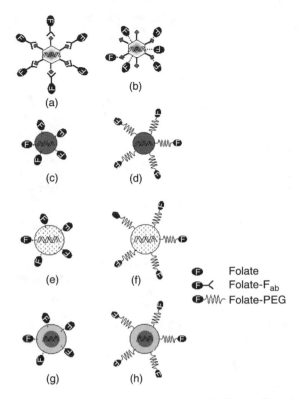

Figure 33.3 Structure of folate-derivatized gene transfer vectors. (A) Folate–F_{ab}–adenovirus, (B) folate–adenovirus, (C) folate polyplexes, (D) folate–PEG polyplexes, (E) folate lipoplexes, (F) folate–PEG lipoplexes, (G) folate lipopolyplexes, and (H) folate–PEG lipopolyplexes.

33.5.2 FR-Targeted Polyplexes

Several cationic polymers have been conjugated to folate. These polymers form tight electrostatic complexes with plasmid DNA. The findings of these studies are briefly summarized below.

33.5.2.1 Folate–Poly(L-lysine) (PLL)

PLL (Figure 33.4A) can efficiently condense plasmid DNA, but lacks endosomal lytic activity, which is important for efficient gene delivery. Gottschalk et al.[56] reported that folate–PLL carrying CMV/β-gal showed relatively low transfection activity in KB cells. However, when the transfection complex was co-incubated with a replication-defective adenovirus, a 1000-fold increase of β-gal activity was observed. These results suggested that endosomalytic activity, which was provided by the viral particles, was essential for efficient DNA delivery.

In a separate study, Mislick et al.[69] synthesized a folate–PLL conjugate and studied its gene transfer properties in the presence of chloroquine. Chloroquine, a lysosomatropic weak base, is known to increase lysosomal pH and possess lysosomalytic activity, similar to that of adenoviral particles. Folate–PLL polyplexes, in the presence of chloroquine (a lysosomotropic agent), exhibited a sixfold higher transfection efficiency than that of nonconjugated PLL in KB cells. The enhancement was blocked by the addition of free folic acid. In addition, the transfection activity of folate–PLL polyplexes decreased in the absence of chloroquine.

More recently, Leamon et al.[70] evaluated the gene transfer properties of folate–PLL conjugates that incorporate a PEG spacer. PLL–PEG–folate polyplexes carrying the pGL3 plasmid were evaluated in FR+ HeLa cervical cancer and IGROV ovarian cancer cells. The highest transfection

Table 33.1 Reported Studies on Targeted Gene Delivery via the FR

Type of Vector	Reporter Gene	Cell Line	Targeting Moiety	Ref.
Viral Vectors				
Adenovirus	pCMVLuc	KB	Folate–F_{ab}	57,68
Adenovirus, retrovirus	β–gal	KB and MDA231	Folate–Ad virus	67
Polyplexes				
PLL	pCMV–β–gal	KB, HeLa, Caco-2, SW620, SKOV3	Folate–PLL	56
	pGL2 (Luc)	KB	Folate–PLL	69
	pGL3 and pSV–β–gal	HeLa and IGROV	Folate–PEG–PLL	70
	PGL3 (Luc)	HeLa	Folate–PEG–PLL	71
PEI	pCMVLuc	KB	Folate–PEI and Folate–PEG–PEI	72
	pCMVLuc	KB, smooth muscle cells, and CT-26	Folate–PEG–PEI	73,74
pDMAEMA copolymer	pCMVLacZ	OVCAR-3	Folate–PEG–polymer	75
Lipoplexes				
C_{14}COrn	pCMVLuc	KB and others	Folate–PEG3400–PE	76
DOPE–RPR209120	PCMVLuc or CAT	M109	Folate–PEG4600–Chol	77
Lipopolyplexes				
DOPC–Chol	pCMVLuc	M109	Folate–cysteine–PEG3400–PE	78
DOPE–CHEMS + PLL	pRSVLuc	KB	Folate–PEG–DSPE	15
C-DOPE–DOPE–Chol or CHEMS + PLL	pCMVlacZ	KB and 24JK-FBP KB, HeLa, MDA231, A549	Folate–PEG–DOPE	79,80
Diolein–CHEMS + PEI	pcDNA3–CMV–Luc	KB	Folate–PEG–DSPE	81

activity was observed with the conjugate that incorporated a PEG 3400 spacer. In addition, transfection activity was concentration dependent and could be blocked by excess-free folate. These observations indicated that PLL–PEG–folate/DNA complexes were taken up via the FR, and that the intramolecular spatial distance between the ligand and the PLL backbone played an important role in optimized transfection.

Finally, Ward et al.[71] reported that surface modification of PLL–DNA complexes with folate–PEG_{800} or folate–PEG_{3400} could enhance transgene expression both in FR^+ HeLa cells and increase the systemic circulation time of the DNA vector in mice.

33.5.2.2 *Folate–Polyethylenimine (PEI)*

In contrast to PLL, PEI (Figure 33.4B) has both DNA condensing and endosomalytic activities due to its buffering capacity at mildly acidic pH. PEI alone forms polyplexes with DNA that exhibit high transfection activities, especially at high nitrogen to phosphate (N/P) ratios. A series of PEI–folate conjugates, either directly linked or linked through a PEG spacer, have been synthesized and evaluated by Guo and Lee.[72] Transfection studies performed in KB cells showed that gene transfer efficiency increased as the N/P ratio increased in all formulations. A greater FR-targeting effect was observed at low N/P ratios. Folate directly attached to PEI did not significantly increase the transfection efficiency. These data suggested that a PEG spacer was required for effective binding of the PEI conjugate to the FR.

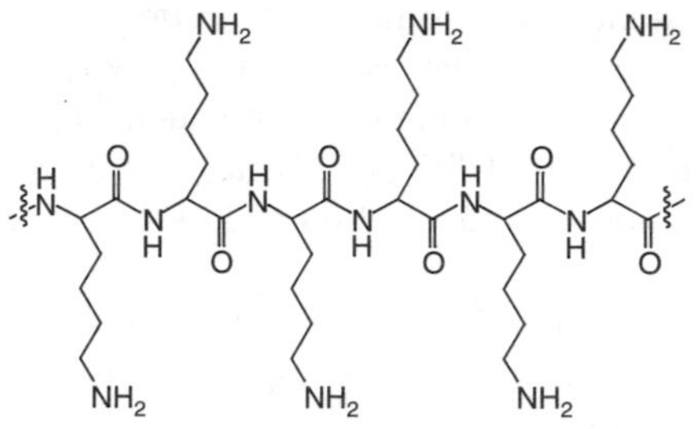

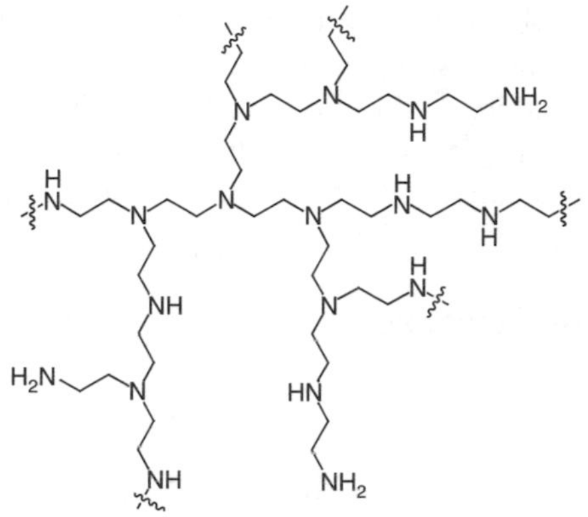

(a) Poly-L-lysine

(b) Polyethylenimine

Figure 33.4 Structure of (A) poly(L-lysine) (PLL) and (B) polyethylenimine (PEI).

More recently, Benns et al.[73] reported the synthesis of folate–PEG–folate–PEI by linking folate to both ends of a PEG and then coupling to PEI. At neutral charge (N/P) ratio, polyplexes of this conjugate were less cytotoxic and exhibited higher transfection activity in FR+ colon adenocarcinoma cells but not in FR- cells compared to simple PEI/DNA polyplexes.[74] Free folic acid was able to inhibit the folate-dependent transfection.

33.5.2.3 Folate-pDMAEMA

Van Steenis et al.[75] modified the surface of pDMAEMA polyplexes with folate via a PEG spacer. This resulted in a sharp decrease of the zeta potential and a small increase in particle size of these polyplexes. PEGylated polyplexes with or without folate ligands were less cytotoxic than uncoated polyplexes. *In vitro* transfection of FR+ OVCAR-3 ovarian carcinoma cells was markedly increased compared to nontargeted PEGylated polyplexes, suggesting FR-mediated gene delivery.

In summary, the above reported studies on folate-conjugated polyplexes showed promising transfection result *in vitro*. The incorporation of a PEG spacer and endosomalytic elements appear to be crucial factors for the overall performance of the polyplexes. Further studies are needed to characterize the *in vivo* gene transfer activities of these polyplexes.

33.5.3 FR-Targeted Lipoplexes

Dauty et al.[76] reported a novel vector formulation based on a cationic thiol-detergent, dimerized tetradecyl-ornithinyl-cysteine $(C_{14}COrn)_2$. $(C_{14}COrn)_2$/DNA complexes prepared at N/P = 1.2 were small particles (~30 nm) with transfection efficiency slightly higher than naked DNA. The poor activity was probably a result of low zeta potential (-45 ± 5 mV) of these complexes, which hinders cellular uptake. Increasing the N/P ratio to 4–5 enhanced transfection efficiency to levels comparable to other well-known transfection reagents such as PEI and lipopolyamine. In an attempt to achieve extended circulation time and to target tumors by systemic delivery, 2 mole% of dipalmitoylphosphatidyethanolamine/PEG$_{3400}$/folate (DPPE/PEG$_{3400}$/folate) was incorporated into the formulation. The resulting particles showed good colloidal stability and FR-dependent binding and internalization in FR$^+$ tumor cells.

Hofland et al.[77] evaluated a cationic liposomal formulation comprised of equimolar amounts of RPR209120 (a novel lipopolyamine) and dioleoylphosphatidylethanolamine (DOPE), and 5 mol% folate/PEG/cholesterol (folate/PEG/Chol) (Figure 33.5A). The corresponding lipoplexes showed ~1000-fold higher *in vitro* transfection efficiency than non-targeted-PEG liposomes in M109, an FR$^+$ murine lung carcinoma cell line. *In vivo* biodistribution studies of these lipoplexes in BALB/c mice carrying subcutaneous M109 tumors showed similar uptake of the FR-targeted and the nontargeted control lipoplexes in most organs including the tumors except for the lung tissues, in which the FR-targeted group exhibited 50-fold lower uptake than the nontargeted group.

In summary, FR-targeted lipoplexes are attractive vectors given the inherent endosomalytic activity provided by the lipidic components. Their utility for tumor-selective delivery *in vivo* should be further validated in animal model studies.

33.5.4 FR-Targeted Lipopolyplexes

33.5.4.1 LPDI

These vectors are comprised of cationic liposomes complexed with DNA condensed with a polycation. Reddy et al.[78] prepared a liposome-entrapped polycation-condensed DNA vector (LPDI) that incorporated protamine-condensed plasmid DNA and folate/PEG/PE as an FR-targeting ligand. Liposome/protamine/DNA lipopolyplexes with a low concentration (0.01 mole%) of folate–PEG–PE yielded superior transfection activity in ascites cells. Among the tissues in the peritoneal cavity, ascites and solid tumors received most of the injected dose. In both solid tumors and tumor cell-containing ascites, eight- to tenfold more transfection activity was observed in the FR-targeted group than in the nontargeted group. Increasing the folate ligand density resulted in reduced gene transfer activity. This could be due to the steric hindrance effect of PEG, which may inhibit endosomal disruption following internalization of lipoplexes into the FR$^+$ cells.

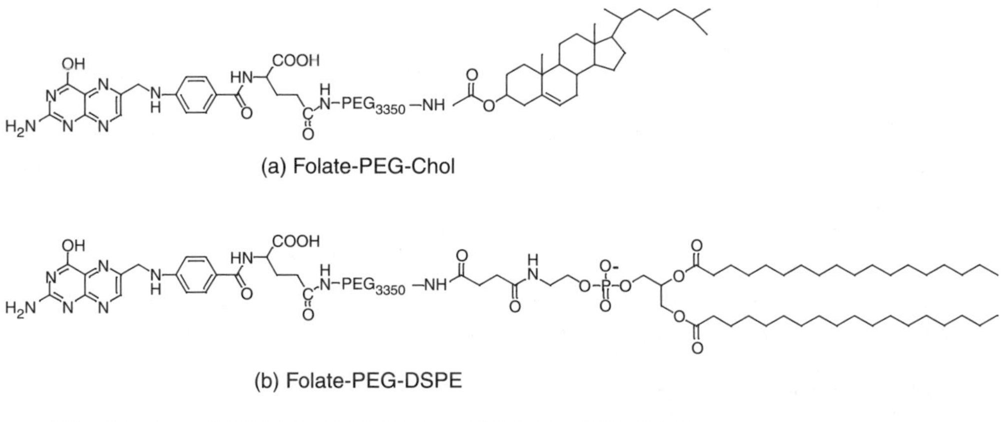

(a) Folate-PEG-Chol

(b) Folate-PEG-DSPE

Figure 33.5 Structure of (A) folate–PEG–Chol and (B) folate–PEG–DSPE.

33.5.4.2 LPDII

These vectors are comprised of anionic liposomes complexed with DNA condensed with a polycation. Lee and Huang[15] have developed an FR-targeted vector, in which DNA was first complexed to PLL at a ratio of 1:0.75 (w/w) and then complexed to folate-coated pH-sensitive anionic liposomes composed of DOPE/cholesteryl hemisuccinate/folate-PEG-DOPE (DOPE/CHEMS/folate-PEG-DOPE) (6:4:0.01 mol/mol) via charge interaction. Transfection efficiency in FR+ KB cells was affected by both the lipid-to-DNA ratio and the lipid composition. Electron micrograph results showed LPDII are spherical particles with a positively stained core enclosed in a lipidic envelope with a mean diameter of 74 +/- 14 nm. For LPDII particles that were positively charged (prepared at low lipid to DNA ratios), transfection and cellular uptake levels were independent of the FR and did not require a pH-sensitive lipid composition. In contrast, for LPDII particles that were negatively charged (prepared at high lipid to DNA ratios), FR-targeted LPDII exhibited significantly higher transfection efficiency than nontargeted LPDII formulations and the transfection of FR-targeted was inhibited by the addition of free folic acid in KB cells. These results indicated that FR-targeting is most efficient with negatively charged vector particles, which lack activity in the absence of cellular targeting.

Reddy et al.[79,80] prepared LPDII-type complexes comprised of a polycation-condensed DNA plasmid associated with a mixture of neutral and anionic lipids supplemented with folate–PEG–DOPE for tumor targeting. N-citraconyl-dioleoylphosphatidylethanolamine (C-DOPE), an acid-labile lipid, is included in liposome formulation for pH-dependent release of endosome-entrapped DNA into the cytoplasm, and a novel plasmid containing a 366 bp segment from SV40 DNA has also been employed to facilitate transport of the plasmid into the nucleus. Transfection studies showed that high molecular weight polymers such as acylated PLL and cationic dendrimers as DNA condensing agents exhibited more significant FR-targeting effect than small cationic molecules such as spermine, spermidine, or gramicidin S. C-DOPE-containing liposomes were stable in neutral pH but became fusogenic at pH 5, which was close to the pH in the endosomes. C-DOPE thus constitutes a novel pH-sensitive liposomal formulation that facilitates the endosomal release of DNA.

Gosselin et al.[81] evaluated the stability and transfection properties of PEI–DNA polyplexes (PEI at 25 kDa) before and after covalent cross-linking with dithiobis(succinimidylpropionate) (DSP) or dimethyl 3,3′-dithiobispropionimidate·2HCl (DTBP), either alone or as a component of LPDII vectors. PEI–DNA complexes could be stabilized by this step at molar ratios greater than or equal to 10:1 (DSP or DTBP:PEI); however, further increasing the ratio of cross-linking agent decreased the transfection efficiency. For preparation of LPDII type vectors, anionic liposomes composed of diolein/CHEMS (6:4 mol/mol), diolein/CHEMS/PEG-distearoylphosphatidylethanolamine (PEG-DSPE) (6:4:0.05 mol/mol), or diolein/CHEMS/folate-PEG-Chol (6:4:0.05 mol/mol) were used. The structures of folate–PEG–Chol and folate–PEG–DSPE are shown in Figure 33.5A and 33.5B, respectively. Electron microscopy studies showed that crosslinked LPDII vectors appeared as roughly spherical aggregated complexes with a rather broad size distribution ranging between 300 and 800 nm. Compared with cross-linked LPDII vectors without PEG–DSPE, inclusion of folate–PEG–Chol increased gene transfer activities 3- to 4-fold at lipid:DNA ratios between 1 and 5 in KB cells, and the transfection activity mediated by LPDII vectors containing folate-PEG-Chol was reduced in the presence of a high concentration of free folate. The vectors that contained covalent stabilization are designed to maintain stability in systemic circulation. The *in vivo* gene transfer properties of this novel formulation remain to be characterized.

33.6 CONCLUSIONS

The numerous formulations developed to target the FR and the exciting preliminary results in cell culture reflect a high level of enthusiasm in the gene delivery community for this tumor targeting strategy. The findings indicated that besides FR-mediated tumor cell uptake, formulation characteristics that provide reduced nonspecific cellular uptake, efficient endosomal escape, increased plasma stability, and prolonged systemic circulation are critical for the specificity and efficiency of gene transfer by receptor targeted vectors. Optimization of these properties is, therefore, urgently needed. Furthermore, key determinants of *in vivo* gene transfect vector performance include its pharmacokinetic property and stability in circulation, and its ability to localize in and distribute within the target tumors. Novel strategies, such as covalent vector stabilization and PEGylation of the vector, might prove critical for the *in vivo* performance of FR-targeted vectors. In addition, the effects of tumor endothelial permeability and the rate of extravascular diffusion within the tumor may have a profound impact on the ability of gene transfer vectors to reach its target tumor cells. Future studies should, therefore, place greater emphasis on the *in vivo* characterization of vector formulations in animal tumor models. Continued efforts in optimization of FR-targeted gene transfer vectors will likely lead to the development of a tumor-specific vehicle for therapeutic gene delivery and promote the advancement of clinical translation of cancer gene therapy.

REFERENCES

1. Dachs, G.U. et al. Targeting gene therapy to cancer: a review, *Oncol. Res.* 9, 313, 1997.
2. Walther, W. and Stein, U., Therapeutic genes for cancer gene therapy, *Mol. Biotechnol.* 13, 21, 1999.
3. Curiel, D.T., Gerritsen, W.R., and Krul, M.R.L., Progress in cancer gene therapy, *Cancer Gene Ther.* 7, 1197, 2000.
4. Meyer, F. and Finer, M., Gene therapy: progress and challenges, *Cell. Mol. Biol.* 47, 1277, 2001.
5. Walther, W. and Stein, U., Viral vectors for gene transfer: a review of their use in the treatment of human diseases, *Drugs* 60, 249, 2000.
6. Seth, P., Adenoviral vectors, in *Cancer Gene Therapy: Past Achievements and Future Challenges*, Habib, N.A., (Ed.), Kluwer Academic/Plenum Publishers, New York, 2000, chap. 2.
7. Barnett, B.G., Crews, C.J., and Douglas, J.T., Targeted adenoviral vectors, *Biochim. Biophys. Acta* 1575, 1, 2002.
8. Lasic, D.D. and Templeton, N.S., Liposomes in gene therapy, *Adv. Drug Deliv. Rev.* 20, 221, 1996.
9. Bilbao, G., et al., Advances in adenoviral vectors for cancer gene therapy, *Expert Opin. Ther. Patents* 7, 1427, 1997.
10. Rolland, A.P., From genes to gene medicines: recent advances in nonviral gene delivery, *Crit. Rev. Ther. Drug Carrier Sys.* 15, 143, 1998.
11. Nishikawa, M. and Huang, L., Nonviral vectors in the new millennium: delivery barriers in gene transfer, *Hum. Gene Ther.* 12, 861, 2001.
12. Wolff, J.A. et al., Direct gene-transfer into mouse muscle *in vivo*, *Science* 247, 1465, 1990.
13. Felgner, P.L. et al., Lipofection: a highly efficient, lipid-mediated DNA- transfection procedure, *Proc. Nat. Acad. Sci. U.S.A.* 84, 7413, 1987.
14. Gao, X. and Huang, L., Potentiation of cationic liposome-mediated gene delivery by polycations, *Biochemistry* 35, 1027, 1996.
15. Lee, R.J. and Huang, L., Folate-targeted, anionic liposome-entrapped polylysine-condensed DNA for tumor cell-specific gene transfer, *J. Biol. Chem.* 271, 8481, 1996.
16. Li, S. et al., Characterization of cationic lipid-protamine-DNA (LPD) complexes for intravenous gene delivery, *Gene Ther.* 5, 930, 1998.
17. Hart, S.L., Synthetic vectors for gene therapy, *Expert Opin. Ther. Patents* 10, 199, 2000.
18. De Smedt, S.C., Demeester, J., and Hennink, W.E., Cationic polymer based gene delivery systems, *Pharm. Res.* 17, 113, 2000.

19. Pouton, C.W. and Seymour, L.W., Key issues in non-viral gene delivery, *Adv. Drug Del. Rev.* 46, 187, 2001.

20. Schatzlein, A.G., Non-viral vectors in cancer gene therapy: principles and progress, *Anti-Cancer Drugs* 12, 275, 2001.

21. Segura, T. and Shea, L.D., Materials for non-viral gene delivery, *Ann. Rev. Mater. Res.* 31, 25, 2001.

22. Merdan, T., Kopecek, J., and Kissel, T., Prospects for cationic polymers in gene and oligonucleotide therapy against cancer, *Adv. Drug Del. Rev.* 54, 715, 2002.

23. Niidome, T. and Huang, L., Gene therapy progress and prospects: Nonviral vectors, *Gene Ther.* 9, 1647, 2002.

24. Wicthoff, C.M. and Middaugh, C.R., Barriers to nonviral gene delivery, *J. Pharm. Sci.* 92, 203, 2003.

25. Lacey, S.W. et al., Complementary-DNA for the folate binding-protein correctly predicts anchoring to the membrane by glycosyl- phosphatidylinositol, *J. Clin. Invest.* 84, 715, 1989.

26. Ratnam, M. et al., Homologous membrane folate binding-proteins in human-placenta: cloning and sequence of a cDNA, *Biochemistry* 28, 8249, 1989.

27. Shen, F. et al., Identification of a novel folate receptor, a truncated receptor, and receptor-type-beta in hematopoietic-cells: cDNA cloning, expression, immunoreactivity, and tissue-specificity, *Biochemistry* 33, 1209, 1994.

28. Shen, F. et al., Folate receptor-type-gamma is primarily a secretory protein due to lack of an efficient signal for glycosylphosphatidylinositol modification: protein characterization and cell-type specificity, *Biochemistry* 34, 5660, 1995.

29. Antony, A.C., The biological chemistry of folate receptors, *Blood* 79, 2807, 1992.

30. Sirotnak, F.M. and Tolner, B., Carrier-mediated membrane transport of folates in mammalian cells, *Annu. Rev. Nutr.* 19, 91, 1999.

31. Weitman, S.D. et al., Distribution of the folate receptor gp38 in normal and malignant-cell lines and tissues, *Cancer Res.* 52, 3396, 1992.

32. Garinchesa, P. et al., Trophoblast and ovarian-cancer antigen-LK26: sensitivity and specificity in immunopathology and molecular-identification as a folate-binding protein, *Am. J. Pathol.* 142, 557, 1993.

33. Ross, J.F. et al., Folate receptor type beta is a neutrophilic lineage marker and is differentially expressed in myeloid leukemia, *Cancer* 85, 348, 1999.

34. Toffoli, G. et al., Overexpression of folate binding protein in ovarian cancers, *Int. J. Cancer* 74, 193, 1997.

35. Antony, A.C., Folate receptors, *Annu. Rev. Nutr.* 16, 501, 1996.

36. Christensen, E.I. et al., Membrane receptors for endocytosis in the renal proximal tubule, *Int. Rev. Cytol.* 180, 237, 1998.

37. Brzezinska, A., Winska, P., and Balinska, M., Cellular aspects of folate and antifolate membrane transport, *Acta Biochim. Pol.* 47, 735, 2000.

38. Molthoff, C.F. et al., Radioimmunotherapy of ovarian cancer with intravenously administered iodine-131 labeled chimeric MOv18 monoclonal antibody, *J. Nuclear Med.* 40, 966, 1999.

39. van Zanten-Przybysz, I. et al., Radioimmunotherapy with intravenously administered i-131-labeled chimeric monoclonal antibody MOv18 in patients with ovarian cancer, *J. Nuclear Med.* 41, 1168, 2000.

40. Kranz, D.M. et al., Conjugates of folate and anti-T-cell-receptor antibodies specifically target folate-receptor-positive tumor-cells for lysis, *Proc. Nat. Acad. Sci. U.S.A.* 92, 9057, 1995.

41. Hwu, P. et al., *In vivo* antitumor-activity of T-cells redirected with chimeric antibody T-cell receptor genes, *Cancer Res.* 55, 3369, 1995.

42. Gabizon, A. et al., Targeting folate receptor with folate linked to extremities of poly(ethylene glycol)-grafted liposomes: *in vitro* studies, *Bioconj. Chem.* 10, 289, 1999.

43. Leamon, C.P. and Low, P.S., Delivery of macromolecules into living cells: a method that exploits folate receptor endocytosis, *Proc. Nat. Acad. Sci. U.S.A.* 88, 5572, 1991.

44. Leamon, C.P. and Low, P.S., Cytotoxicity of momordin-folate conjugates in cultured human cells, *J. Biol. Chem.* 267, 24966, 1992.

45. Leamon, C.P., Pastan, I., and Low, P.S., Cytotoxicity of folate-pseudomonas exotoxin conjugates toward tumor-cells: contribution of translocation domain, *J. Biol. Chem.* 268, 24847, 1993.

46. Atkinson, S.F. et al., Conjugation of folate via gelonin carbohydrate residues retains ribosomal-inactivating properties of the toxin and permits targeting to folate receptor positive cells, *J. Biol. Chem.* 276, 27930, 2001.

47. Lu, J.Y. et al., Folate-targeted enzyme prodrug cancer therapy utilizing penicillin-V amidase and a doxorubicin prodrug, *J. Drug Target.* 7, 43, 1999.

48. Melani, C. et al., Targeting of Interleukin 2 to human ovarian carcinoma by fusion with a single-chain Fv of antifolate receptor antibody, *Cancer Res.* 58, 4146, 1998.

49. Mathias, C.J. et al., Tumor-selective radiopharmaceutical targeting via receptor-mediated endocytosis of gallium-67-deferoxamine-folate, *J. Nuclear Med.* 37, 1003, 1996.

50. Guo, W., Hinkle, G.H., and Lee, R.J., 99mTc-HYNIC-folate: a novel receptor-based targeted radiopharmaceutical for tumor imaging, *J. Nuclear Med.* 40, 1563, 1999.

51. Lee, R.J. and Low, P.S., Folate-mediated tumor-cell targeting of liposome-entrapped doxorubicin *in vitro*, *Biochim. Biophys. Acta* 1233, 134, 1995.

52. Ladino, C.A. et al., Folate-maytansinoids: target-selective drugs of low molecular weight, *Int. J. Cancer* 73, 859, 1997.

53. Konda, S.D. et al., Specific targeting of folate-dendrimer MRI contrast agents to the high affinity folate receptor expressed in ovarian tumor xenografts, *Magn. Reson. Mater. Phys. Biol. Med.* 12, 104, 2001.

54. Lee, R.J. and Low, P.S., Delivery of liposomes into cultured KB cells via folate receptor-mediated endocytosis, *J. Biol. Chem.* 269, 3198, 1994.

55. Wang, S., et al., Delivery of antisense oligodeoxyribonucleotides against the human epidermal growth-factor receptor into cultured KB cells with liposomes conjugated to folate via polyethylene-glycol, *Proc. Nat. Acad. Sci. U.S.A.* 92, 3318, 1995.

56. Gottschalk, S. et al., Folate receptor-mediated DNA delivery into tumor-cells — potosomal disruption results in enhanced gene-expression, *Gene Ther.* 1, 185, 1994.

57. Douglas, J.T. et al., Targeted gene delivery by tropism-modified adenoviral vectors, *Nat. Biotechnol.* 14, 1574, 1996.

58. Wang, S. and Low, P.S., Folate-mediated targeting of antineoplastic drags, imaging agents, and nucleic acids to cancer cells, *J. Control. Release* 53, 39, 1998.

59. Benns, J.M. and Kim, S.W., Tailoring new gene delivery designs for specific targets, *J. Drug Target.* 8, 1, 2000.

60. Sudimack, J. and Lee, R.J., Targeted drug delivery via the folate receptor, *Adv. Drug Del. Rev.* 41, 147, 2000.

61. Leamon, C.P. and Low, P.S., Folate-mediated targeting: from diagnostics to drug and gene delivery, *Drug Discov. Today* 6, 44, 2001.

62. Lu, Y.J. and Low, P.S., Folate-mediated delivery of macromolecular anticancer therapeutic agents, *Adv. Drug Del. Rev.* 54, 675, 2002.

63. Gosselin, M.A. and Lee, R.J., Folate receptor-targeted liposomes as vectors for therapeutic agents, in *Biotechnology Annual Review*, El-Gewely, M.R. (Ed.), Elsevier Science, Amsterdam, 2002, 103.

64. Guo, W.J. et al., Receptor-specific delivery of liposomes via folate-PEG-Chol, *J. Liposome Res.* 10, 179, 2000.

65. Reddy, J.A. and Low, P.S., Folate-mediated targeting of therapeutic and imaging agents to cancers, *Crit. Rev. Ther. Drug Carrier Sys.* 15, 587, 1998.

66. Ward, C.M., Folate-targeted non-viral DNA vectors for cancer gene therapy, *Curr. Opin. Mol. Ther.* 2, 182, 2000.

67. Reddy, J.A., Clapp, D.W., and Low, P.S., Retargeting of viral vectors to the folate receptor endocytic pathway, *J. Control. Release* 74, 77, 2001.

68. Douglas, J.T. and Curiel, D.T., Strategies to accomplish targeted gene delivery to muscle cells employing tropism-modified adenoviral vectors, *Neuromusc. Disord.* 7, 284, 1997.

69. Mislick, K.A. et al., Transfection of folate-polylysine DNA complexes — evidence for lysosomal delivery, *Bioconj. Chem.* 6, 512, 1995.

70. Leamon, C.P., Weigl, D., and Hendren, R.W., Folate copolymer-mediated transfection of cultured cells, *Bioconj. Chem.* 10, 947, 1999.

71. Ward, C.M. et al., Modification of PLL/DNA complexes with a multivalent hydrophilic polymer permits folate-mediated targeting *in vitro* and prolonged plasma circulation *in vivo*, *J. Gene Med.* 4, 536, 2002.

72. Guo, W.J. and Lee, R.J., Receptor-targeted gene delivery via folate-conjugated polyethylenimine, *AAPS Pharmsci.* 1, Article 19, 1999.

73. Benns, J.M. et al., Folate-PEG-folate-graft-polyethylenimine-based gene delivery, *J. Drug Target.* 9, 123, 2001.

74. Benns, J.M., Mahato, R.I., and Kim, S.W., Optimization of factors influencing the transfection efficiency of folate-PEG-folate-graft-polyethylenimine, *J. Control. Release* 79, 255, 2002.

75. van Steenis, J.H. et al., Preparation and characterization of folate-targeted PEG-coated pDMAEMA-based polyplexes, *J. Control. Release* 87, 167, 2003.

76. Dauty, E. et al., Intracellular delivery of nanometric DNA particles via the folate receptor, *Bioconj. Chem.* 13, 831, 2002.

77. Hofland, H.E.J. et al., Folate-targeted gene transfer *in vivo*, *Mol. Ther.* 5, 739, 2002.

78. Reddy, J.A. et al., Folate-targeted, cationic liposome-mediated gene transfer into disseminated peritoneal tumors, *Gene Ther.* 9, 1542, 2002.

79. Reddy, J.A. et al., Optimization of folate-conjugated liposomal vectors for folate receptor-mediated gene therapy, *J. Pharm. Sci.* 88, 1112, 1999.

80. Reddy, J.A. and Low, P.S., Enhanced folate receptor mediated gene therapy using a novel pH-sensitive lipid formulation, *J. Control. Release* 64, 27, 2000.

81. Gosselin, M.A., Guo, W.J., and Lee, R.J., Incorporation of reversibly cross-linked polyplexes into LPDII vectors for gene delivery, *Bioconj. Chem.* 13, 1044, 2002.

CHAPTER 34

Transferrin Receptor-Targeted Gene Delivery Systems

Ralf Kircheis and Ernst Wagner

CONTENTS

0-8493-1934-X/05/$0.00+$1.50
© 2005 by CRC Press LLC

34.1 INTRODUCTION

Nonviral vectors are receiving increasing attention as promising gene delivery vehicles for a broad variety of biomedical applications. While the majority of gene delivery approaches so far has employed adenoviral or retroviral vectors, nonviral vectors are becoming more and more attractive because of important advantages, such as biosafety, stability, low cost, and consistency of production. Moreover, nonviral vectors can be chemically modified with ease, and they offer high flexibility regarding the size of the transgene delivered.[1-3] Nonviral delivery systems which are able to deliver therapeutic genes specifically to the target tissues *in vivo* would be particularly attractive for the development of therapeutics, e.g., in areas of high medical needs such as oncology, cardiovascular disease, and diseases of the central nerve system.

Polyplexes are based essentially on the condensation of negatively charged DNA by electrostatic interactions with polycationic condensing compounds into compact particles, protecting the DNA from degradation.[4] The resulting particles also facilitate the uptake of the DNA into the cells. For efficient transfection the delivery vector must mediate a sequential process that comprises DNA condensation, uptake into the cell, release from the endosomal compartment, migration through the cytoplasm, uptake into the nucleus, and finally decondensation of the DNA for transcription.[5] Furthermore, for *in vivo* application there are additional barriers encountered until the DNA complexes reach the target cells, including anatomical size constraints and nonspecific interaction with biological fluids, extracellular matrix, and nontarget cells. Moreover, the transfection particles must be small enough to enable blood circulation, extravasation, and diffusion through tissues.[6] Recently, strategies have been developed for target-specific gene delivery *in vivo*. Specific recognition and internalization into the target cells aims to avoid undesired gene delivery to nontarget tissues and to diminish toxic side effects.[7] In recent years, effective nonviral delivery systems have been designed that combine mechanisms for DNA compaction, cellular uptake, endosomal release, and, to some extent, nuclear uptake. A variety of polycations, such as polylysine, dendrimers, and polyethylenimines (PEIs) possessing several of these mechanisms have been broadly used for transfection in cell culture as well as in a variety of applications *in vivo*.[8-10] Cell-binding ligands such as transferrin (Tf), integrin-binding RGD motifs, antibodies to surface markers of cells (e.g., anti-CD3), or epithelial growth factor (EGF) have been incorporated into DNA–polycation complexes for receptor-mediated interaction and endocytosis into target cells.[11-14] Combined with other modifications to block undesired, nonspecific interactions with blood components or nontarget cells (see below), the incorporation of target-specific ligands, such as Tf, has been demonstrated to result in targeted gene expression in distant tumors following systemic application of transfection particles through the tail vein in tumor-bearing mice.

34.2 INTRINSIC PROPERTIES OF POLYCATIONS

There are a number of activities essential for efficient transfection that are intrinsic features of polycations, including DNA condensation, electrostatic interaction with cell membrane, release from the endosomal compartment, and migration through the cytoplasm into the nucleus. In contrast, other features, such as specific binding to target cells and shielding from nonspecific interaction in biological fluids and nontarget cells, have to be additionally incorporated into the transfection complexes.

34.2.1 DNA Condensation and Electrostatic Interaction with Cell Membrane

A number of cationic polymers, such as polylysine,[12,13,15] poly(amidoamine) (PAMAM) dendrimers,[9] and polyethylenimines,[10] are able to interact electrostatically via their protonated amine groups with the negatively charged phosphate groups of DNA. Condensation protects the DNA from degradation by nucleases, and results in formation of compact particles, i.e., polyplexes, that can be taken up by cells via natural processes such as adsorptive endocytosis, pinocytosis, and phagocytosis. The extent of condensation between the polycation and DNA depends on the charge ratio, that is, the nitrogen of the polycation to DNA phosphate (N/P) ratio.[4,5] At electroneutrality the complexes are often large and have the tendency to form insoluble aggregates. Compact particles of small size are usually obtained at higher polycation to DNA ratios (i.e., N/P ratios), resulting in complexes with a strong net positive charge.[15]

The positive surface charge of polycation–DNA complexes usually also serves for binding to the cells via electrostatic interactions with the negatively charged cell membrane structures, e.g., sulfated proteoglycans, followed by nonspecific adsorptive endocytosis.[16-19] Interaction with the proteoglycans seems to be essential also for the subsequent complex internalization process.[20] The need for excess positive charge for efficient DNA complexation and cell binding, however, can lead to major problems, particularly for application *in vivo*, as discussed below. In addition to the surface charge, the size of the transfection particles is also a major factor affecting cellular uptake.[21] The optimal size for internalization into particular target cells is dependent on the characteristics of the target cells' preferred mode of uptake (endocytosis, pinocytosis, or phagocytosis); in the same target cell small or large particles can enter different intracellular pathways, resulting in completely different outcomes: transfection or degradation.

The physical properties of the condensed polycation–DNA complexes can be characterized by a number of techniques that measure the size or shape (electrophoretic mobility in agarose gels, electron or atomic force microscopy, laser light scattering), charge (electrophoretic mobility, zeta potential), conformation and condensation (circular dichroism, ethidium bromide exclusion, and centrifugation). It is understood that characterization of essential physical parameters, e.g., surface charge and size of complexes, is essential to understanding the processes involved in gene delivery and to predicting the behavior of the transfection particles in various biological fluids and solutions.

34.2.2 Endosomal Release

Following uptake into the cells, irrespectively whether induced by nonspecific adsorptive endocytosis or receptor-mediated endocytosis (see below), the DNA complexes are in most cases present in endosomes or other internal vesicles, i.e., the material is separated from the cytoplasm by a lipid membrane. After endocytosis, normal cellular trafficking processes direct the endocytosed particles to lysosomes for degradation. The accumulation of DNA complexes in the endosomes and degradation in the endolysosomes strongly limits the efficiency of gene transfer. Furthermore, while condensation of DNA and electrostatic interaction with the cell membrane are common features that are characteristic for most of the polycations; however, there are major differences between different polycations concerning their ability to release the complex from the endosomal compartment.[8-10,22]

34.2.2.1 Polylysine

Poly(L-lysine) was one of the first polycations used as a nonviral vector for gene delivery. Polylysine has well characterized DNA condensing activity. At physiological pH the amino groups of the polylysine are positively charged and interact ionically with the negatively charged phosphate groups of the DNA, forming toroid-like structures.[11-13,15]

Condensed polylysine–DNA complexes interact electrostatically with the negatively charged cell surface and are taken up by absorptive endocytosis. The efficiency of uptake and gene expression can be enhanced by covalent coupling of ligands to the polylysine that can specifically target cells and promote receptor-mediated cellular uptake (see below).[11-13,23-27] Although polylysine can provide uptake of DNA into cells, generally rather low gene expression is found due to a number of factors, such as ineffective endosomal release and inefficient nuclear uptake. Since polylysine is lacking an intrinsic endosomal release activity, the majority of polylysine–DNA complexes are trapped within the endosomes, and will be degraded in the endolysosomes. Different strategies have been developed to increase endosomal release of polylysine–DNA complexes.

The addition of lysosomotropic agents, e.g., chloroquine, into the transfection medium was found to increase transfection efficacy in a number of cell lines.[27] Chloroquine is assumed to accumulate in the vesicles leading to swelling of the endosome by raising the pH. Furthermore it may also help to release the polylysine from the DNA, increasing gene expression due to an enhanced release of partially dissociated DNA into the cytoplasm.[28,29] Similarly, the addition of glycerol was shown to enhance gene expression in a number of cell lines, probably by weakening the endosomal lipid bilayer and allowing polylysine–DNA complex release into the cytoplasm.[30] However, the addition of lysosomotropic agents or glycerol for endosomal release of polylysine–DNA complexes is limited largely to cell culture application.

Another approach developed to enhance endosomal release is the incorporation of viruses that have an intrinsic endosomal disruption mechanism. Replication-deficient and chemically inactivated adenoviruses have been used to enhance release of polylysine–DNA complexes from the endosomal compartment into the cytoplasm.[31-35] While the undesired replicative activity of the virus has been knocked out by genetic deletions of sequences that are essential for viral replication, and additionally by treating with methoxypsoralen plus UV irradiation, the endosomal release activity of the virus capsid is not affected.[36] It has been demonstrated that for optimal endosomal release of the DNA, the viral capsid has to be codelivered with the polylysine–DNA complex. For codelivery the virus can either be covalently coupled to polylysine[33,34] or may be linked to the polylysine, e.g., by means of a biotin-streptavidin bridge.[32] A variety of other viruses such as the avian adenovirus (CELO)[37] and the human rhinovirus[38] have been successfully used to achieve endosomal release of polylysine–DNA complexes. The inclusion of viruses into polyplexes was shown to dramatically enhance endosomal release; however, this approach has the drawback of potentially eliciting an immune response to the virus particle when applied *in vivo*. Therefore, other strategies aside from using the whole virus have selectively employed viral peptide sequences that are responsible for endosomal activity, such as the N-terminal of hemagglutinin 2 subunit,[39,40] rhinovirus VP-1,[38] or synthetic membrane-destabilizing peptides.[39-42] A semi-synthetic transfection system, the so-called adenovirus-enhanced transferrinfection (AVET) system, was designed. This system employed the endosomolytic activity of inactivated adenovirus and the receptor-mediated specific uptake mechanisms of transferrin ligand covalently coupled to polylysine.[13,25,32] This transfection system has shown high transfection efficacy in cell culture and a variety of applications *in vivo*.[43-47] Primary human melanoma cell lines transfected with the gene for the immunostimulatory cytokine IL-2 by using the AVET system[43,47] have been used as autologous or allogeneic tumor vaccines in clinical trials with patients with advanced metastatic melanoma.[48]

34.2.2.2 *Polyethylenimine*

Several polycations have been described that combine both DNA condensation and intrinsic endosomal release activity, thus removing the need for additional endosomolytic agents. Polyethylenimine (PEI) is the compound with the highest charge density, and has a high intrinsic endosomolytic activity because of a strong buffer capacity at biological pHs. PEI is only partially protonated at physiological pH.[10,49] Upon acidification within the endosomes or endolysosomes, PEI is assumed to act as a proton sponge, with the protonation presumably triggering passive chloride ion influx. Proton and chloride ion accumulation is followed by influx of water, causing osmotic swelling and subsequent endosome rupture, thus allowing the escape of the PEI–DNA complexes into the cytosol.[10,49,50] Notably, high levels of gene expression have been found using PEI–DNA or Tf-PEI–DNA complexes in the absence of additional endosomolytic agents like chloroquine or complex-linked adenovirus.[10,50,51] This ability for osmolytic endosomolysis seems to be dependent on particle size with small PEI800kDa or PEI25kDa–DNA complexes formed in water having lower transfection efficacy compared to larger complexes formed at higher ionic strength.[21,52,53] Addition of chloroquine largely restored the impaired transfection efficacy of the small PEI complexes.[21]

Besides the high buffering capacity of PEI, other features, such as the structural flexibility and the molecular weight of the PEI, were also found to be important for transfection efficacy.[22,52] A variety of PEI molecules differing in molecular weight and structure (branched and linear) have been described. Studies comparing PEIs of different molecular weights have shown that high transfection efficacy is characteristic for PEIs with molecular weights greater than or equal to 10 kDa. Low-molecular-weight PEIs (800 Da, or 2 kDa at low charge ratio) showed impaired transfection capacity even though they were able to condense DNA.[50,53,54] Transfection efficacy was restored in these cases by addition of replication-defective adenovirus.[54] Low-molecular-weight PEI at ~2–10 kDa, when applied at higher charge ratios, was found to be effective for DNA delivery in vitro.[55,56] However, it was not effective in receptor-targeted applications and in vivo.[57] PEIs of 10–50 kDa were found to be highly effective; increasing the PEI molecular weight from 20 kDa to 800 kDa, however, seems not to further enhance transfection efficacy.[10,51,52,58-61] Recent studies indicate that 10–25 kDa PEIs may have significant advantages compared to high-molecular-weight PEI (800 kDa) particularly for in vivo applications.[21,46,53,58-61] In comparison with the lower molecular weight PEIs, the 800 kDa PEI has shown high toxicity after systemic application in vivo, probably due to a massive aggregation of erythrocyte.[46,52,59] Furthermore, linear PEI (22 kDa) and branched PEIs (25 kDa and 800 kDa) also showed different behavior in vitro and in vivo applications.[52] The linear PEI showed a particularly high transfection efficacy in a variety of cell types.[52] Moreover, a lower dependency of the transfection efficacy on the cell cycle was found for linear PEI as compared to branched PEI.[62]

Furthermore, high reporter gene expression levels in the lungs were found following systemic application of linear PEI complexes in vivo,[61] which have not been reached with other PEI species.[52] The toxicity profile after systemic application also appears more favorable for linear versus branched PEIs, with the linear PEI showing the lowest tendency for erythrocyte aggregation.[46,52,59]

PEI can be easily modified by coupling of various chemical groups or ligands to the amino groups (see below), making this compound particularly versatile and interesting for further development of nonviral vectors.

34.2.3 Migration through Cytoplasm, Nuclear Entry, and Complex Disassembly

Following uptake into the cell and release from the endosomal compartment, the polyplexes still have to enter the nucleus. The transport of transfected DNA from the cytosol into the nucleus is one of the major limitations for efficient gene transfer by nonviral vectors.[63,64,65] Studies using PEI–DNA complexes that were either microinjectioned into the cytoplasm[64] or exogenously administered have indicated that PEI may have some intrinsic nuclear targeting activity.[66] Alternatively, the polycations may also just provide protection of DNA from nucleases present in the cytosol, increasing the "survival time" of the DNA in the cytoplasm,[67] and leading this way to a higher probability that the complexes finally can enter the nucleus with the next cell division.

The second step, the passage through the nuclear membrane, is also poorly understood so far. However, the fact that intact PEI–DNA complexes have been found in the cell nucleus indicates that a disassembly of the PEI and DNA before nuclear entry might not be necessary.[64,66]

Recent studies[68] have shown that the transfection efficacy of polyplexes is critically dependent on cell division. High transfection efficacy was observed when the complexes were taken up by the cells in the late S or G2 phase, preceding mitosis. Addition of complexes in the G1 phase resulted in low transfection levels, probably due to higher DNA degradation in the cytoplasm before the cells succeeded to enter the next cell cycle.[68] In contrast, transfection with recombinant adenoviruses was not dependent on the cell cycle. Interestingly, also linear PEI was less dependent on the cell cycle in comparison to branched PEIs.[62,68]

The cell cycle dependency decreases the transfection efficacy of nonviral vectors in general. Moreover, while cells under cell culture conditions are constantly dividing, this is much less pronounced for cells within the tissues *in vivo*, making this cell cycle dependency a further limiting factor for gene delivery *in vivo*. The attachment of nuclear localization signal sequences to the polyplexes or the introduction of binding sites for transcription factors in the DNA sequence may help to overcome these barriers.[69-71]

The final step in gene delivery will be the disassembly of the complexes releasing the DNA from the polycation so that the transcription apparatus of the cell can access the DNA for transcription. Efficient disassembly of the polyplexes is an interesting challenge since it represents a diametrically opposite mechanism to the previous steps of gene delivery; premature disassembly of complexes will lead to enhanced DNA degradation and low efficacy of gene delivery, while a delayed or incomplete disassembly will not allow for efficient gene expression.[52]

34.3 RECEPTOR-MEDIATED TRANSFECTION

In order to increase transfection efficiency and improve to target specificity, strategies have been developed to employ specific receptor-mediated cellular uptake mechanism by incorporating cell-binding ligands into the transfection complexes.[7,27,72] The concept of receptor-mediated gene transfer tries to copy well known entry mechanisms which are widely used by viruses and toxins, or which are used for uptake of macromolecules including nutrients (e.g., low-density lipoprotein [LDL], Tf), growth factors and hormones (insulin, vascular endothelial growth factor [VEGF], epidermal growth factor [EGF], FGF) into cells. These entry mechanisms employ natural cellular internalization processes such as receptor-mediated endocytosis, potocytosis, and phagocytosis. As targeting ligands, proteins, peptides, carbohydrates, vitamins, or antibodies to cell surface receptors can be used. The utilization of ligand–receptor binding for uptake into target cells is dependent on (1) the expression level of the receptor on the target cells, (2) the ligand–receptor binding affinity, and (3) whether ligand–receptor binding is followed by internalization. A variety of targeting ligands have been attached covalently or noncovalently to various polycations. Earlier studies on polycation-based receptor-mediated gene delivery systems have used cell-binding ligands covalently coupled to polylysine. This approach was pioneered by the experiments of Wu and Wu,[12,23,26] who demonstrated targeting of asialoorosomucoid-polylysine–DNA complexes to hepatocytes that are known to express asialoglycoprotein receptor. Other ligands widely used for targeting of DNA complexes are Tf, which is overexpressed by proliferating cells such as tumor cells,[13,27,51,72] synthetic galactose-containing ligands for targeting of hepatocytes,[57,73,74] EGF,[14,75,76] basic fibroblast growth factor (bFGF),[77] and antibodies against a variety of cell surface markers.[35,51,78-80] Cell-type-specific gene transfer has been demonstrated using antibodies against the CD3 molecule, which is exclusively expressed on T-lymphocytes or T-cell-derived lymphoid cell lines.[51,79] Although efficient internalisation of ligand-polylysine–DNA complexes was found in most cases, most of the DNA complexes remained trapped in the endolysosomes, resulting in rather moderate transfection efficiencies. Incorporation of additional endosomolytic moieties, such as inactivated adenovirus or membrane-destabilizing peptides, enabled efficient cytoplasmic release of DNA and resulted in high transfection levels.[32,39,40]

In recent years, polycations with intrinsic endosomolytic activity such as PEI[10,49] have been widely used for gene delivery protocols. In parallel, a variety of strategies have already been developed, with the aim to combine the powerful intrinsic transfection activity of these polycations with receptor-mediated cellular uptake mechanism by incorporation of cell-binding ligands.[7,14,51,76,80] In principle, since poly(L-lysine) (PLL) and PEI are both polycationic polyamines, any successful targeting method that has been used for PLL–DNA or other polycation–DNA complexes should be also amenable to PEI–DNA complexes.

Coupling of anti-CD3 antibody to PEI mediates specific gene delivery in the absence of additional endosomal release agents, transferring expression to CD3 positive cells only.[51] A variety of other ligands have been successfully coupled to PEI, including galactose for hepatocyte targeting,[57,74] mannose for enhanced uptake into dendritic cells,[81] EGF for enhanced uptake into epithelial cells[14,76] and integrin–binding RGD peptides.[82] Specific targeting via the integrin-mediated endocytosis pathway by using RGD-PEI–DNA complexes produced up to a 100-fold enhancement in gene expression, which was lost when aspartic acid was replaced by glutamic acid in the targeted peptide sequence (RGD–RGE).[82]

34.3.1 Transferrin-PEI–DNA Complexes

The iron-transport protein Tf has been employed as a targeting ligand for a variety of carriers including PLL and PEI. Coupling of Tf to PLL or PEI can be easily achieved by sodium periodate oxidation of the carbohydrate residues of Tf followed by coupling with the primary amino groups of PLL or PEI.[13,51,59] Alternatively, Tf can be coupled to the distal ends of the polyethylene glycol (PEG), which is used for shielding of the complexes (see below).[83] Coupling of transferrin to PEI (800 kDa) produced up to several-hundred-fold increase in transfection efficiency depending on the cell type targeted (Figure 34.1).[51] Receptor-mediated uptake was demonstrated by (1) competition experiments with excess free ligand, (2) reduced expression for complexes either lacking the targeting ligand or coupled to irrelevant ligands, and (3) enhanced gene expression after up-regulation of the receptor. In many cases, efficient uptake of transferrin-polycation–DNA complexes was found to correlate with high expression levels of the Tf receptor as shown for the Tf receptor-rich erythromyeloid cell line K562.[51,84] Since the Tf receptor is expressed on a broad variety of proliferating cells, Tf receptor-mediated gene transfer is not specific for a certain cell type or tissue type, but rather can be employed for preferential gene delivery to highly proliferating cells, such as tumor cells.[27,51,72,84] Compared to Tf-containing complexes that show increased transfection activity to a broad variety of proliferating cells, coupling of anti-CD3 antibody to PEI resulted in specific gene delivery to CD3 expressing cells only (Figure 34.2).[51]

34.3.2 Specific Ligand-Receptor-Mediated Uptake versus Nonspecific Electrostatic Uptake

The incorporation of specific cell-binding ligands allows for specific ligand-receptor-mediated uptake; additionally, nonspecific electrostatic interactions between the net positively charged polycation–DNA complexes and the negatively charged cell surface can be involved. What mechanism will prevail, the ligand-receptor interaction or the electrostatic interactions will depend on (1) the level of receptor expression on the target cells, (2) the density and amount of negatively charged proteoglycan on the surface of a particular cell type, and (3) the ratio of polycation to DNA in the transfection complex, i.e., the net surface charge of the complex. Increasing Tf-receptor expression on K562 cells by pretreatment with desferrioxamine was shown to increase Tf-mediated uptake into the cells significantly.[72,84] In contrast, cell types such as melanoma cells that possess a high level of negatively charged cell-surface proteoglycans[19] will internalize positively charged PEI–DNA complexes that are much less dependent on receptor-mediated mechanisms. Specific receptor-mediated uptake mechanisms via transferrin or anti-CD3 antibody were found to be most pronounced at low positive or nearly neutral charges of the complexes, whereas nonspecific electrostatic interactions were found particularly at high positive charges.[51]

In addition, the size of the complex can determine whether nonspecific or receptor-mediated uptake mechanisms will prevail. Erbacher and colleagues, using PEI of 25 kDa molecular weight conjugated to integrin-binding RGD peptide, observed receptor-mediated uptake mainly at higher

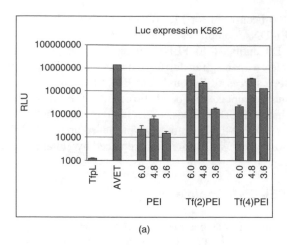

(a)

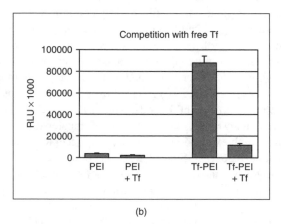

(b)

Figure 34.1   Transferrin-receptor mediated gene delivery. (A) PEI or Tf–PEI conjugates containing covalently coupled transferring at two molar ratios (2:1 and 4:1) were used to complex plasmid DNA (pCMVL, 10 µg) coding for the luciferase reporter gene. DNA complexes were formed at various N/P ratios (3.6; 4.8; 6). DNA complexes (in 0.5 ml HBS) were added to K562 human erythromyeloid leukaemia cells (5×10^5 cells in 1.5 ml RPMI1640, 10% FCS). Transfection medium was removed after 4 h, and luciferase activity was measured 24 h after transfection. Transfection with transferrin-polylysine (Tf-pL)–DNA or adenovirus-enhanced transferrinfection system (AVET) are shown for comparison. Mean and SEM of triplicates are shown. (B) Competitive inhibition with free transferrin. PEI–DNA or Tf-PEI–DNA complexes (N/P ratio 4.8) were used to transfect K562 cells in the absence or presence of 5 mg/ml free Tf. Transfection medium was removed after 3 h, and luciferase activity was measured 24 h after transfection. Mean and SEM of triplicates are shown.

N/P ratios, whereas at neutral charges the formation of large aggregates resulted in a receptor-independent uptake, presumably by phagocytosis.[82] Finally, the amount of DNA complex will be a factor in deciding between nonspecific and receptor-mediated uptake, with the latter being particularly important at low DNA concentrations.[14]

34.4 APPLICATION *IN VIVO*

In the previous section, a variety of intracellular barriers for nonviral gene delivery were discussed. However, following the application of polycation–DNA complexes *in vivo*, additional barriers have to be overcome. For locally applied complexes, the target site, route of access, and chosen therapy will determine the strategy for gene delivery. When the target site is easily accessible, local administration such as intradermal, intramuscular, intratracheal, intraperitoneal,

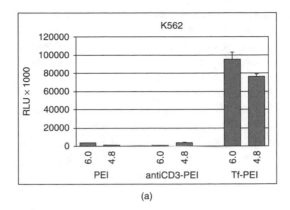

(a)

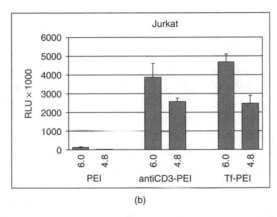

(b)

Figure 34.2 Target-specific gene delivery by cell binding ligands. Human erythromyeloid leukemia cells K562 (CD3 negative) (A) or human T-cell leukemic cells Jurkat E6.1 (CD3 positive) (B) were transfected with PEI–DNA, Tf-PEI–DNA, or anti-CD3-PEI–DNA complexes. DNA complexes (in 0.5 ml HBS) were added to the cells (5×10^5 cells in 1.5ml RPMI1640, 10% FCS). Transfection medium was removed after 4 h, and luciferase activity was measured 24 h after transfection. Mean and SEM of triplicates are shown.

or intratumoral application can be used.[3,44,46,47,85,86] Similarly surgery or catheters can deliver the complexes directly to the target site. Other physical methods can also be used to increase polyplex gene delivery locally, such as continuous application via micropump,[87] aerosol delivery,[88] or jet nebulization.[89]

Local application of polylysine–DNA complexes initially did not show efficient gene transfer. With the addition of endosomal active agents to the polylysine complexes, such as adenoviruses (generating AVET complexes), efficient gene expression was demonstrated in airway epithelium following intratracheal application,[44] nasal epithelium[90] and in tumors after intratumoral application.[46,47] Other cationic polymers, such as PEI, have successfully been used for local gene delivery *in vivo*; mouse brain,[10,58,60] kidney,[91] lung,[92,93] bilary epithelia,[94] fetal mouse liver,[95] carotid artery,[96] and subcutaneous tumors.[46,47,87]

Several groups have described local administration of PEI–DNA complexes to subcutaneous tumors. Generally the low-molecular-weight PEIs (~10–25 kDa) showed higher transfection efficacy and lower toxicity as compared to high-molecular-weight PEI (800 kDa).[46,47,52] Furthermore, intratumoral delivery of linear PEI (22 kDa) into solid or cystic tumor models resulted in transfection of 1% of the cells with the expression persisting for 15 days.[87] The addition of ligands, such as Tf, has been shown to increase the efficiency of gene delivery to subcutaneously growing tumors, with Tf-PEI (800 kDa)–DNA complexes being up to 100 times more efficient than naked DNA, and approximately 10 times more efficient than ligand-free PEI, showing a transfection efficacy similar to that observed with AVET complexes.[46,47]

Once applied, complexes have to diffuse through the solid tumor mass. The diffusion of complexes is dependent on solubility and complex size, which are influenced by the charge ratio and salt conditions at complex formation.[21,52] Intratumoral application of complexes often results in a lot of the injected material refluxing through the needle tract and being lost from the intended site of application; this was partially overcome by administering complexes with a micropump.[87] The injection of a large volume of liquid may help in the diffusion of complexes by weakening the cell-to-cell connections and dispersing the complexes throughout the tissue, via hydrodynamic mechanism, as observed in livers after application of DNA in high injection volumes.[97,98]

The interaction of transfection complexes with extracellular matrix and necrotic tissues can significantly reduce transfection efficiency. Large areas of necrosis are typically found in rapidly growing tumors.[99] Furthermore, some glycosaminoglycans, such as hyaluronic acid or heparin sulphate, can completely inhibit the ability of the complexes to transfect cells by releasing the DNA from complexes, and causing degradation of the DNA.[100] Therefore, strategies that shield complexes from these nonspecific interactions have been developed (see below). However, compared to lipoplexes, polyplexes seem generally to be more resistant to inactivation in biological fluids, such as pulmonary surfactants, and retain their ability to condense DNA.[101]

The activation of the immune system is often a side effect that is not desired, since it can lead to uncontrolled side effects and may prevent repeated application of the vector. Generally the polycations have a lower immunogenicity as compared to viruses, e.g., adenoviruses.[102] However, activation of the complement system by polylysine, PEI, and PAMAM has been observed, and complement activation was found to correlate with the positive surface charge of the polyplexes.[103] The activation of the complement cascade can be avoided by shielding the positive particles with PEG (see below). Stimulation of the immune system by complexes can also be a result of bacterial DNA contamination and lipopolysaccharide (LPS) or CpG islands within plasmid DNA.[104-106] The avoidance of immune stimulation may often be desired. However, when generating a vaccination response to an antigen or induction of an antitumoral effect, these additional nonspecific immunostimulatory activities could potentially be beneficial in eliciting an effective treatment.

34.4.1 Systemic Application

Frequently, the target tissues, e.g., metastatic tumor nodules, cannot be reached via local administration but only via the systemic blood circulation, followed by targeting of the specific site. This route of application must overcome numerous biological and physical obstacles before the intended site can be reached. The physical and colloidal parameters of the transfection complexes such as particle size and surface charge will be critical for successful delivery,[46] playing a role in the transport of complexes through capillaries, in extravasation from the blood vessels into the tissues, and in passing through the inter-tissue space to the target cells.[6] Furthermore, complexes must remain stable and soluble while not aggregating in the blood or unraveling, revealing the DNA to degrading enzymes within the blood or inter-tissue space. The transfection complexes should not interact with components of the blood system, such as plasma, complement, erythrocytes, cells of the RES (e.g., macrophages), extracellular matrix, and other nontarget cells.[6,7]

34.4.1.1 Systemically Delivered Polylysine–DNA Complexes

Using polylysine–DNA complexes or Tf-polylysine–DNA complexes without additional endosomolytic agents in tumor-bearing mice has not been shown to lead to significant gene expression *in vivo*, which is certainly due to the lack of intrinsic activity for endosomal release.

In contrast, optimized AVET complexes have been shown to specifically target gene expression to tumors. Electro-neutral AVET complexes that were systemically administered into Neuro2a tumor-bearing A/J mice by intravenous injection into the tail vein resulted in highly preferential luciferase reporter gene expression in the distant tumor in comparison to very low gene expression found in the major organs.[46]

34.4.1.2 Systemically Delivered PEI–DNA Complexes

Systemically delivered linear low-molecular-weight PEI–DNA complexes (PEI at 22 kDa) resulted in a high gene expression in the lung with lower activity gene expression being observed in other organs (spleen, heart, liver, and kidneys).[61] These complexes rapidly crossed the endothelial cells of the lung capillary bed and gene expression was mainly found in alveolar cells.[107-110] In comparison, branched PEIs (25 and 800 KDa) were less efficient as compared to the linear PEI.[46,52,59,61] Surprisingly, systemic delivery of 22 kDa PEI–DNA complexes in mice bearing lung tumors were found to be less efficient at expressing transgenes as compared to normal lungs.[87] With this type of complex gene delivery is not targeted to a particular cell type but rather is mediated by nonspecific interactions based on the positive surface charges of complexes interacting with components of the blood.[7,59,111] These interactions, however, also cause the particles to be taken out of circulation, for example by nonspecifically binding to erythrocytes and uptake by the RES. The interaction of the polycation with blood proteins, such as albumin, can cause the complex to disassemble, exposing the DNA to degrading enzymes via a similar mechanism as that observed with polylysine.[112] Hence the colloidal and physical parameters of the transfection complexes are critical gene delivery in vivo.[21,46,59,111,113] Among other possible reasons, the toxicity that is associated with PEI polyplexes is thought to arise from the positively charged particles interacting with erythrocytes, causing them to aggregate together and then lodge in the lungs, causing embolism.[46,59,111,114]

34.4.2 Shielding of Polycation–DNA Complexes

Shielding of the surface positive charge of polyplexes has recently been shown to reduce aggregation of the particles, nonspecific interactions with serum components and erythrocytes, or interaction with first-pass organs such as the lung endothelium. This protection from nonspecific interactions has been achieved via a number of methods.

34.4.2.1 PEG Shielding of Polyplexes

The chemical linkage of PEG can shield the positive charge of polycation–DNA complexes and reduces the positive surface charge (ζ-potential) to near neutral charge.[46,83,111] Shielded PEI–DNA complexes when applied in vivo showed reduced interaction with erythrocytes and also reduced toxicity; furthermore, they have increased circulation time before being cleared from the blood stream as compared to unshielded complexes.[46,111] By analogy, PEG-polylysine–DNA shielded complexes were also found to have decreased ζ-potential and reduced toxicity. Furthermore, they had an increased solubility as compared to unshielded polylysine–DNA complexes.[115,116] Copolymers of anionic peptides have been linked to PEG, which stabilizes cationic complexes by inhibiting aggregation and reducing complement activation.[117]

34.4.2.2 Ligand Density

Ligands that are covalently attached to the polycation can be used as shielding agent. Previously, Tf–PEI (800 kDa) complexes were found to have a reduced ζ-potential compared to ligand-free PEI (800 kDa).[46] When the ligand density was increased within the complex, the ζ-potential was reduced to near neutral and the nonspecific interactions with cells (such as erythrocytes) was inhibited. Furthermore, the toxicity of the complexes was reduced when applied in vivo (Figure 34.3).[59]

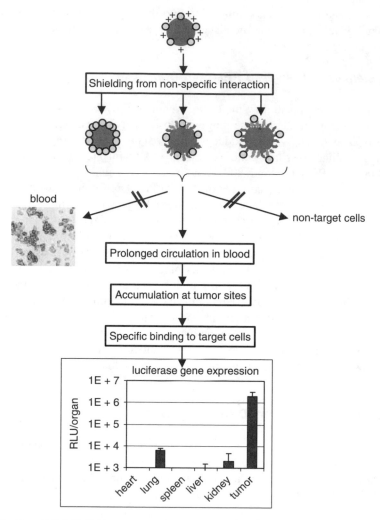

Figure 34.3 Shielding of Tf-PEI–DNA complexes from nonspecific interactions for systemic *in vivo* application. Polycation–DNA complexes with cell-binding ligands (e.g., Tf) can specifically interact with target cells expressing the according receptor. However, at the same time nonspecific interactions with blood components and nontarget cells are possible. Nonspecific interactions can be blocked by shielding the surface charge of the transfection complexes by (1) high ligand density (e.g., high transferrin), (2) PEGylation of ligand-PEI–DNA complexes, or (3) PEGylation with the ligand at the distal end of the PEG coat. Protection from nonspecific interactions enables a prolonged circulation time of the transfection complexes in the blood, leading to accumulation at sites of higher vascular leakiness, e.g., tumor sites, followed by binding and uptake into the target cells.

34.4.2.3 *Poly-N-(2-hydroxypropyl)methacrylamide (pHPMA) Shielding*

Polylysine grafted with pHPMA via a peptide linker had increased solubility with lower toxicity and a negative surface charge. The pHPMA shield reduced the interaction of complexes with blood components.[118]

34.4.2.4 *Poloxamer Shielding*

PEI–DNA complexes shielded with the polyether Pluronic 123 showed reduced lung gene expression with an increased liver expression over unshielded PEI complexes.[119] Local application of DNA shielded with poloxamers increased gene expression in muscle, and furthermore the

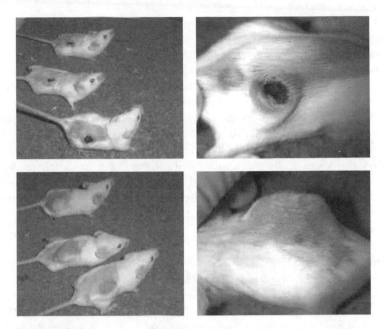

Figure 34.4   (See color insert following page 336) Targeted gene therapy of tumors after systemic application. Surface-shielded Tf-PEI–DNA complexes containing the gene for the highly active cytokine tumor necrosis factor α (TNFα) (upper panel) or the gene for β-galactosidase (control, lower panel) were injected via the tail vein into the systemic blood circulation of mice bearing a subcutaneously growing Neuro2a tumor. Following 4 applications at 2- to 3-day intervals pronounced hemorrhagic tumor necrosis followed by significant decrease in tumor size are found in the TNFα treated mice, whereas no tumor necrosis was found in the control group. Representative pictures are shown.

complexes diffused throughout the tissue more efficiently.[120] Based on these initial reports of shielding approaches, the combination of the above methods will help to generate particles that can have reduced nonspecific interactions while gaining circulation time and increased solubility.

34.4.3   Targeted Gene Expression with Shielded Polyplexes

The development of shielded polycation–DNA complexes for systemic application that can target specific tissues has been achieved with varying success. Examples for targeting polyplexes applicable via systemic application are Tf-PEI–DNA complexes shielded with either PEG[46,83] or by high ligand density.[59] Both delivery systems were shown to target subcutaneous tumors systemically (murine neuroblastoma [Neuro2A]) after application of complexes in the tail vein of syngenic A/J mice. Application of PEGylated Tf-PEI–DNA or high ligand density Tf-PEI–DNA complexes resulted in preferential gene expression in the distant tumors. Furthermore, luciferase gene expression was dramatically reduced in other organs (liver, lung, heart, spleen, and kidney) as compared to unshielded complexes.[46,59,83,121] The targeting of PEGylated-Tf–PEI polyplexes was the first demonstration of tumor-specific targeted gene expression after systemic application.[46] Furthermore, this targeted expression is particularly interesting, as subcutaneous tumors are not directly joined to the main blood supply but via the peripheral blood supply. The reporter gene expression observed in these tumors is probably mediated via two mechanisms. The first is passive targeting due to the shielding of the complexes from nonspecific interactions, which increases the circulation time, allowing the complexes to reach the tumor.[6,111] The second is active targeting of tumor cells via binding of the Tf in the polyplex to the Tf receptor followed by receptor-mediated uptake into the tumor cells (Figure 34.3).[51] However, tumor targeting is only efficient when tumors are well vascularized with a higher permeability as compared to normal vasculature.[99] The size of fenestration within human tumors will be critical for the future development of nonviral particles, as this will put certain size constraints on particle sizes that can be administered.[122,123]

When gene expression was characterized within tumors, often the pattern of gene expression was found as small foci distributed throughout the tumor, with these foci often being close to immature blood vessels.[46,59] The distribution of the stabilized transfection particles was analyzed by southern blotting for DNA distribution in the various organs. Particles that were shielded by increasing the density of transferrin on the polyplex surface were mainly observed in the liver, followed by the tumor. However, particularly in the liver, the majority of DNA was degraded, with almost no gene expression observed in the liver. Histo-chemical analysis of the liver revealed that most of the DNA was taken up by the Kupffer cells within the liver. Southern blotting with the tumor tissue revealed that only ~0.1% of the administered DNA was in tumors, but histo-chemical analysis demonstrated that the DNA was associated mainly with tumor cells and not normal host tissue.[59] The potential of the liver to filter out particles will be a challenge for future vector development to overcome, ensuring that the majority of material administered reaches the targeted organ (e.g., tumor) and is not removed on the first pass through the liver.[124,125]

Recently, surface-shielded Tf-PEI–DNA complexes were used to target the gene coding for a highly active cytokine, tumor necrosis factor α (TNFα), specifically to distant tumors following systemic application in tumor-bearing mice. Irrespective of whether shielding was achieved by incorporation of a high density of Tf in the complex[126,127] or by a PEG coat with Tf covalently coupled to the distal ends of the PEG,[128] gene expression of TNFα was found specifically in the tumor, leading to hemorrhagic tumor necrosis and tumor regression in a variety of tumor models. Importantly, no systemic toxicity — as is usually found with systemic TNFα protein therapy — was detected.

The encouraging potential of targeting genes into tumors is also supported by related findings. Systemic delivery of Tf-targeted liposomes enhanced the gene expression of the p53 gene in head and neck squamous cell carcinoma xenografts, making them more sensitive to radiotherapy.[129] In addition, other targeting ligands have been successfully applied, including EGF[76] and folate acid.[130,131]

34.5 FUTURE DIRECTIONS

Future strategies that combine passive and active targeting with "transcriptional targeting" by using tissue-specific promoters[132,133] should enable a level of target specificity sufficient for the therapeutic use of gene coding for biologically highly active effector molecules, e.g., suicide genes or genes of immunostimulatory cytokines. In addition to target specificity, a sufficiently high transfection efficacy is a further prerequisite for therapeutic efficacy. Future vector development will have to improve further the intracellular steps of the transfection process, including endosomal release and nuclear uptake.[69-71,134] Finally, combination of gene therapy with chemotherapy or radiation can synergize in their therapeautic effects.[135]

34.6 CONCLUSIONS

Therapeutically applicable nonviral gene delivery systems have to perform a number of essential functions, ranging from condensation and protection of DNA to the specific delivery, uptake, and expression of the desired gene. The incorporation of ligands, such as transferrin, endosomal release enhancers, nuclear localization signals, and shielding agents, protecting from nonspecific interactions with blood components and RES and enhancing gene expression, combined with transcriptional and translational regulators, should ensure that the gene of interest is only expressed at the desired target site. An arrangement of these functions in exchangable functional blocks could provide vector systems that are applicable to a broad variety of therapeutic approaches.

REFERENCES

1. Ledley, F.D., Nonviral gene therapy: the promise of genes as pharmaceutical products, *Hum. Gene Ther.* 6, 1129, 1995.
2. Lollo, C.P., Banaszczyk, M.G., and Chiou, H.C., Obstacles and advances in non-viral gene delivery, *Curr. Opin. Mol. Ther.* 2, 136, 2000.
3. Wolff, J.A., Malone, R.W., Williams, P., Chong, P., Acsadi, G., Jani, A., and Felgner, P.L., Direct gene transfer into mouse muscle *in vivo*, *Science* 247, 1465, 1990.
4. Felgner, P.L. et al., Nomenclature for synthetic gene delivery systems, *Hum. Gene Ther.* 8, 511, 1997.
5. Behr, J.P., Gene transfer with synthetic cationic amphiphiles: prospects for gene therapy, *Bioconj. Chem.* 5, 382, 1994.
6. Kircheis, R. and Wagner, E., Polycation–DNA complexes for *in vivo* gene delivery, *Gene Ther. Reg.* 1, 95, 2000.
7. Kircheis, R., Wightman, L., Wagner, E., Design and gene delivery activity of modified polyethylene-imines, *Adv. Drug Del. Rev.* 53, 341, 2001.
8. Huang, C.H., Hung, M.C., and Wagner, E., *Nonviral Vectors for Gene Therapy*, Academic Press, San Diego, 1999.
9. Haensler, J. and Szoka, F.C., Jr., Polyamidoamine cascade polymers mediate efficient transfection of cells in culture, *Bioconj. Chem.* 4, 372, 1993.
10. Boussif, O. et al., A versatile vector for gene and oligonucleotide transfer into cells in culture and *in vivo*: polyethylenimine, *Proc. Nat. Acad. Sci. U.S.A.* 92, 7297, 1995.
11. Zauner, W., Ogris, M., and Wagner, E., Polylysine-based transfection systems utilizing receptor-mediated delivery, *Adv. Drug Del. Rev.* 30, 97, 1998.
12. Wu, G.Y. and Wu, C.H., Receptor-mediated *in vitro* gene transformation by a soluble DNA carrier system, *J. Biol. Chem.* 262, 4429, 1987.
13. Wagner, E. et al., Transferrin-polycation conjugates as carriers for DNA uptake into cells, *Proc. Nat. Acad. Sci. U.S.A.* 87, 3410, 1990.
14. Blessing, T. et al., Different strategies for formation of PEGylated EGF-conjugated PEI/DNA complexes for targeted gene delivery, *Bioconj. Chem.* 12, 529, 2001.
15. Kabanov, A.V. and Kabanov, V.A., DNA complexes with polycations for the delivery of genetic material into cells, *Bioconj. Chem.* 6, 7, 1995.
16. Duncan, R., Pratten, M.K., Lloyd, J.B. Mechanism of polycation stimulation of pinocytosis, *Biochem. Biophys. Acta* 587, 463, 1979.
17. Leonetti, J.P., Degols, G., Lebleu, B., Biological activity of oligonucleotide poly(L-lysine) conjugates: mechanism of cell uptake, *Bioconj. Chem.* 1, 149, 1990.
18. Labat-Moleur, F et al., An electron microscopy study into the mechanism of gene transfer with lipopolyamines, *Gene Ther.* 3, 1010, 1996.
19. Mislick, K.A., and Baldeschwieler, J.D., Evidence for the role of proteoglycans in cation-mediated gene transfer, *Proc. Nat. Acad. Sci. U.S.A.* 93, 12349, 1996.
20. Szoka, F.C., Mechanism of cationic lipoplex gene delivery, 3rd ASGT Meeting, Denver, CO, May 2000; *Mol. Ther.* 1 (5, Part 2), S2, 2000.
21. Ogris, M. et al., The size of DNA/transferrin-PEI complexes is an important factor for gene expression in cultured cells, *Gene Ther.* 5, 1425, 1998.
22. Tang, M.X. and Szoka, F.C., The influence of polymer structure on the interactions of cationic polymers with DNA and morphology of the resulting complex, *Gene Ther.* 4, 823, 1997.
23. McKee, T.D. et al., Preparation of asialoorosomucoid-polylysine conjugates, *Bioconj. Chem.* 5, 306, 1994.
24. Perales, J.C. et al., An evaluation of receptor-mediated gene transfer using synthetic DNA-ligand complexes, *Eur. J. Biochem.* 226, 255, 1994.
25. Wagner, E. et al., Transferrin-polycation-DNA complexes: the effect of polycations on the structure of the complex and DNA delivery to cells, *Proc. Nat. Acad. Sci. U.S.A.* 88, 4255, 1991.
26. Wu, G.Y. and Wu, C.H., Receptor-mediated gene delivery *in vivo*, *J. Biol. Chem.* 263, 14621, 1988.
27. Wagner, E., Curiel, D., and Cotton, M., Delivery of drugs, proteins and genes into cells using transferrin as a ligand for receptor mediated endocytosis, *Adv. Drug Del. Rev.* 14, 113, 1994.
28. Seglen, P.O., Inhibitors of lysosomal function, *Methods Enzymol.* 96, 737, 1993.

29. Erbacher, P. et al., Putative role of chloroquine in gene transfer into a human hepatoma cell line by DNA/lactosylated polylysine complexes, *Exp. Cell. Res.* 225, 186, 1996.

30. Zauner, W. et al., Glycerol and polylysine synergize in their ability to rupture vesicular membranes: a mechanism for increased transferrin-polylysine-mediated gene transfer, *Exp. Cell. Res.* 232, 137, 1997.

31. Curiel, D.T. et al., Adenovirus enhancement of transferrin-polylysine-mediated gene delivery, *Proc. Nat. Acad. Sci. U.S.A.* 88, 8850, 1991.

32. Wagner, E. et al., Coupling of adenovirus to transferrin-polylysine/DNA complexes greatly enhances receptor-mediated gene delivery and expression of transfected genes, *Proc. Nat. Acad. Sci. U.S.A.* 89, 6099, 1992.

33. Cristiano, R.J. et al., Hepatic gene therapy: efficient gene delivery and expression in primary hepatocytes utilizing a conjugated adenovirus-DNA complex, *Proc. Nat. Acad. Sci. U.S.A.* 90, 11548, 1993.

34. Wu, G.Y. et al., Incorporation of adenovirus into a ligand-based DNA carrier system results in retention of original receptor specificity and enhances targeted gene expression, *J. Biol. Chem.* 269, 11542, 1994.

35. Merwin, J.R. et al., CD5-mediated specific delivery of DNA to T lymphocytes: compartmentalization augmented by adenovirus, *J. Immunol. Methods* 186, 257, 1995.

36. Cotten, M. et al., Psoralen treatment of adenovirus particles eliminates virus replication and transcription while maintaining the endosomolytic activity of the virus capsid, *Virology* 205, 254, 1994.

37. Cotten, M. et al., Chicken adenovirus (CELO virus) particles augment receptor-mediated DNA delivery to mammalian cells and yield exceptional levels of stable transformants, *J. Virol.* 67, 3777, 1993.

38. Zauner, W. et al., Rhinovirus-mediated endosomal release of transfection complexes, *J. Virol.* 69, 1085, 1995.

39. Plank, C. et al., The influence of endosome-disruptive peptides on gene transfer using synthetic virus-like gene transfer systems, *J. Biol. Chem.* 269, 12918, 1994.

40. Wagner, E., Application of membrane-active peptides for nonviral gene delivery, *Adv. Drug Del. Rev.* 38, 279, 1999.

41. Parente, R.A., Nir, S., and Szoka, F.C., Jr., Mechanism of leakage of phospholipid vesicle contents induced by the peptide GALA, *Biochemistry* 29, 8720, 1990.

42. Nishikawa, M. et al., Hepatocyte-targeted *in vivo* gene expression by intravenous injection of plasmid DNA complexed with synthetic multi-functional gene delivery system, *Gene Ther.* 7, 548, 2000.

43. Kircheis, R. et al., Cytokine gene-modified tumor cells for prophylactic and therapeutic vaccination: IL-2, IFNgamma, or combination IL-2+IFNgamma, *Cytokines. Cell. Mol. Ther.* 4, 95, 1998.

44. Gao, L et al., Direct *in vivo* gene transfer to airway epithelium employing adenovirus-polylysine-DNA complexes, *Hum. Gene Ther.* 4, 17, 1993.

45. Zatloukal, K. et al., Transferrinfection: a highly efficient way to express gene constructs in eukaryotic cells, *Ann. N.Y. Acad. Sci.* 660, 136, 1992.

46. Kircheis, R. et al., Polycation-based DNA complexes for tumor-targeted gene delivery *in vivo*, *J. Gene Med.* 1, 111, 1999.

47. Wightman, L. et al., Development of transferrin-polycation/DNA based vectors for gene delivery to melanoma cells, *J. Drug Target.* 7, 293, 1999.

48. Schreiber, S. et al., Immunotherapy of metastatic malignant melanoma by a vaccine consisting of autologous interleukin 2-transfected cancer cells: outcome of a phase I study, *Hum. Gene Ther.* 10, 983, 1999.

49. Kichler, A. et al., Polyethylenimine-mediated gene delivery: a mechanistic study, *J. Gene Med.* 3, 135, 2001.

50. Boussif, O., Zanta, M.A., and Behr, J.P., Optimized galenics improve *in vitro* gene transfer with cationic molecules up to a thousand-fold, *Gene Ther.* 3, 1074, 1996.

51. Kircheis, R. et al., Coupling of cell-binding ligands to polyethylenimine for targeted gene delivery, *Gene Ther.* 4, 409, 1997.

52. Wightman, L. et al., Different behavior of branched and linear polyethylenimine for gene delivery *in vitro* and *in vivo*, *J. Gene Med.* 3, 362, 2001.

53. Kichler, A., Behr, J.-P., and Erbacher, P., Polyethylenimines: a family of potent polymers for nucleic acid delivery, in *Nonviral Vectors for Gene Therapy*, Huang, C.H., Hung, M.C., Wagner, E. (Eds.), Academic Press, San Diego, 1999.

54. Baker, A. and Cotten, M., Delivery of bacterial artificial chromosomes into mammalian cells with psoralen-inactivated adenovirus carrier, *Nucleic Acids Res.* 25, 1950, 1997.

55. Fischer, D. et al., A novel non-viral vector for DNA delivery based on low molecular weight, branched polyethylenimine: effect of molecular weight on transfection efficiency and cytotoxicity, *Pharm Res.* 16, 1273, 1999.

56. Bieber, T. and Elsasser H.P., Preparation of a low molecular weight polyethylenimine for efficient cell transfection, *Biotechniques* 30, 74, 2001.

57. Morimoto, K. et al., Molecular weight-dependent gene transfection activity of unmodified and galactosylated polyethyleneimine on hepatoma cells and mouse liver. *Mol Ther.* 7, 254, 2003.

58. Abdallah, B. et al., A powerful nonviral vector for *in vivo* gene transfer into the adult mammalian brain: polyethylenimine, *Hum. Gene Ther.* 7, 1947, 1996.

59. Kircheis, R. et al., Polyethylenimine/DNA complexes shielded by transferrin target gene expression to tumors after systemic application, *Gene Ther.* 8, 28, 2001.

60. Goula, D. et al., Size, diffusibility and transfection performance of linear PEI/DNA complexes in the mouse central nervous system, *Gene Ther.* 5, 712, 1998.

61. Goula, D. et al., Polyethylenimine-based intravenous delivery of transgenes to mouse lung, *Gene Ther.* 5, 1291, 1998.

62. Brunner, S. et al., Overcoming the nuclear barrier: cell cycle independent nonviral gene transfer with linear polyethylenimine or electroporation, *Mol. Ther.* 5, 80, 2002.

63. Zabner, J. et al., Cellular and molecular barriers to gene transfer by a cationic lipid, *J. Biol. Chem.* 8, 839, 1995.

64. Pollard, H.et al., Polyethylenimine but not cationic lipids promote transgene delivery to the nucleus in mammalian cells, *J. Biol. Chem.* 273, 7507, 1998.

65. Fritz, J.D. et al., Gene transfer into mammalian cells using histone-condensed plasmid DNA, *Hum. Gene Ther.* 7, 1395, 1996.

66. Godbey, W.T., Wu, K.K., and Mikos, A.G., Tracking the intracellular path of poly(ethylenimine) DNA complexes for gene delivery, *Proc. Nat. Acad. Sci. U.S.A.* 96, 5177, 1999.

67. Lechardeur, D. et al., Metabolic instability of plasmid DNA in the cytosol: a potential barrier to gene transfer, *Gene Ther.* 6, 482, 1999.

68. Brunner, S. et al., Cell cycle dependence of gene transfer by lipoplex, polyplex and recombinant adenovirus, *Gene Ther.* 7, 401, 2000.

69. Sebestyén, M.G. et al., DNA vector chemistry: the covalent attachment of signal peptides to plasmid DNA, *Nat. Biotechnol.* 16, 80, 1998.

70. Zanta, M.A. et al., Gene delivery: A single nuclear localization signal peptide is sufficient to carry DNA to the cell nucleus, *Proc. Nat. Acad. Sci. U.S.A.* 96, 91, 1999.

71. Brandén, L.J., Mohamed, A.J., and Smith, C.I., A peptide nucleic acid-nuclear localization signal fusion that mediates nuclear transport of DNA, *Nat. Biotechnol.* 17, 784, 1999.

72. Cotten, M., Wagner, E., and Birnstiel, M.L., Receptor-mediated transport of DNA into eukaryotic cells, *Methods Enzymol.* 217, 618, 1993.

73. Plank, C. et al., Gene transfer into hepatocytes using asialoglycoprotein receptor mediated endocytosis of DNA complexed with an artificial tetra-antennary galactose ligand, *Bioconj. Chem.* 3, 533, 1992.

74. Zanta, M.A. et al., *In vitro* gene delivery to hepatocytes with galactosylated polyethylenimine, *Bioconj. Chem.* 8, 841, 1997.

75. Cristiano, R.J. and Roth, J.A., Epidermal growth factor mediated DNA delivery into lung cancer cells via the epidermal growth factor receptor, *Cancer Gene Ther.* 3, 4, 1996.

76. Wolschek, M.F. et al., Specific systemic non-viral gene delivery to human hepatocellular carcinoma xenografts in SCID mice, *Hepatology* 36, 1106, 2002.

77. Sosnowski, B.A. et al., Targeting DNA to cells with basic fibroblast growth factor (FGF2), *J. Biol. Chem.* 271, 33647, 1996.

78. Chen, J. et al., A novel gene delivery system using EGF receptor-mediated endocytosis, *FEBS Lett.* 338, 167, 1994.

79. Buschle, M. et al., Receptor-mediated gene transfer into T-lymphocytes via binding of DNA/CD3 antibody particles to the CD3 T cell receptor complex, *Hum. Gene Ther.* 6, 753, 1995.

80. Erbacher, P. et al., Transfection and physical properties of various saccharide, poly(ethylene glycol), and antibody-derivatized polyethylenimines (PEI), *J. Gene Med.* 1, 210, 1999.

81. Diebold, S.S. et al., Efficient gene delivery into human dendritic cells by adenovirus polyethylenimine and Mannose polyethylenimine transfection, *Hum. Gene Ther.* 10, 775, 1999.

82. Erbacher, P., Remy, J.S. and Behr, J.P., Gene transfer with synthetic virus-like particles via the integrin-mediated endocytosis pathway, *Gene Ther.* 6, 138, 1999.

83. Ogris, M. et al., Tumor-targeted gene therapy: strategies for preparation of ligand-polyethylene glycol-polyethylenimine/DNA complexes, *J. Contr. Release*, 91, 173–181, 2003.

84. Cotten, M. et al., Transferrin-polycation-mediated introduction of DNA into human leukemic cells: stimulation by agents that affect the survival of transfected DNA or modulate transferrin receptor levels, *Proc. Nat. Acad. Sci. U.S.A.* 87, 4033, 1990.

85. Aoki, K. et al., Polyethylenimine-mediated gene transfer into pancreatic tumor dissemination in the murine peritoneal cavity, *Gene Ther.* 8, 508, 2001.

86. Hengge, U.R., Walker, P.S., and Vogel, J.C., Expression of naked DNA in human, pig, and mouse skin, *J. Clin. Invest.* 97, 2911, 1996.

87. Coll, J.L. et al., *In vivo* delivery to tumors of DNA complexed with linear polyethylenimine, *Hum. Gene Ther.* 10, 1659, 1999.

88. Gautam, A. et al., Transgene expression in mouse airway epithelium by aerosol gene therapy with PEI-DNA complexes, *Mol. Ther.* 3, 551, 2001.

89. Rudolph, C., Muller, R.H., and Rosenecker, J., Jet nebulization of PEI/DNA polyplexes: physical stability and *in vitro* gene delivery efficieny, *J. Gene Med.* 4, 66, 2002.

90. Fasbender, A. et al., Complexes of adenovirus with polycationic polymers and cationic lipids increase the efficiency of gene transfer *in vitro* and *in vivo*, *J. Biol. Chem.* 272, 6479, 1997.

91. Boletta, A. et al., Nonviral gene delivery to the rat kidney with polyethylenimine, *Hum. Gene Ther.* 8, 1243, 1997.

92. Ferrari, S. et al., ExGen 500 is an efficient vector for gene delivery to lung epithelial cells *in vitro* and *in vivo*, *Gene Ther.* 4, 1100, 1997.

93. Ferrari, S. et al., Polyethylenimine shows properties of interest for cystic fibrosis gene therapy, *Biochim. Biophys. Acta* 1447, 219, 1999.

94. McKay, T. et al., Selective *in vivo* transfection of murine biliary epithelia using polycation-enhanced adenovirus, *Gene Ther.* 7, 644, 2000.

95. Gharwan, H. et al., Nonviral gene transfer into fetal mouse livers (a comparison between the cationic polymer PEI and naked DNA), *Gene Ther.* 10, 810, 2003.

96. Turunen, M.P. et al., Efficient adventitial gene delivery to rabbit carotid artery with cationic polymer-plasmid complexes, *Gene Ther.* 6, 6, 1999.

97. Liu, F., Song, Y., and Liu, D., Hydrodynamics-based transfection in animals by systemic administration of plasmid DNA, *Gene Ther.* 6, 1258, 1999.

98. Zhang, G., Budker, V., and Wolff, J.A., High levels of foreign gene expression in hepatocytes after tail vein injections of naked plasmid DNA, *Hum. Gene Ther.* 10, 1735, 1999.

99. Smrekar, B. et al., Tissue-dependent factors affect gene delivery to tumors *in vivo*, *Gene Ther.* 10, 1079, 2003.

100. Ruponen, M., Ylä-Herttuala, S., and Urtti, A., Interactions of polymeric and liposomal gene delivery systems with extracellular glycosaminoglycans: physicochemical and transfection studies, *Biochim. Biophys. Acta* 1415, 331, 1999.

101. Ernst, N. et al., Interaction of liposomal and polycationic transfection complexes with pulmonary surfactant, *J. Gene Med.* 1, 331, 1999.

102. Ferkol, T. et al., Immunologic responses to gene transfer into mice via the polymeric immunoglobulin receptor, *Gene Ther.* 3, 669, 1996.

103. Plank, C. et al., Activation of the complement system by synthetic DNA complexes: a potential barrier for intravenous gene delivery, *Hum. Gene Ther.* 7, 1437, 1996.

104. Hartmann, G. and Krieg, A.M., CpG DNA and LPS induce distinct patterns of activation in human monocytes, *Gene Ther.* 6, 893, 1999.

105. Hartmann, G., Weiner, G.J., and Krieg, A.M., CpG DNA: a potent signal for growth, activation, and maturation of human dendritic cells, *Proc. Nat. Acad. Sci. U.S.A.* 96, 9305, 1999.

106. McLachlan, G. et al., Bacterial DNA is implicated in the inflammatory response to delivery of DNA/DOTAP to mouse lungs, *Gene Ther.* 7, 384, 2000.

107. Lemkine, G.F. and Demeneix, B.A., Polyethylenimines for *in vivo* gene delivery, *Curr. Opin. Mol. Ther.* 3, 178, 2001.

108. Bragonzi, A. et al., Comparison between cationic polymers and lipids in mediating systemic gene delivery to the lungs, *Gene Ther.* 6, 1995, 1999.

109. Goula, D. et al., Rapid crossing of the pulmonary endothelial barrier by polyethylenimine/DNA complexes, *Gene Ther.* 7, 499, 2000.

110. Zou, S.M. et al., Systemic linear polyethylenimine (L-PEI)-mediated gene delivery in the mouse, *J. Gene Med.* 2, 128, 2000.

111. Ogris, M. et al., PEGylated DNA/transferrin-PEI complexes: reduced interaction with blood components, extended circulation in blood and potential for systemic gene delivery, *Gene Ther.* 6, 595, 1999.

112. Dash, P.R. et al., Factors affecting blood clearance and *in vivo* distribution of polyelectrolyte complexes for gene delivery, *Gene Ther.* 6, 643, 1999.

113. Erbacher, P. et al., Transfection and physical properties of various saccharide, poly(ethylene glycol), and antibody-derivatized polyethylenimines (PEI), *J. Gene Med.* 1, 210, 1999.

114. Chollet, P. et al., Side-effects of a systemic injection of linear polyethylenimine-DNA complexes, *J. Gene Med.* 4, 84, 2002.

115. Wolfert, M.A. et al., Characterization of vectors for gene therapy formed by self-assembly of DNA with synthetic block co-polymers, *Hum. Gene Ther.* 7, 2123, 1996.

116. Toncheva, V. et al., Novel vectors for gene delivery formed by self-assembly of DNA with poly(L-lysine) grafted with hydrophilic polymers, *Biochim. Biophys. Acta,* 1380, 354, 1998.

117. Finsinger, D. et al., Protective copolymers for nonviral gene vectors: synthesis, vector characterization and application in gene delivery, *Gene Ther.* 7, 1183, 2000.

118. Dash, P.R. et al., Decreased binding to proteins and cells of polymeric gene delivery vectors surface modified with a multivalent hydrophilic polymer and retargeting through attachment of transferring, *J. Biol. Chem.* 275, 3793, 2000.

119. Nguyen, H.K. et al., Evaluation of polyether-polyethyleneimine graft copolymers as gene transfer agents, *Gene Ther.* 7, 126, 2000.

120. Lemieux, P. et al., A combination of poloxamers increases gene expression of plasmid DNA in skeletal muscle, *Gene Ther.* 7, 986, 2000.

121. Hildebrandt, I.J. et al., Optical imaging of transferrin targeted PEI/DNA complexes in living subjects, *Gene Ther.* 10, 758, 2003

122. Gerlowski, L.E. and Jain, R.K., Microvascular permeability of normal and neoplastic tissues, *Microvasc. Res.* 31, 288, 1986.

123. McDonald, D.M. and Choyke, P.L., Imaging of angiogenesis: from microscope to clinic, *Nat. Med.* 9, 713, 2003.

124. Mahato, R.I. et al., Physicochemical and pharmacokinetic characteristics of plasmid DNA/cationic liposome complexes, *J. Pharmacol. Sci.* 84, 1267, 1995.

125. Liu, F., Qi, H., Huang, L., and Liu, D., Factors controlling the efficiency of cationic lipid-mediated transfection *in vivo* via intravenous administration, *Gene Ther.* 4, 517, 1997.

126. Kircheis, R. et al., Tumor-targeted gene delivery: an attractive strategy to use highly active effector molecules in cancer treatment, *Gene Ther.* 9, 731, 2002.

127. Kircheis, R. et al., Tumor-targeted gene delivery of tumor necrosis factor-alpha induces tumor necrosis and tumor regression without systemic toxicity, *Cancer Gene Ther.* 9, 673, 2002.

128. Kursa, M. et al., Novel shielded transferrin-polyethylene glycol-polyethylenimine/DNA complexes for systemic tumor-targeted gene transfer, *Bioconj. Chem.* 14, 222, 2003.

129. Xu, L. et al., Transferrin-liposome-mediated systemic p53 gene therapy in combination with radiation results in regression of human head and neck cancer xenografts, *Hum. Gene Ther.* 10, 2941, 1999.

130. Hofland, H.E. et al., Folate-targeted gene transfer *in vivo*, *Mol. Ther.* 5, 739, 2002.

131. Rait, A.S. et al., Tumor-targeting, systemically delivered antisense HER-2 chemosensitizes human breast cancer xenografts irrespective of HER-2 levels, *Mol. Med.* 8, 475, 2002.

132. Vile, R.G. and Hart, I.R., *In vitro* and *in vivo* targeting of gene expression to melanoma cells, *Cancer Res.* 53, 962, 1993.

133. Dachs, G.U. et al., Targeting gene expression to hypoxic tumor cells, *Nature Med.* 3, 515, 1997.
134. Vacik, J. et al., Cell-specific nuclear import of plasmid DNA, *Gene Ther.* 6, 1006, 1999.
135. Wagner, E., Kircheis, R. and Walker, S.F., Targeted nucleic acid delivery into tumors: new avenues for cancer therapy, *Biomed. Pharmaco. Ther.* 58, 152, 2004.

CHAPTER **35**

Gene Delivery to the Lungs

Berma M. Kinsey, Charles L. Densmore, and Frank M. Orson

CONTENTS

35.1 INTRODUCTION

35.1.1 Gene Delivery to the Lung

A number of pulmonary diseases, including cystic fibrosis, α-1-antitrypsin deficiency, lung cancer, asthma and pulmonary alveolitis, and fibrosis are all candidates for pulmonary gene delivery.

0-8493-1934-X/05/$0.00+$1.50
© 2005 by CRC Press LLC

Genes, in the form of DNA that expresses the desired product, can be delivered to the lung by routes such as intravenous or airway administration, and various agents have been used to deliver genes. Viruses, especially adenovirus, have been much used as gene delivery agents, but they suffer from the drawback that they elicit undesirable immune responses precluding repeated dosing, which will undoubtedly be necessary for many applications. In view of this, the most attractive vehicle for gene delivery that emerges is plasmid DNA, since it is easy to engineer and produce, stable, and nontoxic. Simple delivery of so-called "naked" plasmid DNA has been tried, but it has generally not proved satisfactory due to insufficient gene expression. Cationic lipid–DNA complexes were used very early on as agents for lung gene delivery by all routes, as they protect plasmid DNA from degradation and give better gene expression than DNA alone. Unfortunately, these agents suffer from toxicity problems, and enthusiasm for their use has waned.[1,2] In this review, we will focus on cationic polymer–DNA complexes as agents for gene delivery to the lung. The use of these versatile agents, particularly in the area of gene delivery for cancer therapy,[3] has increased significantly in the last several years.

35.1.2 Polycations

Polycations are synthetic or semisynthetic molecules containing multiple amino groups. They include polyethylenimine (PEI), poly(L-lysine) (PLL), polyarginine, dendrimers, peptides, and chitosan. They all bind DNA and protect it from degradation, and they have all been used for gene delivery to the lung with varying degrees of success. Polycation–DNA complexes have been dubbed "polyplexes" in distinction from cationic lipid–DNA complexes or "lipoplexes" (liposomes). For different applications, the ratio of the nitrogen atoms in the polycation to the phosphate groups in the DNA (N/P ratio) can range from 4 to 20. Polycations can be derivatized through their amino groups with targeting moieties such as peptides and saccharides, with protective groups such as polyethylene glycol or hydrophobic molecules, and fluorescent or radioactive molecules for detection purposes. Gene delivery to specific components of pulmonary tissue with polyplexes may also be affected in part by the delivery method: intravenous infusion, administration as a liquid suspension or spray directly into the airways, or aerosol deposition. These methods will be discussed in turn.

35.2 INTRAVENOUS GENE DELIVERY

35.2.1 Introduction

The earliest polycations used for gene delivery were polypeptides of lysine and argenine. However, intravenous delivery of uncomplexed PLL or polyarginine was reported to cause bronchial hyper-responsiveness,[4] and intravenously delivered PLL–DNA is rapidly cleared from the circulation by the liver,[5] most likely through clearance after binding to serum albumin.[6] Recently, polycationic membrane-destabilizing peptides complexed to DNA were reported to give significant gene expression in the mouse lung. However, they were less efficient than other synthetic vectors used for intravenous (IV) administration, and there was some mortality associated with their use.[7] The most widely used polycation for IV delivery to the lung is PEI in its various forms. In 1995, PEI was reported to be a useful vector for gene delivery *in vitro* and *in vivo*.[8] PEI has a large buffering capacity attributable to the fact that every third atom is a protonatable nitrogen. It is believed that DNA bound to this highly charged cationic polymer is significantly protected from the low pH environment of the lysosomes, and that the "proton sponge" property of PEI swells the endosomes and allows efficient escape of the DNA. Although the mechanisms for PEI-mediated transfection are not fully understood, there is evidence that PEI and PEI–DNA complexes undergo nuclear localization more effectively than lipoplexes.[9] Recent work has shown that PEI–DNA nanoparticles arc carried actively on microtubles to the perinuclear region within minutes of cellular entry, in a manner similar to the trafficking of the most efficient viruses.[10] PEI is available in branched and linear forms, with molecular weights ranging from 200 D to 1 MD. PEI–DNA complexes can be prepared for *in vivo* use in water or in buffers such as PBS or HEPES or in 5% glucose. PEI is easy to derivatize using a variety of methods, making it feasible to alter various properties of the PEI polyplex, particularly the ability to bind to specific cell surface receptors.[11,12]

35.2.2 Intravenous Gene Delivery by Polyethylenimine

PEI–DNA complexes cross the endothelial barrier rapidly, and gene expression is seen in the respiratory epithelial cells, as well as endothelial and interstitial cells.[13,14] Early experiments indicated that 22 kD linear PEI–DNA complexes formed in 5% glucose gave the highest expression when compared to 25 kD branched PEI, various liposomes, and salt-containing formulations (Table 35.1).[15,16] Additionally, these studies showed that PEI–DNA complexes formed with 125 µg DNA at an N/P ratio of 6:1 caused 50% mortality in the mice, whereas ratios up to 15:1 using only 50 µg DNA gave increased gene expression with no mortality. Later studies confirmed these results, but found further that less than 50 µg DNA gave essentially no gene expression, while older animals and certain mouse strains were more sensitive to the toxic effects of the linear PEI.[17]

Several other parameters have been considered in order to optimize the conditions for PEI–DNA gene delivery to the lung. It was found that injection of 50 µg DNA complexed with 22 kD linear PEI at an N/P ratio of 10:1 in a large volume (400 µl) of 5% glucose gave the highest gene expression in the lung on a per mg of protein basis as compared to other organs.[18] Another study compared gene expression after delivery of PEI–DNA complexes consisting of 800 kD or 25 kD branched PEI or 22 kD linear PEI under salt and salt-free conditions.[19] All of the conditions tested revealed that the lung had the highest expression on a per organ basis. PEI800 was the highest of the three when delivered in salt (0.5X HBS), while linear PEI22 in 5% glucose was the highest overall (Table 35.2).

Table 35.1 Gene Delivery to the Lung by Various Agents[a]

	N/P Ratio	RLU/mg Protein
DNA alone	—	20
Linear 22 kD PEI	15	30,000,000
Branched 25 kD PEI	15	1,000
DOTMA–DOPE	5	100
DOTAP	4.8	1,000,000
GL67–DOPE	4	8,000

[a] Adapted from Bragonzi et al.,[15] 50 µg DNA delivered in 600-700 µL 5% glucose.

Table 35.2 Gene Delivery to the Lung in Salt or Salt-free Solutions[a]

	Solution	RLU/Organ
Branched 800 kD PEI	0.5X HEPES	700,000
Branched 25 kD PEI	0.5X HEPES	70,000
Linear 22 kD PEI	0.5X HEPES	30,000
Branched 800 kD PEI	5% glucose	70,000
Branched 25 kD PEI	5% glucose	8,000
Linear 22 kD PEI	5% glucose	3,000,000

[a] Adapted from Wightman et al.[19] 50 µg DNA in 250 µl of solution.

Other work looking at the amount of DNA delivered to organs showed that delivery of DNA by 25 kD branched PEI–DNA in 5% glucose was at least an order of magnitude more efficient than delivery of naked DNA, and that the DNA was quite persistent.[20] At 24 h, the liver had the highest amount of DNA per organ, followed by the lung and kidney. No inflammation was seen in these experiments after 3 weekly injections of 50 μg DNA complexed with PEI, but twice a week doses for 3 weeks induced signs of inflammation in the liver.

35.2.3 Ternary Complexes with Polyethylenimine

A recent study again compared the efficacy of 25 kD branched PEI–DNA and 22 kD linear PEI–DNA complexes as delivery agents, but this time with the addition of a polyanion, polyacrylic acid, in order to reduce the negative charge of the complexes along with the toxicity,[21] reminiscent of what was done in the case of cationic lipids.[22] The ternary complexes were made using 50 μg DNA, high N/P ratios (30–60:1), and various amounts of polyacrylic acid. The complexes incorporating branched PEI were much more toxic and less efficient than those with linear PEI, while cationic lipid ternary complexes were nontoxic but quite inefficient. "Chasing" the administration of optimized linear PEI ternary complexes with 1.5 g of polyacrylic acid maintained gene expression while cutting down on the toxicity as seen by histological screening.

We have developed a formulation consisting of 750 kD PEI, DNA, and protein (albumin or other proteins) in PBS, which gives efficient gene expression using only 1 μg of DNA (Figure 35.1).[23] Gene expression dropped by a factor of 100 4 days after administration, but could still be detected for 7 days. When injections were given every other day for 20 days, the level of expression achieved 2 days after one injection was maintained. There was no evidence of toxicity following the repeated injections, and toxicity studies involving more prolonged gene administration are ongoing. Branched PEIs with nominal molecular weights of 750 kD and 70 kD were found to be equivalent in transfection efficiency, while PEIs with molecular weights of less than 70 kD, including linear forms, were not functional in this system using low plasmid doses (Figure 35.2). There is some uncertainty about the molecular weight of PEI polymers, depending on what method is used for determination. For example, nominally 800 kD and 70 kD samples actually appeared to have weight average molecular weights of 155 kD and 133 kD, respectively, when measured by gel filtration.[24]

There has been a widespread perception that PEI is too toxic for clinical applications, especially as most of the applications that have been envisioned, whether for gene therapy or immunization, would require repeated dosing because the gene being delivered is very unlikely to be permanently incorporated into cells. We feel that toxicity concerns will be obviated by demonstrating that satisfactory and wide-spread gene expression in the lungs can be safely maintained over repeated doses by using small amounts of PEI and DNA in ternary complexes with proteins and that PEI will thus prove to be a very useful agent for IV gene delivery to this important organ.

35.2.4 Polyethylenimine Conjugated to Other Molecules

Conjugation of targeting moieties to PEI has led to some success, but the targeting usually involved a malignant tumor rather than normal lung tissue. In one case, however, PEI was conjugated to an anti-platelet endothelial cell adhesion molecule-1 (anti-PECAM, or CD321) antibody in order to specifically target lung vasculature.[25] It was found that the 22 kD linear PEI conjugate was more effective and less toxic than the 25 kD branched polymer conjugate, and that antibody conjugation increased lung gene expression about 20-fold. Pretreatment with the anti-inflammatory agent dexamethasone further increased gene expression fourfold. Biodistribution studies done at 30 min postinjection showed that lung uptake of the targeted complexes was increased while at the same time liver uptake was decreased. Furthermore, production of the inflammatory cytokine tumor

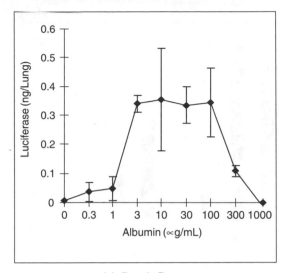

(a). Protein Dose.

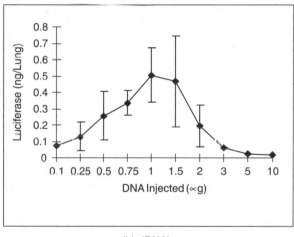

(b). (DNA).

Figure 35.1 Influence of albumin concentration and plasmid dose on luciferase expression. (A) Mice were injected with 1 µg of pCMV-Luc complexed with 45 nM of PEI N with albumin (N/P ratio of 15:1) over a range of albumin concentrations in a total volume of 200 µl of PBS. Two days after injection, the mice were sacrificed and lung extracts were tested for luciferase activity, expressed as ng of luciferase in the total lung extract. (B) Mice were injected with the indicated dose of pCMV-Luc in 200 µl of PBS with quantities of PEI–HSA appropriate for a 15:1 N/P ratio. Two days after injection, the mice were sacrificed and lung extracts were tested for luciferase activity, expressed as ng of luciferase in the total lung extract. (From Orson, F.M., et al., Gene delivery to the lung using protein/polyethylenimine/plasmid complexes, *Gene Ther.* 9, 463, 2002. With permission.)

necrosis factor α(TNFα) was substantially reduced by use of the Ab-conjugated PEI as compared to the unconjugated PEI.

Much work has been done on gene delivery using polycations conjugated to transferrin in order to target tumors that overexpress the transferrin receptor.[26] In tumor-free mice, the data showed that expression in normal lung using these agents was somewhat lower than that of the unconjugated PEI. Part of the rationale for the conjugation was that shielding the positively charged PEI–DNA complexes with transferrin molecules would increase the circulation time of the complexes and thereby increase gene expression in the targeted tissue. The same shielding rationale was applied

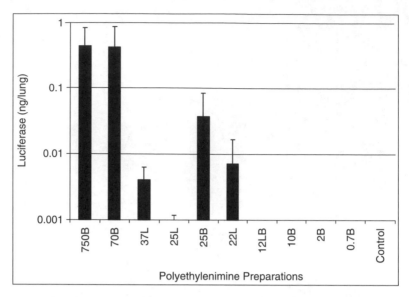

Figure 35.2 Gene expression from plasmid bound in PEI–protein complexes using different sizes of branched and linear PEI forms. Groups of mice (BALB/c) were administered 1 μg pCMV-Luc bound in PEI–protein complexes containing 45 nM of PEI N (N/P ratio of 15:1) in a total volume of 200 μl of PBS. One day after injection, the mice were sacrificed and lung extracts were tested for luciferase activity, expressed in the figure as ng of luciferase in the total lung extract.

to polyethylene glycol molecules. Adding anionic peptides conjugated to polyethylene glycol (PEG) to positively charged PEI–DNA complexes did not seem to change gene expression in the lung compared to PEI–DNA alone,[27] but again, direct conjugation of shielding PEG molecules to PEI resulted in substantially lower transfection in the lungs and concomitantly higher transfection in the kidneys or liver.[28]

Somewhat along the same lines, glycosylation of PEI has been accomplished in order to target various saccharide receptors of interest, but *in vitro* work suggested that glycosylated PEI–DNA complexes might have lessened utility for lung gene transfer.[29] In confirmation, we found in our laboratories that while there was increased transfection using glycosylated PEIs *in vitro*, there was essentially no gene expression following intravenous or aerosol delivery using these PEI derivatives (unpublished results).

35.2.5 Gene Delivery by Macroaggregated Albumin–PEI Particles

In our laboratory, we have developed a system using macroaggregated albumin–PEI conjugates (MAA–PEI) for gene delivery to the lung.[30] Radiolabeled MAA particles have been used for many years as a safe lung imaging agent. We found that gene expression by our MAA–PEI particles loaded with up to 5 μg DNA was limited to the lung, and no toxicity was observed. Similar responses in several different strains of mice were observed, although C57BL/6 mice had roughly 50% of the response of others, an example of which is shown in Figure 35.3. Gene expression from MAA–PEI-bound plasmid encoding an antigenic protein elicited both pulmonary mucosal and systemic immune responses following IV administration. The responses compared favorably with those obtained after other gene delivery techniques (Figure 35.4).

35.2.6 Cancer Gene Therapy by Intravenous Gene Delivery

A plasmid containing the gene coding for the antiangiogenesis factor endostatin was delivered IV to mice who carried lung fibrosarcoma tumors.[31] The plasmid was complexed either to linear

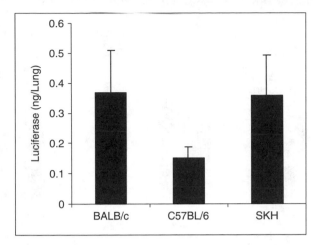

Figure 35.3 Gene expression from MAA–PEI-bound plasmid administered IV to different strains of mice. Groups of mice from the designated strains were injected with 1 μg of pCMV-Luc complexed with MAA–PEI containing 45 nM of PEI N (N/P ratio of 15:1) in a total volume of 200 μl of PBS. Two days after injection, the mice were sacrificed and lung extracts were tested for luciferase activity, expressed in the figure as ng of luciferase in the total lung extract. Similar responses to those of BALB/c and SKH-HR1 were observed in separate experiments with the CD-1 outbred strain.

22 kD PEI or to a cationic lipid vector, GL67–dioleoyl phosphatidylethanolamine (GL67–DOPE). There was essentially no difference in the level of expression of the endostatin gene between the two vectors. The delivery of the gene resulted in a reduced number and weight of the tumors, and prolonged survival of the mice.

35.3 DELIVERY OF GENES TO THE AIRWAYS

35.3.1 Introduction

Two delivery techniques have been used to foster gene expression in the airways of the lung: fluid instillation into the airway by direct intratracheal (IT) injection or intranasal (IN) instillation with aspiration of the fluid and aerosol inhalation. The simplest and most practical experimental method would be delivery of naked DNA by the intranasal route. However, gene expression is low and much of the DNA may not reach the lungs. Mice represent by far the largest majority of the animals used for gene expression studies, and although mice are obligate nose breathers, some effort must be made to ensure that the dose is inhaled and not swallowed or sneezed and coughed out. Direct delivery to the airway has also been done by IT injection or cannulation, usually involving surgical exposure of the trachea. A microspray device has been developed for IT delivery in mice and monkeys,[32,33] but it has not yet been used with polycation–DNA complexes. Aerosol gene delivery has been successfully used on groups of mice with a variety of lipoplexes and polyplexes. This method has the advantage of being well tolerated and noninvasive, and may be more readily adapted for clinical applications.

35.3.2 Fluid Instillation of Polyplexes into the Airway

IT and IN delivery of naked DNA to the upper airways has achieved some success,[34] but lung surfactant inhibits gene expression.[35] When dendrimers (globular branched polyamidoamines) were complexed with DNA and administered IT or IN, there was a lower gene expression as compared with uncomplexed DNA (the opposite of what was found after IV delivery).[36] In these studies, no

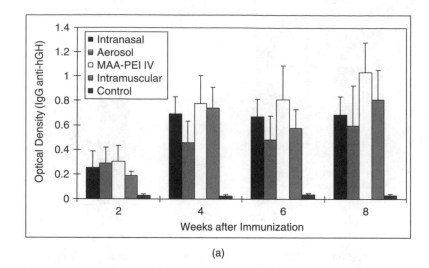

(a)

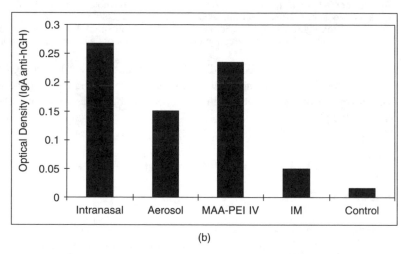

(b)

Figure 35.4   Humoral immune responses to genetic immunization via different routes. Groups of five mice were exposed to pCMV-hGH using alternate methods of immunization, including intranasal, aerosol, MAA–PEI particles, and intramuscular. Immune responses in serum were monitored at biweekly intervals, as detailed by Orson et al.[30] and Densmore et al.[49] After 8 weeks, the mice were sacrificed, and the lungs were lavaged with 1 ml of phosphate-buffered saline. Enzyme-linked immunosorbent assays (ELISAs) for hGH specific IgG antibodies (serum, panel a) and mucosal IgA antibodies (bronchoalveolar lavage fluid, panel b) were performed with a 1:500 dilution of serum for each mouse and the neat lavage fluid for pools of mice in each group, respectively. The data represent the net optical density for each group with the background in wells without serum subtracted (0.1 OD units).

reporter gene expression was found in any other organ. In other studies, fractured dendrimers were found to give very low gene expression in the lung.[37] Chitosan, a linear biodegradable polysaccha-ride containing amino groups, has been considered promising for lung gene delivery in several models. Reasonable gene expression was demonstrated following IT administration of chito-san–DNA[38] and delivery of chitosan–DNA in the form of a dry powder was found to give higher gene expression than delivery of naked DNA or chitosan–DNA in water solution.[39] IN delivery of chitosan–DNA complexes expressing epitopes of respiratory syncytial virus caused a significant reduction of viral load in the lungs of immunized mice.[40] Nevertheless, by far the majority of work done with airway delivery has been with polyethylenimine (PEI).

Soon after the discovery of PEI as an efficient gene delivery agent, it was used to transfect lung tissue by IT injection in rabbits, using 22 kD linear PEI.[41] No lung toxicity was observed at doses up to 700 μg DNA (N/P of 5). Subsequent experiments by the same group showed that rabbits treated daily for 7 days showed areas of inflammation in the lungs, but no increased inflammation after an additional 14 and 21 consecutive days.[42] When mice were injected IT with PEI–DNA, there was little difference between the linear 22 kD PEI and the 25 kD branched PEI, but they were statistically better than the two cationic lipids tested (DOTAP and GL-67).[14] In these experiments, the level of gene expression for the linear 22 kD PEI following IV injection was two orders of magnitude greater than IT, and 25 kD branched PEI performed very poorly. Another group found that 25 kD PEI gave good gene expression via IT cannula intubation, especially when the PEI–DNA complexes were prepared in water rather than 25 mM HEPES buffer (pH 7.4).[37] Several studies have been done in rats, administering the gene complex by IT intubation to one lung and using the other lung as a control.[43,44] Several gene transfer agents were compared, and linear 22 kD PEI–DNA prepared in 5% glucose was favored, although all the tested agents impaired lung function somewhat as compared to naked DNA.

35.3.3 Fluid Phase Airway Gene Delivery with Poly(ethylenimine)–Poly(ethylene Glycol)

It was reported that conjugation of linear PEI and PEG allowed higher concentrations of PEI–DNA complexes to be prepared, and IN instillation of 100 μg DNA did give significant levels of transgene expression.[45] Another group found that IT administration of PEI–DNA complexes to which were added PEG–anionic peptide conjugates that bound electrostatically to the complexes resulted in substantially *less* gene expression as compared to the use of unprotected PEI–DNA.[27]

35.3.4 Fluid Phase Airway Gene Delivery by Ternary PEI–TAT Peptide–DNA Complexes

Recently, gene delivery by the arginine-rich motif of the HIV-1 TAT protein (TAT peptide) was demonstrated in cell culture. The dimer repeat of the peptide proved to be the most efficient. Ternary complexes of PEI, TAT peptide-2, and DNA were prepared by mixing the peptide and DNA at a 1:1 N/P ratio followed by 25 kD branched PEI at a 10:1 ratio. This formulation mediated a fourfold higher gene expression in the lung over intratracheal instillation of PEI–DNA alone.[46]

35.3.5 Cancer Gene Therapy by Fluid Phase Airway Gene Delivery

The biological effects of genes delivered by airway-administered polyplexes in fluid phase have also been demonstrated in a cancer gene therapy model. When branched 25 kD PEI–DNA complexes expressing murine IL-12 were administered IN twice weekly for 4 weeks to nude mice carrying human osteosarcoma lung metastases, both the number and size of the tumors were greatly diminished.[47] In this case, IL-12 expression was detected in the lung but not in the liver.

35.3.6 Aerosol Gene Delivery to the Lungs by Poly(ethylenimine)

Aerosol delivery has been viewed as an ideal noninvasive mode of delivery to the airway bronchial and alveolar epithelial surfaces. This mode of delivery could directly target the airway surface, thus avoiding many of the problems associated with intravenous delivery, such as interactions with serum proteins, difficulties penetrating endothelial barriers, and accumulation in other tissues after passing through the lungs. A major problem in the use of aerosols for delivery of DNA was previously seen as the rapid loss of plasmid integrity due, in part, to the shear force of jet nebulization. Work in this laboratory and others has shown that complexing plasmid DNA with

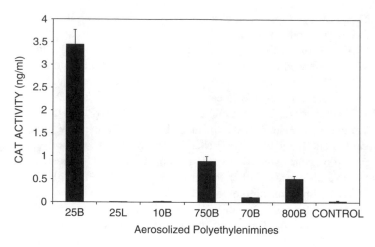

Figure 35.5 Comparison of gene expression following aerosol delivery using different sizes of branched and linear PEI. Two mg of CAT plasmid was complexed with PEI at an N/P ratio of 10:1, and the complex was aerosolized to groups of 5 mice using 5% CO2-in-air. The mice were sacrificed at 24 h, and CAT was quantitated by ELISA. The values show the CAT mean ± SD for each group in ng/ml of extract (1 ml was used for each mouse).

cationic lipids resulted in increased structural stability,[48] but the level of gene expression using these agents was still not satisfactory. Additionally, the inflammatory response resulting from inhalation of lipid-based formulations was found to be unacceptable in patients.[2]

Our laboratory has developed methods for aerosol gene delivery using PEI to stabilize the plasmid DNA to nebulization.[49] Branched 25 kD PEI gave the highest levels of gene expression among the other molecular weights and types of PEI that were tested (Figure 35.5). The PEI–DNA complexes are formed at pH 7.4 in water. This method gives significantly higher levels of gene expression in a mouse model than several cationic lipid formulations which had been optimized for aerosol application and used in human gene therapy clinical trials. Biodistribution studies involving aerosol delivery of PEI-based formulations showed about 20-fold greater transfection in the lungs than in the nose, while expression in nonpulmonary tissues was undetectable.[49] More recently, levels of reporter gene expression 20- to 50-fold higher were achieved by utilizing 5% CO_2 in air for nebulization, by optimizing the N/P ratio and by improving the formulation procedure.[50]

PEI-based formulations also appear to exhibit a high degree of efficiency for gene delivery to the lungs. There was nearly 100% transfection of airway epithelium and lung parenchyma according to immunohistochemistry after delivery of either a reporter gene or the variant tumor suppressor gene p53CD (1-366) (Figure 35.6).[51] Repeated aerosol exposures to PEI–DNA were associated with very low acute toxicity, and the levels of TNFα and interleukin-1 β (IL-1β) induced in bronchoalveolar lavage fluid (BALF) and serum after PEI–DNA aerosol delivery were significantly lower than following IV administration of either PEI–DNA or cationic lipid–DNA complexes or aerosol delivery of liposomes.[52] This implies that there is a lack of the inflammatory response that frequently results from exposure to the unmethylated CpG motifs found in bacterially produced plasmid DNA. The induction of proinflammatory cytokines by plasmid DNA is a serious problem that is associated with lung toxicity and a markedly decreased efficiency of therapies that require repeated gene dosing.[53] The apparent "masking" of this CpG effect by PEI may be responsible for the sustained expression of genes that are delivered via PEI–DNA aerosol.[54] In addition, surfactant, which has been shown to inhibit the transfection efficiency of cationic lipids,[55] does not appear to diminish PEI-mediated transfection.

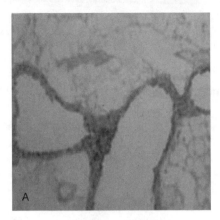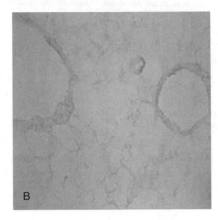

Figure 35.6 Immunohistochemical detection of CAT expression in airway epithelium from aerosol delivery. Mice were treated with 2 mg of CAT plasmid complexed with PEI at an N/P ratio of 10:1 and aerosolized using 5% CO2-in-air for 30 min, and then this same dose was repeated for a total exposure of 4 mg over 60 min. The lungs were harvested 24 h after the aerosol exposure, and inflated with 50% tissue-freezing medium in PBS. Cryosections were cut at 5 μm and then stained and developed using standard methods. (Modified from Reference 51. With permission.)

35.3.7 Cancer Gene Therapy by Aerosol Gene Delivery

We have shown that aerosolized PEI–DNA complexes expressing the human wild-type p53 tumor suppressor factor significantly slowed the growth of aggressive murine melanoma (B16-F10) tumors in the lungs of normal mice and increased their survival.[56] The same effect using either the wild-type p53 gene or a p53 variant was seen in a pulmonary metastasis cancer model using a very slow growing human osteosarcoma derivative in the nude mouse.[57] The different tissues examined in these studies did not reveal any signs of toxicity or inflammation, despite twice weekly aerosol PEI–plasmid exposures for periods of 3 to 8 weeks.

A hypothetical stumbling block for cancer treatment by gene therapy has been the perceived lack of deep penetration of tumor tissue by the delivery agents. Further studies in our melanoma model have suggested that while p53 expression does lead to some (expected) visible apoptosis of tumor cells, suppression of angiogenesis may be the major factor in inhibiting tumor growth.[58] In the osteosarcoma model, aerosol treatment of the mice once a week with a plasmid expressing interleukin-12 (IL-12) also substantially inhibited the growth of the tumors.[61] Both of the foregoing findings indicate that tumor inhibition may be possible even in the absence of gene delivery to every single tumor cell, and the implications for the treatment of lung metastases are promising. Furthermore, our novel approach involving combined aerosol delivery of PEI–p53 and a liposomal chemotherapy agent has proven very effective in the melanoma model.[56]

35.3.8 Genetic Immunization by Aerosol Gene Delivery

We have shown that persistent, high-level antibody responses can be achieved with a single administration of an antigen-expressing plasmid using aerosol or other techniques to deliver the DNA.[49] In this study, we compared the immune response after aerosol, intranasal instillation, and intramuscular modes of delivery of the human growth hormone (hGH) gene to mice. The levels of anti-hGH antibodies measured in serum from these mice by tail bleed at 2-week intervals was indistinguishable between the three groups, and mucosal antibody responses were also induced by the aerosol immunization. Parallel studies at the same time demonstrated that delivery of hGH plasmid bound to IV MAA–PEI–DNA particles elicited similar high level responses for both humoral and cellular immunity.[30] Figure 35.4 shows the humoral responses of these groups of mice systemically (panel A) and mucosally (panel B). Such approaches may have value for respiratory pathogens, and may also be useful for the immunotherapy of cancer and hypersensitivity diseases.

35.4 CONCLUDING REMARKS

The enormous surface area of the lung, with its rich bed of capillaries accessible on a first pass by intravenous injection, and its multitude of alveoli available to aerosols, represents a very tempting target for DNA delivery and gene expression. The IV route of delivery, although more invasive in nature, may be especially valuable in selected clinical applications. Studies need to be done to look at the effectiveness of this method in comparison to delivery via the airways, particularly in terms of the magnitude and persistence of gene expression. For example, it might be that IV gene delivery in combination with delivery via the airways would be particularly efficacious for cancer therapy, while airway delivery alone might be better for gene replacement purposes. For all applications, and especially genetic immunization, the great advantage to gene delivery by polycations is that there is no immune response raised against the gene delivery agent itself, in contrast to viral vectors.[59]

Fluid phase airway administration of polycation–DNA complexes may be useful for genetic immunization protocols, especially when mucosal immunity is desired. It may also be useful in endobronchial cancer therapy when administered locally via bronchoscopy, but it would not seem as useful for applications where robust gene expression in a large area of the lung is needed, such as for gene replacement therapy. IN delivery under anesthesia is a useful experimental technique for rodents, and IN administration may have considerable promise for mucosal immunization, since some of the DNA is taken up in the nasal epithelium, where it can be presented and processed in the nasal associated lymphoid tissue.

Aerosol gene delivery to the lung would seem to be the ideal method for general applications, as aerosol technology for medication delivery is very well developed and accepted for human use. Cancer therapy studies in mouse models have been very promising, and genetic immunization is an application which is gaining acceptance. Further studies on the possible toxic or inflammatory effects of repeated dosing are ongoing. The finding that the level of gene expression can be maintained by repeated dosing has implications not only for cancer therapy, but also for gene replacement therapy in diseases such as cystic fibrosis. Since aerosol technology is clinically well established and accepted, aerosol delivery of polycation–gene formulations could have many applications for infectious diseases, genetic diseases, and asthma, as well as for cancer.

REFERENCES

1. Driskell, R. and Engelhardt, J.F., Current status of gene therapy for inherited lung diseases, *Ann. Rev. Physiol.* 65, 25.1, 2003.
2. Ruiz, F.E. et al., A clinical inflammatory syndrome attributable to aerosolized lipid-DNA administration in cystic fibrosis, *Hum. Gene Ther.* 12, 751, 2001.
3. Merdan, T., Kopecek, J. and Kissel, T., Prospects for cationic polymers in gene and oligonucleotide therapy against cancer, *Adv. Drug Del. Rev.* 54, 715, 2002.
4. Gu, Q. and Lee, L.Y., Hypersensitivity of pulmonary c fibre afferents induced by cationic proteins in the rat, *J. Physiol.* 537 (Pt. 3), 887, 2001.
5. Ward, C.M., Read, M.L. and Seymour, L.W., Systemic circulation of poly(l-lysine)/DNA vectors is influenced by polycation molecular weight and type of DNA: differential circulation in mice and rats and the implications for human gene therapy, *Blood* 97, 2221, 2001.

6. Dash, P.R. et al., Factors affecting blood clearance and *in vivo* distribution of polyelectrolyte complexes for gene delivery, *Gene Ther.* 6, 643, 1999.

7. Rittner, K. et al., New basic membrane-destabilizing peptides for plasmid-based gene delivery *in vitro* and *in vivo*, *Mol. Ther.* 5, 104, 2002.

8. Boussif, O. et al., A versatile vector for gene and oligonucleotide transfer into cells in culture and *in vivo*: polyethylenimine, *PNAS* 92, 7297, 1995.

9. Pollard, H. et al., Polyethylenimine but not cationic lipids promotes transgene delivery to the nucleus in mammalian cells, *J. Biol. Chem.* 273, 7507, 1998.

10. Suh, J., Wirtz, D. and Hanes, J., Efficient active transport of gene nanocarriers to the cell nucleus, *PNAS* 100, 3878, 2003.

11. Kircheis, R., Wightman, L. and Wagner, E., Design and gene delivery activity of modified polyethylenimines, *Adv. Drug Del. Rev.* 53, 341, 2001.

12. Lemkine, G.F. and Demeneix, B.A., Polyethylenimines for *in vivo* gene delivery, *Curr. Opin. Gene Ther.* 3, 178, 2001.

13. Goula, D. et al., Rapid crossing of the pulmonary endothelial barrier by polyethylenimine/DNA complexes, *Gene Ther.* 7, 499, 2000.

14. Bragonzi, A. et al., Biodistribution and transgene expression with nonviral cationic vector/DNA complexes in the lungs, *Gene Ther.* 7, 1753, 2000.

15. Bragonzi, A. et al., Comparison between cationic polymers and lipids in mediating systemic gene delivery to the lungs, *Gene Ther.* 6, 1995, 1999.

16. Goula, D. et al., Polyethylenimine-based intravenous delivery of transgenes to mouse lung, *Gene Ther.* 5, 1291, 1998.

17. Chollet, P. et al., Side-effects of a systemic injection of linear polyethylenimine-DNA complexes, *J. Gene Med.* 4, 84, 2002.

18. Zou, S.M. et al., Systemic linear polyethylenimine (l-pei)-mediated gene delivery in the mouse, *J. Gene Med.* 2, 128, 2000.

19. Wightman, L. et al., Different behavior of branched and linear polyethylenimine for gene delivery *in vitro* and *in vivo*, *J. Gene Med.* 3, 362, 2001.

20. Oh, Y. K. et al., Prolonged organ retention and safety of plasmid DNA administered in polyethylenimine complexes, *Gene Ther.* 8, 1587, 2001.

21. Trubetskoy, V. et al., Recharging cationic DNA complexes with highly charged polyanions for *in vitro* and *in vivo* gene delivery, *Gene Ther.* 10, 261, 2003.

22. Li, S. et al., Characterization of cationic lipip-protamin-DNA (lpd) complexes for intravenous gene delivery, *Gene Ther.* 5, 930, 1998.

23. Orson, F.M. et al., Gene delivery to the lung using protein/polyethylenimine/plasmid complexes, *Gene Ther.* 9, 463, 2002.

24. Godbey, W.T., Wu, K.K., and Mikos, A.G., Poly(ethylenimine) and its role in gene delivery, *J. Control. Release* 60, 149, 1999.

25. Li, S. et al., Targeted gene delivery to pulmonary endothelium by anti-pecam antibody, *Am. J. Physiol. Lung Cell Mol. Physiol.* 278, L504, 2000.

26. Kircheis, R. et al., Tumor targeting with surface-shielded ligand–polycation DNA complexes, *J. Control. Release* 72, 165, 2001.

27. Rudolph, C. et al., Nonviral gene delivery to the lung with copolymer-protected and transferrin-modified polyethylenimine, *BBA* 1573, 75, 2002.

28. Ochietti, B. et al., Altered organ accumulation of oligonucleotides using polyethyleneimine grafted with poly(ethylene oxide) or pluronic as carriers, *J. Drug Targ.* 10, 113, 2002.

29. Fajac, I. et al., Uptake of plasmid/glycosylated polymer complexes and gene transfer efficiency in differentiated airway epithelial cells, *J. Gene Med.* 5, 38, 2003.

30. Orson, F.M. et al., Genetic immunization with lung-targeting macroaggregated polyethyleneimine-albumin conjugates elicits combined systemic and mucosal immune responses, *J. Immunol.* 164, 6313, 2000.

31. Nakashima, Y. et al., Endostatin gene therapy on murine lung metastases model utilizing cationic vector-mediated intravenous gene delivery, *Gene Ther.* 10, 123, 2003.

32. Beck, S.E. et al., Deposition and expression of aerosolized raav vectors in the lungs of rhesus macaques, *Mol. Ther.* 6, 546, 2002.

33. Schughart, K. et al., Solvoplex: a new type of synthetic vector for intrapulmonary gene delivery, *Hum. Gene Ther.* 10, 2891, 1999.

34. Gill, D.R. et al., Increased persistence of lung gene expression using plasmids containing the ubiquitin c or elongation factor 1alpha promoter, *Gene Ther.* 8, 1539, 2001.

35. Raczka, E. et al., The effect of synthetic surfactant exosurf on gene transfer in mouse lung *in vivo*, *Gene Ther.* 5, 1333, 1998.

36. Kukowska-Latallo, J.F. et al., Intravascular and endobronchial DNA delivery to murine lung tissue using a novel, nonviral vector, *Hum. Gene Ther.* 11, 1385, 2000.

37. Rudolph, C. et al., *In vivo* gene delivery to the lung using polyethylenimine and fractured polyamidoamine dendrimers, *J. Gene Med.* 2, 269, 2000.

38. Koping-Hoggard, M. et al., Relationship between the physical shape and the efficiency of oligomeric chitosan as a gene delivery system *in vitro* and *in vivo*, *J. Gene Med.* 5, 130, 2003.

39. Okamoto, H. et al., Pulmonary gene delivery by chitosan-pdna complex powder prepared by a supercritical carbon dioxide process, *J. Pharm. Sci.* 92, 371, 2003.

40. Iqbal, M. et al., Nasal delivery of chitosan-DNA plasmid expressing epitopes of respiratory syncytial virus (rsv) induces protective ctl responses in balb/c mice, *Vaccine* 21, 1478, 2003.

41. Ferrari, S. et al., Exgen 500 is an efficient vector for gene delivery to lung epithelial cells *in vitro* and *in vivo*, *Gene Ther.* 4, 1100, 1997.

42. Ferrari, S. et al., Polyethylenimine shows properties of interest of cystic fibrosis gene therapy, *Biochem. Biophys. Acta.* 1447, 219–225, 1999.

43. Stammberger, U. et al., Non-viral gene delivery to atelectatic and ventilated lungs, *Ann. Thoracic Surg.* 73, 432, 2002.

44. Uduehi, A.N. et al., Effects of linear polyethylenimine and polyethylenimine/DNA on lung function after airway instillation to rat lungs, *Mol. Ther.* 4, 2001.

45. Kichler, A. et al., Intranasal gene delivery with a polyethylenimine-PEG conjugate, *J. Control Release,* 81, 379–388, 2002.

46. Rudolph, C. et al., Oligomers of the arginine-rich motif of the hiv-1 tat protein are capable of transferring plasmid DNA into cells, *J. Biol. Chem.* 278, 11411, 2003.

47. Jia, S.F. et al., Eradication of osteosarcoma lung metastases following intranasal interleukin-12 gene therapy using a nonviral polyethylenimine vector, *Cancer Gene Ther.* 9, 260, 2002.

48. Densmore, C.L. et al., Gene transfer by guanidinium-cholesterol: dioleoylphosphatidyl-ethanolamine liposome-DNA complexes in aerosol, *J. Gene Med.* 1, 251, 1999.

49. Densmore, C.L. et al., Aerosol delivery of robust polyethyleneimine-DNA complexes for gene therapy and genetic immunization, *Mol. Ther.* 1, 180, 2000.

50. Gautam, A. et al., Enhanced gene expression in mouse lung after pei-DNA aerosol delivery, *Mol. Ther.* 2, 63, 2000.

51. Gautam, A. et al., Transgene expression in mouse airway epithelium by aerosol gene therapy with pei-DNA complexes, *Mol. Ther.* 3, 551, 2001.

52. Gautam, A., Densmore, C.L., and Waldrep, J.C., Pulmonary cytokine responses associated with pei-DNA aerosol gene therapy, *Gene Ther.* 8, 254, 2001.

53. Tan, Y. et al., The inhibitory role of cpg immunostimulatory motifs in cationic lipid vector-mediated transgene expression *in vivo*, *Hum. Gene Ther.* 10, 2153, 1999.

54. Densmore, C.L. et al., Sustained gene expression in the lungs following repeated aerosol delivery of pei-plasmid DNA despite the presence of a non-CPG-mediated refractory effect, *Mol. Ther.* 3, S543, 2001.

55. Tsan, M.F., Tsan, G.L., and White, J.E., Surfactant inhibits cationic liposome-mediated gene transfer, *Hum. Gene Ther.* 8, 817, 1997.

56. Gautam, A. et al., Growth inhibition of established b16-f10 lung metastases by sequential aerosol delivery of p53 gene and 9-nitrocamptothecin, *Gene Ther.* 9, 353, 2002.

57. Densmore, C.L. et al., Growth suppression of established human osteosarcoma lung metastases in mice by aerosol gene therapy with pei-p53 complexes, *Cancer Gene Ther.* 8, 619, 2001.

58. Gautam, A. et al., Aerosol delivery of pei-p53 complexes inhibits b16-f10 lung metastases through regulation of angiogenesis, *Cancer Gene Ther.* 9, 28, 2002.

59. Weiss, D.J. Delivery of gene transfer vectors to lung: Obstacles and the role of adjunct techniques for airway administration, *Mol. Ther.* 6, 148, 2002.

60. Gautam, A. et al., Topical gene therapy for pulmonary diseases with PEI-DNA aerosol complexes, in *Methods in Molecular Medicine, Vol. 2: Lung Cancer: Diagnostic and Therapeutic Methods and Reviews*, Driscoll, B. (Ed.), Humana Press, Totowa, New Jersey, 2002, Chap. 33.

61. Jia, S.F. et al., Aerosol gene therapy with PEI:IL-12 eradicates osteosarcoma lung metastases, *Clin. Cancer Res.*, 9, 3462, 2003.

CHAPTER 36

Cutaneous Gene Delivery

James C. Birchall

CONTENTS

36.1 INTRODUCTION

The aim of cutaneous gene therapy is to facilitate the expression of a genetic product in the viable region of skin tissue in order to treat a localized or systemic condition, or to vaccinate against a potential pathogenic disease. The challenge of delivering genes to the skin is a product of the physico-chemical properties of the large hydrophilic DNA molecule, with or without an additional carrier vehicle, and the significant barrier properties of the target tissue. The strategies for overcoming these challenges are diverse and can involve innovative techniques to bypass the physical barrier and formulation methods to maximize the therapeutic potential of the delivered DNA. In certain instances mediating indiscriminate expression of the delivered gene is only a primary goal and restricting expression to a specific location or cell population can support improvements in the treatment strategy.

36.2 DISEASE TARGETS FOR CUTANEOUS GENE THERAPY

The skin is ostensibly an attractive organ for the initial development and clinical administration of both therapeutic and prophylactic genetic medicines. The tissue is readily accessed, is extremely well characterized at both the cellular and the molecular level, has a significant regenerative capacity,

0-8493-1934-X/05/$0.00+$1.50
© 2005 by CRC Press LLC

and is easily monitored directly or by biopsy. Further, the major cell populations that compose the skin components can be readily isolated, grown in an *in vitro* culture environment, and grafted back to the donor following the insertion of a transgene.[1] While this *ex vivo* approach may be inconvenient, painful, and expensive, it does provide the ability to confirm the efficiency of gene transfer prior to delivery back to the patient and avoids the inherent risks involved in administering gene vectors directly to patients.[2] Where the vectors that are employed to mediate the transfer of the DNA are potentially harmful, such as retroviral vectors, this may be of primary importance. Retroviral vectors do not efficiently transfect nondividing cells, and therefore a further advantage of using the *ex vivo* approach with these vectors lies in their ability to transfect very high numbers of cells that are actively dividing in a cell culture environment, prior to returning the graft to the donor.

In contrast to *ex vivo* cutaneous delivery, *in vivo* gene transfer involves the delivery of the gene directly to, or through, the patient's skin, and is therefore potentially less problematic and more convenient. Originally, *in vivo* gene therapy was introduced as a mechanism for the replacement of absent or defective genes implicated in heritable cutaneous disorders, or "genodermatoses."[3-5] An alphabetical list of characterized genodermatoses and their associated aberrant genes is shown in Table 36.1. Although the delivery of a replacement wild-type copy of a dysfunctional gene is a relatively simple theoretical concept, the progression of these treatments to the clinic has been restricted by the inability to accurately target the corrective gene to specific cell types and regulate its expression.[6] In addition, while the correction of inherited abnormalities by transfer of genetic material can be easily achieved in diseases with an autosomal recessive mode of transmission, it becomes more complicated for autosomal dominant inherited diseases where the genetic correction requires inactivation of the mutant allele.[1]

Most current applications of gene therapy, and the majority of clinical trials that have been conducted to date, are directed against the treatment of various forms of malignancies[7] or infectious diseases.[3] It is true that the progression of normal tissue to metastatic tumor may involve numerous separate mutations. In theory these would have to be corrected separately by gene replacement to restore a normal cellular phenotype; however, it has been shown that correction of a single defect may be sufficient to reverse malignancy.[8] In the case of cancer, localized cutaneous malignancies in the skin may also be amenable to alternative gene therapy approaches, such as "suicide gene therapy," whereas systemic cancers may be targeted using a "cancer vaccine." The aim of suicide gene therapy, also known as gene-directed enzyme prodrug therapy (GDEPT),[8] is to endow cancer cells with an exclusive ability to metabolize a nontoxic prodrug into a cytotoxic species. With this approach a tumor-specific promoter sequence may be employed as part of the plasmid DNA (pDNA) construct, specifically to restrict expression of the delivered transgene to tumor cells. Once in place the transgene encodes for an enzyme that will distinctively catalyze the conversion of a prodrug to a more toxic metabolite *in situ*. The most widely used enzyme and prodrug combination that has made the most progress toward the clinic employs the herpes simplex virus thymidine kinase (HSV-TK) gene to convert the prodrug ganciclovir to a more toxic metabolite. Ganciclovir is not a substrate for human enzymes, but is converted into its monophosphate by the enzyme thymidine kinase. The monophosphate is, in turn, converted by endogenous cellular kinases into ganciclovir triphosphate, a toxic metabolite.[8] One advantage of the GDEPT approach is that not all tumor cells would need to express the HSV-TK gene as the active metabolite is able to spread, either via gap junctions or in apoptotic vesicles engulfed by surrounding cells, to adjacent tissue (bystander effect).[9] In cancer vaccination a gene encoding for a protein antigen is introduced into tumor cells to prime and stimulate the host immune system against the tumor.[10] Human clinical trials are currently under way with cancer vaccines against prostate cancer, adenocarcinoma of the breast and colon, melanoma, and several lymphomas.[11]

The effective prophylaxis of infectious disease by genetic vaccination is considered a realistic application for gene therapy. To date, this approach is being investigated for its ability to raise immunity against a variety of pathogens, including human immunodeficiency virus (HIV-1),[12,13]

Table 36.1 Heritable Skin Diseases (Genodermatoses) and Their Implicated Defective Genes[a]

Genodermatosis	Gene(s) Implicated
Albinism	TYR, TYRP-1, OCA2, OA1
Ataxia telangiectasia	ATM
Bannayan-Zonan syndrome	PTEN
Basal cell nevus syndrome	PTC
Chediak-Higashi syndrome	CHS1
Cockayne syndrome	CSA (CKN1), CSB
Congenital erythropoietic porphyria	UROS
Congenital atrichia	HR
Cowden syndrome	PTEN
Darier's disease	ATP2A2
Dyskeratosis congenital	DSC1
Dystrophic epidermolysis bullosa (EB)	COL7A1
EB-muscular dystrophy	PLEC1
EB-pyloric atresia	ITGA6, ITGB4
EB-simplex	KRT5, KRT14
Ectodermal dysplasia/skin fragility	PKP1
Ehlers-Danlos syndrome	COL1A1, COL1A2, ADAMTS2, COL3A1, COL5A1, COL5A2, PLOD, B4GALT7
Epidermolytic hyperkeratosis	KRT1, KRT10
Epidermolytic palmoplantar keratoderma	KRT9
Erythrokeratodermia variabilis	GJB3
Erythropoietic protoporphyria	FECH
Fabry's disease	GLA
Familial cylindromatosis	CYLD
Familial porphyria cutanea tarda	URO-D
Generalized atrophic benign EB	COL17A1
Hailey-Hailey disease	ATP2C1
Haim-Munk syndrome	CTSC
Hereditory hemorrhagic telangiectasia	ENG, ALK 1
Hereditory melanoma	CDK4, CDKN2A
Hermansky-Pudlak syndrome	HPS
Hyperkeratotic cutaneous capillary-venous malformation	KRIT1
Ichthyosis bullosa Siemens	KRT2e
Junctional EB	LAMA3, LAMB3, LAMC2
Lamellar ichthyosis	TGM1
Milroy disease	VEGFR-3
Monilethrix	hHB1, hHB6
Muir-Torre syndrome	hMSH2
Mutilating keratoderma with ichthyosis	LOR
Netherton's syndrome	SPINK5
Neurofibromatosis	NF1, NF2
Nonepidermolytic palmoplantar keratoderma	KRT16, KRT1
Nude	WHN
Occipatal horn syndrome	ATP7A
Pachonychia congenita type ∫	KRT6a, KRT6b, KRT16, KRT17
Palmoplantar keratoderma with deafness	GJB2
Papillon-Lefevre syndrome	CTSC
Peutz-Jeghers	STK11/LKB1
Pseudoxanthoma elasticum	MRP6/ABCC6
Refsum disease	PAHX
Sjörgren-Larsson syndrome	FALDH
Striate palmoplantar keratoderma	DSP, DSG1
Tielz syndrome	MITF
Trichothiodystrophy	XPB, XPD
Tuberous sclerosis	TSC1, TSC2
Variegate porphyria	PPO
Vohwinkel's syndrome	GJB2
Waardenburg syndrome	PAX3

Table 36.1 Heritable Skin Diseases (Genodermatoses) and Their Implicated Defective Genes[a] (Continued)

Genodermatosis	Gene(s) Implicated
Waardenburg-Hirchsprung disease	EDN3, EDNRB, SOX10
Werner syndrome	WRN
White sponge nevus	KRT4, KRT13
Xeroderma pigmentosum	XPA, XPB, XPC, XPD, XPE, XPF, XPG, hRAD30
X-linked recessive ichthyosis	STS

[a] Adapted from Uitto, J. and Pulkkinen, L., The genodermatoses: candidate diseases for gene therapy, Hum. Gene Ther. 11, 2267, 2000.

malaria,[14] tuberculosis,[15] hepatitis B,[16] hepatitis C,[17] and herpes simplex virus.[18] In genetic vaccination, the patients are immunized by introducing a transgene that can mediate the mucosal expression of foreign antigen and the subsequent induction of an immune response.[19] Cutaneous DNA immunization has received significant attention in recent years as the excellent antigen-presenting capabilities of epidermal Langerhans cells can be exploited to elicit a T-cell mediated immune reaction that can be more efficient and achieved at a lower cost when compared with recombinant protein vaccines.[3] As the immune response is induced by a single gene rather than an entire organism, this approach is also considered to be safer than using live attenuated vaccines.[20] Amongst the many studies in this area, Fynan et al.[19] have demonstrated that DNA vaccines can effectively raise immunity to lethal influenzae challenge in mice and chickens, with delivery of the DNA directly to the epidermis shown to be more effectual than parenteral administration or mucosal application.

With increasing knowledge of skin pathobiology and molecular genetics, the skin will continue to represent an attractive site for localized gene therapy. In the future, the delivery of therapeutic genes to epidermal tissue may also provide potential for the local immune-modulatory DNA-based treatment of dermatological pathologies such as hyper-proliferative dermatoses.[21] The recent elucidation of the molecular events of alopecia[22] also highlights the potential cosmetic interventions for gene therapy targeted to the hair follicle.[23]

36.3 THE SKIN AS A BARRIER TO GENE DELIVERY

Clearly, if cutaneous gene therapy is to succeed, then the transgene must be efficiently transported to the target cell population. The primary role of the skin, however, is to serve as a physical barrier to the invasion of foreign material. Simplistically, the skin can be considered as a structure comprised of three distinct layers — the epidermis, the dermis, and the hypodermis — containing a number of adnexal features such as hair follicles, sebaceous glands, and sweat glands. In humans, the epidermis, the uppermost layer of the skin, is approximately 50–150 µm thick. The surface of the epidermis is comprised of flattened "dead" cells that have lost their nuclei following differentiation from the inner to the outer layer of the epidermis. This nonviable layer is termed the stratum corneum. The stratum corneum, approximately 15–20 µm in thickness, represents the principal barrier to penetration and permeation of substances through the skin. The remainder of the epidermis is a stratified epithelium continually undergoing progressive differentiation (keratinization). Cell types found in the epidermis include keratinocytes, melanocytes, Langerhans cells, and Merkel cells. The differentiated keratinocytes arise from a pool of transient amplifying cells located at the basal layer of the epidermis, which in turn are derived from epidermal stem cells based within and around the hair follicle. Stem cells can be isolated from the epidermis *in vitro* and typically express certain cell surface markers, but these cells are presently impractical to selectively target *in vivo*.[2] The dermis is a connective tissue underlying the epidermis that acts as a protective layer against injuries, provides the skin with elastic properties, and maintains a role in thermal regulation. The

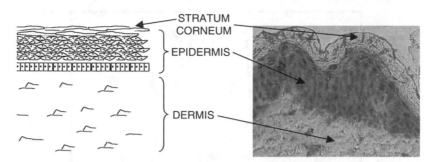

Figure 36.1 (See color insert following page 336) Schematic representation and light photomicrograph of human skin showing the stratum corneum, epidermis, and dermis. Full-thickness human breast skin, obtained from mastectomy with ethical committee approval and informed patient consent, was fixed in 0.5% glutaraldehyde, embedded in OCT medium, frozen, and sectioned with a Leica® CM3050S cryostat. Sectioned tissue (10 μm) was stained with haematoxylin.

main components of the dermis, collagen and elastin, form an amorphous extracellular matrix. The dermis also contains blood and lymphatic vessels, nervous elements, and scattered cells including fibroblasts, mast cells, macrophages, and lymphocytes. A schematic diagram and photomicrograph illustrating the stratum corneum, epidermis, and dermis regions of human skin are displayed in Figure 36.1. The hypodermis is the layer of skin beneath the dermis and is constituted of subcutaneous fat and blood vessels. It maintains skin mobility, supplies energy, and insulates the body.

It has been firmly established that the structural basis of the cutaneous permeability barrier in mammals is the stratum corneum.[24] Therefore, in order to deliver therapeutic compounds to the epidermis or the dermis, delivery strategies must overcome the physical barrier represented by the tightly packed dead cells of the stratum corneum. Traditional transdermal formulation strategies aim to enhance the delivery of small therapeutic molecules, less than 500 mol wt, across the stratum corneum by the paracellular, transcellular, or intracellular routes. However, in order to deliver macromolecular products such as genes and proteins, more innovative and radical methods of drug delivery are required.

36.4 PHYSICAL METHODS FOR ENHANCING GENE DELIVERY TO SKIN

The crudest yet simplest method for bypassing the stratum corneum barrier is to administer the genetic medicine via direct injection. Although alternative pain-free delivery strategies may be of more benefit in the clinic, the ability to deliver genes directly to the epidermis and dermis via a hypodermic needle provides a valuable research tool for studying gene expression in the target tissue. Injection of "naked DNA," DNA that has not been complexed with a carrier vector, into the superficial dermis of porcine skin has shown to mediate the expression of reporter gene β-galactosidase (β-gal) in the middle layer of the epidermis overlaying the injection site with the expressed protein being visualized for up to 3 weeks.[25] The same research group[26] has also studied the expression of naked DNA injected into human, pig, and mouse skin. The authors showed that when both human skin organ cultures, i.e., human skin excised during a surgical procedure, and human grafts on immunocompromised mice are injected with the pCMV-β-gal reporter plasmid, the DNA is expressed in the epidermis in the same way as in pig skin.[25] In contrast, DNA injected into mouse skin is expressed not just in the epidermis but also in the dermis and underlying fat and muscle. Clearly, the delivery of naked DNA via an intradermal needle does not involve intricate formulation. Such factors should not be ignored, however, as the resulting gene expression efficiency can be enhanced if the DNA is delivered in a modified vehicle comprising dextrose or phosphate-buffered saline.[27]

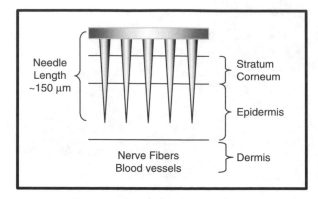

Figure 36.2 Schematic representation of the microneedle delivery concept. The microneedles are designed to penetrate the stratum corneum to facilitate access of molecules to the viable epidermis without impacting on the underlying nerve endings.

The development of microneedles offers a minimally invasive, pain-free alternative to intradermal injection. This approach utilizes standard microfabrication techniques to etch arrays of micron-sized needles that, when inserted into the skin, create conduits for direct and controlled access of molecules across the stratum corneum and subsequent diffusion to the underlying epidermis and dermis. As microneedles are designed to penetrate the stratum corneum without stimulating the pain receptors found in deeper tissue, they are able to circumnavigate the definitive barrier to delivery without causing any pain or discomfort (Figure 36.2). The advantages of utilizing microneedles as opposed to other physical methods include (1) direct and controlled delivery of the formulation; (2) exposure of large surface areas to the delivery agents, of possible importance for skin diseases such as psoriasis which extends to plaques; (3) painless delivery for patients; and (4) the ability to manipulate the formulation that is applied through the microchannels. Microneedle arrays were originally designed to enhance the transdermal delivery of low-molecular-weight drugs to the systemic circulation.[28] Currently the technology is being exploited to investigate the delivery of pDNA to the viable epidermis.[29] The microfabrication process used to create microneedles for localized gene delivery involves a modified Deep Reactive Ion Etching process. At the start of the process a silicon wafer is coated with a positive photosensitive material. A standard high-resolution chromium-plated lithographic mask bearing the appropriate dot array pattern is used during the UV light exposure step to produce a photoresist etch mask. The surface is subsequently etched using a reactive blend of fluorinated and oxygen gases, with those regions protected by the photoresist mask resisting the etching process and leading to the formation of microneedles. The result, following removal of the photoresist, is an array of microneedles, approximately 150 μm in length. Figure 36.3 shows scanning electron micrographs of microneedle arrays prepared using this technique both before (Figure 36.3A) and after (Figure 36.3B) removal of the photoresist. Importantly, the microneedles maintain their integrity following application to full-thickness human skin (Figure 36.3C), allowing for several applications with the same device.

If skin is placed in a water bath under controlled conditions,[30] an epidermal membrane comprising the stratum corneum and viable epidermis can be readily removed and used to determine the penetration and diffusion of materials. Figure 36.4A shows the appearance of human breast epidermal membrane, with epidermis facing uppermost, following application of silicon microneedles. Figure 36.4B shows the microneedle array visualized unremoved from the epidermal membrane. The creation of microchannels in the epidermal membrane with widths varying from approximately 10–15μm to 60–70μm was not unexpected, as the thickness of human skin varies[31] and the pressure applied during the use of the microneedle device may not be evenly distributed. Nevertheless, in all cases the microneedles are clearly seen to cross both the stratum corneum and

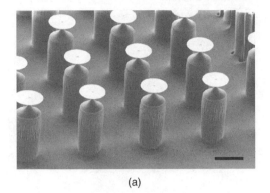

(a)

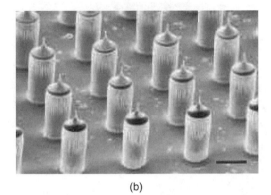

(b)

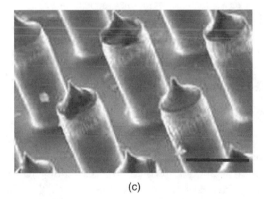

(c)

Figure 36.3 Scanning electron micrographs of silicon microneedles microfabricated using a modified Deep Reactive Ion Etching process. The microfabrication process, as described in Section 36.4, was completed at CCLRC Rutherford Appleton Laboratory (Chilton, Didcot, Oxon, UK). The wafer saw was prepared at the School of Engineering, Cardiff University, Cardiff, UK. (A) Silicon microneedles prior to removal of the photoresist. Bar = 50 μm. (B) Silicon microneedles following removal of the photoresist. Bar = 50 μm. (C) Silicon microneedles following application to full-thickness human skin. Bar = 50 μm.

viable epidermis to facilitate controlled access of molecules to the target region of skin. The microchannels produced using these and other microneedles are demonstrating considerable potential for the delivery of genes and macromolecules to human skin.[29,32]

Ciernik et al.[33] reported the installation and expression of a solution of naked DNA in mice using high-frequency puncturing of skin with oscillating needles (1 cm length, 250 μm diameter). The authors demonstrated that this "tattooing" technique led to enhanced expression of reporter

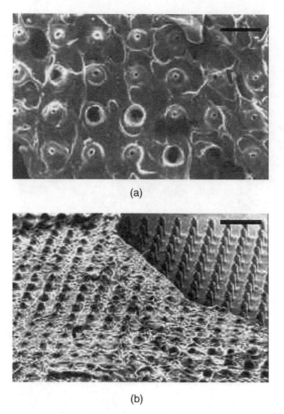

(a)

(b)

Figure 36.4 Scanning electron micrographs of epidermal membrane treated with silicon microneedles. Full-thickness human breast skin, obtained from mastectomy with ethical committee approval and informed patient consent, had all excess adipose tissue removed by blunt dissection. The epidermal membrane, consisting of the stratum corneum and the viable epidermis, was obtained by separation from the dermal layer. The tissue was immersed in distilled water preheated to 60° C for 60 sec, and the upper layers carefully peeled off from the dermal layer using tweezers. Epidermal membranes were treated with microneedles for 30 sec at an approximate pressure of 2 kg/cm². (A) Microneedle-treated epidermal membrane with viable epidermis facing uppermost. Bar = 200 μm. (B) Epidermal membrane with microneedles *in situ*, viable epidermis facing uppermost. Bar = 500 μm.

gene compared with both direct subepidermal injection and topical application. The method was also used to demonstrate the induction of cytotoxic T-lymphocytes using a peptide oligonucleotide. Eriksson et al.[34] used a similar technique, termed "microseeding," to deliver expression plasmids to intact pig skin and partial thickness wounds, and confirmed that the procedure proved more efficient than direct injection and particle-mediated gene transfer.

Electroporation involves the application of short-duration high-intensity electric field pulses to cells and tissues to create transient aqueous pathways within lipid bilayers.[35] Permeability of molecules through cell membranes and tissue can therefore be enhanced by several orders of magnitude. Moreover, the associated local electric field can contribute to transmembrane molecular transport by electrophoresis or electroporation.[36] Electroporation has been used widely to introduce macromolecules, including pDNA, into cells *in vitro*.[37] *In vivo* gene transfer by electroporation has been shown to be proficient in a variety of animal models (for review, see Jaroszeski et al.[38]) and in human skin.[39] Dujardin et al.[35] utilized electroporation to demonstrate the localization of fluorescently labeled plasmid and transient expression of the green fluorescent protein reporter gene in the epidermis of rat skin. The authors noted that the technique did not adversely affect skin viability. Zhang et al.[40] targeted expression of a lacZ reporter gene to various depths of mouse dermis by varying the pulsed electric fields administered by the electrodes and the pressure applied between the electrodes and skin. Glasspool-Malone et al.[41] used the nuclease inhibitor aurintricarboxylic acid (ATA) to enhance the *in vivo* expression mediated by naked pDNA delivered by electroporation. The authors demonstrated the expression of reporter gene in adipocytes, fibroblasts, and dendritic-like cells within the dermis and subdermal tissue, whereas intradermal injection led to a lower level of expression that was restricted to the epidermis.

The use of ballistic devices to propel materials coated onto dense carrier particles through the stratum corneum and into the underlying tissue has been widely reported as a method for enhancing the delivery of drugs through the skin. The commercially available Helios Gene Gun® utilizes a helium cylinder to accelerate DNA-coated gold particles into target cells or tissues. The gold particles are typically around 1 μm in diameter and can penetrate through cell membranes, carrying the bound DNA, typically 0.5–5 μg/mg gold, into the cell cytoplasm. Subsequent dissociation of the DNA from the carrier particles allows the gene to be expressed.[4] Numerous studies have shown the successful delivery and expression of genes using this method, with both reporter genes and therapeutic plasmids having been transported to mammalian cells in culture,[42] oral mucosa,[43] cornea,[44] and animal skin.[3,45–47] The gene gun has demonstrated high levels of local transgene expression[47] with epidermal and dermal implication.[48] While expression of the transgene is usually transient, lasting from a few days up to 4 weeks,[49] Cheng et al.[45] reported sustained luciferase activity in rat dermis one and a half years after *in vivo* particle bombardment.

As the delivery of material by particle bombardment is cell receptor independent, it is clear that genes can be delivered indiscriminately into different mammalian cell populations. The technique is limited, however, by (1) the restricted penetration of the delivered material to the areas beneath the bombarded surface, (2) the relatively uncontrolled depth of penetration of the delivered particles, (3) the necessity to coat the active medicament onto a dense carrier particle and the associated reduction in formulation flexibility, and (4) the quantity of active material that can be loaded in this way.

36.5 FORMULATION STRATEGIES FOR ENHANCING GENE DELIVERY TO SKIN

Whereas the aforementioned physical methods are capable of introducing DNA into the target region of the skin, for the genetic treatment to succeed the DNA must be biologically functional, i.e., it must remain stable and be capable of entering, and translocating through, cells to reach the nucleus for therapeutic gene expression. Gene therapy vectors are therefore used to protect the DNA and encourage cellular uptake, trafficking, and expression. Gene therapy vectors based on viruses are efficient at transfecting cells and encompass the ability to mediate long-term gene expression.[50] There are, however, inherent safety issues associated with viral vectors relating to the stimulation of an immune response, the activation of oncogenes, and the observation of necrosis, cytotoxicity, hyperkeratosis, and acanthosis in mouse skin.[51,52] Synthetic nonviral gene transfer approaches provide a relatively inexpensive, direct, and flexible method for delivering DNA to skin. Nonviral gene vectors are relatively nontoxic in comparison with their viral counterparts and can often be suitable for repeat dosing. Problems associated with nonviral gene transfer approaches, however, include variable levels of expression and the inability to target delivery to specific cell types. Although nonviral gene transfer is not currently suitable for mediating long-term gene expression, as such a strategy would require targeting stem cells and integration into the stem cell genome,[53] nonviral carrier systems such as liposomes, polycations, and polymers have been utilized in cutaneous gene delivery studies.

The interaction of liposomes with skin, and their potential role in promoting the epidermal delivery of therapeutic molecules, has been investigated for a number of years.[54,55] The most exploited nonviral gene delivery strategy involves the use of liposomes comprising cationic lipid species to interact spontaneously with negatively charged DNA to form electrostatic complexes.[56] These lipid–DNA complexes afford the properties of DNA condensation, protection against nuclease-mediated degradation, and improved transgene–cell interactions. Although the work of Cevc et al.[57] provides some evidence for a transporter role for certain lipid-based vesicles, termed transfersomes, in carrying drugs across the skin, it would be expected that an intact stratum corneum would represent an impenetrable barrier for pDNA–liposome structures. Nevertheless, there is increasing data supporting the efficacy of liposomes to deliver drugs to the pilosebaceous unit.[23,58,59] At the junction of the stratum corneum with the hair follicle lays a lipophilic matrix of triglycerides, waxy esters, and cholesterol, which collectively form sebum. This pocket of oily substance, forming the entrance to the pilosebaceous unit, is approximately 70 μm in diameter. It can be envisaged that the physico-chemical properties of lipid–pDNA complexes would allow for partitioning into sebum with subsequent access to cells of the pilosebaceous unit. The immunohistochemistry data of Alexander and Akhurst is intriguing to note: they reported that the topical application of a cationic liposome–pDNA vector to *in vivo* mouse skin in the anagen or active phase of the hair growth cycle resulted in more efficient reporter gene expression than when applied in the telogen phase of the hair growth cycle.[60] This would implicate a significant role in pDNA uptake and expression for cells via the follicular pathway. Indeed, *in vivo* experiments from the laboratory of Weiner demonstrate that topical application of DNA complexed with novel mixtures of nonionic/cationic lipids exhibit transgene expression localized to perifollicular cells.[61] With the expected limited tissue diffusion of such large complexes, observations of significant expression in dermal fibroblasts following topical application of cationic liposome–pDNA vectors[60] may reflect disruption of stratum corneum barrier function as a result of chemical and physical techniques used to remove animal hair prior to the application of gene vectors.[62]

36.6 POLYMER-MEDIATED GENE DELIVERY TO SKIN

The feasibility of using polycationic polymers to mediate the transfection of DNA *ex vivo* in skin was investigated by Nead and McCance.[63] This study showed that 15–20% of primary human keratinocytes could be transfected with reporter pDNA complexed with poly(L-ornithine) when the cells are shocked with the penetration enhancer molecule dimethylsulfoxide (DMSO).

The use of polymer systems to transfer genes into skin *in vivo* as a viable alternative to lipid-based carrier vectors is now receiving increasing attention. Indeed, a recent publication compared the ability of polymers and liposomes in targeting gene delivery to skin cells.[64] The authors directly compared the ability of phospholipid-based cationic liposomes, PEGylated liposomes, nonionic liposomes,[65] nonionic/cationic liposomes, protective interactive noncondensing (PINC) polymers,[66] and propylene glycol:ethanol:water mixtures[67] in delivering luciferase and β-gal reporter genes to cells in the skin of neonate 6-day-old rats. It is interesting to note the authors' conclusion that topically applied nonionic liposomes were the superior vehicle among those tested for the delivery and expression of recombinant genes in skin cells.

Prokop et al.[68] developed and evaluated 40 polymer formulations for use as *in vivo* gene delivery vehicles. The charge density of the polymer–DNA constructs was manipulated by selecting different ratios of polymer-to-plasmid mass as well as altering polymer chemistry. The authors found that, as opposed to *in vitro* observations, favorable *in vivo* luciferase expression was observed using negatively charged polymer–DNA particles, with the Tetronic® polymer series or noncharged polyvinylpyrrolidine formulations proving the most potent carriers. This study also reported the possibility of sustaining gene expression over 7 days and made suggestions for the redesign of improved polymer–plasmid formulations for delivery via particle bombardment.

Alternative approaches utilizing physical polymer formulations to control and enhance the delivery of DNA in skin are receiving increased attention. Li et al.[69] investigated the use of novel biocompatible and biodegradable thermosensitive hydrogels for the controlled delivery of genes to treat skin disorders or facilitate wound healing. Triblock copolymers comprising poly(ethylene glycol-b-[D, L-lactic acid-*co*-glycol acid]-b-ethylene glycol) (PEG–PLGA–PEG) were found to mediate higher levels of luciferase expression in mouse skin when compared to naked pDNA. Importantly, the PEG–PLGA–PEG copolymers displayed low cytotoxicity and a high degree of biocompatibility.

Luu et al.[70] reported an additional method for delivering DNA to skin in a controlled and sustained manner. This approach utilized electrospinning with solutions of poly(lactide-*co*-glycolide) (PLGA) random copolymer and a poly(D,L-lactide)–poly(ethylene glycol) (PLA–PEG) block copolymer to fabricate scaffolds with nanostructured fibers. The method produces unwoven membranes with individual fiber diameters ranging from 50–500 nm. The fibers therefore form a large, interconnected, porous network for gene delivery in tissue engineering. DNA incorporated into the polymer scaffold is released at a rate according to the scaffold composition and morphology, with maximum release at 2 h. Critically, the released DNA was shown to remain structurally intact and biologically functional, mediating reporter gene β-gal expression in cultured MC3T3 cells. Though this system at present has limited bioactivity, the authors aim to further optimize the model with the use of additional transfection enhancement agents and with a combination of genes delivered in a controllable manner.

A similar, previously published concept involves the injection of a biodegradable DNA-containing polymer implant that releases DNA over a 2-month period.[71] In this system, which is based on an injectable biodegradable drug delivery system,[72,73] the aqueous DNA is mixed with PLGA dissolved in glycofurol and injected via a 21-gauge needle. The PLGA solidifies *in situ*, entrapping the DNA within the polymer matrix. When high-molecular-weight PLGA was used in the implant, sustained release of plasmid was observed up to a maximum of 60 days. The released DNA mediated pronounced gene expression when implanted subcutaneously in mice, with maximum luciferase expression observed at 28 days after injection. In addition, a significantly higher level of secreted human placental alkaline phosphatase expression was observed for 67 days after injection when compared with other treatments, i.e., DNA in saline or non-DNA loaded injectable implant followed by DNA in saline. The authors claim that this approach has several advantages over other systems, including ease of administration, high DNA encapsulation efficiency, and an uncomplicated and nontoxic fabrication process. The ability to inject the implant as a solution makes this technology applicable to tissue regeneration in irregular cavities and the sustained release profile may lend the system to the efficient delivery of genetic vaccines.

While the aforementioned uses for polymers as DNA carriers or physical scaffolds for DNA release may have utility for a number of gene therapy applications, the majority of current research has specifically employed polymers as adjuncts in the delivery of genetic vaccines. Intramuscular or subcutaneous delivery of pDNA complexed with PINC polymer systems have been shown to enhance antibody titers to expressed human growth hormone in comparison with pDNA in saline.[74] In DNA vaccination for human immunodeficiency virus (HIV), low viscosity carboxymethylcellulose has an adjunct effect on the specific mucosal antibody, systemic antibody, and cell-mediated immune response following intranasal administration.[75]

Lima et al.[76] coencapsulated a tuberculosis-protective gene construct and trehalose dimicolate into biodegradable PLGA microspheres. The microsphere formulation permitted a tenfold reduction in the DNA dose when compared with naked DNA. Polymeric microencapsulation of pDNA with PLGA has also been investigated as a means of enhancing the stability of genetic vaccines.[77] In a separate study, a variant PLGA microsphere incorporating lipophilic additives was also shown to enhance the B- and T-cell responsiveness to DNA vaccines.[78]

36.7 TARGETING GENE EXPRESSION IN THE SKIN

Persistent gene expression is difficult to achieve in the skin, as the epidermis is a rapidly proliferating and differentiating tissue. Keratinocyte stem cells are the progenitors of all other keratinocytes in the epidermis. Therefore, gene therapy targeted to the keratinocyte stem cells or long-lasting keratinocyte progenitor cells remains an important milestone in the development of a sustainable genetic treatment.[3,79] To this end, recent studies have explored the use of lentiviral[79] and retroviral[79,80] vectors to introduce genes into keratinocyte progenitor cells both *in vitro* and *ex vivo*. The findings suggest that our ability to exploit these cells successfully is currently limited by our considerable lack of understanding of the biological behavior of these cells both *in vivo* and during *in vitro* culturing.

Restricting gene expression to specific cells via specific promoter sequences was initially utilized by cancer researchers to target gene expression to tumor cells. The methodology relies on the fact that promoter sequences that are present in genes responsible for producing certain proteins will implicitly operate in those areas where those proteins are found. Promoter-based gene expression can be used in cutaneous gene therapy to restrict the expression of introduced genes to specific cell layers. For example, promoter sequences that only promote gene expression in the presence of involucrin, a protein produced in the superficial epidermis, have been used to mediate gene expression specifically in differentiating keratinocytes in the skin epidermis.[81-84] Further, promoter sequences that exclusively function in the presence of keratin 5[85-87] and keratin 14[84,85,87] have been shown specifically to enhance gene expression in the differentiated layer of the epidermis.

36.8 SUMMARY

The accessibility and variety of potentially amenable therapeutic targets makes the skin a uniquely attractive somatic tissue for gene therapy. The superficial lipophilic layer of the skin, however, remains a major obstacle in delivering macromolecules to the viable region of the tissue. For intradermal gene therapy to realize its potential, the research community must continue to provide innovative solutions to circumnavigate the barrier properties of the skin, target the appropriate cell population, and promote the stability, efficiency, specificity, and longevity of gene expression once *in situ*. It is essential that research in this area recognizes that the formulation approaches employed may be complex, e.g., tissue-specific gene expression constructs or polymer drug delivery scaffolds, or relatively simple, e.g., simply altering the ionic composition of the delivery vehicle, and acknowledges that no singular delivery system is likely to be superior for all applications. Combining the existing knowledge base with further advances in physical delivery strategies and gene vector technology will progress the concept of cutaneous gene therapy toward clinical reality.

ACKNOWLEDGMENTS

The contributions of Feriel Chabri, Sion Coulman, Marc Pearton, Dr. Chris Gateley, Dr. Alex Anstey, Dr. Chris Allender, and Dr. Keith Brain are gratefully acknowledged. I would also like to thank Dr. Anthony Hann, Cardiff School of Biosciences, for assistance with electron microscopy, and Professor David Barrow, Tyrone Jones, and Kostas Bouris, Cardiff School of Engineering, Cardiff University, for their continued support.

REFERENCES

1. Spirito, F., Meneguzzi, G., Danos, O., and Mezzina, M., Cutaneous gene transfer and therapy: the present and the future, *J. Gene Med.* 3, 21, 2001.
2. Khavari, P.A., Rollman, O., and Vahlquist, A., Cutaneous gene transfer for skin and systemic diseases, *J. Int. Med.* 252, 1, 2002.

3. Lin, M.T.S., Pulkkinen, L., and Uitto, J., Cutaneous gene therapy: principles and prospects, *New Emerg. Ther.* 18, 177, 2000.

4. Lin, M.T.S., Pulkkinen, L., Uitto, J., and Yoon, K., The gene gun: current applications in cutaneous gene therapy, *Int. J. Derm.* 39, 161, 2000.

5. Uitto, J. and Pulkkinen, L., The genodermatoses: candidate diseases for gene therapy, *Hum. Gene Ther.* 11, 2267, 2000.

6. Meneguzzi, G. and Vailly, J., Gene therapy of inherited skin diseases, in *The Skin and Gene Therapy*, Hengge, U. and Volc-Platzer, B. (Eds.), Springer-Verlag Berlin, Heidelberg, 2001, Chap. 7.

7. Hart, I.R. and Vile, R.G., Targeted therapy for malignant melanoma, *Curr. Opin. Oncol.* 6, 221, 1994.

8. McNeish, I.A., Searle, P.F., Young, L.S., and Kerr, D.J., Gene directed enzyme prodrug therapy for cancer, *Adv. Drug Del. Rev.* 26, 173, 1997.

9. Kerr, D.J., Young, L.S., Searle, P.F., and McNeish, I.A., Gene directed enzyme prodrug therapy for cancer, *Adv. Drug. Deliv. Rev.* 26, 173, 1997.

10. Durrant, L., Cancer vaccines, *Anticancer Drugs* 8, 727, 1997.

11. Weiner, D.B. and Kennedy, R.C., Genetic vaccines, *Sci. Am.* 281, 50, 1999.

12. Cohen, A.D., Boyer, J.D., and Weiner, D.B., Modulating the immune response to genetic immunization, *FASEB J.* 12, 1611, 1998.

13. Kim, J.J. and Weiner, D.B., DNA gene vaccination for HIV, *Springer Semin. Immunopathol.* 19, 175, 1997.

14. Hedstrom, R.C., Doolan, D.L., Wang, R., Gardner, M.J., Kumar, A., Sedegah, M., Gramzinski, R.A., Sacci, J.B. Jr., Charoenvit, Y., Weiss, W.R., Margalith, M., Norman, J.A., Hobart, P., and Hoffman, S.L., The development of a multivalent DNA vaccine for malaria, *Springer Semin. Immunopathol.* 19, 147, 1997.

15. Tascon, R.E., Colston, M.J., Ragno, S., Stavropoulos, E., Gregory, D., and Lowrie, D.B., Vaccination against tuberculosis by DNA injection, *Nat. Med.* 2, 888, 1996.

16. Davis, H.L. and Brazolot Millan, C.L., DNA-based immunization against hepatitis B virus, *Springer Semin. Immunopathol.* 19, 195, 1997.

17. Inchauspe, G., Gene vaccination for hepatitis C, *Springer Semin. Immunopathol.* 19, 211, 1997.

18. Bourne, N., Stanberry, L.R., Bernstein, D.I., and Lew, D., DNA immunization against experimental genital herpes simplex virus infection, *J. Infect. Dis.* 173, 800, 1996.

19. Fynan, E.F., Webster, R.G., Fuller, D.H., Haynes, J.R., Santoro, J.C., and Robinson, H.L., DNA vaccines: protective immunizations by parenteral, mucosal and gene-gun inoculations, *Proc. Nat. Acad. Sci. U.S.A.* 90, 11478, 1993.

20. Baca-Estrada, M.E., Foldvari, M., Babiuk, S.L., and Babiuk, L.A., Vaccine delivery: lipid-based delivery systems, *J. Biotech.* 83, 91, 2000.

21. Menter, A., Pathogenesis and genetics of psoriasis, *Cutis* 61, 8, 1998.

22. Ahmed, W., ul Haque, M.F., Brancolini, V., Tsou, H.C., ul Haque, S., Lam, H., Aita, V.M., Owen, J., deBlaquiere, M., Frank, J., Cserhalmi-Friedman, P.B., Leask, A., McGrath, J.A., Peacocke, M., Ahmad, M., Ott, J., and Christiano, A.M., Alopecia universalis associated with a mutation in the human hairless gene, *Science* 279, 720, 1998.

23. Li, L. and Hoffman, R.M., The feasibility of targeted selective gene therapy of the hair follicle, *Nat. Med.* 1, 705, 1995.

24. Menon, G.K. and Elias, P.M., The epidermal barrier and strategies for surmounting it: an overview, in *The Skin and Gene Therapy*, Hengge, U. and Volc-Platzer, B. (Eds.), Springer-Verlag Berlin, Heidelberg, 2001, Chap. 1.

25. Hengge, U.R., Chan, E.F., Foster, R.A., Walker, P.S., and Vogel, J.C., Cytokine gene expression in epidermis with biological effects following injection of naked DNA, *Nat. Genet.* 10, 161, 1995.

26. Hengge, U.R., Walker, P.S., and Vogel, J.C., Expression of naked DNA in human, pig and mouse skin, *J. Clin. Invest.* 97, 2911, 1996.

27. Chesnoy, S. and Huang, L., Enhanced cutaneous gene delivery following intradermal injection of naked DNA in a high ionic strength solution, *Mol. Ther.* 5, 57, 2002.

28. Henry, S., McAllister, D.V., Allen, M.G., and Prausnitz, M.R., Microfabricated microneedles: a novel approach to transdermal drug delivery, *J. Pharm. Sci.* 87, 922, 1998.

29. Chabri, F., Bouris, K., Jones, T., Barrow, D., Hann, A., Allender, C., Brain, K., and Birchall, J., Microfabricated silicon microneedles for non-viral cutaneous gene delivery, *Br. J. Dermatol.* 150, 869–877, 2004.

30. Christophers, E. and Kligman, A., Preparation of isolated sheets of human stratum corneum, *Arch. Dermatol.* 88, 702, 1963.

31. Holbrook, K.A. and Odland, G.F., Regional differences in the thickness (cell layers) of the human stratum corneum: an ultrastructural analysis, *J. Invest. Dermatol.* 62, 415, 1974.

32. McAllister, D.V., Allen, M.G., and Prausnitz, M.R., Microfabricated microneedles for gene an drug delivery, *Ann. Rev. Biomed. Eng.* 2, 289, 2000.

33. Ciernik, I.F., Krayenbühl, B.H., and Carbone, D.P., Puncture-mediated gene transfer to the skin, *Hum. Gene Ther.* 7, 893, 1996.

34. Eriksson, E., Yao, F., Svensjö, T., Winkler, T., Slama, J., Macklin, M.D., Andree, C., McGregor, M., Hinshaw, V., and Swain, W.F., *In vivo* gene transfer to skin and wound by microseeding, *J. Surg. Res.* 78, 85, 1998.

35. Dujardin, N., Van De Smissen, P., and Préat, V., Topical gene transfer into rat skin using electroporation, *Pharm. Res.* 18, 61, 2001.

36. Prausnitz, M.R., Bose, V.G., Langer, R.S., and Weaver, J.C., Electroporation of mammalian skin: a mechanism to enhance transdermal drug delivery, *Proc. Nat. Acad. Sci. U.S.A.* 90, 10504, 1993.

37. Coulberson, A.L., Hud, N.V., LeDoux, J.M., Vilfan, I.D., and Prausnitz, M.R., Gene packaging with lipids, peptides and viruses inhibits transfection by electroporation in vitro, *J. Control. Release* 86, 361, 2003.

38. Jaroszeski, M.J., Gilbert, R., Nicolau, C., and Heller, R., *In vivo* gene delivery by electroporation, *Adv. Drug Del. Rev.* 35, 131, 1999.

39. Zhang, L., Nolan, E., Kreitschitz, S., and Rabussay, D.P., Enhanced delivery of naked DNA to the skin by non-invasive *in vivo* electroporation, *Biochim. Biophys. Acta* 1572, 1, 2002.

40. Zhang, L., Li, L., Hofmann, G.A., and Hoffman, R.M., Depth-targeted efficient gene delivery and expression in the skin by pulsed electric fields: an approach to gene therapy of skin aging and other diseases, *Biochem. Biophys. Res. Comm.* 220, 633, 1996.

41. Glasspool-Malone, J., Somiari, S., Drabick, J.J., and Malone, R.W., Efficient nonviral cutaneous transfection, *Mol. Ther.* 2, 140, 2000.

42. Heiser, W.C., Gene transfer into mammalian cells by particle bombardment, *Anal. Biochem.* 217, 185, 1994.

43. Wang, J., Murakami, T., Hakamata, Y., Ajiki, T., Jinbu, Y., Akasaka, Y., Ohtsuki, M., Nakagawa, H., and Kobayashi, E., Gene gun-mediated oral mucosal transfer of interleukin 12 cDNA coupled with an irradiated melanoma vaccine in a hamster model: successful treatment of oral melanoma and distant skin lesion, *Cancer Gene Ther.* 8, 705, 2001.

44. Tanelian, D.L., Barry, M.A., Johnston, S.A., Le, T., and Smith, G., Controlled gene gun delivery and expression of DNA within the cornea, *Biotechniques* 23, 484, 1997.

45. Cheng, L., Ziegelhoffer, P.R., and Yang, N-S., *In vivo* promoter activity and transgene expression in mammalian somatic tissues evaluated by using particle bombardment, *Proc. Nat. Acad. Sci. U.S.A.* 90, 4455, 1993.

46. Fuller, D.H., Corb, M.M., Barnett, S., Steimer, K., and Haynes, J.R., Enhancement of immunodeficiency virus-specific immune responses in DNA-immunized rhesus macaques, *Vaccine* 15, 924, 1997.

47. Yang, N-S., Burkholder, J., Roberts, B., Martinell, B., and McCabe, D., *In vivo* and *in vitro* gene transfer to mammalian somatic cells by particle bombardment, *Proc. Nat. Acad. Sci. U.S.A.* 87, 9568, 1990.

48. Lu, B., Scott, G., and Goldsmith, L.A., A model for keratinocyte gene therapy: preclinical and therapeutic considerations, *Proc. Assoc. Am. Phys.* 108, 165, 1996.

49. Udvardi, A., Kufferath, I., Grutsch, H., Zatloukal, K., and Volc-Platzer, B., Uptake of exogenous DNA via the skin, *J. Mol. Med.* 77, 744, 1999.

50. Ghazizadeh, S., Harrington, R., and Taichman, L.B., *In vivo* transduction of mouse epidermis with recombinant retroviral vectors: implications for cutaneous gene therapy, *Gene Ther.* 6, 1267, 1999.

51. Lee, R.J. and Huang, L., Lipidic vector systems for gene transfer, *Crit. Rev. Ther. Drug Carr. Syst.* 14, 173, 1997.

52. Lu, B., Federoff, H.J., Wang, Y., Goldsmith, L.A., and Scott, G., Topical application of viral vectors for epidermal gene transfer, *J. Invest. Dermatol.* 108, 803, 1997.

53. Vogel, J.C., Nonviral skin gene therapy, *Hum. Gene Ther.* 11, 2253, 2000.

54. Schaller, M. and Korting, H.C., Interaction of liposomes with human skin: the role of the stratum corneum, *Adv. Drug Deliv. Rev.* 18, 303, 1996.

55. Schmidt, M-H. and Korting, H.C., Therapeutic progress with topical liposome drugs for skin disease, *Adv. Drug Deliv. Rev.* 18, 335, 1996.

56. Tomlinson, E. and Rolland, A.P., Controllable gene therapy: pharmaceutics of non-viral gene delivery systems, *J. Control. Release* 39, 357, 1996.

57. Cevc, G., Blume, G., Schätzlein, A., Gebauer, D., and Paul, A., The skin: a pathway for systemic treatment with patches and lipid-based agent carriers, *Adv. Drug Deliv. Rev.* 18, 349, 1996.

58. Lauer, A.C., Lieb, L.M., Ramachandran, C., Flynn, G.L., and Weiner, N.D., Transfollicular drug delivery, *Pharm. Res.* 12, 179, 1995.

59. Lauer, A.C., Ramachandran, C., Lieb, L.M., Niemiec, S., and Weiner, N.D., Targeted delivery to the pilosebaceous unit via liposomes, *Adv. Drug Deliv. Rev.* 18, 311, 1996.

60. Alexander, M.Y. and Akhurst, R.J., Liposome-mediated gene transfer and expression via the skin, *Hum. Mol. Gen.* 4, 2279, 1995.

61. Niemiec, S.M., Latta, J.M., Ramachandran, C., Weiner, N.D., and Roesller, B.J., Perifollicular trans-genic expression of human interleukin-1 receptor antagonist protein following topical application of novel liposome-plasmid DNA formulations *in vivo*, *J. Pharm. Sci.* 86, 701, 1997.

62. Birchall, J.C., Marichal, C., Campbell, L., Alwan, A., Hadgraft, J., and Gumbleton, M., Gene expression in an intact *ex vivo* skin tissue model following percutaneous delivery of cationic liposome-plasmid DNA complexes, *Int. J. Pharm.* 197, 233, 2000.

63. Nead, M.A. and McCance, D.J., Poly-L-ornithine-mediated transfection of human keratinocytes, *J. Invest. Dermatol.* 105, 668, 1995.

64. Raghavachari, N. and Fahl, W.E., Targeted gene delivery to skin cells in vivo: a comparative study of liposomes and polymers as delivery vehicles, *J. Pharm. Sci.* 91, 615, 2002.

65. Niemic, S.M., Hu, Z., Ramachandran, C., Wallach, D.F.H., and Weiner, N., The effect of dosing volume on the disposition of cyclosporine-a in hairless mouse skin after topical application of a non-ionic liposomal formulation: an *in vitro* diffusion study, *STP Pharma. Sci.* 4, 145, 1994.

66. Mumper, R.J., Wang, J., Klakamp, S.L., Nitta, H., Anwar, K., Tagliaferri, F., and Rolland, A.P., Protective interactive noncondensing (PINC) polymers for enhanced plasmid distribution and expression in rat skeletal muscle, *J. Control. Release* 52, 191, 1998.

67. Weiner, N., Lieb, L., Niemic, S., Ramachandran, C., Hu, Z., and Egbaria, K., Liposomes: a novel delivery system for pharmaceutical and cosmetic applications, *J. Drug Target.* 2, 405, 1994.

68. Prokop, A., Kozlov, E., Moore, W., and Davidson, J.M., Maximising the *in vivo* efficiency of gene transfer by means of nonviral polymeric gene delivery vehicles, *J. Pharm. Sci.* 91, 67, 2002.

69. Li, Z., Ning, W., Wang, J., Choi, A., Lee, P-Y., Tyagi, P., and Huang, L., Controlled gene delivery system based on thermosensitive biodegradable hydrogel, *Pharm. Res.* 20, 884, 2003.

70. Luu, Y.K., Kim, K., Hsiao, B.S., Chu, B., and Hadjiargyrou, M., Development of a nanostructured DNA delivery scaffold via electrospinning of PLGA and PLA-PEG block copolymers, *J. Control. Release* 89, 341, 2003.

71. Eliaz, R.E. and Szoka Jr, F.C., Robust and prolonged gene expression from injectable polymeric implants, *Gene Ther.* 9, 1230, 2002.

72. Eliaz, R.E. and Kost, J., Characterisation of a polymeric PLGA-injectable implant delivery system for the controlled release of proteins, *J. Biomed. Mater. Res.* 50, 388, 2000.

73. Eliaz, R.E., Wallach, D., and Kost, J., Delivery of soluble tumor necrosis factor receptor from *in situ* forming PLGA implants *in vivo*, *Pharm. Res.* 17, 1546, 2000.

74. Anwer, K., Earle, K.A., Shi, M., Wang, J., Mumper, R.J., Proctor, B., Jansa, K., Ledebur, H.C., Davis, S., Eaglstein, W., and Rolland, A.P., Synergistic effect of formulated plasmid and needle-free injection for genetic vaccines, *Pharm. Res.* 16, 889, 1999.

75. Hamajima, K., Sasaki, S., Fukushima, J., Kaneko, T., Xin, K.Q., Kudoh, I., Okuda, K., Intranasal administration of HIV-DNA vaccine formulated with a polymer, carboxymethylcellulose, augments mucosal antibody production and cell-mediated immune response, *Clin. Immunol. Immunopathol.* 88, 205, 1998.

76. Lima, K.M., Santos, S.A., Lima, V.M.F., Coelho-Castelo, A.A.M., Rodrigues Jr, J.M., and Silva, C.L., Single dose of a vaccine based on DNA encoding mycobacterial hsp65 protein plus TDM-loaded PLGA microspheres protects mice against a virulent strain of Mycobacterium tuberculosis, *Gene Ther.* 10, 678, 2003.

77. Walter, E., Moelling, K., Pavlovic, J., and Merkle. H.P., Microencapsulation of DNA using poly(DL-lactide-co-glycolide): stability issues and release characteristics, *J. Control. Release* 61, 361, 1999.

78. McKeever, U., Barman, S., Hao, T., Chambers, P., Song, S., Lunsford, L., Hsu, Y-Y., Roy, K., and Hedley, M.L., Protective immune response elicited in mice by immunization with formulations of poly(lactide-co-glycolide) microparticles, *Vaccine* 20, 1524, 2002.

79. Kuhn, U., Terunuma, A., Pfutzner, W., Foster, R.A., and Vogel, J., *In vivo* assessment of gene delivery to keratinocytes by lentiviral vectors, *J. Virol.* 76, 1496, 2002.

80. Jensen, T.G., Jensen, U.B., and Bolund, L., Production of retroviral vectors I primary keratinocytes after DNA-mediated gene transfer leads to prolonged gene expression, *Acta Derm. Venereol.* 83, 83, 2003.

81. Carroll, J.M. and Taichman, L.B., Characterization of the human involucrin promoter using a transient β-galactosidase assay, *J. Cell Sci.* 103, 925, 1992.

82. Carroll, J.M., Albers, K.M., Garlick, J.A., Harrington, R., and Taichman, L.B., Tissue- and stratum-specific expression of the human involucrin promoter in transgenic mice, *Proc. Nat. Acad. Sci. U.S.A.* 90, 10270, 1993.

83. Ghazizadeh, S., Doumeng, C., and Taichman, L.B., Durable and stratum-specific gene expression in epidermis, *Gene Ther.* 9, 1278, 2002.

84. Lin, M.T.S., Wang, F., Uitto, J., and Yoon, K., Differential expression of tissue-specific promoters by gene gun, *Brit. J. Dermatol.* 144, 34, 2001.

85. Byrne, C., Tainsky, M., and Fuchs, E., Programming gene expression in developing epidermis, *Development* 120, 2369, 1994.

86. Byrne, C. and Fuchs, E., Probing keratinocyte and differentiation specificity of the human K5 promoter *in vitro* and in transgenic mice, *Mol. Cell. Biol.* 13, 3176, 1993.

87. Page, S.M. and Brownlee, G.G., Differentiation-specific enhancer activity in transduced keratinocytes: a model for epidermal gene therapy, *Gene Ther.* 5, 394, 1998.

CHAPTER 37

Enhancement of Wound Repair by Sustained Gene Transfer via Hyaluronan Matrices

Angela P. Kim, Daniel M. Checkla, Don Wen, Philip Dehazya, and Weiliam Chen

CONTENTS

37.1 INTRODUCTION

Gene therapy is a promising method of treatment for various hereditary and acquired illnesses. It is often advantageous over other methods such as recombinant protein therapy. Recombinant proteins may be difficult to purify, can lack stability, and frequently necessitate long durations of

0-8493-1934-X/05/$0.00+$1.50
© 2005 by CRC Press LLC

administration. DNA vectors encoding therapeutic genes, on the other hand, can be isolated in large quantities with relative ease. Furthermore, direct expression of proteins within cells could circumvent many problems typically associated with the production of recombinant proteins, such as improper refolding and subunit assembly, contamination with endotoxin or host cell proteins, and enzymatic degradation during transport to the site of application. A particularly daunting challenge in the evolution of gene therapy is the development of a suitable DNA delivery vehicle. Gene transfer methods such as viral vectors, cationic lipids, and polycations have been investigated; however, many obstacles still have to be surmounted before their clinical potential can be fully realized.

Concerns over safety and regulatory hurdles remain the biggest shortcomings in the use of viral vectors as gene delivery vehicles. These include the potential for existence of replication-competent viruses and the induction of host inflammatory responses.[1-3] Though viral vectors are very efficient in gene transfer, potential inconsistencies could occur between the production batches of viruses; thus, creating guidelines for their therapeutic use has been problematic. The use of materials such as cationic lipids and polycations elicits fewer safety concerns than viral vectors; however, biocompatibility and toxicity issues must still be addressed before they can be effectively utilized for gene delivery. In addition, they may also fail to remain at the site of administration.

Synthetic, semi-synthetic, and natural polymers have the potential to serve as effective gene delivery vehicles that can overcome many of the obstacles involved with the other methods discussed.[1] Their high levels of biocompatibility reduce the problems associated with toxicity and inflammatory response. In addition, they can be manufactured in large quantities at high consistencies and can be formulated into many states, including gels, films, matrices, and microspheres, providing many different routes of administration. Previously, many synthetic and natural polymers such as poly-lactide-*co*-glycolide and collagen have been used as vehicles for DNA delivery.[4,5] However, carbohydrates, as a class of natural biocompatible polymers, have not hitherto been widely explored for DNA delivery.

Hyaluronic acid (HA), or hyaluronan, is a carbohydrate that has the potential to serve as an effective DNA delivery vehicle. HA is a viscous, high-molecular-weight glycosaminoglycan that is present in the skin, the eyes, the synovial fluids of the joints, and other tissues.[6-9] Moreover, it is also a major component of the extracellular matrix that is synthesized in the plasma membrane of fibroblasts and other cells.[6-9] HA is completely evolutionarily conserved, remaining virtually unmodified from species to species. It is also biocompatible, nontoxic, and nonimmunogenic, making it an excellent candidate for designing drug delivery vehicles.[10-12] Likewise, HA might also be used to formulate polymer vehicles for gene delivery.

HA plays an important role in wound healing and is a major component of the second provisional matrix formed during the healing process;[13] therefore, it has unique potential to serve as an effective vehicle to deliver therapeutic genes for the treatment of chronic wounds such as diabetic, venous, and pressure ulcers.[9,11] Chronic wounds do not follow the expected path of normal healing observed in acute wounds and are the leading causes of nontraumatic lower extremity amputations.[14,15] One possible cause of impaired healing is the reduction of growth factor expression, specifically that of platelet-derived growth factor (PDGF).[12,16] Recombinant PDGF protein has already been approved by the FDA to treat chronic neuropathic diabetic ulcers;[17,18] however, the clinical results produced thus far have not been optimal. Ineffective protein delivery and poor retention at the wound site are likely the contributory factors of below-optimal performance. Gene therapy promises to circumvent these limitations. Adenovirus, collagen, and poly-lactide-*co*-glycolide based vehicles have been used to deliver DNA encoding PDGF and have proven effective.[5,19-21] Nonetheless, adenovirus is cytotoxic and collagen has immunogenic potentials. More importantly, none of these vehicles resembles the second provisional matrix found at wound sites.

This study is the first to explore the feasibility of using HA as a vehicle for gene transfer and, in particular, its application to wound healing. By utilizing adipic dihydrazide as a cross-linker that converts water-soluble HA to an insoluble form (thereby entrapping DNA),[22] we have formulated

HA into a matrix with sustained release properties that is also capable of protecting DNA from enzymatic degradation. All reactants are nontoxic and any unreacted residuals can be easily extracted.[22] The HA matrix can be enzymatically degraded by hyaluronidase to release DNA over a prolonged period of time. The resemblance to the second provisional matrix of the scaffold–matrix configuration should aid in fibroblast migration and serve as a foundation for tissue regeneration. PDGF genes introduced by the HA matrix would be expressed by cells within the wound site and promote their proliferation, thus providing an effective therapy for chronic wound healing. We have demonstrated the efficacy of HA matrices to sustain the delivery of DNA capable of promoting expression of active proteins *in vitro* and the ability of PDGF-encoded DNA delivered via HA matrices to enhance cell proliferation in culture and in porcine full-thickness dermal wound models.

37.2 MATERIALS AND METHODS

37.2.1 Plasmid DNA Gene Constructs

A reporter gene construct, $_pCMV_1$ β-galactosidase, was purchased from Clontech (Palo Alto, CA) and a pFLAG–CMV-5a® mammalian expression vector containing a sequence encoding a C-terminal FLAG® fusion protein was purchased from Sigma-Aldrich (St. Louis, MO). Platelet Derived Growth Factor ββ (PDGF-ββ) cDNA was cloned into this vector. The β-galactosidase (β-gal) and PDGF-ββ–FLAG gene constructs were transformed into competent *E. coli* NovaBlue Cells® (Novagen, Madison, WI). Both constructs included sequences encoding ampicillin resistance and cells containing the constructs were selected by using 100 µg/ml carbenicillin (Novagen). Plasmid DNA was isolated in the milligram range using a Qiagen (Valencia, CA) EndoFree Giga Prep® plasmid isolation kit.

37.2.2 DNA–HA Matrix Formulations

β-gal–HA matrices and PDGF-ββ–FLAG–HA matrices with DNA loadings ranging from 0.5% to 2.5% (w/w) were prepared as previously described.[23] Briefly, DNA resuspended in Tris-EDTA (TE) was slowly stirred with a spatula into a 1% HA (Kraeber GMBH & Co., Waldhofstr, Germany) solution for an extended period of time to ensure for homogeneity. One mL of the DNA-HA mixture was deposited into an aluminum mold 1 cm in diameter. The DNA–HA mixture was then snap-frozen on dry ice and lyophilized.

The lyophilized DNA–HA matrices were cross-linked for conversion into insoluble states as previously described.[23] Briefly, they were submerged and incubated for 30 min in 90% N,N-dimethylformamide (DMF) (Sigma) and 10% water containing 1 mg/ml adipic dihydrazide (ADH) (Sigma). Ethyl-3-(3-dimethyl amino) propyl carbodiimide (EDCI) (Sigma) was added to the solution of ADH–DMF at a final concentration of 1.2 mg/ml. The solution was brought to a pH of 5 with 1 N hydrochloric acid to initiate the cross-linking reaction. The incubation times of matrices ranged from 4 to 25 h to impart different degrees of cross-linking. The reaction was terminated by transferring the matrices to a solution of 90% isopropanol (IPA) and 10% water. The matrices were subsequently rinsed with several portions of 90% IPA and 10% water and then washed overnight to remove residual cross-linking reagents. Following dehydration with 100% IPA, the matrices were placed under a vacuum to remove all residual solvents.

37.2.3 Matrix Morphology Assessment

HA matrices with and without DNA were hydrated in 0.02% ethidium bromide solutions. The extent of fluorescence was observed under UV light.

The effect of hyaluronidase degradation on the structure of the DNA–HA matrices was determined. Matrices before and after a 7-day incubation in 10 units/ml hyaluronidase dissolved in phosphate-buffered saline (PBS) (pH of 7.4) were compared by observing scanning electron microscopy (SEM) images.

37.2.4 Assessment of DNA Release Kinetics

DNA–HA matrices with DNA encoding β-gal (0.5% or 2.5% loadings and 5 or 25 h cross-linking times, respectively) were incubated with 10 units/ml hyaluronidase (in pH 7.4 PBS) to compare the release kinetics of different DNA loadings and degrees of cross-linking (previously described[23]). The matrices were incubated at 37°C with gentle agitation and at stipulated time points, they were centrifuged, supernatants were collected, and matrices were replenished with fresh hyaluronidase (at an equivalent volume to that removed). As controls, DNA–HA matrices were incubated in plain PBS. A similar study was conducted in which the DNA–HA matrices were bisected prior to the initial incubation in either hyaluronidase or PBS.

The concentrations of DNA in the release samples were determined using a Picogreen fluorometric assay, described elsewhere,[24] in which Picogreen dye (Molecular Probes, Eugene, OR) was diluted 200-fold in TE buffer and mixed at a 1:1 ratio with each DNA release sample. Fluorescence was measured (480 nm excitation wavelength and 530 nm emission wavelength) and referenced against a standard plot of DNA concentrations ranging from 0 to 1 µg/ml.

37.2.5 Gel Electrophoresis Analyses

DNA release samples were analyzed via gel electrophoresis on a 0.8% agarose gels prepared throughout the release study. Gel electrophoresis was also conducted to compare mixtures of 1% HA solution and DNA (either 0.5% or 2.5% loadings, respectively) with samples of minced DNA–HA matrices (either 0.5% or 2.5% DNA loadings) that were not subjected to release conditions but rather briefly hydrated in TE buffer.

To assess the capability of cross-linked HA to protect DNA from enzymatic degradation, DNA–HA matrices were incubated in fetal bovine serum (FBS) (Mediatech, Inc., Herndon, VA) as previously described.[23] After several washes with TE buffer, the matrices were subjected to hyaluronidase degradation. Samples were collected after 5 days of incubation and were analyzed on an agarose gel.

37.2.6 Transfections Using Release Samples with DNA Encoding β-Galactosidase

COS-1 cells (ATCC, Rockville, MD) amenable to transfection were cultured at 37°C and 5% CO_2. They were sustained in Dulbecco's Modified Eagle Medium (DMEM) supplemented with 10% FBS and 1% penicillin-streptomycin (all of which were supplied by Mediatech). Cells were passaged biweekly.

COS-1 cells were transfected with release samples collected from DNA–HA matrices with either 0.5% or 2.5% β-gal DNA loadings (cross-linked for 5 and 25 h, respectively). Specific activities of β-gal were assessed as described previously.[23] Briefly, cells were seeded in 24 well plates at 2×10^4 cells per well. When 60% confluence was attained, cells were incubated for 18 h with transfection mixtures containing 30 µl of release sample per well using a 1:3 ratio of µg DNA:µl Fugene reagent (Molecular Probes). The transfection mixtures were replaced with DMEM supplemented with 10% FBS. After 3 days of incubation, cells were lysed and collected, and the β-gal enzymatic activities were determined using a High Sensitivity β-Gal Assay Kit (Stratagene, Cedar Creek, TX) at 570 nm on a microplate reader (Dynatech, MR5000, Dynatech Laboratories, Chantilly, VA). Total protein content of the cell lysates was measured using a BCA protein assay reagent kit (Pierce, Rockland, IL) and the enzymatic activity results were reported in units of activity per milligram of protein. Controls for comparing transfection efficiency and enzyme expression were carried out using naked plasmid DNA with concentrations equivalent to those of the release samples.

37.2.7 Transfections Using Release Samples with DNA Encoding PDGF-ββ–FLAG

COS-1 cells were transfected with PDGF-ββ–FLAG release samples collected from DNA–HA matrices that were cross-linked for 4 h and contained either 0.5% or 2% DNA loadings following the same transfection procedures described above. After 3 days of incubation, cells were fixed and immunostained as previously described.[23] Briefly, cells were fixed with a 1:1 ratio of cold acetone to methanol and incubated in a mouse anti-FLAG monoclonal antibody (Sigma) followed by a rabbit anti-mouse Alexa Fluor® 488 labeled secondary antibody (Molecular Probes). Immunofluorescently stained cells were observed at 488 nm excitation and 530 nm emission wavelengths under a fluorescent microscope (Nikon TE300 Eclipse Inverted Microscope, Melville, NY).

A Quantikine® PDGF-ββ receptor-based enzyme-linked immunosorbent assay (ELISA) kit (R & D Systems, Minneapolis, MN) was used to quantitatively measure PDGF-ββ protein expressed by COS-1 cells that were transfected with release samples collected from DNA–HA matrices. The ELISA kit utilized the chimeric PDGF-β receptor, which had specificity to the β subunit, to particularly identify the PDGF-ββ dimeric, disulfide-linked form of PDGF.[25] Along with quantitatively measuring PDGF-ββ expression, the ELISA assay allowed for assessment of whether or not the protein was functionally active with regard to receptor binding. Cell lysates and conditioned media were both analyzed for their PDGF-ββ contents using a microplate reader as previously described.[23] Transfections using β-gal release samples were conducted as controls.

The efficacy of PDGF-ββ–FLAG release samples to promote proliferation of Normal Neonatal Human Dermal Fibroblasts (NNHDF) (Clonetics, Walkersville, MD) was investigated. NNHDF were cultured at 37°C and 5% CO_2 and were grown in FGM-2 BulletKits® (Clonetics) media containing 2% FBS and other supplements (insulin, FGF, and gentamicin) and were passaged biweekly. As described earlier,[23] PDGF-ββ–FLAG-conditioned media containing a reduced FBS content of 0.5% and no supplements were collected from COS-1 cells that were previously transfected with DNA release samples. Cells were fluorescently immunostained to confirm the expression following procedures described above. The recovered conditioned media were diluted in equal volumes of FGM-2 BulletKits® (containing 0.5% FBS and no supplements) and were transferred to NNHDF cells seeded in 96 well plates at 1.5×10^3 cells per well on the previous day. Cells were harvested daily for the following 5 days and were frozen at −70°C. Cells were lysed and assayed for their DNA contents using Hoescht 33342 (Molecular Probes) to indicate proliferation. The lysates were scanned at an excitation wavelength of 360 nm and an emission wavelength of 460 nm using a Cytofluorometer (Cytofluor II, Biosearch, Bedford, MA). Measurements were made in cells per cm^2 by referencing a standard plot of cells seeded at densities between 0 and 5×10^4 cells per well in 96 well plates with a surface area of 0.32 cm^2 per well. Cells transfected with naked plasmid DNA encoding β-gal and untreated cells (no DNA) were used as controls.

37.2.8 *In Vivo* Wound Healing Studies

A porcine model for full-thickness dermal wounds was used for *in vivo* experiments to test the efficacy of HA matrices containing DNA encoding PDGF-ββ to promote enhanced wound healing. Pigs were placed under anesthesia, shaved, and then scrubbed with 70% IPA and betadine solutions. Full-thickness wounds were created using 8 mm biopsy punches. HA matrices and DNA–HA matrices containing DNA encoding β-gal and PDGF-ββ were implanted into the wounds, followed by the application of Tegaderm for closure. The wound sites were harvested after 4 days, and paraffin embedded sections were prepared. The sections were stained with hematoxylin and eosin and their percentages of granulation tissue formation were evaluated following previously described methods.[26] Briefly, sections were coded numerically by a participant blinded to the treatments and digital images were captured under a Nikon Labophot microscope (Melville, NY) with a Sony DXE950P camera (New York, NY). NIH Image Analysis software was used to trace the total wound area and the area covered by granulated tissue. The degree of granulation was expressed as a percentage and calculated using the following formula:

$$\% \text{ granulation } = \text{ granulated tissue area } / \text{ total wound area } \times 100 \qquad (37.1)$$

37.3 RESULTS AND DISCUSSION

37.3.1 Matrix Morphology

Figure 37.1A depicts DNA–HA matrices approximately 1 cm in diameter and 3 mm thick (weighing between 10 and 11 mg). The size and mass of the matrices can be altered accordingly by changing the size of the mold and the volume of 1% HA solution used in the initial steps of matrix formulation.

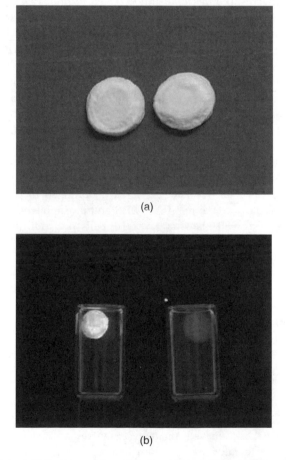

(a)

(b)

Figure 37.1 (A) Cross-linked DNA–HA matrices. (B) Cross-linked HA matrices with DNA (left) and without DNA (right), incubated in 0.02% ethidium bromide. (Adapted from Kim, A. et al., Characterization of DNA-hyaluronan matrix for sustained gene transfer, *J. Control. Release* 90, 81, 2003.)

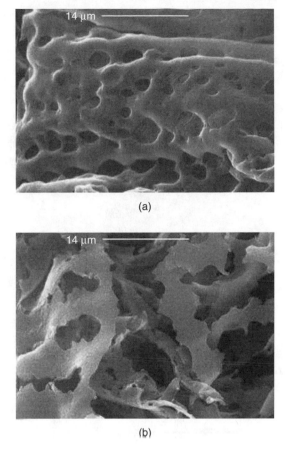

(a)

(b)

Figure 37.2   SEM images of DNA–HA matrices (A) prior to incubation in hyaluronidase solution, and (B) after 7 day incubation in hyaluronidase (10 units/ml). (Adapted from Kim, A. et al., Characterization of DNA-hyaluronan matrix for sustained gene transfer, *J. Control. Release* 90, 81, 2003.)

In Figure 37.1B, the cross-linked DNA–HA matrix on the left and the cross-linked HA matrix on the right (containing no DNA) were incubated in 0.02% ethidium bromide. Fluorescence observed in the matrix containing DNA indicated that the DNA remained entrapped within the matrix.

Figure 37.2A is a SEM image of a DNA–HA matrix before exposure to hyaluronidase and Figure 37.2B is an image of a matrix after 7 days of degradation in hyaluronidase. The DNA–HA matrices were very porous prior to treatment with hyaluronidase and the pores appeared to be smoother and smaller than those of the degraded matrix. Hyaluronidase evidently eroded the matrix at the pore sites by roughening the pore edges and increasing the pore sizes.

37.3.2 Assessment of DNA Release Kinetics

Figure 37.3A depicts the cumulative DNA release kinetics of DNA–HA matrices with 2.5% loadings of DNA encoding β-gal, and cross-linked for either 5 or 25 h, respectively. DNA–HA matrices were either incubated in hyaluronidase solution (10 units/ml in pH 7.4 PBS) or in plain pH 7.4 PBS. During the initial days of release (days 1, 4, and 9), minute amounts of DNA were eluted from matrices incubated in both hyaluronidase solution and PBS. For matrices subjected to hyaluronidase degradation, considerable amounts of DNA were detected beginning on day 16; however, only small amounts of DNA continued to elute from those incubated in PBS. Despite the appearance that less DNA was released into hyaluronidase solution from matrices cross-linked for

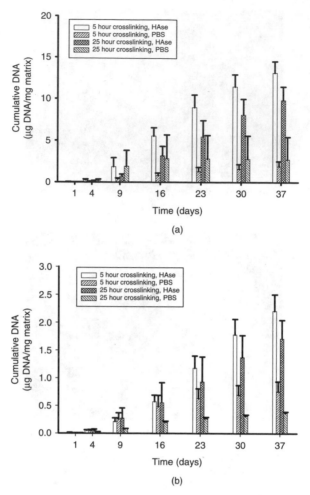

Figure 37.3 Intact DNA–HA matrices incubated in either hyaluronidase (HAse) or PBS solutions. Matrices were cross-linked for either 5 or 25 h. (A) 2.5% loadings of DNA encoding β-galactosidase. (B) 0.5% loadings of DNA encoding β-galactosidase. Each data point represents the average of at least 3 replicates ± SEM. (Modified from Kim, A. et al., Characterization of DNA-hyaluronan matrix for sustained gene transfer, *J. Control. Release* 90, 81, 2003.)

25 h than from those cross-linked for 5 h, there were no statistically significant differences between their DNA levels (p > 0.05 for all data points). The differences between the amounts of DNA released into the PBS from matrices cross-linked for either 5 h or 25 h were also statistically insignificant (p > 0.05 for all data points). Similar results were observed on the DNA release profiles of DNA–HA matrices with 0.5% loadings of DNA encoding β-gal (subjected to either 5 or 25 h of cross-linking) and are displayed in Figure 37.3B. Another release study was conducted on bisected matrices with DNA encoding β-gal, cross-linked for either 5 or 25 h, in order immediately to expose their interiors to the hyaluronidase or PBS solutions. The results are displayed in Figures 37.4A (2.5% DNA loadings) and 37.4B (0.5% DNA loadings). Overall, larger amounts of DNA were released from bisected matrices incubated in hyaluronidase solution as compared to intact matrices (see Figure 37.3). By the completion of the study (day 37), bisected matrices with 2.5% DNA loadings, cross-linked for 5 h, released 66% of their total DNA contents, compared to 53% of the total DNA contents released from the corresponding intact matrices. In addition, substantial release of DNA from bisected matrices occurred at an earlier time point than from intact matrices. For example, when incubated in hyaluronidase solution, considerable levels of DNA were detected

by day 7 for bisected matrices with 2.5% DNA loadings (cross-linked for either 5 or 25 h, respectively). Comparable DNA levels were not detected until day 16 for their intact matrix counterparts. Another distinct characteristic of the release kinetics of bisected DNA–HA matrices was the substantial amounts of DNA released when the matrices were subjected to PBS incubation; whereas only small amounts of DNA were eluted from their intact counterparts. In addition, more DNA appeared to have eluted from matrices subjected to 5 h of cross-linking when incubated in PBS, as compared to those subjected to 25 h of cross-linking, though no statistically significant differences were detected ($p > 0.05$ for all points). The differences between DNA levels released from bisected matrices incubated in hyaluronidase (either 5 or 25 h cross-linking times, respectively) were also statistically insignificant ($p > 0.05$ for all points). Similar results were observed on the release profiles for the DNA-HA matrices with 0.5% loadings of DNA encoding β-gal that were cross-linked for either 5 or 25 h, respectively.

The slow initial release of DNA in both PBS and hyaluronidase of intact matrices and the rapid initial release of DNA upon exposure of the interiors of the bisected matrices could be attributed to different degrees of cross-linking throughout the matrices. This was the result of greater exposure of the matrix exteriors (i.e., the shells) to cross-linking reagent; whereas the interiors (i.e., the cores) were less cross-linked due to the difficulty of cross-linking reagents to penetrate the matrix structure. Thus, the uneven distribution of reagents generated a gradient of decreasing degrees of cross-linking from the exteriors to the interiors of the matrices. Figure 37.5 depicts a cross-sectional matrix model illustrating this heterogeneous cross-linking, in which the darker shades represent the more cross-linked portions (i.e., the exterior) and the lighter shades represent the less cross-linked portions of the core.

When exposed to hyaluronidase solutions, the release of DNA from cross-linked matrices was caused by a combined effect of enzymatic erosion and hydrolysis of the hydrazide linkages, whereas in the absence of hyaluronidase (i.e., plain PBS), the latter was the mechanism of matrix degradation. When the less cross-linked matrix cores were exposed to either hyaluronidase solution or PBS (via bisection), only minimal enzymatic degradation of the matrix structures and/or hydrolysis was sufficient for the release of significant amounts of DNA, thus accounting for the rapid initial release and the overall comparable amounts of DNA released from bisected matrices cross-linked for 5 h (observed in Figure 37.4). In addition, less DNA was released from matrices cross-linked for 25 h when exposed to PBS than their counterparts cross-linked for 5 h because the greater penetration of cross-linking reagents caused their cores to be more resistant to erosion and hydrolysis. Similarly, the exteriors of the matrices were more cross-linked and therefore very resistant to hyaluronidase digestion and hydrolysis. Substantial erosion on the matrices' exterior structures would have to occur before significant amounts of DNA could be released, which resulted in the slow initial release of DNA from intact matrices (see Figure 37.3). Hydrolysis was considerably less efficient than enzymatic degradation in matrix erosion; consequently, larger amounts of DNA were released from the DNA–HA matrices incubated in hyaluronidase solution as compared to those incubated in PBS.

Apparently, the heterogeneous cross-linking pattern of the DNA–HA matrices, in which the cores were less cross-linked than their exteriors, could be advantageous. If the matrices were homogeneously cross-linked, hyaluronidase erosion would cause a gradual decrease in surface area, leading to a continual diminishing of the amounts of DNA released. However, in the presence of gradient cross-linking, the ever-diminishing total surface area for DNA delivery could be compensated for by the ever-increasing rate of DNA elution. This could account for the almost linear rate of DNA release kinetics observed in Figure 37.3, when the intact matrices were incubated in hyaluronidase solution.

37.3.3 Gel Electrophoresis Analyses

DNA–HA mixtures with either 0.5% or 2.5% loadings of DNA encoding β-gal in 1% HA solutions, respectively, were prepared, and electrophoresis was conducted on a 0.8% agarose gel.

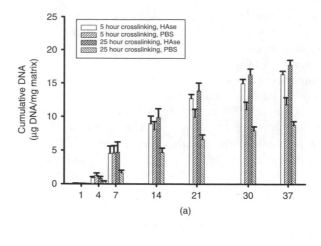

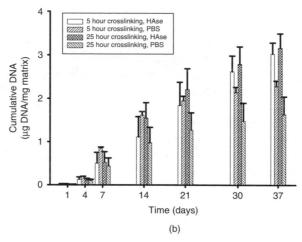

Figure 37.4 Bisected DNA–HA matrices incubated in either hyaluronidase (HAse) or PBS solutions. Matrices were cross-linked for either 5 or 25 h. (A) 2.5% loadings of DNA encoding β-galactosidase. (B) 0.5% loadings of DNA encoding β-galactosidase. Each data point represents the average of at least 3 replicates ± SEM. (Modified from Kim, A. et al., Characterization of DNA-hyaluronan matrix for sustained gene transfer, *J. Control. Release* 90, 81, 2003.)

Figure 37.5 Cross-section of DNA–HA matrix model of cross-linking. Darker shading represents more cross-linked portions and lighter shading represents less cross-linked portions.

The results are depicted in Figure 37.6A. The solutions created intense fluorescence in the wells of the gel caused by trapped DNA. The DNA that did pass through the well appeared in smeared fluorescent trails and did not migrate far. These results suggested that DNA formed a tight association with HA that inhibited it from passing freely through the pores of the agarose gel. HA is heterogeneous, consisting of different molecular weights, and the DNA that remained in the wells

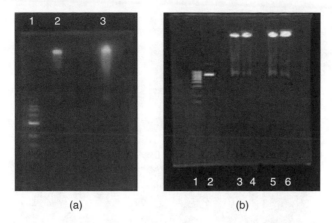

(a)   (b)

Figure 37.6 (A) Gel electrophoresis analyses of DNA–HA solutions: (1) DNA molecular weight marker, (2) 0.5% DNA loading, (3) 2.5% DNA loading. (B) Gel electrophoresis analysis of minced DNA–HA matrices: (1) DNA molecular weight marker, (2) naked plasmid DNA, (3, 4) 0.5% DNA loadings, and (5, 6) 2.5% DNA loadings. (Adapted from Kim, A. et al., Characterization of DNA-hyaluronan matrix for sustained gene transfer, *J. Control. Release,* 90, 81, 2003.)

was likely associated with HA molecules of high molecular weights; whereas the DNA that migrated and formed the fluorescent trails was likely DNA associated with HA molecules of lower molecular weights.

Figure 37.6B displays the gel electrophoresis results of minced cross-linked DNA–HA matrices with either 0.5% (lanes 3 and 4) or 2.5% (lanes 5 and 6) loadings of DNA encoding β-gal that were hydrated in TE. A portion of the DNA from the minced matrices migrated freely along the gel, forming a trail of fluorescence that led to bands located in the regions of supercoiled and uncoiled naked plasmid DNA (lane 2). Also, intense fluorescence was observed in the wells, caused by DNA–HA fragments that were too large to pass through the pores of the gel, yet were still affected by the electrical current and caused well deformation. The fluorescent trails were formed by DNA associated with or entrapped in HA fragments of various molecular weights; whereas the fluorescent bands were likely formed by DNA associated with very small HA fragments of various molecular sizes.

Figure 37.7 depicts the gel electrophoresis results obtained from both intact and bisected DNA–HA matrices. Figure 37.7A shows the representative results using DNA encoding β-gal that was released from intact matrices with 2.5% DNA loadings that were cross-linked for either 5 or 25 h, respectively (see Figure 37.3A for corresponding DNA release profiles). For the initial days of release (days 1 and 4), no visible amounts of DNA appeared on the gels (not shown). Lanes 1, 3, 5, 7, and 9 are the DNA samples released from matrices cross-linked for 5 h; lanes 2, 4, 6, 8, and 10 are the DNA samples released from the matrices cross-linked for 25 h. Visible amounts of DNA began to appear on day 9. Fluorescence was observed in the wells due to the presence of DNA entrapped in the cross-linked HA fragments that were too large to migrate through the gel pores. The amounts of entrapped DNA from matrices subjected to 5 h of cross-linking were less than observed for their counterparts subjected to 25 h of cross-linking. The intensities of fluorescence in the wells increased considerably by day 16. The presence of fluorescent trails indicated that some of the DNA released was entrapped in cross-linked HA fragments, yet their sizes were sufficiently small to migrate through the gel pores. Fluorescent bands likely consisted of free DNA and DNA associated with HA fragments of very low molecular weights.

Figure 37.7B summarizes the gel electrophoresis results using DNA encoding β-gal that was released from bisected DNA–HA matrices with 2.5% DNA loadings (see Figure 37.4A for corresponding DNA release profiles). Lanes 1, 3, 5, 7, 9, and 11 are representative DNA samples from bisected matrices cross-linked for 5 h; lanes 2, 4, 6, 8, 10, and 12 are representative samples from

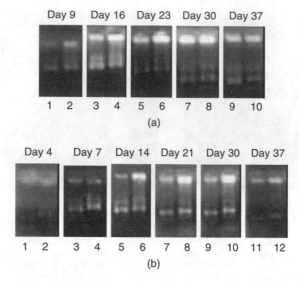

Figure 37.7   Gel electrophoresis analysis of representative DNA release samples from DNA–HA matrices. (A) DNA released from intact DNA–HA matrices, (1, 3, 5, 7, 9) from matrices with 2.5% DNA loadings subjected to 5 h of cross-linking, and (2, 4, 6, 8, 10) from matrices with 0.5% DNA loading subjected to 25 h of cross-linking. (B) DNA released from bisected DNA–HA matrices, (1, 3, 5, 7, 9, 11) from matrices with 2.5% DNA loadings subjected to 5 h of cross-linking, and (2, 4, 6, 8, 10, 12) from matrices with 2.5% DNA loadings subjected to 25 h of cross-linking. (Adapted from Kim, A. et al., Characterization of DNA-hyaluronan matrix for sustained gene transfer, *J. Control. Release*, 90, 81, 2003.)

matrices cross-linked for 25 h. Trace amounts of DNA were detected on day 1 (not shown) and clearly visible amounts of DNA appeared on the gel by day 4. Intense bands were observed by day 7, indicating the release of large amounts of DNA. Considerable amounts of DNA were entrapped by large cross-linked HA fragments (as indicated by the fluorescing wells). However, the fluorescent intensities were substantially less than those observed for intact matrices for comparable samples. Due to the immediate exposure of the less cross-linked interiors of the bisected matrices, more free DNA was released at an earlier time point than that observed for intact matrices. After most of the less cross-linked cores were eroded, the DNA released was primarily from the more cross-linked regions and was more likely entrapped by cross-linked HA fragments, resulting in the intense fluorescence in the wells of samples collected on days 14, 21, and 30. Similar to the observations of intact matrices, the levels of entrapped DNA released from bisected matrices cross-linked for 5 h were less than their counterparts cross-linked for 25 h. These results indicated that matrices cross-linked for longer times (i.e., 25 h) and the more intensively cross-linked portions (i.e., shells) released more entrapped DNA than did matrices cross-linked for shorter times (i.e., 5 h) and the less cross-linked portions (i.e., cores).

Protection studies were conducted to test the potential of the HA matrices to protect the entrapped DNA from enzymatic degradation by incubating DNA–HA matrices in FBS, which has known DNA degrading properties. The matrices were washed several times with PBS and exposed to hyaluronidase degradation. Release samples were collected after 5 days of incubation and gel electrophoresis was conducted. Results are displayed in Figure 37.8. Intact DNA was observed on the gel, concluding that HA provides a mechanism for DNA protection from enzymatic degradation.

37.3.4   β-Galactosidase Release Sample Transfections

β-Gal specific activities were measured for COS-1 cells transfected with release samples from DNA–HA matrices with either 0.5% or 2.5% DNA loadings and cross-linked for 5 or 25 h,

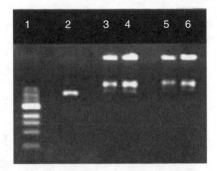

Figure 37.8 Gel electrophoresis analysis of DNA released from DNA–HA matrices previously incubated in serum. (1) DNA molecular weight marker; (2) naked plasmid DNA; (3,4) DNA released from DNA–HA matrices with 2.5% loadings and cross-linked for 5 h; (5,6) DNA released from DNA–HA matrices with 2.5% DNA loadings and cross-linked for 25 h.

respectively. The results are depicted in Figures 37.9 and 37.10. To facilitate comparison, the corresponding DNA concentrations of the release samples used for transfection were presented in the same figures. Transfections using naked plasmid DNA samples with concentrations comparable to those of the release samples were conducted as controls. However, the concentrations of DNA were very high and resulted in excessive cell deaths. Therefore, the naked plasmid DNA solutions were appropriately diluted and resulted in specific activities that mirrored the DNA concentrations, demonstrating the correlation between DNA concentration, transfection efficiency, and protein expression.

Specific enzymatic activity results using release samples from intact DNA–HA matrices with 2.5% DNA loadings are depicted in Figures 37.9A (5 h cross-linking) and 37.9B (25 h cross-linking). For matrices cross-linked for 5 h, a relatively good correlation between enzymatic activities and DNA concentrations was observed. However, despite the comparable DNA levels of samples from matrices cross-linked for 5 h and those cross-linked for 25 h, no trivial correlation between enzymatic activities and DNA concentrations could be deduced for the transfection studies performed using the DNA samples derived from DNA–HA matrices subjected to 25 h of cross-linking. Substantial levels of β-gal activities were not detected until the conclusion of the study (days 30 and 37). There were obvious discrepancies between the amounts of DNA in the samples and their transfection efficiencies, in which the discrepancy was larger for the matrices cross-linked to a greater extent (i.e., 25 h). Similar results were observed when DNA samples released from DNA–HA matrices with 0.5% loadings were used for transfection. These results are depicted in Figures 37.9C (5 h cross-linking) and 37.9D (25 h cross-linking). Correlations between DNA concentrations and enzymatic activities were observed for matrices cross-linked for both 5 and 25 h. However, the β-gal activity levels detected from transfection studies utilizing release samples from matrices cross-linked for 25 h were minute.

A similar study was conducted using DNA released from bisected DNA–HA matrices. The enzymatic activities from transfections of COS-1 cells using DNA samples derived from bisected matrices with 2.5% DNA loadings are depicted in Figures 37.10A (5 h cross-linking) and 37.10B (25 h cross-linking). The levels of β-gal activity for samples from bisected matrices subjected to 5 h of cross-linking measured nearly 8-fold higher than those observed for intact matrices (see Figure 37.9A). In addition, there were strong correlations between enzymatic activities and sample DNA concentrations. Despite their comparable sample DNA concentrations, the enzymatic activities measured from cells transfected with DNA released from matrices cross-linked for 25 h were substantially lower compared to their counterparts cross-linked for 5 h. Similar to the results obtained using intact DNA–HA matrices cross-linked for 25 h (see Figure 37.9B), there was no trivial correlation between enzymatic activities and DNA concentrations. However, for the bisected matrices, substantial amounts of expression were observed at an earlier time point and continued

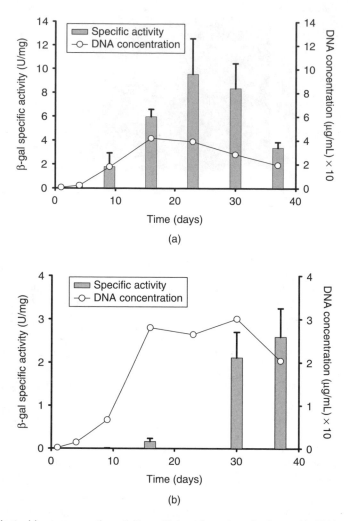

Figure 37.9 β-Galactosidase enzymatic activities obtained from transfections with DNA released from intact DNA–HA matrices. (A) 2.5% DNA loadings, 5 h cross-linking. (B) 2.5% DNA loadings, 25 h cross-linking. (C) 0.5% DNA loadings, 5 h cross-linking. (D) 0.5% DNA loadings, 25 h cross-linking. Each bar represents the average of six replicates ± SEM. Note the variations in scale on the y axes. (Modified from Kim, A. et al., Characterization of DNA-hyaluronan matrix for sustained gene transfer, *J. Control. Release*, 90, 81, 2003.)

throughout the study. Similar results were obtained from β-gal enzymatic activity studies utilizing release samples from bisected DNA–HA matrices that contained 0.5% loadings of DNA encoding β-gal and are displayed in Figures 37.10C (5 h cross-linking) and 37.10D (25 h cross-linking). Similar to the observations for intact matrices, the specific activity levels were minute for matrices cross-linked for 25 h.

As seen previously, gel electrophoresis analyses (see Figure 37.7A) revealed that much of the DNA released from intact DNA–HA matrices was associated with large fragments of HA (DNA that remained trapped in the wells of the agarose gel) and that more DNA released from matrices cross-linked for 25 h was entrapped in HA fragments than from its counterpart cross-linked for 5 h. The discrepancy between DNA concentrations and β-gal enzymatic activities indicated that the entrapped DNA was incapable of transfection. For matrices cross-linked for 25 h, large amounts of DNA were entrapped in HA fragments. However, by day 30 of the release studies (see Figure 37.9B), prolonged enzymatic degradation of HA matrices caused the release of more free and thus

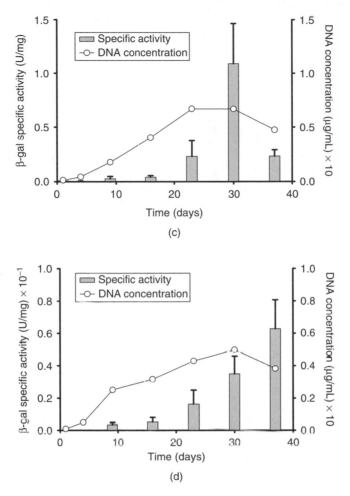

Figure 37.9 (Continued)

transfection-capable DNA (indicated by distinct fluorescent bands), resulting in substantial specific activity levels.

Results from gel electrophoresis showed that less of the DNA released from bisected matrices was entrapped compared to their counterparts that were intact. DNA–HA matrices subjected to 25 h of cross-linking released substantially larger amounts of entrapped DNA than those cross-linked for 5 h. This was due to the immediate exposure of the less cross-linked cores of the bisected matrices to hyaluronidase degradation. The DNA released initially consisted of more free DNA than that associated with HA fragments. Thus, more transfection-capable DNA was released at an earlier time point and in greater quantities, resulting in elevated specific activity levels compared to intact matrices. The matrices cross-linked for 25 h necessitated more extensive hyaluronidase degradation and released less free DNA than matrices cross-linked for 5 h, resulting in a reduction in activity and the discrepancy between β-gal activities and DNA concentrations. Hence, it was apparent that the bisected and less cross-linked matrices (i.e., 5 h) released the least amount of HA-entrapped DNA and resulted in the greatest gene expression over a prolonged period of time.

37.3.5 PDGF-ββ–FLAG Release Sample Transfections

Images from immunofluorescent staining of COS-1 cells transfected with PDGF-ββ–FLAG DNA released from DNA–HA matrices cross-linked for 4 h with either 0.5% or 2% DNA loadings,

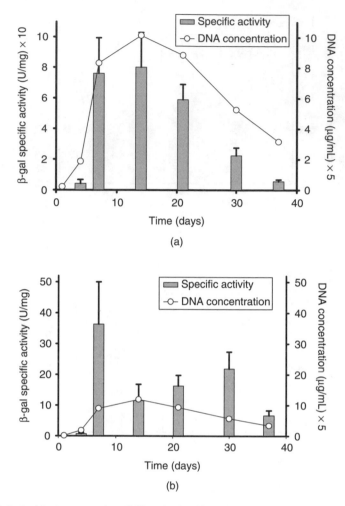

Figure 37.10   β-Galactosidase enzymatic activities obtained from transfections with DNA released from bisected
DNA–HA matrices. (A) 2.5% DNA loadings, 5 h cross-linking. (B) 2.5% DNA loadings, 25 h cross-
linking. (C) 0.5% DNA loadings, 5 h cross-linking. (D) 0.5% DNA loadings, 25 h cross-linking.
Each bar represents the average of six replicates ± SEM. Note the variations in scale on the y
axes. (Modified from Kim, A., et al., Characterization of DNA-hyaluronan matrix for sustained
gene transfer, *J. Control. Release*, 90, 81, 2003.)

respectively, are shown in Figure 37.11. Because an anti-FLAG antibody was utilized, the fluorescent staining represented only the PDGF-ββ–FLAG expressed by the cells and not endogenous
sources of PDGF-ββ. Samples were selected to represent time points at the beginning (day 7),
middle (day 17), and end (day 41) of the study. Cells that were transfected with samples from
matrices with 0.5% DNA loadings are displayed in Figures 37.11A to 37.11C, and cells that were
transfected with DNA released from matrices with 2% DNA loadings are displayed in Figures
37.11D to 37.11F. For both types of matrices, the extent of PDGF-ββ–FLAG expression appeared
to be greatest during the middle of the study and least at the end of the study.

A receptor-based ELISA assay kit was used to measure PDGF-ββ–FLAG expression quantitatively and to test whether it was functionally active (in terms of receptor-binding ability) in COS-1
cells transfected with DNA released from matrices with 2% DNA loadings and cross-linked for 4
h. These results are depicted in Figure 37.12. Control transfections without DNA ("No DNA") and
using plasmid DNA encoding β-gal revealed that the cells produced little or no endogenous PDGF-
ββ. The results of the ELISA assay indicated that the protein was functionally active in that it bound

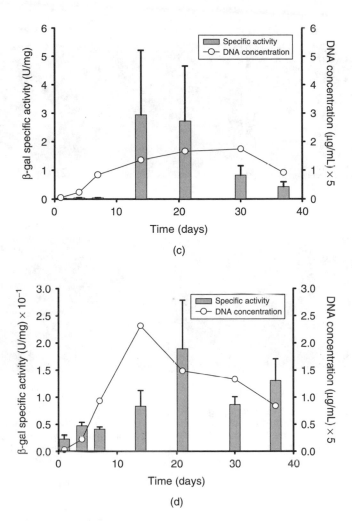

Figure 37.10   (Continued)

to the PDGF-β receptor. Expression was the greatest for samples collected at day 10 of release, compared to the other 2 time points selected (days 1 and 34) and expression was the least at day 1. Only small amounts of PDGF-ββ were detected at day 1 because erosion of the DNA–HA matrices would not have been significant enough to release sufficient amounts of DNA for efficient transfection. Since the DNA construct encoding PDGF-ββ–FLAG contained a membrane binding domain, more of the protein remained within the cell membranes than was secreted into the media and as expected, the levels of PDGF-ββ expression were greater in the lysates than the supernatants.

The PDGF-ββ–FLAG expressed by COS-1 cells was tested for its efficacy in promoting an enhanced proliferation of NNHDF cells. Release samples from matrices cross-linked for 4 h that contained 2% loadings of DNA encoding PDGF-ββ–FLAG were utilized to transfect COS-1 cells. Samples were selected from 3, 10, 17, and 24 days of release. The conditioned media was transferred to NNHDF cells and proliferation was measured using Hoescht 33342. The results are depicted in Figure 37.13. PDGF-ββ–FLAG promoted a statistically significant increase in proliferation compared to the negative controls of naked DNA encoding β-gal and of no DNA ($p < 0.05$ for all time points). Proliferation appeared to have reached a plateau by day 3, likely due to depletion of nutrients in the media with the serum content fourfold less than the standard NNHDF growth media. Naked plasmid DNA transfections using DNA encoding PDGF-ββ–FLAG were also conducted and

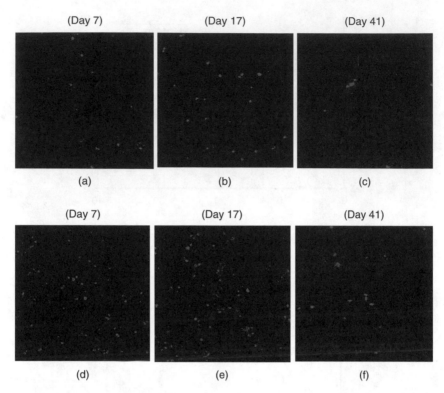

Figure 37.11 **(See color insert following page 336)** Immunofluorescent staining of COS-1 cells transfected with DNA encoding PDGF-ββ–FLAG that was released from DNA–HA matrices, with 0.5% DNA loadings at (A) day 7, (B) day 17, and (C) day 41 of release, or with 2.5% DNA loadings at (A) day 7, (B) day 17, and (C) day 41 of release. (Adapted from Kim, A., et al., Characterization of DNA-hyaluronan matrix for sustained gene transfer, *J. Control. Release,* 90, 81, 2003.)

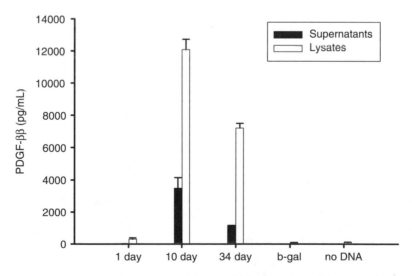

Figure 37.12 Receptor-based ELISA assay of COS-1 cells transfected with PDGF-ββ–FLAG DNA released from DNA–HA matrices cross-linked for 4 h with 2% DNA loadings. (Adapted from Kim, A. et al., Characterization of DNA-hyaluronan matrix for sustained gene transfer, *J. Control. Release,* 90, 81, 2003.)

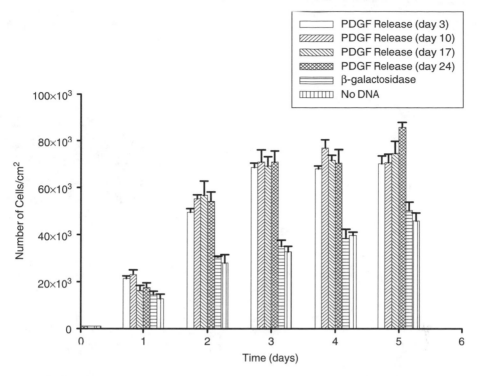

Figure 37.13 Proliferation of NNHDF cells treated with conditioned media of COS-1 cells transfected with release samples from matrices containing DNA encoding PDGF-ββ–FLAG. (Adapted from Kim, A. et al., Characterization of DNA-hyaluronan matrix for sustained gene transfer, *J. Control. Release,* 90, 81, 2003.)

proliferation levels were substantially higher than those detected using release sample DNA for transfections (data not shown).

37.3.6 *In Vivo* Wound Healing

Full thickness wounds were created on the backs of pigs. Plain HA matrices and DNA–HA matrices containing DNA encoding either PDGF-ββ or β-gal were implanted into the wounds. The general experimental setup for wound sites is presented in Figure 37.14. Four days after implantation, animals were sacrificed and wounds were excised. Paraffin embedded tissue sections were prepared and stained with hematoxylin and eosin. Histological evaluation was conducted via microscopy. Measurements of percent granulation tissue within the wounds indicated different degrees of granulation tissue formation between wounds treated with matrices containing PDGF-ββ compared to both β-gal–HA matrices and plain HA matrices ($p < 0.05$). These results are summarized in Table 37.1. Another group of animals was sacrificed after 11 days of matrix implantation. A representative histology section is depicted in Figure 37.15. Evidently, the HA matrix was integrated into the healed wound. Therefore, the HA matrix not only served as a platform for the delivery of a bioactive agent (i.e., PDGF) but also became a scaffold for tissue regeneration.

37.4 CONCLUSIONS

This study has concluded that HA matrices serve as effective delivery systems for DNA. They had the capability to protect DNA from enzymatic degradation and allowed for DNA release over

A

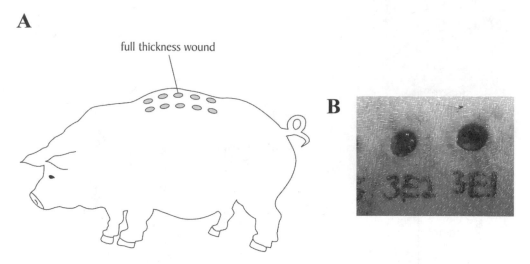

B

Figure 37.14   Full-thickness wounds: (A) sites on porcine dorsum, and (B) their appearances.

Table 37.1 Efficacy in Pigs Full-Thickness Dermal Wounds

DNA in Matrix	PDGF-ββ-Flag (CMV Promoter)	β-Gal (CMV Promoter)	None
% Granulation (AVG ± SD)	72.0 ± 13.5	46.1 ± 7.5	55.8 ± 3.4
p-value	N/A	0.003	0.014

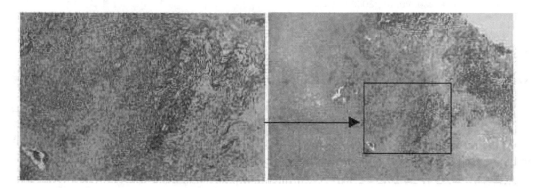

Figure 37.15   Hematoxylin- and eosin-stained paraffin embedded section of a full-thickness wound treated with PDGF-ββ matrix at 40 X (right) and at 100 X (left).

a prolonged period of time. The release of DNA can be modulated by changing either the DNA loadings or the cross-linking times. Assessment of DNA release profiles, gel electrophoresis, and β-gal specific activity results indicated that matrices that were cross-linked more extensively caused a delay in release of DNA and that much of the DNA released was entrapped in fragments of HA, leaving little available for transfection. This could serve as a mechanism to control the onset of gene transfection.

Immunofluorescent staining indicated that the DNA encoding PDGF-ββ–FLAG that was released was expressed, and semi-quantitatively suggested that most expression occurred in cells transfected with samples collected at a time point in the middle of the study. An ELISA assay provided a quantitative analysis of cytokine expressed in both conditioned media and cell lysates, in which more expression of the membrane bound PDGF-ββ–FLAG was observed in lysates. The

ELISA also indicated that the protein was functionally active, having the ability to bind the PDGF-β receptor. The expressed PDGF-ββ–FLAG was efficacious in promoting human dermal fibroblast proliferation. *In vivo* studies also revealed the efficacy of the DNA–HA matrix to promote an increased rate of wound healing in a pig full-thickness wound model.

ACKNOWLEDGMENTS

This study was funded by National Institutes of Health (NIH) grant AG17778 and partially supported by NIH grant HL65175.

REFERENCES

1. Luo, D. and Saltzman, W.M., Synthetic DNA delivery systems, *Nat. Biotechnol.* 18, 33, 2000.
2. Anderson, W.F., Human gene therapy, *Nature* 392 (6679 suppl.), 25, 1998.
3. Eming, S.A., Morgan, J.R., and Berger, A., Gene therapy for tissue repair: approaches and prospects, *Br. J. Plast. Surg.* 50, 491, 1997.
4. Shea, L.D. et al., DNA delivery from polymer matrices for tissue engineering, *Biotechnology* 17, 551, 1999.
5. Tyrone, J.W. et al., Collagen-embedded platelet-derived growth factor DNA plasmid promotes wound healing in a dermal ulcer model, *J. Surg. Res.* 93, 230, 2000.
6. Goa, K.L and Benfield, P., Hyaluronic acid: a review of its pharmacology and use as a surgical aid in ophthalmology, and its therapeutic potential in joint disease and wound healing, *Drugs* 47, 536, 1994.
7. Laurent, T.C. and Fraser, J., Hyaluronan, *FASEB J.* 6, 2397, 1992.
8. Toole, B.P., Hyaluronan and its binding proteins, the hyaladherins, *Curr. Opin. Cell Biol.* 2, 839, 1990.
9. Chen, W.Y. and Abatangelo, G., Functions of hyaluronan in wound repair, *Wound Repair Regen.* 7, 79, 1999
10. Balazs, E.A. and Denlinger, J.L., Clinical uses of hyaluronan, *Ciba Found. Symp.* 143, 265, 1989.
11. Steed, D.L, The role of growth factors in wound healing, *Surg. Clin. North Am.* 77, 575, 1997.
12. Edmonds, M. et al., New treatments in ulcer healing and wound infection, *Diabetes Metab. Res. Rev.* 16, S51, 2000.
13. Bronson, R.E., Bertolami, C.N., Siebert, E.P., Modulation of fibroblast growth and glycosaminoglycan synthesis by interleukin-1, *Coll. Relat. Res.* 7, 323, 1987.
14. Nath, C. and Gulati, S.C., Role of cytokines in healing chronic skin wounds, *Acta Haematol.* 99, 175, 1998.
15. Centers for Disease Control, Diabetes-related amputations of lower extremities in the Medicare population — Minnesota, 1993–1995, *MMWR Morb. Mortal. Weekly Rev.* 47, 649, 1998.
16. Beer, H.D., Longaker, M.T., and Werner, S., Reduced expression of PDGF and PDGF receptors during impaired wound healing, *J. Invest. Dermatol.* 109, 132, 1997.
17. Steed, D.L, Clinical evaluation of recombinant human platelet-derived growth factor for the treatment of lower extremity diabetic ulcers, *J. Vasc. Surg.* 21, 71, 1995.
18. Wieman, T.J., Smiell, J.M., and Su, Y., Efficacy and safety of a topical gel formulation of recombinant human platelet-derived growth factor-BB (Becaplermin) in patients with chronic neuropathic diabetic ulcers, *Diabetes Care* 21, 822, 1998.
19. Liechty, K. W. et al., Adenoviral-mediated overexpression of platelet-derived growth factor-B corrects ischemic impaired wound healing, *J. Invest. Dermatol.* 113, 375, 1999.
20. Breitbart, A.S. et al., Treatment of ischemic wounds using cultured dermal fibroblasts transduced retrovirally with PDGF-B and VEGF121 genes, *Ann. Plast. Surg.* 46, 555, 2001.
21. Shea, L.D. et al., DNA delivery from polymer matrices for tissue engineering, *Nat. Biotechnol.* 17, 551, 1999.
22. Vercruysse, K.P. et al., Synthesis and *in vitro* degradation of new polyvalent hydrazide cross-linked hydrogels of hyaluronic acid, *Bioconj. Chem.* 8, 686, 1997.

23. Kim, A. et al., Characterization of DNA-hyaluronan matrix for sustained gene transfer, *J. Control. Release* 90, 81, 2003..

24. Labarca, C., Paigen, K., A simple, rapid, and sensitive DNA assay procedure, *Anal. Biochem.* 102, 344, 1980.

25. Heldin, C.H., Östman, A., Rönnstrand, L, Signal transduction via platelet-derived growth factor receptors, *Biochim. Biophys. Acta* 1378, F79, 1998.

26. Clark, R.F et al., Transient functional expression of αvβ3 on vascular cells during wound repair, *Am. J. Pathol.* 148, 1407, 1996.

CHAPTER 38

Gene Delivery from Tissue Engineering Matrices

Zain Bengali, Christopher B. Rives, and Lonnie D. Shea

CONTENTS

38.1 INTRODUCTION

The combination of gene therapy with tissue engineering offers the potential to direct progenitor cell proliferation and differentiation into a functional tissue replacement.[1–4] The approaches to direct progenitor cell function can be classified into three basic categories: conductive, inductive, and cell transplantation.[5] In the conductive and inductive approaches, progenitor cells infiltrate from the surrounding tissue, whereas the cell transplantation approach directly delivers progenitor cells into the injury or disease site. Creating an environment with the appropriate cues and stimuli to direct progenitor cells into functional tissues represents a major challenge in the field of tissue engineering. Polymer scaffolds, whether natural, synthetic, or a combination of the two, serve a central role in these approaches and provide the means to regulate the chemical and mechanical stimuli within the tissue microenvironment.

The design of tissue engineering matrices attempts to mimic the numerous functions of the natural extracellular matrix (ECM), including providing a support for cell adhesion, organizing cells into structures, and serving as a reservoir for growth factors. Several fundamental requirements have been identified from the variety of materials and tissue systems that have been examined. These basic requirements include being biodegradable and biocompatible, having sufficient mechanical integrity, and having the ability to provide a suitable environment for new tissue

0-8493-1934-X/05/$0.00+$1.50
© 2005 by CRC Press LLC

formation that can integrate with the surrounding tissue.[5] Matrices implanted at a particular site must function to create and maintain a space for tissue formation, and should be resorbed or degraded at a rate that is comparable to that of new tissue formation. For synthetic polymers (e.g., copolymers of lactide and glycolide [PLG]), matrices are typically highly porous, and thus can initially support nutrient transport by diffusion from the surrounding tissue and also allow cell infiltration from the surrounding tissue. For matrices fabricated from natural polymers such as collagen or hyaluronic acid, cells can migrate through the scaffold by means of specific cellular interactions with the natural polymer and their ability to degrade the matrix. Cellular infiltration from the surrounding tissue is important for integration of the engineered tissue with the host, and also for development of a vascular network throughout the tissue to supply necessary metabolites once the tissue has developed. The basic properties of the scaffolds can be augmented to provide specific cellular cues. For example, adhesion molecules, peptides, and extracellular matrix proteins can be immobilized to the biomaterial to regulate the cellular interactions with the matrix.[6-8] Growth factors can be released from the matrix, which can then bind to cell surface receptors and initiate a variety of cellular processes.[8,9] The appropriate combination of signals provided by the matrix must coordinate the pattern of gene expression to initiate the cellular processes that lead to tissue formation.

Gene therapy approaches provide a mechanism to directly alter gene expression within a developing tissue. The delivery of DNA has been used to obtain localized and prolonged expression of tissue inductive factors that can stimulate tissue regeneration. For wound healing and tissue regeneration applications, transient expression of the encoded gene may be desirable such that expression subsides when the tissue has formed. The delivery strategies have typically employed direct injection of the viral or nonviral vector, the gene gun, or the transplantation of cells that are genetically engineered *ex vivo*. Using DNA encoding for tissue inductive factors, the transfected or engineered cells serve as bioreactors for the localized production of the proteins.[2] Gene delivery may provide more opportunities for sustained, localized production of tissue inductive factors relative to controlled release protein systems. Additionally, inducible vectors could be employed to regulate the quantity of protein produced after delivery.[10]

The combination of biomaterial and drug delivery technologies has been applied to create polymeric matrices capable of delivering DNA for applications in tissue engineering and regeneration. These matrices can be employed with either the inductive or cell transplantation approaches and may overcome some of the challenges associated with delivery by direct injection, which include clearance from the delivery site and degradation or inactivation of the vector. Matrix-mediated delivery for gene therapy can result in the DNA vector and the transgene product being retained within the wound site with decreased likelihood of vector dissemination to normal tissue. The matrix can distribute the DNA throughout the desired three-dimensional space of the injury site, rather than only at the site of injection. This chapter reports on the approaches for fabrication of tissue engineering matrices, the ability of these matrices to effect gene transfer and promote tissue formation, and the challenges that remain. The approaches have been subdivided into matrix-based release, in which encapsulated DNA is released as a soluble factor, or solid phase delivery, whereby the vector interacts with the matrix.

38.2 MATRIX-BASED RELEASE

Tissue engineering matrices for DNA delivery have been proposed to enhance gene transfer *in vivo* by (1) delaying clearance from the desired tissue, (2) protecting DNA from degradation, (3) providing sustained delivery to maintain the vector at effective levels within the target tissue, and (4) extending opportunities for internalization. Many of these properties have been observed for controlled release systems that deliver proteins.[11] The enhancement of gene transfer may increase the number of cells expressing the transgene along with the extent of transgene expression, while minimizing the quantity of vector used. Polymer encapsulation strategies can shield the vector against degradation, clearance, and an immune response. Drug release into the tissue can occur rapidly, as in a bolus delivery, or can occur over an extended period of time, which may affect the local concentration and cellular internalization. Alternatively, the DNA may be retained *in situ*. For rapid release, levels would be expected to quickly rise and decline as the DNA is cleared or degraded. For sustained delivery, the concentration may be maintained within an appropriate range by adjusting the release rate (e.g., through the polymer choice). Additionally, sustained delivery may reduce the number of dosages or the required cumulative dose.[11]

38.2.1 Nonviral Delivery

Plasmid-based approaches provide the flexibility to design synthetic vectors that alter the surface charge density, reduce vector degradation, target the vector to cellular receptors, and generally package the vector for efficient internalization and intracellular trafficking. Moderate doses of naked DNA for gene transfer are generally regarded as safe, as high doses may stimulate an immune response from nonmethylated CpG sequences on bacterially derived plasmids;[2] however, the transfection efficiency can be relatively low, which has motivated development of synthetic vectors and improved delivery systems. Naked plasmids are relatively large molecules (10^3–10^4 base pairs in length) with effective hydrodynamic diameters in excess of 100 nm.[12] The plasmids have a negative surface charge density, with zeta potentials ranging from −30 mV to −70 mV,[13] due to the abundance of phosphate groups, which can limit internalization into a cell that also has a net negative surface charge density.[12] To reduce the negative surface charge and decrease the effective diameter, plasmids can be complexed with cationic lipids (e.g., DOTAP–DOPE) or cationic polymers (e.g., poly[L-lysine] [PLL] and polyethylenimine [PEI]).[13] These cationic agents self-assemble with DNA to form less negatively charged particles, which can protect the DNA from degradation and facilitate cellular interactions by reducing the size and negative charge density. These cationic polymers and lipids can be modified with functional groups, thus altering the physical properties of the complexes,[14-16] decreasing interactions with serum components,[16,17] and targeting specific cellular receptors.[18-21]

The delivery of naked DNA and DNA complexes has been employed *in vivo* for applications such as DNA vaccines,[22,23] systemic protein delivery,[24] cancer treatment,[25,26] and more recently, wound healing and tissue regeneration.[2,27] Intradermal injections of 100 µg of plasmid encoding keratinocyte growth factor-1 (KGF-1) were used to enhance wound healing; however, multiple injections were required to achieve prolonged transgene expression in a consistent manner.[28] A single injection of 100 µg naked DNA encoding for vascular endothelial growth factor (VEGF) enhanced healing of esophageal ulcers.[29] Alternatively, the gene gun has been used to deliver DNA encoding for transforming growth factor β1 (TGF-β1) as a means of enhancing wound healing.[27] Under the appropriate conditions of injection volume and speed, intravascular delivery of naked DNA can be an effective gene delivery methodology, with the capacity to transfect 5–20% of hepatocytes in the liver[30] and 20% of myofibers within some areas of the muscle.[31] Intravenous injection of DNA complexes, rather than naked DNA, results in increased levels of DNA and longer retention times; however, proper dosing is required to prevent toxicity.[32,33]

38.2.1.1 Hydrogels

Naked DNA delivery from collagen matrices has been employed to promote tissue formation by transfecting invading fibroblasts. Transfected fibroblasts within DNA-loaded collagen scaffolds, also termed gene-activated matrices (GAMs), subsequently act as bioreactors for localized production of tissue inductive factors. Matrices were prepared by lyophilization of type I collagen and subsequent immersion in a DNA solution. Collagen is found in many tissues and has been used as a matrix for numerous tissue engineering applications, including bone, skin, nerve, and cartilage.[34-36]

The collagen serves as a scaffold for the migration of repair cells into the matrix, and serves to either retain the DNA within the scaffold[37] or to provide gradual release.[38] Collagen-based release may limit vector degradation, and can induce transgene expression for up to 40 days.[38] For applications in tissue engineering, collagen–DNA constructs have been implanted into an adult rat femur[39] and a canine bone defect model.[37] Matrices loaded with 1 mg of DNA were capable of transfecting cells *in vivo*, which resulted in protein production for up to 3 weeks after implant. For the canine model, however, regeneration required 100 mg of plasmid delivered from the matrix for regeneration.

DNA complexed with cationic lipids or cationic polymers can also be incorporated and released from collagen-based matrices, while maintaining their activity. Naked DNA delivery produces substantial transgene expression *in vivo*, but results in low levels of transgene expression *in vitro*. DNA complexes can transfect cells *in vitro*; thus, the ability to release DNA complexes can extend the applicability of nonviral DNA release matrices to the engineering of tissues *in vitro*. Note, however, that release of DNA complexes may differ significantly from that for naked DNA, due to the different physical properties of complexes relative to naked DNA. Collagen matrices were loaded with DNA by pipetting solutions of naked DNA, PEI–DNA complexes, and lipoplexes onto collagen.[40] *In vitro* release studies demonstrated that naked DNA was rapidly released, that PEI–DNA and lipid complexes were slowly released, and that PEI–DNA complexes with a protective copolymer had intermediate release kinetics. The PEI–DNA complexes with the protective copolymer gave the highest transfection *in vitro* and *in vivo*, with the highest *in vivo* expression occurring at 4 days and measurable quantities observed at 7 days. PLL–DNA complexes encapsulated in a collagen sponge have been implanted into severed rat optic nerves as a means to promote neuron survival and regeneration.[41] The PLL was modified with basic fibroblast growth factor (BFGF) to facilitate internalization and intracellular trafficking. DNA was detected in the retina for up to 3 months. Nerve terminals were observed extending into the collagen, and appeared capable of transporting the DNA by retrograde transport.

Hydrogels based on agarose, hyaluronic acid (HA), chitosan, and fibrin have been employed independently as biomaterials for fabrication of tissue engineering matrices[8,42,43] or as materials to regulate DNA delivery. Agarose gels have provided a sustained release of PLL–DNA complexes to transfect smooth muscle cells *in vitro* with an efficiency less than that obtained by freshly formed complexes, but greater than that obtained with naked DNA.[44] HA matrices have also demonstrated the capacity for sustained release of naked DNA, with release likely occurring following degradation of the matrix.[45] The release rate of naked DNA, some of which may be associated with HA fragments, could be modulated by the extent of cross-linking in the hydrogel. Fibrin sealants have been employed for the delivery of plasmids to promote angiogenesis, with fibrin-based delivery providing similar responses to delivery in PBS solution.[46] Hydrogels based on fibrin, HA, and agarose primarily function to limit the release of DNA; however, hydrogels employing chitosan offer the potential to condense the DNA. Chitosan is a naturally occurring polysaccharide that can form complexes with DNA. Chitosan–DNA complexes exhibit minimal cytotoxicity, can destabilize the lipid bilayer to facilitate internalization,[47] and produce high levels of transfection *in vitro* and *in vivo*.[48,49] Tissue engineering matrices based on these hydrogels, or combinations of these materials, provide a variety of approaches to regulate DNA delivery *in vitro* and *in vivo*.

38.2.1.2 *Synthetic Polymers*

Synthetic polymers used in drug delivery and in tissue engineering can be processed to fabricate matrices capable of controlled DNA delivery. Copolymers of lactide and glycolide (PLG) have been extensively used because they are generally considered to be biocompatible, are FDA approved, and can be designed to degrade over times ranging from a few weeks to more than a year. The ability of matrices fabricated from PLG to deliver cultured cells to a desired site *in vivo* and guide the formation of new tissues with a predefined gross structure has been well documented.[5,50] These polymers enable control over degradation rate and mechanical properties through varying the composition and molecular weight of the polymer, which would allow matrices to be tailored for specific applications. Importantly, PLG has been widely used in the field of drug delivery, to provide controlled release of small molecules, proteins, and DNA.[11,13]

PLG scaffolds have been fabricated which function as a support for cell adhesion, and as vehicles for cell transplantation and plasmid release. A gas foaming and particulate leaching process can be employed to fabricate interconnected open pore structures for controlled release of DNA.[51,52] This process employs carbon dioxide to process a mixture of polymer and porogen, in order to fuse adjacent polymer particles into an interconnected structure. The DNA can be lyophilized with the microspheres[51] or encapsulated within the microspheres.[53,54] Lyophilization of DNA with the microspheres can provide large quantities of incorporated DNA, with relatively rapid release kinetics. Incorporation of DNA into the microspheres provides for a more sustained release relative to the lyophilization method,[53] with the release kinetics dependent on the polymer molecular weight and microsphere size.[54] DNA can be incorporated into polymer microspheres using several approaches.[54-58] Subcutaneous implantation of scaffolds results in transfected cells observed within the scaffold and the tissue immediately adjacent to the scaffold, with protein production sufficient to promote physiological responses.[51] An alternative approach to fabrication of PLG scaffolds for DNA delivery is electrospinning.[59] Electrospinning creates unwoven, nano-fibered membranous structures that release DNA, with maximal release occurring at approximately 2 h.

DNA polyplexes have been encapsulated and released from polymer microspheres, which may enable these microspheres to be fabricated into matrices using an approach such as the gas foaming procedure. PLL–DNA complexes have been incorporated into PLG microspheres using a double emulsion process. DNA is incorporated with efficiencies ranging from 30 to 45%, is released over approximately 35 days, and retains its integrity.[60-62] Alternatively, oligonucleotides (ONs) complexed with PEI have been incorporated and released from PLG microspheres. The release profile of the ON–PEI complexes depended upon the size, loading, and pore structure of the microspheres. The sustained release of ON–PEI complexes resulted in improved intracellular penetration of the delivered vector as compared to uncomplexed DNA.[63,64]

38.2.2 Viral Delivery

The ability of viruses to transfect cells efficiently has been applied to increase the production of tissue inductive factors locally. The most frequently used vectors for gene therapy are adenovirus, adeno-associated virus (AAV), and retrovirus. Adenovirus vectors have high transfection efficiencies; however, they tend to elicit relatively strong immune responses that reduce the efficiency of gene transfer. Adenovirus vectors persist extrachromosomally, which leads to transient expression of the transgene that lasts on the time scale of a few weeks. AAV vectors are effective in gene delivery, but have a relatively small transgene capacity (4–5 kb). Delivered transgenes are also typically retained as extrachromosomal elements, but may integrate randomly into the genome. Retrovirus vectors can provide prolonged expression with relatively low immunogenicity; however, in vivo transfection efficiencies are low, and they only transfect proliferating cells. Retrovirus delivery may lead to stable expression as the gene may be inserted into the chromosome of the host cell.

Ex vivo gene transfer or in vivo injection of viral vectors have been employed for wound healing[65,66] and to promote the regeneration of many tissues, such as skin,[67,68] bone,[69,70] and nerve.[71-74] Many viral vectors employed in tissue regeneration encode for growth factors, such as BMP-2,[75-78] FGF-2,[66,79] BDNF,[41,80] and VEGF.[79,81] Since most growth factors are cleared rapidly from tissues and have a brief half-life,[82,83] localized gene delivery may circumvent this problem through sustained expression of factors. The localized expression of these factors can directly promote cell survival, stimulate proliferation, and induce progenitor cell differentiation down specific pathways.[79,81] Alternatively, viral vectors have been delivered that encode for transcription factors, such as mutant p53 to inhibit tumor growth[84] or Brn-3a to protect neurons from stimuli that induce cell death.[85] The challenges in applying viral vectors to tissue regeneration can include the immune response to the vector or to the cells transduced by the vector. Ectopic bone formation was observed by the injection of a BMP-2 containing adenovirus into the thigh muscles of nude rats. However, bone formation was inhibited in immune competent Sprague Dawley rats.[75] Presence of an immune response against the viral vector may limit the biological activity of the secreted protein.[78,86]

Viruses have been incorporated into both natural and synthetic materials that have been used in fabricating tissue engineering matrices. Incorporation of the virus into collagen gels is accomplished by addition of the virus to a collagen solution at 4°C and subsequent gelation at body temperature. Collagen gels carrying targeted vectors encoding for platelet-derived growth factor B (PDGF-B) demonstrated enhanced transgene expression at significantly lower doses relative to nontargeted vectors.[65] Implantation of adenovirus-loaded collagen gels into skeletal muscle defects resulted in enhanced muscle repair. This repair strategy was able to induce multiple processes involved in regeneration, including angiogenesis, arteriogenesis, and myogenesis.[66] A gelatin sponge matrix for delivery of canarypox virus increased reporter gene expression relative to fluid injection of the virus.[87] Application of this system to express IL-2, IL-12, and tumor necrosis factor α (TNF-α) inhibited tumor growth in heterotopic nodules.

Encapsulation of viral vectors in biomaterials can provide for sustained delivery, with the potential to achieve similar or increased levels of gene transfer. The matrix can serve to locally concentrate the virus, which may increase the efficiency and localize the delivery of the vector. For delivery from collagen, the release of adenovirus depends on the collagen concentration, with minimal release obtained for collagen concentrations above 1%. These adenovirus vectors remained viable *in vivo*, suggesting that the collagen may limit degradation.[38] Coacervates of gelatin (3.5%) and alginate (0.1%) loaded with adenovirus were able to transduce cells at levels that were comparable to equivalent amounts of injected adenovirus.[88] Alternatively, adenovirus vectors were encapsulated and released from synthetic polymers, with viable viruses being released for up to 10 days. Viruses encapsulated into PLG were active *in vitro* and *in vivo*.[89-92]

Encapsulation of viruses within a matrix may also serve to limit the immune response to viral vectors. *In vivo* delivery of viruses results in both cell-mediated and humoral immune responses, which typically prohibit multiple or continuous administration of the vector. Two mechanisms are proposed for this reduced immune response: decreased doses of the vector and shielding of the vector. Large doses of viruses are required when delivery occurs through bolus injection to achieve efficient transfection *in vivo*; matrix-based delivery may reduce the quantity of virus needed for delivery. Alternatively, encapsulation within the material may limit attack by neutralizing antibodies. Adenoviral vectors encapsulated and released from PLG microspheres[89,90] provided gene transfer in mice that were preimmunized to the adenovirus. The encapsulated vectors had 45-fold lower anti-adenovirus titers than those obtained with direct injection of the adenovirus.[90,92] Similarly, encapsulation of adenovirus vectors in alginate beads or collagen minipellets showed significantly greater expression in preimmunized mice than direct delivery of the virus, while reducing the vector-specific immune response.[93]

38.3 SOLID PHASE DELIVERY

Solid phase delivery describes the process of immobilizing DNA to the extracellular matrix, which functions to support cell adhesion and places the DNA directly in the cellular microenvironment. The immobilization of DNA to the matrix may seem counterintuitive given the need for cellular internalization to achieve expression; however, there are natural and synthetic corollaries for this approach. Growth factors can associate with the extracellular matrix, which can function directly from the matrix or upon release.[94-96] Additionally, many viral vectors can associate with the extracellular matrix as a means to facilitate cellular binding and internalization.[97,98] This immobilization to a matrix that supports cell adhesion can overcome the mass transport limitations associated with DNA delivery. Bolus addition of DNA complexes to the culture media overlaying cells results in a fraction of the DNA being internalized.[99,100] Approaches that increase the DNA concentration in the cellular microenvironment have led to an 8.5-fold enhancement in transfection.[101] An additional advantage of immobilized delivery may be the ability to localize gene transfer, for both viral and nonviral vectors. The following paragraphs describe solid phase delivery of viral and nonviral vectors. Many of these approaches have not been directly applied to tissue engineering or regeneration; however, the potential exists to develop these delivery systems for these applications.

38.3.1 Viral Vector Delivery

The design of viral vectors that bind to biomaterials is aimed at localizing delivery and increasing gene transfer. Viral vectors administered to particular regions often diffuse to distant tissues and organs.[77,102-105] For targeting of the liver, almost 30% of injected virus is delivered to nonliver sites.[106] Nonspecific delivery can have detrimental effects, including ectopic gene expression, toxicity of nontarget cells, and eliciting a strong immune response. Modification of the viral vector to bind to a biomaterial can avoid distal transfection side effects through specific cellular exposure, and may enhance the transfection efficiency by concentrating the virus at the delivery site. The strategies for immobilization of the virus involve modification of the material, chemical modification of the virus, or engineering of the virus. It is important to note that the strategies to immobilize the virus to the material must maintain the activity of the virus, which can be compromised by covalent attachment of functional groups or inappropriate binding to a material.

The surface of a biomaterial can allow binding of viruses for immobilized delivery. Polystyrene beads have been used to bind adenovirus vectors, which resulted in elevated levels of infection up to 72 h after transfer.[107] Viruses were not released into media and transgene expression was observed for only those cells in contact with the spheres. Conjugation of an adenoviral vector to polystyrene microspheres yields a fivefold increase in transduction efficiency *in vivo* relative to free vector delivery administered intramuscularly or intravenously.[108] This method also yielded targeting of gene expression to specific tissues by retention of microsphere-bound vectors in the capillary bed.

Alternatively, the matrix can be specifically engineered to bind viruses. Collagen gels were modified with IgG that is specific for the adenovirus hexon, resulting in adenovirus binding upon incubation with the matrix. Transfection was localized to the collagen using the immobilized vectors, compared to disperse transduction found in control conditions.[109] Implantation of an adenovirus-modified gel to the porcine right ventricle resulted in transduction throughout the matrix, differing from that observed by direct injection.[109,110] A similar strategy has been applied to delivery from collagen-coated stents and microcoils, which have applications in the treatment of endothelial and vascular disease, respectively.[111,112]

An alternative to modifying the matrix involves modification of the virus to allow for binding to the matrix, while avoiding inactivation of the virus. Adenovirus vectors have been chemically modified with biotin groups using N-hydroxysuccinimdyl ester (NHS-ester) chemistry to couple biotin to amine groups present on the exterior of the virus. These viral particles can then bind to avidin-conjugated microspheres.[113] This approach was capable of enhancing transgene expression for cells that are not readily transduced by adenovirus. Additionally, infected cells were primarily observed in regions immediately adjacent to the beads. Adenovirus vectors were also attached to paramagnetic beads, allowing for focusing of the vector and localization of transduction.[113] An alternative to chemical modification of the virus involves engineering of functional groups into proteins on the viral shell or on the target cell.[114-116]

38.3.2 Nonviral DNA Delivery

Solid phase delivery of nonviral vectors provides greater versatility in manipulating both the substrate and the vector to enhance transfection. PLL was modified with biotin residues for subsequent complexation with DNA and binding to a neutravidin substrate.[117] Complexes were formed with mixtures of biotinylated and nonbiotinylated PLL. Complexes formed with a greater percentage of biotinylated PLL bound to the substrates at higher densities; however, transfection was greatest when complexes were formed with the lowest percentage of biotinylated PLL. Substrate-mediated delivery of the complexes increased transgene expression 100-fold relative to bolus delivery of similar complexes.[117] Additionally, transfection was observed only in the location to which complexes were bound, suggesting the possibility of spatially regulating DNA delivery. An alternative approach involves adsorption of PEI–DNA complexes to silica nanoparticles.[118,119] Transgene expression by nanoparticle adsorption was comparable to that observed by fluid phase delivery, but with reduced toxicity.

The matrix can also be modified to bind plasmid DNA directly. Matrices formed from PLG or collagen were modified with poly(ε-CBZ-L-lysine) (PCBZL) to present amine groups on the surface, which were subsequently modified with PEI.[120] Incubation of naked DNA with the substrate resulted in binding to the surface and demonstrated the ability to mediate transfection. PLGA and collagen membranes have also been coated with phosphatidyl glycerol (1–5%) to support binding of complexes formed with polyamidoamine (PAMAM) dendrimer.[121] Vectors were slowly released from this scaffold, yielding transfection comparable to solution-based transfection controls. *In vivo* studies demonstrated a six- to eightfold enhancement in transfection relative to naked DNA delivery.

38.4 FUTURE DIRECTIONS AND CHALLENGES

Matrices capable of DNA delivery have demonstrated the potential for localized gene transfer, with sufficient protein production to stimulate physiological responses that lead to tissue formation. The development of these delivery systems will potentially have a broad impact on the ability of clinicians to treat tissue loss or organ failure resulting from disease or trauma. This technology could be applied not only to the inductive approach of guiding tissue regeneration, but also to cell transplantation approaches. Gene delivery is an attractive approach because a variety of cellular processes can be targeted through this single delivery method. Strategies can be developed in which a gene encoding for a secreted protein is delivered and stimulates numerous cells. Alternatively, a gene encoding for an intracellular protein could be delivered, which will control the fate of that individual cell but will likely require greater numbers of transfected cells to be effective. The utility of gene delivery from tissue engineering matrices will increase with continued development of (1) the design parameters of the system and their effects on gene transfer and transgene expression, and (2) an understanding of the biology behind tissue formation and how transgene expression can augment the regenerative process

The studies described herein have illustrated the potential to deliver DNA in several approaches, and the continued examination of the mechanism underlying gene transfer may identify molecular-level design parameters. Numerous studies have examined the mechanism by which vectors traffic through the cell and into the nucleus for subsequent expression.[15,99,122] Prior to cellular internalization, however, the vector may interact with the polymer matrix and other components in the extracellular milieu. These interactions can influence vector release from the matrix, stability, transport through the extracellular space, and, ultimately, internalization and trafficking. Correlating the stability and transport of the vector with the amount and duration of transgene expression and the number and distribution of transfected cells will ultimately lead to an understanding of the molecular design for the matrix and vector.

The application of these matrices will also require a more thorough understanding of the biological requirements for tissue regeneration. Cellular signaling within the scaffold will depend on various factors, such as implant location and cell types present. The expression of transgenes, encoding for growth factors or transcription factors, will likely influence cellular processes such as proliferation, differentiation, and migration. The integration of the transgene expression profile with other design components that influence gene expression must be considered for its cumulative effect on tissue formation. While the proposed studies have illustrated the potential for extending the production of growth factors locally, the delivery strategies may also be adapted to control delivery on length scales of 10 μm to 1 mm. The ability to regulate expression of one or more factors in time and space may be a critical piece to the engineering of complex tissue architectures, such as those found in vascular networks and the nervous system.

REFERENCES

1. Griffith, L.G. and Naughton, G., Tissue engineering: current challenges and expanding opportunities, *Science* 295 (5557), 1009–1014, 2002.
2. Bonadio, J., Genetic approaches to tissue repair, *Ann. N.Y. Acad. Sci.* 961, 58–60, 2002.
3. Richardson, T.P., Murphy, W.L., and Mooney, D.J., Polymeric delivery of proteins and plasmid DNA for tissue engineering and gene therapy, *Crit. Rev. Eukaryot. Gene Expr.* 11 (1–3), 47–58, 2001.
4. Lauffenburger, D.A. and Schaffer, D.V., The matrix delivers, *Nat. Med.* 5 (7), 733–734, 1999.
5. Murphy, W.L. and Mooney, D.J., Controlled delivery of inductive proteins, plasmid DNA and cells from tissue engineering matrices, *J. Periodontal Res.* 34 (7), 413–419, 1999.
6. Rowley, J.A., Madlambayan, G., and Mooney, D.J., Alginate hydrogels as synthetic extracellular matrix materials, *Biomaterials* 20 (1), 45–53, 1999.
7. Hubbell, J.A., Bioactive biomaterials, *Curr. Opin. Biotechnol.* 10 (2), 123–129, 1999.
8. Sakiyama-Elbert, S.E. and Hubbell, J.A., Functional biomaterials: design of novel biomaterials, *Ann. Rev. Mater. Res.* 31, 183–201, 2001.
9. Saltzman, W.M. and Olbricht, W.L., Building drug delivery into tissue engineering, *Nat. Rev. Drug Discov.* 1 (3), 177–186, 2002.
10. Ye, X. et al., Regulated delivery of therapeutic proteins after *in vivo* somatic cell gene transfer, *Science* 283 (5398), 88–91, 1999.
11. Langer, R., Drug delivery and targeting, *Nature* 392 (6679 Suppl.), 5–10, 1998.
12. Ledley, F.D., Pharmaceutical approach to somatic gene therapy, *Pharm. Res.* 13 (11), 1595–1614, 1996.
13. Segura, T. and Shea, L.D., Materials for non-viral gene delivery, *Ann. Rev. Mater. Res.* 31, 25–46, 2001.
14. Adami, R.C. et al., Stability of peptide-condensed plasmid DNA formulations, *J. Pharm. Sci.* 87 (6), 678–683, 1998.
15. Zabner, J. et al., Cellular and molecular barriers to gene transfer by a cationic lipid, *J. Biol. Chem.* 270 (32), 18997–9007, 1995.
16. Blessing, T. et al., Different strategies for formation of pegylated EGF-conjugated PEI/DNA complexes for targeted gene delivery, *Bioconj. Chem.* 12 (4), 529–537, 2001.
17. Pedroso de Lima, M.C. et al., Cationic liposomes for gene delivery: from biophysics to biological applications, *Curr. Med. Chem.* 10 (14), 1221–1231, 2003.
18. Gottschalk, S. et al., Folate receptor mediated DNA delivery into tumor cells: potosomal disruption results in enhanced gene expression, *Gene Ther.* 1 (3), 185–191, 1994.
19. Wagner, E. et al., Transferrin-polycation conjugates as carriers for DNA uptake into cells, *Proc. Nat. Acad. Sci. U.S.A.* 87 (9), 3410–3414, 1990.
20. Harbottle, R.P. et al., An RGD-oligolysine peptide: a prototype construct for integrin-mediated gene delivery, *Hum Gene Ther.* 9 (7), 1037–1047, 1998.
21. Schaffer, D.V. and Lauffenburger, D.A., Targeted synthetic gene delivery vectors, *Curr. Opin. Mol. Ther.* 2 (2), 155–161, 2000.
22. Kumar, M. et al., Intranasal gene transfer by chitosan-DNA nanospheres protects BALB/c mice against acute respiratory syncytial virus infection, *Hum. Gene Ther.* 13 (12), 1415–1425, 2002.

23. Donnelly, J., Berry, K., and Ulmer, J.B., Technical and regulatory hurdles for DNA vaccines, *Int. J. Parasitol.* 33 (5–6), 457–467, 2003.

24. Ochiya, T. et al., New delivery system for plasmid DNA *in vivo* using atelocollagen as a carrier material: the minipellet, *Nat. Med.* 5 (6), 707–710, 1999.

25. Mincheff, M. et al., Naked DNA and adenoviral immunizations for immunotherapy of prostate cancer: a phase I/II clinical trial, *Eur. Urol.* 38 (2), 208–217, 2000.

26. Nishitani, M. et al., Cytokine gene therapy for cancer with naked DNA, *Mol. Urol.* 4 (2), 47–50, 2000.

27. Andree, C. et al., *In vivo* transfer and expression of a human epidermal growth factor gene accelerates wound repair, *Proc. Nat. Acad. Sci. U.S.A.* 91 (25), 12188–12192, 1994.

28. Byrnes, C.K. et al., Success and limitations of a naked, plasmid transfection protocol for keratinocyte growth factor-1 to enhance cutaneous wound healing, *Wound Rep. Regen.* 9 (5), 341–346, 2001.

29. Baatar, D. et al., Esophageal ulceration triggers expression of hypoxia-inducible factor-1 alpha and activates vascular endothelial growth factor gene: implications for angiogenesis and ulcer healing, *Am. J. Pathol.* 161 (4), 1449–1457, 2002.

30. Budker, V. et al., Hypothesis: naked plasmid DNA is taken up by cells *in vivo* by a receptor-mediated process, *J. Gene Med.* 2 (2), 76–88, 2000.

31. Herweijer, H. and Wolff, J.A., Progress and prospects: naked DNA gene transfer and therapy, *Gene Ther.* 10 (6), 453–458, 2003.

32. Li, S. and Huang, L., *In vivo* gene transfer via intravenous administration of cationic lipid-protamine-DNA (LPD) complexes, *Gene Ther.* 4 (9), 891–900, 1997.

33. Oh, Y.K. et al., Prolonged organ retention and safety of plasmid DNA administered in polyethylenimine complexes, *Gene Ther.* 8 (20), 1587–1592, 2001.

34. Pieper, J.S. et al., Crosslinked type II collagen matrices: preparation, characterization, and potential for cartilage engineering, *Biomaterials* 23 (15), 3183–3192, 2002.

35. Verdu, E. et al., Alignment of collagen and laminin-containing gels improve nerve regeneration within silicone tubes, *Restor. Neurol. Neurosci.* 20 (5), 169–179, 2002.

36. Wisser, D. and Steffes, J., Skin replacement with a collagen based dermal substitute, autologous keratinocytes and fibroblasts in burn trauma, *Burns* 29 (4), 375–380, 2003.

37. Bonadio, J. et al., Localized, direct plasmid gene delivery in vivo: prolonged therapy results in reproducible tissue regeneration, *Nat. Med.* 5 (7), 753–759, 1999.

38. Ochiya, T. et al., Biomaterials for gene delivery: atelocollagen-mediated controlled release of molecular medicines, *Curr. Gene Ther.* 1 (1), 31–52, 2001.

39. Fang, J. et al., Stimulation of new bone formation by direct transfer of osteogenic plasmid genes, *Proc. Nat. Acad. Sci. U.S.A.* 93 (12), 5753–5758, 1996.

40. Scherer, F. et al., Nonviral vector loaded collagen sponges for sustained gene delivery *in vitro* and *in vivo*, *J. Gene Med.* 4 (6), 634–643, 2002.

41. Berry, M. et al., Sustained effects of gene-activated matrices after CNS injury, *Mol. Cell. Neurosci.* 17 (4), 706–716, 2001.

42. Bellamkonda, R., Ranieri, J.P., and Aebischer, P., Laminin oligopeptide derivatized agarose gels allow three-dimensional neurite extension *in vitro*, *J. Neurosci. Res.* 41 (4), 501–509, 1995.

43. Madihally, S.V. and Matthew, H.W., Porous chitosan scaffolds for tissue engineering, *Biomaterials* 20 (12), 1133–1142, 1999.

44. Meilander, N.J. et al., Sustained release of plasmid DNA using lipid microtubules and agarose hydrogel, *J. Control. Release* 88 (2), 321–331, 2003.

45. Kim, A. et al., Characterization of DNA-hyaluronan matrix for sustained gene transfer, *J. Control. Release* 90 (1), 81–95, 2003.

46. Jozkowicz, A. et al., Delivery of high dose VEGF plasmid using fibrin carrier does not influence its angiogenic potency, *Int. J. Artif. Org.* 26 (2), 161–169, 2003.

47. Fang, N. et al., Interactions of phospholipid bilayer with chitosan: effect of molecular weight and pH, *Biomacromolecules* 2 (4), 1161–1168, 2001.

48. Leong, K.W. et al., DNA-polycation nanospheres as non-viral gene delivery vehicles, *J. Control. Release* 53 (1–3), 183–193, 1998.

49. Mao, H.Q. et al., Chitosan-DNA nanoparticles as gene carriers: synthesis, characterization and transfection efficiency, *J. Control. Release* 70 (3), 399–421, 2001.

50. Putnam, A.J. and Mooney, D.J., Tissue engineering using synthetic extracellular matrices, *Nat. Med.* 2 (7), 824–826, 1996.

51. Shea, L.D. et al., DNA delivery from polymer matrices for tissue engineering, *Nat. Biotechnol.* 17 (6), 551–554, 1999.

52. Harris, L.D., Kim, B.S., and Mooney, D.J., Open pore biodegradable matrices formed with gas foaming, *J. Biomed. Mater. Res.* 42 (3), 396–402, 1998.

53. Nof, M. and Shea, L.D., Drug-releasing scaffolds fabricated from drug-loaded microspheres, *J. Biomed. Mater. Res.* 59 (2), 349–356, 2002.

54. Jang, J.H. and Shea, L.D., Controllable delivery of non-viral DNA from porous scaffolds, *J. Control. Release* 86 (1), 157–168, 2003.

55. Hedley, M.L., Curley, J., and Urban, R., Microspheres containing plasmid-encoded antigens elicit cytotoxic T-cell responses, *Nat. Med.* 4 (3), 365–368, 1998.

56. Ando, S. et al., PLGA microspheres containing plasmid DNA: preservation of supercoiled DNA via cryopreparation and carbohydrate stabilization, *J. Pharm. Sci.* 88 (1), 126–130, 1999.

57. Hsu, Y.Y., Hao, T., and Hedley, M.L., Comparison of process parameters for microencapsulation of plasmid DNA in poly(D,L-lactic-co-glycolic) acid microspheres, *J. Drug Target.* 7 (4), 313–323, 1999.

58. Luo, D. et al., Controlled DNA delivery systems, *Pharm. Res.* 16 (8), 1300–1308, 1999.

59. Luu, Y.K. et al., Development of a nanostructured DNA delivery scaffold via electrospinning of PLGA and PLA-PEG block copolymers, *J. Control. Release* 89 (2), 341–353, 2003.

60. Capan, Y. et al., Influence of formulation parameters on the characteristics of poly(D, L-lactide-co-glycolide) microspheres containing poly(L-lysine) complexed plasmid DNA, *J. Control. Release* 60 (2–3), 279–286, 1999.

61. Capan, Y. et al., Stability of poly(L-lysine)-complexed plasmid DNA during mechanical stress and DNase I treatment, *Pharm. Dev. Technol.* 4 (4), 491–498, 1999.

62. Capan, Y. et al., Preparation and characterization of poly (D,L-lactide-co-glycolide) microspheres for controlled release of poly(L-lysine) complexed plasmid DNA, *Pharm. Res.* 16 (4), 509–513, 1999.

63. De Rosa, G. et al., Long-term release and improved intracellular penetration of oligonucleotide-polyethylenimine complexes entrapped in biodegradable microspheres, *Biomacromolecules* 4 (3), 529–536, 2003.

64. De Rosa, G. et al., A new delivery system for antisense therapy: PLGA microspheres encapsulating oligonucleotide/polyethyleneimine solid complexes, *Int. J. Pharm.* 254 (1), 89–93, 2003.

65. Chandler, L.A. et al., FGF2-Targeted adenovirus encoding platelet-derived growth factor-B enhances *de novo* tissue formation, *Mol. Ther.* 2 (2), 153–160, 2000.

66. Doukas, J. et al., Delivery of FGF genes to wound repair cells enhances arteriogenesis and myogenesis in skeletal muscle, *Mol. Ther.* 5 (5 Pt. 1), 517–527, 2002.

67. Ghazizadeh, S. and Taichman, L.B., Virus-mediated gene transfer for cutaneous gene therapy, *Hum. Gene Ther.* 11 (16), 2247–2251, 2000.

68. Ghazizadeh, S., Doumeng, C., and Taichman, L.B., Durable and stratum-specific gene expression in epidermis, *Gene Ther.* 9 (19), 1278–1285, 2002.

69. Tsuda, H. et al., Efficient BMP2 gene transfer and bone formation of mesenchymal stem cells by a fiber-mutant adenoviral vector, *Mol. Ther.* 7 (3), 354–365, 2003.

70. Huard, J. et al., Gene therapy and tissue engineering for sports medicine, *J. Gene Med.* 5 (2), 93–108, 2003.

71. Brooks, A.I. et al., Nerve growth factor somatic mosaicism produced by herpes virus-directed expression of cre recombinase, *Nat. Biotechnol.* 15 (1), 57–62, 1997.

72. Conner, J.M. et al., Nontropic actions of neurotrophins: subcortical nerve growth factor gene delivery reverses age-related degeneration of primate cortical cholinergic innervation, *Proc. Nat. Acad. Sci. U.S.A.* 98 (4), 1941–1946, 2001.

73. Choi-Lundberg, D.L. et al., Dopaminergic neurons protected from degeneration by GDNF gene therapy, *Science* 275 (5301), 838–841, 1997.

74. Davidson, B.L. and Breakefield, X.O., Viral vectors for gene delivery to the nervous system, *Nat. Rev. Neurosci.* 4 (5), 353–364, 2003.

75. Alden, T.D. et al., *In vivo* endochondral bone formation using a bone morphogenetic protein 2 adenoviral vector, *Hum. Gene Ther.* 10 (13), 2245–2253, 1999.

76. Baltzer, A.W. et al., Genetic enhancement of fracture repair: healing of an experimental segmental defect by adenoviral transfer of the BMP-2 gene, *Gene Ther.* 7 (9), 734–739, 2000.

77. Gelse, K. et al., Fibroblast-mediated delivery of growth factor complementary DNA into mouse joints induces chondrogenesis but avoids the disadvantages of direct viral gene transfer, *Arthritis Rheum.* 44 (8), 1943–1953, 2001.

78. Musgrave, D.S. et al., Adenovirus-mediated direct gene therapy with bone morphogenetic protein-2 produces bone, *Bone* 24 (6), 541–547, 1999.

79. Matsuoka, N. et al., Adenovirus-mediated gene transfer of fibroblast growth factor-2 increases BrdU-positive cells after forebrain ischemia in gerbils, *Stroke* 34 (6), 1519–1525, 2003.

80. Jia, Q. et al., Expression of brain-derived neurotrophic factor in the central nervous system of mice using a poliovirus-based vector, *J. Neurovirol.* 8 (1), 14–23, 2002.

81. Galeano, M. et al., Adeno-associated viral vector-mediated human vascular endothelial growth factor gene transfer stimulates angiogenesis and wound healing in the genetically diabetic mouse, *Diabetologia* 46 (4), 546–555, 2003.

82. Kang, R. et al., Gene therapy for arthritis: principles and clinical practice, *Biochem. Soc. Trans.* 25 (2), 533–537, 1997.

83. Levick, J.R., A method for estimating macromolecular reflection by human synovium, using measurements of intra-articular half lives, *Ann. Rheum. Dis.* 57 (6), 339–344, 1998.

84. Lang, F.F. et al., Adenovirus-mediated p53 gene therapy for human gliomas, *Neurosurgery* 45 (5), 1093–1104, 1999.

85. Smith, M.D. et al., The POU domain transcription factor Brn-3a protects cortical neurons from apoptosis, *Neuroreport* 12 (15), 3183–3188, 2001.

86. Okubo, Y. et al., Osteoinduction by bone morphogenetic protein-2 via adenoviral vector under transient immunosuppression, *Biochem. Biophys. Res. Comm.* 267 (1), 382–387, 2000.

87. Siemens, D.R. et al., Viral vector delivery in solid-state vehicles: gene expression in a murine prostate cancer model, *J. Nat. Cancer Inst.* 92 (5), 403–412, 2000.

88. Kalyanasundaram, S. et al., Coacervate microspheres as carriers of recombinant adenoviruses, *Cancer Gene Ther.* 6 (2), 107–112, 1999.

89. Davidson, B.L., Hilfinger, J.M., and Beer, S.J., Extended release of adenovirus from polymer microspheres: potential use in gene therapy for brain tumors, *Adv. Drug Del. Rev.* 27 (1), 59–66, 1997.

90. Beer, S.J. et al., Poly (lactic-glycolic) acid copolymer encapsulation of recombinant adenovirus reduces immunogenicity *in vivo*, *Gene Ther.* 5 (6), 740–746, 1998.

91. Beer, S.J., Hilfinger, J.M., and Davidson, B.L., Extended release of adenovirus from polymer microspheres: Potential use in gene therapy for brain tumors, *Adv. Drug Del. Rev.* 27 (1), 59–66, 1997.

92. Matthews, C. et al., Poly-L-lysine improves gene transfer with adenovirus formulated in PLGA microspheres, *Gene Ther.* 6 (9), 1558–1564, 1999.

93. Sailaja, G. et al., Encapsulation of recombinant adenovirus into alginate microspheres circumvents vector-specific immune response, *Gene Ther.* 9 (24), 1722–1729, 2002.

94. Kuhl, P.R. and Griffith-Cima, L.G., Tethered epidermal growth factor as a paradigm for growth factor-induced stimulation from the solid phase, *Nat. Med.* 2 (9), 1022–1027, 1996. (Published erratum appears in *Nat. Med.* 3 (1), 93, 1997.)

95. Sakiyama-Elbert, S.E., Panitch, A., and Hubbell, J.A., Development of growth factor fusion proteins for cell-triggered drug delivery, *FASEB J.* 15 (7), 1300–1302, 2001.

96. Zisch, A.H. et al., Covalently conjugated VEGF-fibrin matrices for endothelialization, *J. Control. Release* 72 (1–3), 101–113, 2001.

97. Williams, D.A., Retroviral-fibronectin interactions in transduction of mammalian cells, *Ann. N.Y. Acad. Sci.* 872, 109–113, discussion 113–114, 1999.

98. Proctor, R.A., Fibronectin: a brief overview of its structure, function, and physiology, *Rev. Infect. Dis.* 9 Suppl. 4, S317–321, 1987.

99. Varga, C.M., Hong, K., and Lauffenburger, D.A., Quantitative analysis of synthetic gene delivery vector design properties, *Mol. Ther.* 4 (5), 438–446, 2001.

100. Tseng, W.C., Haselton, F.R., and Giorgio, T.D., Transfection by cationic liposomes using simultaneous single cell measurements of plasmid delivery and transgene expression, *J. Biol. Chem.* 272 (41), 25641–25647, 1997.

101. Luo, D. and Saltzman, W.M., Enhancement of transfection by physical concentration of DNA at the cell surface, *Nat. Biotechnol.* 18 (8), 893–895, 2000.

102. Wirtz, S., Galle, P.R., and Neurath, M.F., Efficient gene delivery to the inflamed colon by local administration of recombinant adenoviruses with normal or modified fibre structure, *Gut* 44 (6), 800–807, 1999.

103. Fechner, H. et al., Trans-complementation of vector replication versus Coxsackie-adenovirus-receptor overexpression to improve transgene expression in poorly permissive cancer cells, *Gene Ther.* 7 (22), 1954–1968, 2000.

104. Barbara, G. et al., Interleukin 10 gene transfer prevents experimental colitis in rats, *Gut* 46 (3), 344–349, 2000.

105. Schellingerhout, D. et al., Quantitation of HSV mass distribution in a rodent brain tumor model, *Gene Ther.* 7 (19), 1648–1655, 2000.

106. Huard, J. et al., The route of administration is a major determinant of the transduction efficiency of rat tissues by adenoviral recombinants, *Gene Ther.* 2 (2), 107–115, 1995.

107. Cavanagh, H.M. et al., Cell contact dependent extended release of adenovirus by microparticles *in vitro*, *J. Virol. Methods* 95 (1–2), 57–64, 2001.

108. Mah, C. et al., Improved method of recombinant AAV2 delivery for systemic targeted gene therapy, *Mol. Ther.* 6 (1), 106–112, 2002.

109. Levy, R.J. et al., Localized adenovirus gene delivery using antiviral IgG complexation, *Gene Ther.* 8 (9), 659–667, 2001.

110. Klugherz, B.D. et al., Gene delivery from a DNA controlled-release stent in porcine coronary arteries, *Nat. Biotechnol.* 18 (11), 1181–1184, 2000.

111. Klugherz, B.D. et al., Gene delivery to pig coronary arteries from stents carrying antibody-tethered adenovirus, *Hum. Gene Ther.* 13 (3), 443–454, 2002.

112. Abrahams, J.M. et al., Endovascular microcoil gene delivery using immobilized anti-adenovirus antibody for vector tethering, *Stroke* 33 (5), 1376–1382, 2002.

113. Pandori, M., Hobson, D., and Sano, T., Adenovirus-microbead conjugates possess enhanced infectivity: a new strategy for localized gene delivery, *Virology* 299 (2), 204–212, 2002.

114. Parrott, M.B. and Barry, M.A., Metabolic biotinylation of recombinant proteins in mammalian cells and in mice, *Mol. Ther.* 1 (1), 96–104, 2000.

115. Parrott, M.B. and Barry, M.A., Metabolic biotinylation of secreted and cell surface proteins from mammalian cells, *Biochem. Biophys. Res. Comm.* 281 (4), 993–1000, 2001.

116. Lee, J.H. et al., Engineering novel cell surface receptors for virus-mediated gene transfer, *J. Biol. Chem.* 274 (31), 21878–21884, 1999.

117. Segura, T. and Shea, L.D., Surface-tethered DNA complexes for enhanced gene delivery, *Bioconj. Chem.* 13 (3), 621–629, 2002.

118. Kneuer, C. et al., A nonviral DNA delivery system based on surface modified silica- nanoparticles can efficiently transfect cells *in vitro*, *Bioconj. Chem.* 11 (6), 926–932, 2000.

119. Manuel, W.S., Zheng, J.I., and Hornsby, P.J., Transfection by polyethyleneimine-coated microspheres, *J. Drug Target.* 9 (1), 15–22, 2001.

120. Zheng, J., Manuel, W.S., and Hornsby, P.J., Transfection of cells mediated by biodegradable polymer materials with surface-bound polyethyleneimine, *Biotechnol. Prog.* 16 (2), 254–257, 2000.

121. Bielinska, A.U. et al., Application of membrane-based dendrimer/DNA complexes for solid phase transfection *in vitro* and *in vivo*, *Biomaterials* 21 (9), 877–887, 2000.

122. Ledley, T.S. and Ledley, F.D., Multicompartment, numerical model of cellular events in the pharmacokinetics of gene therapies, *Hum. Gene Ther.* 5 (6), 679–691, 1994.

CHAPTER 39

Gene Therapy Stents for In-Stent Restenosis

Ilia Fishbein, Itay Perlstein, and Robert Levy

CONTENTS

0-8493-1934-X/05/$0.00+$1.50
© 2005 by CRC Press LLC

39.1 DELIVERING GENE THERAPY WITH A STENT

A paradigm shift has taken place over the past decade with the emergence of stent angioplasty as a major therapeutic advance for treating coronary artery disease. Coronary stenting typically utilizes balloon expandable metallic meshwork stents (Figures 39.1A and 39.1B) that provide immediate relief to arteries obstructed by either progressive arteriosclerosis or thrombosis or both. A second wave of novel technology is now emerging in this field with the advent of drug eluting stents to prevent the development of in-stent restenosis (ISR). ISR is a complex disease process (see below) in which the angioplasty-injured artery develops a secondary obstructive injury response (Figure 39.1C). At this time laboratory investigations are actively exploring the next frontier for stent development, which will involve molecular therapy constructs configured on stents that will utilize either gene therapy vectors or cellular therapies with genetically engineered cells. The use of stents as delivery systems for molecular therapy constructs offers the greatest promise of all of the advances thus far, since it can enable not only therapies to address ISR, but treatment of the underlying vascular disease, and even downstream disease processes that may impact on individual patients. This chapter will approach the subject of gene delivery stents by reviewing the challenges posed by stent angioplasty and how they can be addressed with relevant candidate genes and delivery systems.

39.2 IN-STENT RESTENOSIS: THE SCOPE OF THE PROBLEM

The prognosis of patients with advanced coronary artery disease has dramatically improved since the percutaneous coronary angioplasty (PCA) was introduced in clinical practice in the late 1970s.[1] Presently over 1.5 million percutaneous coronary interventional (PCI) procedures are performed worldwide each year.[2] Currently more than 75% of PCI involve stent deployment.[2] The exponential increase in stent use during the last 5 years has been accompanied by an increasing awareness of PCI's "Achilles heel," coronary restenosis. Restenosis, a renarrowing of angioplastied arterial segments, compromises results achieved by PCI in 30 to 50% of patients in a time frame of 3 to 6 months.[3] Animal[4] and clinical[5] studies have demonstrated that restenosis is a combination of neointimal formation and arterial remodeling, with the latter being responsible for roughly 65% of lumen loss.[5] Rigid stents deployed at the PCI sites virtually eliminate inward remodeling[6] and maintain stable external dimensions of the artery. Elimination of remodeling due to stenting was shown to decrease the restenosis rate significantly in the vast majority of patients referred for coronary revascularization.[7] However, in about 20% of patients, restenosis develops despite the presence of a stent,[8] since, in contrast to remodeling, the neointimal component is actually exaggerated by stenting.[9]

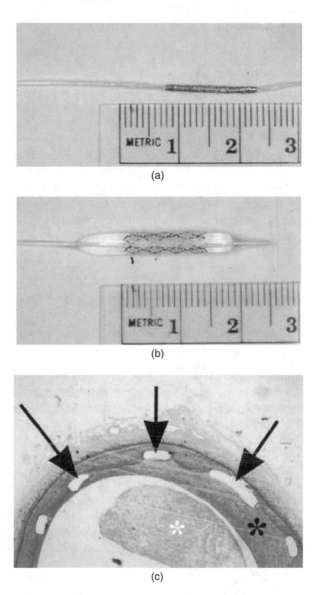

Figure 39.1 Stent angioplasty: Crown® coronary stent (Cordis, Warren, NJ) crimped over an angioplasty balloon (A) prior to deployment and (B) following mock deployment. (C) Porcine coronary in-stent restenosis lesion. Neointimal tissue (black asterisk) covering the stent struts (arrows), and large thrombus (white asterisk) result in partial blockage of the lumen. (Hematoxylin and eosin staining; 50 × magnification.)

Since the pathophysiology and potential treatment of restenosis after balloon angioplasty and after stenting are not identical, the vascular renarrowing in the stented arterial segments was recently categorized as ISR, an independent nosologic entity.[10] The exact sequence of events leading to the ISR is not completely known yet. Moreover, it probably differs among patients and even between lesions in the same patient. However, most researchers in this field agree that several interrelated processes triggered by vascular trauma contribute to the neointimal formation and lumen loss after stenting.[10,11]

39.2.1 ISR Mechanisms: Platelet Aggregation, Thrombosis, and Thrombus Organization

Stents are routinely deployed under a pressure of 10–16 atm. Forceful stent expansion leads to endothelial denudation, internal elastic lamina fractures, and penetration of stent struts into the media. Exposure of subendothelial thrombogenic material such as collagen and fibronectin to blood causes immediate platelet accumulation, especially in proximity to stent wires. Engagement of the IIb/IIIa integrin receptor with RGD sequences present in fibronectin and vitronectin promotes platelet aggregation and activation with local release of ADP, thromboxane A2, and serotonin. Those compounds recruit new platelets to the injury site, establishing a positive feedback loop.

Mural and parietal thrombi incorporating red blood cells, neutrophils, and platelets are a consistent finding in postmortem specimens of stented coronary segments from patients who died less than 2 weeks after stent deployment.[12] Since activated platelets are a rich source of smooth muscle cell chemoattractants and mitogens such as platelet-derived growth factor (PDGF) and basic fibroblast growth factor (BFGF), mural thrombus might play the role of a scaffold, facilitating repopulation of the thrombus by the smooth muscle cells (SMCs).

39.2.2 ISR Mechanisms: SMC Proliferation, Migration, and Matrix Synthesis

Mitogenic growth factors at the site of stent angioplasty are derived from several cellular sources:

- Direct crushing injury of plaque and healthy media
- Activated platelets attached to denuded subendothelium
- Circulating cytokines with direct proliferative effects on SMCs (AII, thrombin, endothelin), which are normally inactive due to an intact endothelium
- Monocytes transmigrating from blood to the vessel wall and converting to macrophages

Pro-proliferative signaling is activated immediately after stenting and sustained by the cytokines and growth factors released from the blood-borne and resident cells invading the stented arterial segment. Collectively, mitogens acting via protein tyrosine kinase and G protein-coupled receptors elicit inward signaling circuits, which culminate in nuclear activation of c-fos, c-jun, and c-myc with a subsequent G0–G1 shift of quiescent medial SMCs. Human ISR samples demonstrate increased proliferation indices that correlate with the severity of vascular trauma.[13,14] It is important to note that increased proliferation is almost exclusively seen in the medial and neointimal cells located near stent struts, while between struts proliferation is close to zero.[15]

Growth-factor-gradient-directed luminal migration of SMCs and adventitial myofibroblasts across the internal elastic lamina represents a crucial step in neointimal formation. This process requires appropriate cellular signaling in the context of the coordinated modification of the extra-cellular matrix (ECM).[16] The growth factor gradient is maintained by the chemotactic products expressed in the innermost neointima by blood-borne platelets and monocytes. The synthetic activity of SMCs within the developing neointima differs from that of their medial counterparts. Neointimal SMCs produce smaller amounts of contractile proteins at the expense of a marked increase in production and secretion of ECM proteins and glycosaminoglycans. The matrix substance consti-tutes about 90% of the typical restenotic lesion volume with the rest composed of SMCs and macrophages.

39.2.3 Treatment of In-Stent Restenosis

By the time the importance of ISR was recognized, an extensive body of experimental and clinical data related to the treatment of the post-angioplasty restenosis had already accumulated.

Therefore, the first trials for ISR prevention revisited already existing antirestenotic protocols, although in slightly modified settings. The logic behind this approach was straightforward: since arterial narrowing after stenting is solely a consequence of neointimal accumulation, the drugs that inhibit SMC proliferation and synthetic activity are more likely to be effective for ISR than for post-angioplasty restenosis. The clinical success rate of those initial ISR studies was found to be disappointingly low, despite encouraging results in animal experiments.[10] Specifically, only probucol,[17] statins,[18] and troglitazone[19] have shown modest efficacy for ISR prevention in selected patients. The most plausible reason for the discrepancy between the experimental results and human restenosis therapeutic response after systemic drug administration is the higher doses used in the animal studies, which are either hazardous or poorly tolerated in humans.

Radiation-emitting stents present a promising approach for ISR prevention.[20] Ionizing radiation is known to disrupt DNA replication selectively in actively dividing cells, rendering them nonproliferative. Both β- and γ-irradiation sources have been explored for vascular applications.[21] Although several studies (WRIST, SCRIPPS, START) demonstrated impressive results at shorter time points (up to 6 months), this effect was far less apparent at longer follow-up periods, suggesting that vascular irradiation delays rather than prevents in-stent restenosis.[22] Furthermore, use of ^{32}P-β-emitting stents was associated with an unusually high incidence of stent-edge restenosis, or so-called "candy wrapper" lesions.[23] The optimal total dose of irradiation remains unknown as yet. A therapeutic window for radioactive stents appears to be narrow, since low doses (< 10 Gy) might actually promote SMC proliferation, while higher doses (> 30 Gy) applied to mediastinum for the treatment of Hodgkin's lymphoma were shown to be associated with rapid progression of coronary atherosclerosis.[24]

39.3 STENT COMPOSITION AND DESIGN

Currently about 30 companies worldwide manufacture more than 60 different stent devices. They fall into two major categories, balloon-expandable and self-expandable, and are designed in tubular, ring, multi-design, coil, and mesh configurations.[25] Stent material and surface treatment are important determinants of the arterial reaction to stent deployment.[25,26] The majority of stents are manufactured from 316 L stainless steel. Some devices are made of other alloys, such as nitinol, gold, tantalum, and cobalt-chromium alloy. The metal of an expandable stent must exhibit elasticity, radial strength, biocompatibility, and corrosion resistance. Some metallic stents have nonmetallic surface coatings to improve biocompatibility, such as the silicon carbide-coated Tenax stent.

39.3.1 Coated Stents

The use of stent coatings pursues two main goals: to enhance biocompatibility and to provide a controlled release matrix for anti-restenosis drug release. Ideal intravascular devices should not elicit either thrombus formation or a sustained foreign body response; i.e., ideally these materials should be biocompatible. Unfortunately, 316 L stainless steel and nitinol, the alloys most commonly used for stent manufacture, do not exhibit these properties. Therefore, without proper anticoagulation and anti-platelet therapy, the majority of stents are found to be occluded shortly after deployment.

A significant effort has been made to improve stent biocompatibility by engineering surface polymer coatings.[27-31] After deployment the polymer layer located at the interface between the stent and vascular wall shields the artery from the metallic surface and "passivates" the stented site. However, the overwhelming majority of coatings proved to cause more severe biocompatibility problems than bare alloy stents. This problem is apparent with both biodegradable (polylactic acid, polyhydroxybutyrate valerate, polyorthoester[31]) and nonbiodegradable (polyurethane, silicone, polyethylene terephthalate[31]) compounds. The only exception among the biostable polymers is a phospholipid, phosphorylcholine, that has demonstrated a minimal inflammatory response when cross-linked onto the surface of a stainless steel stent.[32] Although phosphorylcholine seems to decrease protein binding and platelet adhesion to a stent, a reduction of in-stent restenosis was not demonstrated.[33] It is of interest that the biocompatibility of biodegradable polymers apparently depends inversely on their degradation rate, since low-molecular-weight polylactic acid coatings (rapidly-degrading) were associated with much stronger local toxicity than their high-molecular-weight (slow-degrading) counterparts.[34] Additionally, the inflammatory response might not be necessarily directed to the polymer per se, but could be caused by manufacturing related characteristics (surface roughness, porosity, presence of contaminants, etc.).[26]

39.3.2 Drug-Eluting Stents

The relative lack of success with stent coatings for arterial wall passivation has shifted the interests of researchers to the paradigm of using a polymer matrix on stent surfaces for drug incorporation and sustained release. Development of drug-eluting stents[35-37] is the most important recent addition to the antirestenosis therapeutic strategy. After validation in the porcine coronary stent angioplasty model,[38,39] several drugs configured on stent surfaces have shown very clear benefits in human clinical trials.[40-45] The drugs studied for the stent coatings possess primarily antiproliferative, antimigration and anti-inflammatory effects, thus matching the core motifs of restenosis pathophysiology. Two of the drug-eluting devices, rapamycin- and taxol-coated stents,[46] have recently received FDA approval for marketing in the U.S., while several other devices incorporating either dexamethasone, actinomycin, batimastat, everolimus, or estradiol are in advanced stages of the approval process.[47]

Rapamycin was introduced into clinical practice in early 1990s as an immunosuppresant, and was later shown to have very powerful antiproliferative[48] and antimigration[49] activity for vascular SMCs. A landmark clinical trial, RAVEL,[50,51] completed in 2002, has demonstrated zero restenosis rate at 6 months time point in human subjects treated with rapamycin stents. However, these clinical results were obtained in a selected group of patients with single *de novo* lesions in large-caliber coronary arteries, which are less prone to restenosis. Indeed, when the same treatment was applied to more complicated cases in patients enrolled in the SIRIUS trial, a 9.2% restenosis rate at 6 months time point was reported.[47] A very similar fate faced the taxol-coated stents developed by Boston Scientific. The initial enthusiasm triggered by the TAXUS I trial[44] that showed zero restenosis incidence in a similarly biased patient population became somewhat subdued when the taxol stent was tried in the "real world" study (ELUTES[52]), which demonstrated 5% stenosis rate at a 12 month follow-up. Several additional coated stent clinical studies with derivatives of rapamycin and taxol, as well as with other antiproliferative, antimigratory, and antiinflammatory compounds (dexamethasone, actinomycin D, batimostat, estradiol) are currently under way.[47] Despite the fact that drug-coated stents have revolutionized the field of interventional cardiology, their long-term safety and efficacy have not been sufficiently studied as yet. Some specific concerns are related to the inhibition of re-endothelialization[53-55] by the drugs with combined negative effects on proliferation and migration, polymer-induced inflammation,[54] accelerated atherosclerosis at sites adjacent to deployed stent[56] and delayed stent thrombosis.[53] Furthermore, late malapposition of stents observed in 21% of intravascular ultrasound studies of patients treated under the RAVEL protocol,[51] though clinically silent is a potential concern.

39.4 MOLECULAR THERAPEUTICS FOR IN-STENT RESTENOSIS: CRITICAL ISSUES

ISR represents an advantageous setting for local gene therapy since coronary arteries are accessible via percutaneous catheterization, a pathological process in the vessel wall is localized and has a definite onset. Moreover, the main disease processes contributing to ISR are identified and studied in sufficient detail to allow for a rational approach for gene therapy. However, dense anatomical structures, the presence of blood flow, and functional properties of the cells constituting the mammalian arterial wall make vascular tissue a more difficult target for gene therapy than many other organs, such as liver or lungs. Nonetheless, many examples of successful reporter and therapeutic gene expression have been published over the last 15 years of research in this field.[57-59] It should be noted that all of the following examples used local delivery catheters rather than gene delivery stents, as discussed below.

39.4.1 Plasmid-Mediated Gene Transfer

The first *in vivo* attempts to transfer plasmid DNA into the arterial wall demonstrated the feasibility of this approach. However, the percentage of transfected cells and the total amount of synthesized transgene product remains low.[60,61] Moreover, transfection achieved with naked DNA is very short-lived and abates within several days. The main reason for poor efficiency of naked DNA is a lack of special mechanisms allowing the DNA molecule to traverse plasma and nuclear membranes, and high sensitivity to enzymatic degradation by nucleases both inside and outside cells. Augmentation of plasmid-mediated transfection based on pressure/volume overload employed in hydrodynamic techniques[62] is rarely applicable to vascular gene delivery, with the exception of *ex vivo* treatment of vascular grafts prior to bypass surgery.[63,64]

The efficiency of plasmid transfection might be improved by the complexation of DNA with cationic lipids. Association with lipids provides DNA with limited stability against degradation and facilitates cell uptake due to nonspecific electrostatic interactions. Recently, Armeanu et al. screened a panel of commercially available transfection agents for their utility in arterial gene delivery and demonstrated that optimized formulation based on the spermine derivative DOCSPER is superior to other transfectants in pig coronary arteries.[65] However, in most cases, liposome-mediated transfection does not reach clinically significant levels, mainly because of inadequate endosomal escape and inability to pass through the nuclear membrane of nondividing cells.

Liposomes incorporating fusiogenic elements of the Sendai/HVJ virus consistently demonstrate higher vascular transfection levels than plain liposomes.[66,67] This augmentation of transfection was tracked to the ability of fusiogenic liposomes to attain direct entry to cytoplasm, bypassing the endosomal/lysosomal degradative pathway.[68] Also, incorporation of targeting and anchoring peptides recognizing integrin receptors[69] and ECM glycosaminoglycans[70] significantly augments the efficiency of liposomal vascular gene transfer.

39.4.2 Replication-Defective Adenovirus-Mediated Gene Transfer

Adenoviral vectors (Ad) that have been engineered to be replication defective are currently the most powerful tools for vascular gene transfer.[71,72] Transgene expression achieved with recombinant Ad in *in vivo* models of vascular injury is 2 to 3 orders of magnitude higher than with liposomal or retroviral vectors.[72] Recombinant Ad are nonpathogenic per se, and can be easily produced and stored in high titers (up to 10^{13} particles per ml). Unlike retroviral vectors, Ad is able to transduce both proliferating and nondividing cells, which is an apparent advantage in the setting of angioplastied and stented human vessels, where the majority of cells remain quiescent. The genome of Ad is easily manipulated. These vectors are also large enough to accommodate relatively large transgenes (up to 37 kb).

The most important shortcoming of Ad as a gene vector is the intensive cellular and humoral immune response against virus proteins and transgene products, which usually precludes repetitive vector administration and renders Ad-based gene expression short-lived.[73,74] Since immune-mediated inflammation might contribute to vascular pathology via cytokine release and activation of SMCs, many efforts have been made to generate less immunogenic Ad vectors. In particular, modification of capsid proteins with polyethylene glycol,[75] as well as overexpression of NO synthase,[76] were shown to mitigate Ad-triggered immune attack. Additionally, development of third generation "gutless" Ad with a unique safety profile and prolonged vascular expression of transgene[77,78] might prove to be the main avenue to overcome Ad immunogeneity.

Another disadvantage of Ad vectors is widespread tropism secondary to the ubiquitous prevalence of the Ad receptor, known as the Coxsackie adenovirus receptor (CAR), in epithelial tissues. This problem has only a minor impact on biodistribution of locally delivered Ad. However, it might preclude systemic virus administration. The Ad promiscuity problem has been partially resolved by the ablation of native virus tropism ("deknobbed Ad") in combination with the retargeting of the vector using antibodies or ligands to the receptor abundantly expressed in the target tissue.[79] The use of the tissue-specific promoters restricting transgene expression to SMCs[80] or endothelium[81] also bestows a degree of freedom in configuring Ad vectors for vascular applications.

39.4.3 Adeno-Associated-Virus-Mediated Gene Transfer

The excellent safety profile of adeno-associated virus (AAV) rooted into their lack of pathogenicity in humans and mild-to-nil inflammatory response[82] makes this viral vector an increasingly popular tool in vascular biology.[72,83] A unique advantage of AAV is a high rate of human genome integration into a defined site on chromosome 19.[84] To date no oncogenic transformation has been linked to this specific recombination phenomenon.

However, the small size of the recombinant gene cassette (4.7 kb) that might be accommodated into this vector strongly limits the choice of the possible therapeutic transgenes. AAV is also dependent for production and packaging on replicating Ad, which makes amplification cumbersome. High titers are unachievable, and this poses a problem for AAV purification from the admixture with Ad.[72]

39.4.4 Retrovirus- and Lentivirus-Mediated Gene Transfer

Historically, retroviruses were the first gene vectors used for vascular gene transfer.[85,86] Unlike AAV, they integrate into the human genome stochastically, and thus may disrupt cellular functions critical to maintenance of normal homeostasis, resulting in oncogenic transformation. Moreover, retroviruses infect only cells in mitosis, which is a minor population even in a severely injured arterial wall. However, this problem may be avoided in retroviruses pseudotyped with a coating that includes the G-protein of vesicular stomatitis virus.[87] These shortcomings of retroviral systems have resulted in relatively limited current use of retroviruses in vascular research, almost exclusively limited to *ex vivo* transduction of cells or whole vessels with subsequent reimplantation *in vivo*.[88] Lenti-viruses are RNA viruses with comparable properties to retroviruses. However, lenti-viruses can transduce nondividing cells. These vectors have not been studied as yet for use with ISR.

39.4.5 Antisense-Based Strategies

Conceptually plasmid- and virus-mediated gene therapy is intended to introduce a new functionality or modify an existing genetic function on transcription level. In contrast, antisense oligonucleotides,[89] ribozymes,[90] DNAzyme,[91] decoy constructs,[92] and siRNA[93] may disrupt an inappropriately enhanced function by inhibition of translation. Short nucleotide sequences are able to enter cells both *in vitro* and *in vivo*. Their cellular uptake might be further enhanced by complexation with cationic lipids or polymers. No specific immune response is elicited to antisense sequences, which makes repetitive administration possible. Successful use of antisense constructs, transcription factor decoys, ribozymes, and DNAzymes for restenosis prevention in animal models has been described.[58,59,72,90,92,94,95] No published study has yet examined the utility of siRNA for treating vascular injury.

39.4.6 Cell-Mediated Gene Therapy

The concept of using genetically modified cells as a gene therapy strategy was studied early in the evolution of gene therapy. In general, *in vitro* transfection and transduction methods combined with a positive selection of stably modified cells could bestow a much higher yield and uniformity of transgene expression than the best examples of *in vivo* gene vector delivery. *Ex vivo* modified cells (allogenic or autologous) might be expanded *in vitro* to the needed amount and transplanted in the patient either surgically or using an interventional device such as a catheter or stent. Recent insights in the differentiation and plasticity of stem cells[96] could allow for a potentially unlimited supply of cellular material regardless of the difficulties in procuring some cell types. Currently cell-based gene therapy protocols are investigated most intensively in orthopedic[97] and cardiovascular[98-100] applications, as well as in cancer[101] and diabetes research.[102]

39.5 GENE THERAPY MOLECULAR TARGETS IN RESTENOSIS

Excessive accumulation of neointimal tissue in ISR is a net result of several interrelated processes occurring in injured arterial walls. SMC proliferation, inward migration, local thrombosis, inflammation, and the increase of ECM significantly contribute to restenotic lesion development. The multifaceted nature of the pathologic process in the injured artery provides numerous targets where gene therapeutic interventions might be applied (Table 39.1). On the other hand, this multiplicity of pro-restenotic processes makes it very unlikely that the manipulation of any single gene will completely eliminate ISR.

Intuitively, therapeutic interventions that reduce the SMCs' accumulation in the neointima, which is a common denominator of ISR, could prove to be most effective for prevention of restenosis. Indeed, antithrombotic gene therapy is able to prevent thrombus formation following stenting and thus avert thrombus repopulation of organized thrombus by the activated SMCs. However, this approach would probably be ineffective for the prevention of SMC proliferation in the media, where the main stimulus for SMC proliferation is stent-induced vascular trauma and sustained cellular activation as a part of the foreign body reaction. In contrast, expression of antiproliferative genes in theory will negate neointimal SMC proliferation irrespective of the origin of involved cells.

39.5.1 Antiproliferative Gene Therapy

Gene therapy approaches to inhibit inappropriate proliferative activity in vasculoproliferative disorders have been extensively investigated.[57-59,72] These studies employed either destruction of cells entering the S-phase of the cell cycle (a "cytotoxic" approach), or an inhibition of cell cycle entry and progression (the "cytostatic" approach). The most cited example of the cytotoxic strategy is the local intramural expression of the herpes simplex virus thymidine kinase isozyme that converts a relatively nontoxic prodrug, ganciclovir, into a highly potent phosphate derivative, which interrupts DNA synthesis in dividing cells, thus killing them. This enzyme/prodrug combination was shown to be effective in rat,[103] rabbit,[104] and pig[105] arterial injury models, causing significant (up to 80%) decrease in neointimal mass. Similarly, local vascular expression of cytosine deaminase with subsequent systemic administration of 5-fluorocytosine decreased restenosis indices in a porcine coronary angioplasty model.[106] Notably, both combinations of a gene and prodrug possess marked bystander effects due to an excess of the activated prodrug from the destroyed cells and its uptake by the neighbor cells that do not express converting enzyme. Another powerful cytotoxic approach is based on the transduction of Fas ligand (FasL) to the cell populations contributing to neointimal formation.[107,108] Engagement of FasL with Fas receptor (CD 95) ubiquitously expressed on the surface of cells within the arterial wall leads to the apoptosis of the target cells and reduction of

Table 39.1 Candidate Antirestenotic Genes for Stent-Based Delivery

Mode of Action	Gene	Vector	Model	Restenosis Inhibition	Refs.
Antiproliferative (cytotoxic)	Thymidine kinase	Ad	Rat carotid	46%	103
	Thymidine kinase	Ad	Rabbit carotid	49%	104
	Thymidine kinase	Ad	Pig coronary	87%	105
	Cytosine deaminase	Ad	Rabbit femoral	45%	106
	FasL	Ad	Rat carotid	73%	107
	FasL	Ad	Rat carotid	58%	108
Antiproliferative (cytostatic)	cdc2, cdk2	ODN/HVJ	Rat carotid	> 90%	112
	cdc2, cyclin B	ODN/HVJ	Rat carotid	60%	113
	p21	Ad	Rat carotid	49%	114
	p27	Ad	Pig femoral	51%	115
	p16–p27	Ad	Pig coronary	66%	116
	p53	Ad	Rat carotid	95%	118
	p53	Plasmid/HVJ	Rabbit carotid	83%	119
	Nonphosphorylatable Rb	Ad	Rat carotid, porcine femoral	42%, 47%	121
	Rb/E2F chimera	Ad	Rat carotid	37%	122
	E2F decoy	ODN/HVJ	Rat carotid	70%	123
	Truncated PKG	Ad	Pig coronary[a]	35%	124
	PCNA	Ribozyme	Pig coronary[a]	28%	90
	Egr-1	DNAzyme	Pig coronary[a]	30%	95
	Dominant-negative H-ras	Ad	Rat carotid	82%	126
	Gax homeobox	Ad	Rabbit iliac[a]	56%	127, 202
	GATA homeobox	Ad	Rat carotid	50%	128
	IFN-β	Ad	Pig coronary	24%	129
Antimigration	PDGF receptor β	ODN	Rat carotid	41%	130
	TIMP-1	Ad	Rabbit aortic	68%	131
	TIMP-1	Ad	Rat carotid	40%	132
	PAI-1	Ad	Mouse carotid	86%	133
Antithrombotic	Hirudin	Ad	Rat carotid	35%	134
	TFPI	Plasmid/HVJ	Rabbit iliac	54%	135
	TFPI	Ad	Rabbit carotid	43%	136
	Prostacyclin synthase	Plasmid/HVJ	Rat carotid	30%	137
Re-endothelialization	VEGF	Plasmid	Rabbit femoral	80%	139
	VEGF	Plasmid	Rabbit carotid	45%	140
	VEGF	Ad	Rabbit aortic	40%	141
	VEGF	Plasmid	Rabbit iliac[a]	58%	142
Mixed mechanism	NOS-3	Plasmid/HVJ	Rat carotid	70%	158
	NOS-3	Ad	Pig coronary	26%	159
	NOS-3	Ad	Rat carotid	72%	160
	NOS-2	Ad	Pig coronary[a]	37%	162
	NOS-2	Plasmid	Pig coronary[a]	45%	163

[a] Indicates studies that employed stenting of injured arterial segments.

Abbreviations: Ad: replication defective adenovirus; ODN: oligodeoxynucleotides; HVJ: hemagglutinating virus of Japan.

the neointimal mass. This strategy partially recapitulates the natural history of the healing response after angioplasty and stenting, since apoptosis has been recognized as an important factor determining benign vascular remodeling after trauma.[109]

The mammalian cell cycle is very intricately controlled, allowing fine-tuning of proliferation rates as a function of both intracellular and extracellular microenvironment dynamics.[110,111] The central motif of this regulation is the interaction between cyclin dependent kinases (CDK) and their cognate cyclin counterparts, which drive the cell through several checkpoints, after which cell division is imminent.[110] Therefore, the core machinery of the cell cycle became a pivotal target of cytostatic gene therapy strategies. Local vascular downregulation of two kinases, CDK2 and CDC2,[112,113] as well as a cyclin B1,[113] with antisense oligodeoxynucleotides (ODN) resulted in a 60% reduction of the neointima to media ratio in a rat carotid angioplasty model. A similar effect was observed following local delivery of adenoviral constructs driving overexpression of naturally occurring CDK inhibitors p21[114] and p27.[115] A chimeric p16–p27 fusion construct was recently shown to be especially active resulting in threefold reduction of neointimal thickness in angioplastied porcine coronaries.[116] The tumor suppressor gene product p53 is able to inactivate cyclin–CDK complexes in G1 phase via a p21 functional enhancement. Notably, deficient p53 activity in human medial SMCs might create a predisposition for clinical restenosis.[117] Overexpression of p53 in balloon-injured rat arteries using adenoviral-vector-[118] or Sendai-virus-facilitated liposomal gene transfer[119] significantly decreased neointimal hyperplasia. Since upregulation of p53 expression also causes SMCs apoptosis,[120] this molecular target might combine the potential of both cytostatic and cytotoxic gene therapies.

Another important checkpoint protein, Rb, in its active (hypophosphorylated) form binds and sequesters the transcription factor E2F. When Rb is phosphorylated in response to cyclin–CDK-mediated signal propagation, it changes conformation and releases E2F. The latter translocates into the nucleus, where it activates promoters of immediate early genes, essential for the completion of proliferation. The gene for the truncated (nonphosphorylatable) form of Rb delivered using Ad vector to injured rat and pig arteries[121] was proven to be an effective antirestenotic construct. Also, Ad transfer of a E2F–Rb fusion chimera with constitutive repression of E2F activity decreased neointimal formation in the rat carotid angioplasty model.[122] Inactivation of E2F was also achieved by the transfection of short decoy sequences that are identical to the sequence E2F naturally recognizes in the promoters of target genes. This strategy proved to be an effective antirestenotic approach in animal models.[123] Moreover, it was assessed in the PREVENT clinical trial for the *ex vivo* pressure-facilitated transfection of autologous grafts during bypass surgery, and showed very promising results at a 1-year time point.[64]

Neointimal formation after vascular injury is associated with attenuated signaling through the protein kinase G (PKG) pathway, which is regulated by the local level of cyclic GMP. Ad transduction of SMCs with a constitutively active form of PKG reduced serum-induced migration and upregulated serum-deprivation-triggered apoptosis in SMCs.[124] Catheter-mediated intramural delivery of this Ad construct in stented pig coronaries resulted in a 35% reduction of luminal stenosis 4 weeks following stenting.[124] Chimeric DNA–RNA hammerhead ribozymes specific to the proliferating cell nuclear antigen (PCNA), a cofactor for the DNA polymerase, catheter-delivered to the stented porcine coronaries, also moderately reduced thickness of neointima overlaying struts 4 weeks after stent placement.[90]

Early growth response factor 1 (Egr-1) is a zinc-finger transcription factor involved in transactivation of the processes linking injuring stimuli (shear stress, mechanical trauma, hypoxia, and oxidative stress) with the expression of effector molecules, such as PDGF, FGF-2, and thrombospondin.[125] DNAzyme-targeting human Egr-1 mRNA was able to cleave Egr-1 transcripts and totally shut off activation of Egr-1 in response to mitogenic stimulation *in vitro*. Intracoronary catheter delivery of Erg-1–DNAzyme simultaneous with stenting resulted in a 30% reduction of ISR in a pig model.[95] The proto-oncogene H-ras is one of the key control elements in several mitotic pathways. A dominant negative mutant of H-ras was engineered into recombinant Ad and intraluminally delivered into denuded rat carotid arteries. A 15-min exposure to the dominant negative H-ras construct led to an 82% inhibition of ensuing neointima.[126] Gax[127] and Gata-6[128] are homeobox genes that function as transcription factors for growth arrest genes. Their increased activity correlates with a quiescent phenotype in vascular SMCs. Adenovirus-driven expression of those two homeobox genes in the balloon-injured vessel wall resulted in a 50 to 70% decrease of neointimal mass in rabbit and rat models, respectively. Interferon β, a cytokine with SMC antiproliferative effects *in vitro* was assessed in pig coronary artery stent angioplasty and demonstrated moderate effectiveness (a 24% reduction of the neointima-to-media ratio).[129]

39.5.2 Antimigration Gene Therapy

Since the vast majority of SMCs in normal and atherosclerotic arteries reside in the medial layer, no significant injury-triggered neointimal thickening is possible without directed centripetal migration of SMCs. Key elements of this process include activation of specific cellular signaling pathways, reorganization of cytoskeletal fibers, and local changes in ECM composition. Migration-related signaling is mainly induced by the growth factors (PDGF, NGF) acting via cell membrane receptors with intrinsic tyrosine kinase activity. Indeed, local delivery of an antisense construct for the PDGF receptor β inhibited neointimal formation in a rat carotid angioplasty model.[130]

The physical characteristics of ECM in the quiescent adult arterial wall makes directed movement of cells difficult primarily due to the dense network of collagen and elastin. Balloon/stent injury shifts the synthetic pattern of SMCs to increase synthesis and secretion of ECM-modifying proteases (metalloproteinases and plasminogen activators). These enzymes partially digest ECM, facilitating the traversal of SMCs through the vessel wall layers. Therefore, interference with matrix remodeling provides an attractive way to control the vascular hyperplastic response. To this end, local[131] and systemic[132] administration of Ad vectors encoding a gene for the tissue inhibitor of metalloproteinases (TIMP-1) decreased restenosis in rodent models. Similarly, Ad transfer of plasminogen activator inhibitor (PAI-1) to PAI-1 deficient mice reduced the hyperplastic neointimal response.[133]

39.5.3 Antithrombotic Gene Therapy

Intramural and parietal thrombi are consistently formed in stented arterial segments. Though nonoccluding, these thrombi help promote SMC infiltration and significantly contribute to the size of ISR lesions. Therefore, gene therapy to inhibit the clotting cascade or to promote lysis of the formed thrombus was explored along with antiproliferative and antimigration approaches. A leach-derived anticoagulant, hirudin, locally expressed in rat carotid artery using an Ad vector resulted in 35% decrease of neointimal formation after angioplasty, without systemic anticoagulation.[134] A similar level of neointimal reduction was achieved with the tissue factor pathway inhibitor over-expressed in injured rabbit arteries using HVJ–liposomal transfection[135] or Ad vector.[136]

The propensity of activated platelets to adhere to dysfunctional endothelium is critically dependent on the ratio between prostacyclin and thromboxane A2. This ratio might be shifted due to the prevalence of antiadhesive factors as a result of a local increase of cyclooxygenase activity. Indeed, adenoviral-vector-derived prostacyclin synthase, a cyclooxygenase-family enzyme, both prevented thrombus formation and reduced neointimal hyperplasia in a rat carotid angioplasty model.[137]

39.5.4 Gene Therapy for Re-Endothelization

Recently, significant efforts have been made to promote vascular healing via accelerated restoration of a functional endothelial layer at the site of vascular trauma.[138] Once the endothelium has overgrown the angioplastied and stented arterial segment and the local balance of antithrombotic and antiproliferative factors is reestablished, the SMCs regain a quiescent phenotype, thereby stabilizing lesions. During angiogenesis, as well as following injury, endothelial cell proliferation is primarily controlled by the availability of vascular endothelial growth factor (VEGF) and basic fibroblast growth factor (BFGF-2). However, BFGF-2, in addition to being an endothelial mitogen, upregulates SMC proliferation as well, which is an apparent disadvantage of this molecule in restenosis settings. On the other hand, VEGF is endothelium specific, and thus was widely studied as a putative enhancer of re-endothelialization.

A reduction of neointimal tissue that correlated with an accelerated time course of endothelial recovery was shown for rabbit vasculature after intraluminal[139] or perivascular[140] delivery of VEGF plasmid DNA. Qualitatively similar findings were observed in the rabbit aortic angioplasty model with an Ad–VEGF construct.[141] Notably, the efficacy of VEGF overexpression was also confirmed in an experimental model involving stenting.[142] Moreover, several clinical trials examined VEGF-based gene therapy with plasmids[143] and Ad vectors[144] for peripheral and coronary arterial disease. Those studies (all uncontrolled) consistently demonstrated safety, feasibility, and in some cases possible therapeutic effects in terms of clinical and angiographic restenosis prevention.

39.5.5 Nitric-Oxide-Related Genes: Pleiotropic Therapeutic Effects

The multifaceted nature of the mechanisms underlying ISR hinders the ultimate success of therapy. However, another aspect is the fact that some genes have pleiotropic actions on different related elements of ISR pathogenesis. For example, experimental success with plasminogen modulation discussed above in the context of matrix remodeling might be partially related to the parietal thrombus dissolution due to fibrinolytic effects. Nitric oxide (NO) and enzymes involved in its synthesis (NOS 1, 2, and 3) represent the most versatile pleiotropic modulators of the processes central to restenosis lesion evolution.[145] Indeed, NO inhibits proliferation[146,147] and migration[148] of SMC, induces apoptosis,[149] and possesses marked antiplatelet effects.[150] Moreover, NO induces phenotype modulation of SMCs in injured arteries,[151,152] thus decreasing ECM elaboration.[153] Local production of NO is markedly decreased after balloon injury.[154] Its restoration by NO donors[155,156] or by the enhanced dietary uptake of NO synthesis precursors[155,157] was shown to have an antirestenotic effect. Due to its unique characteristics, NO upregulation is currently the most studied molecular target in gene therapy of ISR. Clear therapeutic efficacy of focal NO synthase overexpression has been demonstrated with liposomal[158] or Ad[159,160] vectors for NOS isoforms in rat and pig models. Recently these results were corroborated in stented rat[161] and pig[162,163] arteries with both nonviral and viral vectors. It is of interest that the beneficial effects of increased physical activity on ISR formation demonstrated in rat model studies was apparently mediated by NO synthesis and was completely reversible by cotreatment with NO synthesis inhibitors.[161]

39.6 GENE DELIVERY STENTS

The idea of using stents as gene delivery platforms poses a number of important challenges, both biological and technical.

First, the surface area of the majority of stent devices is less than 1 cm^2. Thus, the most obvious limitation is the amount of gene vector that can be accommodated on the artery-contacting surface of the stent. Therefore, gene stents require dense surface packaging of highly effective vectors.

Second, since gene vectors cannot be loaded on a bare metallic stent, some type of coating is needed to contain vector particles. Both natural and synthetic polymers were intensively studied for medical applications over the last decades and a few coated stents were approved for human use.[164,165] However, vascular application of polymer coatings poses additional requirements because of direct contact with blood flow and the high propensity for vasculature implants to develop localized inflammatory reactions and thrombi. Indeed, a seminal study of van Giesen at al.[31] demonstrated that the presence of thin polymer coatings on metallic stents fabricated from a variety of biodegradable and nondegradable polymers caused acute and chronic inflammatory reactions with a persistent foreign body response.

Third, gene therapeutics configured on the stent should not compromise the mechanical properties of the stent. Nonuniform coatings were shown to prevent an optimal opposition of stent struts to the vessel wall, especially when eccentric plaque is present.[25]

Fourth, the release rate of genetic material from the stent platform should be protracted enough to match the natural course of the pathogenesis of restenosis. In-stent restenosis lesions are usually completely developed by 6 months. However, the main processes critical to the development of restenosis are activated earlier (0–30 days). Thus, a practical aim would be to provide the release of effective amounts of genetic material during the first month after stenting.

Fifth, polymer-based gene coatings should not increase the thrombogenicity of stent surfaces. Currently, acute thrombosis is a relatively rare complication of stenting due to the aggressive use of antiplatelet agents and anticoagulants. Stent thrombus is a concern because of the potential for life-threatening emergencies and because thrombus can serve as a scaffold for proliferating SMCs. Thus, the material used for stent coating has to be extensively characterized for potential thrombogenicity.

Lastly, stent surface modifications should not preclude the ingrowth of endothelium, since endothelialization is needed to help restore the local balance of antiproliferative, antimigratory, and antithrombotic factors, thus making the stented segment less prone to the neointimal narrowing.

Relatively few studies have explored the strategy of arterial gene delivery using stents as a platform for vector immobilization. From a theoretical standpoint, immobilization of gene therapy vectors on the stent has a number of important advantages. First, this approach can achieve high concentrations of vector–DNA at sites where transgene activity is most desirable,[166] while diminishing distal spread of the vector. Animal and clinical studies have consistently shown that mural thrombosis, SMC activation, and proliferation occur primarily near stent struts.[15] Thus, a relatively small amount of stent-immobilized gene vectors strategically placed at the interface of tissue and implant might be sufficient to produce clinically significant transfection of invading cells, rendering them quiescent. Moreover, in that circumstance, the shearing effect of the bloodstream should not be a critical determinant of vector persistence in the treated vascular segment, since the gene vector will be physically shielded by the deployed stent itself. Furthermore, immobilization hypothetically decreases hematogenic distal dissemination of vectors with subsequent decrease of their exposure to the antigen presenting cells in spleen and bone marrow. Hence, the extent of the immune response elicited by the immobilized gene vector is presumably lower than in the case of the matched amount of systemically administered free vector. The studies that have investigated gene delivery stents thus far vary regarding the type of vector (cells, adenovirus, naked DNA), the type of polymer coating (gelatin, PLGA, etc.) and the way in which the vector is attached (bulk immobilization versus surface tethering) (Table 39.2). At present, the majority of gene delivery stent studies have employed easily traceable reporter genes, rather than therapeutic gene constructs relevant to the pathophysiology of ISR.

For several years, our laboratory has been involved in stent-based gene delivery platforms. Several strategies employing either plasmids or adenoviral vectors, as well as a number of different polymer coatings, were investigated *in vitro* and in animal experiments.

39.6.1 Plasmid Delivery from a Stent: Bulk Immobilization

Our laboratory was the first to demonstrate plasmid DNA delivery from a stent platform.[167] We examined a biodegradable polymer, polylactic-polyglycolic acid (PLGA), as a stent coating with DNA dispersed in the bulk of the PLGA. To formulate the emulsion coating, GFP plasmid in Tris-EDTA buffer was emulsified into a PLGA solution in chloroform using vortexing on the ice. Stent coating was achieved by multiple cycles of dipping stents into emulsion and evaporation to dryness. In a typical formulation, 1 mg of DNA was incorporated in 8 to 9 mg of PLGA. Using rhodamine-labeled DNA we documented a uniform distribution of plasmid DNA material throughout the coating, which was present both on the stent struts and between them in the form of a thin web-like film. No fragmentation of PLGA was noted following forceful stent deployment (Figure 39.2A). *In vitro* dissolution studies demonstrated a marked burst effect with 50% of the DNA load released after 1 h. The rest of the plasmid DNA was released according to first order kinetics over 10 days (Figure 39.2B). Released DNA retained an intact structure (Figure 39.2C) and was able to transfect SMCs when complexed with lipofectamine. When DNA–PLGA-coated stent wires were placed in SMC culture, marked localized transfection (7.9%) of cells proximal to the rod was observed. This

Table 39.2 Vectors, Transgenes, and Polymer Coatings Examined for Stent Delivery

Vector	Study	Model	Transgene	Polymer	Immobilization	Arterial Expression
Plasmid	Klugherz et al.[167]	Pig coronary	GFP	PLGA	Bulk	1.14% (media)
	Perlstein et al.[168]	Pig coronary	GFP	Collagen, PLGA	Bulk	10.4% (neointima)
	Takahashi et al.[172]	Rabbit iliac	GFP, β-Gal, luciferase	Polyurethane	Bulk	> 3-fold increase over control
	Feldman et al.[173]	Rabbit iliac	β-Gal	NA	NA	ND
Adenovirus	Levy et al.[174]	—	GFP, TK	Collagen	Surface	ND
	Klugherz et al.[175]	Pig coronary	GFP	Collagen	Surface	17% (neointima)
	Abrahams et al.[176]	Rat carotid	GFP	Collagen	Surface	13.3% of cells in thrombus
	Yuan at al.[177]	Pig carotid	β-Gal	Gelatin	Bulk	38.6% (media)
	Ye at al.[179,180]	Rabbit iliac	β-Gal	PLLA–PCL	Bulk	Moderate (media)
	Rajasubramanian et al.[178]	—	β-Gal	PLLA–PCL	Bulk	ND
	Nakayama et al.[182]	Rabbit carotid	β-Gal	Gelatin	Bulk	Moderate
Endothelial cells	Dichek et al.[184]	Sheep coronary	tPA	Fibrinogen	Surface	ND
	Flugelman et al.[185]	Sheep coronary	β-Gal	Fibrinogen	Surface	ND
	Scott et al.[186]	Sheep coronary	Simlan virus 40 large T antigen	Fibrinogen	Surface	ND
	Eton et al.[88]	Pig iliac	β-Gal	Polyurethane	Bulk	Moderate
FPC	Shirota et al.[191]	Hybrid tissue	—	Gelatin	Surface	ND
SMCs	Panetta et al.[189]	Pig coronary	GFP	Fibronectin	Surface	High

Abbreviations: TK: thymidine kinase; PLLA–PCL: poly(L-lactic) acid/poly-e-caprolactone blend; NA: not applicable; ND: not determined.

finding is intriguing since naked DNA is a poor transfection agent for cultured SMC. We hypothesize that direct contact of cell membranes with DNA molecules entrapped in the PLGA matrix facilitates transfection. Pig angioplasty studies with our DNA-delivery stents showed transfection rates of 1.14% of total cells in the stented segment with preferential transfection of the media, and to a lesser extent of neointima and adventitia (Figures 39.2D and 39.2E). Distal spread of vector DNA to the lungs was detected by PCR in two of six animals.

39.6.1.1 *Denatured Collagen-Coated Stents Demonstrate Facilitated Plasmid Transfection*

Although DNA–PLGA stents demonstrated promise, the relatively low levels of transfection achieved in the target neointima layer and the fact that chronic inflammatory activity due to foreign body response was reported in arteries exposed to PLGA[31,34] prompted us to explore alternative formulations for stent coating. To this end, a formulation based on heat-denatured collagen for plasmid DNA delivery was investigated.[168] The rationale for the use of denatured rather than native collagen for stent coating is twofold. Native collagen gels, because of their fragile nature, proved to be unsuitable for stent coatings; these preparations broke down upon stent deployment with immediate shedding of the native collagen coating. Our *in vitro* experiments have also demonstrated a dramatic increase in transfection in cells grown on denatured collagen that was not detected with native collagen substrates ($18.3 \pm 1.8\%$ versus $1.0 \pm 0.1\%$, respectively) (Figure 39.3A–39.3C).[168]

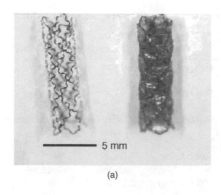

(a)

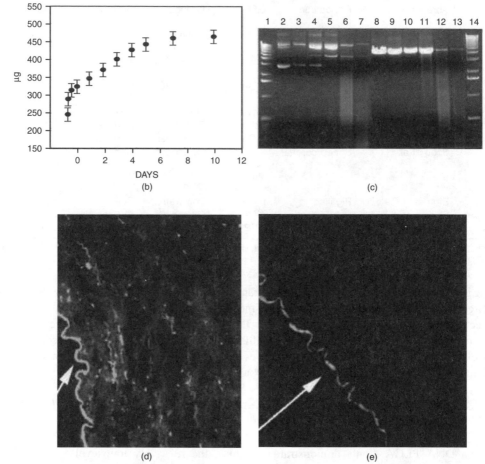

(b)   (c)

(d)   (e)

Figure 39.2 **(See color insert following page 336)** DNA delivery stent: DNA–PLGA micro-emulsion coated stent. (A) Balloon-expanded stents comparing uncoated (left) and DNA–PLGA coated (right). The latter incorporates Nile Red dye for better visualization of the coating. (B) Cumulative release of DNA from coated stents in Tris-EDTA at 37° C, quantified by a thiazole orange fluorescence assay. (C) Agarose gel electrophoresis confirms intact plasmid DNA release from coated stents. Original plasmid (lane 2); samples obtained at 1, 3, 24, and 72 h, and 10 days (lanes 3–7); same as lanes 2–7, but after *Bag*II digestion (lanes 8–13); DNA ladders (lanes 1 and 14). (D, E) Photomicrographs of the stented segments of pig coronary arteries explanted 7 days following stent deployment. The animals were stented with (D) DNA–PLGA coated stents or (E) control PLGA-coated stents. Punctuate GFP fluorescence, seen predominantly in the medial layer of D, is absent in E (FITC; 200 X magnification). (From Klugherz, B.D., Jones, P.L., Cui, X. et al., Gene delivery from a DNA controlled-release stent in porcine coronary arteries, *Nat. Biotechnol.* 18, 1181, 2000. With permission.)

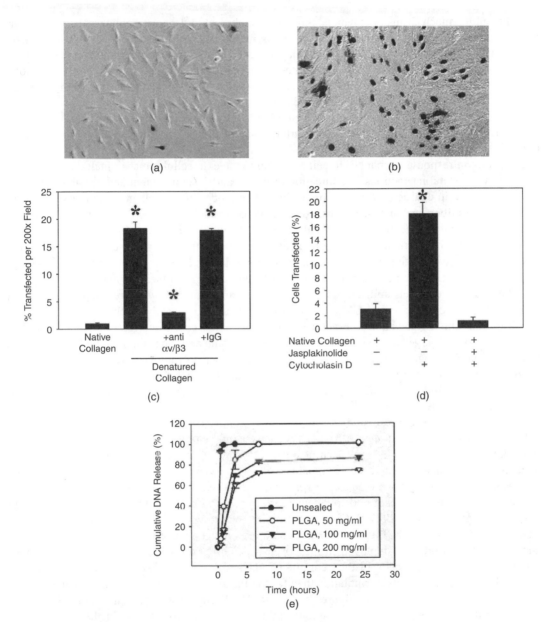

Figure 39.3 **(See color insert)** DNA delivery stent coated with denatured collagen for facilitated plasmid transfection. (A) Smooth muscle cells (A10) grown on native collagen substrates assume an elongated morphology and exhibit poor transfection rates, as indicated by the paucity of β-gal-expressing cells (dark blue). (B) A10 cultures grown on denatured collagen are denser, and show a higher transfection efficiency than (A). (A and B: Bright-field photomicrographs of X-gal-stained cultures, 200 X magnification.) (C) Increased transfection in cells grown on denatured collagen versus native collagen is completely blocked by the specific anti-αvβ3 integrin antibody. Control antibody (nonspecific IgG) had no effect on transfection. (D) Cells grown on native collagen after the treatment with F-actin depolymerization agent cytocholasin-D demonstrate a significant transfection rate increase. Inhibition of the cytocholasin-D effect with jasplakinolide results in a low transfection rate, comparable to that observed for native collagen (*p < 0.001). (E) Cumulative plasmid DNA release from denatured-collagen-coated stents shows controlled release of DNA load over 24 h as a function of the PLGA concentration in the sealant PLGA layer. (From Perlstein, I., Connolly, J.M., Cui, X. et al., DNA delivery from an intravascular stent with a denatured collagen-polylactic-polyglycolic acid-controlled release coating: mechanisms of enhanced transfection, *Gene Ther.* 10, 1420, 2003. With permission.)

The unique mechanisms of enhanced transfection inherent to denatured collagen-based systems were shown to occur via $\alpha v \beta 3$ integrin receptor interactions and associated cytoskeletal actin modifications. The relevance of denatured collagen stent coatings for vascular transfection is verified by the finding of injury triggered $\alpha v \beta 3$ integrin upregulation in SMCs constituting restenotic lesions.[169,170] The transfection enhancement effect of denatured collagen was completely blocked with an anti-$\alpha v3$ integrin antibody (Figure 39.3C). On the other hand, supplementation of native collagen substrates in cell culture with tenascin-C (TN-C), a protein recognized by the $\alpha v3$ integrin receptor, resulted in a 33-fold increase in arterial smooth muscle cell transfection compared to control.[168]

A common response to changes in cell adhesion to the extracellular matrix includes alterations in actin cytoskeletal interactions. Modification of the G- and F-actin content and distribution might also have a role in enhancing transfection. Cells grown on denatured collagen had marked F-actin enriched stress fibers and intense perinuclear G-actin, compared to those grown on native collagen, which demonstrated F-actin enriched focal adhesions without perinuclear G-actin localization.[168] Cytochalasin-D, an F-actin depolymerizing agent, caused significantly increased SMC transfection in cells cultivated on native collagen compared to control cells (18.0 + 1.8% versus 3.02 + 0.9%) (Figure 39.3D), indicating that actin-related cytoskeletal changes influence transfection.[168]

The cytoskeletal levels and ratio of F- and G-actin are dynamically altered depending on the phase of the cell cycle and the ECM composition. Since DNA is negatively charged, we speculate that high amounts of polyanionic F-actin in the cytoskeleton could interfere with plasmid DNA transfection on a charge basis, since DNA is also polyanionic and thus could compete for cationic lipids via ion exchange, as well as electrically repelling other polyanions. Furthermore, G-actin is an inhibitor of nuclease-I.[171] Thus, switching the cytoskeleton balance in favor of G-actin or modifying the cellular actin distribution might enhance DNA transfection based on decreased plasmid destruction.

The stent coating for these studies was formulated using carbodiimide cross-linked denatured collagen deposited on the stent surface by repeated coating and drying cycles. In typical preparations, a total of 1 mg of collagen was admixed with 500 µg of DNA and was applied as a coating on a 15 mm Crown stent. For vascular implants a thin additional sealant layer of PLGA was formed on the surface of the collagen–DNA matrix since cell culture experiments have shown that the plasmid release rate from denatured collagen is too rapid, with virtually all of the DNA load released in less than 30 min (Figure 39.3E). On the other hand, the presence of the PLGA sealant prolonged DNA release beyond 24 h (Figure 39.3E). Denatured, collagen coated, PLGA-sealed stents containing GFP plasmid were deployed in pig coronaries. GFP expression levels assessed as a percentage of GFP-positive cells in the neointima 7 days after stenting were 9-fold higher for the collagen–DNA coated stents' formulation than for the PLGA–DNA stents[167] described above (10.4 ± 1.23% versus 1.2% ± 0.7%) (Figure 39.4A–39.4C). Furthermore, denatured collagen-PLGA coated stents contained only half the amount of DNA that was formulated in the PLGA–DNA coated stents (500 µg and 1000 µg, respectively).

39.6.1.2 Other DNA Delivery Stents

Recently Takahashi et al.[172] reported arterial transfection using a stent with DNA immobilized within polyurethane, a biostable polymer. The polyurethane coating was formulated by a dipping technique and was 35 µm thick. Depending on the concentration of DNA dispersed within the polyurethane, 40 to 96 µg of DNA was loaded onto the stent. A burst effect during the first 24 h was responsible for the release of 60% of the DNA payload in Tris-EDTA buffer (pH 7.4). Stents loaded with GFP plasmid were deployed into rabbit iliac arteries. Seven days following implantation the authors observed multiple fluorescent foci in the media of stented segments. Using immunofluorescence staining with cell-type-specific antibodies, the researchers demonstrated that both SMCs and macrophages expressed the transgene.

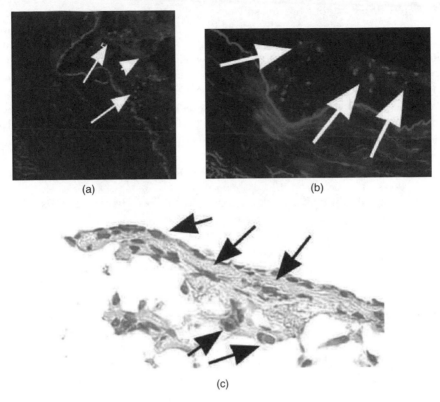

(a)   (b)

(c)

Figure 39.4 **(See color insert)** DNA delivery stent (pig coronary artery model): DNA–denatured-collagen coating with a thin PLGA sealant layer. (A and B) Fluorescent photomicrographs showing arterial GFP expression 7 days after denatured-collagen-coated stent deployment. The sites of enhanced neointimal GFP expression are indicated by the arrows (FITC; 200 X magnification). (C) Photomicrograph of the section parallel to one shown in (B), stained by the GFP immunohistochemistry protocol (VIP; 200 X magnification). (From Perlstein, I., Connolly, J.M., Cui, X. et al., DNA delivery from an intravascular stent with a denatured collagen-polylactic-polyglycolic acid-controlled release coating: mechanisms of enhanced transfection, *Gene Ther.* 10, 1420, 2003. With permission.)

A conceptually different approach for stent-based gene therapy was suggested by Feldman et al.[173] These authors used a microneedle-studded stent, enabling focal delivery of plasmid DNA into media and adventitia of healthy and atherosclerotic rabbit arteries. Microneedles spaced 40 to 100 μm apart were 20 to 140 μm in length with a tip diameter of 0.4 μm, which is in the range of micropipettes used for DNA injection into individual cells. The microneedles were each loaded with 6.7 ± 0.6 μg DNA that was delivered to SMCs in culture with an efficacy similar to that of calcium-phosphate-mediated transfection. No detailed investigation of *in vivo* transfection mechanisms was attempted in this pilot study.

39.6.2 Adenovirus-Mediated Gene Transfer from Stents: The Tethering Strategy

A common denominator of the techniques for plasmid immobilization described above is a bulk incorporation of the vector in a DNA controlled-release-matrix layer adsorbed on the stent surface. The release of the embedded plasmid DNA vectors is driven by the combined effect of vector diffusion and, with biodegradable polymers, matrix degradation. Recently, our group described a novel approach for genetic vector immobilization based on the use of high affinity ligands to tether a vector to a substrate.[174] Theoretically, vector surface tethering is advantageous compared to bulk incorporation in terms of the decreased total amount of genetic material per stent, and potentially provides better control of vector release kinetics. Furthermore, adenoviruses and other viral vectors do not tolerate organic solvents, unlike DNA. In line with this idea, recombinant replication-defective Ad was successfully immobilized in a collagen–avidin gel using biotinylated antihexon antibody for Ad binding. In cell culture this system was shown to enable site-specific localization of either Ad–β-galactosidase or Ad–thymidine kinase (plus ganciclovir).[174] Pursuing this approach we tested a panel of commercially available anti-AdV antibodies, comparing their potential to serve as affinity ligands for Ad tethering. A monoclonal antiknob antibody demonstrated the highest affinity and best results in terms of immobilization density and minimal distal spread of gene vector. The encouraging *in vitro* results with affinity ligand-mediated immobilization of Ad prompted us to use the same technique in a pig coronary stenting model.[175]

39.6.2.1 *Stent-Tethered Ad Using Antiknob Antibodies*

The stents were collagen-coated using carbodiimide to cross-link collagen fibers to stabilize a coating. Surface collagen was activated using a bifunctional cross-linker, N-succinimidyl 3-(2-pyridyldithio) propionate (SPDP). The stents were then reacted with antiknob F(ab)' fragments with subsequent immobilization of recombinant adenoviruses driving expression of green fluorescent protein (AdGFP). The loading capacity of the antiknob antibody-derivatized collagen films was determined by incubations with progressively increasing Ad titers. The maximal amount of Ad associated with collagen-antibody formulation was shown to be 2.5×10^{10} particles per 1 mg of the film (Figures 39.5B and 39.5C). Incubation of the collagen–Ad films in physiologic buffer demonstrated high affinity retention of transduction-competent Ad in films with slow release of 30% of bound virus into the buffer over 20 days. In contrast, when Ad was passively associated with the collagen films without antibody as an affinity adapter, 80% of the Ad payload was released into the incubation buffer after 7 days. In SMC culture both collagen–Ad films and collagen-coated Ad-tethered stents demonstrated intensive though site-specific transgene expression within 100 μm of the film or the stent wire borders (Figure 39.5A and 39.5D).

39.6.2.2 *Transduction Using Ad-Delivery Stents in the Pig Coronary Model*

To ascertain *in vivo* transduction activity of stent-immobilized Ad, the Ad-GFP tethered stents were deployed into pig coronaries.[175] Vessels were harvested for analysis 1 week after stenting. Fluorescent microscopy revealed widespread, though not uniform, GFP expression, which was only minimal in control segments treated with the stents conjugated with nonspecific F(ab)' prior to incubation with Ad-GFP (Figures 39.5E and 39.5F). GFP expression was confirmed by immuno-histochemistry. The morphometry study demonstrated a 17% transduction rate in neointima with a lower percentage of GFP-positive cells in media and adventitia (7% and 1%, respectively) (Figures 39.5E and 39.5F). The actual extent of GFP-positive cells in neointima was probably even higher, since a fraction of the transduced cells was found to be permanently attached to the stent surface and was thus lost during specimen processing. Since the neointima contributes to ISR, this inward gradient of transduction is highly desirable, and probably is a result of direct physical contact of immobilized Ad with innermost elements of the vessel wall. The biodistribution of Ad vector delivered from the stent platform[175] in pigs was studied by PCR analyses of tissue samples using GFP-specific primers. The tissues collected for the analysis were stented segments of the artery, a segment of the manipulated artery distal to the implanted stent, and a segment of nonstented artery, as well as myocardium, liver, kidney, and lung. A GFP signal was consistently found only in the arterial fragment underlying the stent. However, no GFP DNA was observed in adjacent coronary segments and distal organs, even after 35 amplification cycles.

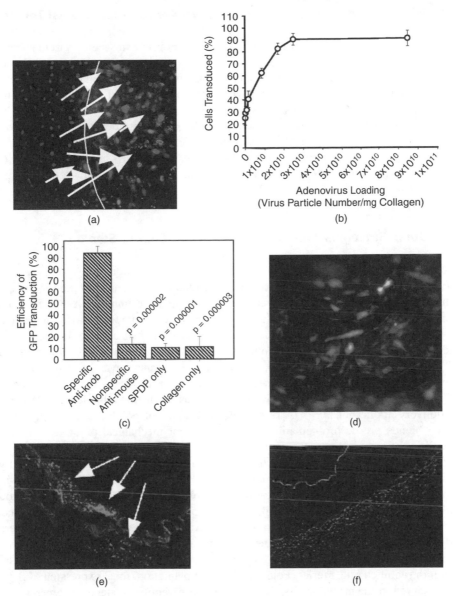

Figure 39.5 **(See color insert)** Adenovirus delivery stent. (A) Strict spatial restriction of transduction by Ad–GFP immobilized on the surface of collagen film using antiknob antibody as an affinity ligand. The A10 cells were DAPI-counterstained. The collagen film's border is delineated by white lines; the arrows point to transduced cells (Merged DAPI/FITC image, 200 X magnification). (B) Viral loading determined by transduction efficiency following exposure of antiknob derivatized collagen films to successively higher levels of Ad–GFP demonstrates asymptotic transduction at 2.5 x 10^{10} viral particles/mg derivatized collagen. (C) Specificity of antibody affinity for vector tethering determined by the extent of transduction in cell culture. Properly derivatized collagen films containing antiknob F(ab)'2 are compared to control films containing either a nonspecific antibody, films reacted with SPDP cross-linker alone, or collagen films that were not exposed to the derivatization procedure. (D) Antibody tethered Ad–GFP delivery from collagen-coated stents in A10 cell culture results in abundant spatially localized GFP expression (FITC, 200 X magnification). (E) Pig coronary photomicrograph demonstrating arterial wall transduction following stenting with collagen-immobilized antiknob tethered adenovirus. Positive GFP expression (arrows) compared with (F) the control stent-treated artery demonstrating typical levels of autofluorescence (FITC, 400 X magnification). Klugherz, B.D., Song, C., DeFelice, S. et al., Gene delivery to pig coronary arteries from stents carrying antibody-tethered adenovirus, *Hum. Gene Ther.* 13, 443, 2002. With permission.)

39.6.2.3   Microcoil-Ad Delivery Using Antiknob Antibody for Arterial Transduction

To prove the clinical versatility of Ad immobilization on the surface of vascular devices, we examined the utility of Ad-mediated gene transfer from a microcoil platform.[176] Microcoils are used clinically for the treatment of intracranial aneurysms. Platinum coil segments were coated with collagen and consecutively conjugated with antknob F(ab)' and Ad–GFP according to the procedure described above for the use of stents.[175] Coil placement into confluent SMC cultures (A10 cells) resulted in highly localized transduction. Coils with antibody-tethered adenovirus were implanted in isolated carotid segments in rats in simulated aneurysms. Highly efficient transduction (13%) of cells invading the intraluminal thrombus was observed 7 days after surgery. The leukocyte origin of the majority of the GFP-positive cells was confirmed by immunofluorescence studies. No transduction of the vessel wall elements or distal spread of transgene was observed at any time following microcoil implantation.

39.6.2.4   Other Strategies Employing Adenovirus-Eluting Stents

Several other groups have investigated focal arterial transduction using stents to deliver Ad vectors. Yuan at al.[177] have reported incorporation of recombinant LacZ adenovirus into a gelatin coating built up on stent surface by the repetitive cycles of stent immersion into gelatin solution and air-drying. The presence of an undisclosed cross-linker led to the formation of a stable polymer layer (5–10 µm) with bulk incorporated adenovirus vector. One week after deployment into porcine coronaries, 38.6% of medial SMC were found to be β-galactosidase-positive. No spread of vector to distal arterial segments, myocardium, liver, lungs, or kidneys was observed. However, the sensitivity of the histochemical method used for atopic β-galactosidase expression is questionable.

A bioengineering group from the University of Texas fabricated a biodegradable porous stent for gene delivery applications.[178-180] A resorbable stent made of polylactic acid–polycaprolactone (PLA–PCL) blends proved problematic due to the inferior mechanical properties in comparison with metallic devices.[25,181] Incorporation of polyethylene glycol into the outermost layers of PLA–PCL stents by an acid swelling technique increased the porosity and hydrophilicity of the polymer and made possible absorption of aqueous viral suspensions (10^6–10^9 pfu) following 15 to 60 min exposure of porous stent to the suspension of adenovirus. However, the aqueous virus suspension within the stents quickly eluted, with 85% of vector load lost after a 15 min incubation *in vitro* in tissue culture medium. The stent-delivered viral vectors retained viability and were shown to transduce cells effectively *in vitro*. The mechanical properties of the stents precluded their percutaneous deployment. Therefore the polymeric stents loaded with AdLacZ were surgically implanted into rabbit carotid arteries. Four days after implantation, focal expression of β-galactosidase was observed in the media, and occasionally in the adventitia of stented segments. Occluding thrombus was observed in the majority of samples. Additionally, in animals that were implanted with stents with higher adenovirus loadings, as many as 70% of hepatocytes expressed β-galactosidase due to the widespread biodistribution of the aqueous viral suspension contained within the stent.

Recently, Nakayama and coauthors[182] described an adenovirus-loaded stent coating based on the photopolymerization of a styrenated derivative of a gelatin with a photoactivator, carboxylated camphorquinone. Both reagents were added to AdLacZ suspension (10^9 pfu/ml), and gold stents were dip-coated and photoactivated for a few minutes with the formation of a hydrogel layer on the surface of stents. Transgene expression in organ culture experiments, and *in vivo* in a rabbit carotid stent–angioplasty model, was demonstrated 1 and 3 weeks after stent deployment. Unfortunately, only luminal aspects of the stented segments were studied; therefore, no conclusions can be made regarding a prevalence of the transduced cells in the media and adventitia. The distal spread of adenoviral vector was not examined in this study.

39.6.3 Cell-Seeded Stents

Stent seeding with cells, either native or genetically modified to express an advantageous transgene, is an interesting alternative to the immobilization of genetic vectors per se. However, virtually none of the stent-based cell therapies studied thus far have appropriately addressed the interaction of cells with the stent during the course of deployment. For example, the force of deployment and the dramatic changes in dimensions of the stents' shape at the time of deployment must result in considerable loss of seeded cells, and yet this has not been investigated. Furthermore, the fate of compressed cells trapped between the stent struts and the arterial wall also has not been properly studied. In addition, covered stents could conceivably be used to deliver cell-seeded stent grafts. However, the seeding in this design configuration would need to be on the luminal side of the stent, and thus would be in contact with the expanding balloon, surely resulting in extensive cellular loss. The use of a self-deploying stent could partially solve this problem, but nevertheless the dramatic changes in dimensions with stent deployment would result in a discontinuous endothelial covering, which would require weeks to months for regrowth of an intact continuous endothelium.

39.6.3.1 Endothelial Cells

The concept of cell-coated stents has emerged from the realization of the pivotal role of endothelium in the healing of stent-induced vascular damage. Neointima that forms in stented rabbit iliac segments is more extensive when the stent is deployed in the segment that was denuded of endothelium prior to stenting.[183] Therefore, initial prototype experiments[184-186] investigated a protocol that involved retrieving sheep peripheral veins, establishing and expanding primary endothelial culture, transducing retrovirus with reporter or therapeutic genes, seeding modified autologous endothelial cells on a stent, and deploying the stent into the coronary circulation. About 30% of the cells immobilized on the stent were lost during deployment.[185] However, the majority of endothelial cells were shown to persist for at least 4 hours, especially on the lateral aspects of the stent.[186] The compromised persistence of seeded cells may have been partially related to the nature of therapeutic gene used in the cited studies. Its product, tissue type plasminogen activator, is a potent protease, which might impair fragile binding of the seeded cells to the stent and ECM. Eton et al.[88] examined long-term survival of stent-immobilized genetically engineered endothelial cells. Syngenic endothelial cells expressing β-galactosidase were suffused into polyurethane stent graft, expanded for 3 days in culture, and delivered with deployed stents into pig iliac arteries. Numerous β-galactosidase positive cells embedded into newly developed neointima were identified 1 month after deployment.

39.6.3.2 Smooth Muscle Cells

SMCs were previously investigated for the delivery of therapeutic genes into the balloon-injured vasculature.[187,188] Unlike endothelial cells, SMCs can repopulate medial and intimal layers and impact neighbor "native" SMCs by the paracrine secretion of antiproliferative, antimigration, and/or antifibrotic proteins. Recently, Panetta and colleagues reported on fabrication, *in vitro* behavior, and *in vivo* activities of a custom-designed mesh stent seeded with genetically modified SMCs.[189] The stents were coated with fibronectin, and porcine SMCs with stable overexpression of GFP were seeded at a high density onto the mesh. The stents were deployed in pig coronaries, and the pigs were terminated 4 weeks after the procedure. The seeded cells were identified by the GFP fluorescence and their number was not significantly changed compared to the predeployment time point. Surprisingly, stent-delivered SMCs did not induce neointimal hyperplasia in comparison with fibrinogen-coated control stents.

39.6.3.3 Stem Cells

Recent progress in the isolation and *ex vivo* expansion of endothelial progenitor cells (EPCs) from bone marrow and peripheral blood[190] may provide an alternative cellular source for stent seeding. Shirota et al. have examined EPC association with photocured gelatin-coated metallic and polyurethane stents.[191] EPCs were isolated from canine peripheral blood and expanded *in vitro*. Seeded EPC formed confluent monolayers on the stent surface. The EPC-seeded stents were then deployed in artificial vascular tissue, engineered using SMCs grown on tubular collagen sheets. After 7 days in culture, EPC migrated from the stent into the hybrid medial tissue, proliferated, and endothelialized the luminal surface of the artificial vessel.

39.7 FUTURE DIRECTIONS

There is strong evidence[192] that the abrupt compliance mismatch that exists at the junctions between stent ends and the adjacent free arterial wall disturbs local hemodynamics and results in shear stresses that promote atherosclerosis[193] and restenosis.[194] These stent-induced alterations are greatly reduced by minimizing the compliance mismatch between the stent and nonstented segment, using the recently developed compliance-matching stent.[195]

Stent-angioplastied arteries in general demonstrate no significant changes in arterial diameter after 6 months following stent deployment. Thus, after this time the presence of a stent is not necessary and might even be harmful due to the continuing foreign body reaction. Thus, a biodegradable stent that could be resorbed by this time could circumvent the long-term mechanically related adverse effects noted by the Virmanis group[13] in human autopsied stented arteries. The main advantage of biodegradable stents is their disappearance at the later stages of vascular healing. The manufacture and processing of biodegradable stents is compatible with gene vector immobilization, both as bulk incorporation[178-180] and as surface modification. However, radial strength and elastic expansion characteristics of biodegradable stents cannot be compared with those parameters of metallic devices.[25,181] Several polymer compounds were examined as potential materials for biolabile stents.[25,180,196] The majority of biodegradable devices were not biocompatible. Thus, only a high-molecular-weight PLLA, the Igaki-Tamai stent, was tested in a clinical trial[197] and showed restenosis at a rate in the range typical for metal devices. Notably, this stent was heated to 70° C prior to deployment to make polymer malleable. It also typically takes 18 to 24 months for a PLLA stent to biodegrade completely. Another feasible solution for the unnecessary lifelong presence of a stent in the healed artery is the development of a removable coronary stent, which could be extracted. Removable stents are widely used for the bronchial[198] and urologic[199] stenting.

Gene delivery stents have potential indications beyond the current clinical scenarios wherein stenting is performed. Along these lines, stents might be considered as a convenient platform for conventional and genetic drugs that could be brought in proximity to a diseased artery and left in place without compromise of blood flow. Since forceful expansion of a stent may not be necessary in these circumstances, only minimal neointimal proliferation might be expected. Potential clinical settings that may exploit the benefits of the "delivery only" gene stents range from low-density lipoprotein (LDL) receptor transfer in patients with familial hypercholesterolemia to fibrosis-promoting gene therapy in patients with arterial aneurysms. In contrast to the restenosis setting, gene therapy for the modification of underlying vascular diseases does not have a distinct time frame and thus might require indefinite administration (as in the case of cell therapy). Potential exhaustion of the vector loaded onto this type of stent could be approached by the catheter-mediated recharge of the device. The generic technology of rechargeable devices for the sustained release of therapeutics is well established[200] and might be adjusted to fit this specific application.

39.8 CONCLUSIONS

A merging of two concepts, gene therapy and controlled release drug delivery, led to the development of gene delivery stents. Although still in its infancy, the field of gene stents has provided evidence for the safety and potential efficacy of gene-loaded vascular devices. Much work remains to be done in identifying optimal genes for ISR therapy, preferred vectors, and ways of configuring vectors on the stent. Recent clinical experience with rapamycin- and taxol-coated stents has raised expectations in interventional cardiology. Thus, in order to present a viable alternative, gene stents must be optimized in terms of polymers employed for the binding and incorporation of genetic material. In addition to excellent mechanical properties and release profiles, the polymer of choice has to fit the stringent biocompatibility requirements of vascular tissue. To this end, immobilization techniques based on vector tethering rather than on bulk incorporation could be advantageous.

Finally, the potential use of gene stents is not restricted solely to cardiovascular applications and might be envisioned in different medical settings where scaffolds are introduced to provide a patency of viscera. Bronchial, biliary, and urologic stents are some examples. In these settings, stents loaded with gene vectors would be able to modify a neoplastic, inflammatory, or infective process by focal delivery of respective gene constructs. A proof-of-concept study that demonstrated a tracheal expression of β-galactosidase driven by the stent-tethered adenoviral vector was recently reported by our group.[201]

ACKNOWLEDGMENTS

The authors thank Ms. Jennifer LeBold and Mrs. Jeanne Connolly for their help in preparing this chapter. The authors also acknowledge research support from the National Institutes of Health (HL 72108, HL 64388, HL 07915), the American Heart Association (Dr. Perlstein), the Cystic Fibrosis Foundation (Dr. Fishbein), the Nanotechnology Institute of Pennsylvania (Dr. Levy), the Cordis Corporation (Dr. Levy), and the William J. Rashkind Endowment of the Children's Hospital of Philadelphia (Dr. Levy).

REFERENCES

1. Gruntzig, A.R., Senning, A., and Siegenthaler, W.E., Nonoperative dilatation of coronary-artery stenosis: percutaneous transluminal coronary angioplasty, *N. Engl. J. Med.* 301, 61, 1979.
2. American Heart Association, *2002 Heart and Stroke Statistical Update*, 2001.
3. Puma, J.A., Sketch, M.H., Jr., Tcheng, J.E. et al., Percutaneous revascularization of chronic coronary occlusions: an overview, *J. Am. Coll. Cardiol.* 26, 1, 1995.
4. Post, M.J., Borst, C., and Kuntz, R.E., The relative importance of arterial remodeling compared with intimal hyperplasia in lumen renarrowing after balloon angioplasty: a study in the normal rabbit and the hypercholesterolemic Yucatan micropig, *Circulation* 89, 2816, 1994.
5. Mintz, G.S., Popma, J.J., Pichard, A.D. et al., Arterial remodeling after coronary angioplasty: a serial intravascular ultrasound study, *Circulation* 94, 35, 1996.
6. Hoffmann, R., Mintz, G.S., Dussaillant, G.R. et al., Patterns and mechanisms of in-stent restenosis: a serial intravascular ultrasound study, *Circulation* 94, 1247, 1996.
7. Serruys, P.W., de Jaegere, P., Kiemeneij, F. et al., A comparison of balloon-expandable-stent implantation with balloon angioplasty in patients with coronary artery disease, Benestent Study Group, *N. Engl. J. Med.* 331, 489, 1994.
8. Macaya, C., Serruys, P.W., Ruygrok, P. et al., Continued benefit of coronary stenting versus balloon angioplasty: one-year clinical follow-up of Benestent trial, Benestent Study Group, *J. Am. Coll. Cardiol.* 27, 255, 1996.

9. Ueda, Y., Nanto, S., Komamura, K. et al., Elastic recoil and intimal thickening after coronary stenting, *J. Inter. Cardiol.* 8, 137, 1995.

10. Lowe, H.C., Oesterle, S.N., and Khachigian, L.M., Coronary in-stent restenosis: current status and future strategies, *J. Am. Coll. Cardiol.* 39, 183, 2002.

11. Bennett, M.R. and O'Sullivan, M., Mechanisms of angioplasty and stent restenosis: implications for design of rational therapy, *Pharmacol. Ther.* 91, 149, 2001.

12. Grewe, P.H., Deneke, T., Machraoui, A. et al., Acute and chronic tissue response to coronary stent implantation: pathologic findings in human specimen, *J. Am. Coll. Cardiol.* 35, 157, 2000.

13. Farb, A., Sangiorgi, G., Carter, A.J. et al., Pathology of acute and chronic coronary stenting in humans, *Circulation* 99, 44, 1999.

14. Moreno, P.R., Palacios, I.F., Leon, M.N. et al., Histopathologic comparison of human coronary in-stent and post-balloon angioplasty restenotic tissue, *Am. J. Cardiol.* 84, 462, 1999.

15. Yutani, C., Ishibashi-Ueda, H., Suzuki, T. et al., Histologic evidence of foreign body granulation tissue and *de novo* lesions in patients with coronary stent restenosis, *Cardiology* 92, 171, 1999.

16. Schwartz, S.M., Smooth muscle migration in atherosclerosis and restenosis, *J. Clin. Invest.* 99, 2814, 1997.

17. Wakeyama, T., Ogawa, H., Iida, H. et al., Effects of candesartan and probucol on restenosis after coronary stenting, *Circ. J.* 67, 519, 2003.

18. Walter, D.H., Schachinger, V., Elsner, M. et al., Statin therapy is associated with reduced restenosis rates after coronary stent implantation in carriers of the Pl(A2)allele of the platelet glycoprotein IIIa gene, *Eur. Heart J.* 22, 587, 2001.

19. Takagi, T., Akasaka, T., Yamamuro, A. et al., Troglitazone reduces neointimal tissue proliferation after coronary stent implantation in patients with non-insulin dependent diabetes mellitus: a serial intra-vascular ultrasound study, *J. Am. Coll. Cardiol.* 36, 1529, 2000.

20. Lew, R., Ajani, A., and Waksman, R., Review of intracoronary radiation for in-stent restenosis, *J. Invasive Cardiol.* 15 Suppl. A, 2A, 2003.

21. French, M.H. and Faxon, D.P., Update on radiation for restenosis, *Rev. Cardiovasc. Med.* 3, 1, 2002.

22. Holmes, D.R., Jr., In-stent restenosis, *Rev. Cardiovasc. Med.* 2, 115, 2001.

23. Albiero, R., Adamian, M., Kobayashi, N. et al., Short- and intermediate-term results of (32)P radio-active beta-emitting stent implantation in patients with coronary artery disease: the Milan dose-response study, *Circulation* 101, 18, 2000.

24. Reinders, J.G., Heijmen, B.J., Olofsen-van Acht, M.J. et al., Ischemic heart disease after mantlefield irradiation for Hodgkin's disease in long-term follow-up, *Radiother. Oncol.* 51, 35, 1999.

25. Regar, E., Sianos, G., and Serruys, P.W., Stent development and local drug delivery, *Br. Med. Bull.* 59, 227, 2001.

26. van Beusekom, H.M., Schwartz, R.S., and van der Giessen, W.J., Synthetic polymers, *Semin. Interv. Cardiol.* 3, 145, 1998.

27. al-Lamee, K., Whelan, D.M., van der Giessen, W.J. et al., Surface modification of stents for improving biocompatibility, *Med. Device Technol.* 11, 12, 2000.

28. Bertrand, O.F., Sipehia, R., Mongrain, R. et al., Biocompatibility aspects of new stent technology, *J. Am. Coll. Cardiol.* 32, 562, 1998.

29. De Scheerder, I.K., Wilczek, K.L., Verbeken, E.V. et al., Biocompatibility of biodegradable and nonbiodegradable polymer-coated stents implanted in porcine peripheral arteries, *Cardiovasc. Interv. Radiol.* 18, 227, 1995.

30. De Scheerder, I.K., Wilczek, K.L., Verbeken, E.V. et al., Biocompatibility of polymer-coated oversized metallic stents implanted in normal porcine coronary arteries, *Atherosclerosis* 114, 105, 1995.

31. van der Giessen, W.J., Lincoff, A.M., Schwartz, R.S. et al., Marked inflammatory sequelae to implantation of biodegradable and nonbiodegradable polymers in porcine coronary arteries, *Circulation* 94, 1690, 1996.

32. Whelan, D.M., van der Giessen, W.J., Krabbendam, S.C. et al., Biocompatibility of phosphorylcholine coated stents in normal porcine coronary arteries, *Heart* 83, 338, 2000.

33. Boland, J.L., Corbeij, H.A., van der Giessen, W. et al., Multicenter evaluation of the phosphorylcho-line-coated biodivYsio stent in short *de novo* coronary lesions: the SOPHOS study, *Int. J. Cardiovasc. Interv.* 3, 215, 2000.

34. Lincoff, A.M., Furst, J.G., Ellis, S.G. et al., Sustained local delivery of dexamethasone by a novel intravascular eluting stent to prevent restenosis in the porcine coronary injury model, *J. Am. Coll. Cardiol.* 29, 808, 1997.

35. Fattori, R. and Piva, T., Drug-eluting stents in vascular intervention, *Lancet* 361, 247, 2003.

36. Curcio, A., Torella, D., Coppola, C. et al., Coated stents: a novel approach to prevent in-stent restenosis, *Ital. Heart J.* 3 Suppl. 4, 16S, 2002.

37. Schwartz, R.S., Edelman, E.R., Carter, A. et al., Drug-eluting stents in preclinical studies: recommended evaluation from a consensus group, *Circulation* 106, 1867, 2002.

38. Hong, M.K., Kornowski, R., Bramwell, O. et al., Paclitaxel-coated Gianturco-Roubin II (GR II) stents reduce neointimal hyperplasia in a porcine coronary in-stent restenosis model, *Coron. Artery Dis.* 12, 513, 2001.

39. Suzuki, T., Kopia, G., Hayashi, S. et al., Stent-based delivery of sirolimus reduces neointimal formation in a porcine coronary model, *Circulation* 104, 1188, 2001.

40. Park, S.J., Shim, W.H., Ho, D.S. et al., A paclitaxel-eluting stent for the prevention of coronary restenosis, *N. Engl. J. Med.* 348, 1537, 2003.

41. Tanabe, K., Serruys, P.W., Grube, E. et al., TAXUS III Trial: in-stent restenosis treated with stent-based delivery of paclitaxel incorporated in a slow-release polymer formulation, *Circulation* 107, 559, 2003.

42. Hong, M.K., Mintz, G.S., Lee, C.W. et al., Paclitaxel coating reduces in-stent intimal hyperplasia in human coronary arteries: a serial volumetric intravascular ultrasound analysis from the Asian Paclitaxel-Eluting Stent Clinical Trial (ASPECT), *Circulation* 107, 517, 2003.

43. Sousa, J.E., Costa, M.A., Sousa, A.G. et al., Two-year angiographic and intravascular ultrasound follow-up after implantation of sirolimus-eluting stents in human coronary arteries, *Circulation* 107, 381, 2003.

44. Grube, E., Silber, S., Hauptmann, K.E. et al., TAXUS I: six- and twelve-month results from a randomized, double-blind trial on a slow-release paclitaxel-eluting stent for *de novo* coronary lesions, *Circulation* 107, 38, 2003.

45. Degertekin, M., Serruys, P.W., Foley, D.P. et al., Persistent inhibition of neointimal hyperplasia after sirolimus-eluting stent implantation: long-term (up to 2 years) clinical, angiographic, and intravascular ultrasound follow-up, *Circulation* 106, 1610, 2002.

46. Oberhoff, M., Herdeg, C., Baumbach, A. et al., Stent-based antirestenotic coatings (sirolimus/paclitaxel), *Catheter Cardiovasc. Interv.* 55, 404, 2002.

47. Granada, J.F., Kaluza, G.L., and Raizner, A., Drug-eluting stents for cardiovascular disorders, *Curr. Atheroscler. Rep.* 5, 308, 2003.

48. Marx, S.O., Jayaraman, T., Go, L.O. et al., Rapamycin-FKBP inhibits cell cycle regulators of proliferation in vascular smooth muscle cells, *Circ. Res.* 76, 412, 1995.

49. Poon, M., Marx, S.O., Gallo, R. et al., Rapamycin inhibits vascular smooth muscle cell migration, *J. Clin. Invest.* 98, 2277, 1996.

50. Regar, E., Serruys, P.W., Bode, C. et al., Angiographic findings of the multicenter randomized study with the sirolimus-eluting Bx velocity balloon-expandable stent (RAVEL): sirolimus-eluting stents inhibit restenosis irrespective of the vessel size, *Circulation* 106, 1949, 2002.

51. Serruys, P.W., Degertekin, M., Tanabe, K. et al., Intravascular ultrasound findings in the multicenter, randomized, double-blind RAVEL (RAndomized study with the sirolimus-eluting VElocity balloon-expandable stent in the treatment of patients with de novo native coronary artery Lesions) trial, *Circulation* 106, 798, 2002.

52. Gershlick, A.H., DeScheerder, I., and Chevalier, B., Local drug delivery to inhibit coronary artery restenosis: data from the ELUTES (Evaluation of Paclitaxel Eluting Stent) clinical trial, *Circulation* 104 (Suppl. II), 416, 2001.

53. Liistro, F. and Colombo, A., Late acute thrombosis after paclitaxel eluting stent implantation, *Heart* 86, 262, 2001.

54. Farb, A., Heller, P.F., Shroff, S. et al., Pathological analysis of local delivery of paclitaxel via a polymer-coated stent, *Circulation* 104, 473, 2001.

55. Guba, M., von Breitenbuch, P., Steinbauer, M. et al., Rapamycin inhibits primary and metastatic tumor growth by antiangiogenesis: involvement of vascular endothelial growth factor, *Nat. Med.* 8, 128, 2002.

56. Honda, Y., Grube, E., de La Fuente, L.M. et al., Novel drug-delivery stent: intravascular ultrasound observations from the first human experience with the QP2-eluting polymer stent system, *Circulation* 104, 380, 2001.

57. George, S.J. and Baker, A.H., Gene transfer to the vasculature: historical perspective and implication for future research objectives, *Mol. Biotechnol.* 22, 153, 2002.

58. Janssens, S.P., Applied gene therapy in preclinical models of vascular injury, *Curr. Atheroscler. Rep.* 5, 186, 2003.

59. Quarck, R. and Holvoet, P., Restenosis and gene therapy, *Expert Opin. Biol. Ther.* 1, 79, 2001.

60. Leclerc, G., Gal, D., Takeshita, S. et al., Percutaneous arterial gene transfer in a rabbit model: efficiency in normal and balloon-dilated atherosclerotic arteries, *J. Clin. Invest.* 90, 936, 1992.

61. Chapman, G.D., Lim, C.S., Gammon, R.S. et al., Gene transfer into coronary arteries of intact animals with a percutaneous balloon catheter, *Circ. Res.* 71, 27, 1992.

62. Budker, V., Zhang, G., Danko, I. et al., The efficient expression of intravascularly delivered DNA in rat muscle, *Gene Ther.* 5, 272, 1998.

63. von der Leyen, H.E., Braun-Dullaeus, R., Mann, M.J. et al., A pressure-mediated nonviral method for efficient arterial gene and oligonucleotide transfer, *Hum. Gene Ther.* 10, 2355, 1999.

64. Mann, M.J., Whittemore, A.D., Donaldson, M.C. et al., *Ex vivo* gene therapy of human vascular bypass grafts with E2F decoy: the PREVENT single-centre, randomized, controlled trial, *Lancet* 354, 1493, 1999.

65. Armeanu, S., Pelisek, J., Krausz, E. et al., Optimization of nonviral gene transfer of vascular smooth muscle cells *in vitro* and *in vivo*, *Mol. Ther.* 1, 366, 2000.

66. Mann, M.J., Morishita, R., Gibbons, G.H. et al., DNA transfer into vascular smooth muscle using fusigenic Sendai virus (HVJ)-liposomes, *Mol. Cell Biochem.* 172, 3, 1997.

67. Dzau, V.J., Mann, M.J., Morishita, R. et al., Fusigenic viral liposome for gene therapy in cardiovascular diseases, *Proc. Nat. Acad. Sci. U.S.A.* 93, 11421, 1996.

68. Kaneda, Y., Saeki, Y., and Morishita, R., Gene therapy using HVJ-liposomes: the best of both worlds? *Mol. Med. Today* 5, 298, 1999.

69. Lestini, B.J., Sagnella, S.M., Xu, Z. et al., Surface modification of liposomes for selective cell targeting in cardiovascular drug delivery, *J. Control. Release* 78, 235, 2002.

70. Nah, J.W., Yu, L., Han, S. et al., Artery wall binding peptide-poly(ethylene glycol)-grafted-poly(L-lysine)-based gene delivery to artery wall cells, *J. Control. Release* 78, 273, 2002.

71. Lee, S.W., Trapnell, B.C., Rade, J.J. et al., *In vivo* adenoviral vector-mediated gene transfer into balloon-injured rat carotid arteries, *Circ. Res.* 73, 797, 1993.

72. Baek, S. and March, K.L., Gene therapy for restenosis: getting nearer the heart of the matter, *Circ. Res.* 82, 295, 1998.

73. Jooss, K. and Chirmule, N., Immunity to adenovirus and adeno-associated viral vectors: implications for gene therapy, *Gene Ther.* 10, 955, 2003.

74. Liu, Q. and Muruve, D.A., Molecular basis of the inflammatory response to adenovirus vectors, *Gene Ther.* 10, 935, 2003.

75. Croyle, M.A., Chirmule, N., Zhang, Y. et al., "Stealth" adenoviruses blunt cell-mediated and humoral immune responses against the virus and allow for significant gene expression upon readministration in the lung, *J. Virol.* 75, 4792, 2001.

76. Szelid, Z., Sinnaeve, P., Vermeersch, P. et al., Preexisting antiadenoviral immunity and regional myocardial gene transfer: modulation by nitric oxide, *Hum. Gene Ther.* 13, 2185, 2002.

77. Wen, S., Graf, S., and Dichek, D.A., A helper-dependent adenoviral vector improves arterial wall gene transfer, *Mol. Ther.* 7, S241, 2003.

78. Belalcazar, L.M., Merched, A., Carr, B. et al., Long-term stable expression of human apolipoprotein A-I mediated by helper-dependent adenovirus gene transfer inhibits atherosclerosis progression and remodels atherosclerotic plaques in a mouse model of familial hypercholesterolemia, *Circulation* 107, 2726, 2003.

79. Hong, S.S., Magnusson, M.K., Henning, P. et al., Adenovirus stripping: a versatile method to generate adenovirus vectors with new cell target specificity, *Mol. Ther.* 7, 692, 2003.

80. Ribault, S., Neuville, P., Mechine-Neuville, A. et al., Chimeric smooth muscle-specific enhancer/promoters: valuable tools for adenovirus-mediated cardiovascular gene therapy, *Circ. Res.* 88, 468, 2001.

81. Nicklin, S.A., Reynolds, P.N., Brosnan, M.J. et al., Analysis of cell-specific promoters for viral gene therapy targeted at the vascular endothelium, *Hypertension* 38, 65, 2001.

82. Sun, J.Y., Anand-Jawa, V., Chatterjee, S. et al., Immune responses to adeno-associated virus and its recombinant vectors, *Gene Ther.* 10, 964, 2003.

83. Nicklin, S.A., Buening, H., Dishart, K.L. et al., Efficient and selective AAV2-mediated gene transfer directed to human vascular endothelial cells, *Mol. Ther.* 4, 174, 2001.

84. Kotin, R.M., Linden, R.M., and Berns, K.I., Characterization of a preferred site on human chromosome 19q for integration of adeno-associated virus DNA by non-homologous recombination, *EMBO J.* 11, 5071, 1992.

85. Nabel, E.G., Plautz, G., Boyce, F.M. et al., Recombinant gene expression *in vivo* within endothelial cells of the arterial wall, *Science* 244, 1342, 1989.

86. Nabel, E.G., Plautz, G., and Nabel, G.J., Site-specific gene expression *in vivo* by direct gene transfer into the arterial wall, *Science* 249, 1285, 1990.

87. Friedmann, T. and Yee, J.K., Pseudotyped retroviral vectors for studies of human gene therapy, *Nat. Med.* 1, 275, 1995.

88. Eton, D., Terramani, T.T., Wang, Y. et al., Genetic engineering of stent grafts with a highly efficient pseudotyped retroviral vector, *J. Vasc. Surg.* 29, 863, 1999.

89. Kutryk, M.J., Foley, D.P., van den Brand, M. et al., Local intracoronary administration of antisense oligonucleotide against c-myc for the prevention of in-stent restenosis: results of the randomized investigation by the Thoraxcenter of antisense DNA using local delivery and IVUS after coronary stenting (ITALICS) trial, *J. Am. Coll. Cardiol.* 39, 281, 2002.

90. Frimerman, A., Welch, P.J., Jin, X. et al., Chimeric DNA-RNA hammerhead ribozyme to proliferating cell nuclear antigen reduces stent-induced stenosis in a porcine coronary model, *Circulation* 99, 697, 1999.

91. Khachigian, L.M., Fahmy, R.G., Zhang, G. et al., c-Jun regulates vascular smooth muscle cell growth and neointima formation after arterial injury: inhibition by a novel DNA enzyme targeting c-Jun, *J. Biol. Chem.* 277, 22985, 2002.

92. Morishita, R., Aoki, M., and Kaneda, Y., Decoy oligodeoxynucleotides as novel cardiovascular drugs for cardiovascular disease, *Ann. N.Y. Acad. Sci.* 947, 294, 2001.

93. Hannon, G.J., RNA interference, *Nature* 418, 244, 2002.

94. Gunn, J., Holt, C.M., Francis, S.E. et al., The effects of oligonucleotides to c-myb on vascular smooth muscle cell proliferation and neointima formation after porcine coronary angioplasty, *Circ. Res.* 80, 520, 1997.

95. Lowe, H.C., Fahmy, R.G., Kavurma, M.M. et al., Catalytic oligodeoxynucleotides define a key regulatory role for early growth response factor-1 in the porcine model of coronary in-stent restenosis, *Circ. Res.* 89, 670, 2001.

96. Bianco, P. and Robey, P.G., Stem cells in tissue engineering, *Nature* 414, 118, 2001.

97. Laurencin, C.T., Attawia, M.A., Lu, L.Q. et al., Poly(lactide-*co*-glycolide)/hydroxyapatite delivery of BMP-2- producing cells: a regional gene therapy approach to bone regeneration, *Biomaterials* 22, 1271, 2001.

98. Orlic, D., Hill, J.M., and Arai, A.E., Stem cells for myocardial regeneration, *Circ. Res.* 91, 1092, 2002.

99. Parikh, S.A. and Edelman, E.R., Endothelial cell delivery for cardiovascular therapy, *Adv. Drug Del. Rev.* 42, 139, 2000.

100. Roell, W., Fan, Y., Xia, Y. et al., Cellular cardiomyoplasty in a transgenic mouse model, *Transplantation* 73, 462, 2002.

101. Harrington, K., Alvarez-Vallina, L., Crittenden, M. et al., Cells as vehicles for cancer gene therapy: the missing link between targeted vectors and systemic delivery? *Hum. Gene Ther.* 13, 1263, 2002.

102. Xu, R., Li, H., Tse, L.Y. et al., Diabetes gene therapy: potential and challenges, *Curr. Gene Ther.* 3, 65, 2003.

103. Guzman, R.J., Hirschowitz, E.A., Brody, S.L. et al., *In vivo* suppression of injury-induced vascular smooth muscle cell accumulation using adenovirus-mediated transfer of the herpes simplex virus thymidine kinase gene, *Proc. Nat. Acad. Sci. U.S.A.* 91, 10732, 1994.

104. Simari, R.D., San, H., Rekhter, M. et al., Regulation of cellular proliferation and intimal formation following balloon injury in atherosclerotic rabbit arteries, *J. Clin. Invest.* 98, 225, 1996.

105. Ohno, T., Gordon, D., San, H. et al., Gene therapy for vascular smooth muscle cell proliferation after arterial injury, *Science* 265, 781, 1994.

106. Harrell, R.L., Rajanayagam, S., Doanes, A.M. et al., Inhibition of vascular smooth muscle cell proliferation and neointimal accumulation by adenovirus-mediated gene transfer of cytosine deaminase, *Circulation* 96, 621, 1997.

107. Luo, Z., Sata, M., Nguyen, T. et al., Adenovirus-mediated delivery of Fas ligand inhibits intimal hyperplasia after balloon injury in immunologically primed animals, *Circulation* 99, 1776, 1999.

108. Mano, T., Luo, Z., Suhara, T. et al., Expression of wild-type and noncleavable Fas ligand by tetracycline-regulated adenoviral vectors to limit intimal hyperplasia in vascular lesions, *Hum. Gene Ther.* 11, 1625, 2000.

109. Perlman, H., Maillard, L., Krasinski, K. et al., Evidence for the rapid onset of apoptosis in medial smooth muscle cells after balloon injury, *Circulation* 95, 981, 1997.

110. Bicknell, K.A., Surry, E.L., and Brooks, G., Targeting the cell cycle machinery for the treatment of cardiovascular disease, *J. Pharm. Pharmacol.* 55, 571, 2003.

111. Sriram, V. and Patterson, C., Cell cycle in vasculoproliferative diseases: potential interventions and routes of delivery, *Circulation* 103, 2414, 2001.

112. Morishita, R., Gibbons, G.H., Ellison, K.E. et al., Intimal hyperplasia after vascular injury is inhibited by antisense cdk 2 kinase oligonucleotides, *J. Clin. Invest.* 93, 1458, 1994.

113. Morishita, R., Gibbons, G.H., Kaneda, Y. et al., Pharmacokinetics of antisense oligodeoxyribonucleotides (cyclin B1 and CDC 2 kinase) in the vessel wall *in vivo*: enhanced therapeutic utility for restenosis by HVJ-liposome delivery, *Gene* 149, 13, 1994.

114. Chen, D., Krasinski, K., Sylvester, A. et al., Downregulation of cyclin-dependent kinase 2 activity and cyclin A promoter activity in vascular smooth muscle cells by p27(KIP1), an inhibitor of neointima formation in the rat carotid artery, *J. Clin. Invest.* 99, 2334, 1997.

115. Tanner, F.C., Boehm, M., Akyurek, L.M. et al., Differential effects of the cyclin-dependent kinase inhibitors p27(Kip1), p21(Cip1), and p16(Ink4) on vascular smooth muscle cell proliferation, *Circulation* 101, 2022, 2000.

116. Tsui, L.V., Camrud, A., Mondesire, J. et al., p27-p16 fusion gene inhibits angioplasty-induced neointimal hyperplasia and coronary artery occlusion, *Circ. Res.* 89, 323, 2001.

117. Speir, E., Modali, R., Huang, E.S. et al., Potential role of human cytomegalovirus and p53 interaction in coronary restenosis, *Science* 265, 391, 1994.

118. Scheinman, M., Ascher, E., Levi, G.S. et al., P53 gene transfer to the injured rat carotid artery decreases neointimal formation, *J. Vasc. Surg.* 29, 360, 1999.

119. Yonemitsu, Y., Kaneda, Y., Tanaka, S. et al., Transfer of wild-type p53 gene effectively inhibits vascular smooth muscle cell proliferation *in vitro* and *in vivo*, *Circ. Res.* 82, 147, 1998.

120. Scheinman, M., Ascher, E., Kallakuri, S. et al., P53 gene transfer to the injured rat carotid artery promotes apoptosis, *Surgery* 126, 863, 1999.

121. Chang, M.W., Barr, E., Seltzer, J. et al., Cytostatic gene therapy for vascular proliferative disorders with a constitutively active form of the retinoblastoma gene product, *Science* 267, 518, 1995.

122. Wills, K.N., Mano, T., Avanzini, J.B. et al., Tissue-specific expression of an anti-proliferative hybrid transgene from the human smooth muscle alpha-actin promoter suppresses smooth muscle cell proliferation and neointima formation, *Gene Ther.* 8, 1847, 2001.

123. Morishita, R., Gibbons, G.H., Horiuchi, M. et al., A gene therapy strategy using a transcription factor decoy of the E2F binding site inhibits smooth muscle proliferation *in vivo*, *Proc. Nat. Acad. Sci. U.S.A.* 92, 5855, 1995.

124. Sinnaeve, P., Chiche, J.D., Gillijns, H. et al., Overexpression of a constitutively active protein kinase G mutant reduces neointima formation and in-stent restenosis, *Circulation* 105, 2911, 2002.

125. Silverman, E.S. and Collins, T., Pathways of Egr-1-mediated gene transcription in vascular biology, *Am. J. Pathol.* 154, 665, 1999.

126. Ueno, H., Yamamoto, H., Ito, S. et al., Adenovirus-mediated transfer of a dominant-negative H-ras suppresses neointimal formation in balloon-injured arteries *in vivo*, *Arterioscler. Thromb. Vasc. Biol.* 17, 898, 1997.

127. Maillard, L., Van Belle, E., Smith, R.C. et al., Percutaneous delivery of the gax gene inhibits vessel stenosis in a rabbit model of balloon angioplasty, *Cardiovasc. Res.* 35, 536, 1997.

128. Mano, T., Luo, Z., Malendowicz, S.L. et al., Reversal of GATA-6 downregulation promotes smooth muscle differentiation and inhibits intimal hyperplasia in balloon-injured rat carotid artery, *Circ. Res.* 84, 647, 1999.

129. Stephan, D., San, H., Yang, Z.Y. et al., Inhibition of vascular smooth muscle cell proliferation and intimal hyperplasia by gene transfer of beta-interferon, *Mol. Med.* 3, 593, 1997.

130. Cohen-Sacks, H., Najajreh, Y., Tchaikovski, V. et al., Novel PDGFbetaR antisense encapsulated in polymeric nanospheres for the treatment of restenosis, *Gene Ther.* 9, 1607, 2002.

131. Turunen, M.P., Puhakka, H.L., Koponen, J.K. et al., Peptide-retargeted adenovirus encoding a tissue inhibitor of metalloproteinase-1 decreases restenosis after intravascular gene transfer, *Mol. Ther.* 6, 306, 2002.

132. Furman, C., Luo, Z., Walsh, K. et al., Systemic tissue inhibitor of metalloproteinase-1 gene delivery reduces neointimal hyperplasia in balloon-injured rat carotid artery, *FEBS Lett.* 531, 122, 2002.

133. Carmeliet, P., Moons, L., Lijnen, R. et al., Inhibitory role of plasminogen activator inhibitor-1 in arterial wound healing and neointima formation: a gene targeting and gene transfer study in mice, *Circulation* 96, 3180, 1997.

134. Rade, J.J., Schulick, A.H., Virmani, R. et al., Local adenoviral-mediated expression of recombinant hirudin reduces neointima formation after arterial injury, *Nat. Med.* 2, 293, 1996.

135. Yin, X., Yutani, C., Ikeda, Y. et al., Tissue factor pathway inhibitor gene delivery using HVJ-AVE liposomes markedly reduces restenosis in atherosclerotic arteries, *Cardiovasc. Res.* 56, 454, 2002.

136. Zoldhelyi, P., Chen, Z.Q., Shelat, H.S. et al., Local gene transfer of tissue factor pathway inhibitor regulates intimal hyperplasia in atherosclerotic arteries, *Proc. Nat. Acad. Sci. U.S.A.* 98, 4078, 2001.

137. Todaka, T., Yokoyama, C., Yanamoto, H. et al., Gene transfer of human prostacyclin synthase prevents neointimal formation after carotid balloon injury in rats, *Stroke* 30, 419, 1999.

138. Losordo, D.W., Isner, J.M., and Diaz-Sandoval, L.J., Endothelial recovery: the next target in restenosis prevention, *Circulation* 107, 2635, 2003.

139. Asahara, T., Chen, D., Tsurumi, Y. et al., Accelerated restitution of endothelial integrity and endothelium-dependent function after phVEGF165 gene transfer, *Circulation* 94, 3291, 1996.

140. Laitinen, M., Zachary, I., Breier, G. et al., VEGF gene transfer reduces intimal thickening via increased production of nitric oxide in carotid arteries, *Hum. Gene Ther.* 8, 1737, 1997.

141. Hiltunen, M.Ö., Laitinen, M., Turunen, M.P. et al., Intravascular adenovirus-mediated VEGF-C gene transfer reduces neointima formation in balloon-denuded rabbit aorta, *Circulation* 102, 2262, 2000.

142. Van Belle, E., Tio, F.O., Chen, D. et al., Passivation of metallic stents after arterial gene transfer of phVEGF165 inhibits thrombus formation and intimal thickening, *J. Am. Coll. Cardiol.* 29, 1371, 1997.

143. Isner, J.M., Pieczek, A., Schainfeld, R. et al., Clinical evidence of angiogenesis after arterial gene transfer of phVEGF165 in patient with ischaemic limb, *Lancet* 348, 370, 1996.

144. Hedman, M., Hartikainen, J., Syvanne, M. et al., Safety and feasibility of catheter-based local intracoronary vascular endothelial growth factor gene transfer in the prevention of postangioplasty and in-stent restenosis and in the treatment of chronic myocardial ischemia: phase II results of the Kuopio Angiogenesis Trial (KAT), *Circulation* 107, 2677, 2003.

145. George, S.E., Nitric oxide and restenosis: opportunities for therapeutic intervention, *Coron. Artery Dis.* 10, 295, 1999.

146. Groves, P.H., Banning, A.P., Penny, W.J. et al., The effects of exogenous nitric oxide on smooth muscle cell proliferation following porcine carotid angioplasty, *Cardiovasc. Res.* 30, 87, 1995.

147. Garg, U.C. and Hassid, A., Nitric oxide-generating vasodilators and 8-bromo-cyclic guanosine monophosphate inhibit mitogenesis and proliferation of cultured rat vascular smooth muscle cells, *J. Clin. Invest.* 83, 1774, 1989.

148. Sarkar, R., Meinberg, E.G., Stanley, J.C. et al., Nitric oxide reversibly inhibits the migration of cultured vascular smooth muscle cells, *Circ. Res.* 78, 225, 1996.

149. Pollman, M.J., Yamada, T., Horiuchi, M. et al., Vasoactive substances regulate vascular smooth muscle cell apoptosis: countervailing influences of nitric oxide and angiotensin II, *Circ. Res.* 79, 748, 1996.

150. Nong, Z., Hoylaerts, M., Van Pelt, N. et al., Nitric oxide inhalation inhibits platelet aggregation and platelet-mediated pulmonary thrombosis in rats, *Circ. Res.* 81, 865, 1997.

151. Yan, Z. and Hansson, G.K., Overexpression of inducible nitric oxide synthase by neointimal smooth muscle cells, *Circ. Res.* 82, 21, 1998.

152. Lincoln, T.M., Dey, N.B., Boerth, N.J. et al., Nitric oxide–cyclic GMP pathway regulates vascular smooth muscle cell phenotypic modulation: implications in vascular diseases, *Acta Physiol. Scand.* 164, 507, 1998.

153. Rizvi, M.A. and Myers, P.R., Nitric oxide modulates basal and endothelin-induced coronary artery vascular smooth muscle cell proliferation and collagen levels, *J. Mol. Cell Cardiol.* 29, 1779, 1997.

154. Jeremy, J.Y., Rowe, D., Emsley, A.M. et al., Nitric oxide and the proliferation of vascular smooth muscle cells *Cardiovasc. Res.* 43, 580, 1999.

155. Kalinowski, M., Alfke, H., Bergen, S. et al., Comparative trial of local pharmacotherapy with L-arginine, r-hirudin, and molsidomine to reduce restenosis after balloon angioplasty of stenotic rabbit iliac arteries, *Radiology* 219, 716, 2001.

156. Rolland, P.H., Bartoli, J.M., Piquet, P. et al., Local delivery of NO-donor molsidomine post-PTA improves haemodynamics, wall mechanics and histomorphometry in atherosclerotic porcine SFA, *Eur. J. Vasc. Endovasc. Surg.* 23, 226, 2002.

157. Schwarzacher, S.P., Lim, T.T., Wang, B. et al., Local intramural delivery of l-arginine enhances nitric oxide generation and inhibits lesion formation after balloon angioplasty, *Circulation* 95, 1863, 1997.

158. von der Leyen, H.E., Gibbons, G.H., Morishita, R. et al., Gene therapy inhibiting neointimal vascular lesion: *in vivo* transfer of endothelial cell nitric oxide synthase gene, *Proc. Nat. Acad. Sci. U.S.A.* 92, 1137, 1995.

159. Varenne, O., Pislaru, S., Gillijns, H. et al., Local adenovirus-mediated transfer of human endothelial nitric oxide synthase reduces luminal narrowing after coronary angioplasty in pigs, *Circulation* 98, 919, 1998.

160. Janssens, S., Flaherty, D., Nong, Z. et al., Human endothelial nitric oxide synthase gene transfer inhibits vascular smooth muscle cell proliferation and neointima formation after balloon injury in rats, *Circulation* 97, 1274, 1998.

161. Indolfi, C., Torella, D., Coppola, C. et al., Physical training increases eNOS vascular expression and activity and reduces restenosis after balloon angioplasty or arterial stenting in rats, *Circ. Res.* 91, 1190, 2002.

162. Wang, K., Kessler, P.D., Zhou, Z. et al., Local adenoviral-mediated inducible nitric oxide synthase gene transfer inhibits neointimal formation in the porcine coronary stented model, *Mol. Ther.* 7, 597, 2003.

163. Muhs, A., Heublein, B., Schletter, J. et al., Preclinical evaluation of inducible nitric oxide synthase lipoplex gene therapy for inhibition of stent-induced vascular neointimal lesion formation, *Hum. Gene Ther.* 14, 375, 2003.

164. Rabkin, E., and Schoen, F.J., Cardiovascular tissue engineering, *Cardiovasc. Pathol.* 11, 305, 2002.

165. Griffith, L.G., Polymeric biomaterials, *Acta Material.* 48, 263, 2000.

166. Ettenson, D.S. and Edelman, E.R., Local drug delivery: an emerging approach in the treatment of restenosis, *Vasc. Med.* 5, 97, 2000.

167. Klugherz, B.D., Jones, P.L., Cui, X. et al., Gene delivery from a DNA controlled-release stent in porcine coronary arteries, *Nat. Biotechnol.* 18, 1181, 2000.

168. Perlstein, I., Connolly, J.M., Cui, X. et al., DNA delivery from an intravascular stent with a denatured collagen-polylactic-polyglycolic acid-controlled release coating: mechanisms of enhanced transfection, *Gene Ther.* 10, 1420, 2003.

169. Srivatsa, S.S., Fitzpatrick, L.A., Tsao, P.W. et al., Selective alpha v beta 3 integrin blockade potently limits neointimal hyperplasia and lumen stenosis following deep coronary arterial stent injury: evidence for the functional importance of integrin alpha v beta 3 and osteopontin expression during neointima formation, *Cardiovasc. Res.* 36, 408, 1997.

170. Stouffer, G.A., Hu, Z., Sajid, M. et al., Beta3 integrins are upregulated after vascular injury and modulate thrombospondin- and thrombin-induced proliferation of cultured smooth muscle cells, *Circulation* 97, 907, 1998.

171. Lazarides, E. and Lindberg, U., Actin is the naturally occurring inhibitor of deoxyribonuclease I, *Proc. Nat. Acad. Sci. U.S.A.* 71, 4742, 1974.

172. Takahashi, A., Palmer-Opolski, M., Smith, R.C. et al., Transgene delivery of plasmid DNA to smooth muscle cells and macrophages from a biostable polymer-coated stent, *Gene Ther.* 10, 1471, 2003.

173. Feldman, M.D., Sun, B., Koci, B.J. et al., Stent-based gene therapy, *J. Long Term Eff. Med. Implants* 10, 47, 2000.

174. Levy, R.J., Song, C., Tallapragada, S. et al., Localized adenovirus gene delivery using antiviral IgG complexation, *Gene Ther.* 8, 659, 2001.

175. Klugherz, B.D., Song, C., DeFelice, S. et al., Gene delivery to pig coronary arteries from stents carrying antibody-tethered adenovirus, *Hum. Gene Ther.* 13, 443, 2002.

176. Abrahams, J.M., Song, C., DeFelice, S. et al., Endovascular microcoil gene delivery using immobilized anti-adenovirus antibody for vector tethering, *Stroke* 33, 1376, 2002.

177. Yuan, J., Gao, R., Shi, R. et al., Intravascular local gene transfer mediated by protein-coated metallic stent, *Chin. Med. J. (Engl.)* 114, 1043, 2001.

178. Rajasubramanian, G., Meidell, R.S., Landau, C. et al., Fabrication of resorbable microporous intravascular stents for gene therapy applications, *ASAIO J.* 40, M584, 1994.

179. Ye, Y.W., Landau, C., Willard, J.E. et al., Bioresorbable microporous stents deliver recombinant adenovirus gene transfer vectors to the arterial wall, *Ann. Biomed. Eng.* 26, 398, 1998.

180. Ye, Y.W., Landau, C., Meidell, R.S. et al., Improved bioresorbable microporous intravascular stents for gene therapy, *ASAIO J.* 42, M823, 1996.

181. Schlun, M., Martin, H., Grabow, N. et al., Design strategy for balloon-expandable stents made of biodegradable polymers using finite element analysis, *Biomed. Tech. (Berl.)* 47 Suppl. 1 Pt. 2, 831, 2002.

182. Nakayama, Y., Ji-Youn, K., Nishi, S. et al., Development of high-performance stent: gelatinous photogel-coated stent that permits drug delivery and gene transfer, *J. Biomed. Mater. Res.* 57, 559, 2001.

183. Rogers, C., Parikh, S., Seifert, P. et al., Endogenous cell seeding: remnant endothelium after stenting enhances vascular repair, *Circulation* 94, 2909, 1996.

184. Dichek, D.A., Neville, R.F., Zwiebel, J.A. et al., Seeding of intravascular stents with genetically engineered endothelial cells, *Circulation* 80, 1347, 1989.

185. Flugelman, M.Y., Virmani, R., Leon, M.B. et al., Genetically engineered endothelial cells remain adherent and viable after stent deployment and exposure to flow *in vitro*, *Circ. Res.* 70, 348, 1992.

186. Scott, N.A., Candal, F.J., Robinson, K.A. et al., Seeding of intracoronary stents with immortalized human microvascular endothelial cells, *Am. Heart J.* 129, 860, 1995.

187. Gomes, D., Louedec, L., Plissonnier, D. et al., Endoluminal smooth muscle cell seeding limits intimal hyperplasia, *J. Vasc. Surg.* 34, 707, 2001.

188. Scott-Burden, T., Tock, C.L., Schwarz, J.J. et al., Genetically engineered smooth muscle cells as linings to improve the biocompatibility of cardiovascular prostheses, *Circulation* 94, II235, 1996.

189. Panetta, C.J., Miyauchi, K., Berry, D. et al., A tissue-engineered stent for cell-based vascular gene transfer, *Hum. Gene Ther.* 13, 433, 2002.

190. Szmitko, P.E., Fedak, P.W., Weisel, R.D. et al., Endothelial progenitor cells: new hope for a broken heart. Hematopoietic, vascular and cardiac fates of bone marrow-derived stem cells, *Circulation* 107, 3093, 2003.

191. Shirota, T., Yasui, H., Shimokawa, H. et al., Fabrication of endothelial progenitor cell (EPC)-seeded intravascular stent devices and *in vitro* endothelialization on hybrid vascular tissue, *Biomaterials* 24, 2295, 2003.

192. Rachev, A., Manoach, E., Berry, J. et al., A model of stress-induced geometrical remodeling of vessel segments adjacent to stents and artery/graft anastomoses, *J. Theor. Biol.* 206, 429, 2000.

193. Paszkowiak, J.J. and Dardik, A., Arterial wall shear stress: observations from the bench to the bedside, *Vasc. Endovasc. Surg.* 37, 47, 2003.

194. Ward, M.R., Tsao, P.S., Agrotis, A. et al., Low blood flow after angioplasty augments mechanisms of restenosis: inward vessel remodeling, cell migration, and activity of genes regulating migration, *Arterioscler. Thromb. Vasc. Biol.* 21, 208, 2001.

195. Berry, J.L., Manoach, E., Mekkaoui, C. et al., Hemodynamics and wall mechanics of a compliance matching stent: *in vitro* and *in vivo* analysis, *J. Vasc. Interv. Radiol.* 13, 97, 2002.

196. Colombo, A. and Karvouni, E., Biodegradable stents: "fulfilling the mission and stepping away," *Circulation* 102, 371, 2000.

197. Tamai, H., Igaki, K., Kyo, E. et al., Initial and 6-month results of biodegradable poly-l-lactic acid coronary stents in humans, *Circulation* 102, 399, 2000.

198. Rafanan, A.L. and Mehta, A.C., Stenting of the tracheobronchial tree, *Radiol. Clin. North Am.* 38, 395, 2000.

199. Marks, L.S., Ettekal, B., Cohen, M.S. et al., Use of a shape-memory alloy (nitinol) in a removable prostate stent, *Tech. Urol.* 5, 226, 1999.
200. Santini, J.T., Jr., Cima, M.J., and Langer, R., A controlled-release microchip, *Nature* 397, 335, 1999.
201. Kruklitis, R.J., Fishbein, I., Singhal, S. et al., Stent-mediated gene transfer: a novel and efficient means of gene delivery to the tracheobronchial tree, *Mol. Ther.* 7, S121, 2003.
202. Maillard, L., Van Belle, E., Tio, F.O. et al., Effect of percutaneous adenovirus-mediated Gax gene delivery to the arterial wall in double-injured atheromatous stented rabbit iliac arteries, *Gene Ther.* 7, 1353, 2000.

CHAPTER 40

Gene Delivery Using BioMEMS

Krishnendu Roy

CONTENTS

40.1 INTRODUCTION

Microelectromechanical systems (MEMS) have attracted tremendous interest in recent years. The miniaturization of devices containing traditional electrical and mechanical components to the micro and meso scales has revolutionized technology and opened up new directions in areas such as chemistry and biology. Microfabrication techniques and MEMS devices started as a tool for the semiconductor industry and have developed into one of the most significant technological advances of the past century. In the past few years, microfabrication concepts have proven successful in creating a variety of diagnostic and therapeutic devices, e.g., lab-on-a-chip systems, microelectrodes, surgical devices, cardiovascular stents, and so on. Collectively, efforts to integrate microfabrication concepts into solving biological and medical problems have been termed BioMEMS. These techniques have been widely studied in various areas of biology and medicine including cellular manipulations, biomaterials, tissue engineering, drug delivery, and so on, and have shown exciting promise. It is at this juncture that we will try to evaluate the prospect of BioMEMS for use in gene therapy applications. Although the use of MEMS in gene delivery is somewhat analogous to that of drug delivery, there are some fundamental differences which must be addressed before applying MEMS-based drug delivery ideas for gene therapy applications. In this chapter we will

0-8493-1934-X/05/$0.00+$1.50
© 2005 by CRC Press LLC

address the applicability of BioMEMS to drug and gene therapy and provide current concepts and future directions in the field. The reader should keep in mind that only limited research has been done in the specific area of gene delivery using microfabrication, but enough relevant concepts have been evaluated for us to gain basic insights into the promise of BioMEMS for gene therapy.

40.2 DRUG AND GENE DELIVERY SYSTEMS

The field of drug delivery has made tremendous advances in the past several decades. Traditional pharmaceutical formulations — such as suspensions, tablets, capsules, and injections — have become integrated with our daily health care. Over the last 20 years or so, polymer-controlled delivery via different routes has become the primary research focus in the field. Numerous delivery systems have proven effective and some of those are already on the market today. Novel degradable and nondegradable polymers, tailored for specific applications, have opened new directions in targeted and efficient drug delivery. Another recent focus in drug delivery has been in the area of stimuli-sensitive release of bioactive molecules. Physical and chemical changes of polymer properties in response to physiological stimuli have been used to design triggered delivery systems that could be highly effective in reducing systemic toxicity, controlling drug dosage, and administering drugs only where and when necessary. Research in these areas has primarily focused on hydrogels and their stimuli responsive properties, but newer systems, especially microfabricated systems, are garnering increased interest.

Gene therapy started in the late 1980s with the use of viruses as the primary carriers for plasmid DNA. With the realization that inefficient delivery is the fundamental bottleneck to the ultimate clinical success of genetic therapy, the field has gone through some fundamental changes over the last decade. It is only in the last few years that the promise of gene-based therapeutics has started to become reality. However, effective gene-based human therapeutics still awaits significant development in delivery technologies. Pharmaceutical and polymer scientists have made tremendous contributions to gene therapy, and numerous nonviral, polymer, and lipid-based delivery concepts have been reported in recent years. However, some fundamental limitations of drug and gene delivery must be addressed before we can move beyond the current dogma of diffusion- and degradation-controlled delivery systems that has dominated the field for decades. These limitations include our inability to do the following:

- To accurately and reproducibly control *in vivo* release kinetics
- To produce controlled-release delivery devices on a large scale and with little polydispersity
- To fabricate integrated multiple-drug delivery devices that can achieve a truly combinatorial release kinetics
- To combine real time sensing with stimuli-triggered delivery of the bioactive molecule

Microfabrication technology and BioMEMS have the unique potential to address all of these issues. In this chapter we will present some of the work that has been done in these areas and its potential in addressing the specific problems in gene therapy.

As mentioned earlier, the state of the art in drug delivery research has been polymer-controlled delivery. In the last 20 years or so, numerous research papers have described the design and synthesis of new polymers and lipid molecules, the fabrication of novel delivery devices (particles, liposomes, matrices, reservoirs, etc.), and the use of these devices in delivering drugs and genes to different regions of the body. The previous chapters have discussed in great detail the state of the art in polymer-controlled gene therapy. The reader is referred to numerous review articles and book chapters on lipid-based delivery systems.[1-5] The fundamental theme in most of these delivery methods has been the uptake of DNA-carrying vehicles into cells and the release of nucleic acids through diffusion, degradation, or uncoupling of the polymer. Most of these intracellular transport, uncoupling, and delivery processes are still poorly understood. Concepts of stimuli-triggered gene delivery, especially sensor-controlled delivery of drugs and genes, are still in their infancy, but nevertheless, the field is gradually moving toward a new direction and MEMS is playing a significant role.

40.3 BARRIERS IN GENE DELIVERY

Most of the current work in MEMS-based delivery has focused on the delivery of small molecule drugs or proteins and peptides. Although these concepts are generally applicable to gene delivery also, there are some specific concerns that need to be addressed. These include the following.

First, the genetic material needs to be protected from extracellular nucleases during transport inside the body until it reaches the target tissue or cell.

Second, in gene therapy and genetic vaccination it is fundamentally critical to achieve efficient targeting to specific cells. Therefore, general systemic delivery is often inadequate to generate effective therapeutic effects. Although some drug delivery applications would benefit from efficient tissue targeting, this is a key challenge for the ultimate success of gene therapy.

Third, genes must be delivered into the cell nucleus in order for the corresponding protein to be expressed. Either the MEMS device itself or its released content must be able to at least traverse the cell membrane.

This chapter will focus on MEMS-based drug delivery systems that have been reported so far and will emphasize their limitations and advantages for gene therapy applications.

40.4 BIOMEMS AND DRUG DELIVERY: CURRENT RESEARCH

40.4.1 Microfabricated Microneedles

One of the first and most successful applications of microfabrication technology in drug delivery is the use of microneedles. Several groups have worked on fabricating these needles, primarily for delivering drugs and vaccines across the skin.[6-11] The idea of painless microneedle insertion into the epidermal or dermal region of the skin seems simple and naïve, but research in the past several years has proven that the concept is highly applicable, practical, and successful. Transdermal drug delivery is an extremely attractive mode because of high patient compliance, avoidance of hepatic first-pass metabolism, and repeated and prolonged applicability of the drug. However, the skin is designed to protect against passive penetration and hence is a major barrier for the delivery of bioactive molecules. Transdermal patches that are now on the market use permeation enhancers or iontophoresis to deliver molecules across the stratum corneum and the epidermis into the blood-vessel-rich dermal layer. Unfortunately, these techniques are largely inapplicable for high-molecular-weight drugs such as proteins, peptides, or DNA. Microneedles provide a unique means of physically disrupting the skin barrier and directly delivering the drug into the skin. The basic idea is to keep the needles small enough so that the nerve endings in the dermal layer of the skin do not get stimulated, hence making the process essentially painless. In the context of gene therapy, efficient delivery into the skin could provide a tremendous advantage, especially in the area of genetic immunization.

Cutaneous delivery of drugs or macromolecules, especially vaccines, is significant for three fundamental reasons.

First, skin is a highly immuno-active organ and naturally consists of a large population of antigen-presenting dendritic cells (Langerhan's cells). This immune-surveillance property ensures that antigenic moieties that penetrate the epithelial barrier are rapidly "intercepted" and transported to the local lymph nodes to initiate a potent protective immune response. Hence, from a basic science point of view, cutaneous immunization provides significant advantages over other routes.

Second, skin is the most readily accessible organ. Needle-free delivery into the skin can be a highly patient-compliant, cost-effective, and self-applicable modality for vaccination that would be particularly useful for children and older adults. The worldwide transdermal market is close to $3 billion but consists of only 10 small molecule drugs.[11] The tremendous economic impact of efficient cutaneous delivery and vaccination is therefore easily conceivable.

Third, from a public health standpoint, cutaneous delivery promises to be a safe, easy to apply, and accessible immunization strategy. The World Health Organization (WHO) has identified the urgent need for needle-free vaccinations for its infectious diseases program. The public-health burden posed by the unsafe use of needles, especially in third world countries, is of epidemic proportions[12] and often precludes the use of mass immunization strategies. The WHO has proposed transcutaneous and transmucosal delivery as potential alternative routes for effective mass vaccination. However, strategies that require specialized devices and skilled personnel, e.g., gene gun or Powderject® delivery devices, could restrict the accessibility of vaccines to a privileged few. Self-applicable, patch-type cutaneous immunization can potentially reach the most underprivileged populations and save millions of lives every year.

Microfabricated needles were originally developed for intracellular microinjection of DNA and proteins.[13,14] Glass microcapillaries, produced by bulk silicon micromachining, have been used for microinjection into plant cells. Microfabrication technology has also been employed to fabricate microprobes for injection into neural tissues. These design concepts have been later translated into microneedle arrays for transcutaneous drug and gene delivery.[7]

The outermost layer of the skin, the stratum corneum, is a major barrier to the penetration of drugs and macromolecules into the skin, and is devoid of any nerve endings. Hence, physical disruption of the layer with needles that are long enough to penetrate across the stratum corneum but also short enough not to stimulate the deeper nerve endings of the dermis would provide an essentially painless method of delivering drugs and genes into the skin. Based on this concept, the first-generation microneedles were designed as solid silicon microprojections that were coated with the desired drug.[7] These were fabricated using a deep reactive ion etching process in which a chromium mask, deposited on a silicon wafer, was etched out using a plasma of fluorine and oxygen, leaving behind solid projections (or spikes) of silicon. The second-generation needles were designed to have a hole (channel) through the silicon projections for "injecting" (i.e., flowing) the drug through the needles.[6] These were fabricated by a coupled deep-etching process, wherein the channel is deep-etched on a silicon wafer, followed by the reactive ion etching process described earlier.

Solid microneedle arrays have been developed containing 400 needles (20 x 20) in a single array. These needles have a base diameter of 80 μm and a projection (height) of 150 μm, with a radius of curvature less than 1 μm. Hollow needles have similar dimensions with 5 to 70 μm channels. These needles have been evaluated in skin penetration studies demonstrating that they can effectively pass through the stratum corneum into the epidermal layer, with a majority of the needle tips remaining intact (Figure 40.1). In drug penetration experiments the solid needle arrays increased skin permeation 10,000-fold over background while hollow needles were able to increase permeation ~100,000-fold. Studies in humans have shown that the needles are essentially painless and could provide a highly attractive and effective drug and gene delivery system.

In a recent study, Mikszta et al.[9] demonstrated that microprojection or microneedle arrays provide more enhanced and reproducible immune response following cutaneous genetic immunization in mice (Figure 40.2). Furthermore, in a human safety trial,[9] a significant increase in transepithelial water loss was reported with minimal skin irritation and negligible perception (Figure 40.3). The microprojections remained intact after skin penetration and the observed perception was less than during application of ECG preparation pads.

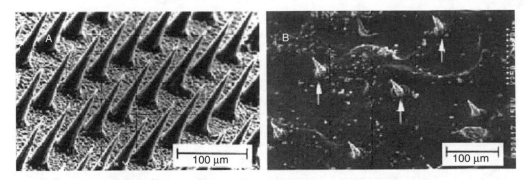

Figure 40.1   (A) Scanning electron micrograph of solid microfabricated microneedles. (B) Microneedles inserted into the skin epidermis. The tissue is viewed from the underside after needle insertion. This demonstrates that the microneedles penetrated into the skin without damaging the tips. Arrows indicate some of the microneedle tips. (From Henry, S., McAllister, D.V., Allen, M.G., and Prausnitz, M.R., *J. Pharm. Sci.* 87, 922–925, 1998. With permission.)

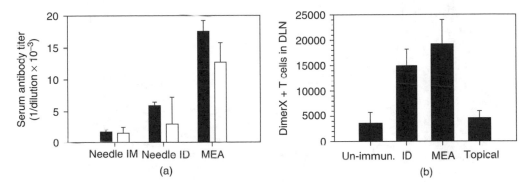

Figure 40.2   *In vivo* immunization in mice with microneedles (microenhancer arrays [MEA]). (A) Serum antibody (IgG1 [closed bars] and IgG2a [open bars]) response after third immunization with MEA carrying plasmid DNA encoding for hepatitis B surface antigen. MEA administered DNA vaccine clearly generates stronger humoral response compared to needle delivery. (B) Antigen-specific T-cells in MEA immunized and control groups of mice. MEA-based delivery generates stronger cellular immune response compared to conventional immunization methods. (From Mikszta, J.A., Alarcon, J.B., Brittingham, J.M., Sutter, D.E., Pettis, R.J., and Harvey, N.G., *Nat. Med.* 8, 415–419, 2002. With permission.)

40.4.2 Integrated Delivery Systems

Liepmann et al. reported an integrated MEMS-based drug reconstitution and delivery system for insulin delivery.[15] This DARPA-supported project is designed to develop a self-contained drug storage and on-demand delivery module that is fully microfabricated. This would include power supply, control systems, microvalves, pumps, needles, and injectors, all contained within a 4-mm thick credit card size device. The device is designed to reconstitute and deliver dry (lyophilized) drugs (antibiotics or insulin) for 24 to 48 h. The possibility of a closed loop sensor-controlled design could be ideal for diabetics or pain management therapy.

Some of the basic components of this integrated system have already been developed, specifically those involving microneedles and microfluidic concepts. Molded polysilicon needles have been fabricated for the delivery of bioactive molecules and have been designed with various geometries and sizes. One of the novel concepts emerging out of this project is a MEMS-based microfluidic device that uses bubbles as actuators. This concept essentially involves a device with no moving parts and a planar geometry leading to low cost and ease of fabrication. Instead of

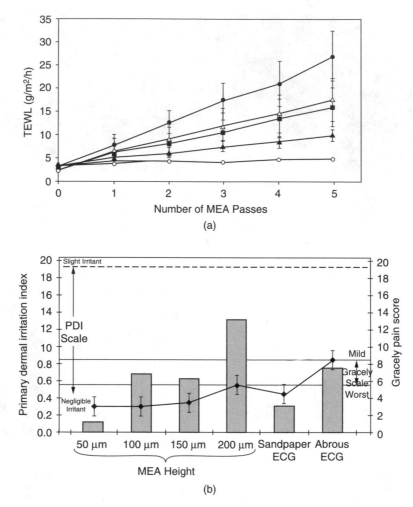

Figure 40.3 Human clinical safety and efficacy of microneedles (microenhancer arrays [MEA]). (A) Transepithelial water loss (TEWL) as a function of MEA projection height and number of passes. Data represent mean ± SEM. MEA 200 μm; MEA 150 μm; MEA 100 μm; MEA 50 μm; fibrous ECG pad; sandpaper ECG pad. (B) Skin irritation and perception scores. Irritiation is plotted as grey bars. Perception is plotted as Gracey scale (closed symbols). The clinical trials were with 4 males and 4 females. (From Mikszta, J.A., Alarcon, J.B., Brittingham, J.M., Sutter, D.E., Pettis, R.J., and Harvey, N.G., *Nat. Med.* 8, 415–419, 2002. With permission.)

thermally generating bubbles to control fluid flow through a microchannel, the researchers developed electrolysis bubbles, which provided a significantly low energy option for closing and opening a MEMS microvalve. This type of valve is a key component of any delivery device involving MEMS concepts because it is absolutely critical to control the kinetics and amount of flow of the drug or gene in the device with high accuracy. This bubble-activated valve could be coupled to a constant-pressure microreservoir containing the drug that would act as a pump for accurate on-demand delivery.

Another integrated delivery device incorporating the concept of "artificial muscles" has been reported by Madou and colleagues.[16] This design, which is being commercialized by ChipRx, Inc., incorporates a drug reservoir coupled to a biosensor. The sensor output is coupled to a hydrogel-based drug reservoir, which could be activated by the sensor and would then "squeeze" the drug out. The key component here is a polymer hydrogel that swells or shrinks in response to a stimulus and thereby closes or opens a microvalve.[16] This type of microactuator (the so-called artificial muscle) allows for a controlled delivery mechanism using MEMS concepts. The researchers used poly(hydroxyl ethyl methacrylate) (PHEMA) gels blended with polyaniline, an electricity-conducting polymer. The blends were shaped into different valve configurations, e.g., sphincter-like configuration, plunger configuration, and tube configuration. The shrinking and swelling of the hydrogel–polyaniline blend were evaluated following electrochemical actuation and effective valve closing and opening were observed.[16]

These devices each exemplify an ideal MEMS system for drug delivery that can also be adopted for delivering gene therapy formulations, especially intramuscular delivery of naked DNA or parenteral delivery of a constant dose of DNA-lipid–DNA-polycation complexes. These types of devices, in their current configuration, would not be suitable for direct intracellular delivery of nucleic acids but would still be an excellent option for the delivery of genetic materials for specific applications.

40.4.3 Microchip-Based Delivery Systems

A BioMEMS drug delivery system capable of incorporating multiple drugs and triggered pulsatile release profiles has been developed by Langer and colleagues.[17,18] The initial prototype involves a 17×17 mm "chip" that contains 34 drug reservoirs that can be loaded with the same drug or different drugs. The device can be miniaturized down to a few mm in size and can be implanted in the body. The 17 mm size device could hold about 1000 reservoirs for a truly automated, prolonged, pulsatile release of multiple drugs.

These devices were fabricated using MEMS techniques, including photolithography, reactive ion etching, chemical vapor deposition, and E-beam evaporation. The reservoir volumes were ~25 nl. One end of the reservoirs was sealed with gold anodes, chosen for gold's unique dissolution properties upon application of an electric field. In addition, gold is a biocompatible material widely used in medical applications. The gold anodes, in presence of chloride ions, were selectively dissolved using a potential of +1.04 volts to release model drugs (Figure 40.4). It has been demonstrated that pulsatile release of multiple drugs can be obtained in a controlled fashion using appropriate triggering sequences. This device is currently under commercial development (Micro-CHIPS, Inc.) and awaits human clinical trials.

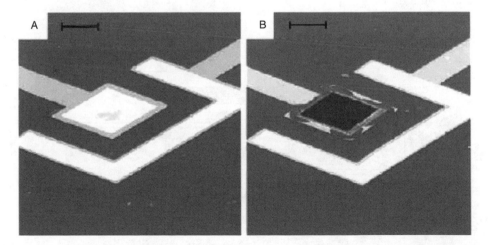

Figure 40.4 Electrochemical dissolution of the gold anode membrane to initiate release from a reservoir. (A, B) Scanning electron micrographs of the gold membrane anode covering a drug reservoir before (A) and after (B) the application of +1.04 V for several seconds in phosphate-buffered saline (PBS). (Scale bar = 50 μm.) (From Santini, J.T., Jr., Cima, M.J., and Langer, R., *Nature* 397, 335–338, 1999. With permission.)

40.4.4 Delivery from Immunoisolated Genetically Modified Cells

One concept for long-term, sustained delivery of therapeutic proteins to diseased tissues is the use of microencapsulation of genetically modified cells in immunoisolated devices. Similar concepts have been widely used to deliver insulin from encapsulated allogeneic or xenogeneic pancreatic beta cells.[19-21] The use of nonautologous cells necessitates immunoisolation in order to prevent rejection of the transplant by the host immune system. A major advantage of such a technique is that the release of therapeutics is controlled by true physiological stimuli in a manner identical to normal tissue. However, the principal limitation in this research has been the fabrication of truly immunoisolated devices that can, at the same time, efficiently deliver the therapeutic molecule. Research has primarily focused on polymer-based encapsulation techniques in which the cells are entrapped in a polymer matrix using physical or chemical cross-linking techniques. In these processes, it is extremely difficult accurately to control the pore sizes of the matrix/membrane that could provide optimal isolation from antibodies and immune complexes while adequately delivering the therapeutic molecule. Novel methods that could produce highly controlled membranes with small, uniform porosities on a large scale would have a tremendous impact in the field of cell-based gene therapy.

Recently, BioMEMS have been applied to construct such immunoisolation devices for efficient cell encapsulation.[22-24] The key to such design is the fabrication of small monodisperse pores that are large enough for the diffusion of the therapeutic molecule but small enough to prevent the entry of antibodies and immune complexes into the device. Desai and colleagues reported such a design (Figure 40.5) using bulk and surface micromachining techniques. This microfabricated biocapsule consisted of two separate membranes bonded together with cells entrapped inside the cavity.[23] The wafer surface consisted of a micromachined, porous (as low as 15 nm pore size) membrane filter that prevented immunoglobulins (> 15 nm) from entering the device while allowing the drug and nutrients (small molecules, proteins, or peptides, < 6 nm) to pass through. The final biocapsule was 1100 μm in thickness, 4 × 4 mm in lateral dimension, 10.4 mm² in membrane area, 10 μl in cavity volume, and 9 μm in membrane thickness. Pancreatic beta cells, suspended in an alginate matrix, were loaded in a half-capsule; the remaining half of the capsule was glued on in order to achieve a full encapsulation. *In vitro* and *in vivo* studies showed cell viability, effective immunoisolation, and insulin diffusion.[23]

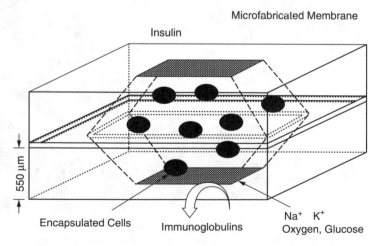

Figure 40.5 Schematic diagram of a microfabricated immunoisolation biocapsule as proposed by Desai et al. (From Desai, T.A., Hansford, D.J., and Ferrari, M., *Biomol. Eng.* 17, 23–36, 2000. With permission.)

Figure 40.6 Microfabricated PMMA particles for oral delivery of drugs. (From Tao, S.L., Lubeley, M.W., and Desai, T.A., *J. Control. Release* 88, 215–228, 2003. With permission.)

In the context of gene therapy, this type of design could be easily incorporated to entrap genetically modified cells, e.g., cells secreting nerve growth factors for diseases like Parkinson's or implants delivering anti-inflammatory molecules from genetically modified cells for long-term pain management, as in the case of cancer patients.

40.4.5 Microfabricated Particles for Oral Drug Delivery

Recently, microfabrication of oral drug delivery particles has been developed.[25,26] The first-generation particles involved silicon wafer processing,[25] while subsequent designs have involved polymethyl methacrylate (PMMA) particles suitable for *in vivo* use.[26] The basic concept is to create a planar particle design with a higher surface-contact area than spherical particles and to achieve increased adhesion to the mucosal surface of the gastrointestinal tract. Furthermore, by microfabricating these particles, one would achieve a monodisperse population with uniform size. In addition, multiple drug reservoirs can be fabricated within a single particle, thereby delivering several drugs within the same vehicle.

The microparticles are fabricated using several consecutive lithography procedures followed by surface modification and lectin conjugation. The PMMA surface is modified using aminolysis followed by avidin–biotin conjugation chemistry to attach mucoadhesive biomolecules such as lectins. These ligands ensure that the device attaches to the intestinal mucosa and allows the drug microcontainers to dissolve and release the drug at an appropriate location. The area of drug release can be controlled by the polymer properties (e.g., pH dependent polymer dissolution). Figure 40.6 shows PMMA particles fabricated using this concept.[26]

These lectin-modified microfabricated particles were evaluated *in vitro* in a Caco-2 human intestinal epithelium cell culture model. The devices showed enhanced cell-adhesive properties when modified with plant lectins. These particles are currently undergoing further evaluation in collaboration with iMEDD, Inc.[26]

40.5 BIOCOMPATIBILITY OF MEMS MATERIALS

Current MEMS devices pose some fundamental concerns in terms of their ultimate clinical applicability. As discussed in this chapter, most of the microfabrication concepts have only been demonstrated using silicon and its derivatives (silicon oxide, silicon nitride, etc.) or with metals such as gold. Only recently, some applications have been reported that use polymeric materials (e.g., PMMA). Although gold has been used for some *in vivo* applications for a long time (e.g., tooth replacement and gene-gun-based delivery) and is known to be fairly biocompatible, the compatibility of silicon materials has been a serious concern in the biomaterials community. In addition, the issue of biofouling needs to be evaluated, especially for applications involving sensor-integrated delivery. Recently, Langer and colleagues performed a comprehensive study with MEMS materials and demonstrated that these substances are biocompatible in an *in vivo* animal model.[27]

The researchers evaluated *in vivo* inflammatory and wound-healing responses in an animal model using the cage-implant method. Several MEMS materials — gold, silicon nitride (SiN), silicon dioxide (SiO_2), silicon (Si), and the photoresist SU-8, which are widely used in microfabrication — were implanted subcutaneously in rodents using a stainless steel cage. Empty cages served as negative controls. At different time points, inflammatory responses were evaluated histologically and by using leukocyte counts. Biofouling was studied using scanning electron microscopy. It was concluded that gold, SiN, SiO_2, Su08, and Si were all biocompatible. In addition, gold, SiN, SiO_2, and Su-8 demonstrated reduced biofouling.

Taken together, this provides the first comprehensive look at the biocompatibility of materials that have traditionally been used in the semiconductor industry. It is quite exciting that most of the materials were quite compatible and showed reduced biofouling. This would imply that existing MEMS techniques could be translated into medical device manufacturing even before polymeric material-based MEMS processes have evolved.

40.6 POTENTIAL OF MEMS IN GENE DELIVERY APPLICATIONS

Throughout this chapter we have discussed current research in microfabricated drug delivery devices with only a few examples specifically evaluating DNA delivery. In a broad sense, most of these technologies could be applicable, to some extent, for the delivery of nucleic acids either as "naked" DNA or as DNA in a secondary vehicle; e.g., in polycation-DNA complexes, lipid DNA complexes, or micro- and nanoparticles. In addition to the issues of storage stability (or bioactivity) of nucleic acids in microfabricated devices, for the latter applications the physical stability of the secondary vehicles inside a MEMS device (e.g., how long lipid–DNA complexes or polymer–DNA complexes are stable inside a microchip-type device) is a critical concern that needs to be addressed. In spite of these significant hurdles, the tremendous progress made in recent years in transforming concepts from the semiconductor industry into biomedical applications suggests that BioMEMS-based gene therapy will definitely be a reality in the near future.

REFERENCES

1. Anwer, K., Meaney, C., Kao, G., Hussain, N., Shelvin, R., Earls, R.M., Leonard, P., Quezada, A., Rolland, A.P., and Sullivan, S.M., Cationic lipid-based delivery system for systemic cancer gene therapy. *Cancer Gene Ther.* 7, 1156–1164, 2000.
2. Chesnoy, S. and Huang, L., Structure and function of lipid-DNA complexes for gene delivery. *Ann. Rev. Biophys. Biomol. Struct.* 29, 27–47, 2000.
3. Ilies, M.A., Seitz, W.A., and Balaban, A.T., Cationic lipids in gene delivery: principles, vector design and therapeutical applications. *Curr. Pharm. Des.* 8, 2441–2473, 2002.
4. Pedroso de Lima, M.C., Simoes, S., Pires, P., Faneca, H., and Duzgunes, N., Cationic lipid-DNA complexes in gene delivery: from biophysics to biological applications. *Adv. Drug Del. Rev.* 47, 277–294, 2001.
5. Zhdanov, R.I., Podobed, O.V., and Vlassov, V.V., Cationic lipid-DNA complexes-lipoplexes-for gene transfer and therapy. *Bioelectrochemistry* 58, 53–64, 2002.

6. Henry, S., McAllister, D.V., Allen, M.G., and Prausnitz, M.R., Microfabricated microneedles: A novel approach to transdermal drug delivery. *J. Pharm. Sci.* 88, 948, 1999.
7. Henry, S., McAllister, D.V., Allen, M.G., and Prausnitz, M.R., Microfabricated microneedles: a novel approach to transdermal drug delivery. *J. Pharm. Sci.* 87, 922–925, 1998.
8. McAllister, D.V., Allen, M.G., and Prausnitz, M.R., Microfabricated microneedles for gene and drug delivery. *Ann. Rev. Biomed. Eng.* 2, 289–313, 2000.
9. Mikszta, J.A., Alarcon, J.B., Brittingham, J.M., Sutter, D.E., Pettis, R.J., and Harvey, N.G., Improved genetic immunization via micromechanical disruption of skin-barrier function and targeted epidermal delivery. *Nat. Med.* 8, 415–419, 2002.
10. Prausnitz, M.R., Overcoming skin's barrier: the search for effective and user-friendly drug delivery. *Diabetes Technol. Ther.* 3, 233–236, 2001.
11. Barry, B.W., Novel mechanisms and devices to enable successful transdermal drug delivery. *Eur. J. Pharm. Sci.* 14, 101–114, 2001.
12. Immunisation, W. H. O. E. P. o. Reducing the risk of unsafe injections in immunisation programs: the role of injection equipment. 1996.
13. Trimmer, W., Ling, P., Chin, C.-K., Orton, P., Gaugler, R., Hashmi, S., Hashmi, G., Brunett, B., and Reed, M., Injection of DNA into plant and animal tissues with micromechanical piercing structures. in *Proceedings IEEE Micro Electro Mechanical Systems.* 1995.
14. Hashmi, S., Ling, P., Hashmi, G., Reed, M., Gaugler, R., Trimmer, W., Genetic transformation of nematodes using arrays of micromechanical piercing structures. *Biotechniques* 19, 766–770, 1995.
15. Liepmann, D., Pisano, A.P., and Sage, B., Microelectromechanical systems technology to deliver insulin. *Diabetes Technol. Ther.* 1, 469–476, 1999.
16. Low, L.-M., Seetharaman, S., He, K.-Q., and Madou, M.J., Microactuators toward microvalves for responsive controlled drug delivery. *Sens. Actuat. B: Chem.* 67, 149–160, 2000.
17. Santini, J.T., Jr., Cima, M.J., and Langer, R., A controlled-release microchip. *Nature* 397, 335–338, 1999.
18. Santini, J.T., Jr., Richards, A.C., Scheidt, R.A., Cima, M.J., and Langer, R.S., Microchip technology in drug delivery. *Ann. Med.* 32, 377–379, 2000.
19. Zhou, D., Kintsourashvili, E., Mamujee, S., Vacek, I., and Sun, A.M., Bioartificial pancreas: alternative supply of insulin-secreting cells. *Ann. N.Y. Acad. Sci.* 875, 208–218, 1999.
20. Lanza, R.P., Beyer, A.M., Staruk, J.E., and Chick, W.L., Biohybrid artificial pancreas. Long-term function of discordant islet xenografts in streptozotocin diabetic rats. *Transplantation* 56, 1067–1072, 1993.
21. Pollok, J.M., Lorenzen, M., Kolln, P.A., Torok, E., Kaufmann, P.M., Kluth, D., Bohuslavizki, K.H., Gundlach, M., and Rogiers, X., In vitro function of islets of Langerhans encapsulated with a membrane of porcine chondrocytes for immunoisolation. *Dig. Surg.* 18, 204–210, 2001.
22. Desai, T.A., Microfabrication technology for pancreatic cell encapsulation. *Expert Opin. Biol. Ther.* 2, 633–646, 2002.
23. Desai, T.A., Hansford, D.J., and Ferrari, M., Micromachined interfaces: new approaches in cell immunoisolation and biomolecular separation. *Biomol. Eng.* 17, 23–36, 2000.
24. Desai, T.A., Chu, W.H., Tu, J.K., Beattie, G.M., Hayek, A., and Ferrari, M., Microfabricated immunoisolating biocapsules. *Biotechnol. Bioeng.* 57, 118–120, 1998.
25. Ahmed, A., Bonner, C., and Desai, T.A., Bioadhesive microdevices with multiple reservoirs: a new platform for oral drug delivery. *J. Control. Release* 81, 291–306, 2002.
26. Tao, S.L., Lubeley, M.W., and Desai, T.A., Bioadhesive poly(methyl methacrylate) microdevices for controlled drug delivery. *J. Control. Release* 88, 215–228, 2003.
27. Voskerician, G., Shive, M.S., Shawgo, R.S., von Recum, H., Anderson, J.M., Cima, M.J., and Langer, R., Biocompatibility and biofouling of MEMS drug delivery devices. *Biomaterials* 24, 1959–1967, 2003.

Index